Wireless Communications Over Rapidly Time-Varying Channels

Wireless Communications Over Rapidly Time-Varying Channels

Edited by

Franz Hlawatsch

Gerald Matz

ELSEVIER

AMSTERDAM • BOSTON • HEIDELBERG • LONDON
NEW YORK • OXFORD • PARIS • SAN DIEGO
SAN FRANCISCO • SINGAPORE • SYDNEY • TOKYO

Academic Press is an imprint of Elsevier

Academic Press is an imprint of Elsevier
The Boulevard, Langford Lane, Kidlington, Oxford OX5 1GB, UK
30 Corporate Drive, Suite 400, Burlington, MA 01803, USA

First edition 2011

British Library Cataloguing in Publication Data
A catalogue record for this book is available from the British Library.

Library of Congress Cataloging-in-Publication Data
A catalog record for this book is available from the Library of Congress.

ISBN: 978-0-323-16579-2

For information on all Academic Press publications visit our web site at *www.books.elsevier.com*

Printed and bound in USA
11 12 13 14 15 10 9 8 7 6 5 4 3 2 1

Contents

Preface

Wireless communications has become a field of enormous scientific and economic interest. Recent success stories include 2G and 3G cellular voice and data services (e.g., GSM and UMTS), wireless local area networks (WiFi/IEEE 802.11x), wireless broadband access (WiMAX/IEEE 802.16x), and digital broadcast systems (DVB, DAB, DRM). On the physical layer side, traditional designs typically assume that the radio channel remains constant for the duration of a data block. However, researchers and system designers are increasingly shifting their attention to channels that may vary within a block. In addition to time dispersion caused by multipath propagation, these rapidly time-varying channels feature frequency dispersion resulting from the Doppler effect. They are, thus, often referred to as being "doubly dispersive."

Historically, channels with time variation and frequency dispersion were first considered mostly in the context of ionospheric and tropospheric communications and in radio astronomy. The theoretical foundations of rapidly time-varying channels were established by Bello, Gallager, Kailath, Kennedy, and others in the sixties of the twentieth century. More recently, rapidly time-varying channels have become important in novel application scenarios with potentially high economic relevance and societal impact.

- User mobility, a source of significant Doppler frequency shifts, is an essential factor in today's cellular and broadband access systems. An extreme example is given by radio access links for high-speed trains. Channels with rapid time variation are also encountered in car-to-car and car-to-infrastructure communications, which are becoming increasingly important.
- In advanced wireless networks, nodes may cooperate to achieve spatial diversity gains in a distributed manner. An example is the base station cooperation option (also known as network MIMO or cooperative multipoint transmission) in 3GPP Long Term Evolution. In such systems, the carrier frequency offsets of different nodes accumulate and, together with mobility-induced Doppler frequency shifts, result in channels with rapid time variation.
- In underwater acoustic communications, the relative Doppler shifts are potentially much larger than in terrestrial radio systems because the speed of sound is much smaller than the speed of light. Furthermore, the smaller propagation speed of acoustic waves results in larger propagation delays. Underwater channels are, therefore, instances of particularly harsh doubly dispersive channels.

Rapid channel variations induced by Doppler shifts provide an extra dimension that offers additional gains. At the same time, doubly dispersive channels pose tough design challenges and necessitate the use of sophisticated methods to combat the detrimental effects of the channel and to realize the additional gains. Thus, understanding the fundamental properties of doubly dispersive channels and the resulting design paradigms will become essential know-how in the future wireless arena.

This book explains the system-theoretic and information-theoretic foundations of doubly dispersive channels and describes the current state of the art in algorithm and system design. It is intended to present a comprehensive and coherent discussion of the challenges and developments in the field, which will help researchers and engineers understand and develop future wireless communication technologies. Contributed by leading experts, the individual chapters of this book address the most important aspects of the theory and methodology of wireless communications over rapidly time-varying channels. Wireless transceiver design and modern techniques such as iterative turbo-style detection, multicarrier (OFDM) modulation, and multiantenna (MIMO) processing are given special attention.

In the introductory chapter, Chapter 1, we discuss the properties and mathematical characterization of doubly dispersive channels. Further topics addressed include propagation effects, system-theoretic aspects, stochastic channel characterizations, parsimonious channel models, and measurement principles.

Chapter 2, by G. Durisi, V. Morgenshtern, H. Bölcskei, U. Schuster, and S. Shamai, discusses information-theoretic aspects of random time-varying channels, including MIMO channels. This chapter focuses on noncoherent channel capacity (i.e., channel capacity in the absence of channel state information) in the large-bandwidth and high-SNR regimes.

Chapter 3, by E. Viterbo and Y. Hong, addresses the design of channel codes for fast-fading channels, using methods from algebraic number theory and lattice theory. The sphere decoder is discussed as an efficient means to recover the transmitted code words.

Chapter 4, by G. Leus, Z. Tang, and P. Banelli, considers the estimation of rapidly time-varying channels in single-carrier and multicarrier communication systems. A block-based approach is adopted that builds on a basis expansion model for the channel and the transmission of dedicated pilot (training) symbols.

Chapter 5, by M. Dong, B. M. Sadler, and L. Tong, complements Chapter 4 by discussing training designs for the estimation of time-varying channels. The optimization of the number, placement, and power of pilot symbols is studied for various system configurations (single carrier, multicarrier, multiantenna) and performance criteria.

Chapter 6, by P. Schniter, S.-J. Hwang, S. Das, and A. P. Kannu, presents equalization techniques for doubly dispersive channels. Both coherent and noncoherent detection are addressed, using linear and tree-search methods, iterative approaches, and joint detection-estimation schemes.

Chapter 7, by L. Rugini, P. Banelli, and G. Leus, is dedicated to orthogonal frequency division multiplex (OFDM) transmissions over time-varying channels. This chapter discusses methods for equalizing intercarrier interference and for channel estimation and comments on the relevance of these methods to existing standards.

Chapter 8, by C. Dumard, J. Jaldén, and T. Zemen, considers a multiuser system employing multiple antennas and a multicarrier CDMA transmission format. An iterative (turbo) receiver is developed, which performs estimation of the time-varying channels, multiuser separation, and channel decoding, with complexity reductions due to Krylov subspace and sphere decoding techniques.

The final chapter, Chapter 9, by A. Papandreou-Suppappola, C. Ioana, and J. J. Zhang, discusses wideband channels that are more suitably characterized in terms of Doppler scaling than in terms of Doppler shifts. Theoretical considerations and advanced receiver designs are exemplified by an underwater acoustic communication system.

We would like to thank all people who contributed to this book in one way or another. We are especially grateful to the chapter authors for their expertise and hard work, and for accepting the constraints of a predefined, common notation. We thank Tim Pitts of Elsevier for inviting us to edit this book. Tim and his colleagues—Melanie Benson, Susan Li, Melissa Read, and Naomi Robertson—provided much appreciated assistance during the various stages of this project. Finally, we acknowledge support by the Austrian Science Fund (FWF) under Grants S10603 (Statistical Inference) and S10606 (Information Networks) within the National Research Network SISE.

Franz Hlawatsch
Gerald Matz

About the Editors

Franz Hlawatsch received the Dipl.-Ing., Dr. techn., and Univ.-Dozent (habilitation) degrees in electrical engineering/signal processing from Vienna University of Technology, Vienna, Austria, in 1983, 1988, and 1996, respectively. Since 1983, he has been with the Institute of Telecommunications, Vienna University of Technology, as an associate professor. During 1991–1992, as a recipient of an Erwin Schrödinger Fellowship, he spent a sabbatical year with the Department of Electrical Engineering, University of Rhode Island, Kingston, RI, USA. In 1999, 2000, and 2001, he held one-month visiting professor positions with INP–ENSEEIHT/TéSA (Toulouse, France) and IRCCyN (Nantes, France). He (co)authored a book, a review paper that appeared in the *IEEE Signal Processing Magazine*, about 180 refereed or invited scientific papers and book chapters, and three patents. He coedited three books. His research interests include signal processing for wireless communications, statistical signal processing, and compressive signal processing. Prof. Hlawatsch was a Technical Program Co-Chair of EUSIPCO 2004 and has served on the technical committees of numerous international conferences. From 2003 to 2007, he served as an associate editor for the IEEE Transactions on Signal Processing. He is currently serving as an associate editor for the IEEE Transactions on Information Theory. From 2004 to 2009, he was a member of the IEEE Signal Processing for Communications Technical Committee. He is coauthor of a paper that won an IEEE Signal Processing Society Young Author Best Paper Award.

Gerald Matz received the Dipl.-Ing. and Dr. techn. degrees in electrical engineering in 1994 and 2000, respectively, and the Habilitation degree for communication systems in 2004, all from Vienna University of Technology,Vienna, Austria. Since 1995, he has been with the Institute of Telecommunications, Vienna University of Technology, where he currently holds a tenured position as associate professor. From March 2004 to February 2005, he was on leave as an Erwin Schrödinger Fellow with the Laboratoire des Signaux et Systèmes, Ecole Supérieure d'Electricité, France. During summer 2007, he was a guest researcher with the Communication Theory Lab at ETH Zurich, Switzerland. He has directed or actively participated in several research projects funded by the Austrian Science Fund (FWF), the Vienna Science and Technology Fund (WWTF), and the European Union. He has published more than 140 papers in international journals, conference proceedings, and edited books. His research interests include wireless communications, statistical signal processing, and information theory. Prof. Matz serves as a member of the IEEE Signal Processing Society (SPS) Technical Committee on Signal Processing for Communications and Networking and of the IEEE SPS Technical Committee on Signal Processing Theory and Methods. He was an associate editor for the IEEE Transactions of Signal Processing (2006–2010), for the IEEE Signal Processing Letters (2004–2008), and for the EURASIP journal Signal Processing (2007–2010). He was a Technical Program Co-Chair of EUSIPCO 2004 and has been on the Technical Program Committee of numerous international conferences. In 2006, he received the Kardinal Innitzer Most Promising Young Investigator Award.

Contributing Authors

Paolo Banelli, *Università di Perugia* (Perugia, Italy)

Helmut Bölcskei, *ETH Zurich* (Zurich, Switzerland)

Sibasish Das, *Qualcomm Inc.* (San Diego, CA, USA)

Min Dong, *University of Ontario Institute of Technology* (Oshawa, Ontario, Canada)

Charlotte Dumard, *FTW Forschungszentrum Telekommunikation Wien* (Vienna, Austria)

Giuseppe Durisi, *Chalmers University of Technology* (Gothenburg, Switzerland)

Franz Hlawatsch, *Vienna University of Technology* (Vienna, Austria)

Yi Hong, *Monash University* (Clayton, Melbourne, Australia)

Sung-Jun Hwang, *Qualcomm Inc.* (Santa Clara, CA, USA)

Cornel Ioana, *National Polytechnic Institute of Grenoble* (Grenoble, France)

Joakim Jaldén, *Royal Institute of Technology (KTH)* (Stockholm, Sweden)

Arun P. Kannu, *Indian Institute of Technology* (Madras, Chennai, India)

Geert Leus, *Delft University of Technology* (Delft, The Netherlands)

Gerald Matz, *Vienna University of Technology* (Vienna, Austria)

Veniamin I. Morgenshtern, *ETH Zurich* (Zurich, Switzerland)

Antonia Papandreou-Suppappola, *Arizona State University* (Tempe, AZ, USA)

Luca Rugini, *Università di Perugia* (Perugia, Italy)

Brian M. Sadler, *Army Research Laboratory* (Adelphi, MD, USA)

Philip Schniter, *Ohio State University* (Columbus, OH, USA)

Ulrich G. Schuster, *Robert Bosch GmbH* (Stuttgart, Germany)

Shlomo Shamai (Shitz), *Technion–Israel Institute of Technology* (Haifa, Israel)

Zijian Tang, *TNO Defence, Security and Safety* (The Hague, The Netherlands)

Lang Tong, *Cornell University* (Ithaca, NY, USA)

Emanuele Viterbo, *Monash University* (Clayton, Melbourne, Australia)

Thomas Zemen, *FTW Forschungszentrum Telekommunikation Wien* (Vienna, Austria)

Jun Jason Zhang, *Arizona State University* (Tempe, AZ, USA)

Notations and Symbols

Basic Notations

$\mathbb{R}, \mathbb{C}, \mathbb{Z}$	real/complex/integer numbers
j	$\sqrt{-1}$
x^*	complex conjugation
z^{-1}	unit delay
$\frac{\mathrm{d}}{\mathrm{d}t}x(t)$, $x'(t)$	differentiation/derivative
$\langle x, y\rangle$	inner product
$\\|x\\| = \sqrt{\langle x, x\rangle}$	norm
$E_x = \\|x\\|^2$	energy
$(x * y)(t)$	continuous-time convolution
$(x * y)[n]$	discrete-time convolution
$\mathrm{Re}\{\cdot\}$, $\mathrm{Im}\{\cdot\}$	real part, imaginary part
$L^2(\mathbb{R})$	space of square-integrable functions on $\mathbb{R}$
$l^2(\mathbb{Z})$	space of square-summable functions on $\mathbb{Z}$
$\triangleq$	definition
$\lceil x \rceil$	smallest integer not less than x
$\lfloor x \rfloor$	largest integer not greater than x

Basic Symbols

t	continuous time
f	frequency (Hz)
τ	continuous time-delay
ν	Doppler frequency (Hz)
α	continuous timescale parameter
n	discrete time
m	discrete time-delay
θ	normalized frequency
l	discrete frequency
ξ	normalized Doppler frequency
d	discrete Doppler frequency
κ	discrete scale parameter
$\Delta t, \Delta n$, etc.	lag/difference
$x(t)$	continuous-time signal
$x[n]$	discrete-time signal

$X(f)$	continuous-time Fourier transform
$X(\theta)$	discrete-time Fourier transform
$X[l]$	discrete Fourier transform (DFT)
$X(z)$	z-transform
f_s, T_s	sampling frequency/period
$s(t), s[n]$	transmit signal
$w(t), w[n]$	white channel noise
$r(t), r[n]$	receive signal
$g(t), \gamma(t)$	transmit/receive pulse
$b[i]$	information bits
$c[j]$	coded bits
$L_c[j]$	log-likelihood ratio
$a[k], a[k,l]$	transmit symbols
$y[k], y[k,l]$	demodulated symbols
$z[k], z[k,l]$	noise after demodulation
$p[k], p[k,l]$	pilot symbols

Transceiver and Channel Parameters

T_0, N	block length
B	transmit bandwidth (Hz)
f_c	carrier frequency (Hz)
T	symbol duration
R	symbol rate
F	subcarrier frequency spacing
T_b	bit duration
R_b	bit rate
Π	interleaver
k	symbol time index
l	subcarrier frequency index
$\mathscr{A}$	symbol alphabet
N_a	size (cardinality) of symbol alphabet
$\mathscr{P}$	set of pilot locations k or (k,l)
N_p	number of pilot symbols
K	number of transmitted symbols
L	number of subcarriers
N_g	guard interval length
L_{CP}	cyclic prefix length

$M_{\mathrm{T}}, M_{\mathrm{R}}$	number of transmit/receive antennas
U	number of users
ρ	signal-to-noise ratio
I	basis expansion model order
$\tau_{\max}, M$	maximum delay (channel length)
$\nu_{\max}, D$	maximum Doppler frequency
A_s	Doppler scale spread
T_{c}	coherence time
F_{c}	coherence bandwidth
T_{s}	stationarity time
F_{s}	stationarity bandwidth

Channel and Signal Representations

$\mathbf{H}$	channel operator
$h(t,\tau), h[n,m]$	time-varying impulse response
$H(f,\nu), H[l,d]$	frequency-domain impulse response
$L_{\mathbf{H}}(t,f), L_{\mathbf{H}}[n,l]$	(time-frequency/time-varying) transfer function
$S_{\mathbf{H}}(\tau,\nu), S_{\mathbf{H}}[m,d]$	(delay-Doppler) spreading function
$F_{\mathbf{H}}(\tau,\alpha), F_{\mathbf{H}}[m,\kappa]$	wideband spreading function
$C_{\mathbf{H}}(\tau,\nu), C_{\mathbf{H}}[m,d]$	scattering function
$B_{\mathbf{H}}(\tau,\alpha), B_{\mathbf{H}}[m,\kappa]$	wideband scattering function
$R_{\mathbf{H}}(\Delta t,\Delta f), R_{\mathbf{H}}[\Delta n,\Delta l]$	time-frequency correlation function
$A_x(\Delta t,\Delta f)$	ambiguity function
$A_{xy}(\Delta t,\Delta f)$	cross-ambiguity function

Special Functions and Signals

$\ln(\cdot)$	natural logarithm
$\log(\cdot), \log_{10}(\cdot)$	base-10 logarithm
$\log_2(\cdot)$	binary logarithm
$e^t, \exp(t)$	exponential function
$e^{j2\pi ft}$	complex sinusoid
$\sin(\cdot), \cos(\cdot), \tan(\cdot), \cot(\cdot)$	trigonometric functions
$\sinh(\cdot), \cosh(\cdot), \tanh(\cdot), \coth(\cdot)$	hyperbolic functions
$u(t), u[n]$	unit step
$1_{\mathcal{T}}(t), 1_{\mathcal{N}}[n]$	indicator function

$\delta(t)$	Dirac impulse
$\delta[n]$	unit sample
δ_{ij}	Kronecker delta
$\mathrm{sign}(\cdot)$	sign function
$\mathrm{sinc}(t) = \frac{\sin(t)}{t}$	sinc function

Vectors, Matrices, and Operators

$\mathbf{x}$	(column) vector
x_i, $[\mathbf{x}]_i$	ith element of vector $\mathbf{x}$
$\Vert\mathbf{x}\Vert$	norm of $\mathbf{x}$
$\Vert\mathbf{x}\Vert_{\mathbf{W}}$	weighted norm of $\mathbf{x}$
$\mathbf{A}$, (a_{ij})	matrix
a_{ij}, $[\mathbf{A}]_{ij}$	element of matrix $\mathbf{A}$
$\det\{\mathbf{A}\}$	determinant
$\mathrm{tr}\{\mathbf{A}\}$	trace
$\dim\{\mathbf{x}\}$	dimension
$\mathbf{A}^{-1}$	inverse
$\mathbf{A}^{\#}$	pseudo-inverse
$\mathbf{x}^T$, $\mathbf{A}^T$	transpose
$\mathbf{x}^H$, $\mathbf{A}^H$	Hermitian transpose
$\mathbf{I}$	identity matrix
$\mathbf{W}$	DFT matrix
$\mathbf{U}\boldsymbol{\Sigma}\mathbf{V}^H$	singular value decomposition
$\mathbf{U}\boldsymbol{\Lambda}\mathbf{U}^H$	eigenvalue decomposition
H	linear operator
H^{-1}	inverse of H
H^*	adjoint
$\mathsf{H}^{\#}$	pseudo-inverse
I	identity operator

Probability, Random Variables, and Random Processes

$\Pr\{\mathscr{E}\}$	probability of event $\mathscr{E}$
$f(x)$	probability density function (pdf)
$p(x)$	probability mass function (pmf)
$f(x\vert y)$, $p(x\vert y)$	conditional pdf/pmf
$F(x)$	cumulative distribution function

$\mathsf{E}\{\cdot\}$	expectation
$\phi_x(s)$	characteristic function
μ_x, $\boldsymbol{\mu}_x$, $\mu_x(t)$, $\mu_x[n]$	mean (vector, signal)
$\sigma_x^2 = \text{var}\{x\}$	variance
$C_x(t_1,t_2)$, $C_x[n_1,n_2]$	covariance function
$C_{xy}(t_1,t_2)$, $C_{xy}[n_1,n_2]$	cross-covariance function
$R_x(t_1,t_2)$, $R_x[n_1,n_2]$	correlation function
$R_{xy}(t_1,t_2)$, $R_{xy}[n_1,n_2]$	cross-correlation function
$\mathbf{C}_x$, $\mathbf{C}_{xy}$	(cross-)covariance matrix
$\mathbf{R}_x$, $\mathbf{R}_{xy}$	(cross-)correlation matrix
$c_x(\Delta t)$, $c_x[\Delta n]$	autocovariance (stationary case)
$c_{xy}(\Delta t)$, $c_{xy}[\Delta n]$	cross-covariance (stationary case)
$r_x(\Delta t)$, $r_x[\Delta n]$	autocorrelation (stationary case)
$r_{xy}(\Delta t)$, $r_{xy}[\Delta n]$	cross-correlation (stationary case)
$P_x(f)$, $P_x(\theta)$, $P_x[l]$	power spectral density
$P_{xy}(f)$, $P_{xy}(\theta)$, $P_{xy}[l]$	cross power spectral density
$\hat{a}$	estimate of a

Abbreviations

2D	two-dimensional
3G	third-generation
4G	fourth-generation
AoA	angle of arrival
AoD	angle of departure
APP	a posteriori probability
AR	autoregressive
ARMA	autoregressive moving average
AUV	autonomous underwater vehicle
AWGN	additive white Gaussian noise
BCJR	Bahl, Cocke, Jelinek, Raviv
BEM	basis expansion model
BER	bit error rate
BLUE	best linear unbiased estimator
BPSK	binary phase-shift keying
CCE	critically sampled complex exponential
CCF	channel correlation function
CDMA	code-division multiple access
CE	complex exponential
CE-BEM	complex exponential BEM
CFO	carrier frequency offset
CIR	channel impulse response
CM	constant modulus
CP	cyclic prefix
CP-OFDM	cyclic-prefixed orthogonal frequency division multiplexing
CP-SCM	cyclic-prefixed single-carrier modulation
CRB	Cramér-Rao bound
CS	critically sampled
CSI	channel state information
DAB	digital audio broadcasting
DFE	decision feedback equalization
DFT	discrete Fourier transform
DKL-BEM	discrete Karhunen-Loève BEM
DPS	discrete prolate spheroidal
DPS-BEM	discrete prolate spheroidal BEM

DPSS	discrete prolate spheroidal sequence
DSSS	direct sequence spread spectrum
DVB	digital video broadcasting
EFDOM	equivalent frequency-domain OFDM model
EM	expectation-maximization
EMB	Bayesian EM
EXT	extrinsic probability
FDE	frequency-domain equalization
FDKD	frequency-domain Kronecker delta
FIR	finite impulse response
FM	frequency-modulated
FT	Fourier transform
GLRT	generalized likelihood ratio test
GSM	Global System for Mobile Communications
IAI	inter-antenna interference
IBI	interblock interference
ICI	intercarrier interference
i.i.d.	independent and identically distributed
ISI	intersymbol interference
JPG	jointly proper Gaussian
KF	Kalman filter
LLR	log likelihood ratio
LMMSE	linear minimum mean-squared error
LS	least squares
LSF	local scattering function
LTE	Long Term Evolution
LTI	linear time-invariant
LTV	linear time-varying
LZF	linear zero-forcing
MA	moving average
MAI	multiple-access interference
MAP	maximum a posteriori
MAPSD	maximum a posteriori sequence detection
MBAE	minimum band approximation error
MC	multicarrier
MC-CDMA	multicarrier CDMA
MEP	minimum error probability
MF	matched filter

MIMO	multiple-input multiple-output
ML	maximum likelihood
MLSD	maximum likelihood sequence detection
MMSE	minimum mean-squared error
MSE	mean-squared error
NMMSE	normalized MMSE
NZP	non-zero padding
OFDM	orthogonal frequency division multiplexing
OFDMA	orthogonal frequency division multiple access
OOB	out-of-band
PAM	pulse amplitude modulation
PAPR	peak-to-average power ratio
PAT	pilot-assisted transmission
P-BEM	polynomial BEM
pdf	probability density function
PIC	parallel interference cancelation
PN	pseudo-noise
PS	pulse-shaped
PSD	power spectral density
PSK	phase-shift keying
PSP	per-survivor processing
QAM	quadrature amplitude modulation
QPSK	quadrature phase-shift keying
RF	radio frequency
RHS	right-hand side
RMS	root-mean-square
RPP	regular periodic placement
SCM	single-carrier modulation
SD	sequence detection
SF	spreading function
SFBC	space-frequency block coding
SFC	space-frequency coding
SFM	scattering function matrix
SFN	single-frequency network
SIC	successive interference cancelation
SINR	signal-to-interference-and-noise ratio
SIR	signal-to-interference ratio
SISO	single-input single-output

SNR	signal-to-noise ratio
SOE	sum of exponentials
SOVA	soft-output Viterbi algorithm
STBC	space-time block coding
STC	space-time coding
STFBC	space-time-frequency block coding
STFC	space-time-frequency coding
STFT	short-time Fourier transform
SVD	singular value decomposition
TDKD	time-domain Kronecker delta
TDM	time-division multiplexing
TDMS	time-division multiplexed switching
TF	time-frequency
TFAR	time-frequency AR
TFR	time-frequency representation
TR	transmitted reference
TV	time-varying
US	uncorrelated scattering
UWB	ultra wideband
VSSO	vector-state scalar-observation
WCDMA	wideband CDMA
WH	Weyl-Heisenberg
WLAN	wireless local area network
WSS	wide-sense stationary
WSSUS	wide-sense stationary uncorrelated scattering
ZF	zero-forcing
ZP	zero padding
ZP-SCM	zero-padded single-carrier modulation

CHAPTER

1 Fundamentals of Time-Varying Communication Channels

Gerald Matz, Franz Hlawatsch
Vienna University of Technology, Vienna, Austria

1.1 INTRODUCTION

Wireless communication systems, i.e., systems transmitting information via electromagnetic (radio) or acoustic (sound) waves, have become ubiquitous. In many of these systems, the transmitter or the receiver is mobile. Even if both link ends are static, scatterers – i.e., objects that reflect, scatter, or diffract the propagating waves – may move with significant velocities. These situations give rise to time variations of the wireless channel due to the Doppler effect. Nonideal local oscillators are another source of temporal channel variations, even in the case of wireline channels. Because of their practical relevance, *linear time-varying (LTV) channels* have attracted considerable interest in the fields of signal processing, communications, propagation, information theory, and mathematics. In their most general form, LTV channels are also referred to as *time-frequency (TF) dispersive* or *doubly dispersive*, as well as *TF selective* or *doubly selective*.

In this chapter, we discuss the fundamentals of wireless channels from a signal processing and communications perspective. In contrast to existing textbooks (e.g., Jakes, 1974; Molisch, 2005; Parsons, 1992; Vaughan & Bach Andersen, 2003), our focus will be on LTV channels. Many of the theoretical foundations of LTV channels were laid in the 1950s and 1960s. Zadeh (1950) proposed a "system function" that characterizes an LTV system in a joint TF domain. Driven by increasing interest in ionospheric channels, Kailath complemented Zadeh's work by introducing a dual system function, discussing sampling models, and addressing measurement issues (Kailath, 1959, 1962). A related discussion focusing on the concept of duality (an important notion in TF analysis) was provided by Gersho (1963). In a seminal paper on random LTV channels. Bello (1963) introduced the assumption of *wide-sense stationary uncorrelated scattering* (WSSUS), which has been used almost universally since. The estimation of channel statistics was addressed by Gallager (1964) and a few years later by Gaarder (1968). A fairly comprehensive coverage of the modeling of and communication over random LTV channels was provided by Kennedy (1969). Information-theoretic aspects of LTV channels were addressed in Biglieri, Proakis and Shamai (1998) and in Gallager (1968) (see also Chapter 2).

This chapter provides a review of this early work and a discussion of several more recent results. In Section 1.2, we summarize the most important physical aspects of LTV channels. Some basic tools for a deterministic description of LTV channels are discussed in Section 1.3, while the statistical description of random LTV channels is considered in Section 1.4. Section 1.5 is devoted to the important class of underspread channels and their properties. Parsimonious channel models are reviewed in Section 1.6.

Finally, Section 1.7 discusses the measurement of LTV channels and of their statistics. Throughout this chapter, we will consider noise-free systems since our focus is on the signal distortions caused by LTV wireless channels and not on noise effects. The equivalent complex baseband representation of signals and systems (channels) will be used in most cases.

1.2 THE PHYSICS OF TIME-VARYING CHANNELS

In this section, we briefly describe some of the physical phenomena associated with wireless channels. We shall concentrate on radio channels, although much of our discussion is also relevant to acoustic channels. The term "wireless channel" will be understood as an abstraction of all effects on the transmit signal caused by the transmission. This typically includes the effects of antennas and radio-frequency front ends in addition to the propagation environment affecting the electromagnetic waves.

1.2.1 Wave Propagation

In wireless communications, information is transmitted by radiating a modulated electromagnetic wave at a certain carrier frequency by means of a transmit antenna and picking up energy of the radiated wave by means of a receive antenna. The behavior of the radio waves is determined by the propagation environment according to Maxwell's equations. For most scenarios of interest, solving Maxwell's equations is infeasible (even if the propagation environment is completely known, which rarely happens in practice). This is due to the fact that except for free-space propagation, the wave interacts with dielectric or conducting objects. These interactions are usually classified as reflection, transmission, scattering, and diffraction. We will follow the prevailing terminology and refer to interacting objects simply as "scatterers," without distinguishing between the different types of interaction. While the behavior of radio waves strongly depends on the carrier frequency f_c (or, equivalently, wavelength λ_c), there are a number of common phenomena that lead to a high-level characterization valid for all wireless channels.

1.2.2 Multipath Propagation and Time Dispersion

The presence of multiple scatterers (buildings, vehicles, hills, and so on) causes a transmitted radio wave to propagate along several different paths that terminate at the receiver. Hence, the receive antenna picks up a superposition of multiple attenuated copies of the transmit signal. This phenomenon is referred to as *multipath propagation*. Due to different lengths of the propagation paths, the individual multipath components experience different delays (time shifts). The receiver thus observes a temporally smeared-out version of the transmit signal. Even though the medium itself is not physically dispersive (in the sense that different frequencies propagate with different velocities), such channels are termed *time-dispersive*. The following example considers a simple idealized scenario.

Example 1.1

Consider two propagation paths in a static environment. The receive signal in the equivalent complex baseband domain is given by

$$r(t) = h_1\, s(t-\tau_1) + h_2\, s(t-\tau_2).$$

Here, $h_p = |h_p|e^{j\varphi_p}$ and τ_p are, respectively, the complex attenuation factor and delay associated with the pth path. The magnitude of the Fourier transform $R(f) \triangleq \int_{-\infty}^{\infty} r(t)\,e^{-j2\pi ft}dt$ of the receive signal follows as

$$|R(f)| = |S(f)|\sqrt{|h_1|^2 + |h_2|^2 + 2|h_1||h_2|\cos(2\pi(\tau_1 - \tau_2)f - (\varphi_1 - \varphi_2))}.$$

■

As can be seen from Example 1.1, a time-dispersive channel has a multiplicative effect on the transmit signal in the frequency domain (this, of course, is a basic equivalence in Fourier analysis). Therefore, time-dispersive channels are *frequency-selective* in the sense that different frequencies are attenuated differently; see Fig. 1.1 for illustration. These differences in attenuation become more severe when the difference of the path delays is large and the difference between the path attenuations is small.

Multipath propagation is not the only source of time dispersion. Further potential sources are transmitter and receiver imperfections, such as transmit/receive pulses not satisfying the Nyquist criterion, imperfect timing recovery, or sampling jitter. In the following example, we consider an equivalent discrete-time baseband representation that includes pulse amplitude modulation (PAM), analog-to-digital and digital-to-analog conversion, and demodulation.

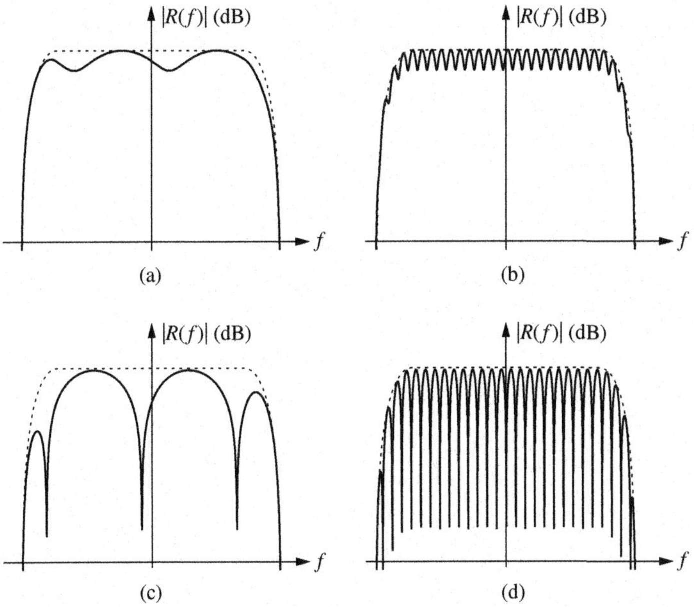

FIGURE 1.1

Illustration of frequency selectivity in a receive spectrum (thick solid line) for a two-path channel and a raised-cosine transmit spectrum (thin dotted line): (a) small $|\tau_1 - \tau_2|$ and $|h_1| \gg |h_2|$, (b) large $|\tau_1 - \tau_2|$ and $|h_1| \gg |h_2|$, (c) small $|\tau_1 - \tau_2|$ and $|h_1| \approx |h_2|$, and (d) large $|\tau_1 - \tau_2|$ and $|h_1| \approx |h_2|$.

Example 1.2

Consider a single propagation path with complex attenuation factor h and delay 0 and a digital PAM system with symbol period T and transmit and receive pulses whose convolution yields a Nyquist pulse $p(t)$. Assuming a timing error ΔT, the received sequence in the equivalent discrete-time (symbol-rate) baseband domain equals

$$r[k] = \sum_{l=-\infty}^{\infty} h_l a[k-l], \quad \text{with} \quad h_k = hp(kT - \Delta T),$$

where $a[k]$ denotes the sequence of transmit symbols. Note that in spite of a single propagation path, there is significant temporal dispersion unless $\Delta T = 0$ (in which case $h_k = 0$ for $k \neq 0$).

Although multipath propagation has traditionally been viewed as a transmission impairment, nowadays there is a tendency to consider it as beneficial since it provides additional degrees of freedom that are known as *delay diversity* or *frequency diversity* and that can be exploited to realize diversity gains or, in the context of multiantenna systems, even multiplexing gains (Tse & Viswanath, 2005).

1.2.3 Doppler Effect and Frequency Dispersion

In many wireless systems, the transmitter, receiver, and/or scatterers are moving. In such situations, the emitted wave is subject to the Doppler effect and hence experiences frequency shifts. We first restrict our discussion to a simple scenario with a static transmitter, no scatterers, and a receiver moving with velocity υ. In this case, a purely sinusoidal carrier wave of frequency f_c is observed by the receiver as a sinusoidal wave of frequency

$$\left(1 - \frac{\upsilon\cos(\phi)}{c_0 + \upsilon\cos(\phi)}\right) f_c \approx \left(1 - \frac{\upsilon\cos(\phi)}{c_0}\right) f_c, \tag{1.1}$$

where ϕ is the angle of arrival of the wave relative to the direction of motion of the receiver and c_0 is the speed of light. The above approximation on the right-hand side of (1.1) holds for the practically predominant case $\upsilon \ll c_0$. For a general transmit signal $s(t)$ with Fourier transform $S(f)$, one can then show the following expressions of the receive signal (h is the complex attenuation factor):

$$R(f) = hS(\alpha f), \quad r(t) = \frac{h}{\alpha} s\left(\frac{t}{\alpha}\right), \quad \text{with} \quad \alpha = 1 - \frac{\upsilon\cos(\phi)}{c_0}. \tag{1.2}$$

This shows that the Doppler effect results in a temporal/spectral scaling (i.e., compression or dilation).

In many practical cases, the transmit signal is effectively band-limited around the carrier frequency f_c, i.e., $S(f)$ is effectively zero outside a band $[f_c - B/2, f_c + B/2]$, where $B \ll f_c$. The approximation $\alpha f = f - \frac{\upsilon\cos(\phi)}{c_0} f \approx f - \frac{\upsilon\cos(\phi)}{c_0} f_c$ (whose accuracy increases with decreasing normalized bandwidth B/f_c) then implies

$$R(f) \approx hS(f - \nu), \quad r(t) \approx hs(t)e^{j2\pi\nu t}, \quad \text{with} \quad \nu = \frac{\upsilon\cos(\phi)}{c_0} f_c. \tag{1.3}$$

Here, the Doppler effect essentially results in a frequency shift, with the Doppler shift frequency ν being proportional to both the velocity υ and the carrier frequency f_c. The relations (1.2) and (1.3) are often referred to as wideband and narrowband Doppler effect, respectively, even though the "narrowband" approximation (1.3) holds true also for systems usually considered wideband by communication engineers (e.g., for a WLAN at carrier frequency $f_c = 2.4$ GHz and with bandwidth $B = 20$ MHz, there is $B/f_c = 8.3 \cdot 10^{-3}$).

In the general case of multipath propagation and moving transmitter, receiver, and/or scatterers, the received multipath components (echoes) experience different Doppler shifts since the angles of arrival/departure and the relative velocities associated with the individual multipath components are typically different. Hence, the transmit signal is spread out in the frequency domain – it experiences *frequency dispersion.*

Example 1.3

Consider two propagation paths with equal delay τ_0 but different Doppler frequencies ν_1 and ν_2. Here, the Fourier transform of the receive signal is obtained as

$$R(f) = \left[h_1 S(f-\nu_1) + h_2 S(f-\nu_2)\right] e^{-j2\pi\tau_0 f}, \tag{1.4}$$

where $h_p = |h_p| e^{j\varphi_p}$ denotes the complex attenuation factor of the pth path. The magnitude of the receive signal follows as

$$|r(t)| = |s(t-\tau_0)| \sqrt{|h_1|^2 + |h_2|^2 + 2|h_1||h_2| \cos(2\pi(\nu_1-\nu_2)(t-\tau_0) + (\varphi_1 - \varphi_2))}. \tag{1.5}$$

While (1.4) illustrates the frequency dispersion, (1.5) shows that the Doppler effect leads to time-varying multiplicative modifications of the transmit signal in the time domain. Thus, channels involving Doppler shifts are also referred to as being *time-selective.* With the replacements $R(f) \to r(t)$, $f \to t$, $\tau_1 \to \nu_1$, and $\tau_2 \to \nu_2$, Fig. 1.1 can also be viewed as an illustration of time selectivity. Depending on the system architecture, time selectivity may be viewed as a transmission impairment or as a beneficial effect offering *Doppler diversity* (also termed *time diversity*).

Apart from the Doppler effect due to mobility, imperfect local oscillators are another cause of frequency dispersion because they result in carrier frequency offsets (i.e., different and possibly time-varying carrier frequencies at the transmitter and receiver), oscillator drift, and phase noise.

Example 1.4

Consider a static scenario with line-of-sight propagation without multipath, i.e., the transmit and receive signals in the (complex) bandpass domain are related as $r(t) = hs(t)$. The transmitter uses a perfect local oscillator, i.e., $s(t) = s_B(t) e^{j2\pi f_c t}$ ($s_B(t)$ denotes the baseband signal). The local oscillator at the receiver is characterized by $m(t) = p(t) e^{-j2\pi(f_c+\Delta f)t}$, where Δf is a carrier frequency offset and $p(t)$ models phase noise effects that broaden the oscillator's line spectrum. In this case, the received baseband signal $r_B(t) = r(t)m(t)$ and its Fourier transform are given by

$$r_B(t) = h s_B(t) p(t) e^{-j2\pi \Delta f t}, \qquad R_B(f) = h \int_{-\infty}^{\infty} P(\nu - \Delta f) S_B(f-\nu)\, d\nu.$$

Clearly, in spite of the static scenario (no Doppler effect), the transmit signal experiences temporal selectivity and frequency dispersion.

■

1.2.4 Path Loss and Fading

Wireless channels are characterized by severe fluctuations in the receive power, i.e., in the strength of the electromagnetic field at the receiver position. The receive power is usually modeled as a combination of three phenomena: path loss, large-scale fading, and small-scale fading.

The *path loss* describes the distance-dependent power decay of electromagnetic waves. Let us model the attenuation factor as $d^{-\beta}$, where d is the distance the wave has traveled and β denotes the path loss exponent, which is typically assumed to lie between 2 and 4. The path loss in decibels is then obtained as $P_{\mathrm{L}} = 10\beta \log_{10}(d)$.

Two receivers located at the same distance d from the transmitter may still experience significantly different receive powers if the radio waves have propagated through different environments. In particular, obstacles like buildings or dense vegetation can block or attenuate propagation paths and result in *shadowing* and *absorption loss*, respectively. This type of fading is referred to as *large-scale fading* since its effect on the receive power is constant within geographic regions whose dimensions are on the order of $10\lambda_{\mathrm{c}} \cdots 100\lambda_{\mathrm{c}}$, i.e., large relative to the wavelength λ_{c}. Experimental evidence indicates that for many systems, large-scale fading can be accurately modeled as a random variable with log-normal distribution (Molisch, 2005).

Finally, constructive and destructive interference of field components corresponding to different propagation paths causes receive power fluctuations within "small" regions whose dimensions are on the order of a few wavelengths. This *small-scale fading* can vary over several decades and is usually modeled stochastically by channel coefficients with Gaussian distribution. The magnitude of the channel coefficients is then Rayleigh distributed for zero-mean channel coefficients and Rice distributed for nonzero-mean channel coefficients. Rician fading is often assumed when there exists a line of sight.

Path loss and large-scale fading change only gradually and are relevant to the link budget and average receive signal-to-noise ratio (SNR); they are often combated by a feedback loop performing power control at the transmitter. In contrast, small-scale fading causes the receive power to fluctuate so rapidly that adjusting the transmit power is infeasible. Hence, small-scale fading has a direct impact on system performance (capacity, error probability, and so on). The main approach to mitigating small-scale fading is the use of diversity techniques (in time, frequency, or space).

1.2.5 Spatial Characteristics

In a multipath scenario, the *angle of departure* (AoD) of a propagation path indicates the direction in which the planar wave corresponding to that path departs from the transmitter. Similarly, the *angle of arrival* (AoA) indicates from which direction the wave arrives at the receiver. AoA and AoD are spatial channel characteristics that can be measured using an antenna array at the respective link end. The angular resolution of an antenna array is determined by the number of individual antennas, their arrangement, and their distance. The transformation between the array signal vector and the angular domain is based on the array steering vector. In the case of a uniform linear array, this transformation is a discrete Fourier transform (Tse & Viswanath, 2005).

1.3 DETERMINISTIC DESCRIPTION

We next discuss some basic deterministic[1] characterizations of LTV channels. We consider a wireless system operating at carrier frequency f_c. We will generally describe this system in the equivalent complex baseband domain for simplicity (an exception being the wideband system considered in Section 1.3.2). The LTV channel will be viewed and denoted as a linear operator (Naylor & Sell, 1982) $\mathbf{H}$ that acts on the transmit signal $s(t)$ and yields the receive signal $r(t) = (\mathbf{H}s)(t)$.

1.3.1 Delay-Doppler Domain – Spreading Function

As mentioned before, the physical effects underlying LTV channels are mainly multipath propagation and the Doppler effect. Hence, a physically meaningful and intuitive characterization of LTV channels is in terms of time delays and Doppler frequency shifts. Let us first assume an LTV channel $\mathbf{H}$ with P discrete propagation paths. The receive signal $r(t) = (\mathbf{H}s)(t)$ is here given by

$$r(t) = \sum_{p=1}^{P} h_p\, s(t-\tau_p)\, e^{j2\pi \nu_p t}, \tag{1.6}$$

where h_p, τ_p, and ν_p denote, respectively, the complex attenuation factor, time delay, and Doppler frequency associated with the pth path. Equation (1.6) models the effect of P discrete specular scatterers (ideal point scatterers). This expression can be generalized to a continuum of scatterers as (Bello, 1963; Molisch, 2005; Proakis, 1995)

$$r(t) = \int_{-\infty}^{\infty}\int_{-\infty}^{\infty} S_{\mathbf{H}}(\tau,\nu)\, s(t-\tau)\, e^{j2\pi\nu t}\, d\tau\, d\nu. \tag{1.7}$$

The weight function $S_{\mathbf{H}}(\tau,\nu)$ is termed the *(delay-Doppler) spreading function* of the LTV channel $\mathbf{H}$ since it describes the spreading of the transmit signal in time and frequency. The value of the spreading function $S_{\mathbf{H}}(\tau,\nu)$ at a given delay-Doppler point (τ,ν) characterizes the overall complex attenuation and scatterer reflectivity associated with all paths of delay τ and Doppler ν, and it describes how the delayed and Doppler-shifted version $s(t-\tau)\,e^{j2\pi\nu t}$ of the transmit signal $s(t)$ contributes to the receive signal $r(t)$. Thus, the spreading function expresses the channel's *TF dispersion characteristics*. As such, it is a generalization of the impulse response of time-invariant systems, which describes the time dispersion. An example is shown in Fig. 1.2. Note that (1.6) is reobtained as a special case of (1.7) for

$$S_{\mathbf{H}}(\tau,\nu) = \sum_{p=1}^{P} h_p\, \delta(\tau-\tau_p)\, \delta(\nu-\nu_p). \tag{1.8}$$

[1]We call these characterizations "deterministic" because they do not assume a stochastic model of the channel; however, for a random channel, they are themselves random, i.e., nondeterministic – see Section 1.4.

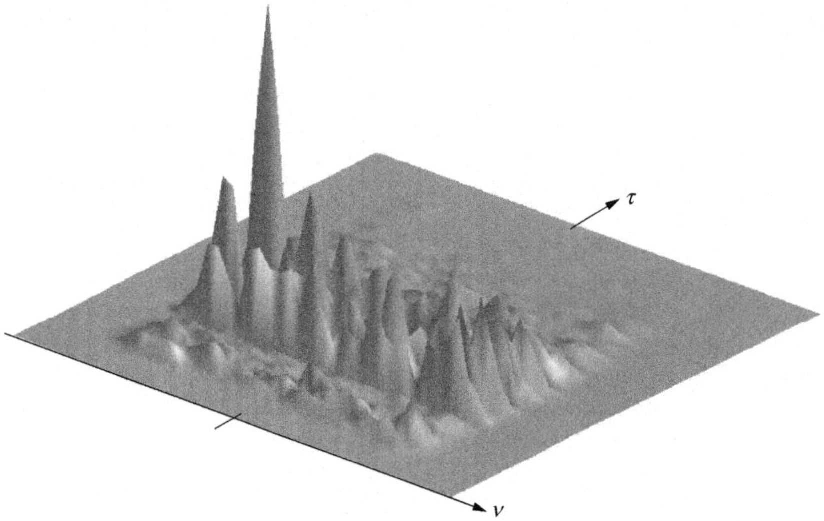

FIGURE 1.2

Example of a spreading function (magnitude).

A dual representation of the channel's TF dispersion in the frequency domain, again in terms of the spreading function $S_{\mathbf{H}}(\tau,\nu)$, is

$$R(f) = \int_{-\infty}^{\infty}\int_{-\infty}^{\infty} S_{\mathbf{H}}(\tau,\nu)\, S(f-\nu)\, \mathrm{e}^{-j2\pi\tau(f-\nu)}\, d\tau\, d\nu.$$

We can also obtain a representation of the TF dispersion in a joint TF domain. We will use the *short-time Fourier transform* (STFT) of a signal $x(t)$, which is a linear TF signal representation defined as $X^{(g)}(t,f) \triangleq \int_{-\infty}^{\infty} x(t')\, g^*(t'-t)\, \mathrm{e}^{-j2\pi f t'}\, dt'$, where $g(t)$ denotes a normalized analysis window (Flandrin, 1999; Hlawatsch & Boudreaux-Bartels, 1992; Nawab & Quatieri, 1988). The STFT of the receive signal in (1.7) can be expressed as

$$R^{(g)}(t,f) = \int_{-\infty}^{\infty}\int_{-\infty}^{\infty} S_{\mathbf{H}}(\tau,\nu)\, S^{(g)}(t-\tau, f-\nu)\, \mathrm{e}^{-j2\pi\tau(f-\nu)}\, d\tau\, d\nu. \tag{1.9}$$

Apart from the phase factor, this is the two-dimensional (2D) convolution of the STFT of the transmit signal with the spreading function of the channel. This again demonstrates that the spreading function describes the channel's TF dispersion.

Example 1.5

For a two-path channel with delays τ_1, τ_2 and Doppler frequencies ν_1, ν_2, the spreading function is given by

$$S_{\mathbf{H}}(\tau,\nu) = h_1\, \delta(\tau-\tau_1)\, \delta(\nu-\nu_1) + h_2\, \delta(\tau-\tau_2)\, \delta(\nu-\nu_2).$$

Inserting this expression into (1.7) yields

$$r(t) = h_1\, s(t-\tau_1)\, e^{j2\pi\nu_1 t} + h_2\, s(t-\tau_2)\, e^{j2\pi\nu_2 t}.$$

We note that Examples 1.1 and 1.3 are essentially reobtained as special cases with $\nu_1 = \nu_2 = 0$ and $\tau_1 = \tau_2 = \tau_0$, respectively.

■

Writing (1.7) as $r(t) = \int_{-\infty}^{\infty} \big[\int_{-\infty}^{\infty} S_{\mathbf{H}}(\tau,\nu)\, s(t-\tau)\, d\tau\big] e^{j2\pi\nu t} d\nu = \int_{-\infty}^{\infty} r_\nu(t)\, d\nu$, we see that the LTV channel $\mathbf{H}$ can be viewed as a continuous (infinitesimal) parallel connection of systems parameterized by the Doppler frequency ν. The output signals of these systems are given by

$$r_\nu(t) = \tilde{r}_\nu(t)\, e^{j2\pi\nu t} \quad \text{with} \quad \tilde{r}_\nu(t) = (S_{\mathbf{H}}(\cdot,\nu) * s)(t) = \int_{-\infty}^{\infty} S_{\mathbf{H}}(\tau,\nu)\, s(t-\tau)\, d\tau.$$

Thus, each system consists of a time-invariant filter with impulse response $S_{\mathbf{H}}(\tau,\nu)$, followed by a modulator (mixer) with frequency ν.

For a time-invariant channel with impulse response $h(\tau)$, the spreading function equals $S_{\mathbf{H}}(\tau,\nu) = h(\tau)\delta(\nu)$ so that (1.7) reduces to the convolution of $s(t)$ with $h(\tau)$. This correctly indicates the absence of Doppler shifts (frequency dispersion). In the dual case of a channel without frequency selectivity, i.e., $r(t) = \tilde{h}(t)\, s(t)$, there is $S_{\mathbf{H}}(\tau,\nu) = \widetilde{H}(\nu)\delta(\tau)$ with $\widetilde{H}(\nu) = \int_{-\infty}^{\infty} \tilde{h}(t)\, e^{-j2\pi\nu t} dt$, which correctly indicates the absence of time dispersion.

1.3.2 Delay-Scale Domain – Delay-Scale Spreading Function

Whereas for narrowband systems ($B/f_c \ll 1$), the Doppler effect can be represented as a frequency shift, it must be characterized by a time-scaling (compression/dilation) in the case of *(ultra)wideband* systems. Relation (1.6) is here replaced by (Molisch, 2005)

$$r(t) = \sum_{p=1}^{P} a_p \frac{1}{\sqrt{\alpha_p}}\, s\!\left(\frac{t-\tau_p}{\alpha_p}\right), \quad \text{with} \quad \alpha_p = 1 - \frac{v\cos(\phi_p)}{c_0}.$$

Generalizing to a continuum of scatterers, we obtain

$$r(t) = \int_{-\infty}^{\infty} \int_{0}^{\infty} F_{\mathbf{H}}(\tau,\alpha) \frac{1}{\sqrt{\alpha}}\, s\!\left(\frac{t-\tau}{\alpha}\right) d\tau\, d\alpha. \tag{1.10}$$

Here, $F_{\mathbf{H}}(\tau,\alpha)$ denotes the *delay-scale spreading function* of the LTV channel $\mathbf{H}$ (Margetts, Schniter, & Swami, 2007; Ye & Papandreou-Suppappola, 2003). Like (1.7), expression (1.10) can represent any LTV channel, but it is most efficient (parsimonious) for wideband channels. The delay-scale description of LTV channels will be discussed in more detail in Chapter 9.

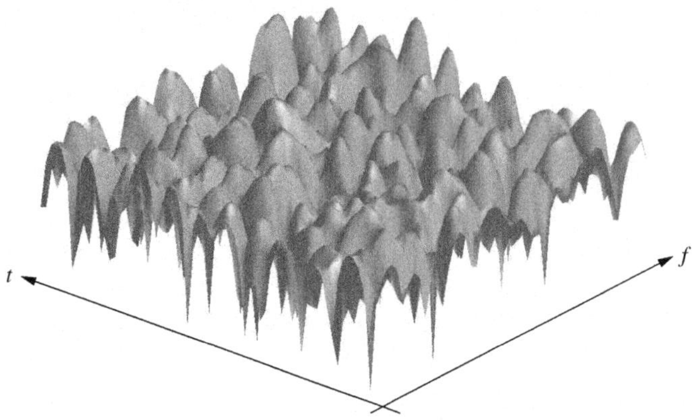

FIGURE 1.3

Example of a TF transfer function (magnitude, in decibel).

1.3.3 Time-Frequency Domain – Time-Varying Transfer Function

As explained in Section 1.2, time-dispersiveness corresponds to frequency selectivity, and frequency-dispersiveness corresponds to time selectivity. The joint TF selectivity of an LTV channel is characterized by the *TF (or time-varying) transfer function* (Bello, 1963; Zadeh, 1950)

$$L_{\mathbf{H}}(t,f) \triangleq \int_{-\infty}^{\infty}\int_{-\infty}^{\infty} S_{\mathbf{H}}(\tau,\nu)\, e^{j2\pi(t\nu - f\tau)}\, d\tau\, d\nu. \tag{1.11}$$

This 2D Fourier transform relation between shift (dispersion) domain and weight (selectivity) domain extends the 1D Fourier transform relation $H(f) = \int_{-\infty}^{\infty} h(\tau)e^{-j2\pi f\tau}\,d\tau$ of time-invariant channels to the time-varying case. According to (1.11), the TF echoes described by $S_{\mathbf{H}}(\tau,\nu)$ correspond to TF fluctuations of $L_{\mathbf{H}}(t,f)$, which are a TF description of small-scale fading. For *underspread* channels (to be defined in Section 1.5), the TF transfer function $L_{\mathbf{H}}(t,f)$ can be interpreted as the channel's complex attenuation factor at time t and frequency f, and it inherits many properties of the transfer function (frequency response) defined in the time-invariant case. An example is shown in Fig. 1.3.

Example 1.6

We reconsider the two-path channel from Example 1.5. Using (1.11), it can be shown that the squared magnitude of the channel's TF transfer function is given by

$$|L_{\mathbf{H}}(t,f)|^2 = |h_1|^2 + |h_2|^2 + 2|h_1||h_2|\cos(2\pi[t(\nu_1-\nu_2) - f(\tau_1-\tau_2)] + \varphi_1 - \varphi_2).$$

Clearly, this channel is TF selective in the sense that $L_{\mathbf{H}}(t,f)$ fluctuates with time and frequency. The rapidity of fluctuation with time is proportional to the "Doppler spread" $|\nu_1-\nu_2|$, whereas the rapidity of fluctuation with frequency is proportional to the "delay spread" $|\tau_1-\tau_2|$.

Inserting (1.11) into (1.7) and developing the integrals with respect to τ and ν leads to the channel input–output relation

$$r(t) = \int_{-\infty}^{\infty} L_{\mathbf{H}}(t,f)\, S(f)\, e^{j2\pi tf}\, df. \tag{1.12}$$

In spite of its apparent similarity to the relation $r(t) = \int_{-\infty}^{\infty} H(f)\, S(f)\, e^{j2\pi ft}\, df$ valid for time-invariant channels, (1.12) has to be interpreted with care. Specifically, (1.12) is not a simple inverse Fourier transform since $L_{\mathbf{H}}(t,f)\, S(f)$ also depends on t.

For the special case of a time-invariant channel (no frequency dispersion), the TF transfer function reduces to the frequency response, i.e., $L_{\mathbf{H}}(t,f) = H(f)$, and (1.12) corresponds to $R(f) = H(f)\, S(f)$. This correctly reflects the channel's pure frequency selectivity. In the dual case of a channel without time dispersion, the TF transfer function simplifies according to $L_{\mathbf{H}}(t,f) = \tilde{h}(t)$, and (1.12) thus reduces to the relation $r(t) = \tilde{h}(t)\, s(t)$, which describes the channel's pure time selectivity.

1.3.4 Time-Delay Domain – Time-Varying Impulse Response

While the spreading function was motivated by a specific physical model (multipath propagation, Doppler effect), it actually applies to any LTV system. To see this, we develop (1.7) as

$$r(t) = \int_{-\infty}^{\infty} \underbrace{\left[\int_{-\infty}^{\infty} S_{\mathbf{H}}(\tau,\nu)\, e^{j2\pi t\nu}\, d\nu \right]}_{h(t,\tau)} s(t-\tau)\, d\tau = \int_{-\infty}^{\infty} h(t,\tau)\, s(t-\tau)\, d\tau, \tag{1.13}$$

where $h(t,\tau) = \int_{-\infty}^{\infty} S_{\mathbf{H}}(\tau,\nu)\, e^{j2\pi t\nu}\, d\nu$ is the *(time-varying) impulse response* of the LTV channel $\mathbf{H}$. An example is depicted in Fig. 1.4. Defining the *kernel* of $\mathbf{H}$ as $k_{\mathbf{H}}(t,t') \triangleq h(t,t-t')$, (1.13) can be

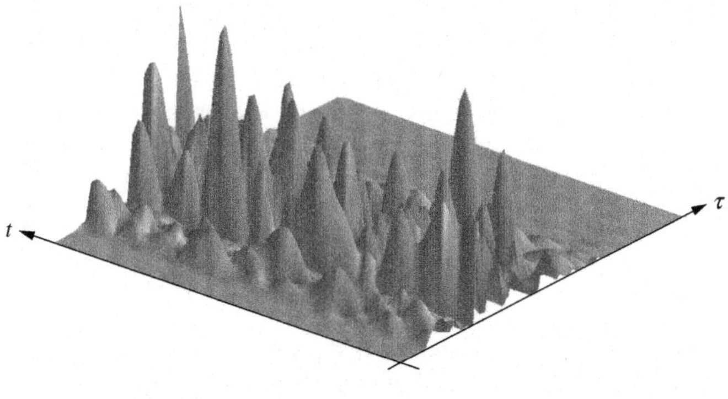

FIGURE 1.4

Example of a time-varying impulse response (magnitude).

rewritten as

$$r(t) = \int_{-\infty}^{\infty} k_{\mathbf{H}}(t, t')\, s(t')\, dt', \tag{1.14}$$

which is the integral representation of a linear operator (Naylor & Sell, 1982). This shows that the input–output relation (1.7) is completely general, i.e., any LTV system (channel) can be characterized in terms of its spreading function. The spreading function and TF transfer function can be written in terms of the impulse response $h(t, \tau)$ as

$$S_{\mathbf{H}}(\tau, \nu) = \int_{-\infty}^{\infty} h(t, \tau)\, e^{-j2\pi \nu t}\, dt, \tag{1.15}$$

$$L_{\mathbf{H}}(t, f) = \int_{-\infty}^{\infty} h(t, \tau)\, e^{-j2\pi f \tau}\, d\tau. \tag{1.16}$$

From (1.13) and (1.14), it follows that for a transmit signal that is an impulse, i.e., $s(t) = \delta(t - t_0)$, the receive signal equals $r(t) = k(t, t_0) = h(t, t - t_0)$. The impulse response can also be interpreted in terms of a "continuous tapped delay line": for a fixed tap (delay) τ, $h(t, \tau)$ as a function of t describes the time-varying tap weight function that multiplies the delayed transmit signal $s(t - \tau)$. In the special case of a time-invariant channel, $h(t, \tau)$ simplifies to a function of τ only, and for a frequency-nonselective channel, it simplifies as $h(t, \tau) = \tilde{h}(t)\, \delta(\tau)$ with some $\tilde{h}(t)$.

Example 1.7

A popular model of an LTV channel with specular scattering is specified in terms of the impulse response as

$$h(t, \tau) = \sum_{p=1}^{P} h_p\, e^{j2\pi \nu_p t}\, \delta(\tau - \tau_p(t)). \tag{1.17}$$

Apart from the time dependence of the delays $\tau_p(t)$, this is just the inverse Fourier transform of (1.8) with respect to ν. The time-varying delays $\tau_p(t)$ account for a "delay drift" that is due to changing path lengths caused by the movement of the transmitter, receiver, and/or scatterers. However, these changes are much slower than the phase fluctuations resulting from small-scale fading (which are described by the exponential functions in (1.17)).

1.3.5 Extension to Multiantenna Systems

Consider a *multiple-input multiple-output* (MIMO) wireless system with M_{T} transmit antennas and M_{R} receive antennas (Bölcskei, Gesbert, Papadias, & van der Veen, 2006; Paulraj, Nabar & Gore, 2003). The signals emitted by the jth transmit antenna and captured by the ith receive antenna will be denoted by $s_j(t)$ and $r_i(t)$, respectively. Each receive signal $r_i(t)$ is a superposition of distorted versions of all transmit signals, i.e.,

$$r_i(t) = \sum_{j=1}^{M_T} (\mathsf{H}_{ij} s_j)(t), \quad i = 1, \ldots, M_R, \tag{1.18}$$

where H_{ij} denotes the LTV channel between transmit antenna j and receive antenna i. (For a survey of MIMO channel modeling aspects, see Almers et al., 2007 and references therein.) Defining the length-M_T transmit signal vector $\mathbf{s}(t) = (s_1(t) \cdots s_{M_T}(t))^T$ and the length-M_R receive vector $\mathbf{r}(t) = (r_1(t) \cdots r_{M_R}(t))^T$, all input–output relations (1.18) can be combined as

$$\mathbf{r}(t) = (\underline{\mathsf{H}}\mathbf{s})(t), \quad \text{with} \quad \underline{\mathsf{H}} \triangleq \begin{pmatrix} \mathsf{H}_{11} & \cdots & \mathsf{H}_{1M_T} \\ \vdots & \ddots & \vdots \\ \mathsf{H}_{M_R 1} & \cdots & \mathsf{H}_{M_R M_T} \end{pmatrix}. \tag{1.19}$$

The delay-Doppler, TF, and time-delay characterizations of single-antenna channels can be easily generalized to the MIMO case. Let us define the $M_R \times M_T$ matrices $\mathbf{S_H}(\tau, \nu)$, $\mathbf{L_H}(t, f)$, and $\mathbf{H}(t, \tau)$ whose (i,j)th elements equal the delay-Doppler spreading function, TF transfer function, and impulse response of H_{ij}, respectively. Then,

$$\begin{aligned} \mathbf{r}(t) &= \int_{-\infty}^{\infty}\int_{-\infty}^{\infty} \mathbf{S_H}(\tau, \nu)\, \mathbf{s}(t-\tau)\, e^{j2\pi \nu t}\, d\tau\, d\nu \\ &= \int_{-\infty}^{\infty} \mathbf{H}(t, \tau)\, \mathbf{s}(t-\tau)\, d\tau, \end{aligned} \tag{1.20}$$

with

$$\mathbf{S_H}(\tau, \nu) = \int_{-\infty}^{\infty} \mathbf{H}(t, \tau)\, e^{-j2\pi \nu t}\, dt. \tag{1.21}$$

Furthermore,

$$\begin{aligned} \mathbf{L_H}(t, f) &= \int_{-\infty}^{\infty}\int_{-\infty}^{\infty} \mathbf{S_H}(\tau, \nu)\, e^{j2\pi(t\nu - f\tau)}\, d\tau\, d\nu \\ &= \int_{-\infty}^{\infty} \mathbf{H}(t, \tau)\, e^{-j2\pi f\tau}\, d\tau. \end{aligned} \tag{1.22}$$

The main difference between these relations and their single-antenna counterparts is the *spatial resolution* offered by the aperture of the antenna arrays. The transmit array and receive array make it possible to resolve to a certain extent the AoD and AoA, respectively, of the individual paths. This spatial or directional resolution is essentially determined by the array steering vectors that define a transformation to the angular domain (Sayeed, 2002; Tse & Viswanath, 2005).

Example 1.8

Assuming uniform linear arrays at the transmitter and the receiver with antenna separation $\Delta_T\lambda_c$ and $\Delta_R\lambda_c$, respectively, the array steering vectors for given AoD ϕ and AoA ψ are (Molisch, 2005)

$$\mathbf{a}_T(\phi) = \frac{1}{\sqrt{M_T}} \begin{pmatrix} 1 \\ e^{-j2\pi\Delta_T\cos(\phi)} \\ \vdots \\ e^{-j2\pi(M_T-1)\Delta_T\cos(\phi)} \end{pmatrix}, \quad \mathbf{a}_R(\psi) = \frac{1}{\sqrt{M_R}} \begin{pmatrix} 1 \\ e^{-j2\pi\Delta_R\cos(\psi)} \\ \vdots \\ e^{-j2\pi(M_R-1)\Delta_R\cos(\psi)} \end{pmatrix}.$$

Further assuming purely specular scattering with P paths, where each path has its distinct AoD ϕ_p and AoA ψ_p, the matrix-valued spreading function is given by (cf. (1.8))

$$\mathbf{S_H}(\tau,\nu) = \sum_{p=1}^{P} \mathbf{H}_p\,\delta(\tau-\tau_p)\,\delta(\nu-\nu_p), \quad \text{with} \quad \mathbf{H}_p = h_p\,\mathbf{a}_R(\psi_p)\,\mathbf{a}_T^T(\phi_p).$$

Note that here the MIMO matrices $\mathbf{H}_p = h_p\,\mathbf{a}_R(\psi_p)\,\mathbf{a}_T^T(\phi_p)$ describing the individual paths in the delay-Doppler domain (i.e., determining $\mathbf{S_H}(\tau,\nu)$ at the corresponding delay-Doppler points (τ_p,ν_p)) all have rank equal to one. The TF transfer function is obtained as

$$\mathbf{L_H}(t,f) = \sum_{p=1}^{P} \mathbf{H}_p\,e^{j2\pi(t\nu_p - f\tau_p)};$$

it involves a superposition of all matrices $\mathbf{H}_p$ at each TF point (t,f) and hence in general will have full rank everywhere, provided that more than $\min\{M_T, M_R\}$ paths have sufficiently distinct AoA/AoD (rich scattering). We note that because of the finite aperture of the antenna arrays ($M_T\Delta_T$ and $M_R\Delta_R$), no more than, respectively, M_T and M_R orthogonal directions can be effectively resolved in the angular domain (Sayeed, 2002; Tse & Viswanath, 2005).

The spatial/angular dispersion of MIMO channels – i.e., the mixing of the signals emitted from all transmit antennas – can be viewed as an inconvenience necessitating spatial equalization. However, spatial dispersion actually provides additional degrees of freedom that can be exploited to realize *spatial diversity*. This diversity is analogous to the delay diversity due to time dispersion and the Doppler diversity due to frequency dispersion.

1.4 STOCHASTIC DESCRIPTION

A complete deterministic characterization of LTV channels (e.g., based on Maxwell's equations) is infeasible in virtually all scenarios of practical relevance. Even if such a characterization were possible, it would only apply to a specific environment, whereas wireless systems need to be designed for a wide variety of operating conditions. This motivates stochastic characterizations, which consider an LTV channel as a random quantity whose statistics describe common properties of an underlying *ensemble* of wireless channels.

We will restrict our discussion to the common case of Rayleigh fading, where the channel's system functions $S_{\mathbf{H}}(\tau,\nu)$, $L_{\mathbf{H}}(t,f)$, and $h(t,\tau)$ are 2D complex Gaussian random processes with zero mean.

For Rayleigh fading, the stochastic characterization of a channel reduces to the specification of its second-order statistics.

1.4.1 WSSUS Channels

1.4.1.1 The WSS, US, and WSSUS Properties

The second-order statistics of the 2D system functions of an LTV channel (spreading function, TF transfer function, and impulse response) generally depend on four variables. In his seminal paper, Bello (1963) provided a simplified description in terms of only two variables by introducing the assumption of *wide-sense stationary uncorrelated scattering* (WSSUS). The WSSUS property is also discussed, e.g., in Matz and Hlawatsch (2003); Molisch (2005); Proakis (1995).

A random LTV channel is said to feature *uncorrelated scattering* (US) if different channel taps (delay coefficients) are uncorrelated (Bello, 1963), i.e.,

$$\mathsf{E}\{h(t,\tau)\,h^*(t',\tau')\} = r_h'(t,t';\tau)\,\delta(\tau-\tau'),$$

with some correlation function $r_h'(t,t';\tau)$. Note that different taps can be interpreted as belonging to different scatterers. Furthermore, a channel is said to be *wide-sense stationary* (WSS) if the channel taps are jointly wide-sense stationary with respect to the time variable t (Bello, 1963), i.e.,

$$\mathsf{E}\{h(t,\tau)\,h^*(t',\tau')\} = \tilde{r}_h(t-t';\tau,\tau'),$$

with some correlation function $\tilde{r}_h(\Delta t;\tau,\tau')$. Combining the WSS and US properties, we obtain the WSSUS property

$$\mathsf{E}\{h(t,\tau)\,h^*(t',\tau')\} = r_h(t-t';\tau)\,\delta(\tau-\tau'). \tag{1.23}$$

This shows that the second-order statistics of a WSSUS channel are fully described by the 2D function $r_h(\Delta t;\tau)$, which is a correlation function in the time-difference variable Δt.

1.4.1.2 Scattering Function and TF Correlation Function

US channels have uncorrelated delay coefficients, and it can be shown that for WSS channels, different Doppler frequency coefficients are uncorrelated. Taken together, this implies that the spreading function $S_{\mathbf{H}}(\tau,\nu)$ of a WSSUS channel is a 2D *white* (but nonstationary) process, i.e.,

$$\mathsf{E}\big\{S_{\mathbf{H}}(\tau,\nu)\,S_{\mathbf{H}}^*(\tau',\nu')\big\} = C_{\mathbf{H}}(\tau,\nu)\,\delta(\tau-\tau')\,\delta(\nu-\nu'). \tag{1.24}$$

The rationale here is that each delay-Doppler pair (τ,ν) corresponds to a scatterer with reflectivity $S_{\mathbf{H}}(\tau,\nu)$, and the reflectivities of any two distinct scatterers (i.e., scatterers with different delay τ or different Doppler ν) are uncorrelated. The mean intensity of the 2D white spreading function process, $C_{\mathbf{H}}(\tau,\nu) \geq 0$, is known as the channel's *scattering function* (Bello, 1963; Matz & Hlawatsch, 2003; Molisch, 2005). The scattering function characterizes the average strength of scatterers with delay τ and Doppler frequency ν, and thus, it provides a statistical characterization of the TF dispersion produced by a WSSUS channel. A *wideband scattering function* based on the delay-scale spreading function $F_{\mathbf{H}}(\tau,\alpha)$ in (1.10) can be defined in a similar manner (see Balan, Poor, Rickard, & Verdu, 2004; Margetts et al., 2007; Ye & Papandreou-Suppappola, 2003; and Chapter 9).

By definition, WSS channels are stationary in time; furthermore, it can be shown that US channels are stationary in frequency. It follows that the statistics of a WSSUS channel do not change with time or frequency, and hence, the TF transfer function $L_{\mathbf{H}}(t,f)$ is a 2D *stationary* process, i.e.,

$$\mathsf{E}\{L_{\mathbf{H}}(t,f)L_{\mathbf{H}}^{*}(t',f')\} = R_{\mathbf{H}}(t-t',f-f'). \tag{1.25}$$

Here, $R_{\mathbf{H}}(\Delta t,\Delta f)$ denotes the channel's *TF correlation function*. The stationarity of $L_{\mathbf{H}}(t,f)$ as expressed by the above equation is consistent with the fact that the spreading function $S_{\mathbf{H}}(\tau,\nu)$, which is the 2D Fourier transform of $L_{\mathbf{H}}(t,f)$, is a white process.

Using the inverse of the Fourier transform relation (1.11) in (1.24), we obtain a similar Fourier transform relation between the TF correlation function $R_{\mathbf{H}}(\Delta t,\Delta f)$ and the scattering function $C_{\mathbf{H}}(\tau,\nu)$:

$$C_{\mathbf{H}}(\tau,\nu) = \int_{-\infty}^{\infty}\int_{-\infty}^{\infty} R_{\mathbf{H}}(\Delta t,\Delta f)\,\mathrm{e}^{-j2\pi(\nu\Delta t-\tau\Delta f)}\,\mathrm{d}\Delta t\,\mathrm{d}\Delta f. \tag{1.26}$$

By inspecting (1.25) and (1.26), it is seen that the scattering function is the 2D *power spectral density* of the 2D stationary process $L_{\mathbf{H}}(t,f)$. This observation suggests the use of spectrum estimation techniques to measure the scattering function (Kay & Doyle, 2003; see also Section 1.7.4).

1.4.1.3 ***Statistical Input–Output Relations***

The scattering function is a statistical characterization of the TF dispersion produced by a WSSUS channel. This interpretation can be made more explicit by considering the *Rihaczek spectra* (Flandrin, 1999; Matz & Hlawatsch, 2006) of transmit signal $s(t)$ and receive signal $r(t)$. The Rihaczek spectrum of a (generally nonstationary) random process $x(t)$ with correlation function $R_x(t,t') = \mathsf{E}\{x(t)x^{*}(t')\}$ is defined as

$$\overline{\Gamma}_x(t,f) \triangleq \int_{-\infty}^{\infty} R_x(t,t-\Delta t)\,\mathrm{e}^{-j2\pi f\Delta t}\,\mathrm{d}\Delta t.$$

With some precautions, $\overline{\Gamma}_x(t,f)$ can be interpreted as a mean energy distribution of $x(t)$ over the TF plane. Thus, it generalizes the power spectral density of stationary processes.

Starting from (1.7), it can be shown that

$$\overline{\Gamma}_r(t,f) = \int_{-\infty}^{\infty}\int_{-\infty}^{\infty} C_{\mathbf{H}}(\tau,\nu)\,\overline{\Gamma}_s(t-\tau,f-\nu)\,\mathrm{d}\tau\,\mathrm{d}\nu. \tag{1.27}$$

This means that the TF energy spectrum of the receive signal is a superposition of TF-translated versions of the TF energy spectrum of the transmit signal, weighted by the corresponding values of the scattering function. The "statistical input–output relation" (1.27) thus represents a second-order statistical analogue of the deterministic linear input–output relation (1.9).

Performing a 2D Fourier transform of the convolution relation (1.27) yields

$$\bar{A}_r(\Delta t,\Delta f) = R_{\mathbf{H}}(\Delta t,\Delta f)\bar{A}_s(\Delta t,\Delta f), \tag{1.28}$$

where $\bar{A}_x(\Delta t,\Delta f) \triangleq \int_{-\infty}^{\infty} R_x(t,t-\Delta t)\,\mathrm{e}^{-j2\pi t\Delta f}\,\mathrm{d}t$, the 2D Fourier transform of $\overline{\Gamma}_x(t,f)$, is the *expected ambiguity function* of a random process $x(t)$. The simple multiplicative input–output relation (1.28) is the basis of certain methods for estimating the scattering function (Artés, Matz, & Hlawatsch, 2004; Gaarder, 1968). Specifically, $\bar{A}_s(\Delta t, \Delta f)$ is known by design, and $\bar{A}_r(\Delta t, \Delta f)$ can be estimated from the receive signal. According to (1.28), an estimate of the TF correlation function $R_{\mathbf{H}}(\Delta t, \Delta f)$ can then be obtained by a (regularized) division of the estimate of $\bar{A}_r(\Delta t, \Delta f)$ by $\bar{A}_s(\Delta t, \Delta f)$, and an estimate of the scattering function $C_{\mathbf{H}}(\tau, \nu)$ is finally obtained by a 2D Fourier transform according to (1.26). Further details of this approach are provided in Section 1.7.4.

1.4.1.4 Delay and Doppler Profiles, Time and Frequency Correlation Functions

In some situations, only the delays or only the Doppler shifts of a WSSUS channel are of interest. An example is the exploitation of delay diversity in *orthogonal frequency division multiplexing* (OFDM) systems – see Chapter 7 – by (pre)coding across tones; here, the channel's Doppler characteristics are irrelevant. In such cases, the 2D descriptions provided by the scattering function $C_{\mathbf{H}}(\tau, \nu)$ or TF correlation function $R_{\mathbf{H}}(\Delta t, \Delta f)$ may be too detailed, and it is sufficient to use one of the "marginals" of the scattering function. These marginals are defined as

$$c_{\mathbf{H}}^{(1)}(\tau) \triangleq \int_{-\infty}^{\infty} C_{\mathbf{H}}(\tau, \nu)\, d\nu, \qquad c_{\mathbf{H}}^{(2)}(\nu) \triangleq \int_{-\infty}^{\infty} C_{\mathbf{H}}(\tau, \nu)\, d\tau, \tag{1.29}$$

and termed *delay power profile* and *Doppler power profile*, respectively. The name "delay power profile" for $c_{\mathbf{H}}^{(1)}(\tau)$ is motivated by the relation $c_{\mathbf{H}}^{(1)}(\tau) = \mathsf{E}\{|h(t, \tau)|^2\}$, which shows that $c_{\mathbf{H}}^{(1)}(\tau)$ is the mean power of the channel tap with delay τ (this does not depend on t since WSSUS channels are stationary with respect to time). A similar relation and interpretation hold for $c_{\mathbf{H}}^{(2)}(\nu)$. Because of (1.26), the (1D) Fourier transforms of the delay power profile $c_{\mathbf{H}}^{(1)}(\tau)$ and Doppler power profile $c_{\mathbf{H}}^{(2)}(\nu)$ are given by the *frequency correlation function* and *time correlation function* defined, respectively, as

$$r_{\mathbf{H}}^{(1)}(\Delta f) \triangleq R_{\mathbf{H}}(0, \Delta f) = \mathsf{E}\{L_{\mathbf{H}}(t, f) L_{\mathbf{H}}^*(t, f - \Delta f)\},$$

$$r_{\mathbf{H}}^{(2)}(\Delta t) \triangleq R_{\mathbf{H}}(\Delta t, 0) = \mathsf{E}\{L_{\mathbf{H}}(t, f) L_{\mathbf{H}}^*(t - \Delta t, f)\}.$$

The 1D channel statistics discussed above can be used to formulate statistical input–output relations for stationary or white input (transmit) processes. For a stationary transmit signal with power spectral density $P_s(f)$, it can be shown that the receive signal produced by a WSSUS channel is stationary as well and its power spectral density and correlation function are respectively given by

$$P_r(f) = \int_{-\infty}^{\infty} c_{\mathbf{H}}^{(2)}(\nu) P_s(f - \nu)\, d\nu, \qquad r_r(\Delta t) = r_{\mathbf{H}}^{(2)}(\Delta t)\, r_s(\Delta t).$$

Dual relations involving $c_{\mathbf{H}}^{(1)}(\tau)$ and $r_{\mathbf{H}}^{(1)}(\Delta f)$ hold for white transmit signals. Furthermore, similar relations exist for cyclostationary processes (Matz & Hlawatsch, 2003).

1.4.1.5 Global Channel Parameters

For many design and analysis tasks in wireless communications, only global channel parameters are relevant. These parameters summarize important properties of the scattering function and TF correlation function such as overall strength, position, and extension (spread).

As discussed in Section 1.2.4, the path loss (average power attenuation) is important, e.g., for link budget considerations. In the case of WSSUS channels, the path loss is equal to the volume of the scattering function or, equivalently, the maximum amplitude of the TF correlation function. That is, we have $P_{\mathrm{L}} = -10\log_{10}(\rho_{\mathbf{H}}^2)$ with

$$\begin{aligned}\rho_{\mathbf{H}}^2 &= \int_{-\infty}^{\infty}\int_{-\infty}^{\infty} C_{\mathbf{H}}(\tau,\nu)\,\mathrm{d}\tau\,\mathrm{d}\nu \\ &= \int_{-\infty}^{\infty} c_{\mathbf{H}}^{(1)}(\tau)\,\mathrm{d}\tau = \int_{-\infty}^{\infty} c_{\mathbf{H}}^{(2)}(\nu)\,\mathrm{d}\nu \\ &= R_{\mathbf{H}}(0,0) = \mathsf{E}\{|L_{\mathbf{H}}(t,f)|^2\}.\end{aligned}$$

Note that the last expression, $\mathsf{E}\{|L_{\mathbf{H}}(t,f)|^2\}$, does not depend on (t,f) due to the TF stationarity of WSSUS channels.

Further useful parameters are the *mean delay* and *mean Doppler shift* of a WSSUS channel, which are defined by the first moments (centers of gravity)

$$\bar{\tau} \triangleq \frac{1}{\rho_{\mathbf{H}}^2}\int_{-\infty}^{\infty}\int_{-\infty}^{\infty} \tau\, C_{\mathbf{H}}(\tau,\nu)\,\mathrm{d}\tau\,\mathrm{d}\nu = \frac{1}{\rho_{\mathbf{H}}^2}\int_{-\infty}^{\infty} \tau\, c_{\mathbf{H}}^{(1)}(\tau)\,\mathrm{d}\tau, \tag{1.30a}$$

$$\bar{\nu} \triangleq \frac{1}{\rho_{\mathbf{H}}^2}\int_{-\infty}^{\infty}\int_{-\infty}^{\infty} \nu\, C_{\mathbf{H}}(\tau,\nu)\,\mathrm{d}\tau\,\mathrm{d}\nu = \frac{1}{\rho_{\mathbf{H}}^2}\int_{-\infty}^{\infty} \nu\, c_{\mathbf{H}}^{(2)}(\nu)\,\mathrm{d}\nu. \tag{1.30b}$$

In particular, $\bar{\tau}$ describes the distance-dependent mean propagation delay. For physical channels, causality implies that $C_{\mathbf{H}}(\tau,\nu) = 0$ for $\tau < 0$ and consequently that $\bar{\tau} \geq 0$. Assuming that the receiver's timing recovery unit locks to the center of gravity of the delay power profile, the subsequent receiver stages will see an equivalent channel $\tilde{\mathbf{H}}$ where the mean delay $\bar{\tau}$ is split off. That is, $\mathbf{H} = \tilde{\mathbf{H}}\mathbf{D}_{\bar{\tau}}$ where $\mathbf{D}_{\bar{\tau}}$ is a pure time-delay operator acting as $(\mathbf{D}_{\tau}s)(t) = s(t-\bar{\tau})$ and $\tilde{\mathbf{H}}$ is an equivalent channel whose mean delay is zero. In many treatments, the equivalent channel $\tilde{\mathbf{H}}$ is considered even though this is not stated explicitly. Similar considerations apply to the mean Doppler shift $\bar{\nu}$, which can also be split off by frequency offset compensation techniques, resulting in an equivalent channel with mean Doppler shift equal to zero.

The mean delay $\bar{\tau}$ and mean Doppler shift $\bar{\nu}$ describe the overall location of the scattering function $C_{\mathbf{H}}(\tau,\nu)$ in the (τ,ν) plane. The extension of $C_{\mathbf{H}}(\tau,\nu)$ about $(\bar{\tau},\bar{\nu})$ can be measured by the *delay spread* and *Doppler spread*, which are defined as the root-mean-square (RMS) widths of delay power profile $c_{\mathbf{H}}^{(1)}(\tau)$ and Doppler power profile $c_{\mathbf{H}}^{(2)}(\nu)$, respectively:

$$\sigma_{\tau} \triangleq \frac{1}{\rho_{\mathbf{H}}}\sqrt{\int_{-\infty}^{\infty}\int_{-\infty}^{\infty} (\tau-\bar{\tau})^2\, C_{\mathbf{H}}(\tau,\nu)\,\mathrm{d}\tau\,\mathrm{d}\nu} = \frac{1}{\rho_{\mathbf{H}}}\sqrt{\int_{-\infty}^{\infty} (\tau-\bar{\tau})^2\, c_{\mathbf{H}}^{(1)}(\tau)\,\mathrm{d}\tau}, \tag{1.31a}$$

$$\sigma_{\nu} \triangleq \frac{1}{\rho_{\mathbf{H}}}\sqrt{\int_{-\infty}^{\infty}\int_{-\infty}^{\infty} (\nu-\bar{\nu})^2\, C_{\mathbf{H}}(\tau,\nu)\,\mathrm{d}\tau\,\mathrm{d}\nu} = \frac{1}{\rho_{\mathbf{H}}}\sqrt{\int_{-\infty}^{\infty} (\nu-\bar{\nu})^2\, c_{\mathbf{H}}^{(2)}(\nu)\,\mathrm{d}\nu}. \tag{1.31b}$$

Sometimes, it is more convenient to work with the reciprocals of Doppler spread and delay spread,

$$T_c \triangleq \frac{1}{\sigma_\nu}, \qquad F_c \triangleq \frac{1}{\sigma_\tau}, \tag{1.32}$$

which are known as the *coherence time* and *coherence bandwidth*, respectively. These two parameters can be used to quantify the duration and bandwidth within which the channel is approximately constant (or, at least, strongly correlated). This interpretation is supported by two arguments. First, it can be shown that the curvatures in the Δt and Δf directions of the squared magnitude of the TF correlation function $R_{\mathbf{H}}(\Delta t, \Delta f)$ at the origin are inversely proportional to the squared coherence bandwidth and squared coherence time, respectively. This corresponds to the following second-order Taylor series approximation of $|R_{\mathbf{H}}(\Delta t, \Delta f)|^2$ about the origin:

$$|R_{\mathbf{H}}(\Delta t, \Delta f)|^2 \approx \rho_{\mathbf{H}}^4 \left[1 - \left(\frac{2\pi \Delta t}{T_c}\right)^2 - \left(\frac{2\pi \Delta f}{F_c}\right)^2\right].$$

(The first-order terms vanish since $|R_{\mathbf{H}}(\Delta t, \Delta f)|^2 \le |R_{\mathbf{H}}(0,0)|^2 = \rho_{\mathbf{H}}^4$, i.e., $|R_{\mathbf{H}}(\Delta t, \Delta f)|^2$ assumes its maximum at the origin.) Thus, within durations $|\Delta t|$ smaller than T_c and bandwidths $|\Delta f|$ smaller than F_c, the channel will be strongly correlated. In addition, it can be shown that

$$\frac{1}{\rho_{\mathbf{H}}^2} \mathsf{E}\{|L_{\mathbf{H}}(t+\Delta t, f+\Delta f) - L_{\mathbf{H}}(t,f)|^2\} \le 2\pi \left[\left(\frac{\Delta t}{T_c}\right)^2 + \left(\frac{\Delta f}{F_c}\right)^2\right]. \tag{1.33}$$

This implies that within TF regions of duration $|\Delta t|$ smaller than T_c and bandwidth $|\Delta f|$ smaller than F_c, the channel is approximately constant (in the mean-square sense). More specifically, within the local *ε-coherence region* $\mathscr{B}_c^\varepsilon(t,f) \triangleq [t, t+\varepsilon T_c] \times [f, f+\varepsilon F_c]$, the RMS error of the approximation $L_{\mathbf{H}}(t+\Delta t, f+\Delta f) \approx L_{\mathbf{H}}(t,f)$ is of order ε. An illustration of this coherence region will be shown in Fig. 1.8 in Section 1.4.3.3.

We finally illustrate the characterization of WSSUS channels with a simple example.

Example 1.9

A popular WSSUS channel model uses a separable scattering function $C_{\mathbf{H}}(\tau,\nu) = \frac{1}{\rho_{\mathbf{H}}^2} c_{\mathbf{H}}^{(1)}(\tau) c_{\mathbf{H}}^{(2)}(\nu)$ with an exponential delay power profile and a so-called Jakes Doppler power profile (Molisch, 2005; Proakis, 1995):

$$c_{\mathbf{H}}^{(1)}(\tau) = \begin{cases} \frac{\rho_{\mathbf{H}}^2}{\tau_0} e^{-\tau/\tau_0}, & \tau \ge 0, \\ 0, & \tau < 0, \end{cases} \qquad c_{\mathbf{H}}^{(2)}(\nu) = \begin{cases} \frac{\rho_{\mathbf{H}}^2}{\pi\sqrt{\nu_{\max}^2 - \nu^2}}, & |\nu| < \nu_{\max}, \\ 0, & |\nu| > \nu_{\max}. \end{cases}$$

Here, τ_0 is a delay parameter, and $\nu_{\max}$ is the maximum Doppler shift. The exponential delay profile is motivated by the exponential decay of receive power with path length (which is proportional to delay), and the Jakes Doppler profile results from the assumption of uniformly distributed AoA (Jakes, 1974). Note that $c_{\mathbf{H}}^{(1)}(\tau)$ ignores the fundamental propagation delay of the first (and, in this case, also strongest) multipath component. The corresponding TF correlation function is separable as well, i.e., $R_{\mathbf{H}}(\Delta t, \Delta f) = \frac{1}{\rho_{\mathbf{H}}^2} r_{\mathbf{H}}^{(2)}(\Delta t) r_{\mathbf{H}}^{(1)}(\Delta f)$, with time correlation function and frequency correlation function given by

$$r_{\mathbf{H}}^{(2)}(\Delta t) = \rho_{\mathbf{H}}^2 J_0(2\pi \nu_{\max} \Delta t), \qquad r_{\mathbf{H}}^{(1)}(\Delta f) = \frac{\rho_{\mathbf{H}}^2}{1 + j2\pi \tau_0 \Delta f}.$$

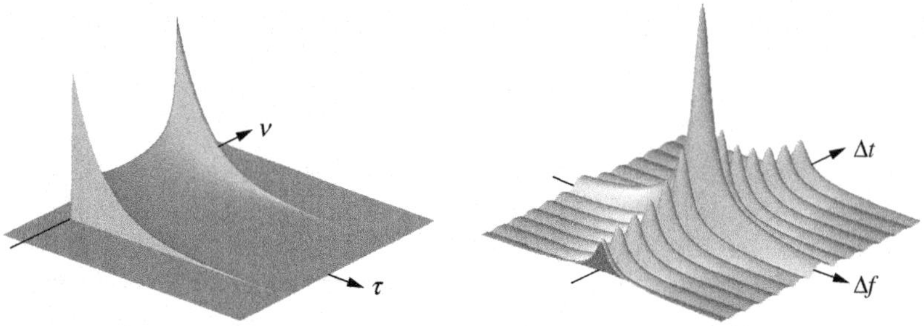

FIGURE 1.5

WSSUS channel following the Jakes-exponential model: scattering function (left) and magnitude of TF correlation function (right).

Here, $J_0(\cdot)$ denotes the zeroth-order Bessel function of the first kind. The scattering function and TF correlation function for this WSSUS channel are depicted in Fig. 1.5.

Assuming a delay parameter $\tau_0 = 10\,\mu\text{s}$ and maximum Doppler $\nu_{max} = 100$ Hz, we obtain the mean delay $\bar{\tau} = \tau_0 = 10\,\mu\text{s}$ and mean Doppler $\bar{\nu} = 0$ Hz. Furthermore, the delay spread and Doppler spread follow as

$$\sigma_\tau = \tau_0 = 10\,\mu\text{s}, \qquad \sigma_\nu = \frac{\nu_{max}}{\sqrt{2}} = 70.71\text{ Hz},$$

and the corresponding coherence time and coherence bandwidth are

$$T_c = \frac{\sqrt{2}}{\nu_{max}} = 14.14\text{ ms}, \qquad F_c = \frac{1}{\tau_0} = 100\text{ kHz}.$$

For a narrowband system with bandwidth 10 kHz and frame duration 1 ms, the bound (1.33) implies that the mean-square difference between any two values of $L_{\mathbf{H}}(t, f)$ within the frame duration and transmit band is at most roughly 1.5%. Thus, the TF transfer function can be assumed constant, i.e., $L_{\mathbf{H}}(t, f) \approx h$. Within one frame, the input–output relation (1.12) then simplifies to $r(t) \approx hs(t)$, a model known as block flat fading. ■

1.4.2 Extension to Multiantenna Systems

We next outline the extension of the WSSUS property to multiple-antenna (MIMO) channels, mostly following Matz (2006). The main difference from the single-antenna case is the need for joint statistics of the individual links.

1.4.2.1 Scattering Function Matrix and Space-Time-Frequency Correlation Function Matrix

Extending Bello (1963) (see also Section 1.4.1), we call a MIMO channel WSSUS if all $M_T M_R$ elements of the TF transfer function matrix $\mathbf{L_H}(t, f)$ in (1.22) are *jointly* (wide-sense) stationary. Defining the length-$M_T M_R$ vector $\mathbf{l_H}(t, f) \triangleq \text{vec}\{\mathbf{L_H}(t, f)\}$, this condition can be written as (cf. (1.25))

$$\mathsf{E}\{\mathbf{l_H}(t, f)\mathbf{l_H}^H(t', f')\} = \mathbf{R_H}(t - t', f - f'). \tag{1.34}$$

Here, the $M_T M_R \times M_T M_R$ matrix $\mathbf{R_H}(\Delta t, \Delta f)$ is referred to as the *space-time-frequency correlation function matrix* of the channel. This matrix-valued function describes the correlation of the transfer functions $L_{\mathsf{H}_{ij}}(t, f)$ and $L_{\mathsf{H}_{i'j'}}(t', f')$ of any two component channels H_{ij} and $\mathsf{H}_{i'j'}$ at time lag $t - t' = \Delta t$ and frequency lag $f - f' = \Delta f$.

Equivalently, the MIMO-WSSUS property expresses the fact that the elements of the spreading function matrix $\mathbf{S_H}(\tau, \nu)$ in (1.21) are *jointly* white (cf. (1.24)), i.e.,

$$\mathsf{E}\{\mathbf{s_H}(\tau,\nu)\,\mathbf{s}_\mathbf{H}^H(\tau',\nu')\} = \mathbf{C_H}(\tau,\nu)\,\delta(\tau-\tau')\,\delta(\nu-\nu'),$$

with $\mathbf{s_H}(\tau,\nu) \triangleq \mathrm{vec}\{\mathbf{S_H}(\tau,\nu)\}$. The $M_T M_R \times M_T M_R$ matrix $\mathbf{C_H}(\tau,\nu)$ will be referred to as the *scattering function matrix* (SFM). The SFM is nonnegative definite for all (τ,ν). It summarizes the mean spatial characteristics and strength of all scatterers with delay τ and Doppler ν. The SFM and the space-time-frequency correlation function matrix are related via a 2D Fourier transform (cf. (1.26)):

$$\mathbf{C_H}(\tau,\nu) = \int_{-\infty}^{\infty}\int_{-\infty}^{\infty} \mathbf{R_H}(\Delta t, \Delta f)\,\mathrm{e}^{-j2\pi(\nu\Delta t - \tau\Delta f)}\,\mathrm{d}\Delta t\,\mathrm{d}\Delta f.$$

Thus, recalling (1.34), the SFM $\mathbf{C_H}(\tau,\nu)$ can be interpreted as the 2D power spectral density matrix of the 2D stationary multivariate random process $\mathbf{l_H}(t, f)$.

1.4.2.2 Canonical Decomposition

While the SFM provides a decorrelated representation with respect to delay and Doppler, it still features spatial correlations. For a spatially decorrelated representation, consider the (τ,ν)-dependent eigendecomposition of the SFM,

$$\mathbf{C_H}(\tau,\nu) = \sum_{i=1}^{M_R}\sum_{j=1}^{M_T} \lambda_{ij}(\tau,\nu)\,\mathbf{u}_{ij}(\tau,\nu)\,\mathbf{u}_{ij}^H(\tau,\nu).$$

Here, $\lambda_{ij}(\tau,\nu) \geq 0$ and $\mathbf{u}_{ij}(\tau,\nu)$ denote the eigenvalues and eigenvectors of $\mathbf{C_H}(\tau,\nu)$, respectively (we use 2D indexing for later convenience). For each (τ,ν), the $M_T M_R$ vectors $\mathbf{u}_{i,j}(\tau,\nu)$ form an orthonormal basis of $\mathbb{C}^{M_T M_R}$. Using the $M_R \times M_T$ matrix form of this basis, $\mathbf{U}_{ij}(\tau,\nu) \triangleq \mathrm{unvec}\{\mathbf{u}_{ij}(\tau,\nu)\}$, the channel's spreading function can be expanded as

$$\mathbf{S_H}(\tau,\nu) = \sum_{i=1}^{M_R}\sum_{j=1}^{M_T} \alpha_{ij}(\tau,\nu)\,\mathbf{U}_{ij}(\tau,\nu), \tag{1.35}$$

with the random coefficients $\alpha_{ij}(\tau,\nu) \triangleq \mathbf{u}_{ij}^H(\tau,\nu)\,\mathbf{s_H}(\tau,\nu)$. It can be shown that these coefficients are orthogonal with respect to delay, Doppler, and space, i.e.,

$$\mathsf{E}\{\alpha_{ij}(\tau,\nu)\,\alpha_{i'j'}^*(\tau',\nu')\} = \lambda_{ij}(\tau,\nu)\,\delta_{i,i'}\,\delta_{j,j'}\,\delta(\tau-\tau')\,\delta(\nu-\nu').$$

The expansion (1.35) entails the following representation of the MIMO channel (see (1.19) and (1.20)):

$$\mathbf{r}(t) = (\underline{\mathbf{H}}\mathbf{s})(t) = \int_{-\infty}^{\infty}\int_{-\infty}^{\infty}\sum_{i=1}^{M_R}\sum_{j=1}^{M_T} \alpha_{ij}(\tau,\nu)\,\mathbf{U}_{ij}(\tau,\nu)\,\mathbf{s}(t-\tau)\,\mathrm{e}^{j2\pi\nu t}\,\mathrm{d}\tau\,\mathrm{d}\nu. \tag{1.36}$$

It is seen that the eigenvector matrices $\mathbf{U}_{ij}(\tau,\nu)$ describe the spatial characteristics of deterministic atomic MIMO channels associated with delay τ and Doppler frequency ν. The expansion (1.36) is "doubly orthogonal" since for any (τ,ν), the matrices $\mathbf{U}_{ij}(\tau,\nu)$ are (deterministically) orthonormal and the coefficients $\alpha_{ij}(\tau,\nu)$ are stochastically orthogonal. Thus, (1.36) represents any MIMO-WSSUS channel as a superposition of deterministic atomic MIMO channels weighted by uncorrelated scalar random coefficients. In this representation, the channel transfer effects (space-time-frequency dispersion/selectivity) are separated from the channel stochastics.

Example 1.10

For spatially i.i.d. MIMO-WSSUS channels, the SFM is given by $\mathbf{C}_{\mathbf{H}}(\tau,\nu) = C(\tau,\nu)\mathbf{I}$, i.e., all component channels $\mathbf{H}_{ij}$ are independent WSSUS channels with identical scattering function $C(\tau,\nu)$. In this case, $\lambda_{ij}(\tau,\nu) = C(\tau,\nu)$ and $\mathbf{U}_{ij}(\tau,\nu) = \mathbf{e}_i\mathbf{e}_j^H$, with $\mathbf{e}_i$ denoting the ith unit vector. Here, the action of the atomic channels is $\mathbf{U}_{ij}(\tau,\nu)\mathbf{s}(t) = s_j(t)\,\mathbf{e}_i$, which means that for the atomic channels, the ith receive antenna observes the signal emitted by the jth transmit antenna. Since due to the i.i.d. assumption the individual spatial links are independent, there is $\alpha_{ij}(\tau,\nu) = \mathbf{u}_{ij}^H(\tau,\nu)\,\mathbf{s}_{\mathbf{H}}(\tau,\nu) = S_{\mathbf{H}_{ij}}(\tau,\nu)$, and (1.36) simplifies to

$$\mathbf{r}(t) = \int_{-\infty}^{\infty}\int_{-\infty}^{\infty}\sum_{i=1}^{M_\mathrm{R}}\sum_{j=1}^{M_\mathrm{T}} S_{\mathbf{H}_{ij}}(\tau,\nu)\, s_j(t-\tau)\,e^{j2\pi\nu t}\,\mathbf{e}_i\,d\tau\,d\nu.$$

The i.i.d. MIMO-WSSUS model is an extremely simple model in that it is already completely decorrelated in all domains. Slightly more complex – but more realistic – MIMO-WSSUS models are discussed in the next example.

Example 1.11

An extension of the flat-fading MIMO model of Weichselberger, Herdin, Özcelik, & Bonek (2006) assumes that the spatial eigenmodes (but not necessarily the associated powers $\lambda_{ij}(\tau,\nu)$) are separable, i.e., $\mathbf{U}_{ij}(\tau,\nu) = \mathbf{v}_i(\tau,\nu)\,\mathbf{w}_j^H(\tau,\nu)$. Here, the spatial modes $\mathbf{v}_i(\tau,\nu)$ and $\mathbf{w}_j(\tau,\nu)$ can be interpreted as transmit and receive beamforming vectors, respectively. The resulting simplified version of (1.35) can be written as

$$\mathbf{S}_{\mathbf{H}}(\tau,\nu) = \mathbf{V}(\tau,\nu)\,\boldsymbol{\Sigma}(\tau,\nu)\,\mathbf{W}^H(\tau,\nu), \tag{1.37}$$

with the deterministic matrices $\mathbf{V}(\tau,\nu) = (\mathbf{v}_1(\tau,\nu)\cdots\mathbf{v}_{M_\mathrm{R}}(\tau,\nu))$ (dimension $M_\mathrm{R}\times M_\mathrm{R}$) and $\mathbf{W}(\tau,\nu) = (\mathbf{w}_1(\tau,\nu)\cdots\mathbf{w}_{M_\mathrm{T}}(\tau,\nu))$ (dimension $M_\mathrm{T}\times M_\mathrm{T}$) and the random matrix $\boldsymbol{\Sigma}(\tau,\nu)$ (dimension $M_\mathrm{R}\times M_\mathrm{T}$) given by $[\boldsymbol{\Sigma}(\tau,\nu)]_{ij} = \alpha_{ij}(\tau,\nu)$. In this context, the average powers $\lambda_{ij}(\tau,\nu)$ are referred to as *coupling coefficients* since they describe how strongly the spatial transmit modes $\mathbf{w}_j(\tau,\nu)$ and spatial receive modes $\mathbf{v}_i(\tau,\nu)$ are coupled on average.

A special case of the above model for uniform linear arrays is obtained by assuming that the spatial modes equal the array steering vectors, i.e., $\mathbf{V}(\tau,\nu) = \mathbf{F}_{M_\mathrm{R}}$ and $\mathbf{W}(\tau,\nu) = \mathbf{F}_{M_\mathrm{T}}$, where $\mathbf{F}_N$ denotes the $N\times N$ DFT matrix. This model is known in the literature as the *virtual MIMO model* (Sayeed, 2002).

Another simplification of (1.37) is obtained by assuming that also the SFM eigenvalues are spatially separable, i.e., $\lambda_{ij}(\tau,\nu) = \kappa_i(\tau,\nu)\,\mu_j(\tau,\nu)$. This corresponds to a WSSUS extension of the so-called *Kronecker model* (Kermoal, Schumacher, Pedersen, Mogensen, & Frederiksen, 2002).

An analysis of channel measurements in Matz (2006) revealed that for a channel having a dominant scatterer with delay τ_0 and Doppler frequency ν_0, the SFM at $(\tau,\nu) = (\tau_0,\nu_0)$, $\mathbf{C_H}(\tau_0,\nu_0)$, has a single dominant eigenvalue (i.e., effective rank one), and the same holds true for the associated atomic MIMO channel matrix $\mathbf{U}_{11}(\tau_0,\nu_0)$. That is, $\mathbf{U}_{11}(\tau_0,\nu_0) = \mathbf{v}_1(\tau_0,\nu_0)\,\mathbf{w}_1^H(\tau_0,\nu_0)$, where the spatial signatures $\mathbf{v}_1(\tau_0,\nu_0)$ and $\mathbf{w}_1(\tau_0,\nu_0)$ essentially capture the AoA and AoD, respectively, associated with that dominant scatterer. It follows that in the delay-Doppler domain, a rank-one Kronecker model sufficiently characterizes the channel, i.e., $\mathbf{S_H}(\tau_0,\nu_0) = \alpha_{11}(\tau_0,\nu_0)\,\mathbf{v}_1(\tau_0,\nu_0)\,\mathbf{w}_1^H(\tau_0,\nu_0)$. Note, however, that the Kronecker model is not necessarily applicable to the TF transfer function matrix $\mathbf{L_H}(t,f)$. This is because the spatial averaging effected by the Fourier transform (1.22) will generally build up full-rank matrices.

1.4.3 Non-WSSUS Channels

The WSSUS assumption greatly simplifies the statistical characterization of LTV channels. However, it is satisfied by practical wireless channels only approximately within certain time intervals and frequency bands. A similarly simple and intuitive framework for *non-WSSUS* channels is provided next, following to a large extent Matz (2005). This framework includes WSSUS channels as a special case.

A fundamental property of WSSUS channels is the fact that different scatterers (delay-Doppler components) are uncorrelated, i.e., the spreading function $S_{\mathbf{H}}(\tau,\nu)$ is a white process. In practice, this property will not be satisfied because channel components that are close to each other in the delay-Doppler domain often result from the same physical scatterer and will hence be correlated. In addition, filters, antennas, and windowing operations at the transmit and/or receive side are often viewed as part of the channel; they cause some extra time and frequency dispersion that results in correlations of the spreading function of the overall channel.

Example 1.12

Consider a channel with a single specular scatterer with delay τ_0, Doppler shift ν_0, and random reflectivity h. The transmitter uses a filter with impulse response $g(\tau)$, and the receiver multiplies the receive signal by a window $\gamma(t)$. It can be shown that the spreading function of the effective channel (including transmit filter and receiver window) is given by

$$S_{\mathbf{H}}(\tau,\nu) = h\, g(\tau - \tau_0)\,\Gamma(\nu - \nu_0),$$

where $\Gamma(\nu)$ is the Fourier transform of $\gamma(t)$. Clearly, the spreading function exhibits correlations in a neighborhood of (τ_0,ν_0) that is determined by the effective duration of $g(\tau)$ and the effective bandwidth of $\gamma(t)$.

An alternative view of non-WSSUS channels builds on the TF transfer function, which is no longer TF stationary. The physical mechanisms causing $L_{\mathbf{H}}(t,f)$ to be nonstationary include shadowing, delay and Doppler drift due to mobility, and changes in the propagation environment. These effects occur at a much larger scale than small-scale fading.

Example 1.13

Consider a receiver approaching the transmitter with changing speed $v(t)$ so that their distance decreases according to $d(t) = d_0 - \int_0^t v(t')\,dt'$. In this case, the transmit signal is delayed by $\tau_1(t) = d(t)/c_0$ and Doppler shifted by $\nu_1(t) = f_c v(t)/c_0$. The impulse response and TF transfer function are here given by $h(t,\tau) = h e^{j2\pi\nu_1(t)t}\delta(\tau - \tau_1(t))$ and $L_{\mathbf{H}}(t,f) = h e^{j2\pi(\nu_1(t)t - \tau_1(t)f)}$, respectively. The correlation function of $L_{\mathbf{H}}(t,f)$ can be shown to depend explicitly on time. Hence, the channel is temporally nonstationary.

1.4.3.1 *Local Scattering Function and Channel Correlation Function*

In Matz (2005), the *local scattering function* (LSF) was introduced as a physically meaningful second-order statistic that extends the scattering function $C_{\mathbf{H}}(\tau,\nu)$ of WSSUS channels to the case of non-WSSUS channels. The LSF is defined as a 2D Fourier transform of the 4D correlation function of the TF transfer function $L_{\mathbf{H}}(t,f)$ or the spreading function $S_{\mathbf{H}}(\tau,\nu)$ with respect to the lag variables, i.e.,

$$\mathcal{C}_{\mathbf{H}}(t,f;\tau,\nu) \triangleq \int_{-\infty}^{\infty}\int_{-\infty}^{\infty} \mathsf{E}\{L_{\mathbf{H}}(t,f+\Delta f)L^*_{\mathbf{H}}(t-\Delta t,f)\}\, e^{-j2\pi(\nu\Delta t-\tau\Delta f)}\, d\Delta t\, d\Delta f$$

$$= \int_{-\infty}^{\infty}\int_{-\infty}^{\infty} \mathsf{E}\{S_{\mathbf{H}}(\tau,\nu+\Delta\nu)S^*_{\mathbf{H}}(\tau-\Delta\tau,\nu)\}\, e^{j2\pi(t\Delta\nu-f\Delta\tau)}\, d\Delta\tau\, d\Delta\nu.$$

For WSSUS channels, $\mathcal{C}_{\mathbf{H}}(t,f;\tau,\nu) = C_{\mathbf{H}}(\tau,\nu)$ (cf. (1.26)). It was shown in Matz (2005) that the LSF describes the power of multipath components with delay τ and Doppler shift ν occurring at time t and frequency f. This interpretation can be supported by the following channel input–output relation extending (1.27):

$$\overline{\Gamma}_r(t,f) = \int_{-\infty}^{\infty}\int_{-\infty}^{\infty} \mathcal{C}_{\mathbf{H}}(t,f-\nu;\tau,\nu)\,\overline{\Gamma}_s(t-\tau,f-\nu)\,d\tau\,d\nu,$$

where, as before, $\overline{\Gamma}_x(t,f)$ denotes the Rihaczek spectrum of a random process $x(t)$. This relation shows that the LSF $\mathcal{C}_{\mathbf{H}}(t,f;\tau,\nu)$ describes the TF energy shifts from $(t-\tau,f)$ to $(t,f+\nu)$.

The LSF is a channel statistic that reveals the nonstationarities (in time and frequency) of a channel via its dependence on t and f. A dual second-order channel statistic that is better suited for describing the channel's delay-Doppler correlations (in addition to TF correlations) is provided by the *channel correlation function* (CCF) defined as

$$\mathcal{R}_{\mathbf{H}}(\Delta t,\Delta f;\Delta\tau,\Delta\nu) \triangleq \int_{-\infty}^{\infty}\int_{-\infty}^{\infty} \mathsf{E}\{L_{\mathbf{H}}(t,f+\Delta f)L^*_{\mathbf{H}}(t-\Delta t,f)\}\, e^{-j2\pi(t\Delta\nu-f\Delta\tau)}\, dt\, df$$

$$= \int_{-\infty}^{\infty}\int_{-\infty}^{\infty} \mathsf{E}\{S_{\mathbf{H}}(\tau,\nu+\Delta\nu)S^*_{\mathbf{H}}(\tau-\Delta\tau,\nu)\}\, e^{j2\pi(\nu\Delta t-\tau\Delta f)}\, d\tau\, d\nu.$$

The CCF can be shown to characterize the correlation of multipath components separated by Δt in time, by Δf in frequency, by $\Delta\tau$ in delay, and by $\Delta\nu$ in Doppler. It generalizes the TF correlation function $R_{\mathbf{H}}(\Delta t, \Delta f)$ of WSSUS channels to the non-WSSUS case: for WSSUS channels, $\mathscr{R}_{\mathbf{H}}(\Delta t, \Delta f; \Delta\tau, \Delta\nu) = R_{\mathbf{H}}(\Delta t, \Delta f)\,\delta(\Delta\tau)\,\delta(\Delta\nu)$, which correctly indicates the absence of delay and Doppler correlations. The CCF is symmetric and assumes its maximum at the origin. It is related to the LSF via a 4D Fourier transform,

$$\mathscr{C}_{\mathbf{H}}(t, f; \tau, \nu) = \int_{-\infty}^{\infty}\int_{-\infty}^{\infty}\int_{-\infty}^{\infty}\int_{-\infty}^{\infty} \mathscr{R}_{\mathbf{H}}(\Delta t, \Delta f; \Delta\tau, \Delta\nu)\, \mathrm{e}^{-j2\pi(\nu\Delta t - \tau\Delta f - t\Delta\nu + f\Delta\tau)} \mathrm{d}\Delta t\, \mathrm{d}\Delta f\, \mathrm{d}\Delta\tau\, \mathrm{d}\Delta\nu,$$

in which time t and Doppler lag $\Delta\nu$ are Fourier dual variables, and so are frequency f and delay lag $\Delta\tau$. Once again, this indicates that delay-Doppler correlations manifest themselves as TF nonstationarities (and vice versa).

1.4.3.2 *Reduced-Detail Channel Descriptions*

Several less detailed channel statistics for non-WSSUS channels can be obtained as marginals of the LSF or as cross sections of the CCF. Of specific interest are the *average scattering function*,

$$\bar{C}_{\mathbf{H}}(\tau, \nu) \triangleq \int_{-\infty}^{\infty}\int_{-\infty}^{\infty} \mathscr{C}_{\mathbf{H}}(t, f; \tau, \nu)\, \mathrm{d}t\, \mathrm{d}f = \mathsf{E}\{|S_{\mathbf{H}}(\tau, \nu)|^2\}$$

and its Fourier dual,

$$\mathscr{R}_{\mathbf{H}}(\Delta t, \Delta f; 0, 0) = \int_{-\infty}^{\infty}\int_{-\infty}^{\infty} \bar{C}_{\mathbf{H}}(\tau, \nu)\, \mathrm{e}^{j2\pi(\nu\Delta t - \tau\Delta f)}\, \mathrm{d}\tau\, \mathrm{d}\nu,$$

which characterize the global delay-Doppler dispersion and TF correlations much in the same way as the scattering function $C_{\mathbf{H}}(\tau, \nu)$ and TF correlation function $R_{\mathbf{H}}(\Delta t, \Delta f)$ of WSSUS channels. Dual channel statistics describing particularly the nonstationarities and delay-Doppler correlations of non-WSSUS channels are given by the *TF-dependent path loss*

$$\rho^2_{\mathbf{H}}(t, f) \triangleq \int_{-\infty}^{\infty}\int_{-\infty}^{\infty} \mathscr{C}_{\mathbf{H}}(t, f; \tau, \nu)\, \mathrm{d}\tau\, \mathrm{d}\nu = \mathsf{E}\{|L_{\mathbf{H}}(t, f)|^2\}$$

and its Fourier dual

$$\mathscr{R}_{\mathbf{H}}(0, 0; \Delta\tau, \Delta\nu) = \int_{-\infty}^{\infty}\int_{-\infty}^{\infty} \rho^2_{\mathbf{H}}(t, f)\, \mathrm{e}^{-j2\pi(t\Delta\nu - f\Delta\tau)}\, \mathrm{d}t\, \mathrm{d}f.$$

In addition, it is possible to define TF-dependent delay and Doppler power profiles,

$$c^{(1)}_{\mathbf{H}}(t, f; \tau) \triangleq \int_{-\infty}^{\infty} \mathscr{C}_{\mathbf{H}}(t, f; \tau, \nu)\, \mathrm{d}\nu, \qquad c^{(2)}_{\mathbf{H}}(t, f; \nu) \triangleq \int_{-\infty}^{\infty} \mathscr{C}_{\mathbf{H}}(t, f; \tau, \nu)\, \mathrm{d}\tau,$$

whose usefulness straightforwardly generalizes from the WSSUS case (see (1.29)).

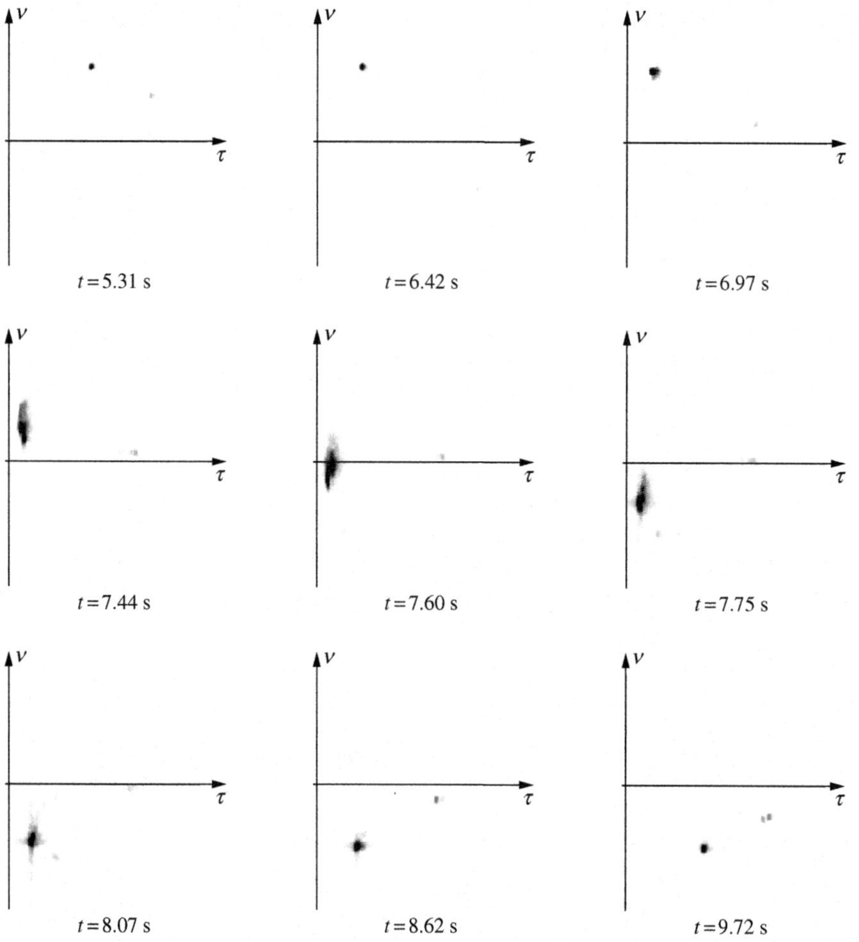

FIGURE 1.6

Temporal snapshots of the LSF estimate $\hat{\mathscr{C}}_{\mathbf{H}}(t, f_c; \tau, \nu)$ for a car-to-car channel.

Example 1.14

We consider measurement data of a mobile radio channel for car-to-car communications. The channel measurements were recorded during 10s at $f_c = 5.2$ GHz with the transmitter and receiver located in two cars that moved in opposite directions on a highway. (We note that channel sounding is addressed in Section 1.7.) Details of the measurement campaign are described in Paier et al. (2009), and the measurement data are available at http://measurements.ftw.at. Figure 1.6 shows nine snapshots of the estimated LSF $\hat{\mathscr{C}}_{\mathbf{H}}(t, f; \tau, \nu)$ at different time instants t and with frequency f fixed to f_c (see Section 1.7.4 for an estimator of the LSF). The figure depicts three successive phases: during phase I (top row), the cars approach each other; during

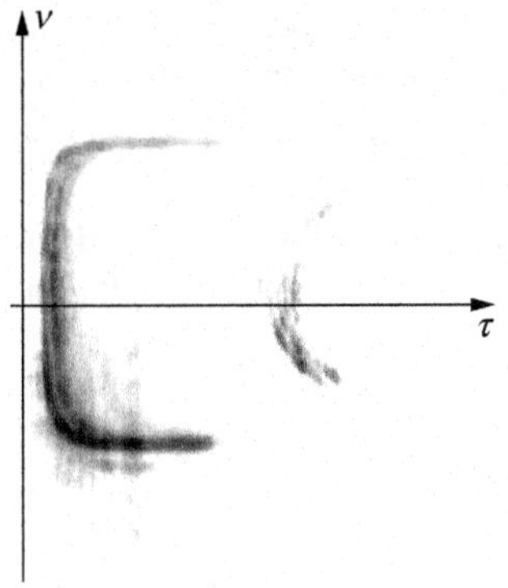

FIGURE 1.7

Estimate of the average LSF $\bar{C}_{\mathbf{H}}(\tau,\nu)$ for the car-to-car channel.

phase II (middle row), they drive by each other; and during phase III (bottom row), they move away from each other. At each time instant, the LSF is seen to consist of only a small number of dominant components. These components correspond to (1) the direct path between the two cars, (2) a path involving a reflection by a building located sideways of the highway, and (3) further paths corresponding to reflections by other vehicles on the highway. The direct path has a large delay and large positive Doppler frequency during phase I, a small delay and near-zero Doppler frequency during phase II, and a large delay and large negative Doppler frequency during phase III. Similar observations apply to the other multipath components.

Figure 1.7 shows an estimate of the average LSF $\bar{C}_{\mathbf{H}}(\tau,\nu)$. While this representation indicates the maximum delay and Doppler frequency, it suggests a continuum of scatterers, and thus fails to indicate that at each time instant there are only a few dominant multipath components.

■

1.4.3.3 *Global Channel Parameters*

As in the WSSUS case, it is desirable to be able to characterize non-WSSUS channels in terms of a few global scalar parameters. In particular, the *transmission loss* is given by

$$\begin{aligned}
\mathcal{E}_{\mathbf{H}}^2 &\triangleq \int_{-\infty}^{\infty}\int_{-\infty}^{\infty}\int_{-\infty}^{\infty}\int_{-\infty}^{\infty} \mathcal{C}_{\mathbf{H}}(t,f;\tau,\nu)\,\mathrm{d}t\,\mathrm{d}f\,\mathrm{d}\tau\,\mathrm{d}\nu \\
&= \int_{-\infty}^{\infty}\int_{-\infty}^{\infty} \bar{C}_{\mathbf{H}}(\tau,\nu)\,\mathrm{d}\tau\,\mathrm{d}\nu \\
&= \int_{-\infty}^{\infty}\int_{-\infty}^{\infty} \rho_{\mathbf{H}}^2(t,f)\,\mathrm{d}t\,\mathrm{d}f \\
&= \mathrm{E}\{\|\mathbf{H}\|^2\},
\end{aligned}$$

where $\|\mathbf{H}\|^2 \triangleq \int_{-\infty}^{\infty}\int_{-\infty}^{\infty} |k_{\mathbf{H}}(t,t')|^2 \, dt \, dt'$. The transmission loss quantifies the mean received energy for a normalized stationary and white transmit signal. Furthermore, non-WSSUS versions of the mean delay $\bar{\tau}$, mean Doppler $\bar{\nu}$, delay spread σ_τ, and Doppler spread σ_ν can be defined by replacing the scattering function $C_{\mathbf{H}}(\tau,\nu)$ and path loss $\rho_{\mathbf{H}}^2$ in the respective WSSUS-case definitions (1.30), (1.31) with the average scattering function $\bar{C}_{\mathbf{H}}(\tau,\nu)$ and transmission loss $\mathcal{E}_{\mathbf{H}}^2$, respectively. Time-dependent or frequency-dependent versions of these parameters can also be defined. As an example, we mention the time-dependent delay spread given by

$$\sigma_\tau(t) \triangleq \frac{1}{\rho_{\mathbf{H}}(t)} \sqrt{\int_{-\infty}^{\infty}\int_{-\infty}^{\infty}\int_{-\infty}^{\infty} [\tau - \bar{\tau}(t)]^2 \mathcal{C}_{\mathbf{H}}(t,f;\tau,\nu) \, df \, d\tau \, d\nu}$$

with

$$\rho_{\mathbf{H}}^2(t) \triangleq \int_{-\infty}^{\infty} \rho_{\mathbf{H}}^2(t,f) \, df, \qquad \bar{\tau}(t) \triangleq \frac{1}{\rho_{\mathbf{H}}^2(t)} \int_{-\infty}^{\infty}\int_{-\infty}^{\infty}\int_{-\infty}^{\infty} \tau \, \mathcal{C}_{\mathbf{H}}(t,f;\tau,\nu) \, df \, d\tau \, d\nu.$$

As in the case of WSSUS channels, a coherence time T_c and a coherence bandwidth F_c can be defined as the reciprocal of Doppler spread σ_ν and delay spread σ_τ, respectively (Matz, 2005). These coherence parameters can be combined into a local *ε-coherence region* $\mathcal{B}_c^\varepsilon(t,f) \triangleq [t, t+\varepsilon T_c] \times [f, f+\varepsilon F_c]$. It can then be shown that the TF transfer function is approximately constant within $\mathcal{B}_c^\varepsilon(t,f)$ in the sense that the normalized RMS error of the approximation $L_{\mathbf{H}}(t',f') \approx L_{\mathbf{H}}(t,f)$ is maximally of order ε for all $(t',f') \in \mathcal{B}_c^\varepsilon(t,f)$.

While much of the above discussion involved concepts familiar from WSSUS channels, delay-Doppler correlations and channel nonstationarity are phenomena specific to non-WSSUS channels. The amount of delay correlation and Doppler correlation can be measured by the following moments of the CCF:

$$\overline{\Delta\tau} \triangleq \frac{1}{\|\mathcal{R}_{\mathbf{H}}\|_1} \int_{-\infty}^{\infty}\int_{-\infty}^{\infty}\int_{-\infty}^{\infty}\int_{-\infty}^{\infty} |\Delta\tau| \, |\mathcal{R}_{\mathbf{H}}(\Delta t, \Delta f; \Delta\tau, \Delta\nu)| \, d\Delta t \, d\Delta f \, d\Delta\tau \, d\Delta\nu, \tag{1.38a}$$

$$\overline{\Delta\nu} \triangleq \frac{1}{\|\mathcal{R}_{\mathbf{H}}\|_1} \int_{-\infty}^{\infty}\int_{-\infty}^{\infty}\int_{-\infty}^{\infty}\int_{-\infty}^{\infty} |\Delta\nu| \, |\mathcal{R}_{\mathbf{H}}(\Delta t, \Delta f; \Delta\tau, \Delta\nu)| \, d\Delta t \, d\Delta f \, d\Delta\tau \, d\Delta\nu. \tag{1.38b}$$

These parameters quantify the delay lag and Doppler lag spans within which there are significant correlations. Delay-Doppler correlations correspond to channel nonstationarity in the (dual) TF domain. The amount of (non-)stationarity can be measured in terms of a *stationarity time* and a *stationarity bandwidth* that are respectively defined as

$$T_s \triangleq \frac{1}{\overline{\Delta\nu}}, \qquad F_s \triangleq \frac{1}{\overline{\Delta\tau}}. \tag{1.39}$$

These two stationarity parameters can be combined into a local *ε-stationarity region* $\mathcal{B}_s^\varepsilon(t,f) \triangleq [t, t+\varepsilon T_s] \times [f, f+\varepsilon F_s]$. It can be shown that the LSF is approximately constant within $\mathcal{B}_s^\varepsilon(t,f)$ in the sense that the normalized error magnitude of the approximation $\mathcal{C}_{\mathbf{H}}(t',f';\tau,\nu) \approx \mathcal{C}_{\mathbf{H}}(t,f;\tau,\nu)$

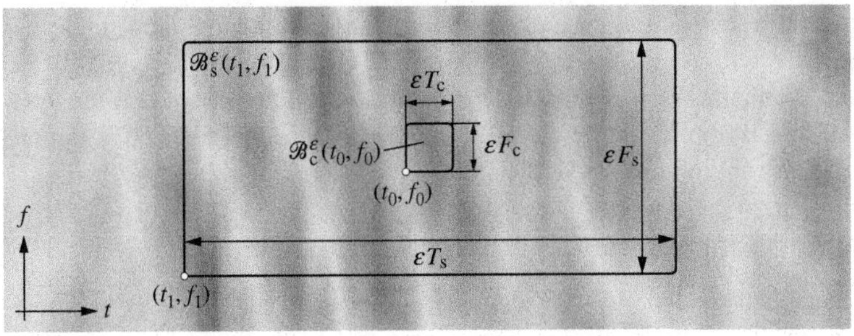

FIGURE 1.8

Illustration of coherence region $\mathscr{B}_c^\varepsilon(t_0,f_0)$ and stationarity region $\mathscr{B}_s^\varepsilon(t_1,f_1)$; the gray-shaded background corresponds to the magnitude of the channel's TF transfer function.

is maximally of order ε for all $(t',f') \in \mathscr{B}_s^\varepsilon(t,f)$. The stationarity region essentially quantifies the duration and bandwidth within which the channel can be approximated with good accuracy by a WSSUS channel. The relevance of the stationarity region to wireless system design was discussed in Matz (2005). For example, the ratio of the size of the stationarity region and the size of the coherence region, $T_S F_S/(T_C F_C)$, is crucial for the operational meaning of ergodic capacity. An illustration of the stationarity region and the coherence region is provided in Fig. 1.8.

1.5 UNDERSPREAD CHANNELS

So far, some of the system functions (e.g., TF transfer function and LSF) have been defined only formally without addressing their theoretical justification or practical applications.

Example 1.15

If a pure carrier signal $s(t) = e^{j2\pi f_c t}$ is transmitted over a linear time-invariant (purely time-dispersive/frequency-selective) channel with impulse response $h(\tau)$, the receive signal is given by the transmit signal multiplied by a complex factor, i.e.,

$$r(t) = H(f_c)\, e^{j2\pi f_c t}, \tag{1.40}$$

where $H(f) = \int_{-\infty}^{\infty} h(\tau)\, e^{-j2\pi f\tau}\, d\tau$ is the conventional channel transfer function. In mathematical language, complex exponentials are eigenfunctions of time-invariant channels.

For an LTV channel with spreading function $S_{\mathbf{H}}(\tau,\nu)$, we have

$$r(t) = L_{\mathbf{H}}(t, f_c)\, e^{j2\pi f_c t}, \tag{1.41}$$

where $L_{\mathbf{H}}(t,f) = \int_{-\infty}^{\infty}\int_{-\infty}^{\infty} S_{\mathbf{H}}(\tau,\nu)\, e^{j2\pi(t\nu - f\tau)}\, d\tau\, d\nu$ is the channel's TF transfer function. In spite of the formal similarity to (1.40), the receive signal in (1.41) is not a complex exponential in general; the time-dependence of the complex factor $L_{\mathbf{H}}(t, f_c)$ may result in strong amplitude and frequency modulation.

Thus, the interpretation of the TF transfer function $L_{\mathbf{H}}(t, f)$ as a "transfer function" appears to be a doubtful matter. In this section, we will argue that this problem and related ones can be resolved using the concept of *underspread* channels. We will introduce two different underspread notions: one is based on the *dispersion spread* and is applicable to any LTV channel, while the other builds upon the *correlation spread* and is only relevant to non-WSSUS channels.

1.5.1 Dispersion-Underspread Property

A difficulty with general LTV channels is the fact that they can cause arbitrarily large joint TF spreads, i.e., arbitrarily severe TF dispersion. The amount of TF dispersion can be quantified by measuring the spread (extension, effective support) of a suitable delay-Doppler representation of the channel (the spreading function for a given channel realization, the scattering function for WSSUS channels, the LSF for non-WSSUS channels). In the following, we will focus on WSSUS channels and, thus, the scattering function. For a discussion of the formulation and implications of the dispersion-underspread property for individual channel realizations and non-WSSUS channels, we refer to Matz and Hlawatsch (1998) and Matz (2005), respectively.

The spread of the scattering function $C_{\mathbf{H}}(\tau, \nu)$ of a WSSUS channel can be measured in different ways. If $C_{\mathbf{H}}(\tau, \nu)$ has a compact support $\mathscr{S}_{\mathbf{H}}$, i.e., $C_{\mathbf{H}}(\tau, \nu) = 0$ for $(\tau, \nu) \neq \mathscr{S}_{\mathbf{H}}$, a simple measure of its spread is given by the area of its support region, $|\mathscr{S}_{\mathbf{H}}|$. An even simpler – but also less accurate – measure of dispersion spread in the compact-support case is given by the area of the smallest rectangle circumscribing $\mathscr{S}_{\mathbf{H}}$, i.e.,

$$d_{\mathbf{H}} \triangleq 4\tau_{\max}\nu_{\max}, \tag{1.42}$$

where $\tau_{\max} \triangleq \max\{|\tau| : (\tau, \nu) \in \mathscr{S}_{\mathbf{H}}\}$ and $\nu_{\max} \triangleq \max\{|\nu| : (\tau, \nu) \in \mathscr{S}_{\mathbf{H}}\}$ are the channel's maximum delay and maximum Doppler, respectively.

A limitation of support-based measures of dispersion spread is the fact that the scattering function of most channels is not compactly supported. We could then use the notion of an "effective support," but this notion presents some arbitrariness. An alternative is provided by moments or, more generally, weighted integrals of the scattering function. The quantities most commonly used in this context are the delay spread σ_τ and Doppler spread σ_ν defined in (1.31), whose product

$$\sigma_{\mathbf{H}} \triangleq \sigma_\tau \sigma_\nu$$

quantifies the dispersion spread without requiring a compact support of $C_{\mathbf{H}}(\tau, \nu)$.

The *dispersion-underspread* property expresses the fact that the channel's dispersion spread is "small." Usually, this constraint is formulated as (Matz & Hlawatsch, 2003; Molisch, 2005; Proakis, 1995)

$$d_{\mathbf{H}} \leq 1 \quad \text{or} \quad \sigma_{\mathbf{H}} \leq 1, \tag{1.43}$$

depending on which measure of dispersion spread is being used. A channel that does not satisfy (1.43) is termed *overspread*. Sometimes, the stronger condition $d_{\mathbf{H}} \ll 1$ or $\sigma_{\mathbf{H}} \ll 1$ is imposed. The underspread conditions (1.43) imply that the channel does not cause *both* strong time dispersion and strong frequency dispersion. An alternative formulation is in terms of the channel's coherence parameters (see (1.32)):

$$T_c F_c \geq 1, \qquad T_c \geq \sigma_\tau, \qquad F_c \geq \sigma_\nu.$$

Each of these three inequalities is strictly equivalent to the second inequality in (1.43). The first inequality states that (the TF transfer function of) an underspread channel has to remain constant within a TF region of area (much) larger than 1; thus, the channel cannot be *both* strongly time-selective and strongly frequency-selective. The second inequality states that the channel stays approximately constant over a time period that is larger than its delay spread, and the third inequality states that the channel stays approximately constant over a frequency band that is larger than its Doppler spread. Note that none of the above conditions requires the channel to be "slowly time-varying." In fact, the time variation (Doppler spread) can be arbitrarily large – i.e., the coherence time can be arbitrarily small – as long as the impulse response is sufficiently short.

Real-world wireless (radio) channels are virtually always underspread. This is because the delay and Doppler of any propagation path are both inversely proportional to the speed of light c_0, i.e., $\tau_{\max} = d_{\max}/c_0$ and $\nu_{\max} = v_{\max} f_c/c_0$, where $d_{\max}$ and $v_{\max}$ denote the maximum path length and maximum relative velocity, respectively. It then follows from (1.42) that $d_{\mathbf{H}} = 4 d_{\max} v_{\max} f_c/c_0^2$. Since $1/c_0^2 \approx 10^{-17}\,\mathrm{s}^2/\mathrm{m}^2$, violation of (1.43) would require that the product $4 d_{\max} v_{\max} f_c$ is on the order of $10^{17}\,\mathrm{m}^2/\mathrm{s}^2$, which is practically impossible for wireless (radio) communications. However, the situation can be different for underwater acoustic channels (see Chapter 9).

Example 1.16

Reconsider the WSSUS channel with exponential power profile and Jakes Doppler profile from Example 1.9. Here, $\sigma_\tau = 10\,\mu\mathrm{s}$ and $\sigma_\nu = 70.71$ Hz or equivalently $T_c = 14.14$ ms and $F_c = 100$ kHz. It follows that $\sigma_{\mathbf{H}} = 7.1 \cdot 10^{-4}$ and $T_c F_c = 1.4 \cdot 10^3$, which shows that this channel is strongly underspread. Indeed, the channel stays approximately constant over several milliseconds, which is much larger than the effective impulse response duration of some tens of microseconds.

For an underspread channel, it follows from (1.11) that the TF transfer function $L_{\mathbf{H}}(t,f)$ is a *smooth* function. Indeed, its Fourier transform – the spreading function – has a small "delay-Doppler bandwidth." This interpretation is furthermore supported by (1.33), which can be shown to imply

$$\frac{1}{\rho_{\mathbf{H}}^2} \mathrm{E}\left\{\left|L_{\mathbf{H}}(t+\Delta t, f+\Delta f) - L_{\mathbf{H}}(t,f)\right|^2\right\} \leq 4\pi\,|\Delta t|\,|\Delta f|\,\sigma_{\mathbf{H}}.$$

This means that the normalized TF transfer function varies very little within TF regions of area $|\Delta t|\,|\Delta f| \ll 1/\sigma_{\mathbf{H}}$. For a channel that is more underspread (smaller $\sigma_{\mathbf{H}}$), the TF region within which $L_{\mathbf{H}}(t,f)$ stays approximately constant is larger.

An alternative view of the channel's TF variation is obtained by studying the TF correlation function $R_{\mathbf{H}}(\Delta t, \Delta f)$. By working out the square on the left-hand side of (1.33), one can show that

$$\frac{|R_{\mathbf{H}}(\Delta t, \Delta f)|}{\rho_{\mathbf{H}}^2} \geq 1 - \pi\left[(\Delta t)^2 \sigma_\nu^2 + (\Delta f)^2 \sigma_\tau^2\right].$$

Thus, $R_{\mathbf{H}}(\Delta t, \Delta f)$ has a slow decay, which means that $L_{\mathbf{H}}(t,f)$ features significant correlations over TF regions whose area is on the order of $1/(\sigma_\tau^2 \sigma_\nu^2)$.

In addition to the smoothness and correlation of the TF transfer function, the underspread property has several other useful consequences; some of them are discussed in Sections 1.5.3–1.5.5. Further results in a similar spirit can be found in Matz (2005) and Matz and Hlawatsch (1998, 2003).

1.5.2 Correlation-Underspread Property

For WSSUS channels, the TF transfer function $L_{\mathbf{H}}(t,f)$ is a 2D stationary process whose (delay-Doppler) power spectrum is given by the scattering function $C_{\mathbf{H}}(\tau,\nu)$. Even though this no longer holds true for non-WSSUS channels, one would intuitively expect, e.g., that the LSF $\mathcal{C}_{\mathbf{H}}(t,f;\tau,\nu)$ of a non-WSSUS channel is a TF-dependent (delay-Doppler) power spectrum. While such an interpretation does not hold in general, it holds in an approximate sense for the practically important class of correlation-underspread channels.

We will first introduce a global measure of the amount of correlation of a non-WSSUS channel via the extension (effective support) of the CCF $\mathcal{R}_{\mathbf{H}}(\Delta t,\Delta f;\Delta\tau,\Delta\nu)$. Specifically, the amount of correlation in time and frequency is measured by the coherence time T_{c} and coherence bandwidth F_{c} defined in (1.32), respectively, and the amount of correlation in delay and Doppler is measured by the moments $\overline{\Delta\tau}$ and $\overline{\Delta\nu}$ defined in (1.38), respectively. Based on these parameters, we define the *correlation spread* as

$$c_{\mathbf{H}} \triangleq T_{\mathrm{c}} F_{\mathrm{c}} \overline{\Delta\tau}\,\overline{\Delta\nu}.$$

The correlation spread characterizes the effective support of the CCF. Note that for a WSSUS channel, $\overline{\Delta\tau} = \overline{\Delta\nu} = 0$ and hence $c_{\mathbf{H}} = 0$.

An LTV channel is called *correlation underspread* if its correlation spread is (much) less than one, i.e., if

$$c_{\mathbf{H}} \leq 1. \tag{1.44}$$

This condition requires that the channel does not exhibit both strong TF correlations and strong delay-Doppler correlations. It is a 2D extension of the underspread notion for 1D nonstationary processes as described, e.g., in Matz and Hlawatsch (2006). The correlation-underspread property for non-WSSUS channels is satisfied by most practical wireless (radio) channels.

Using $T_{\mathrm{c}} = 1/\sigma_\nu$ and $F_{\mathrm{c}} = 1/\sigma_\tau$, the correlation-underspread property (1.44) can be alternatively written as

$$\overline{\Delta\tau}\,\overline{\Delta\nu} \leq \sigma_\tau\sigma_\nu = \sigma_{\mathbf{H}}.$$

That is, the amount of delay-Doppler correlation has to be smaller than the amount of delay-Doppler dispersion. This requires that only delay-Doppler components that are close to each other in the delay-Doppler plane are correlated. Yet another interpretation is obtained by using (1.39), which gives

$$T_{\mathrm{c}} F_{\mathrm{c}} \leq T_{\mathrm{s}} F_{\mathrm{s}}.$$

This means that the channel's stationarity region (whose area is measured by $T_{\mathrm{s}}F_{\mathrm{s}}$) is larger than the channel's coherence region (whose area is measured by $T_{\mathrm{c}}F_{\mathrm{c}}$). The equivalent requirement that the channel statistics change at a much slower rate than the channel's transfer characteristics is fundamental and indispensable for any transceiver technique that is based on channel statistics (in fact, it is tacitly assumed in most of the relevant publications).

Example 1.17

Consider a mobile WiMAX system operating at carrier frequency $f_{\mathrm{c}} = 3.5$ GHz. We assume that the user's speed is around $v = 30$ m/s and the path length between base station and user is around $d = 6$ km. This corresponds to a Doppler spread of $\sigma_\nu = v f_c / c_0 = 350$ Hz and a delay spread of $\sigma_\tau = d/c_0 = 20$ μs. The dispersion spread is thus obtained as $\sigma_{\mathbf{H}} = 7 \cdot 10^{-3}$, which shows that the channel is dispersion-underspread. The coherence parameters are $T_c = 2.9$ ms and $F_c = 50$ kHz.

Reflections from the same physical object (e.g., a building) are an important source of correlated delay-Doppler components. Assuming objects of width $w = 30$ m and maximum angular spread (viewed from the user's position) of $\delta = 2°$ yields the moments $\overline{\Delta\tau} = w/c_0 = 0.1$ μs and $\overline{\Delta\nu} = 2\sigma_\nu \sin(\delta/2) = 12.2$ Hz.

Thus, for this scenario, we obtain a correlation spread of $c_{\mathbf{H}} = 1.7 \cdot 10^{-4}$, and hence, the channel is strongly correlation-underspread. This can also be concluded from the relation $\overline{\Delta\tau}\,\overline{\Delta\nu} = 1.2 \cdot 10^{-6} \ll \sigma_{\mathbf{H}} = 7 \cdot 10^{-3}$. The stationarity parameters are $T_s = 82$ ms and $F_s = 10$ MHz. Comparing with the coherence parameters T_c, F_c given above, we also have $T_c F_c = 1.4 \cdot 10^2 \ll T_s F_s = 8.3 \cdot 10^5$.

■

1.5.3 Approximate Eigenrelation

The eigendecomposition of a linear system (channel) $\mathbf{H}$ is of fundamental importance. The *eigenfunctions* $u_k(t)$ and *eigenvalues* λ_k are defined by the eigenequation[2]

$$(\mathbf{H}u_k)(t) \equiv \int_{-\infty}^{\infty} k_{\mathbf{H}}(t,t')\, u_k(t')\, \mathrm{d}t' = \lambda_k u_k(t). \tag{1.45}$$

For normal channels ($\mathbf{H}\mathbf{H}^* = \mathbf{H}^*\mathbf{H}$, where $\mathbf{H}^*$ denotes the adjoint of $\mathbf{H}$ (Naylor & Sell, 1982)), the (normalized) eigenfunctions $u_k(t)$ constitute an orthonormal basis, i.e., any square-integrable signal can be expanded into the $u_k(t)$ and $\langle u_k, u_l \rangle = \delta_{kl}$. Together with (1.45), this implies the operator diagonalization (Naylor & Sell, 1982)

$$\langle \mathbf{H}u_k, u_l \rangle = \langle \lambda_k u_k, u_l \rangle = \lambda_k \langle u_k, u_l \rangle = \lambda_k \delta_{kl}$$

and, also, the following decomposition of the channel $\mathbf{H}$ or its kernel $k_{\mathbf{H}}(t,t')$:

$$\mathbf{H} = \sum_k \lambda_k u_k \otimes u_k^*, \qquad k_{\mathbf{H}}(t,t') = \sum_k \lambda_k u_k(t)\, u_k^*(t').$$

Here, $u_k \otimes u_k^*$ denotes the rank-one projection operator with kernel $u_k(t)\, u_k^*(t')$.

In the context of digital communications, the channel eigendecomposition is useful for modulator design. Modulating the data symbols a_k onto the eigenfunctions $u_k(t)$, in the sense that the eigenfunctions constitute the "transmit pulses," yields the transmit signal $s(t) = \sum_k a_k u_k(t)$. The resulting receive signal (ignoring noise) is $r(t) = \sum_k a_k (\mathbf{H}u_k)(t) = \sum_k a_k \lambda_k u_k(t)$. Performing demodulation by correlating with the transmit pulses, i.e., projecting $r(t)$ onto the eigenfunctions, gives $\langle r, u_l \rangle = \lambda_l a_l$. Thus, all symbols are recovered free of mutual interference. Note that this scheme requires knowledge of the channel eigenfunctions at the transmitter.

In the limiting case of a time-invariant channel, the index k becomes a frequency variable, the eigenfunctions are given by the complex exponentials, and the eigenvalues correspond to the

[2] We do not consider singular value decompositions since underspread channels can be shown (Matz & Hlawatsch, 1998) to be approximately normal, and for normal systems, the singular value decomposition is equivalent to the eigendecomposition (Naylor & Sell, 1982).

channel's frequency response. The eigendecomposition here possesses an intuitive physical interpretation (in terms of the physically meaningful quantity "frequency") and a computationally attractive structure. With LTV channels, on the other hand, the set of eigenfunctions is no longer independent of $\mathbf{H}$, and it provides neither a physical interpretation nor a computationally attractive structure. However, following Matz and Hlawatsch (1998, 2003), we will now show that these problems can be resolved in an approximate manner for the class of dispersion-underspread WSSUS channels. (Similar results, not discussed here because of space restrictions, apply to non-WSSUS channels.)

1.5.3.1 Approximate Eigenfunctions

We will first argue that at the output of an underspread WSSUS channel, a transmit signal that is well TF-localized is received almost undistorted. Consider an underspread WSSUS channel $\mathbf{H}$ whose mean delay and mean Doppler are zero, i.e., $\bar{\tau} = \bar{\nu} = 0$ (the case of nonzero $\bar{\tau}$ and $\bar{\nu}$ will be discussed later). The transmit signal is constructed as a TF-shifted version of a given unit-energy pulse $g(t)$, i.e., $s(t) = g(t - t_0)\, e^{j2\pi f_0 t}$, where $g(t)$ is well TF-localized about time 0 and frequency 0 and t_0 and f_0 are arbitrary (thus, $s(t)$ is well TF-localized about time t_0 and frequency f_0). It is then possible to show the approximations (see, e.g., Matz & Hlawatsch, 1998 and Chapter 2)

$$r(t) = (\mathbf{H}s)(t) \approx \langle \mathbf{H}s, s\rangle\, s(t) \approx L_{\mathbf{H}}(t_0, f_0)\, s(t). \tag{1.46}$$

Hence, up to small errors, the effect of an underspread channel on a well TF-localized transmit signal is merely a (complex-valued) amplitude scaling, with the scaling factor being approximately equal to $L_{\mathbf{H}}(t_0, f_0)$, i.e., the value of the TF transfer function at the TF location of the transmit signal. The approximation error $r(t) - L_{\mathbf{H}}(t_0, f_0)\, s(t)$ in (1.46) can be bounded as

$$\begin{aligned} \|r - L_{\mathbf{H}}(t_0, f_0)\, s\| &\le \|r - \langle \mathbf{H}s, s\rangle\, s\| + \|\langle \mathbf{H}s, s\rangle\, s - L_{\mathbf{H}}(t_0, f_0)\, s\| \\ &= \|r - \langle r, s\rangle\, s\| + \big|\langle \mathbf{H}s, s\rangle - L_{\mathbf{H}}(t_0, f_0)\big|, \end{aligned} \tag{1.47}$$

where we subtracted/added the term $\langle \mathbf{H}s, s\rangle\, s(t)$, applied the triangle inequality, and used $\|s\| = 1$. We will next quantify the two terms in this bound in the mean-square sense.

With regard to the first term, it can be shown that

$$\mathsf{E}\{\|r - \langle r, s\rangle\, s\|^2\} = \int_{-\infty}^{\infty}\int_{-\infty}^{\infty} C_{\mathbf{H}}(\tau, \nu)\big[1 - |A_g(\tau, \nu)|^2\big]\, d\tau\, d\nu, \tag{1.48}$$

with the *ambiguity function* (Flandrin, 2003; Hlawatsch & Boudreaux-Bartels, 1992; Woodward, 1953)

$$A_g(\tau, \nu) \triangleq \int_{-\infty}^{\infty} g(t)\, g^*(t - \tau)\, e^{-j2\pi\nu t}\, dt.$$

We note that the ambiguity function satisfies

$$|A_g(\tau, \nu)| \le |A_g(0, 0)| = \|g\|^2 = 1. \tag{1.49}$$

In order for the right-hand side in (1.48) to be small, since both $C_{\mathbf{H}}(\tau, \nu)$ and $1 - |A_g(\tau, \nu)|^2$ are non-negative, there must be $|A_g(\tau, \nu)|^2 \approx 1$ on the effective support of $C_{\mathbf{H}}(\tau, \nu)$. For a well TF-localized pulse $g(t)$, consistent with (1.49), $|A_g(\tau,\nu)|^2$ decreases when moving away from the origin, and thus $|A_g(\tau,\nu)|^2 \approx 1$ only in a small region about the origin. Hence, the above requirement means that $C_{\mathbf{H}}(\tau,\nu) \approx 0$ outside this region. In other words, the effective support of $C_{\mathbf{H}}(\tau,\nu)$ has to be concentrated about the origin, i.e., $\mathbf{H}$ has to be an underspread channel.

This line of reasoning can be made more explicit by using the second-order Taylor series expansion of $|A_g(\tau,\nu)|^2$. To simplify the discussion, we assume $g(t)$ to be real-valued and even-symmetric. We then obtain

$$|A_g(\tau,\nu)|^2 \approx 1 - 2\pi^2 \left(B_g^2\tau^2 + D_g^2\nu^2\right). \tag{1.50}$$

Here, we have used the facts that the first-order and mixed second-order terms are zero because of the even symmetry of $|A_g(\tau,\nu)|^2$ (i.e., $|A_g(-\tau,-\nu)|^2 = |A_g(\tau,\nu)|^2$), and that the curvatures of $|A_g(\tau,\nu)|^2$ at the origin are determined by the RMS bandwidth B_g and RMS duration D_g of $g(t)$:

$$-\frac{1}{4\pi^2}\frac{\partial^2 |A_g(\tau,\nu)|^2}{\partial \tau^2}\bigg|_{(0,0)} = B_g^2 \triangleq \int_{-\infty}^{\infty} f^2\, |G(f)|^2\, df,$$

$$-\frac{1}{4\pi^2}\frac{\partial^2 |A_g(\tau,\nu)|^2}{\partial \nu^2}\bigg|_{(0,0)} = D_g^2 \triangleq \int_{-\infty}^{\infty} t^2\, |g(t)|^2\, dt.$$

Inserting (1.50) into (1.48) gives

$$\mathsf{E}\left\{\|r - \langle r,s\rangle\, s\|^2\right\} \approx 2\pi^2 \rho_{\mathbf{H}}^2 \left(\sigma_\tau^2 B_g^2 + \sigma_\nu^2 D_g^2\right). \tag{1.51}$$

This expression is minimized by temporally scaling $g(t)$ such that its duration and bandwidth are balanced according to

$$\frac{D_g}{B_g} = \frac{\sigma_\tau}{\sigma_\nu}, \tag{1.52}$$

in which case

$$\mathsf{E}\left\{\|r - \langle r,s\rangle\, s\|^2\right\} \approx 4\pi^2 \rho_{\mathbf{H}}^2 \sigma_{\mathbf{H}}\, B_g D_g. \tag{1.53}$$

We conclude from this final expression that $r(t) \approx \langle r,s\rangle\, s(t)$ in the mean-square sense if $g(t)$ is well TF-localized (i.e., $B_g D_g$ is small) and $\mathbf{H}$ is underspread (i.e., $\sigma_{\mathbf{H}}$ is small). Hence, under these conditions, $s(t) = g(t-t_0)\, e^{j2\pi f_0 t}$ is an *approximate eigenfunction* of a WSSUS channel. In contrast to the exact eigenfunctions of $\mathbf{H}$, the approximate eigenfunctions $g(t-t_0)\, e^{j2\pi f_0 t}$ are highly structured because they are TF translates of a single prototype function $g(t)$; they do not depend on the specific channel realization, and their parameters t_0, f_0 have an immediate physical interpretation.

We note that the above results and interpretations remain valid even in the case $\bar{\tau} \neq 0$, $\bar{\nu} \neq 0$, provided that the eigenequation is relaxed as

$$(\mathbf{H}s)(t) = \lambda\, s(t-\bar{\tau})\, e^{j2\pi \bar{\nu} t}.$$

This still preserves the shape of the transmit signal but allows for a time delay and frequency shift.

Next, we consider the second term in the bound (1.47).

1.5.3.2 Approximate Eigenvalues

The final approximation in (1.46), $r(t) = (\mathbf{H}s)(t) \approx L_{\mathbf{H}}(t_0, f_0)\, s(t)$, suggests that $L_{\mathbf{H}}(t_0, f_0)$, i.e., the TF transfer function evaluated at the TF location of $s(t)$, plays the role of an approximate eigenvalue. This extends a similar interpretation of the frequency response of time-invariant channels (see (1.40)).

A qualitative argument corroborating this interpretation is as follows. The above approximation implies $\langle r, s\rangle = \langle \mathbf{H}s, s\rangle \approx L_{\mathbf{H}}(t_0, f_0)$. Using $s(t) = g(t - t_0)\, e^{j2\pi f_0 t}$, one can show

$$\langle r, s\rangle = \langle \mathbf{H}s, s\rangle = \int_{-\infty}^{\infty}\int_{-\infty}^{\infty} L_{\mathbf{H}}(t, f)\, \Gamma_g^*(t - t_0, f - f_0)\, dt\, df, \tag{1.54}$$

where $\Gamma_g(t, f) \triangleq \int_{-\infty}^{\infty} g(t)\, g^*(t - \tau)\, e^{-j2\pi f\tau}\, d\tau$ is the *Rihaczek distribution* of the function $g(t)$ (Flandrin, 1999; Hlawatsch & Boudreaux-Bartels, 1992). We note that $\int_{-\infty}^{\infty}\int_{-\infty}^{\infty} \Gamma_g(t, f)\, dt\, df = \|g\|^2 = 1$. Therefore, in order for the weighted integral in (1.54) to be approximately equal to $L_{\mathbf{H}}(t_0, f_0)$, $L_{\mathbf{H}}(t, f)$ has to be effectively constant in a neighborhood of (t_0, f_0) that is determined by the effective support of $\Gamma_g(t - t_0, f - f_0)$. We thus conclude that the approximation $\langle r, s\rangle \approx L_{\mathbf{H}}(t_0, f_0)$ will be better for a smaller support of $\Gamma_g(t - t_0, f - f_0)$ (better TF concentration of $g(t)$) and a smaller rate of variation of $L_{\mathbf{H}}(t, f)$ (smaller dispersion spread of $\mathbf{H}$).

To support this argument by a quantitative result, we consider the second term in the bound (1.47), $|\langle r, s\rangle - L_{\mathbf{H}}(t_0, f_0)|$, in the mean-square sense. Indeed, the mean-square difference between $\langle r, s\rangle$ and $L_{\mathbf{H}}(t_0, f_0)$ can be shown to equal

$$\mathrm{E}\{|\langle r, s\rangle - L_{\mathbf{H}}(t_0, f_0)|^2\} = \int_{-\infty}^{\infty}\int_{-\infty}^{\infty} C_{\mathbf{H}}(\tau, \nu)\, |1 - A_g(\tau, \nu)|^2\, d\tau\, d\nu.$$

This expression is very similar to (1.48), and thus, similar conclusions can be made. That is, in order for $\langle r, s\rangle - L_{\mathbf{H}}(t_0, f_0)$ to be small in the mean-square sense, it is necessary that $s(t)$ (and hence $g(t)$) is well TF-localized and $\mathbf{H}$ is underspread. Similarly to the previous subsection, a second-order Taylor series expansion of $A_g(\tau, \nu)$ (analogous to (1.50)) can be used to make this conclusion more concrete. Specifically, the approximation $|1 - A_g(\tau, \nu)|^2 \approx 2\pi^2(B_g^2\tau^2 + D_g^2\nu^2)$ entails

$$\mathrm{E}\{|\langle r, s\rangle - L_{\mathbf{H}}(t_0, f_0)|^2\} \approx 2\pi^2 \rho_{\mathbf{H}}^2 \left(\sigma_\tau^2 B_g^2 + \sigma_\nu^2 D_g^2\right).$$

The interpretation of this latter approximation is analogous to that of (1.51).

Example 1.18

For the WSSUS channel of Example 1.9, we found $\sigma_\tau = 10\ \mu\text{s}$, $\sigma_\nu = 70.71$ Hz, and $\sigma_{\mathbf{H}} = 7.1 \cdot 10^{-4}$. Consider the Gaussian pulse $g(t) = \frac{1}{\sqrt{\sqrt{2\pi}\, D_g}} \exp\left(-\left(\frac{t}{2D_g}\right)^2\right)$ with RMS duration D_g and RMS bandwidth $B_g = 1/(4\pi D_g)$. The rule (1.52) requires that $D_g/B_g = 4\pi D_g^2$ is equal to $\sigma_\tau/\sigma_\nu = \sqrt{2} \cdot 10^{-7}$, which yields

$$D_g = 106\ \mu\text{s}, \quad B_g = 750\ \text{Hz}.$$

With this choice, the mean-square error (normalized by $\rho_{\mathbf{H}}^2$) of the approximate eigenrelation $(\mathbf{H}s)(t) \approx L_{\mathbf{H}}(t_0, f_0)s(t)$ with $s(t) = g(t - t_0)e^{j2\pi f_0 t}$ is on the order of (cf. (1.53))

$$4\pi^2 \sigma_{\mathbf{H}} B_g D_g = 2.2 \cdot 10^{-3}.$$

■

1.5.4 Time-Frequency Sampling

In many problems like transceiver design and performance evaluation, sampling models for WSSUS channels play an important role. In this section, we consider 2D uniform sampling of the channel's TF transfer function $L_{\mathbf{H}}(t, f)$. This is particularly important for systems employing OFDM (Bingham, 1990), e.g., for OFDM-based channel sounding (see de Lacerda, Cardoso, Knopp, Gesbert, & Debbah, 2007; Suzuki, Thi Van Anh Tran, & Collings, 2007; and Section 1.7.2). Without loss of generality, we assume $\bar{\tau} = \bar{\nu} = 0$ (note that a bulk delay and Doppler shift can be easily split off from $\mathbf{H}$).

Let us consider the representation of a WSSUS channel $\mathbf{H}$ by the samples of $L_{\mathbf{H}}(t, f)$ taken on the uniform rectangular sampling lattice (kT, lF) with $k, l \in \mathbb{Z}$. From the samples $L_{\mathbf{H}}(kT, lF)$, we may attempt to reconstruct $L_{\mathbf{H}}(t, f)$ by means of the following interpolation:

$$\hat{L}_{\mathbf{H}}(t, f) = \sum_{k=-\infty}^{\infty} \sum_{l=-\infty}^{\infty} L_{\mathbf{H}}(kT, lF) \operatorname{sinc}\left(\frac{\pi}{T}(t - kT)\right) \operatorname{sinc}\left(\frac{\pi}{F}(f - lF)\right). \tag{1.55}$$

This interpolation is consistent with the samples in that $\hat{L}_{\mathbf{H}}(kT, lF) = L_{\mathbf{H}}(kT, lF)$. In order to have perfect reconstruction, i.e., $\hat{L}_{\mathbf{H}}(t, f) = L_{\mathbf{H}}(t, f)$ for all t, f in the mean-square sense, the channel's scattering function $C_{\mathbf{H}}(\tau, \nu)$ must have a compact support contained within the rectangular area $[-\tau_{\max}, \tau_{\max}] \times [-\nu_{\max}, \nu_{\max}]$ with $\tau_{\max} \leq 1/(2F)$ and $\nu_{\max} \leq 1/(2T)$. (Using different sampling lattices and reconstruction kernels, it may be possible to slightly relax these conditions.) Note that for decreasing dispersion spread $d_{\mathbf{H}} = 4\tau_{\max}\nu_{\max}$, the number of required transfer function samples per second per Hertz decreases.

If the above conditions are not satisfied (i.e., if the sampling lattice is too coarse or if the scattering function does not have compact support), the reconstructed TF transfer function $\hat{L}_{\mathbf{H}}(t, f)$ will differ from $L_{\mathbf{H}}(t, f)$ due to aliasing. However, it can be shown that the normalized mean-square reconstruction error is bounded as

$$\frac{1}{\rho_{\mathbf{H}}^2} \mathsf{E}\left\{\left|\hat{L}_{\mathbf{H}}(t, f) - L_{\mathbf{H}}(t, f)\right|^2\right\} \leq 2\left(\sigma_\tau^2 F^2 + \sigma_\nu^2 T^2\right).$$

Even though this bound is sometimes rather loose, it yields some useful insights. First, it shows that the mean-square reconstruction error will be small if σ_τ and σ_ν are small relative to the reciprocals $1/F$ and $1/T$ of the sampling periods F and T, respectively, i.e., if the channel is underspread. Second, the bound is minimized by choosing the sampling lattice such that

$$\frac{T}{F} = \frac{\sigma_\tau}{\sigma_\nu}. \tag{1.56}$$

In this case, we obtain

$$\frac{1}{\rho_{\mathbf{H}}^2} \mathsf{E}\left\{\left|\hat{L}_{\mathbf{H}}(t, f) - L_{\mathbf{H}}(t, f)\right|^2\right\} \leq 4TF\sigma_{\mathbf{H}}, \tag{1.57}$$

which shows that good reconstruction is guaranteed for underspread channels (small $\sigma_{\mathbf{H}}$) and sufficiently dense sampling (small TF). Note, in particular, that $4TF\sigma_{\mathbf{H}} \ll 1$ or equivalently $4TF \ll 1/\sigma_{\mathbf{H}}$ ensures small aliasing errors: together with (1.56), $4TF \ll 1/\sigma_{\mathbf{H}}$ implies $2T \ll 1/\sigma_\nu$ and $2F \ll 1/\sigma_\tau$.

Example 1.19

The example WSSUS channel considered previously has an exponential delay profile with $\sigma_\tau = \tau_0 = 10\,\mu\text{s}$ and a Jakes Doppler profile with $\sigma_\nu = 70.71$ Hz ($\nu_{\max} = 100$ Hz), and thus a dispersion spread of $\sigma_{\mathbf{H}} = 7.1 \cdot 10^{-4}$. For this channel, temporal aliasing is avoided if $T \leq 1/(2\nu_{\max}) = 5$ ms. Let us choose $T = 0.5$ ms. The optimum choice of F according to (1.56) is then $F = 3.5$ kHz. With these parameters, our upper bound on the normalized mean-square reconstruction error in (1.57) is obtained as $4TF\sigma_{\mathbf{H}} = 5 \cdot 10^{-3}$.

1.5.5 Approximate Karhunen–Loève Expansion

Non-WSSUS channels are complicated objects because they exhibit two kinds of TF variation:

1. their transfer characteristics, characterized by $L_{\mathbf{H}}(t,f)$, depend on time and frequency;
2. their second-order statistics, described by $\mathcal{C}_{\mathbf{H}}(t,f;\tau,\nu)$, are TF-dependent as well.

However, we have seen in Section 1.5.2 that for correlation-underspread channels, the rate of TF variation of $\mathcal{C}_{\mathbf{H}}(t,f;\tau,\nu)$ is much smaller than that of $L_{\mathbf{H}}(t,f)$.

To simplify the characterization of a non-WSSUS channel $\mathbf{H}$, it is desirable to decouple the channel's transfer characteristics (TF selectivity) and the channel's randomness. A canonical way of doing this is provided by the *Karhunen–Loève expansion* (Chow & Venetsanopoulos, 1974; Kennedy, 1969; Van Trees, 1992)

$$\mathbf{H} = \sum_k \eta_k \mathbf{G}_k. \tag{1.58}$$

Here, the $\mathbf{G}_k$ are "atomic" LTV channels that are deterministic and *orthonormal* in the sense that $\langle \mathbf{G}_k, \mathbf{G}_l \rangle \triangleq \int_{-\infty}^{\infty}\int_{-\infty}^{\infty} k_{\mathbf{G}_k}(t,t')\, k_{\mathbf{G}_l}^*(t,t')\, dt\, dt' = \delta_{kl}$, and the η_k are random coefficients that are *uncorrelated*. The uncorrelatedness of the η_k is due to an appropriate choice of the $\mathbf{G}_k$. Assuming $\mathbf{H}$ to be zero-mean for simplicity, the η_k are zero-mean as well, and thus, their uncorrelatedness is equivalent to statistical orthogonality, i.e., $\mathrm{E}\{\eta_k \eta_l^*\} = 0$ for $k \neq l$.

In spite of this remarkable double-orthogonality property (deterministic orthogonality of the $\mathbf{G}_k$ and statistical orthogonality of the η_k), the Karhunen–Loève expansion is difficult to use in practice since the atomic channels $\mathbf{G}_k$ are generally unstructured, not localized with respect to time and frequency or delay and Doppler, and dependent on the channel's second-order statistics. An important exception is given by WSSUS channels. For these, a continuous version of (1.58) reads

$$\mathbf{H} = \int_{-\infty}^{\infty}\int_{-\infty}^{\infty} S_{\mathbf{H}}(\tau,\nu)\, \mathbf{S}_{\tau,\nu}\, d\tau\, d\nu,$$

where $\mathbf{S}_{\tau,\nu}$ denotes the *TF shift operators* defined by $(\mathbf{S}_{\tau,\nu} s)(t) = s(t-\tau)e^{j2\pi\nu t}$. Here, the delay-Doppler integration variable (τ,ν) replaces the summation index k, the TF shift operators $\mathbf{S}_{\tau,\nu}$ play the role of the atomic channels $\mathbf{G}_k$, and the spreading function $S_{\mathbf{H}}(\tau,\nu)$ takes the place of the coefficients

η_k. The atomic channels $\mathbf{S}_{\tau,\nu}$ are orthogonal, and the coefficients $S_{\mathbf{H}}(\tau,\nu)$ are uncorrelated (because the spreading function of a WSSUS channel is white). In addition, the $\mathbf{S}_{\tau,\nu}$ are structured, physically meaningful, perfectly localized with respect to delay and Doppler, and they do not depend on the channel's second-order statistics.

We will now develop an *approximate* Karhunen–Loève expansion for the broad class of correlation-underspread non-WSSUS channels. This expansion uses highly structured atomic channels that are well localized *both* in time and frequency and in delay and Doppler. Consider a deterministic underspread prototype channel $\mathbf{G}$ that is normalized such that $\|\mathbf{G}\|^2 = \int_{-\infty}^{\infty}\int_{-\infty}^{\infty} |k_{\mathbf{G}}(t,t')|^2\, dt\, dt' = 1$, and whose TF transfer function $L_{\mathbf{G}}(t,f)$ is localized about the origin and smooth so that also the spreading function $S_{\mathbf{G}}(\tau,\nu)$ is concentrated about the origin. This channel acts like a filter that passes only signal components localized about the origin of the TF plane. Based on $\mathbf{G}$, we define

$$\mathbf{G}_{t,f}^{\tau,\nu} \triangleq \mathbf{S}_{t,f+\nu}\, \mathbf{G}\mathbf{S}_{t-\tau,f}^{*}, \tag{1.59}$$

which is a continuously parameterized set of deterministic dispersion-underspread channels that pick up transmit signal components localized about the TF point $(t-\tau,f)$ and shift them to the TF point $(t,f+\nu)$. That is, $\mathbf{G}_{t,f}^{\tau,\nu}$ performs a TF shift by the delay-Doppler pair (τ,ν) on signals localized about the TF point (t,f) and suppresses all other signals (see Fig. 1.9 for an illustration). However, rather than the continuous parameterization in (1.59), our approximate Karhunen–Loève expansion uses the discrete parameterization $\tilde{\mathbf{G}}_{k,l}^{m,d} \triangleq \mathbf{G}_{kT_0,lF_0}^{m\tau_0,d\nu_0}$, where T_0, F_0, τ_0, and ν_0 are sampling periods chosen such that $\{\tilde{\mathbf{G}}_{k,l}^{m,d}\}$ is a *frame* (Gröchenig, 2001) for the space of square-integrable channels (this presupposes, in particular, that $T_0F_0\tau_0\nu_0 \le 1$).

With these assumptions, any square-integrable channel $\mathbf{H}$ (i.e., $\|\mathbf{H}\| < \infty$) can be decomposed as

$$\mathbf{H} = \sum_k \sum_l \sum_m \sum_d \eta_{k,l}^{m,d}\, \tilde{\mathbf{G}}_{k,l}^{m,d}. \tag{1.60}$$

Here, the coefficients are given by $\eta_{k,l}^{m,d} = \langle \mathbf{H}, \mathbf{\Gamma}_{k,l}^{m,d} \rangle$, with $\mathbf{\Gamma}$ denoting a prototype channel that is *dual* to $\mathbf{G}$ (Gröchenig, 2001). The expansion (1.60) constitutes a *Gabor expansion* (Gröchenig, 2001) of the

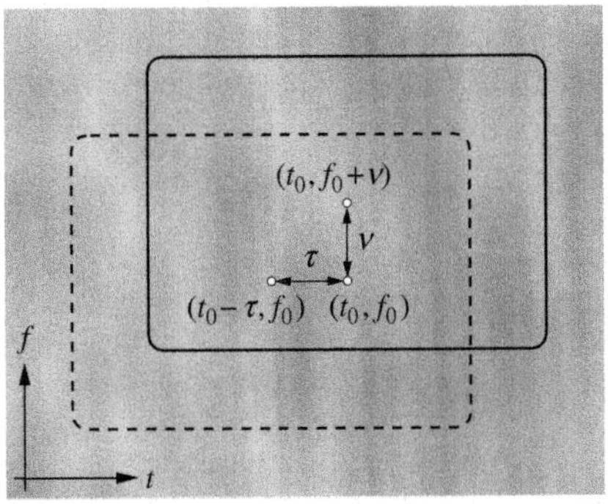

FIGURE 1.9

Action of the atomic channel $\mathbf{G}_{t_0,f_0}^{\tau,\nu}$: signal energy is picked up in the region centered about $(t_0-\tau,f_0)$ (dashed line) and shifted by (τ,ν) to the region centered about $(t_0,f_0+\nu)$ (solid line).

channel $\mathbf{H}$. It is formally similar to the Karhunen–Loève expansion (1.58) but uses a different parameterization and a set of channels with a strong mathematical structure and an intuitive interpretation. Note that (1.60) indeed decouples the transfer characteristics and randomness of the channel $\mathbf{H}$: the atomic channels $\tilde{\mathbf{G}}_{k,l}^{m,d}$ are deterministic and capture the TF selectivity/dispersiveness of $\mathbf{H}$, whereas the coefficients $\eta_{k,l}^{m,d}$ are random and capture the (nonstationary) channel statistics.

In order for (1.60) to constitute an approximate Karhunen–Loève expansion, the coefficients $\eta_{k,l}^{m,d}$ need to be effectively uncorrelated. Indeed, it can be shown that if $\mathbf{H}$ is correlation-underspread and if $\mathbf{G}$ and T_0, F_0, τ_0, ν_0 are suitably chosen, then

$$\mathsf{E}\left\{\eta_{k,l}^{m,d}\left(\eta_{k',l'}^{m',d'}\right)^*\right\} \approx \mathscr{C}_{\mathbf{H}}(kT_0, lF_0; m\tau_0, d\nu_0)\, \delta_{kk'}\delta_{ll'}\delta_{mm'}\delta_{dd'}.$$

This also reveals that the mean power of $\eta_{k,l}^{m,d}$ is approximately equal to the LSF evaluated at $t = kT_0$, $f = lF_0$, $\tau = m\tau_0$, and $\nu = d\nu_0$.

A bound showing that the coefficients $\eta_{k,l}^{m,d}$ are approximately uncorrelated for correlation-underspread channels is provided in Matz (2005). Here, we only consider the approximation $\mathsf{E}\{|\eta_{k,l}^{m,d}|^2\} \approx \mathscr{C}_{\mathbf{H}}(kT_0, lF_0; m\tau_0, d\nu_0)$, whose error can be bounded as

$$\left|\mathsf{E}\{|\eta_{k,l}^{m,d}|^2\} - \mathscr{C}_{\mathbf{H}}(kT_0, lF_0; m\tau_0, d\nu_0)\right|$$
$$\leq \int_{-\infty}^{\infty}\int_{-\infty}^{\infty}\int_{-\infty}^{\infty}\int_{-\infty}^{\infty} \left|\mathscr{R}_{\mathbf{H}}(\Delta t, \Delta f; \Delta\tau, \Delta\nu)\right| \left|1 - \mathscr{R}_{\Gamma}(\Delta t, \Delta f; \Delta\tau, \Delta\nu)\right| \mathrm{d}\Delta t\, \mathrm{d}\Delta f\, \mathrm{d}\Delta\tau\, \mathrm{d}\Delta\nu.$$

For this bound to be small, we require $\mathscr{R}_{\Gamma}(\Delta t, \Delta f; \Delta\tau, \Delta\nu) \approx 1$ on the effective support of the channel's CCF $\mathscr{R}_{\mathbf{H}}(\Delta t, \Delta f; \Delta\tau, \Delta\nu)$. Since $\mathscr{R}_{\Gamma}(\Delta t, \Delta f; \Delta\tau, \Delta\nu) \leq \mathscr{R}_{\Gamma}(0,0;0,0) = 1$ and $\mathscr{R}_{\Gamma}(\Delta t, \Delta f; \Delta\tau, \Delta\nu)$ tends to decrease when moving away from the origin, this requirement can only be satisfied if the effective support of $\mathscr{R}_{\mathbf{H}}(\Delta t, \Delta f; \Delta\tau, \Delta\nu)$ is small, i.e., if the channel $\mathbf{H}$ is correlation-underspread. This argument is largely analogous to the one we provided in the context of (1.48), and it can be made more precise in a similar manner, i.e., by means of a second-order Taylor expansion of $\mathscr{R}_{\Gamma}(\Delta t, \Delta f; \Delta\tau, \Delta\nu)$.

We conclude that (1.60) provides a decomposition of correlation-underspread channels into simple, well-structured deterministic channels; these channels are weighted by effectively uncorrelated random coefficients whose mean power is approximately given by samples of the LSF.

1.6 PARSIMONIOUS CHANNEL MODELS

The complete mathematical description of LTV channels is rather complex. Characterizing T seconds of a transmission – i.e., how T seconds of the receive signal depend on T seconds of the transmit signal – requires a $T \times T$ section of the channel's kernel $k_{\mathbf{H}}(t, t')$ (see (1.14)). Thus, the complexity of characterization grows quadratically with the duration of transmission. Fortunately, most practical channels feature some additional structure, such as the underspread property described in the previous section, which simplifies the description in the sense that a smaller number of parameters is sufficient to model the channel's behavior. Several such parsimonious (low-dimensional, low-rank) representations of LTV channels have been proposed and found to be useful in many applications like channel estimation and equalization (see Chapters 4–8).

For simplicity, we will discuss parsimonious channel representations in a discrete-time setting where the channel's input–output relation reads

$$r[n] = \sum_{m=0}^{M-1} h[n,m]\, s[n-m]. \tag{1.61}$$

Here, $s[n]$, $r[n]$, and $h[n,m]$ are sampled versions of, respectively, $s(t)$, $r(t)$, and $h(t,\tau)$ in (1.13) (the sampling frequency $f_s = 1/T_s$ is assumed to be larger than $B + \nu_{\max}$, where B is the transmit bandwidth); furthermore, $M = \lceil \tau_{\max}/T_s \rceil$ is the number of discrete channel taps, i.e., the maximum discrete-time delay.

1.6.1 Basis Expansion Models

A popular class of low-rank channel models uses an expansion with respect to time n of each tap of the channel impulse response $h[n,m]$ into a basis $\{u_i[n]\}_{i=0,\ldots,I-1}$ (Giannakis & Tepedelenlioğlu, 1998; Tsatsanis & Giannakis, 1996), i.e.,

$$h[n,m] = \sum_{i=0}^{I-1} c_i[m]\, u_i[n]. \tag{1.62}$$

This *basis expansion model* (BEM) is motivated by the observation that the temporal (n) variation of $h[n,m]$ is usually rather smooth due to the channel's limited Doppler spread, and hence, $\{u_i[n]\}_{i=0,\ldots,I-1}$ can be chosen as a small set of smooth functions. In most cases, the BEM (1.62) is considered only within a finite interval, hereafter assumed to be $[0, N-1]$ without loss of generality. The ith coefficient for the mth tap in (1.62) is given by

$$c_i[m] = \langle h[\cdot, m], \tilde{u}_i \rangle = \sum_{n=0}^{N-1} h[n,m]\, \tilde{u}_i^*[n],$$

where $\{\tilde{u}_i[n]\}_{i=0,\ldots,I-1}$ is the biorthogonal basis for the span of $\{u_i[n]\}_{i=0,\ldots,I-1}$ (i.e., $\langle u_i, \tilde{u}_{i'} \rangle = \delta_{ii'}$ for all i, i'). In particular, choosing complex exponentials and polynomials for the $u_i[n]$ results in Fourier and Taylor series, respectively. These two BEMs were essentially proposed already in Bello (1963). The usefulness of (1.62) is due to the fact that the complexity of characterizing $h[n,m]$ on the interval $[0, N-1]$ is reduced from N^2 to $MI \ll N^2$ numbers. However, it is important to note that in most practical cases, an extension of the time interval will require a proportional increase in the BEM model order (i.e., $I \propto N$).

Inserting the basis expansion (1.62) into (1.61) results in

$$r[n] = \sum_{m=0}^{M-1} \sum_{i=0}^{I-1} c_i[m]\, u_i[n]\, s[n-m] = \sum_{i=0}^{I-1} u_i[n] \underbrace{\sum_{m=0}^{M-1} c_i[m]\, s[n-m]}_{\tilde{r}_i[n]}.$$

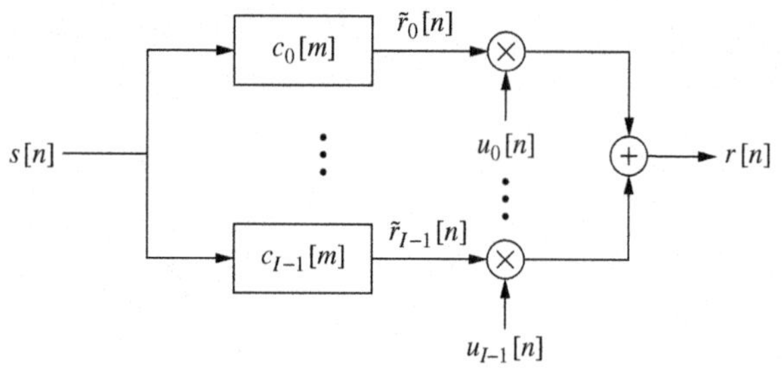

FIGURE 1.10

BEM-induced block diagram of an LTV channel.

Hence, the channel can be viewed as a bank of I time-invariant filters (convolutions) with impulse responses $c_i[m]$ whose outputs $\tilde{r}_i[n]$ are multiplied by the basis functions $u_i[n]$ and added. This parallel connection of "convolution-multiplication" branches is depicted in Fig. 1.10.

By taking length-N discrete Fourier transforms (DFT) of (1.62) with respect to m or n, we obtain the following expressions for the discrete time-varying transfer function (cf. (1.16)) and the discrete spreading function (cf. (1.15)):

$$L_{\mathbf{H}}[n,l] \triangleq \sum_{m=0}^{N-1} h[n,m]\,\mathrm{e}^{-j2\pi\frac{lm}{N}} = \sum_{i=0}^{I-1} C_i[l]\,u_i[n], \tag{1.63}$$

$$S_{\mathbf{H}}[m,d] \triangleq \sum_{n=0}^{N-1} h[n,m]\,\mathrm{e}^{-j2\pi\frac{dn}{N}} = \sum_{i=0}^{I-1} c_i[m]\,U_i[d]. \tag{1.64}$$

Here, l and d denote discrete frequency and discrete Doppler, respectively, and $C_i[l]$ and $U_i[d]$ denote the DFT of $c_i[m]$ and $u_i[n]$, respectively. Since the spreading function of practical wireless channels is concentrated about the origin, (1.64) suggests that the functions $U_i[d]$ have to be concentrated about the origin as well and, hence, that the basis functions $u_i[n]$ have to be smooth.

The expressions (1.62), (1.63), and (1.64) imply that the $N \times N$ matrices $\mathbf{H}$, $\mathbf{L}$, and $\mathbf{S}$ with respective elements $[\mathbf{H}]_{n+1,m+1} = h[n,m]$, $[\mathbf{L}]_{n+1,l+1} = L_{\mathbf{H}}[n,l]$, and $[\mathbf{S}]_{m+1,d+1} = S_{\mathbf{H}}[m,d]$ are sums of I dyadic products, and hence, their rank is (at most) I. For a given channel matrix, the best low-rank approximation is given by the dominant modes of the singular value decomposition (SVD) of that matrix. Unfortunately, the singular vectors depend explicitly on the channel realization and lack a computationally convenient structure. BEMs can thus be viewed as low-rank approximations of the channel's SVD using a fixed and nicely structured set of functions.

1.6.1.1 *Complex Exponential (Fourier) Basis*

The BEM most often employed in practice uses a basis of complex exponentials (Giannakis & Tepedelenlioğlu, 1998; Tsatsanis & Giannakis, 1996). This can be motivated by considering the inverse DFT of (1.64), i.e.,

$$h[n,m] = \frac{1}{N} \sum_{d=0}^{N-1} S_{\mathbf{H}}[m,d]\,\mathrm{e}^{j2\pi\frac{nd}{N}}.$$

Assuming $S_{\mathbf{H}}[m,d]=0$ for $|d|>D$, with D denoting the maximum discrete Doppler shift, results in the so-called (critically sampled) complex exponential (CE) BEM. Here, the model order equals $I=2D+1$, and the basis functions and coefficients are given by $u_i[n]=\mathrm{e}^{j2\pi\frac{(i-D)n}{N}}$ and $c_i[m]=\frac{1}{N}S_{\mathbf{H}}[m,i-D]$, respectively. Unfortunately, the assumption of a maximum discrete Doppler shift is poorly justified even if the underlying continuous channel has a compactly supported spreading function. This is because the multiplicative temporal windowing (i.e., time limitation to the interval $[0,N-1]$) leading to (1.64) corresponds to a convolution in the Doppler domain. This causes Doppler leakage and thus requires a rather large D to achieve satisfactory modeling accuracy. An alternative interpretation is obtained by noting that the uniformly spaced discrete Doppler frequencies d/N (i.e., Doppler resolution $1/N$) usually do not coincide with the actual Doppler frequencies of the continuous channel. In the time (n) domain, this problem manifests itself as a Gibbs (or ringing) phenomenon, which degrades the quality of the CE-BEM especially near the interval boundaries.

To mitigate this Doppler resolution/Gibbs phenomenon problem, oversampling has been proposed (Thomas & Vook, 2000). The basis functions are here given by $\{u_i[n]=\mathrm{e}^{j2\pi\frac{(i-qD)n}{qN}}\}_{i=0,\ldots,I-1}$, where $I=2qD+1$, with oversampling factor $q\in\mathbb{N}$. An even more general BEM based on complex exponentials reads

$$h[n,m]=\sum_{i=0}^{I-1}c_i[m]\,\mathrm{e}^{j2\pi\xi_i n}, \tag{1.65}$$

where $\{\xi_i\}_{i=0,\ldots,I-1}$, with $\xi_i\in(-1/2,1/2)$, is an arbitrary set of normalized Doppler frequencies. The critically sampled case and the oversampled case with uniform Doppler spacing are reobtained for $\xi_i=(i-D)/N$ and $\xi_i=(i-qD)/(qN)$, respectively. A BEM with nonuniform Doppler spacing is potentially able to achieve better modeling accuracy, provided that the "active" Doppler frequencies ξ_i are appropriately determined. Another modification of the CE-BEM that improves modeling accuracy uses windowing techniques (Leus, 2004; Schniter, 2004).

1.6.1.2 Polynomial Basis

Polynomial BEMs correspond to a Taylor series expansion (Borah & Hart, 1999; Tomasin, Gorokhov, Yang, & Linnartz, 2005). More specifically, the continuous-time channel impulse response $h(t,\tau)$ can be approximated about any time instant t_0 as

$$h(t_0+t,\tau)\approx\sum_{i=0}^{I-1}c_i(t_0,\tau)\,t^i, \quad \text{with} \quad c_i(t_0,\tau)=\frac{1}{i!}\left.\frac{\partial^i h(t,\tau)}{\partial t^i}\right|_{t=t_0}.$$

For underspread channels that evolve smoothly with time, the magnitude of the Taylor series coefficients $c_i(t_0,\tau)$ quickly decreases for increasing i so that a small order I is typically sufficient to approximate the channel within a short time (t) interval. The discrete-time (sampled) version of the polynomial basis expansion model on the interval $[0,N-1]$, using $t_0=0$, is then given by

$$h[n,m]=\sum_{i=0}^{I-1}c_i[m]\left(\frac{n}{N}\right)^i, \tag{1.66}$$

where $c_i[m]=c_i(0,mT_s)\,(NT_s)^i$.

An alternative to the monomials t^i underlying (1.66) is given by Legendre polynomials. Here, $u_i[n] = P_i\left(\frac{2n}{N} - 1\right)$, where $P_i(t)$ is the continuous-time Legendre polynomial of degree i defined as (Abramowitz & Stegun, 1965)

$$P_i(t) = \frac{1}{2^i i!} \frac{\mathrm{d}^i}{\mathrm{d}t^i}(t^2 - 1)^i.$$

Even though all polynomial BEMs with identical model order I are mathematically equivalent in the sense that their basis functions span the same subspace, Legendre polynomials are numerically more stable than monomials. In particular, they have the advantage of being orthogonal on the interval $(-1, 1)$.

Because for a given model order I, polynomials tend to vary less with time than complex exponentials, a polynomial BEM is advantageous over a CE-BEM for low Doppler spreads.

1.6.1.3 Slepian Basis

An approach to combating the Doppler leakage occurring in the CE-BEM (see Zemen & Mecklenbräuker, 2005 and Chapter 8), is to replace the complex exponential basis functions by truncated versions of discrete prolate spheroidal sequences (DPSSs) (Slepian, 1978). DPSSs are functions that are band-limited and maximally time-concentrated in the sense of having minimum energy outside a prescribed time interval $[0, N-1]$. For a given temporal blocklength N and a given maximum normalized Doppler frequency $\xi_{\max}$, they are the solutions $u_i[n]$ to the eigenvalue problem

$$\sum_{n'=0}^{N-1} \frac{\sin(2\pi\xi_{\max}(n-n'))}{\pi(n-n')} u_i[n'] = \lambda_i u_i[n], \quad n \in \mathbb{Z}.$$

The DPSSs $u_i[n]$ form an orthogonal basis on $[0, N-1]$ and an orthonormal basis on $\mathbb{Z}$. The eigenvalues λ_i indicate the percentage of energy of $u_i[n]$ within the interval $[0, N-1]$. In fact, for a BEM, only the samples in the interval $[0, N-1]$ are used; these finite-length sequences have been termed *Slepian sequences*. Although originally proposed in Zemen and Mecklenbräuker (2005) for the (flat fading) channel coefficients of individual OFDM subcarriers, the same idea is applicable in the time-delay domain on a per-tap basis in the sense of (1.62). Slepian sequences usually yield a better modeling accuracy than complex exponential sequences, provided the maximum Doppler frequency $\xi_{\max}$ is known with sufficient accuracy. Interestingly, as $\xi_{\max}$ approaches 0, the Slepian sequences asymptotically coincide with the Legendre polynomials.

1.6.2 Parsimonious WSSUS Models

Parsimonious models for the second-order channel statistics are of practical interest, even though in the WSSUS case the statistics do not change with time. Applications of such models include linear minimum mean-square error methods for channel estimation (see Chapters 4, 5, and 7) and channel simulation.

1.6.2.1 AR, MA, and ARMA Models

A flexible statistical channel model is the *autoregressive moving average* (ARMA) model (Baddour & Beaulieu, 2001; Schafhuber, Matz, & Hlawatsch, 2001; Tsatsanis & Giannakis, 1997). For a (discrete-time) WSSUS channel, the channel taps $h[n,m]$ are stationary processes with respect to time n and uncorrelated for different delays m. An ARMA model with respect to n is then defined separately for each tap (m):

$$h[n,m] = -\sum_{\Delta n=1}^{N_{AR}} a_m[\Delta n]h[n-\Delta n,m] + \sum_{\Delta n=0}^{N_{MA}-1} b_m[\Delta n]e_m[n-\Delta n], \quad m=0,\ldots,M-1. \tag{1.67}$$

Here, the $e_m[n]$ are normalized, stationary, and white innovations processes that are mutually independent for different m, N_{AR} and N_{MA} denote the autoregressive (AR) and moving average (MA) model orders, respectively, and $\{a_m[\Delta n]\}_{\Delta n=1,\ldots,N_{AR}}$ and $\{b_m[\Delta n]\}_{\Delta n=0,\ldots,N_{MA}-1}$ are the (deterministic, time-invariant) parameters of the AR and MA part, respectively. For simplicity, we here assume identical ARMA model orders for all channel taps; the extension to m-dependent orders is straightforward.

The power spectral densities of the stationary tap processes constitute the scattering function according to

$$C_{\mathbf{H}}(m,\xi) = \left|\frac{B_m(\xi)}{A_m(\xi)}\right|^2. \tag{1.68}$$

Here, ξ is the normalized Doppler frequency and

$$A_m(\xi) \triangleq \sum_{\Delta n=0}^{N_{AR}} a_m[\Delta n]e^{-j2\pi\xi\Delta n}, \qquad B_m(\xi) \triangleq \sum_{\Delta n=0}^{N_{MA}-1} b_m[\Delta n]e^{-j2\pi\xi\Delta n},$$

with the convention that $a_m[0] \triangleq 1$. Hence, the second-order channel statistics are characterized by $M(N_{AR}+N_{MA})$ complex parameters. The *autoregressive* and *moving average* channel models are obtained as special cases of (1.67) for $N_{MA}=1$ and $N_{AR}=0$, respectively.

There is a rich literature on the properties of ARMA models and on methods for estimating the ARMA parameters $\{a_m[\Delta n], b_m[\Delta n]\}$ and model orders N_{AR}, N_{MA} (Kay, 1988; Stoica & Moses, 1997). These results can be applied to ARMA channel models. However, a difficulty relates to the extremely small effective bandwidth (i.e., maximum normalized Doppler frequency ξ_{max}) of the channel tap processes, which is due to the fact that the sampling frequency f_s is typically orders of magnitude larger than physical Doppler frequencies. Consider for example a mobile broadband transmission with terminal velocity 30 m/s (108 km/h), carrier frequency $f_c = 5$ GHz, and bandwidth $B = 20$ MHz. The maximum Doppler frequency is obtained as $\nu_{max} = 500$ Hz, and for critical sampling ($f_s = B + \nu_{max}$), the maximum normalized Doppler frequency is $\xi_{max} = \nu_{max}/f_s = 2.5 \cdot 10^{-4}$. This means that the tap processes are extremely narrowband, a fact causing difficulties with conventional ARMA design methods. This problem was tackled in Schafhuber et al. (2001) via a multirate approach: a "subsampled" ARMA model designed for an intermediate sampling frequency that is close to the maximum Doppler frequency is followed by an optimum multistage interpolator in order to match the actual system sampling frequency.

Example 1.20

A *first-order Gauss–Markov model* is often used for theoretical considerations and for Monte-Carlo simulations. This is a pure AR channel model with model order $N_{AR} = 1$, i.e.,

$$h[n,m] = -a_m[1]h[n-1,m] + b_m[0]e_m[n],$$

where $|a_m[1]| < 1$. Setting $a_m[1] = A_m e^{j\varphi_m}$ with $0 \le A_m < 1$ and using (1.68), we obtain for the scattering function

$$C_{\mathbf{H}}(m,\xi) = \frac{|b_m[0]|^2}{|1 + a_m[1]e^{-j2\pi\xi}|^2} = \frac{|b_m[0]|^2}{1 + A_m^2 + 2A_m\cos(2\pi\xi - \varphi_m)}.$$

It is seen that $C_{\mathbf{H}}(m,\xi)$ has peaks at the normalized Doppler frequencies $\xi = \frac{\varphi_m}{2\pi} \pm \frac{1}{2}$. These peaks are more pronounced when A_m is closer to 1.

■

1.6.2.2 *Other Parametric WSSUS Models*

Besides ARMA models, there are several other parametric models for the second-order statistics of a WSSUS channel. A simple and frequently used model, previously considered in Example 1.9, is given by the following (continuous-time) scattering function that has a separable structure with exponential delay power profile and Jakes Doppler power profile (Molisch, 2005; Proakis, 1995):

$$C_{\mathbf{H}}(\tau,\nu) = \begin{cases} \frac{\rho_{\mathbf{H}}^2}{\pi\tau_0} e^{-\tau/\tau_0} \frac{1}{\sqrt{\nu_{\max}^2 - \nu^2}}, & |\nu| < \nu_{\max} \\ 0, & |\nu| > \nu_{\max}. \end{cases}$$

This model involves only three parameters: the path gain $\rho_{\mathbf{H}}^2$, the RMS delay spread $\sigma_\tau = \tau_0$, and the maximum Doppler $\nu_{\max}$. The Jakes Doppler profile results from the assumption of a uniform distribution of the azimuth (i.e., the AoA in the horizontal plane).

A very simple WSSUS channel model is given by the brick-shaped scattering function

$$C_{\mathbf{H}}(\tau,\nu) = \begin{cases} \frac{\rho_{\mathbf{H}}^2}{2\tau_{\max}\nu_{\max}}, & (\tau,\nu) \in [0,\tau_{\max}] \times [-\nu_{\max},\nu_{\max}] \\ 0, & \text{else.} \end{cases}$$

This model results in uniform delay and Doppler profiles. There are again three parameters: path gain $\rho_{\mathbf{H}}^2$, maximum delay $\tau_{\max}$, and maximum Doppler $\nu_{\max}$. A brick-shaped scattering function is convenient for theoretical calculations, and it often corresponds to the worst-case statistics for given $\tau_{\max}$ and $\nu_{\max}$ (see Li, Cimini, & Sollenberger, 1998 and Section 2.3.4.4). We note that a uniform Doppler profile is obtained by assuming a jointly uniform distribution of azimuth and elevation, i.e., uniform AoA in all three dimensions (Molisch, 2005).

1.6.3 Parsimonious Non-WSSUS Models

Due to their explicit dependence on time and frequency, the statistics of non-WSSUS channels are more difficult to characterize than those of WSSUS channels. In particular, the LSF and related statistical descriptions of non-WSSUS channels (see Section 1.4.3) are 4D functions. However, the channel statistics change at a much lower rate than the channel itself. This fact can be exploited by parsimonious parametric models for non-WSSUS channels.

1.6.3.1 *Non-WSSUS ARMA Models*

The ARMA channel model (1.67) can be extended to non-WSSUS channels. In order to obtain correlated and nonstationary tap processes, we consider the following joint vector ARMA model with time-varying ARMA parameter matrices:

$$\mathbf{h}[n] = -\sum_{\Delta n=1}^{N_{\mathrm{AR}}} \mathbf{A}[n,\Delta n]\mathbf{h}[n-\Delta n] + \sum_{\Delta n=0}^{N_{\mathrm{MA}}-1} \mathbf{B}[n,\Delta n]\mathbf{e}[n-\Delta n]. \tag{1.69}$$

Here, the vector $\mathbf{h}[n] \triangleq (h[n,0] \cdots h[n,M-1])^T$ contains the M channel taps at time n, $\mathbf{e}[n] \triangleq (e_0[n] \cdots e_{M-1}[n])^T$ is an innovations vector, and $\mathbf{A}[n,\Delta n]$ and $\mathbf{B}[n,\Delta n]$ denote the time-varying AR and MA parameter matrices, respectively. As before, the innovations processes $e_m[n]$ are normalized, stationary, and white, as well as mutually independent for different m. A pure AR version of this model was previously considered in Jachan and Matz (2005). For (temporally) stationary channels, whose taps are still allowed to be correlated, the AR and MA parameter matrices do not depend on time n. This case was previously considered in Tsatsanis, Giannakis, and Zhou (1996). If the tap processes are nonstationary but mutually uncorrelated, the AR and MA parameter matrices become diagonal.

In the general case, (1.69) involves $\mathcal{O}(N(N_{\mathrm{AR}}+N_{\mathrm{MA}})M^2)$ scalar parameters for a block of duration N. This is not really parsimonious, although more parsimonious than a general innovations representation, which would require $\mathcal{O}(N^2M^2)$ scalar parameters. To reduce the description complexity of (1.69), we can expand the ARMA parameter matrices into a low-dimensional basis, similarly to the BEM considered in Section 1.6.1 for the channel itself. Furthermore, exploiting the fact that only closely spaced channel taps are significantly correlated, we can impose a band structure on the ARMA parameter matrices. These two measures reduce the number of scalar parameters to $\mathcal{O}(I(N_{\mathrm{AR}}+N_{\mathrm{MA}})MD)$, where I and D denote the number of basis functions and the number of nonzero diagonals, respectively. We note that nonstationary vector AR processes and corresponding parameter and order estimation methods are discussed in more detail in Jachan, Matz, and Hlawatsch (2009).

1.6.3.2 BEM Statistics

Another parsimonious description of the channel statistics can be based on a channel BEM as given by (1.62). Since the basis $\{u_i[n]\}$ is deterministic, only the statistics of the MI BEM coefficients $c_i[m]$ need to be characterized. From (1.62), we obtain

$$\mathsf{E}\{h[n,m]h^*[n',m']\} = \sum_{i=0}^{I-1}\sum_{i'=0}^{I-1} \mathsf{E}\{c_i[m]c_{i'}^*[m']\}u_i[n]u_{i'}^*[n'].$$

This shows that the correlation function of $h[n,m]$ is completely characterized by the correlation function of $c_i[m]$. The description complexity is thus reduced from $(MN)^2$ to only $(MI)^2$.

For uncorrelated channel taps, $\mathsf{E}\{c_i[m]c_{i'}^*[m']\} = 0$ for $m \neq m'$ so that the channel statistics are described by only MI^2 parameters. In addition, it is often reasonable to assume that two BEM coefficients $c_i[m], c_{i'}[m]$ are uncorrelated even for the same m so that

$$\mathsf{E}\{c_i[m]c_{i'}^*[m']\} = \mathsf{E}\{|c_i[m]|^2\}\delta_{mm'}\delta_{ii'}. \tag{1.70}$$

Here, only the MI powers $\mathsf{E}\{|c_i[m]|^2\}$ need to be specified. The channel statistics corresponding to (1.70) are still non-WSSUS in general. The only exception is a CE-BEM, i.e., $u_i[n] = \exp(j2\pi\xi_i n)$ as in (1.65). Here, (1.70) corresponds to a discrete-time version of a WSSUS channel (cf. (1.23)), i.e.,

$$\mathsf{E}\{h[n,m]h^*[n',m']\} = r_h[n-n';m]\delta_{mm'}, \tag{1.71}$$

with

$$r_h[\Delta n; m] = \sum_{i=0}^{I-1} \mathsf{E}\{|c_i[m]|^2\} e^{j2\pi\xi_i \Delta n}.$$

The associated (discrete-time) scattering function is given by

$$C_{\mathbf{H}}(m,\xi) = \sum_{i=0}^{I-1} \mathsf{E}\{|c_i[m]|^2\} \delta(\xi - \xi_i).$$

1.7 MEASUREMENT

The measurement of wireless channels is vital for the design, simulation, and performance evaluation of broadband wireless systems. In this section, therefore, we review some of the techniques that can be used to obtain measurements of channel realizations (a task also known as *channel sounding*) and of channel statistics. We assume a transmission phase that is exclusively dedicated to channel measurement, without simultaneous data transmission. However, some of the insights gained apply also to training-based channel estimation during data transmission (see Chapters 4, 5, and 7). Throughout this section, we consider a noise-free scenario for simplicity; this can be justified by the fact that channel sounders usually operate in the high-SNR regime.

1.7.1 Spread-Spectrum-Like Channel Sounding

We first describe how the impulse response or transfer function of an LTV channel can be measured by means of wideband channel sounders using spreading sequences. Popular types of channel sounders are pseudo-noise sequence correlation sounders, swept time-delay cross-correlation sounders, and chirp sounders (Cullen, Fannin, & Molina, 1993; Fannin, Molina, Swords, & Cullen, 1991; Parsons, Demery, & Turkmani, 1991). These sounders employ correlation/pulse-compression techniques that are motivated by the identification of time-invariant systems. As we will demonstrate, the application of such techniques to LTV systems or channels results in systematic measurement errors (Matz et al., 2002). These errors must be kept small by an appropriate choice of the sounder parameters.

1.7.1.1 Idealized Impulse-Train Sounder

A conceptual basis for the above-mentioned correlation/pulse-compression techniques is provided by a very simple channel sounder (Kailath, 1962) that uses as the transmit (sounding) signal an impulse train of period T:

$$s(t) = \Delta(t) \triangleq \sum_{k=-\infty}^{\infty} \delta(t - kT). \tag{1.72}$$

The resulting channel output is a superposition of slices of the kernel $k_{\mathbf{H}}(t,t')$ in (1.14), i.e.,

$$r(t) = (\mathbf{H}\Delta)(t) = \int_{-\infty}^{\infty} k_{\mathbf{H}}(t,t')\,\Delta(t')\,dt' = \sum_{k=-\infty}^{\infty} k_{\mathbf{H}}(t,kT). \tag{1.73}$$

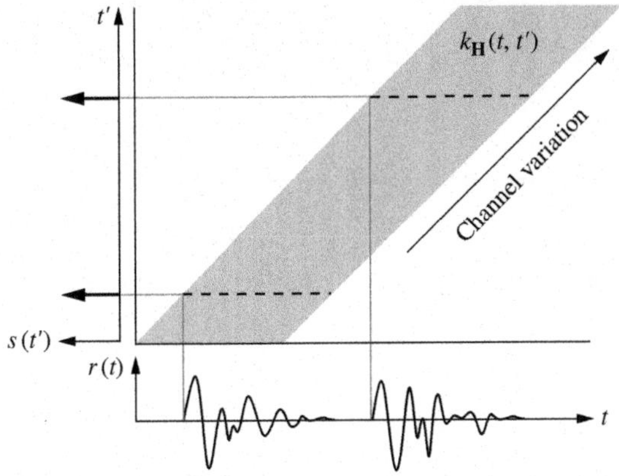

FIGURE 1.11

Schematic illustration of idealized impulse-train channel sounding.

This channel sounding principle is illustrated in Fig. 1.11. Let $p(\tau)$ denote a rectangular window of length T (i.e., $p(\tau) = 1$ for $0 \leq \tau < T$ and $p(\tau) = 0$ otherwise), and assume that the channel's maximum delay satisfies $\tau_{\max} \leq T$ (i.e., $k_{\mathbf{H}}(t,t') = 0$ for $t - t' \notin [0,T)$). We then obtain with (1.73)

$$r(\tau + kT)p(\tau) = \sum_{k'=-\infty}^{\infty} k_{\mathbf{H}}(\tau + kT, k'T)p(\tau) = k_{\mathbf{H}}(\tau + kT, kT) = \tilde{h}(kT, \tau), \tag{1.74}$$

where $\tilde{h}(t,\tau) \triangleq k_{\mathbf{H}}(t + \tau, t)$ is a channel impulse response that is dual to $h(t,\tau) = k_{\mathbf{H}}(t, t - \tau)$. Relation (1.74) shows that the kth block of the receive signal $r(t)$ (the kth "channel snapshot") is a slice of $\tilde{h}(t,\tau)$ at $t = kT$. Note, however, that it is *not* a slice of the usual channel impulse response $h(t,\tau)$. If the channel's maximum Doppler shift satisfies $\nu_{\max} \leq 1/(2T)$, the complete dual impulse response can be recovered by interpolating between successive measurements (Kailath, 1962):

$$\tilde{h}(t,\tau) = (\Upsilon r)(t,\tau) \triangleq \sum_{k=-\infty}^{\infty} r(\tau + kT)p(\tau)\,\mathrm{sinc}\Big(\frac{\pi}{T}(t - kT)\Big). \tag{1.75}$$

From $\tilde{h}(t,\tau)$, the usual impulse response can then be obtained as $h(t,\tau) = \tilde{h}(t - \tau, \tau)$. Note that the two conditions for this sounding method to work are $\tau_{\max} \leq T$ and $\nu_{\max} \leq 1/(2T)$. Combining them gives the *underspread* condition $\tau_{\max}\nu_{\max} \leq 1/2$ (see Section 1.5.1).

1.7.1.2 Generic Correlative Channel Sounder Model

While being conceptually simple, the idealized impulse-train sounder is impractical because the impulse-train sounding signal $\Delta(t)$ has a prohibitively large – theoretically infinite – crest factor (or peak-to-average power ratio). This disadvantage can be avoided by the use of correlation/pulse-compression techniques. Figure 1.12 shows a generic *correlative channel sounder*. The sounding signal

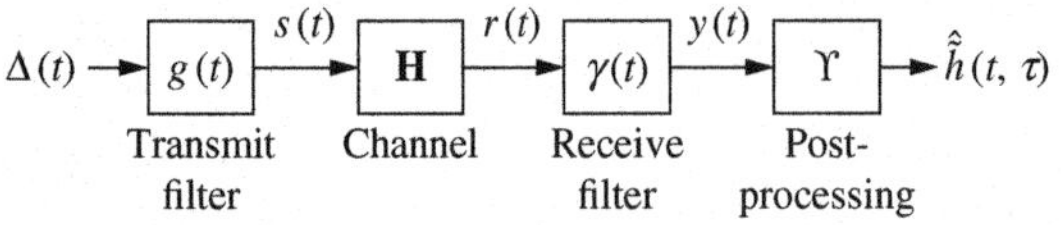

FIGURE 1.12

Correlative channel sounder model.

transmitted over the channel is given by

$$s(t) = (\Delta * g)(t) = \sum_{k=-\infty}^{\infty} g(t-kT), \tag{1.76}$$

where T is the sounding period and $g(t)$ is a sounding pulse or, equivalently, the impulse response of a time-invariant transmit (pulse-shaping) filter. The receive signal $r(t) = (\mathbf{H}s)(t)$ is passed through a time-invariant receive filter with impulse response $\gamma(t)$, which yields the signal $y(t) = (r * \gamma)(t)$. Finally, an estimate $\hat{\tilde{h}}(t,\tau)$ of the channel's dual impulse response $\tilde{h}(t,\tau)$ is obtained by forming the snapshots $\hat{\tilde{h}}(kT,\tau) = y(\tau + kT)p(\tau)$ as in (1.74) and interpolating them (cf. (1.75)):

$$\hat{\tilde{h}}(t,\tau) = (\Upsilon y)(t,\tau). \tag{1.77}$$

We note that the idealized impulse-train channel sounder is a special case of the generic correlative channel sounder with transmit and receive filters $g(t) = \gamma(t) = \delta(t)$.

Correlative channel sounding is based on the assumption that the order of the channel and the receive filter (see Fig. 1.12) can be approximately reversed (in the theoretical analysis) so that the signal $y(t)$ at the output of the receive filter $\gamma(t)$ can be approximately expressed as

$$y(t) \approx (\mathbf{H}\tilde{\Delta})(t), \quad \text{with} \quad \tilde{\Delta}(t) \triangleq \sum_{k=-\infty}^{\infty} (g * \gamma)(t-kT).$$

Hence, the overall scheme can be interpreted as if the channel were sounded using the *virtual sounding signal* $\tilde{\Delta}(t)$. The transmit and receive filters are designed such that[3] $(g * \gamma)(t) \approx \delta(t)$, since then the virtual sounding signal approximately equals an impulse train, i.e., $\tilde{\Delta}(t) \approx \Delta(t)$, and thus, the sounder approximates the idealized impulse-train channel sounder, even though the actual sounding signal in (1.76) can have a small crest factor.

For a time-invariant channel, the system with channel and receive filter interchanged is equivalent to the original system. For an LTV channel, however, this equivalence holds only approximately at best, and this fact entails a systematic error in the sounding result. This measurement error grows with the maximum Doppler frequency of the channel and with the duration of the impulse response of the receive filter (Matz et al., 2002).

[3]Since only a band-limited part (with bandwidth B) of the channel is measured and since practical transmit and receive filters are causal, one should rather require $(g * \gamma)(t) \approx B\,\mathrm{sinc}(\pi B(t-t_0))$ with t_0 sufficiently large. However, here we use $t_0 = 0$ and $B \to \infty$ for simplicity.

Correlative channel sounding can also be represented in the TF domain. A definition of the frequency response of an LTV channel that is dual to the TF transfer function $L_{\mathbf{H}}(t,f)$ in (1.16) is given by the *frequency-dependent modulation function* (Bello, 1963)

$$\tilde{L}_{\mathbf{H}}(t,f) \triangleq \int_{-\infty}^{\infty} \tilde{h}(t,\tau)\,\mathrm{e}^{-j2\pi f\tau}\,\mathrm{d}\tau.$$

An estimate of $\tilde{L}_{\mathbf{H}}(t,f)$ at time $t=kT$ can be obtained by Fourier transforming $\hat{\tilde{h}}(kT,\tau)=y(\tau+kT)p(\tau)$ with respect to τ:

$$\hat{\tilde{L}}_{\mathbf{H}}(kT,f) = \int_{-\infty}^{\infty} \hat{\tilde{h}}(kT,\tau)\,\mathrm{e}^{-j2\pi f\tau}\,\mathrm{d}\tau = \int_{-\infty}^{\infty} y(\tau+kT)\,p(\tau)\,\mathrm{e}^{-j2\pi f\tau}\,\mathrm{d}\tau. \tag{1.78}$$

Hence, $\hat{\tilde{L}}_{\mathbf{H}}(kT,f)$ can be viewed as the short-time Fourier transform (Flandrin, 1999; Hlawatsch & Boudreaux-Bartels, 1992; Nawab & Quatieri, 1988) of $y(t)$ with analysis window $p(t)$, sampled at $t=kT$. An estimate of $\tilde{L}_{\mathbf{H}}(t,f)$ for arbitrary t can be obtained either by interpolating $\hat{\tilde{L}}_{\mathbf{H}}(kT,f)$ or by Fourier transforming the interpolated impulse response estimate $\hat{\tilde{h}}(t,\tau)$ in (1.77).

1.7.1.3 Sounder Types

As mentioned above, the design of correlative channel sounders aims at achieving $(g*\gamma)(t)\approx\delta(t)$, since then the virtual sounding signal $\tilde{\Delta}(t)$ approximates an impulse train $\Delta(t)$. We next discuss two common designs.

- *PN-sequence sounder*: While all sounders within the correlative sounding model employ correlation techniques, the term "correlation sounder" usually refers to sounders whose transmit and receive filters are based on a binary pseudo-noise (PN) sequence $b_i \in \{-1,1\}$ according to

$$g(t) = \sum_{i=1}^{N} b_i\,c(t-iT_c), \qquad \gamma(t) = g(NT_c - t) = \sum_{i=i}^{N} b_{N-i+1}\,c(-(t-iT_c)).$$

  Here, $c(t)$ is a chip pulse, and T_c is the chip duration; note that $NT_c \le T$. For N sufficiently large, $g(t)$ and $\gamma(t)$ induce a virtual sounding signal $\tilde{\Delta}(t)$ that approximates an impulse train within the measurement bandwidth $1/(2T_c)$. We note that sounding sequences b_i other than PN sequences and receive filters that are not time-reversed replicas of the transmit filter have also been proposed (e.g., Schwarz, Martin, & Schüßler, 1993; Thomä et al., 2000). Furthermore, there exists a practical modification of the PN sequence sounder, termed *swept time-delay cross-correlator*, that trades delay resolution against channel tracking capability (Cox, 1972; Rappaport, 1996).
- *Chirp sounder*: As an alternative to PN sequences, chirp sounders use chirp signals, which have a similarly low crest factor. The transmit and receive filters are given by (Salous, Nikandrou, & Bajj, 1998)

$$g(t) = \mathrm{e}^{j2\pi B(t^2/T_g - t)}, \quad \gamma(t) = g^*(T_g - t) = \mathrm{e}^{-j2\pi B(t^2/T_g - t)}, \quad \text{for } 0 \le t < T_g,$$

where $T_g \le T$. These filters again induce a virtual sounding signal $\tilde{\Delta}(t)$ that approximates an impulse train within the measurement band. This approximation can be improved by using nonrectangular (e.g., Gaussian) envelopes for $g(t)$ and $\gamma(t)$ (Skolnik, 1984).

1.7.1.4 Measurement Errors

For LTV channels, the correlative sounding techniques described above suffer from several types of systematic measurement errors (Matz et al., 2002), in addition to the random effects of measurement noise.

- As mentioned previously, the commutation argument underlying correlative channel sounding – i.e., obtaining a virtual sounding signal by swapping channel and receive filter – can hold true for LTV channels only in an approximate manner. The resulting *commutation error* increases with the Doppler spread of the channel and with the length of the receive filter.
- Practically realizable transmit and receive filters have imperfect pulse-compression (correlation) properties, i.e., $(g * \gamma)(t) \neq \delta(t)$ and hence $\tilde{\Delta}(t) \neq \Delta(t)$. This leads to a *pulse-compression error*, even in the case of a time-invariant channel. This error tends to be particularly pronounced for transmit and receive filters of small length. Usually, one attempts to reduce pulse-compression errors via back-to-back calibration, where transmitter and receiver are connected by a short cable, and the response $(g * \gamma)(t)$ is recorded for later equalization of the channel measurements. However, back-to-back calibration tends to increase the measurement noise, and for the case of LTV channels, it is also affected by systematic errors (Matz et al., 1999).
- For channels with maximum delay $\tau_{\max} > T$, successive channel snapshots will overlap (aliasing in the delay domain), and for channels with maximum Doppler shift $\nu_{\max} > 1/(2T)$, the channel snapshots do not track the channel variations sufficiently fast (aliasing in the Doppler domain). These two error mechanisms are combined in the *aliasing error*, which grows with the channel's dispersion spread and can be minimized by an appropriate choice of the sounding period T. The aliasing error vanishes if and only if (Kailath, 1962; Matz et al., 2002)

$$\tau_{\max} \le T \le \frac{1}{2\nu_{\max}}.$$

  As remarked previously, this in turn presupposes $\tau_{\max}\nu_{\max} \le 1/2$, i.e., a dispersion-underspread channel (see Section 1.5.1).
- Finally, the measured function $\hat{\tilde{h}}(t,\tau)$ is typically used as an estimate of the impulse response $h(t,\tau)$, even though it is actually an estimate of the dual impulse response $\tilde{h}(t,\tau)$. This *misinterpretation error* again grows with the channel's dispersion spread. However, it can be easily avoided by converting $\hat{\tilde{h}}(t,\tau)$ to an estimate of $h(t,\tau)$ according to $\hat{h}(t,\tau) = \hat{\tilde{h}}(t-\tau,\tau)$.

A quantitative analysis of these systematic errors along with a corresponding optimization of relevant sounder parameters has been performed in Matz et al. (2002).

1.7.2 Multicarrier Channel Sounding

The channel sounders described in the previous subsection resemble a spread spectrum communication system; the spreading is used to achieve favorable pulse compression/correlation properties. Next, we discuss channel sounding techniques that resemble multicarrier communication systems such as OFDM systems. Such sounders are described, e.g., in Suzuki et al. (2007); de Lacerda et al. (2007).

1.7.2.1 Multicarrier Basics

The basic principle of multicarrier systems – see Chapters 2 and 7 for details – is to split the transmit band (bandwidth B) into L subcarriers and transmit L symbol streams $a[k,l]$, $l = 0,\dots,L-1$ in parallel over these subcarriers. This results in a transmit signal of the form

$$s(t) = \sum_k \sum_{l=0}^{L-1} a[k,l]\, g_{k,l}(t), \quad \text{with} \quad g_{k,l}(t) \triangleq g(t-kT)\, e^{j2\pi lFt},$$

where k and l denote the symbol (time) index and the subcarrier index, respectively, T is the symbol duration, $F = B/L$ is the frequency separation of the subcarriers, and $g(t)$ is a transmit pulse. From the channel output $r(t)$, the receiver calculates the following estimate of the transmit symbols $a[k,l]$:

$$y[k,l] = \langle r, \gamma_{k,l} \rangle = \int_{-\infty}^{\infty} r(t)\, \gamma_{k,l}^*(t)\, dt, \quad \text{with} \quad \gamma_{k,l}(t) \triangleq \gamma(t-kT)\, e^{j2\pi lFt},$$

where $\gamma(t)$ is a receive pulse. In the absence of noise and channel distortions, i.e., when $r(t) = s(t)$, there is $y[k,l] = a[k,l]$ if and only if the *biorthogonality condition* (Gröchenig, 2001) $\langle g_{k',l'}, \gamma_{k,l} \rangle = \delta_{kk'}\delta_{ll'}$ is satisfied. A necessary condition for biorthogonality is $TF \geq 1$. The most prominent example of a biorthogonal system is given by the rectangular pulses

$$g(t) = \begin{cases} \sqrt{F}, & 0 \leq t \leq T, \\ 0, & \text{else}, \end{cases} \qquad \gamma(t) = \begin{cases} \sqrt{F}, & 0 \leq t \leq 1/F, \\ 0, & \text{else}. \end{cases} \tag{1.79}$$

This corresponds to a classical OFDM system with a cyclic prefix (CP) of duration $T_{cp} = T - 1/F$.

1.7.2.2 Measurement Principle

When using a multicarrier system for channel sounding, the symbols $a[k,l]$ are actually training symbols known by the receiver. The measurement principle of multicarrier channel sounders is motivated by the approximate eigenrelation discussed in Section 1.5.3 (see (1.46)),

$$(\mathbf{H} g_{k,l})(t) \approx L_{\mathbf{H}}(kT, lF)\, g_{k,l}(t). \tag{1.80}$$

This approximation is accurate if the transmit pulse $g(t)$ is well TF-localized and the channel is sufficiently dispersion-underspread. We note that for a time-invariant channel and CP-OFDM (see (1.79)), the approximation is exact, i.e., $(\mathbf{H} g_{k,l})(t) = H(lF)\, g_{k,l}(t)$, provided that the channel's maximum delay satisfies $\tau_{max} \leq T_{cp}$. Inserting the approximation (1.80) in $r(t) = (\mathbf{H}s)(t) = \sum_k \sum_{l=0}^{L-1} a[k,l]\, (\mathbf{H} g_{k,l})(t)$, we obtain for the demodulated symbols

$$y[k,l] = \langle r, \gamma_{k,l} \rangle \approx \sum_{k'} \sum_{l'=0}^{L-1} a[k',l']\, L_{\mathbf{H}}(k'T, l'F)\, \langle g_{k',l'}, \gamma_{k,l} \rangle = a[k,l]\, L_{\mathbf{H}}(kT, lF), \tag{1.81}$$

where the biorthogonality property $\langle g_{k',l'}, \gamma_{k,l}\rangle = \delta_{kk'}\delta_{ll'}$ was used in the last step. Based on (1.81), an estimate of the channel transfer function at time kT and frequency lT ($l = 0,\ldots,L-1$) is obtained as

$$\hat{L}_{\mathbf{H}}(kT, lF) = \frac{y[k,l]}{a[k,l]}. \tag{1.82}$$

Still assuming a dispersion-underspread channel, it follows from our discussion in Section 1.5.4 that an estimate of the overall transfer function $\hat{L}_{\mathbf{H}}(t,f)$ can be calculated by lowpass interpolation of the estimate $\hat{L}_{\mathbf{H}}(kT, lF)$ according to (1.55). An estimate of the impulse response $h(t,\tau)$ can then be obtained from (1.82) via the Fourier series

$$\hat{h}(kT,\tau) = \sum_{l=0}^{L-1} \hat{L}_{\mathbf{H}}(kT, lF)\, e^{j2\pi lF\tau}, \qquad 0 \le \tau \le \frac{1}{F},$$

and subsequent interpolation with respect to t.

1.7.2.3 *Measurement Errors*

The measurement principle described above hinges fundamentally on the approximation (1.80). This approximation becomes less accurate for a larger channel dispersion spread $\sigma_{\mathbf{H}}$ and for a larger time-bandwidth product of the transmit pulse $g(t)$ (we assume that the RMS duration and RMS bandwidth of $g(t)$ are chosen according to (1.52)). The approximation error in (1.80) manifests itself as *intersymbol interference* (ISI) and *intercarrier interference* (ICI), which are not described by (1.81). ISI and ICI represent the main limitations of multicarrier channel sounding. In CP-OFDM, ISI is avoided by an appropriate choice of the CP duration T_{cp}; however, the rectangular pulses used in CP-OFDM result in increased ICI, especially for fast time-varying channels (see, e.g., Matz, Schafhuber, Gröchenig, Hartmann, & Hlawatsch, 2007).

Since (1.82) only provides estimates of samples of $L_{\mathbf{H}}(t,f)$ at $t = kT$ and $f = lF$, aliasing errors may arise (see Section 1.5.4). This is similar to spread-spectrum-like sounders. Assuming the optimum choice $T/F = \sigma_\tau/\sigma_\nu$ (see (1.56)), the aliasing errors can be bounded in terms of $\sigma_{\mathbf{H}}$ and TF according to (1.57).

1.7.3 Extension to Multiantenna Systems

The channel sounding techniques described above can be extended to multiantenna systems (i.e., MIMO systems) by means of a procedure known as *time-division multiplexed switching* (TDMS). Here, the individual spatial channels are measured successively by activating one antenna pair at a time. This can be done by physically moving the antennas of a single-antenna channel sounder (e.g., using a step motor), thereby creating a virtual multiantenna system. Alternatively, a single RF chain may be switched electronically between multiple antennas.

For time-invariant channels, theoretically, this measurement principle does not introduce any additional errors. In contrast, for LTV channels, the channel impulse response changes between successive measurements; thus, pretending that these measurements were taken simultaneously leads to errors. In practice, such errors arise even in static scenarios since oscillator drift and phase noise result in noticeable time variations of the effective channel (see Example 1.4). The resulting errors can be significant in the multiantenna setup (Baum & Bölcskei, 2009).

Example 1.21

Consider a 2×2 MIMO system with a rank-one flat-fading time-invariant channel, i.e., $\mathbf{H} = h\binom{1}{1}\binom{1}{1}^T = h\begin{pmatrix}1 & 1\\ 1 & 1\end{pmatrix}$. Using a TDMS-based channel sounder with oscillator phase noise (but no other errors) results in the channel estimate $\hat{\mathbf{H}} = h\begin{pmatrix}e^{j\phi_1} & e^{j\phi_2}\\ e^{j\phi_3} & e^{j\phi_4}\end{pmatrix}$, where the matrix entries $e^{j\phi_i}$ describe phase rotations that are due to the oscillator phase noise in the four successive measurement periods. Clearly, the measured channel will be typically full-rank even though the actual channel has rank one.

A detailed analysis of the sounding errors resulting from antenna switching or displacement is provided in Baum and Bölcskei (2009). It is shown that the errors become larger for increasing temporal separation of successive spatial measurements. Furthermore, they typically lead to an increased rank of the measured MIMO channel matrix and, in turn, to over-estimation of the MIMO channel capacity.

Switching errors can be avoided by using a dedicated RF chain for each antenna and separating the spatial channels in the frequency domain rather than in the time domain (e.g., by dedicating each subcarrier in a multicarrier channel sounder to a single antenna pair).

1.7.4 Measurement of Second-Order Statistics

Second-order channel statistics are important for many tasks in transceiver design and performance evaluation, such as the design of channel estimators, precoders and beamformers, as well as channel simulation. In the following, we summarize some nonparametric and parametric estimators of the second-order channel statistics. We will again consider a noise-free transmission for simplicity.

1.7.4.1 Nonparametric Estimation

In the case of WSSUS channels, we are interested in estimating the scattering function $C_{\mathbf{H}}(\tau, \nu)$ or the TF correlation function $R_{\mathbf{H}}(\Delta t, \Delta f)$ (see (1.26)). An estimator of the scattering function that is based on the statistical input–output relation (1.28) was proposed in Gaarder (1968). For a deterministic transmit signal $s(t)$, (1.28) becomes $\bar{A}_r(\Delta t, \Delta f) = R_{\mathbf{H}}(\Delta t, \Delta f) A_s(\Delta t, \Delta f)$, where $\bar{A}_r(\Delta t, \Delta f)$ is the expected ambiguity function of $r(t)$ as defined in Section 1.4.1 and $A_s(\Delta t, \Delta f) = \int_{-\infty}^{\infty} s(t)\, s^*(t - \Delta t)\, e^{-j2\pi t \Delta f} dt$ is the ambiguity function of $s(t)$ previously used in Section 1.5.3. Hence, the TF correlation function can be expressed as $R_{\mathbf{H}}(\Delta t, \Delta f) = \bar{A}_r(\Delta t, \Delta f)/A_s(\Delta t, \Delta f)$. An estimate of $R_{\mathbf{H}}(\Delta t, \Delta f)$ can then be calculated as

$$\hat{R}_{\mathbf{H}}(\Delta t, \Delta f) = \frac{\hat{A}_r(\Delta t, \Delta f)}{A_s(\Delta t, \Delta f)}, \tag{1.83}$$

where $\hat{A}_r(\Delta t, \Delta f)$ is an estimate of $\bar{A}_r(\Delta t, \Delta f)$. To obtain this latter estimate, a finite-duration sounding signal $s(t)$ is successively transmitted Q times, resulting in Q receive signals $r_q(t)$ (we assume that the different arrival times of these signals have been compensated so that all $r_q(t)$ are temporally aligned). The estimated expected ambiguity function $\hat{A}_r(\Delta t, \Delta f)$ is taken to be the average of the ambiguity functions of the $r_q(t)$:

$$\hat{A}_r(\Delta t, \Delta f) = \frac{1}{Q} \sum_{q=1}^{Q} A_{r_q}(\Delta t, \Delta f).$$

Finally, an estimate of the scattering function $C_{\mathbf{H}}(\tau,\nu)$ is derived from $\hat{R}_{\mathbf{H}}(\Delta t,\Delta f)$ according to (1.26), i.e.,

$$\hat{C}_{\mathbf{H}}(\tau,\nu) = \int_{-\infty}^{\infty}\int_{-\infty}^{\infty} \hat{R}_{\mathbf{H}}(\Delta t,\Delta f)\,\mathrm{e}^{-j2\pi(\nu\Delta t-\tau\Delta f)}\,\mathrm{d}\Delta t\,\mathrm{d}\Delta f.$$

A problem of this simple estimator is due to the following facts. On the one hand, the area of the effective support of $A_s(\Delta t,\Delta f)$ equals one. This follows from the identity

$$\int_{-\infty}^{\infty}\int_{-\infty}^{\infty} |A_s(\Delta t,\Delta f)|^2 \mathrm{d}\Delta t\,\mathrm{d}\Delta f = |A_s(0,0)|^2$$

known as the *radar uncertainty principle* (Wilcox, 1991). On the other hand, the area of the effective support of $R_{\mathbf{H}}(\Delta t,\Delta f)$ is roughly equal to the product of coherence time and coherence bandwidth $T_c F_c = 1/\sigma_{\mathbf{H}}$, which, due to the underspread property, is usually much larger than one. Consequently, the "equalization" with $1/A_s(\Delta t,\Delta f)$ performed in (1.83) involves divisions by values near zero, which result in numerical errors and noise enhancement.

An improved version of the estimator described above can be obtained by exploiting the underspread property[4] (Artés et al., 2004). For an underspread channel, the scattering function $C_{\mathbf{H}}(\tau,\nu)$ is effectively supported in a region $[0,\tau_{\max}]\times[-\nu_{\max},\nu_{\max}]$, with $\tau_{\max}\nu_{\max}\ll 1$. We can thus apply the sampling theorem to the 2D Fourier transform of $C_{\mathbf{H}}(\tau,\nu)$, which is the TF correlation function $R_{\mathbf{H}}(\Delta t,\Delta f)$ (see (1.26)). Specifically, the samples $R_{\mathbf{H}}(i\Delta T, j\Delta F)$ completely determine $C_{\mathbf{H}}(\tau,\nu)$ provided that $\Delta T \le 1/(2\nu_{\max})$ and $\Delta F \le 1/\tau_{\max}$. Therefore, the TF correlation function needs to be estimated only at the lattice points $(i\Delta T, j\Delta F)$, i.e., (1.83) reduces to $\hat{R}_{\mathbf{H}}(i\Delta T, j\Delta F) = \hat{A}_r(i\Delta T, j\Delta F)/A_s(i\Delta T, j\Delta F)$. Since the radar uncertainty principle does not prevent us from achieving $|A_s(i\Delta T, j\Delta F)| \approx |A_s(0,0)|$ by an appropriate design of the sounding signal $s(t)$, divisions by near-zero numbers can be avoided.

Regarding the design of $s(t)$, it is interesting to note that the ambiguity function of the impulse train $\Delta(t) = \sum_{k=-\infty}^{\infty}\delta(t-kT)$ in (1.72) equals $A_\Delta(\Delta t,\Delta f) = \sum_{k=-\infty}^{\infty}\sum_{l=-\infty}^{\infty}\delta(\Delta t-kT,\Delta f-l/T)$. While $\Delta(t)$ is impractical as an actual sounding signal, we can approximate it by means of the correlative channel sounding principle explained in Section 1.7.1, i.e., by using a sounding signal of the form $s(t) = \sum_{k=-\infty}^{\infty} g(t-kT)$ and a receive filter $\gamma(t)$. If $g(t)$ and $\gamma(t)$ are appropriately designed, then, according to Section 1.7.1, the output of the receive filter can be approximated as

$$y(t) \approx (\mathbf{H}\tilde{\Delta})(t),$$

where $\tilde{\Delta}(t) = \sum_{k=-\infty}^{\infty}(g*\gamma)(t-kT)$ is a reasonable approximation to $\Delta(t)$. Based on this approximation, an estimate of the sampled TF correlation function can be obtained by evaluating (1.83) at the lattice points $(iT, j/T)$, with $A_s(iT, j/T)$ replaced by $A_{\tilde{\Delta}}(iT, j/T)$ and $\hat{A}_r(iT, j/T)$ replaced by $\hat{A}_y(iT, j/T)$:

$$\hat{R}_{\mathbf{H}}\left(iT,\frac{j}{T}\right) = \frac{\hat{A}_y\left(iT,\frac{j}{T}\right)}{A_{\tilde{\Delta}}\left(iT,\frac{j}{T}\right)}. \tag{1.84}$$

[4]The development in Artés et al. (2004) is based on discretized signals and channel representations. We here describe the main ideas in a continuous-time setup.

The main advantage over (1.83) is that we can achieve $A_{\tilde{\Delta}}(iT,j/T) \approx \text{const.}$ and thus avoid divisions by near-zero numbers in (1.84), by a suitable design of $g(t)$ and $\gamma(t)$. A 2D Fourier transform of $\hat{R}_{\mathbf{H}}(iT,j/T)$ finally gives an estimate of $C_{\mathbf{H}}(\tau,\nu)$ within the delay-Doppler region $[0,T] \times [-1/(2T),1/(2T)]$ of area 1. In the underspread case considered, this region contains the effective support of $C_{\mathbf{H}}(\tau,\nu)$, provided T is suitably chosen. A detailed bias-variance analysis of this estimator is provided in Artés et al. (2004), along with modifications that use a regularized division or allow for data-driven operation during an ongoing data transmission.

It is interesting to observe that for an idealized impulse-train sounding signal, the discrete-time version of the scattering function estimator described above can be equivalently obtained by first measuring the TF transfer function (cf. (1.78)) and then computing the 2D periodogram (i.e., magnitude-squared 2D Fourier transform) of the measured TF transfer function (Artés et al., 2004). This is not surprising since, as discussed in Section 1.4.1, the scattering function is the 2D power spectral density of the TF transfer function (which is 2D stationary in the WSSUS case). More generally, we can obtain a scattering function estimate by applying any 2D spectrum estimator to a measured TF transfer function. A parametric 2D spectrum estimator will be considered next.

1.7.4.2 Parametric Estimation

Parametric estimation of the scattering function of a WSSUS channel can be based on the discrete-time ARMA model described in Section 1.6. For simplicity, we will here consider only the special case of the AR model (Kay & Doyle, 2003). This model is given by (1.67) with $N_{\mathrm{MA}} = 1$, i.e., the channel impulse response is represented as

$$h[n,m] = -\sum_{\Delta n=1}^{N_{\mathrm{AR}}} a_m[\Delta n]h[n-\Delta n,m] + b_m[0]e_m[n], \quad m=0,\ldots,M-1, \tag{1.85}$$

where $e_m[n]$ is normalized stationary white noise. According to (1.68), the (discrete-time) scattering function is given by

$$C_{\mathbf{H}}(m,\xi) = \frac{|b_m[0]|^2}{\left|\sum_{\Delta n=0}^{N_{\mathrm{AR}}} a_m[\Delta n]\mathrm{e}^{-j2\pi\xi\Delta n}\right|^2}, \quad m=0,\ldots,M-1, \tag{1.86}$$

with $a_m[0]=1$. Estimating $C_{\mathbf{H}}(m,\xi)$ thus reduces to estimating the MN_{AR} AR parameters $a_m[\Delta n]$ (where $m=0,\ldots,M-1$ and $\Delta n = 1,\ldots,N_{\mathrm{AR}}$) and the M innovations variances $|b_m[0]|^2$ and, possibly, the AR model order N_{AR}.

The classical approach to estimating the AR parameters (Kay, 1988; Scharf, 1991; Stoica & Moses, 1997) capitalizes on the fact that the recursive structure (1.85) of the channel tap process $h[n,m]$ is inherited by the correlation function $r_h[\Delta n,m] = \mathsf{E}\{h[n,m]h^*[n-\Delta n,m]\}$ (see (1.71)), i.e.,

$$r_h[\Delta n,m] = -\sum_{\Delta n'=1}^{N_{\mathrm{AR}}} a_m[\Delta n']r_h[\Delta n-\Delta n',m], \quad \text{for} \quad \Delta n > 0. \tag{1.87}$$

For m fixed and $\Delta n = 1,\ldots,N_{\mathrm{AR}}$, (1.87) constitutes a system of N_{AR} linear equations in the N_{AR} AR parameters $a_m[\Delta n]$. These equations are known as the *Yule–Walker equations.* Since the convolution structure of (1.87) entails a Toeplitz structure of the system matrix, the Yule–Walker equations can be solved efficiently by means of the Levinson algorithm (Kay, 1988; Scharf, 1991; Stoica & Moses, 1997). For a practical estimator of the $a_m[\Delta n]$, the correlation $r_h[\Delta n,m]$ is replaced by the sample estimate $\hat{r}_h[\Delta n, m] = \sum_n \hat{h}[n,m]\hat{h}^*[n-\Delta n, m]$, where $\hat{h}[n,m]$ is a measured impulse response obtained via channel sounding (see Sections 1.7.1 and 1.7.2) and the summation with respect to n is over the measurement interval. The innovations variances $|b_m[0]|^2$ can be estimated as

$$\widehat{|b_m[0]|^2} = \hat{r}_h[0,m] + \sum_{\Delta n=1}^{N_{\mathrm{AR}}} \hat{a}_m[\Delta n]\hat{r}_h[-\Delta n, m].$$

Finally, inserting the parameter estimates $\hat{a}_m[\Delta n]$ and $\widehat{|b_m[0]|^2}$ into (1.86) yields an estimate of the scattering function.

1.7.4.3 Non-WSSUS Case

The estimation of the LSF of a non-WSSUS channel (see Section 1.4.3) can be viewed as a nonstationary spectral estimation problem. Accordingly, a nonparametric LSF estimator proposed in Matz (2003) computes local multiwindow periodograms (cf. Thomson, 1982) of a measured channel transfer function $\hat{L}_{\mathbf{H}}(t,f)$, i.e.,

$$\hat{C}_{\mathbf{H}}(t,f;\tau,\nu) = \sum_{i=1}^{I} \left| \int_{-\infty}^{\infty}\int_{-\infty}^{\infty} \hat{L}_{\mathbf{H}}(t',f')\, u_i(t'-t)\, v_i(f'-f)\, e^{-j2\pi(\nu t' - \tau f')}\, dt'\, df' \right|^2 .$$

Here, $u_i(t)$ and $v_i(f)$, $i=1,\ldots,I$, are suitably chosen window functions (e.g., prolate spheroidal wave functions). The number of individual periodograms, I, affects the bias and variance of the estimator; a larger I yields increased bias and reduced variance. A more detailed discussion of this estimator is provided in Matz (2003).

A parametric LSF estimator can be based on the time-varying ARMA model (1.69) (Jachan & Matz, 2005). We will consider only the AR part with $\mathbf{B}[n,0]=\mathbf{I}$ for simplicity. It is convenient to represent the time-varying AR coefficient matrices $\mathbf{A}[n,\Delta n]$ over a block of duration N in terms of a $(2K_{\mathrm{AR}}+1)$-dimensional exponential basis, i.e., $\mathbf{A}[n,\Delta n] = \sum_{\Delta k=-K_{\mathrm{AR}}}^{K_{\mathrm{AR}}} \tilde{\mathbf{A}}[\Delta n, \Delta k] e^{j2\pi\frac{n\Delta k}{N}}$, $n=0,\ldots,N-1$, with $2K_{\mathrm{AR}}+1 \ll N$ (Jachan & Matz, 2005; Jachan et al., 2009). We then obtain from (1.69) the following noisy recursion for the channel vector $\mathbf{h}[n] = (h[n,0]\cdots h[n,M-1])^T$:

$$\mathbf{h}[n] = -\sum_{\Delta n=1}^{N_{\mathrm{AR}}}\sum_{\Delta k=-K_{\mathrm{AR}}}^{K_{\mathrm{AR}}} \tilde{\mathbf{A}}[\Delta n,\Delta k]\mathbf{h}[n-\Delta n]e^{j2\pi\frac{n\Delta k}{N}} + \mathbf{e}[n], \quad n=0,\ldots,N-1.$$

This is autoregressive not only in time (time-delayed copies of $\mathbf{h}[n]$) but also in frequency (frequency-shifted copies of $\mathbf{h}[n]$). This *time-frequency AR* (TFAR) structure is inherited by the second-order statistics of $\mathbf{h}[n]$. Specifically, the expected matrix-valued ambiguity function of $\mathbf{h}[n]$,

$$\bar{\mathbf{A}}[\Delta n,\Delta k] \triangleq \mathsf{E}\left\{\sum_{n=0}^{N-1} \mathbf{h}[n]\mathbf{h}^H[n-\Delta n]e^{-j2\pi\frac{n\Delta k}{N}}\right\}, \tag{1.88}$$

can be shown to satisfy the recursion (Jachan & Matz, 2005; Jachan et al., 2009)

$$\bar{\mathbf{A}}[\Delta n,\Delta k] = -\sum_{\Delta n'=1}^{N_{\mathrm{AR}}}\sum_{\Delta k'=-K_{\mathrm{AR}}}^{K_{\mathrm{AR}}} \tilde{\mathbf{A}}[\Delta n',\Delta k']\bar{\mathbf{A}}[\Delta n-\Delta n',\Delta k-\Delta k']e^{j2\pi\frac{\Delta n'(\Delta k'-\Delta k)}{N}}, \tag{1.89}$$

for $\Delta n = 1, \ldots, N_{\mathrm{AR}}$ and $\Delta k = -K_{\mathrm{AR}}, \ldots, K_{\mathrm{AR}}$. These "TF Yule–Walker equations" constitute a system of $N_{\mathrm{AR}}\,(2K_{\mathrm{AR}}+1)$ linear equations in the $N_{\mathrm{AR}}\,(2K_{\mathrm{AR}}+1)$ TFAR parameter matrices $\tilde{\mathbf{A}}[\Delta n, \Delta k]$. For a correlation-underspread channel (see Section 1.5.2), the phase factor $e^{j2\pi \Delta n'(\Delta k' - \Delta k)/N}$ in (1.89) can be approximated by 1. With this approximation, the TF Yule–Walker equations exhibit a two-level block-Toeplitz structure and can hence be solved efficiently by means of a vector version of the Wax-Kailath algorithm (Wax & Kailath, 1983) (see Jachan & Matz, 2005; Jachan et al., 2009 for algorithmic details).

For a practical estimator of the TFAR parameter matrices $\tilde{\mathbf{A}}[\Delta n, \Delta k]$, the expected matrix ambiguity function $\bar{\mathbf{A}}[\Delta n, \Delta k]$ is replaced by a sample estimate that is obtained by dropping the expectation in (1.88) and substituting an estimate $\hat{\mathbf{h}}[n]$ for $\mathbf{h}[n]$. The resulting TFAR parameter estimates $\hat{\tilde{\mathbf{A}}}[\Delta n, \Delta k]$ can finally be used to compute an estimate of the LSF, based on an LSF expression that is similar in spirit to (1.86) (Jachan & Matz, 2005; Jachan et al., 2009).

1.8 CONCLUSION

In this chapter, we provided a survey of many of the concepts and tools that have been developed during the past six decades for characterizing, modeling, and measuring time-frequency dispersive channels. Our treatment was motivated by the twofold goal of presenting some fundamentals that may be helpful for understanding the subsequent chapters of this book and, more generally, of providing a convenient entry point into the rich literature on rapidly time-varying wireless channels.

Based on considerations of the physical propagation mechanisms, we discussed various deterministic and stochastic channel descriptions. Regarding the latter, both WSSUS and non-WSSUS channels were treated. Some emphasis was placed on the practically most relevant class of underspread channels and on important consequences of the underspread property. We reviewed parsimonious channel models, concentrating on the particularly useful basis expansion and ARMA-type models. Finally, we briefly discussed several techniques for measuring time-varying channels and their statistics. For some of these topics, extensions to multiantenna (MIMO) channels were presented. We note that while our discussion focused on terrestrial wireless channels, much of it applies to other channels as well (e.g., underwater and satellite channels).

Acknowledgment

This work was supported by the FWF under Grant S10603 (Statistical Inference) and Grant S10606 (Information Networks) within the National Research Network SISE and by the WWTF under Grant MA 44 (MOHAWI).

References

Abramowitz, M., & Stegun, I. (1965). *Handbook of mathematical functions*. New York, NY.

Almers, P., Bonek, E., Burr, A., Czink, N., Debbah, M., Degli-Esposti, V., et al. (2007). Survey of channel and radio propagation models for wireless MIMO systems. *EURASIP Journal on Wireless Communications and Networking*, Vol. 2007, Article ID 19070.

Artés, H., Matz, G., & Hlawatsch, F. (2004). Unbiased scattering function estimators for underspread channels and extension to data-driven operation. *IEEE Transactions on Signal Processing, 52*(5), 1387–1402.

Baddour, K. E., & Beaulieu, N. C. (2001, November). Autoregressive models for fading channel simulation. In *Proceedings of the IEEE GLOBECOM-2001* (pp. 1187–1192). San Antonio, TX.

Balan, R., Poor, H. V., Rickard, S., & Verdu, S. (2004, September). Time-frequency and time-scale canonical representations of doubly spread channels. In *Proceedings of EUSIPCO–04*: (pp. 445–448). Vienna, Austria.

Baum, D., & Bölcskei, H. (2009). Information-theoretic analysis of MIMO channel sounding. Submitted.

Bello, P. A. (1963). Characterization of randomly time-variant linear channels. *IEEE Transactions on Communications System, 11*, 360–393.

Biglieri, E., Proakis, J., & Shamai (Shitz) S. (1998). Fading channels: Information-theoretic and communications aspects. *IEEE Transactions on Information Theory, 44*(6), 2619–2692.

Bingham, J. A. C. (1990). Multicarrier modulation for data transmission: An idea whose time has come. *IEEE Communications Magazine, 28*(5), 5–14.

Bölcskei, H., Gesbert, D., Papadias, C. B., & van der Veen, A.-J. (Eds.), (2006). *Space-time wireless systems: From array processing to MIMO communications*. New York, NY: Cambridge University Press.

Borah, D. K., & Hart, B. D. (1999). Frequency-selective fading channel estimation with a polynomial time-varying channel model. *IEEE Transactions on Communications, 47*, 862–873.

Chow, S.-K., & Venetsanopoulos, A. N. (1974). Optimal on-off signaling over linear time-varying stochastic channels. *IEEE Transactions on Information Theory, 20*(5), 602–609.

Cox, D. (1972). Delay Doppler characteristics of multipath propagation in a suburban mobile radio environment. *IEEE Transactions on Antennas and Propagation, 20*, 625–635.

Cullen, P. J., Fannin, P. C., & Molina, A. (1993). Wide-band measurement and analysis techniques for the mobile radio channel. *IEEE Transactions on Vehicular Technology, 42*(4), 589–603.

de Lacerda, R., Cardoso, L. S., Knopp, R., Gesbert, D., & Debbah, M. (2007, October). EMOS platform: Real-time capacity estimation of MIMO channels in the UMTS-TDD band. In *Proceedings of the International Symposium on Wireless Communication Systems (ISWCS)*, Trondheim, Norway.

Fannin, P. C., Molina, A., Swords, S. S., & Cullen, P. J. (1991). Digital signal processing techniques applied to mobile radio channel sounding. *Proceedings of IEE-F, 138*(5), 502–508.

Flandrin, P. (1999). *Time-frequency/time-scale analysis*. San Diego, CA: Academic Press.

Flandrin, P. (2003). Ambiguity function. In B. Boashash (Ed.), *Time-frequency signal analysis and processing: A comprehensive reference*, chap. 5.1, (pp. 160–167). Oxford, UK: Elsevier.

Gaarder, N. T. (1968). Scattering function estimation. *IEEE Transactions on Information Theory, 14*(5), 684–693.

Gallager, R. G. (1964, April). *Characterization and measurement of time- and frequency-spread channels*. (Tech. Rep. No. 352). Cambridge, MA: M.I.T. Lincoln Lab.

Gallager, R. G. (1968). *Information theory and reliable communication*. New York: Wiley.

Gersho, A. (1963). Characterization of time-varying linear systems. *Proceedings of IEEE, 51*(1), 238.

Giannakis, G. B., & Tepedelenlioğlu, C. (1998). Basis expansion models and diversity techniques for blind identification and equalization of time-varying channels. *Proceedings of IEEE, 86*(10), 1969–1986.

Gröchenig, K. (2001). *Foundations of time-frequency analysis*. Boston, MA: Birkhäuser.

Hlawatsch, F., & Boudreaux-Bartels, G. F. (1992). Linear and quadratic time-frequency signal representations. *IEEE Signal Processing Magazine, 9*(2), 21–67.

Jachan, M., & Matz, G. (2005, June). Nonstationary vector AR modeling of wireless channels. In *Proceedings of IEEE SPAWC-05* (pp. 648–652). New York, NY.

Jachan, M., Matz, G., & Hlawatsch, F. (2009). Vector time-frequency AR models for nonstationary multivariate random processes. *IEEE Transactions on Signal Processing, 57*(12), 4646–4659.

Jakes, W. C. (1974). *Microwave mobile communications*. New York, NY: Wiley.

Kailath, T. (1959, May). *Sampling models for linear time-variant filters.* (Tech. Rep. No. 352). Cambridge, MA: M.I.T. Research Lab. of Electronics.

Kailath, T. (1962). Measurements on time-variant communication channels. *IEEE Transactions on Information Theory*, *8*(5), 229–236.

Kay, S. M. (1988). *Modern spectral estimation.* Englewood Cliffs, NJ: Prentice Hall.

Kay, S. M., & Doyle, S. B. (2003). Rapid estimation of the range-Doppler scattering function. *IEEE Transactions on Signal Processing*, *51*(1), 255–268.

Kennedy, R. S. (1969). *Fading dispersive communication channels.* New York, NY: Wiley.

Kermoal, J. P., Schumacher, L., Pedersen, K. I., Mogensen, P. E., & Frederiksen, F. (2002). A stochastic MIMO radio channel model with experimental validation. *IEEE Journal on Selected Areas in Communications*, *20*(6), 1211–1226.

Leus, G. (2004, September). On the estimation of rapidly time-varying channels. In *Proceedings of EUSIPCO 2004* (pp. 2227–2230). Vienna, Austria.

Li, Y., Cimini, L., & Sollenberger, N. (1998). Robust channel estimation for OFDM systems with rapid dispersive fading channels. *IEEE Transactions on Communications*, *46*(7), 902–915.

Matz, G. (2003, June). Doubly underspread non-WSSUS channels: Analysis and estimation of channel statistics. In *Proceedings of IEEE SPAWC-03* (pp. 190–194). Rome, Italy.

Matz, G. (2005). On non-WSSUS wireless fading channels. *IEEE Transactions on Wireless Communications*, *4*(5), 2465–2478.

Matz, G. (2006, October/November). Characterization and analysis of doubly dispersive MIMO channels. In *Proceedings of the 40th Asilomar Conference on Signals, Systems, and Computers*: (pp. 946–950). Pacific Grove, CA.

Matz, G., & Hlawatsch, F. (1998). Time-frequency transfer function calculus (symbolic calculus) of linear time-varying systems (linear operators) based on a generalized underspread theory. *Journal of Mathematical Physics, Special Issue on Wavelet and Time-Frequency Analysis*, *39*(8), 4041–4071.

Matz, G., & Hlawatsch, F. (2003). Time-frequency characterization of random time-varying channels. In B. Boashash, (Ed.), *Time-frequency signal analysis and processing: A comprehensive reference*, Chap. 9.5 (pp. 410–419). Oxford, UK: Elsevier.

Matz, G., & Hlawatsch, F. (2006). Nonstationary spectral analysis based on time-frequency operator symbols and underspread approximations. *IEEE Transactions on Information Theory*, *52*(3), 1067–1086.

Matz, G., Molisch, A. F., Steinbauer, M., Hlawatsch, F., Gaspard, I., & Artés, H. (1999, September). Bounds on the systematic measurement errors of channel sounders for time-varying mobile radio channels. In *Proceedings of IEEE VTC-99 Fall* (pp. 1465–1470). Amsterdam, The Netherlands.

Matz, G., Molisch, A. F., Hlawatsch, F., Steinbauer, M., & Gaspard, I. (2002). On the systematic measurement errors of correlative mobile radio channel sounders. *IEEE Transactions on Communications*, *50*(5), 808–821.

Matz, G., Schafhuber, D., Gröchenig, K., Hartmann, M., & Hlawatsch, F. (2007). Analysis, optimization, and implementation of low-interference wireless multicarrier systems. *IEEE Transactions on Wireless Communications*, *6*(5), 1921–1931.

Margetts, A. R., Schniter, P., & Swami, A. (2007). Joint scale-lag diversity in wideband mobile direct sequence spread spectrum systems. *IEEE Transactions on Wireless Communications*, *6*(12), 4308–4319.

Molisch, A. F. (2005). *Wireless communications.* Chichester, UK: Wiley.

Nawab, S. H., & Quatieri, T. F. (1988). Short-time Fourier transform. In J. S. Lim and A. V. Oppenheim, (Eds.), *Advanced topics in signal processing*, Chap. 6 (pp. 289–337). Englewood Cliffs, NJ: Prentice Hall.

Naylor, A. W., & Sell, G. R. (1982). *Linear operator theory in engineering and science* (2nd ed.). New York, NY: Springer.

Paier, A., Karedal, J., Czink, N., Dumard, C., Zemen, T., Tufvesson, F., et al. (2009). Characterization of vehicle-to-vehicle radio channels from measurements at 5.2 GHz. *Wireless Personal Communications*, *50*, 19–32.

Parsons, J. D. (1992). *The mobile radio propagation channel*. London, UK: Pentech Press.

Parsons, J. D., Demery, D. A., & Turkmani, A. M. D. (1991, October). Sounding techniques for wideband mobile radio channels: A review. *Proceedings of the IEE-I, 138*(5), 437–446.

Paulraj, A., Nabar, R. U., & Gore, D. (2003). *Introduction to space-time wireless communications*. Cambridge, UK: Cambridge University Press.

Proakis, J. G. (1995). *Digital communications* (3rd ed.). New York, NY: McGraw-Hill.

Rappaport, T. S. (1996). *Wireless communications: Principles & practice*. Upper Saddle River, NJ: Prentice Hall.

Salous, S., Nikandrou, N., & Bajj, N. F. (1998). Digital techniques for mobile radio chirp sounders. *IEE Proceedings, Communications, 145*(3), 191–196.

Sayeed, A. M. (2002). Deconstructing multiantenna fading channels. *IEEE Transactions on Signal Processing, 50*(10), 2563–2579.

Schafhuber, D., Matz, G., & Hlawatsch, F. (2001, August). Simulation of wideband mobile radio channels using subsampled ARMA models and multistage interpolation. In: *Proceedings of the 11th IEEE Workshop on Statistical Signal Processing* (pp. 571–574). Singapore.

Scharf, L. L. (1991). *Statistical signal processing*. Reading, MA: Addison Wesley.

Schniter, P. (2004). Low-complexity equalization of OFDM in doubly-selective channels. *IEEE Transactions on Signal Processing, 52*(4), 1002–1011.

Schwarz, K., Martin, U., & Schüßler, H. W. (1993). Devices for propagation measurements in mobile radio channels. In: *Proceedings of IEEE PIMRC-93* (pp. 387–391). Yokohama, Japan.

Skolnik, M. (1984). *Radar handbook*. New York, NY: McGraw-Hill.

Slepian, D. (1978). Prolate spheroidal wave functions, Fourier analysis, and uncertainty—V: The discrete case. *Bell System Technical Journal, 57*(5), 1371–1430.

Stoica, P., & Moses, R. (1997). *Introduction to spectral analysis*. Englewood Cliffs, NJ: Prentice Hall.

Suzuki, H., Thi Van Anh Tran, & Collings, I. B. (2007). Characteristics of MIMO-OFDM channels in indoor environments. *EURASIP Journal on Wireless Communications and Networking*, Vol. 2007, Article ID 19728.

Thomä, R., Hampicke, D., Richter, A., Sommerkorn, G., Schneider, A., Trautwein, U., et al. (2000). Identification of time-variant directional mobile radio channels. *IEEE Transactions on Instrumentation and Measurement, 49*(2), 357–364.

Thomas, T. A., & Vook, F. W. (2000, March). Multi-user frequency-domain channel identification, interference suppression, and equalization for time-varying broadband wireless communications. In *Proceedings of the IEEE Sensor Array and Multichannel Signal Processing Workshop* (pp. 444–448). Boston, MA.

Thomson, D. J. (1982). Spectrum estimation and harmonic analysis. *Proceedings of the IEEE, 70*(9), 1055–1096.

Tomasin, S., Gorokhov, A., Yang, H., & Linnartz, J. P. (2005). Iterative interference cancellation and channel estimation for mobile OFDM. *IEEE Transactions on Wireless Communications, 4*(1), 238–245.

Tsatsanis, M. K., & Giannakis, G. B. (1996). Modeling and equalization of rapidly fading channels. *International Journal of Adaptive Control and Signal Processing, 10*, 159–176.

Tsatsanis, M. K., & Giannakis, G. B. (1997). Subspace methods for blind estimation of time-varying FIR channels. *IEEE Transactions on Signal Processing, 45*(12), 3084–3093.

Tsatsanis, M. K., Giannakis, G. B., & Zhou, G. (1996). Estimation and equalization of fading channels with random coefficients. *Signal Processing, 53*(2–3), 211–229.

Tse, D., & Viswanath, P. (2005). *Fundamentals of wireless communication*. New York, NY: Cambridge University Press.

Van Trees, H. L. (1992). *Detection, estimation, and modulation theory, Part III: Radar-sonar signal processing and Gaussian signals in noise*. Malabar, FL: Krieger.

Vaughan, R., & Bach Andersen, J. (2003). *Channels, propagation and antennas for mobile communications*. London, UK: IEE Press.

Wax, M., & Kailath, T. (1983). Efficient inversion of Toeplitz-block Toeplitz matrix. *IEEE Transactions on Acoustics, Speech, and Signal Processing*, *31*(5), 1218–1221.

Weichselberger, W., Herdin, M., Özcelik, H., & Bonek, E. (2006). A stochastic MIMO channel model with joint correlation of both link ends. *IEEE Transactions on Wireless Communications*, *5*(1), 90–100.

Wilcox, C. H. (1991). The synthesis problem for radar ambiguity functions. In R. E. Blahut, W. Miller, Jr., and C. H. Wilcox, (Eds.), *Radar and sonar*, Part 1. (pp. 229–260). New York, NY: Springer.

Woodward, P. M. (1953). *Probability and information theory with application to radar*. London, UK: Pergamon Press.

Ye, J., & Papandreou-Suppappola, A. (2003, September). Wideband time-varying channels with time-scale diversity and estimation. In *Proceedings of IEEE SSP-03* (pp. 50–53). St. Louis, MO.

Zadeh, L. A. (1950). Frequency analysis of variable networks. *Proceedings of the IRE*, *38*(3), 291–299.

Zemen, T., & Mecklenbräuker, C. F. (2005). Time-variant channel estimation using discrete prolate spheroidal sequences. *IEEE Transactions on Signal Processing*, *53*(9), 3597–3607.

CHAPTER

Information Theory of Underspread WSSUS Channels

2

Giuseppe Durisi[1], Veniamin I. Morgenshtern[2], Helmut Bölcskei[2], Ulrich G. Schuster[3], Shlomo Shamai (Shitz)[4]

[1]*Chalmers University of Technology, Gothenburg, Sweden*

[2]*ETH Zurich, Switzerland*

[3]*Robert Bosch GmbH, Stuttgart, Germany*

[4]*Technion – Israel Institute of Technology, Haifa, Israel*

2.1 THE ROLE OF A SYSTEM MODEL

2.1.1 A Realistic Model

In this chapter, we are interested in the *ultimate limit* on the rate of reliable communication through Rayleigh-fading channels that satisfy the wide-sense stationary (WSS) and uncorrelated scattering (US) assumptions and are underspread (Bello, 1963; Kennedy, 1969). Therefore, the natural setting is an information-theoretic one, and the performance metric is channel *capacity* (Cover & Thomas, 1991; Gallager, 1968).

The family of Rayleigh-fading underspread WSSUS channels (reviewed in Chapter 1) constitutes a good model for real-world wireless channels: their stochastic properties, like amplitude and phase distributions match channel measurement results (Schuster, 2009; Schuster & Bölcskei, 2007). The Rayleigh-fading and the WSSUS assumptions imply that the stochastic properties of the channel are fully described by a two-dimensional power spectral density (PSD) function, often referred to as *scattering function* (Bello, 1963). The underspread assumption implies that the scattering function is highly concentrated in the delay-Doppler plane.

To analyze wireless channels with information-theoretic tools, a *system model*, not just a *channel model*, needs to be specified. A system model is more comprehensive than a channel model because it defines, among other parameters, the transmit-power constraints and the channel knowledge available at the transmitter and the receiver. The choice of a realistic system model is crucial for the insights and guidelines provided by information theory to be useful for the design of practical systems. Two important aspects need to be accounted for by a model that aims at being realistic:

1. *Neither the transmitter nor the receiver knows the realization of the channel:* In most wireless systems, channel state information (CSI) is acquired by allocating part of the available resources to channel estimation. For example, pilot symbols can be embedded into the data stream, as explained in Chapters 4 and 5, to aid the receiver in the channel-estimation process. From an information-theoretic perspective, pilot-based channel estimation is just a special case of coding.

Wireless Communications Over Rapidly Time-Varying Channels

Hence, the rate achievable with training-based schemes cannot exceed the capacity in the absence of CSI at transmitter and receiver.

We refer to the setting where no CSI is available at transmitter and receiver, but both know the statistics of the channel, as the *noncoherent setting* (Sethuraman, Hajek, & Narayanan, 2005; Sethuraman, Wang, Hajek, & Lapidoth, 2009; Zheng & Tse, 2002), in contrast to the *coherent setting*, where a genie provides the receiver with perfect CSI (Biglieri, Proakis, & Shamai (Shitz), 1998, Section III.B). Furthermore, we denote capacity in the noncoherent setting as *noncoherent capacity*.

2. *The peak power of the transmit signal is limited:* Every power amplifier has finite gain and every mobile transmitter has limited battery resources. In addition, regulatory bodies often constrain the admissible radiated power. Hence, in a realistic system model, a peak constraint should be imposed on the transmit signal.

Motivated by these two aspects, we provide an information-theoretic analysis of Rayleigh-fading underspread WSSUS channels in the noncoherent setting, under the additional assumption that the transmit signal is peak-constrained.

2.1.2 A Brief Literature Survey

The noncoherent capacity of fading channels is notoriously difficult to characterize analytically, even for simple channel models (Abou-Faycal, Trott, & Shamai (Shitz), 2001; Marzetta & Hochwald, 1999). Most of the results available in the literature pertain to either the *large-bandwidth* or the *high-SNR* regime. In the following, large-bandwidth regime refers to the case where the average power P is fixed and the bandwidth B is large, so that the SNR, which is proportional to P/B, is small. High-SNR regime refers to the case of fixed B and large P and, hence, large SNR. In this section, we briefly review the relevant literature; two elements will emerge from this review:

1. The modeling aspects we identified in the previous section (WSSUS, noncoherent setting, peak constraint) are *fundamental* in that the capacity is highly sensitive to these aspects.
2. In spite of the large number of results available in the literature, several questions of practical engineering relevance about the design of wireless systems operating over fading channels are still open.

Large-bandwidth (low-SNR) regime

The noncoherent capacity of fading channels has been a subject of investigation in information theory for several decades. The first contributions, which date back to the sixties (Gallager, 1968; Kennedy, 1969; Pierce, 1966; Viterbi, 1967) (see Biglieri et al., 1998 for a more complete list of references), mainly deal with the characterization of the asymptotic behavior of noncoherent capacity in the large-bandwidth limit.

The results in Gallager (1968), Kennedy (1969), Pierce (1966) and Viterbi (1967) illustrate the sensitivity of noncoherent capacity to the presence of a peak constraint. Specifically, the outcome of this analysis is the following rather surprising result: in the large-bandwidth limit, the noncoherent capacity of a fading channel coincides with that of an additive white Gaussian noise (AWGN) channel with the same receive power. However, the signaling schemes that achieve the noncoherent capacity of a fading channel in the wideband limit have unbounded peak power – a result recently formalized by Verdú (2002). Hence, these signaling schemes are not practical.

If a peak constraint is imposed on the transmit signal, AWGN channel capacity cannot be achieved in the infinite-bandwidth limit, and the actual behavior of noncoherent capacity in the wideband regime depends on the specific form of the peak constraint (Durisi, Schuster, Bölcskei, & Shamai (Shitz), 2010; Médard & Gallager, 2002; Subramanian & Hajek, 2002; Telatar & Tse, 2000; Viterbi, 1967). In particular, when the transmit signal is subject to a peak constraint both in time and frequency (as in most practical systems), noncoherent capacity vanishes as the bandwidth grows large (Durisi et al., 2010; Médard & Gallager, 2002; Subramanian & Hajek, 2002; Telatar & Tse, 2000), provided that the number of independent diversity branches of the channel scales linearly with bandwidth,[1] which is the case for WSSUS channels. Intuitively, under a peak constraint on the transmit signal, the receiver is no longer able to resolve the channel uncertainty as the bandwidth, and, hence, the number of independent diversity branches, increases. This result implies that, for a large class of fading channels, the corresponding noncoherent capacity has a global maximum at a certain finite bandwidth, commonly referred to as the *critical bandwidth*. Computing this critical bandwidth is obviously of great practical interest. Moreover, it is important to understand the role played by the spatial degrees of freedom provided by multiple antennas at the transmitter and/or the receiver: can they be used to increase capacity, or do they merely lead to a rate loss because of the increase in channel uncertainty?

High-SNR regime

Characterizing noncoherent capacity in the high-SNR regime is of great practical interest for systems that operate over narrow frequency bands. As no closed-form expressions for the noncoherent capacity are known, not even for memoryless channels, one typically resorts to analyzing the asymptotic behavior of capacity as SNR goes to infinity. Differently from the large-bandwidth regime, where capacity results are robust with respect to the underlying channel model, in the high-SNR regime the capacity behavior is highly sensitive to the *fine details* of the channel model (Lapidoth, 2005; Lapidoth & Moser, 2003; Liang & Veeravalli, 2004; Marzetta & Hochwald, 1999; Telatar, 1999; Zheng & Tse, 2002). The following results support this claim.

In the coherent setting, capacity grows logarithmically with SNR (Telatar, 1999); logarithmic growth also holds in the noncoherent setting for *block-fading* channels (Marzetta & Hochwald, 1999; Zheng & Tse, 2002).[2] An alternative – more general – approach to modeling the time variation of a

[1]The independent diversity branches of a fading channel are sometimes referred to as *stochastic degrees of freedom* (Schuster and Bölcskei, 2007). We will not use this convention here. Instead, we will use the term *degrees of freedom* to refer to signal-space dimensions. Information-theoretic analyses of wireless channels for which the number of independent diversity branches scales sublinearly with bandwidth, can be found, for example, in Porrat, Tse, and Nacu (2007); Raghavan, Hariharan, and Sayeed (2007).

[2]The block-fading model is the simplest model that captures the time variation of a wireless channel. In this model, the channel is taken to be constant over a given time interval, called *block*, and assumed to change independently from one such block to the next. The independence assumption across blocks can be justified, e.g., for systems employing frequency hopping or if the data symbols are interleaved.

wireless channel is to assume that the fading process is stationary.[3] Surprisingly, if the fading process is stationary, the noncoherent capacity does not necessarily grow logarithmically with SNR: other scaling behaviors are possible (Lapidoth, 2005). For example, consider two stationary discrete-time Rayleigh-fading channels subject to additive Gaussian noise. The fading process of the first channel has PSD equal to $1/\Delta$ on the interval $[-\Delta/2, \Delta/2]$, where $0 < \Delta < 1$, and 0 else. The fading process of the second channel has PSD equal to $(1-\varepsilon)/\Delta$ on the interval $[-\Delta/2, \Delta/2]$ and $\varepsilon/(1-\Delta)$ else ($0 < \varepsilon < 1$). These two channels have completely different high-SNR capacity behavior, no matter how small ε is: the noncoherent capacity of the first channel grows logarithmically in SNR, with *pre-log* factor equal to $1-\Delta$ (Lapidoth, 2005); the noncoherent capacity of the second channel grows double-logarithmically in SNR (Lapidoth & Moser, 2003).

Such a result is unsatisfactory from an engineering point of view because the support of a PSD cannot be determined through measurements (measurement noise is one of the reasons, another one is the finite time duration of any physically meaningful measurement process). In other words, capacity turns out to be highly sensitive to a parameter – the measure of the support of the PSD – that has, in the words of Slepian (1976), "... no direct meaningful counterparts in the real world ...". Such a dependency of the capacity behavior on fine details of the channel model suggests that the stationary model is not *robust* in the high-SNR regime. An engineering-relevant problem is then to establish the SNR value at which this lack of robustness starts to manifest itself.

2.1.3 Capacity Bounds Answering Engineering-Relevant Questions

The purpose of this chapter is to present tight upper and lower bounds on the noncoherent capacity of Rayleigh-fading underspread WSSUS channels. On the basis of these bounds, answers to the following engineering-relevant questions can be given:

1. How does the noncoherent capacity of this class of channels differ from the corresponding coherent capacity and from the capacity of an AWGN channel with the same receive power?
2. How much bandwidth and how many antennas should be used to maximize capacity?
3. How robust is the Rayleigh-fading WSSUS underspread channel model? More specifically, at which SNR values does the noncoherent capacity start being sensitive to the fine details of the channel model?

The capacity bounds presented in this chapter make use of information-theoretic tools recently developed in Guo, Shamai (Shitz), & Verdú (2005); Lapidoth (2005); Lapidoth & Moser (2003); Sethuraman et al. (2009). One of the difficulties we shall encounter is to adapt these tools to the continuous-time setting considered in this chapter. Harmonic analysis plays a fundamental role in this respect: it provides an effective method for converting a general continuous-time channel into a discretized channel that can be analyzed using standard information-theoretic tools. The discretization is accomplished by transmitting and receiving on a highly structured set of signals, similar to what is done in pulse-shaped (PS) orthogonal frequency-division multiplexing (OFDM) systems (Kozek & Molisch, 1998). This ensures that the resulting discretized channel inherits the statistical properties of

[3] Stationarity in time, together with the assumption that scatterers corresponding to paths of different delays are uncorrelated, is the fundamental feature of the WSSUS model we focus on in this chapter.

the underlying continuous-time channel (in particular, stationarity), a fact that is crucial for the ensuing analysis. As a byproduct, our results yield a novel information-theoretic criterion for the design of PS-OFDM systems (see Matz et al. (2007) for a recent overview on this topic).

2.2 A DISCRETIZED SYSTEM MODEL

2.2.1 The Channel Model

2.2.1.1 *Continuous-Time Input–Output Relation*

The input–output (I/O) relation of a general continuous-time stochastic linear time-varying (LTV) channel **H** can be written as

$$r(t) = \int_{-\infty}^{\infty} h(t,\tau)s(t-\tau)\mathrm{d}\tau + w(t). \tag{2.1}$$

Here, $s(t)$ is the (stochastic) input signal, whose realizations can be taken as elements of the Hilbert space $L^2(\mathbb{R})$ of square-integrable functions over the real line $\mathbb{R}$; $r(t)$ is the output signal; and $w(t)$ is a zero-mean unit-variance proper AWGN random process. Finally, the time-varying channel impulse response $h(t,\tau)$ is a zero-mean jointly proper Gaussian (JPG) random process that satisfies the WSSUS assumption (Bello, 1963), i.e., it is stationary in time t and uncorrelated in delay τ:

$$\mathsf{E}\{h(t,\tau)h^*(t',\tau')\} \triangleq r_h(t-t',\tau)\delta(\tau-\tau').$$

As a consequence of the JPG and WSSUS assumptions, the *time-delay correlation function* $r_h(t,\tau)$ fully characterizes the channel statistics. The JPG assumption is empirically supported for narrowband channels (Vaughan & Bach Andersen, 2003); even ultrawideband (UWB) channels with bandwidth up to several Gigahertz can be modeled as Gaussian-distributed (Schuster & Bölcskei, 2007). The WSSUS assumption is widely used in wireless channel modeling (Bello, 1963; Biglieri et al., 1998; Kennedy, 1969; Matz & Hlawatsch, 2003; Proakis, 2001; Vaughan & Bach Andersen, 2003). It is in good agreement with measurements of tropospheric scattering channels (Kennedy, 1969), and it provides a reasonable model for many types of mobile radio channels (Cox, 1973a,b; Jakes, 1974), at least when observed over a limited time duration and limited bandwidth (Bello, 1963). A more detailed review of the WSSUS channel model can be found in Chapter 1.

Often, it will be convenient to describe **H** in domains other than the time-delay domain. The *time-varying transfer function* $L_{\mathbf{H}}(t,f) = \mathbb{F}_{\tau\to f}\{h(t,\tau)\}$ and the *delay-Doppler spreading function* $S_{\mathbf{H}}(\tau,\nu) = \mathbb{F}_{t\to\nu}\{h(t,\tau)\}$ can be used for this purpose. As a consequence of the WSSUS assumption, $L_{\mathbf{H}}(t,f)$ is WSS both in time t and in frequency f, and $S_{\mathbf{H}}(\tau,\nu)$ is uncorrelated in delay τ and Doppler ν:

$$\mathsf{E}\{L_{\mathbf{H}}(t,f)L_{\mathbf{H}}^*(t',f')\} \triangleq R_{\mathbf{H}}(t-t',f-f'). \tag{2.2}$$

$$\mathsf{E}\{S_{\mathbf{H}}(\tau,\nu)S_{\mathbf{H}}^*(\tau',\nu')\} \triangleq C_{\mathbf{H}}(\tau,\nu)\delta(\tau-\tau')\delta(\nu-\nu'). \tag{2.3}$$

The function $R_{\mathbf{H}}(t,f)$ is usually referred to as the channel *time-frequency correlation function*, and $C_{\mathbf{H}}(\tau,\nu)$ is called the channel *scattering function*. The two functions are related by the two-dimensional

Fourier transform $C_{\mathbf{H}}(\tau,\nu) = \mathbb{F}_{t\to\nu,f\to\tau}\{R_{\mathbf{H}}(t,f)\}$. We assume throughout that

$$\int_{-\infty}^{\infty}\int_{-\infty}^{\infty} C_{\mathbf{H}}(\tau,\nu)d\tau d\nu = 1 \tag{2.4}$$

for simplicity.

2.2.1.2 The Underspread Assumption

Almost all WSSUS channels of practical interest are *underspread*, i.e., they have a scattering function $C_{\mathbf{H}}(\tau,\nu)$ that is highly concentrated around the origin of the delay-Doppler plane (Bello, 1963).

A mathematically precise definition of the underspread property is available for the case when $C_{\mathbf{H}}(\tau,\nu)$ is compactly supported within a rectangle in the delay-Doppler plane. In this case, a WSSUS channel is said to be underspread if the support area of $C_{\mathbf{H}}(\tau,\nu)$ is much less than one (Durisi et al., 2010; Kozek, 1997). In practice, it is not possible to determine through channel measurements whether $C_{\mathbf{H}}(\tau,\nu)$ is compactly supported or not. Hence, in the terminology introduced in Section 2.1.2, the measure (area) of the support of the scattering function is a fine detail of the channel model. To investigate the sensitivity of noncoherent capacity to this fine detail, and assess the robustness of the Rayleigh-fading WSSUS model, we replace the compact-support assumption by the following physically more meaningful assumption: $C_{\mathbf{H}}(\tau,\nu)$ has a small fraction of its total volume supported outside a rectangle of an area that is much smaller than 1. More precisely, we have the following definition:

Definition 2.1 Let $\tau_0,\nu_0 \in \mathbb{R}_+, \varepsilon \in [0,1]$, and let $\mathscr{H}(\tau_0,\nu_0,\varepsilon)$ be the set of all Rayleigh-fading WSSUS channels **H** with scattering function $C_{\mathbf{H}}(\tau,\nu)$ that satisfies

$$\int_{-\nu_0}^{\nu_0}\int_{-\tau_0}^{\tau_0} C_{\mathbf{H}}(\tau,\nu)d\tau d\nu \geq 1-\varepsilon.$$

We say that the channels in $\mathscr{H}(\tau_0,\nu_0,\varepsilon)$ are underspread if $\Delta_{\mathbf{H}} \triangleq 4\tau_0\nu_0 \ll 1$ and $\varepsilon \ll 1$.

Remark 2.1 Definition 2.1 is inspired by Slepian's treatment of $L^2(\mathbb{R})$ signals that are approximately limited in time and frequency (Slepian, 1976). Note that $\varepsilon = 0$ in Definition 2.1 yields the compact-support underspread definition of Durisi et al. (2010) and Kozek (1997). An alternative definition of the underspread property, also not requiring that $C_{\mathbf{H}}(\tau,\nu)$ is compactly supported, was used in Matz and Hlawatsch (2003). The definition in Matz and Hlawatsch (2003) is in terms of moments of the scattering function.

Typical wireless channels are (highly) underspread, with most of the volume of $C_{\mathbf{H}}(\tau,\nu)$ supported within a rectangle of area $\Delta_{\mathbf{H}} \approx 10^{-3}$ for typical land-mobile channels, and as low as 10^{-7} for some indoor channels with restricted mobility of the terminals (Hashemi, 1993; Parsons, 2000; Rappaport, 2002). In the remainder of the chapter, we refer to $\Delta_{\mathbf{H}}$ as the *channel spread*.

2.2.1.3 Continuous-Time Channel Model and Capacity

The goal of this chapter is to provide a characterization of the capacity of the continuous-time channel with I/O relation (2.1), under the assumptions that:

1. Neither the transmitter nor the receiver knows the realizations of $h(t,\tau)$, but both are aware of the channel statistics. For the Rayleigh-fading WSSUS channel model, knowledge of the channel statistics amounts to knowledge of the channel scattering function.
2. The input signal $s(t)$ is subject to a bandwidth constraint, an average-power constraint, and a peak-power constraint.

A formal definition of the capacity of the continuous-time channel (2.1) can be found in Durisi, Morgenshtern, and Bölcskei (2010). This definition is along the lines of Wyner's treatment of the capacity of a bandlimited continuous-time AWGN channel (Gallager, 1968; Wyner, 1966). The key element in this definition is the precise specification of the set of constraints (approximate time duration, bandwidth, average power, ...) that are imposed on the input signal $s(t)$. For the sake of simplicity of exposition, we refrain from presenting this definition here (the interested reader is referred to Durisi, Morgenshtern, and Bölcskei (2009); Durisi, Morgenshtern, and Bölcskei (2011)). We take, instead, a somewhat less rigorous approach, which has the advantage of simplifying the exposition drastically and better illustrates the harmonic analysis aspects of the problem: we first discretize the channel I/O relation, and then impose a set of constraints on the resulting discretized input signal. These constraints "mimic" the ones that are natural to impose on the underlying continuous-time input signal. We emphasize that all the results stated in the next sections can be made precise in an information-theoretic sense following the approach in Durisi et al. (2010), and Durisi, Morgenshtern, and Bölcskei (2011).

2.2.2 Discretization of the Continuous-Time Input–Output Relation

Different ways to discretize LTV channels have been proposed in the literature (some are reviewed in Chapter 1); not all the induced discretized I/O relations are, however, equally well suited for an information-theoretic analysis. Stationarity of the discretized system functions and statistically independent noise samples are some of the desiderata regarding the discretization step.

The most common approach to the discretization of random LTV channels is based on sampling (Bello, 1963; Médard, 2000; Médard & Gallager, 2002), often combined with a basis expansion model (BEM) (see Chapter 1 for a detailed discussion). A different approach, particularly well suited for information-theoretic analyses (Gallager, 1968; Wyner, 1966), is based on a channel operator eigendecomposition or, more generally, singular-value decomposition. We briefly review these two approaches and discuss their shortcomings when used for the problem we are dealing with in this chapter. These shortcomings will motivate us to pursue a different approach detailed in Section 2.2.2.3.

2.2.2.1 Sampling and Basis Expansion

Under the assumption that $C_{\mathbf{H}}(\tau,\nu)$ is compactly supported in ν over the interval $[-\nu_0,\nu_0]$, and the input signal $s(t)$ is strictly bandlimited with bandwidth B (i.e., $S(f) \triangleq \mathbb{F}\{s(t)\} = 0$ for $|f| > B$), the I/O relation (2.1) can be discretized by means of the sampling theorem (see, for example, Artés, Matz, and Hlawatsch (2004) for a detailed derivation). The resulting discretized I/O relation is given by

$$r\left(\frac{n}{f_o}\right) = \frac{1}{f_i}\sum_{m=-\infty}^{\infty} h_c\left(\frac{n}{f_o},\frac{m}{f_i}\right) s\left(\frac{n}{f_o}-\frac{m}{f_i}\right) + w\left(\frac{n}{f_o}\right), \tag{2.5}$$

where $f_o = 2(B+\nu_0), f_i = 2B$, and

$$h_c(t,\tau) \triangleq \int_{-\infty}^{\infty} h(t,z)\frac{\sin[2\pi B(\tau - z)]}{\pi(\tau - z)}dz. \tag{2.6}$$

The discretized I/O relation (2.5) can be further simplified if the input signal is oversampled, i.e., if f_i is chosen to be equal to f_o. In this case, (2.5) can be rewritten as

$$r[n] = \sum_{m=-\infty}^{\infty} h_c[n,m]s[n-m] + w[n]. \tag{2.7}$$

One evident limitation of the sampling approach just discussed is the compact-support assumption on $C_{\mathbf{H}}(\tau,\nu)$ (cf. Definition 2.1). Furthermore, as a consequence of (2.6), $h_c[n,m]$ does not inherit the US property of $h(t,\tau)$, a fact that makes the information-theoretic analysis of (2.7) more involved. Finally, the apparently harmless oversampling step imposes an implicit constraint on the set of the input-signal samples; this constraint is hard to account for in an information-theoretic analysis. More specifically, if the input-signal samples are chosen in an arbitrary way, the resulting continuous-time input signal may have a bandwidth as large as $B+\nu_0$, rather than just B.

Often, the sampling approach is combined with a BEM. In the most common form of the BEM, the basis consists of complex exponentials and $h_c[n,m]$ is given by

$$h_c[n,m] = \sum_{i=0}^{I-1} c_i[m]e^{j2\pi\nu_i n}, \tag{2.8}$$

where $\{\nu_i\}_{i=0}^{I-1}$ is a set of Doppler frequencies. For general WSSUS underspread channels, the validity of the modeling assumption underlying (2.8) is difficult to assess, and information-theoretic results obtained on the basis of (2.8) might lack generality.

2.2.2.2 Discretization through Channel Eigendecomposition

As discussed in Chapter 1, the kernel $k_{\mathbf{H}}(t,t') \triangleq h(t,t-t')$ associated with a general LTV channel $\mathbf{H}$ (under the assumption that the channel operator $\mathbf{H}$ is normal and compact (Durisi et al., 2010; Naylor & Sell, 1982)) can be decomposed as (Naylor & Sell, 1982, Theorem 6.14.1)

$$k_{\mathbf{H}}(t,t') = \sum_{k=0}^{\infty} \lambda_k u_k(t) u_k^*(t'), \tag{2.9}$$

where λ_k and $u_k(t)$ are the channel eigenvalues and eigenfunctions, respectively. The set $\{u_k(t)\}$ is orthonormal and complete in the input space and in the range space of $\mathbf{H}$ (Naylor & Sell, 1982, Theorem 6.14.1). Hence, any input signal $s(t)$ and any output signal $r(t)$ can be expressed, without loss of generality, in terms of its projections onto the set $\{u_k(t)\}$ according to

$$s(t) = \sum_k \underbrace{\langle s(t), u_k(t)\rangle}_{\triangleq s[k]} u_k(t) \tag{2.10}$$

and

$$r(t) = \sum_k \underbrace{\langle r(t), u_k(t) \rangle}_{\triangleq r[k]} u_k(t). \tag{2.11}$$

The decomposition (2.9), together with (2.10) and (2.11), yields a particularly simple I/O relation, which we refer to as *channel diagonalization*:

$$\begin{aligned} r[k] &= \langle (\mathbf{H}s)(t) + w(t), u_k(t) \rangle = \sum_{k'} s[k'] \underbrace{\langle (\mathbf{H}u_{k'})(t), u_k(t) \rangle}_{\lambda_{k'} \delta_{k'k}} + \underbrace{\langle w(t), u_k(t) \rangle}_{\triangleq w[k]} \\ &= \lambda_k s[k] + w[k]. \end{aligned} \tag{2.12}$$

Note that in (2.12), the channel acts on the discretized input through scalar multiplications only. To summarize, it follows from (2.10) and (2.11) that channel diagonalization is achieved by transmitting and receiving on the channel eigenfunctions.

The discretization method just described is used in Wyner (1966) to compute the capacity of bandlimited AWGN channels and in Gallager (1968) to compute the capacity of (deterministic) linear time-invariant (LTI) channels. This approach, however, is not applicable to our setting. Transmitting and receiving on the channel eigenfunctions $\{u_k(t)\}$ requires perfect knowledge of $\{u_k(t)\}$ at the transmitter and the receiver. But as $\mathbf{H}$ is random, its eigenfunctions $u_k(t)$ are (in general) random as well, and, hence, not known at the transmitter and the receiver in the noncoherent setting.

In contrast, if the eigenfunctions of the random channel $\mathbf{H}$ did not depend on the particular realization of $\mathbf{H}$, we could diagonalize $\mathbf{H}$ without knowledge of the channel realizations. This is the case, for example, for LTI channels, where (deterministic) complex sinusoids are always eigenfunctions, independently of the realization of the channel impulse response. This observation is crucial for the ensuing analysis. Specifically, our discretization strategy is based on the following fundamental property of underspread LTV channels (see Chapter 1 for details): the eigenfunctions of a random underspread WSSUS channel can be well approximated by deterministic functions that are well localized in time and frequency.

Before illustrating the details of the discretization step, we would like to point out two difficulties that arise from the approach we are going to pursue. The discretized I/O relation induced by replacing the channel eigenfunctions in (2.10) and (2.11) with approximate eigenfunctions (or, in other words, induced by transmitting and receiving on the approximate channel eigenfunctions) will, in general, not be as simple as (2.12), because of the presence of "off-diagonal" terms. Furthermore, the set of approximate eigenfunctions may not be complete in the input and range spaces of the channel operator $\mathbf{H}$. This results in a loss of dimensions in signal space, i.e., of degrees of freedom, which will need to be accounted for in our information-theoretic analysis.

2.2.2.3 *Discretization through Transmission and Reception on a Weyl-Heisenberg Set*

We accomplish the discretization of the I/O relation (2.1) by transmitting and receiving on the highly structured Weyl-Heisenberg (WH) set of time-frequency shifts of a pulse $g(t)$. We denote such a WH set as $(g(t), T, F) \triangleq \{g_{k,l}(t) = g(t-kT)e^{j2\pi lFt}\}_{k,l\in\mathbb{Z}}$, where $T, F > 0$ are the *grid parameters* of the

WH set (Kozek & Molisch, 1998). The triple $g(t), T, F$ is chosen so that $g(t)$ has unit energy and that the resulting set $(g(t), T, F)$ is orthonormal.[4] Note that we do not require that the set $(g(t), T, F)$ is complete in $L^2(\mathbb{R})$. The time-frequency localization of $g(t)$ plays an important role in our analysis, because the functions in a WH set constructed from a pulse that is well localized in time and frequency are approximate eigenfunctions of underspread WSSUS channels (Kozek, 1997; Matz & Hlawatsch, 1998, 2003).

We consider input signals of the form

$$s(t) = \sum_{k=0}^{K-1} \sum_{l=0}^{L-1} s[k,l] g_{k,l}(t). \tag{2.13}$$

Whenever $g(t)$ is well localized in time and frequency, we can take $D \triangleq KT$ to be the approximate time duration of $s(t)$ and $B \triangleq LF$ to be its approximate bandwidth. Note that the lack of completeness of $(g(t), T, F)$ implies that there exist signals $s(t) \in L^2(\mathbb{R})$ that cannot be written in the form (2.13), even when $K, L \to \infty$. In other words, we may lose degrees of freedom.

The received signal $r(t)$ is projected onto the signal set $\{g_{k,l}(t)\}_{k=0,l=0}^{K-1,L-1}$ to obtain

$$\underbrace{\langle r(t), g_{k,l}(t)\rangle}_{\triangleq r[k,l]} = \underbrace{\langle (\mathbf{H} g_{k,l})(t), g_{k,l}(t)\rangle}_{\triangleq h[k,l]} s[k,l] + \sum_{\substack{k'=0 \\ (k',l')\neq(k,l)}}^{K-1} \sum_{l'=0}^{L-1} \underbrace{\langle (\mathbf{H} g_{k',l'})(t), g_{k,l}(t)\rangle}_{\triangleq z[k',l',k,l]} s[k',l'] + \underbrace{\langle w(t), g_{k,l}(t)\rangle}_{\triangleq w[k,l]}$$

that is

$$r[k,l] = h[k,l]s[k,l] + \sum_{\substack{k'=0 \\ (k',l')\neq(k,l)}}^{K-1} \sum_{l'=0}^{L-1} z[k',l',k,l]s[k',l'] + w[k,l]. \tag{2.14}$$

We refer to the channel with I/O relation (2.14) as the *discretized channel induced by the WH set* $(g(t), T, F)$. The I/O relation (2.14) satisfies the desiderata we listed at the beginning of Section 2.2.2. Specifically, the channel coefficients $h[k,l]$ inherit the two-dimensional stationarity property of the underlying continuous-time system function $L_{\mathbf{H}}(t,f)$ [see (2.2)]. Furthermore, the noise coefficients $w[k,l]$ are i.i.d. $\mathcal{CN}(0,1)$ as a consequence of the orthonormality of $(g(t), T, F)$. These two properties are crucial for the ensuing analysis.

A drawback of (2.14) is the presence of (self-)interference [the second term in (2.14)], which makes the derivation of capacity bounds involved, as will be seen in Section 2.4.

The signaling scheme (2.13) can be interpreted as PS-OFDM (Kozek & Molisch, 1998), where the input symbols $s[k,l]$ are modulated onto a set of orthogonal signals, indexed by discrete time (symbol index) k and discrete frequency (subcarrier index) l. From this perspective, the interference term in (2.14) can be interpreted as intersymbol and intercarrier interference (ISI and ICI). Figure 2.1 provides a qualitative representation of the PS-OFDM signaling scheme.

[4] A systematic approach to constructing orthonormal WH sets is described in Section 2.2.2.6.

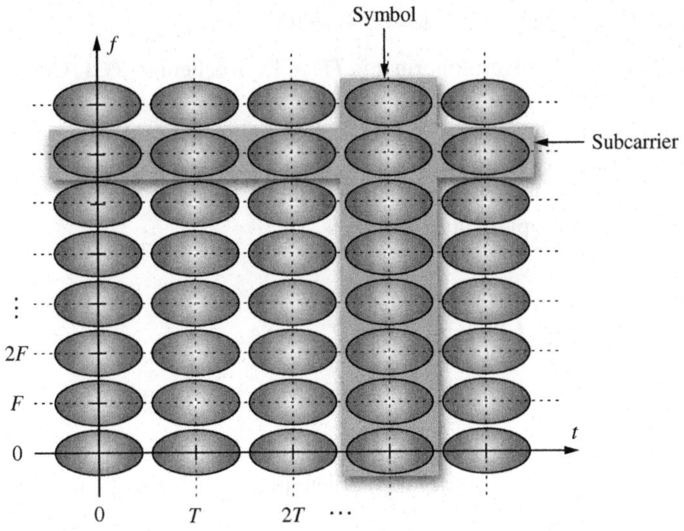

FIGURE 2.1

Pulse-shaped OFDM interpretation of the signaling scheme (2.13). The shaded areas represent the approximate time-frequency support of the pulses $g_{k,l}(t)$.

2.2.2.4 Outline of the Information-Theoretic Analysis

The program pursued in the next sections is to tightly bound the noncoherent capacity of the discretized channel (2.14) under an average-power and a peak-power constraint on the input symbols $s[k,l]$. We refer the interested reader to Durisi et al. (2010) and Durisi, Morgenshtern, and Bölcskei (2011) for a discussion on the relation between the capacity of the discretized channel (2.14) and the capacity of the underlying continuous-time channel (2.1).

The derivation of capacity bounds is made difficult by the presence of the interference term in (2.14). Fortunately, to establish our main results in Sections 2.4 and 2.5, it will be sufficient to compare a trivial capacity upper bound, i.e., the capacity of an AWGN channel with the same receive power as in (2.14), with simple capacity lower bounds obtained by treating the interference term in (2.14) as noise. The corresponding results are of practical interest, as receiver algorithms that take the structure of interference explicitly into account are, in general, computationally expensive. The power of the interference term in (2.14) depends on the choice of the WH set. As it will be shown in Section 2.2.2.7, good time-frequency localization of the pulse $g(t)$ the WH set is generated from is required for the interference term to be small. But good time-frequency localization entails a loss of degrees of freedom.

Before discussing the trade-off between interference minimization and maximization of the number of degrees of freedom, we review some important theoretical results on the construction of WH sets.

2.2.2.5 Orthonormality, Completeness, and Localization

Orthonormality, completeness, and time-frequency localization are desirable properties of the WH set $(g(t), T, F)$. It is, therefore, sensible to ask whether complete orthonormal WH sets generated by a $g(t)$ with prescribed time-frequency localization exist. The answer is as follows:

1. A necessary condition for the set $(g(t), T, F)$ to be orthonormal is $TF \geq 1$ (Gröchenig, 2001, Cor. 7.5.1, Cor. 7.3.2).
2. For $TF = 1$, it is possible to find orthonormal sets $(g(t), T, F)$ that are complete in $L^2(\mathbb{R})$ (Christensen, 2003, Th. 8.3.1). These sets, however, do not exhibit good time-frequency localization, as a consequence of the Balian-Low Theorem (Christensen, 2003, Th. 4.1.1), which states that if $(g(t), T, F)$ is orthonormal and complete in $L^2(\mathbb{R})$, then

$$\left(\int_{-\infty}^{\infty} |tg(t)|^2 dt\right)\left(\int_{-\infty}^{\infty} |fG(f)|^2 df\right) = \infty,$$

   where $G(f) \triangleq \mathbb{F}\{g(t)\}$.
3. For $TF > 1$, it is possible to have orthonormality and good time-frequency localization concurrently, but the resulting set $(g(t), T, F)$ is necessarily incomplete in $L^2(\mathbb{R})$. Lack of completeness entails a loss of degrees of freedom.
4. For $TF < 1$, it is possible to construct WH sets generated by a well-localized $g(t)$, which are also (over)complete in $L^2(\mathbb{R})$. However, as a consequence of overcompleteness, the resulting input signal (2.13) cannot be recovered uniquely at the receiver, even in the absence of noise.

Our choice will be to privilege localization and orthonormality over completeness. The information-theoretic results in Sections 2.4 and 2.5 will show that this choice is sound. In the next two sections, we review the mathematical framework that enables the construction of (noncomplete) orthonormal WH sets that are well localized in time and frequency. We furthermore review a classic criterion for the design of WH sets, which is based on the maximization of the signal-to-interference ratio (SIR) in (2.14). The information-theoretic analysis in Section 2.5 will yield a design criterion that is more fundamental.

2.2.2.6 *Construction of WH Sets*

We next discuss how to construct (noncomplete) WH sets with prescribed time-frequency localization. Frame theory plays a fundamental role in this context.

A WH set $(g(t), T, F)$ is called a WH *frame* for $L^2(\mathbb{R})$ if there exist constants A and B (*frame bounds*) with $0 < A \leq B < \infty$ such that for all $s(t) \in L^2(\mathbb{R})$, we have

$$A\|s(t)\|^2 \leq \sum_k \sum_l |\langle s(t), g_{k,l}(t)\rangle|^2 \leq B\|s(t)\|^2.$$

When $A = B$, the frame is called *tight*. A necessary condition for $(g(t), T, F)$ to be a frame for $L^2(\mathbb{R})$ is $TF \leq 1$ (Gröchenig, 2001, Cor. 7.5.1). For specific pulses, sufficient conditions for the corresponding WH set to be a frame are known. For example, for the Gaussian pulse $g(t) = 2^{1/4} e^{-\pi t^2}$, the condition $TF < 1$ is necessary and sufficient for $(g(t), T, F)$ to be a frame (Gröchenig, 2001, Th. 7.5.3).

A frame $(g(t), T, F)$ for $L^2(\mathbb{R})$ can be transformed into a tight frame $(g^{\perp}(t), T, F)$ for $L^2(\mathbb{R})$ through standard frame-theoretic methods (Christensen, 2003, Th. 5.3.4). Furthermore, the so-obtained $g^{\perp}(t)$ inherits the decay properties of $g(t)$ (see Bölcskei & Janssen, 2000; Matz et al., 2007 for a mathematically precise formulation of this statement).

The key result that makes frame theory relevant for the construction of orthonormal WH sets is the so called *duality theorem* (Daubechies, Landau, & Landau, 1995; Janssen, 1995; Ron & Shen, 1997), which states that $(g(t), T, F)$ with $TF \geq 1$ is an orthonormal WH set if and only if $(g(t), 1/F, 1/T)$ is a tight frame for $L^2(\mathbb{R})$.

The results summarized above are used in the following example to construct an orthonormal WH set that will be used throughout this chapter.

Example 2.1 (Root-raised-cosine WH set)

For later use, we present an example of an orthonormal WH set for the case $T = F = \sqrt{c}$, with $1 \leq c \leq 2$. Let $G(f) = \mathbb{F}\{g(t)\}, \zeta = \sqrt{c}$, and $\mu = c - 1$. We choose $G(f)$ as the (positive) square root of a raised-cosine pulse:

$$G(f) = \begin{cases} \sqrt{\zeta}, & \text{if } \ |f| \leq \frac{1-\mu}{2\zeta} \\ \sqrt{\frac{\zeta}{2}[1+S(f)]}, & \text{if } \ \frac{1-\mu}{2\zeta} \leq |f| \leq \frac{1+\mu}{2\zeta} \\ 0, & \text{otherwise} \end{cases}$$

where $S(f) = \cos\left[\frac{\pi\zeta}{\mu}\left(|f| - \frac{1-\mu}{2\zeta}\right)\right]$. As $(1+\mu)/(2\zeta) = \zeta/2$, the function $G(f)$ has compact support of length $\zeta = \sqrt{c}$. Furthermore, $G(f)$ is real-valued and even, and satisfies

$$\sum_{l=-\infty}^{\infty} G(f - l/\zeta)G(f - l/\zeta - k\zeta) = \zeta\delta_{k0}.$$

By (Christensen, 2003, Th. 8.7.2), we can, therefore, conclude that the WH set $(g(t), 1/\sqrt{c}, 1/\sqrt{c})$ is a tight WH frame for $L^2(\mathbb{R})$, and, by duality, the WH set $(g(t), \sqrt{c}, \sqrt{c})$ is orthonormal. Note that, for $c = 1$ (i.e., $TF = 1$), the pulse $G(f)$ reduces to the rectangular pulse $1_{[-1/2,1/2]}(f)$ and, consequently, $g(t)$ reduces to a *sinc* function, which has poor time localization, as expected from the Balian-Low Theorem. ■

2.2.2.7 *Diagonalization and Loss of Degrees of Freedom*

By choosing $g(t)$ to be well-localized in time and frequency and TF sufficiently large, we can make the variance of the interference term in the I/O relation (2.14) small. The drawback is a loss of degrees of freedom, as formalized next.

Let $A_g(\tau, \nu)$ denote the ambiguity function of $g(t)$, defined as

$$A_g(\tau, \nu) \triangleq \int_{-\infty}^{\infty} g(t)g^*(t-\tau)e^{-j2\pi\nu t}dt.$$

The variance of both $h[k,l] = \langle (\mathbf{H}g_{k,l})(t), g_{k,l}(t)\rangle$ and $z[k',l',k,l] = \langle (\mathbf{H}g_{k',l'})(t), g_{k,l}(t)\rangle$ can be expressed in terms of $A_g(\tau,\nu)$. In fact, as a consequence of the WSSUS property of $\mathbf{H}$, we have that

$$\mathsf{E}\left\{h[k,l]h^*[k',l']\right\} = \int_{-\infty}^{\infty}\int_{-\infty}^{\infty} C_{\mathbf{H}}(\tau,\nu)|A_g(\tau,\nu)|^2 e^{j2\pi[(k-k')T\nu-(l-l')F\tau]}d\tau d\nu \tag{2.15}$$

and, in particular,

$$\sigma_h^2 \triangleq \mathsf{E}\left\{|h[k,l]|^2\right\} = \int_{-\infty}^{\infty}\int_{-\infty}^{\infty} C_{\mathbf{H}}(\tau,\nu)|A_g(\tau,\nu)|^2 d\tau d\nu. \tag{2.16}$$

Because $|A_g(\tau,\nu)|^2 \le \|g(t)\|^4 = 1$ and because the scattering function $C_{\mathbf{H}}(\tau,\nu)$ was assumed to be of unit volume [see (2.4)], we have that $\sigma_h^2 \le 1$. The relation (2.15) implies that $R_{\mathbf{H}}[k,k',l,l'] \triangleq \mathsf{E}\left\{h[k,l]h^*[k',l']\right\} = R_{\mathbf{H}}[k-k',l-l']$, i.e., that $h[k,l]$ is WSS both in discrete time k and discrete frequency l.

The variance of the interference term in (2.14), under the assumption that the $s[k,l]$ are i.i.d. with zero mean and unit variance,[5] can be upper-bounded as follows:

$$\begin{aligned}
\mathsf{E}\left\{\left|\sum_{\substack{k'=0\\(k',l')\neq(k,l)}}^{K-1}\sum_{l'=0}^{L-1} z[k',l',k,l]s[k',l']\right|^2\right\} &= \sum_{\substack{k'=0\\(k',l')\neq(k,l)}}^{K-1}\sum_{l'=0}^{L-1} \mathsf{E}\left\{|z[k',l',k,l]|^2\right\}\\
&\le \sum_{\substack{k'=-\infty\\(k',l')\neq(k,l)}}^{\infty}\sum_{l'=-\infty}^{\infty} \mathsf{E}\left\{|z[k',l',k,l]|^2\right\}\\
&= \sum_{\substack{k'=-\infty\\(k',l')\neq(k,l)}}^{\infty}\sum_{l'=-\infty}^{\infty}\int_{-\infty}^{\infty}\int_{-\infty}^{\infty} \left|A_g(\tau+(k'-k)T,\nu+(l'-l)F)\right|^2 C_{\mathbf{H}}(\tau,\nu)d\tau d\nu\\
&= \sum_{\substack{k=-\infty\\(k,l)\neq(0,0)}}^{\infty}\sum_{l=-\infty}^{\infty}\int_{-\infty}^{\infty}\int_{-\infty}^{\infty} \left|A_g(\tau,\nu)\right|^2 C_{\mathbf{H}}(\tau-kT,\nu-lF)d\tau d\nu\\
&\triangleq \sigma_I^2.
\end{aligned} \tag{2.17}$$

The "infinite-horizon" interference variance σ_I^2 will turn out to be of particular importance in the information-theoretic analysis in Sections 2.4 and 2.5. When $\sigma_I^2 \approx 0$, the I/O relation (2.14) can be well approximated by the following diagonalized I/O relation

$$r[k,l] = h[k,l]s[k,l] + w[k,l]. \tag{2.18}$$

[5]In Sections 2.4 and 2.5, we will use this assumption to obtain capacity lower bounds explicit in σ_I^2.

This simplification eases the derivation of bounds on capacity. As the received signal power in (2.18) is proportional to σ_h^2, it is also desirable to choose WH sets that result in $\sigma_h^2 \approx 1$ (recall that $\sigma_h^2 \leq 1$).

Next, we investigate the design criteria a WH set $(g(t), T, F)$ needs to satisfy so that $\sigma_h^2 \approx 1$ and $\sigma_I^2 \approx 0$. We assume, for simplicity, that $C_{\mathbf{H}}(\tau,\nu)$ is compactly supported within the rectangle $[-\tau_0,\tau_0] \times [-\nu_0,\nu_0]$. Referring to (2.16), (2.17), and to Fig. 2.2, we conclude the following:

1. $\sigma_h^2 \approx 1$ if the ambiguity function of $g(t)$ satisfies $|A_g(\tau,\nu)|^2 \approx 1$ over the support of the scattering function.
2. $\sigma_I^2 \approx 0$ if the ambiguity function of $g(t)$ takes on small values on the rectangles $[-\tau_0 + kT, \tau_0 + kT] \times [-\nu_0 + lF, \nu_0 + lF]$, for $(k,l) \in \mathbb{Z}^2 \setminus \{(0,0)\}$.

For these two conditions to be satisfied, the spread of the channel $\Delta_{\mathbf{H}} = 4\tau_0\nu_0$ needs to be small, and the grid parameters need to be chosen such that the ambiguity function $A_g(\tau,\nu)$ takes on small values outside the solid gray-shaded rectangle centered at the origin in Fig. 2.2. Let D_g and B_g be the root-mean-square duration and the root-mean-square bandwidth, respectively, of the pulse $g(t)$ (see Chapter 1 for a definition of these two quantities). Then, condition 2 above holds if $T > \tau_0 + D_g$ and $F > \nu_0 + B_g$. These two inequalities illustrate the importance of good time-frequency localization of the pulse $g(t)$ (i.e., small D_g and B_g). In fact, large D_g and B_g imply large T and F, and, consequently, a significant loss of degrees of freedom.

For fixed TF, a simple rule on how to choose the grid parameters T and F follows from the observation that for given τ_0 and ν_0, the area of the solid rectangle centered at the origin in Fig. 2.2 is maximized (Kozek, 1997; Kozek & Molisch, 1998; Matz et al., 2007), if

$$\nu_0 T = \tau_0 F. \tag{2.19}$$

This rule is commonly referred to as *grid matching rule*.

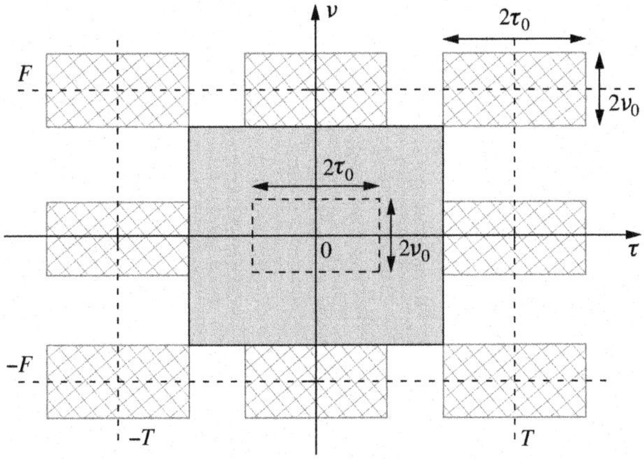

FIGURE 2.2

Support of $C_{\mathbf{H}}(\tau - kT, \nu - lF)$ for some (k,l) pairs. From (2.16), it follows that $\sigma_h^2 \approx 1$ if $|A_g(\tau,\nu)|^2 \approx 1$ over the support of $C_{\mathbf{H}}(\tau,\nu)$. Furthermore, from (2.17) it follows that $\sigma_I^2 \approx 0$ if $|A_g(\tau,\nu)|^2 \approx 0$ outside the area shaded in gray.

Remark 2.2 Whenever T and F are chosen according to the grid matching rule, it is possible to assume, without loss of generality, that $C_{\mathbf{H}}(\tau, \nu)$ in (2.16) and (2.17) is supported on a square rather than a rectangle. A proof of this claim follows from a simple coordinate transformation.

Common approaches for the optimization of WH sets, such as the ones described in Matz et al. (2007), aim at finding – for fixed TF – the pulse $g(t)$ that maximizes the ratio σ_h^2/σ_I^2 (which can be thought of as an SIR). To understand the trade-off between degrees of freedom and SIR, we proceed in a different way, and study how σ_h^2/σ_I^2 varies as a function of TF, for fixed $g(t)$. In Fig. 2.3, we plot a lower bound on σ_h^2/σ_I^2 for the root-raised-cosine WH sets of Example 2.1, as a function of TF and for different channel spreads $\Delta_{\mathbf{H}}$ (see Example 2.2 for more details). As expected, the larger TF, the larger the SIR, but the larger also the loss of degrees of freedom. A common compromise between loss of degrees of freedom and maximization of SIR is to take $TF \approx 1.2$ (Matz et al., 2007).

The limitation of the analysis we just outlined is that, although it sheds light on how σ_h^2/σ_I^2 depends on TF, it does not reveal the influence that σ_h^2, σ_I^2, and TF have on the rate achievable when interference is treated as noise. An information-theoretic analysis of the trade-off between maximization of SIR and minimization of the loss of degrees of freedom is called for. Such an analysis is provided in Section 2.5.

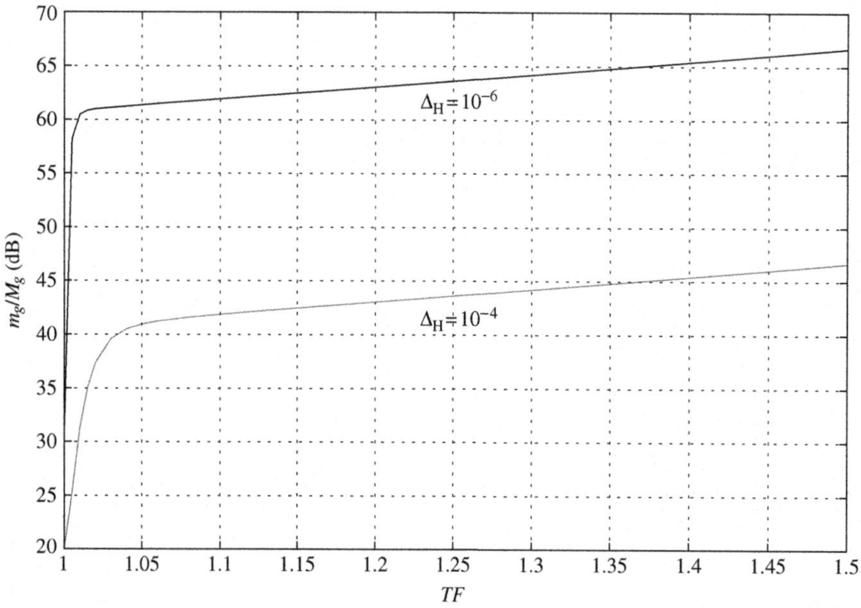

FIGURE 2.3

Trade-off between the product TF of the grid parameters and the signal-to-interference ratio σ_h^2/σ_I^2 for the root-raised-cosine WH sets constructed in Example 2.1. The two cases $\Delta_{\mathbf{H}} = 10^{-4}$ and $\Delta_{\mathbf{H}} = 10^{-6}$ are considered. The ratio m_g/M_g is a lower bound on σ_h^2/σ_I^2, as detailed in Example 2.2.

Example 2.2 (Trade-off between *TF* and σ_h^2/σ_I^2)

Let $\mathscr{D} \triangleq [-\tau_0, \tau_0] \times [-\nu_0, \nu_0]$ and assume that $C_{\mathbf{H}}(\tau, \nu)$ is compactly supported within $\mathscr{D}$ and has unit volume. Then,

$$\sigma_h^2 = \int_{-\infty}^{\infty}\int_{-\infty}^{\infty} C_{\mathbf{H}}(\tau,\nu)|A_g(\tau,\nu)|^2 d\tau d\nu \geq \min_{(\tau,\nu)\in\mathscr{D}} |A_g(\tau,\nu)|^2 \triangleq m_g. \tag{2.20}$$

Furthermore,

$$\begin{aligned}\sigma_I^2 &= \sum_{\substack{k=-\infty \\ (k,l)\neq(0,0)}}^{\infty} \sum_{l=-\infty}^{\infty} \int_{-\infty}^{\infty}\int_{-\infty}^{\infty} |A_g(\tau+kT, \nu+lF)|^2 C_{\mathbf{H}}(\tau,\nu) d\tau d\nu \\ &\leq \max_{(\tau,\nu)\in\mathscr{D}} \sum_{\substack{k=-\infty \\ (k,l)\neq(0,0)}}^{\infty} \sum_{l=-\infty}^{\infty} |A_g(\tau+kT, \nu+lF)|^2 \triangleq M_g. \end{aligned} \tag{2.21}$$

The ratio m_g/M_g is obviously a lower bound on the SIR σ_h^2/σ_I^2, and is easier to compute than σ_h^2/σ_I^2, because it depends on the scattering function only through its support area and not its shape. Under the assumption that T and F are chosen according to the grid matching rule (2.19), it is sufficient to study the ratio m_g/M_g exclusively for the case when the scattering function is compactly supported within a square in the delay-Doppler plane (see Remark 2.2). In Fig. 2.3, the ratio m_g/M_g is plotted as a function of TF and for different values of $\Delta_{\mathbf{H}}$ for the family of root-raised-cosine WH sets defined in Example 2.1. The curves in Fig. 2.3 can be used to determine the value of TF needed to achieve a prescribed SIR for a given channel spread. The ratio m_g/M_g increases significantly when TF is taken only slightly larger than 1. A further increase of TF produces a much less pronounced increase of the ratio m_g/M_g. ■

2.2.2.8 Large-Bandwidth and High-SNR Regimes

The effect of a loss of degrees of freedom on capacity depends on the operating regime of the system. To illustrate this point, let us consider, for simplicity, an AWGN channel and input signals subject to an average-power constraint only. Two operating regimes can be identified, where the impact of a loss of degrees of freedom is drastically different: the *large-bandwidth* (or *power-limited*, or *low-SNR*) regime and the *high-SNR* (or *bandwidth-limited*) regime (Tse & Viswanath, 2005).

In the large-bandwidth regime, capacity is essentially proportional to the receive power, and only mildly sensitive to the number of degrees of freedom. Hence, a loss of degrees of freedom is irrelevant in this regime. In contrast, in the high-SNR regime, capacity grows linearly with the number of degrees of freedom and is only mildly sensitive to the receive power. Therefore, a loss of degrees of freedom is critical in this regime.

One of the main contributions of this chapter is to illustrate that for all SNR values of practical interest, the noncoherent capacity of a Rayleigh-fading underspread WSSUS channel is close to the capacity of an AWGN channel with the same receive power. A key element to establish this result is an appropriate choice of the WH set $(g(t), T, F)$ that is used to discretize the channel. Not surprisingly, this choice turns out to depend on the operating regime of the system. In particular,

1. In the large-bandwidth regime, where capacity is only mildly sensitive to a loss of degrees of freedom, it is sensible to choose $(g(t), T, F)$ so that $\sigma_h^2 \approx 1$ and $\sigma_I^2 \approx 0$, and replace the discretized I/O relation (2.14) by the much simpler diagonalized I/O relation (2.18). In Section 2.3, we study the noncoherent capacity of the diagonalized channel (2.18) in the large-bandwidth regime. Then, in Section 2.4, we assess how well the noncoherent capacity of (2.18) approximates that of (2.14) in this regime. The analysis in these two sections sheds light on the impact of bandwidth, number of antennas, and shape of the scattering function on capacity, and allows us to derive guidelines on the choice of the WH set.
2. In the high-SNR regime, where capacity is sensitive to a loss of degrees of freedom, the choice of $(g(t), T, F)$ that leads to $\sigma_h^2 \approx 1$ and $\sigma_I^2 \approx 0$ may be inadequate, as it could entail a dimension loss that is too high. We, therefore, have to work directly with the I/O relation (2.14), which accounts for ISI/ICI. In Section 2.5, we study the noncoherent capacity of (2.14) in the high-SNR regime. In particular, we address the dependency of noncoherent capacity on the support of the scattering function, and we discuss, from an information-theoretic perspective, the trade-off between maximization of the number of degrees of freedom and maximization of SIR.

2.3 THE LARGE-BANDWIDTH REGIME: DIAGONALIZED I/O RELATION

As a first step toward the characterization of the noncoherent capacity of the channel (2.14) in the large-bandwidth regime, we present in this section tight bounds on the noncoherent capacity of the diagonalized channel (2.18). In Section 2.4, we will then discuss how well the noncoherent capacity of (2.18) approximates that of (2.14). We chose this two-level approach because bounds on the noncoherent capacity of (2.18) are much easier to derive than for (2.14). Furthermore, the information-theoretic tools presented in this section will also turn out useful for the analysis of the noncoherent capacity of (2.14) presented in Section 2.4 and 2.5.

Throughout this section, we shall then focus on the diagonalized I/O relation (2.18), which we recall is obtained by discretizing the continuous-time channel I/O relation (2.1) by means of a WH set $(g(t), T, F)$ for which $\sigma_h^2 \approx 1$ and $\sigma_I^2 \approx 0$. As a consequence, the capacity bounds we shall derive in this section, will depend on $(g(t), T, F)$ only through the grid parameters T and F. A more refined analysis, which is based on the I/O relation with ISI/ICI (2.14) and, hence, leads to bounds explicit in $g(t)$, will be presented in Section 2.4.

We shall assume throughout that the scattering function of the underlying Rayleigh-fading underspread WSSUS channel is compactly supported within $[-\tau_0, \tau_0] \times [-\nu_0, \nu_0]$, and that the grid parameters satisfy the Nyquist condition $T \leq 1/(2\nu_0)$ and $F \leq 1/(2\tau_0)$. These assumptions are not fundamental: they merely serve to simplify the analytic expressions for our capacity bounds.

It is convenient for our analysis to rewrite the I/O relation (2.18) in vector form. As discussed in Section 2.2.2.3, we let $D = KT$ be the approximate duration of the continuous-time input signal $s(t)$ in (2.13) and $B = LF$ be its approximate bandwidth. We denote by $\mathbf{s}$ the KL-dimensional input vector that contains the input symbols $s[k,l]$. The exact way the input symbols $s[k,l]$ are organized into the vector $\mathbf{s}$ is of no concern here, as we will only provide a glimpse of the proof of the capacity bounds. The detailed proof can be found in Durisi et al. (2010). Similarly, the vector $\mathbf{r}$ contains the output-signal samples $r[k,l]$, the vector $\mathbf{h}$ contains the channel coefficients $h[k,l]$, and $\mathbf{w}$ contains the AWGN samples $w[k,l]$. With these definitions, we can now express the I/O relation (2.18) as

$$\mathbf{r} = \mathbf{h} \odot \mathbf{s} + \mathbf{w}, \tag{2.22}$$

where $\odot$ denotes the Hadamard element-wise product.

2.3.1 Power Constraints

We assume that the average power and the peak power of the input signal are constrained as follows:

1. The average power satisfies

$$\frac{1}{T}\mathsf{E}\left\{\|\mathbf{s}\|^2\right\} \leq KP, \tag{2.23}$$

   where P denotes the admissible average power.
2. Among the several ways in which the peak power of the input signal in (2.18) can be constrained (see Durisi et al., 2010 for a detailed discussion), here, we exclusively analyze the case where a joint limitation in time and frequency is imposed, i.e., the case where the amplitude of the input symbols $s[k,l]$ in each time-frequency slot (k,l) is constrained:

$$\frac{1}{T}|s[k,l]|^2 \leq \beta \frac{P}{L}. \tag{2.24}$$

   Here, $\beta \geq 1$ is the *nominal peak-to-average-power ratio* (PAPR). This type of peak constraint is of practical relevance. It models, e.g., a limitation of the radiated peak power in a given frequency band; it also mimics regulatory peak constraints, such as those imposed on UWB systems.

Note that, according to (2.24), the admissible peak power per time-frequency slot goes to zero as the bandwidth [and, hence, L in (2.24)] goes to infinity. An important observation is that the peak constraint in (2.24) depends on the total available bandwidth, rather than the bandwidth that is effectively used by the input signal. As a consequence, it can be shown (Durisi et al., 2010) that the peak constraint (2.24) enforces the use of the total available bandwidth. Consequently, the PAPR of the signal transmitted in each time-frequency slot is effectively limited, and signals with unbounded PAPR, like *flash signals* (Verdú, 2002), are ruled out. Input alphabets commonly used in current systems, like phase-shift keying (PSK) and quadrature-amplitude modulation (QAM), satisfy the constraint (2.24). In contrast, Gaussian inputs, which are often used in information-theoretic analyses, do not satisfy (2.24).

We note that the power constraints (2.23) and (2.24) are imposed on the input symbols $s[k,l]$, rather than on the continuous-time signal $s(t)$. While for the average-power constraint, it is possible to make our analysis more rigorous and let the constraint on $s[k,l]$ follow from an underlying constraint on $s(t)$ (Durisi et al., 2010), the same does not seem to hold true for the peak-power constraint: a limit on the amplitude of $s[k,l]$ does not generally imply a limit on the amplitude of $s(t)$.

2.3.2 Definition of Noncoherent Capacity

Let $\mathcal{Q}$ be the set of probability distributions on $\mathbf{s}$ that satisfy the average-power constraint (2.23) and the peak-power constraint (2.24). The noncoherent capacity of the channel (2.18) is given by (Sethuraman and Hajek, 2005)

$$C \triangleq \lim_{K\to\infty} \frac{1}{KT} \sup_{\mathcal{Q}} I(\mathbf{r};\mathbf{s}), \tag{2.25}$$

where $I(\mathbf{r};\mathbf{s})$ denotes the mutual information between the KL-dimensional output vector $\mathbf{r}$ and the KL-dimensional input vector $\mathbf{s}$ (Cover & Thomas, 1991). The noncoherent capacity C is notoriously hard to characterize analytically. In the next subsections, we present the following bounds on C:

1. An upper bound U_c, which we refer to as coherent-capacity upper bound, that is based on the assumption that the receiver has perfect knowledge of the channel realizations. The derivation of this bound is standard (see, e.g., Biglieri et al., 1998, Sec. III.C.1).
2. An upper bound U_1 that is explicit in the channel scattering function and extends the upper bound (Sethuraman et al., 2009, Prop. 2.2) on the capacity of frequency-flat time-selective channels to general underspread WSSUS channels.
3. A lower bound L_1 that extends the lower bound (Sethuraman et al., 2005, Prop. 2.2) to general underspread WSSUS channels. This bound is explicit in the channel scattering function only for large bandwidth.

Our focus here will be on the engineering insights that can be obtained from the bounds; we will give a flavor of the derivations and refer the reader to Durisi et al. (2010) and Schuster, Durisi, Bölcskei, and Poor (2009) for detailed derivations.

2.3.3 A Coherent-Capacity Upper Bound

The assumptions that the receiver perfectly knows the instantaneous channel realizations and that the input vector $\mathbf{s}$ is subject only to an average-power constraint furnish the following standard capacity upper bound (Biglieri et al., 1998, Sec. III.C.1)

$$U_c(B) \triangleq \frac{B}{TF} \mathsf{E}_h\left\{\ln\left(1 + \frac{PTF}{B}|h|^2\right)\right\}, \tag{2.26}$$

where $h \sim \mathcal{CN}(0,1)$. As the upper bound U_c increases monotonically with bandwidth, this bound does not reflect the noncoherent capacity behavior for large bandwidth accurately. Nevertheless, we shall see in Section 2.3.6, by means of numerical examples, that U_c is quite useful over a large range of bandwidth values of practical interest.

2.3.4 An Upper Bound on Capacity that Is Explicit In $C_{\mathbb{H}}(\tau,\nu)$

In this section, we derive an upper bound on C that will help us to better understand the dependency of noncoherent capacity on bandwidth in the large-bandwidth regime. An important feature of this bound is that it is explicit in the channel scattering function. To simplify the exposition of the key steps used to obtain this bound, we first focus on a very simple channel model.

2.3.4.1 Bounding Idea

Let $h \sim \mathcal{CN}(0,1)$ denote the random gain of a memoryless flat-fading channel with input s, output r, and additive noise $w \sim \mathcal{CN}(0,1)$, i.e., with I/O relation $r = hs + w$. Let $\mathcal{Q}$ denote the set of all probability distributions on s that satisfy the average-power constraint $\mathsf{E}\{|s|^2\} \le P$ and the peak constraint $|s|^2 \le \beta P$. We obtain an upper bound on the capacity $\sup_{\mathcal{Q}} I(r;s)$ along the lines of Sethuraman et al. (2009, Prop. 2.2) (see also Biglieri et al., 1998, p. 2636) as follows:

$$\begin{aligned}
\sup_{\mathcal{Q}} I(r;s) &\overset{(a)}{=} \sup_{\mathcal{Q}} \{I(r;s,h) - I(r;h\,|\,s)\} \\
&\overset{(b)}{\le} \sup_{\mathcal{Q}} \left\{\ln\left(1 + \mathsf{E}\{|s|^2\}\right) - \mathsf{E}\left\{\ln(1+|s|^2)\right\}\right\} \\
&= \sup_{0\le\alpha\le 1} \; \sup_{\substack{\mathcal{Q} \\ \mathsf{E}\{|s|^2\}=\alpha P}} \left\{\ln\left(1 + \mathsf{E}\{|s|^2\}\right) - \mathsf{E}\left\{\ln(1+|s|^2)\right\}\right\} \\
&\overset{(c)}{\le} \sup_{0\le\alpha\le 1} \left\{ \ln(1+\alpha P) - \inf_{\substack{\mathcal{Q} \\ \mathsf{E}\{|s|^2\}=\alpha P}} \mathsf{E}\left\{ \inf_{|u|^2 \le \beta P} \frac{\ln(1+|u|^2)}{|u|^2} |s|^2 \right\}\right\} \\
&\overset{(d)}{=} \sup_{0\le\alpha\le 1} \left\{\ln(1+\alpha P) - \frac{\alpha}{\beta}\ln(1+\beta P)\right\}. \qquad (2.27)
\end{aligned}$$

Here, (a) follows from the chain rule for mutual information (Cover & Thomas, 1991); the inequality (b) results because we take hs as JPG with variance $\mathsf{E}\{|hs|^2\} = \mathsf{E}\{|s|^2\}$. To obtain the inequality (c), we multiply and divide $\ln(1+|s|^2)$ by $|s|^2$ and lower-bound the resulting term $\ln(1+|s|^2)/|s|^2$ by its infimum over all inputs s that satisfy the peak constraint. Finally, (d) results because $\ln(1+|u|^2)/|u|^2$ is monotonically decreasing in $|u|^2$, so that its infimum is achieved for $|u|^2 = \beta P$.

If the supremum in (2.27) is achieved for $\alpha = 1$, the upper bound simplifies to

$$C \le \ln(1+P) - \frac{1}{\beta}\ln(1+\beta P).$$

This bound can be interpreted as the capacity of an AWGN channel with SNR equal to P minus a penalty term that quantifies the capacity loss due to channel uncertainty. The higher the allowed peakness of the input, as measured by the PAPR β, the smaller the penalty.

In spite of its simplicity, the upper bound (2.27) is tight in the low-SNR regime we are interested in (in this section). More precisely, the Taylor-series expansion of the upper bound (2.27) around the point $P = 0$ matches that of capacity up to second order (Sethuraman et al., 2009, Prop. 2.1). In contrast, at high SNR, the upper bound (2.27) exhibits an overly optimistic behavior. In fact, the bound scales logarithmically in P, while the high-SNR capacity scaling for memoryless channels is doubly logarithmic (Lapidoth & Moser, 2003; Taricco & Elia, 1997).

2.3.4.2 The Actual Bound

For the I/O relation of interest in this chapter, namely (2.18), the derivation of a bound similar to (2.27) is complicated by the correlation that $h[k,l]$ exhibits in k and l. A key element in this derivation is the relation between mutual information and minimum mean-square error (MMSE) estimation (Guo et al., 2005), which leads through the classic formula for the infinite-horizon non-causal prediction error for stationary Gaussian processes (Poor, 1994, Eq. (V.D.28)) to a closed-form expression that is explicit in the channel scattering function. The resulting upper bound on C is presented in Theorem 2.1 below. A detailed derivation of this upper bound can be found in Durisi et al. (2010).

Theorem 2.1 The noncoherent capacity of the channel (2.18), under the assumption that the input signal satisfies the average-power constraint (2.23) and the peak-power constraint (2.24), is upper-bounded as $C \le U_1$, where

$$U_1(B) \triangleq \frac{B}{TF} \ln\left(1 + \alpha(B) P \frac{TF}{B}\right) - \alpha(B)\psi(B) \tag{2.28a}$$

with

$$\alpha(B) \triangleq \min\left\{1, \frac{B}{TF}\left(\frac{1}{\psi(B)} - \frac{1}{P}\right)\right\} \tag{2.28b}$$

and

$$\psi(B) \triangleq \frac{B}{\beta} \int_{-\infty}^{\infty}\int_{-\infty}^{\infty} \ln\left(1 + \frac{\beta P}{B} C_{\mathbf{H}}(\tau,\nu)\right) d\tau d\nu. \tag{2.28c}$$

The bound U_1 approaches zero for $B \to \infty$ (Durisi et al., 2010, Sec. III.E), a behavior that is to be expected because the peak constraint (2.24) forces the input signal to be spread out over the total available bandwidth. This behavior is well known (Médard & Gallager, 2002; Subramanian & Hajek, 2002; Telatar & Tse, 2000). However, as the bound (2.28) is explicit in B, it allows us to characterize the capacity behavior also for finite bandwidth. In particular, through a numerical evaluation of U_1 (and of the lower bound we shall derive in Section 2.3.5), it is possible to (coarsely) identify the *critical bandwidth* for which capacity is maximized (see Section 2.3.6).

By means of a Taylor-series expansion, the bound U_1 can be shown to be tight in the large-bandwidth regime. More precisely, the Taylor-series expansion of U_1 around the point $1/B = 0$ matches that of capacity up to first order (Durisi et al., 2010).

2.3.4.3 *Some Simplifications*

Similarly to the very simple memoryless flat-fading channel analyzed in Section 2.3.4.1, if the minimum in (2.28b) were attained for $\alpha(B) = 1$, the first term of the upper bound U_1 in (2.28a) could be interpreted as the capacity of an effective AWGN channel with power P and $B/(TF)$ degrees of freedom, whereas the second term could be seen as a penalty term that characterizes the capacity loss due to channel uncertainty. It turns out, indeed, that $\alpha(B) = 1$ minimizes (2.28b) for virtually all wireless channels and SNR values of practical interest. In particular, a sufficient condition for $\alpha(B) = 1$ is (Durisi et al., 2010)

$$\Delta_{\mathbf{H}} \le \frac{\beta}{3TF} \tag{2.29a}$$

and

$$0 \leq \frac{P}{B} < \frac{\Delta_{\mathbf{H}}}{\beta}\left[\exp\left(\frac{\beta}{2TF\Delta_{\mathbf{H}}}\right) - 1\right]. \tag{2.29b}$$

As virtually all wireless channels are highly underspread, as $\beta \geq 1$, and as, typically, $TF \approx 1.2$ (see Section 2.2.2.7), condition (2.29a) is satisfied in most real-world application scenarios, so that the only relevant condition is (2.29b); but even for large channel spread $\Delta_{\mathbf{H}}$, this condition holds for all SNR values[6] P/B of practical interest. As an example, consider a system with $\beta = 1$ and spread $\Delta_{\mathbf{H}} = 10^{-2}$; for this choice, (2.29b) is satisfied for all SNR values less than 153 dB. As this value is far in excess of the SNR encountered in practical systems, we can safely claim that a capacity upper bound of practical interest results if we substitute $\alpha(B) = 1$ in (2.28a).

2.3.4.4 *Impact of Channel Characteristics*

The spread $\Delta_{\mathbf{H}}$ and the shape of the scattering function $C_{\mathbf{H}}(\tau, \nu)$ are important characteristics of wireless channels. As the upper bound (2.28) is explicit in the scattering function, we can analyze its behavior as a function of $C_{\mathbf{H}}(\tau, \nu)$. We restrict our discussion to the practically relevant case $\alpha(B) = 1$.

Channel spread

For fixed shape of the scattering function, the upper bound U_1 decreases for increasing spread $\Delta_{\mathbf{H}}$. To see this, we define a normalized scattering function $\tilde{C}_{\mathbf{H}}(\tau, \nu)$ supported on a square with unit area, so that

$$C_{\mathbf{H}}(\tau, \nu) = \frac{1}{\Delta_{\mathbf{H}}} \tilde{C}_{\mathbf{H}}\left(\frac{\tau}{2\tau_0}, \frac{\nu}{2\nu_0}\right).$$

By a change of variables, the penalty term can now be written as

$$\begin{aligned}
\psi(B) &= \frac{B}{\beta} \int_{-\infty}^{\infty} \int_{-\infty}^{\infty} \ln\left(1 + \frac{\beta P}{B} C_{\mathbf{H}}(\tau, \nu)\right) d\tau d\nu \\
&= \frac{B\Delta_{\mathbf{H}}}{\beta} \int_{-1/2}^{1/2} \int_{-1/2}^{1/2} \ln\left(1 + \frac{\beta P}{B\Delta_{\mathbf{H}}} \tilde{C}_{\mathbf{H}}(\tau, \nu)\right) d\tau d\nu.
\end{aligned}$$

Because $\Delta_{\mathbf{H}} \ln(1 + \rho/\Delta_{\mathbf{H}})$ is monotonically increasing in $\Delta_{\mathbf{H}}$ for any positive constant $\rho > 0$, the penalty term $\psi(B)$ increases with increasing spread $\Delta_{\mathbf{H}}$. Consequently, as the first term in (2.28a) does not depend on $\Delta_{\mathbf{H}}$, the upper bound U_1 decreases with increasing spread. Because of the Fourier relation $C_{\mathbf{H}}(\tau, \nu) = \mathbb{F}_{t \to \nu, f \to \tau}\{R_{\mathbf{H}}(t, f)\}$, a larger spread implies less correlation in time, frequency, or both; but a channel with less correlation is harder for the receiver to learn; hence, channel uncertainty increases, which ultimately reduces capacity. In a typical system that uses pilot symbols to estimate the channel, a larger spread means that more pilots are required to reliably estimate the channel, so that fewer degrees of freedom are left to transmit data.

[6] Recall that the noise variance was normalized to 1.

Shape of the scattering function

For fixed spread $\Delta_{\mathbf{H}}$, the scattering function that minimizes the upper bound U_1 is the "brick-shaped" scattering function: $C_{\mathbf{H}}(\tau,\nu) = 1/\Delta_{\mathbf{H}}$ for $(\tau,\nu) \in [-\tau_0,\tau_0] \times [-\nu_0,\nu_0]$. This observation follows from Jensen's inequality applied to the penalty term in (2.28c), the normalization of $C_{\mathbf{H}}(\tau,\nu)$ in (2.4), and the fact that a brick-shaped scattering function achieves the resulting upper bound.

First design sketches of a communication system often rely on simple channel parameters like the maximum multipath delay τ_0 and the maximum Doppler shift ν_0. These two parameters completely specify a brick-shaped scattering function, which we just saw to provide the minimum capacity upper bound among all WSSUS channels with a scattering function of prescribed τ_0 and ν_0. Hence, a design on the basis of τ_0 and ν_0 alone is implicitly targeted at the worst-case channel and thus results in a robust design.

2.3.5 A Lower Bound on Capacity

Typically, lower bounds on capacity are easier to obtain than upper bounds because it is sufficient to evaluate the mutual information in (2.25) for an input distribution that satisfies the power constraints (2.23) and (2.24). The main difficulty here is to find input distributions that lead to tight lower bounds. As we are going to show in Section 2.3.6, a good trade-off between analytical tractability and tightness of the resulting bound follows from the choice of i.i.d. zero-mean *constant modulus* input symbols. Constant modulus input symbols are often found in practical systems in the form of PSK constellations, especially in systems designed to operate at low SNR.

2.3.5.1 Bounding Idea

As with the upper bound in Theorem 2.1, we illustrate the main steps in the derivation of the lower bound by considering a simple memoryless channel with I/O relation $r = hu + w$. Here, we pick u to be of zero mean and constant modulus $|u|^2 = P$. A word of warning is appropriate at this point. The choice of transmitting constant-modulus signals on a memoryless fading channel (with the channel not known at the receiver) is a bad one. It is easy to show that the corresponding mutual information $I(r;u)$ (which is a lower bound on the capacity of the memoryless channel) equals zero. The underlying reason is that in constant modulus signals, the information is encoded in the phase of the signal. But for the setting we just described, the conditional probability of r given u depends on the amplitude of u only. Disturbing as this observation might seem, the bounding steps presented below lead to a perfectly sensible lower bound on the capacity of the channel (2.18) we are interested in. In fact, as the discretized channel $h[k,l]$ in (2.18) exhibits correlation both in k and l, the conditional probability of $\mathbf{r}$ given $\mathbf{s}$ [see (2.22)] depends on both the phase and the amplitude of the entries of $\mathbf{s}$. Consequently, as we shall see in Section 2.3.6, we have that $I(\mathbf{r};\mathbf{s}) > 0$ also when constant-modulus inputs are used.

We now derive a lower bound on $I(r;u)$, which not surprisingly will turn out to be negative. We use the chain rule to split the mutual information $I(r;u)$ and obtain

$$\begin{aligned} I(r;u) &= I(r;u,h) - I(r;h\,|\,u) \\ &= I(r;h) + I(r;u\,|\,h) - I(r;h\,|\,u) \\ &\overset{(a)}{\geq} I(r;u\,|\,h) - I(r;h\,|\,u) \end{aligned}$$

$$\stackrel{(b)}{=} I(r;u\,|\,h) - \mathsf{E}\left\{\ln(1+|u|^2)\right\}$$

$$\stackrel{(c)}{=} I(r;u\,|\,h) - \ln(1+P). \tag{2.30}$$

To get the inequality (a), we used the chain rule for mutual information twice, and then dropped the nonnegative term $I(r;h)$. This essentially splits the mutual information into a first component that corresponds to the case when perfect channel knowledge at the receiver is available, and a second component that [like in the upper bound (2.27)] can be interpreted as a penalty term, and quantifies the impact of channel uncertainty. Next, (b) follows because, given the input u, the output r is JPG with variance $1+|u|^2$. Finally, (c) results as u is of constant modulus with $|u|^2 = P$. As expected, the lower bound we arrived at is negative. In fact, the first term in the lower bound (2.30), which is a "coherent" mutual information, is always less than the second term, which is the capacity of an AWGN channel with the same receive power.

If the input signal is subject to a peak-power constraint, with PAPR β strictly larger than 1, we can improve upon (2.30) by *time sharing* (Sethuraman et al., 2005, Cor. 2.1): we let the input signal have squared magnitude γP during a fraction $1/\gamma$ of the total transmission time, where $1 \le \gamma \le \beta$ – that is, we set the channel input s to be $s = \sqrt{\gamma}u$ during this time; for the remaining time, the transmitter is silent, so that the constraint on the average power is satisfied. The resulting bound is

$$\sup_{\mathcal{Q}} I(r;s) \geq \max_{1\le\gamma\le\beta} \frac{1}{\gamma}\left[I(r;\sqrt{\gamma}u\,|\,h) - \ln(1+\gamma P)\right].$$

Because a closed-form expression for the coherent mutual information $I(r;\sqrt{\gamma}u\,|\,h)$ does not exist for constant modulus u, numerical methods are needed for evaluating the lower bound just derived.

2.3.5.2 The Actual Bound

Two difficulties arise when trying to derive a bound similar to (2.30) on the capacity of the channel (2.18): we need to account for the correlation that $h[k,l]$ exhibits in k and l, and we need to compute the limit $K \to \infty$ in (2.25). We choose the input symbols to be i.i.d. of zero mean and constant modulus $|s[k,l]|^2 = PT/K$. The coherent mutual information $I(\mathbf{r};\mathbf{s}\,|\,\mathbf{h})$ is then simply given by $KL \times I(r;s\,|\,h)$, i.e., KL times the coherent mutual information of the scalar memoryless Rayleigh-fading channel we analyzed previously.

In the limit $K \to \infty$, the penalty term $I(\mathbf{r};\mathbf{h}\,|\,\mathbf{s})$ is explicit in the matrix-valued spectrum $\mathbf{P}_h(\theta)$ of the multivariate channel process $\{\mathbf{h}[k] \triangleq \left(h[k,0]\;\; h[k,1]\;\; \cdots\;\; h[k,L-1]\right)^T\}$

$$\mathbf{P}_h(\theta) \triangleq \sum_{k=-\infty}^{\infty} \mathbf{R}_h[k]e^{-j2\pi k\theta}, \quad |\theta| \le \frac{1}{2}, \tag{2.31}$$

where $\mathbf{R}_h[k] \triangleq \mathsf{E}\left\{\mathbf{h}[k'+k]\mathbf{h}^H[k']\right\}$. This result follows from a generalization of Szegö's theorem on the asymptotic distribution of Toeplitz matrices (see Durisi et al., 2010 for further details).

The final lower bound on capacity is stated in the following theorem. As before, a detailed derivation can be found in Durisi et al. (2010).

Theorem 2.2 The noncoherent capacity of the channel (2.18), under the assumption that the input signal satisfies the average-power constraint (2.23) and the peak-power constraint (2.24), is lower-bounded as $C \geq L_1$, where

$$L_1(B) = \max_{1 \leq \gamma \leq \beta} \left\{ \frac{B}{\gamma TF} I(r; \sqrt{\gamma} u \,|\, h) - \frac{1}{\gamma T} \int_{-1/2}^{1/2} \ln\det \left\{ \mathbf{I}_L + \frac{\gamma PTF}{B} \mathbf{P}_h(\theta) \right\} d\theta \right\}. \tag{2.32}$$

2.3.5.3 Some Approximations

The lower bound (2.32) differs from the upper bound in Theorem 2.1 in two important aspects. (i) The first term inside the braces in (2.32) cannot be expressed in closed form but needs to be evaluated numerically because of the constant modulus signaling assumption. (ii) The penalty term depends on the scattering function only indirectly through $\mathbf{P}_h(\theta)$, which again complicates the evaluation of the lower bound. The reason for the complicated structure are edge effects caused by the finite bandwidth $B = LF$. In the large-bandwidth regime, however, we can approximate the penalty term by exploiting the asymptotic equivalence between Toeplitz and circulant matrices (Gray, 2005). This yields

$$L_1(B) \approx L_a(B) \triangleq \max_{1 \leq \gamma \leq \beta} \left\{ \frac{B}{\gamma TF} I(r; \sqrt{\gamma} u \,|\, h) - \frac{B}{\gamma} \int_{-\infty}^{\infty} \int_{-\infty}^{\infty} \ln\left(1 + \frac{\gamma P}{B} C_{\mathbf{H}}(\tau, \nu)\right) d\tau d\nu \right\}. \tag{2.33}$$

Furthermore, we can replace the mutual information $I(r; \sqrt{\gamma} u \,|\, h)$ in (2.33) by its first-order Taylor series expansion for $B \to \infty$ (Verdú, 2002, Th. 14), to obtain the approximation

$$L_a(B) \approx L_{aa}(B) \triangleq \max_{1 \leq \gamma \leq \beta} \left\{ P - \frac{\gamma P^2 TF}{B} - \frac{B}{\gamma} \int_{-\infty}^{\infty} \int_{-\infty}^{\infty} \ln\left(1 + \frac{\gamma P}{B} C_{\mathbf{H}}(\tau, \nu)\right) d\tau d\nu \right\}. \tag{2.34}$$

Both L_a and L_{aa} are no longer true lower bounds, yet they agree with L_1 in (2.32) for large B. More details on how well L_a and L_{aa} approximate L_1 can be found in Durisi et al. (2010).

2.3.6 A Numerical Example

We evaluate the bounds presented previously for the following set of practically relevant system parameters:

1. Brick-shaped scattering function with maximum delay $\tau_0 = 0.5\,\mu\text{s}$, maximum Doppler shift $\nu_0 = 5\,\text{Hz}$, and corresponding spread $\Delta_{\mathbf{H}} = 4\tau_0\nu_0 = 10^{-5}$.
2. Grid parameters $T = 0.35\,\text{ms}$ and $F = 3.53\,\text{kHz}$, so that $TF \approx 1.25$ and $\nu_0 T = \tau_0 F$, as suggested by the design rule (2.19).
3. Receive power normalized with respect to the noise spectral density $P/(1\,\text{W/Hz}) = 2.42 \cdot 10^7 \text{s}^{-1}$.

These parameter values are representative of several different types of systems. For example:

1. An IEEE 802.11a system with transmit power of 200 mW, path loss of 118 dB, and receiver noise figure (Razavi, 1998) of 5 dB; the path loss is rather pessimistic for typical indoor link distances and includes the attenuation of the signal, e.g., by a concrete wall.
2. A UWB system with transmit power of 0.5 mW, path loss of 77 dB, and receiver noise figure of 20 dB.

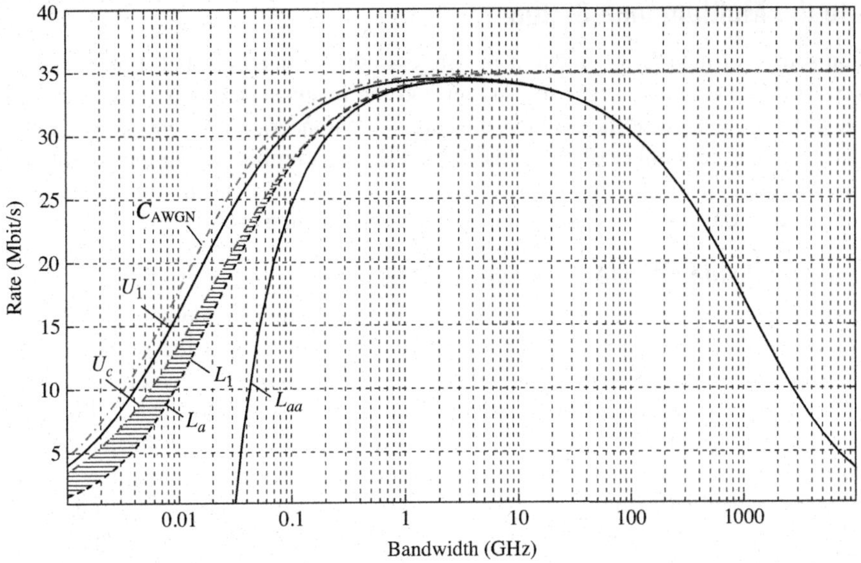

FIGURE 2.4

The coherent-capacity upper bound U_c in (2.26), the upper bound U_1 in (2.28), as well as the lower bound L_1 in (2.32) (evaluated for a QPSK input alphabet) and its large-bandwidth approximations L_a and L_{aa} in (2.33) and (2.34), respectively, for $\beta = 1$ and a brick-shaped scattering function with spread $\Delta_{\mathsf{H}} = 10^{-5}$. The noncoherent capacity is confined to the hatched area. The AWGN-capacity upper bound (2.35) is also plotted for reference.

As illustrated in Fig. 2.4, the upper bound U_1 and the lower bound L_1 (which is evaluated for a QPSK input alphabet) take on their maximum at a large but finite bandwidth; above this *critical bandwidth*, additional bandwidth is detrimental, and the capacity approaches zero as the bandwidth increases further. In this regime, the rate gain resulting from the additional degrees of freedom is offset by the resources required to resolve channel uncertainty. In particular, we can see from Fig. 2.4 that many current wireless systems operate well below the critical bandwidth.

For bandwidth values smaller than the critical bandwidth, L_1 comes quite close to the coherent-capacity upper bound U_c; we show in Section 2.4.2 that this gap can be further reduced by a more sophisticated choice of the input distribution.

We finally note that, for a large range of bandwidth values of practical interest, both U_1 and L_1 (and, hence, the noncoherent capacity of (2.18)) are close to the capacity of an AWGN channel with the same receive power (a trivial upper bound on (2.25)), given by[7]

$$C_{\mathrm{AWGN}}(B) = B\ln\left(1 + \frac{P}{B}\right). \tag{2.35}$$

This observation will be helpful in Section 2.4, where we quantify how well the noncoherent capacity of the diagonalized channel (2.18) approximates that of the channel with ISI/ICI (2.14) in the large-bandwidth regime.

[7] In the remainder of the chapter, we will refer to C_{AWGN} as AWGN-capacity upper bound.

2.3.7 Extension to the Multiantenna Setting

Bandwidth is the main source of degrees of freedom in wireless communications, but it also results in channel uncertainty. Multiple antennas at the transmitter and the receiver can be used to increase the number of degrees of freedom even further. Hence, questions of engineering relevance are (i) how channel uncertainty grows with the number of antennas, (ii) if there is a regime where the additional *spatial* degrees of freedom are detrimental just as the degrees of freedom associated with bandwidth are, and (iii) what is the impact on capacity of spatial correlation across antennas.

2.3.7.1 Modeling Multiantenna Channels – A Formal Extension

To answer the questions above, we need an appropriate model for multiantenna underspread WSSUS channels. As the accurate modeling of multiantenna LTV channels goes beyond the scope of this chapter, we limit our analysis to a multiantenna channel model that is a *formal* extension of the single-input single-output (SISO) model used so far. More precisely, we extend the diagonalized SISO I/O relation (2.18) to a multiple-input multiple-output (MIMO) I/O relation with M_{T} transmit antennas, indexed by q, and M_{R} receive antennas, indexed by r. We assume that all *component channels*, which we denote by $h_{r,q}[k,l]$, are identically distributed, although not necessarily statistically independent.

For a given slot (k,l), we arrange the corresponding component channels $h_{r,q}[k,l]$ in an $M_{\mathrm{R}} \times M_{\mathrm{T}}$ matrix $\mathbf{H}[k,l]$ with entries $[\mathbf{H}[k,l]]_{r,q} = h_{r,q}[k,l]$. The diagonalized I/O relation of the MIMO channel is then given by

$$\mathbf{r}[k,l] = \mathbf{H}[k,l]\mathbf{s}[k,l] + \mathbf{w}[k,l], \tag{2.36}$$

where, for each slot (k,l), $\mathbf{w}[k,l]$ is the M_{R}-dimensional noise vector, $\mathbf{s}[k,l]$ is the M_{T}-dimensional input vector, and $\mathbf{r}[k,l]$ is the output vector of dimension M_{R}. We allow for spatial correlation according to the separable (Kronecker) correlation model (Chuah, Tse, Kahn, & Valenzuela, 2002; Kermoal, Schumacher, Pedersen, Mogensen, & Frederiksen, 2002), i.e.,

$$\mathsf{E}\{h_{r,q}[k'+k,l'+l]h^*_{r',q'}[k',l']\} = B[r,r']A[q,q']R_{\mathbb{H}}[k,l].$$

The $M_{\mathrm{T}} \times M_{\mathrm{T}}$ matrix $\mathbf{A}$ with entries $[\mathbf{A}]_{q,q'} = A[q,q']$ is called the *transmit correlation matrix*, and the $M_{\mathrm{R}} \times M_{\mathrm{R}}$ matrix $\mathbf{B}$ with entries $[\mathbf{B}]_{r,r'} = B[r,r']$ is the *receive correlation matrix*. We normalize $\mathbf{A}$ and $\mathbf{B}$ so that $\mathrm{tr}\{\mathbf{A}\} = M_{\mathrm{T}}$ and $\mathrm{tr}\{\mathbf{B}\} = M_{\mathrm{R}}$. Finally, we let $\sigma_0 \geq \sigma_1 \geq \cdots \geq \sigma_{M_{\mathrm{T}}-1}$ be the eigenvalues of $\mathbf{A}$, and $\lambda_0 \geq \lambda_1 \geq \cdots \geq \lambda_{M_{\mathrm{R}}-1}$ the eigenvalues of $\mathbf{B}$.

A detailed discussion and formal description of the MIMO extension just outlined can be found in Schuster (2009) and Schuster et al. (2009).

2.3.7.2 Capacity Bounds for MIMO Channels in the Large-Bandwidth Regime

Upper and lower bounds on the capacity of the MIMO channel (2.36), under the average-power constraint

$$\frac{1}{T}\mathsf{E}\left\{\sum_{k=0}^{K-1}\sum_{l=0}^{L-1}\|\mathbf{s}[k,l]\|^2\right\} \leq KP \tag{2.37}$$

and the peak-power constraint

$$\frac{1}{T}\|\mathbf{s}[k,l]\|^2 \leq \beta\frac{P}{L} \tag{2.38}$$

can be obtained using the same techniques as in the SISO case. The resulting upper and lower bounds are presented in Theorems 2.3 and 2.4 below. A detailed derivation of these bounds can be found in Schuster (2009) and Schuster et al. (2009).

2.3.7.3 The Upper Bound

Theorem 2.3 The noncoherent capacity of the channel (2.36), under the assumption that the input signal satisfies the average-power constraint (2.37) and the peak-power constraint (2.38), is upper-bounded as $C \leq U_1^{\text{mimo}}$, where

$$U_1^{\text{mimo}}(B) \triangleq \sup_{0 \leq \alpha \leq \sigma_0} \sum_{r=0}^{M_{\mathrm{R}}-1} \left[\frac{B}{TF} \ln\left(1 + \alpha\lambda_r \frac{PTF}{B}\right) - \alpha\psi_r(B) \right] \tag{2.39a}$$

with

$$\psi_r(B) \triangleq \frac{B}{\sigma_0 \beta} \int_{-\infty}^{\infty} \int_{-\infty}^{\infty} \ln\left(1 + \frac{\sigma_0 \lambda_r \beta P}{B} C_{\mathbb{H}}(\tau, \nu)\right) \mathrm{d}\tau \mathrm{d}\nu. \tag{2.39b}$$

Similar to the single-antenna case (see Section 2.3.4.3), it can be shown that for virtually all SNR values of practical interest, the supremum in (2.39a) is attained for $\alpha = \sigma_0$. Hence, the upper bound can be interpreted as the capacity of a set of M_{R} parallel AWGN channels with $B/(TF)$ degrees of freedom and received power $\sigma_0 \lambda_r P$, minus a penalty term that quantifies the capacity loss because of channel uncertainty. The observations on the impact of the shape and spread of the scattering function made in Section 2.3.4.4 remain valid.

2.3.7.4 The Lower Bound

Theorem 2.4 The noncoherent capacity of the channel (2.36), under the assumption that the input signal satisfies the average-power constraint (2.37) and the peak-power constraint (2.38), is lower-bounded as

$$C(B) \geq \max_{1 \leq Q \leq M_{\mathrm{T}}} L_1^{\text{mimo}}(B, Q),$$

where

$$L_1^{\text{mimo}}(B, Q) = \max_{1 \leq \gamma \leq \beta} \left\{ \frac{B}{\gamma TF} I(\tilde{\mathbf{r}}; \sqrt{\gamma}\,\tilde{\mathbf{s}} \,|\, \tilde{\mathbf{H}}) - \frac{1}{\gamma T} \sum_{q=0}^{Q-1} \sum_{r=0}^{M_{\mathrm{R}}-1} \int_{-1/2}^{1/2} \ln\det\left(\mathbf{I}_L + \sigma_q \lambda_r \frac{\gamma PTF}{QB} \mathbf{P}_h(\theta)\right) \mathrm{d}\theta \right\}$$

with

$$\tilde{\mathbf{r}} = \mathbf{\Lambda}^{1/2} \tilde{\mathbf{H}} \mathbf{\Sigma}^{1/2} \tilde{\mathbf{s}} + \tilde{\mathbf{w}}.$$

Here, $\tilde{\mathbf{s}}$ is a M_{T}-dimensional vector whose first Q elements are i.i.d. of zero mean and constant modulus $|[\tilde{\mathbf{s}}]_q|^2 = PT/(QL)$, and the remaining $M_{\mathrm{T}} - Q$ elements are equal to zero. Both the $M_{\mathrm{R}} \times M_{\mathrm{T}}$ matrix $\tilde{\mathbf{H}}$ and the M_{R}-dimensional vector $\tilde{\mathbf{w}}$ have i.i.d. *JPG* entries of zero mean and unit variance. Finally, $\mathbf{\Sigma} = \operatorname{diag}(\sigma_0\ \sigma_1 \cdots \sigma_{M_{\mathrm{T}}-1})^T$ and $\mathbf{\Lambda} = \operatorname{diag}(\lambda_0\ \lambda_1 \cdots \lambda_{M_{\mathrm{R}}-1})^T$.

The lower bound L_1^{mimo} needs to be optimized with respect to the number of active *transmit eigenmodes* Q (i.e., the number of eigenmodes of the transmit correlation matrix $\mathbf{A}$ being signaled over, see Schuster et al. (2009) for more details). Note that when the channel is spatially uncorrelated at the transmitter, Q simply denotes the number of active transmit antennas.

It can be shown that, at very large bandwidth, it is optimal to signal along the strongest eigenmode only (Schuster et al., 2009), a scheme often referred to as *rank-one statistical beamforming* or *eigen-beamforming* (Verdú, 2002). In particular, when the channel is spatially uncorrelated at the transmitter, at very large bandwidth, it is optimal to use a single transmit antenna, an observation previously made in Sethuraman et al. (2009) for frequency-flat time-selective channels. At intermediate bandwidth values, the number of transmit antennas to use (for uncorrelated component channels at the transmitter side), or the number of transmit eigenmodes to signal over (if correlation is present), can be determined by numerical evaluation of the bounds, as shown in Section 2.3.8.

2.3.8 Numerical Examples

For a 3×3 MIMO system, we show in Figs 2.5–2.7 plots of the upper bound U_1^{mimo}, and – for a number of active transmit eigenmodes Q ranging from 1 to 3 – plots of the lower bound L_1^{mimo} and of a large-bandwidth approximation of L_1^{mimo} denoted as L_{aa}^{mimo}.[8]

Parameter settings

All plots are obtained for receive power normalized with respect to the noise spectral density $P/(1\,\text{W/Hz}) = 1.26 \cdot 10^8\,\text{s}^{-1}$; this corresponds, e.g., to a transmit power of 0.5 mW, thermal noise level at the receiver of -174 dBm/Hz, free-space path loss corresponding to a distance of 10 m, and a rather conservative receiver noise figure of 20 dB. Furthermore, we assume that the scattering function is brick-shaped with $\tau_0 = 5\,\mu\text{s}$, $\nu_0 = 50\,\text{Hz}$, and corresponding spread $\Delta_{\mathbf{H}} = 10^{-3}$. We also set $\beta = 1$. For this set of parameter values, we analyze three different scenarios: a spatially uncorrelated channel, spatial correlation at the receiver only, and spatial correlation at the transmitter only.

Spatially uncorrelated channel

Figure 2.5 shows the upper bound U_1^{mimo} and the lower bound L_1^{mimo} for the spatially uncorrelated case. For comparison, we also plot a standard capacity upper bound U_c^{mimo} obtained for the coherent setting and with input subject to an average-power constraint only (Tse & Viswanath, 2005). We can observe that U_c^{mimo} is tighter than our upper bound U_1^{mimo} for small bandwidth; this holds true in general, as for small bandwidth and spatially uncorrelated channels, $U_1^{\text{mimo}}(B) \approx [(M_R B)/(TF)] \ln(1 + PTF/B)$, which is the Jensen upper bound on U_c^{mimo}.

For small and medium bandwidth, the lower bound L_1^{mimo} increases with Q and comes surprisingly close to the coherent capacity upper bound U_c^{mimo} for $Q = 3$.

[8] The analytic expression for L_{aa}^{mimo}, which is similar to (2.34), can be found in Schuster et al. (2009), Eq. (19).

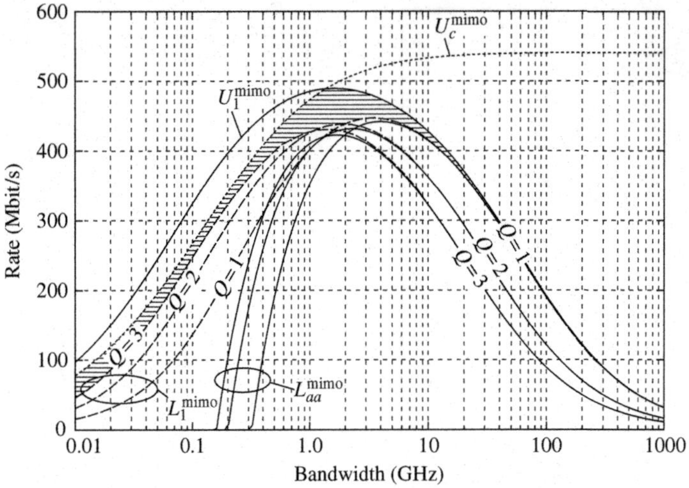

FIGURE 2.5

Upper and lower bounds on the noncoherent capacity of a spatially uncorrelated MIMO underspread WSSUS channel with $M_T = M_R = 3, \beta = 1$, and $\Delta_H = 10^{-3}$. The bounds confine the noncoherent capacity to the hatched area.

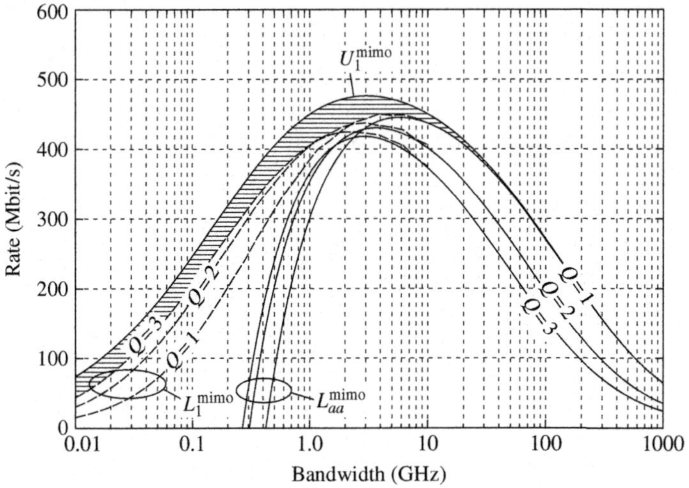

FIGURE 2.6

Upper and lower bounds on the noncoherent capacity of a MIMO underspread WSSUS channel that is spatially uncorrelated at the transmitter but correlated at the receiver with eigenvalues of the receive correlation matrix given by $\{2.6, 0.3, 0.1\}$; $M_T = M_R = 3, \beta = 1$, and $\Delta_H = 10^{-3}$. The bounds confine the noncoherent capacity to the hatched area.

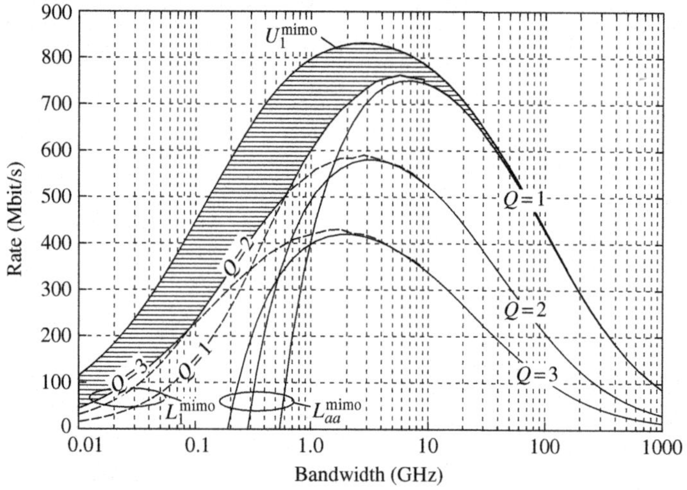

FIGURE 2.7

Upper and lower bounds on the noncoherent capacity of a MIMO underspread WSSUS channel that is spatially correlated at the transmitter with eigenvalues of the transmit correlation matrix $\{1.7, 1.0, 0.3\}$ and uncorrelated at the receiver; $M_T = M_R = 3$, $\beta = 1$, and $\Delta_H = 10^{-3}$. The bounds confine the noncoherent capacity to the hatched area.

As for the SISO case, when the bandwidth exceeds a certain critical bandwidth, both U_1^{mimo} and L_1^{mimo} start to decrease, because the rate gain due to the additional degrees of freedom is offset by the increase in channel uncertainty. The same argument holds in the wideband regime for the degrees of freedom provided by multiple transmit antennas: in this regime, using a single transmit antenna is optimal (Schuster et al., 2009).

Impact of receive correlation

Figure 2.6 shows the same bounds as before but evaluated for spatial correlation at the receiver according to a correlation matrix with eigenvalues $\{2.6, 0.3, 0.1\}$ and a spatially uncorrelated channel at the transmitter. The curves in Fig. 2.6 are very similar to the ones shown in Fig. 2.5 for the spatially uncorrelated case, yet they are shifted toward higher bandwidth and the maximum is lower. In general, receive correlation reduces capacity at small bandwidth but is beneficial at large bandwidth (Schuster et al., 2009).

Impact of transmit correlation

We evaluate the same bounds once more, but this time for spatial correlation at the transmitter according to a correlation matrix with eigenvalues $\{1.7, 1.0, 0.3\}$ and a spatially uncorrelated channel at the receiver. The corresponding curves are shown in Fig. 2.7. The maximum of both the upper and the lower bound is higher than the corresponding maxima in the previous two examples. This rate increase at large bandwidth is caused by the power gain due to statistical beamforming (Schuster et al., 2009). The impact of transmit correlation at small bandwidth is more difficult to characterize because the distance between upper and lower bound is larger than for the spatially uncorrelated case.

An observation of practical importance is that both U_1^{mimo} and L_1^{mimo} are rather flat over a large range of bandwidth values around their respective maxima. Further numerical results (not presented here) point at the following: (i) for smaller values of the channel spread Δ_{H}, these maxima broaden and extend toward higher bandwidth; (ii) an increase in β increases the gap between upper and lower bounds.

2.4 THE LARGE-BANDWIDTH REGIME: I/O RELATION WITH INTERFERENCE

So far, we based our analysis on the diagonalized I/O relation (2.18). The goal of this section is to determine how well the noncoherent capacity of (2.18) approximates that of the channel with ISI and ICI (2.14) in the large-bandwidth regime. The presence of interference makes the derivation of tight capacity bounds (in particular, upper bounds) technically challenging. Nevertheless, our analysis will be sufficient to establish that for a large range of bandwidth values of practical interest, the presence of interference does not change the noncoherent capacity behavior significantly, whenever the channel is underspread.

To establish this result, we derive a lower bound on the capacity of (2.14) by treating interference as noise. We then show that, whenever the channel is underspread, this lower bound, evaluated for an appropriately chosen root-raised-cosine WH set (see Example 2.1), is close for a large range of bandwidth values of practical interest both to the lower bound L_1 we derived in the previous section (for the diagonalized I/O relation (2.18)) and to the AWGN-capacity upper bound C_{AWGN}.

To get an I/O relation in vector form, similar to the one we worked with in the previous section, we arrange the intersymbol and intercarrier interference terms $\{z[k',l',k,l]\}$ in (2.14) in a $KL \times KL$ matrix $\mathbf{Z}$ with entries

$$[\mathbf{Z}]_{(l'+k'L),(l+kL)} = \begin{cases} z[k',l',k,l], & \text{if } (k',l') \neq (k,l) \\ 0, & \text{otherwise.} \end{cases}$$

This definition, together with the definitions in Section 2.3, allows us to compactly express (2.14) as

$$\mathbf{r} = \mathbf{h} \odot \mathbf{s} + \mathbf{Z}\mathbf{s} + \mathbf{w}.$$

For a given WH set, the noncoherent capacity C of the induced discretized channel (2.14) is defined as in (2.25). We next derive a lower bound on C. As in Section 2.3, we first illustrate the main steps in the derivation of the lower bound using a simplified setting.

2.4.1 A Lower Bound on Capacity

2.4.1.1 The Bounding Idea

We consider, for simplicity, a block-fading channel with block length 2 and I/O relation

$$\begin{aligned} r[1] &= hs[1] + zs[2] + w[1] \\ r[2] &= hs[2] + zs[1] + w[2], \end{aligned} \tag{2.40}$$

where $h, w[1], w[2] \sim \mathcal{CN}(0,1)$ are mutually independent, while $z \sim \mathcal{CN}(0,\sigma_z^2)$ is independent of the noise samples $w[1]$ and $w[2]$, but not necessarily independent of h. For this setting, the interference matrix $\mathbf{Z}$ reduces to

$$\mathbf{Z} = \begin{bmatrix} 0 & z \\ z & 0 \end{bmatrix}$$

and the I/O relation (2.40) can be recast as

$$\mathbf{r} = h\mathbf{s} + \mathbf{Z}\mathbf{s} + \mathbf{w}.$$

Let $\mathcal{Q}$ be the set of probability distributions on $\mathbf{s}$ satisfying the average-power constraint $\mathsf{E}\{|s[i]|^2\} \leq P$ and the peak constraint $|s[i]|^2 \leq \beta P$, for $i = 1,2$. To obtain a lower bound on noncoherent capacity, which is given by[9] $(1/2)\sup_{\mathcal{Q}} I(\mathbf{r};\mathbf{s})$, we compute the mutual information for a specific probability distribution in $\mathcal{Q}$. In particular, we take a probability distribution for which $s[1]$ and $s[2]$ are i.i.d. and $\mathsf{h}(s[1]) > -\infty$, where $\mathsf{h}(\cdot)$ denotes differential entropy. We denote by ρ_s the average power of $s[1]$ and $s[2]$, i.e., $\rho_s \triangleq \mathsf{E}\{|s[1]|^2\} = \mathsf{E}\{|s[2]|^2\} \leq P$. The presence of the interference term in (2.40) makes the derivation of a capacity lower bound involved even for this simple setting; bounding steps more sophisticated than the ones we used in Section 2.3.5.1 are needed. As in (2.30), we first use the chain rule and the nonnegativity of mutual information to split $I(\mathbf{r};\mathbf{s})$ as

$$\begin{aligned} I(\mathbf{r};\mathbf{s}) &= I(\mathbf{r};h) + I(\mathbf{r};\mathbf{s}\,|\,h) - I(\mathbf{r};h\,|\,\mathbf{s}) \\ &\geq I(\mathbf{r};\mathbf{s}\,|\,h) - I(\mathbf{r};h\,|\,\mathbf{s}). \end{aligned} \tag{2.41}$$

We next bound each of the two terms on the right-hand side (RHS) of (2.41).

A lower bound on the first term

To lower-bound the first term on the RHS of (2.41), we use the chain rule for differential entropy, the independence of $s[1]$ and $s[2]$ and the fact that conditioning reduces entropy (Cover & Thomas, 1991, Sec. 9.6):

$$\begin{aligned} I(\mathbf{r};\mathbf{s}\,|\,h) &= \mathsf{h}(\mathbf{s}\,|\,h) - \mathsf{h}(\mathbf{s}\,|\,\mathbf{r},h) \\ &= \mathsf{h}(s[1]) + \mathsf{h}(s[2]) - \mathsf{h}(s[1]\,|\,\mathbf{r},h) - \mathsf{h}(s[2]\,|\,\mathbf{r},h,s[1]) \\ &\geq \mathsf{h}(s[1]) + \mathsf{h}(s[2]) - \mathsf{h}(s[1]\,|\,r[1],h) - \mathsf{h}(s[2]\,|\,r[2],h) \\ &= I(r[1];s[1]\,|\,h) + I(r[2];s[2]\,|\,h) \\ &= 2I(r[1];s[1]\,|\,h). \end{aligned} \tag{2.42}$$

The mutual information $I(r[1];s[1]\,|\,h)$ is still difficult to evaluate because of the presence of the interference term $zs[2]$. A simple way to deal with this issue is to treat the interference as noise, as shown

[9] The factor 1/2 arises because of the normalization with respect to the block length (2 in this case).

below (Lapidoth, 2005, App. I):

$$\begin{aligned}
I(r[1];s[1]\,|\,h) &= \mathsf{h}(s[1]) - \mathsf{h}(s[1]\,|\,r[1],h)\\
&\overset{(a)}{=} \mathsf{h}(s[1]) - \min_{\delta}\mathsf{h}(s[1]-\delta r[1]\,|\,r[1],h)\\
&\overset{(b)}{\geq} \mathsf{h}(s[1]) - \min_{\delta}\mathsf{h}(s[1]-\delta r[1]\,|\,h)\\
&\overset{(c)}{\geq} \mathsf{h}(s[1]) - \min_{\delta}\mathsf{E}_h\left\{\ln\left(\pi e\,\mathsf{E}\left\{|s[1]-\delta r[1]|^2\,|\,h\right\}\right)\right\}\\
&\overset{(d)}{=} \mathsf{h}(s[1]) - \mathsf{E}_h\left\{\ln\left(\pi e\frac{\rho_s(\sigma_z^2\rho_s+1)}{|h|^2\rho_s+\sigma_z^2\rho_s+1}\right)\right\}\\
&= \mathsf{E}_h\left\{\ln\left(1+\frac{|h|^2\rho_s}{\sigma_z^2\rho_s+1}\right)\right\} - \left[\ln(\pi e\rho_s) - \mathsf{h}(s[1])\right].
\end{aligned} \tag{2.43}$$

In (a) we used the fact that the differential entropy is invariant to translations, in (b) that conditioning reduces entropy, (c) follows because the Gaussian distribution maximizes the differential entropy, and (d) follows by choosing δ to be the coefficient of the linear estimator of $s[1]$ (from $r[1]$) that minimizes the mean square error. Note that the term $\ln(\pi e\rho_s) - \mathsf{h}(s[1])$ is always positive, and can be interpreted as a penalty term that quantifies the rate loss due to the peak constraint. In fact, for any distribution for which $\mathsf{E}\{|s[1]|^2\} = \rho_s$, we have $\mathsf{h}(s[1]) \leq \ln(\pi e\rho_s)$, with equality if and only if $s[1] \sim \mathcal{CN}(0,\rho_s)$. But the complex Gaussian distribution does not belong to $\mathcal{Q}$ because it violates the peak constraint.

An upper bound on the second term

We next upper-bound the second term on the RHS of (2.41). Let $\mathbf{w}^{(1)} \sim \mathcal{CN}(\mathbf{0},\alpha\mathbf{I}_2)$ and $\mathbf{w}^{(2)} \sim \mathcal{CN}(\mathbf{0},(1-\alpha)\mathbf{I}_2)$, where $0<\alpha<1$, be two independent JPG vectors. Furthermore, let $\mathbf{r}^{(1)} = h\mathbf{s} + \mathbf{w}^{(1)}$ and $\mathbf{r}^{(2)} = \mathbf{Z}\mathbf{s} + \mathbf{w}^{(2)}$. By the data-processing inequality and the chain rule for mutual information, we have that

$$\begin{aligned}
I(\mathbf{r};h\,|\,\mathbf{s}) &\leq I(\mathbf{r}^{(1)},\mathbf{r}^{(2)};h\,|\,\mathbf{s})\\
&= I(\mathbf{r}^{(1)};h\,|\,\mathbf{s}) + I(\mathbf{r}^{(2)};h\,|\,\mathbf{s},\mathbf{r}^{(1)}).
\end{aligned} \tag{2.44}$$

The first term on the RHS of (2.44) can be upper-bounded according to

$$I(\mathbf{r}^{(1)};h\,|\,\mathbf{s}) = \mathsf{E}\left\{\ln\left(1+\frac{\|\mathbf{s}\|^2}{\alpha}\right)\right\} \leq \ln\left(1+\frac{\rho_s}{\alpha}\right).$$

For the second term on the RHS of (2.44), we proceed as follows:

$$
\begin{aligned}
I(\mathbf{r}^{(2)};h\,|\,\mathbf{s},\mathbf{r}^{(1)}) &= \mathsf{h}(\mathbf{r}^{(2)}\,|\,\mathbf{s},\mathbf{r}^{(1)}) - \mathsf{h}(\mathbf{r}^{(2)}\,|\,\mathbf{s},\mathbf{r}^{(1)},h) \\
&\overset{(a)}{=} \mathsf{h}(\mathbf{r}^{(2)}\,|\,\mathbf{s},\mathbf{r}^{(1)}) - \mathsf{h}(\mathbf{r}^{(2)}\,|\,\mathbf{s},h) \\
&\overset{(b)}{\leq} \mathsf{h}(\mathbf{r}^{(2)}\,|\,\mathbf{s}) - \mathsf{h}(\mathbf{r}^{(2)}\,|\,\mathbf{s},h,\mathbf{Z}) \\
&\overset{(c)}{=} \mathsf{h}(\mathbf{r}^{(2)}\,|\,\mathbf{s}) - \mathsf{h}(\mathbf{r}^{(2)}\,|\,\mathbf{s},\mathbf{Z}) \\
&= I(\mathbf{r}^{(2)};\mathbf{Z}\,|\,\mathbf{s}) \\
&= \mathsf{E}\left\{\ln\left(1+\frac{\sigma_z^2}{1-\alpha}\|\mathbf{s}\|^2\right)\right\} \\
&\leq \ln\left(1+\frac{\sigma_z^2\rho_s}{1-\alpha}\right).
\end{aligned}
$$

Here, (a) holds because $\mathbf{r}^{(1)}$ and $\mathbf{r}^{(2)}$ are conditionally independent given $\mathbf{s}$ and h, in (b) we used that conditioning reduces entropy, and (c) follows because $\mathbf{r}^{(2)}$ and h are conditionally independent given $\mathbf{Z}$. Combining the two bounds, we get

$$
I(\mathbf{r};h\,|\,\mathbf{s}) \leq \ln\left(1+\frac{\rho_s}{\alpha}\right) + \ln\left(1+\frac{\rho_s\sigma_z^2}{1-\alpha}\right).
$$

Furthermore, as the bound holds for any $\alpha \in (0,1)$, we have

$$
I(\mathbf{r};h\,|\,\mathbf{s}) \leq \inf_{0<\alpha<1}\left\{\ln\left(1+\frac{\rho_s}{\alpha}\right) + \ln\left(1+\frac{\rho_s\sigma_z^2}{1-\alpha}\right)\right\}. \tag{2.45}
$$

Completing the bound

To get the final bound, we first insert (2.43) into (2.42), and then (2.42) and (2.45) into (2.41):

$$
\begin{aligned}
\sup_{\mathcal{Q}} I(\mathbf{r};\mathbf{s}) \geq 2\left[\mathsf{E}_h\left\{\ln\left(1+\frac{|h|^2\rho_s}{\sigma_z^2\rho_s+1}\right)\right\} - \left(\ln(\pi e\rho_s) - \mathsf{h}(s[1])\right)\right] \\
- \inf_{0<\alpha<1}\left\{\ln\left(1+\frac{\rho_s}{\alpha}\right) + \ln\left(1+\frac{\rho_s\sigma_z^2}{1-\alpha}\right)\right\}.
\end{aligned} \tag{2.46}
$$

The bound just obtained can be tightened by maximizing it over the set of probability distributions on $s[1]$ that satisfy the average-power constraint $\mathsf{E}\{|s[1]|^2\} \le P$ and the peak-power constraint $|s[1]|^2 \le \beta P$.

2.4.1.2 The Actual Bound

The application of the bounding steps just illustrated to the channel (2.14), is made difficult by the correlation exhibited by $h[k,l]$ and $z[k',l',k,l]$. We deal with this difficulty as follows. As in Section 2.3.5.2, the stationarity of $h[k,l]$ in k allows us to use Szegö's theorem on the eigenvalue distribution of Toeplitz matrices to obtain a bound explicit in the matrix-valued spectral density $\mathbf{P}_h(\theta)$ defined in (2.31). The statistical properties of $z[k',l',k,l]$ are captured through the interference variance σ_I^2 defined in (2.17), which replaces σ_z^2 in (2.46). More details on these bounding steps can be found in Durisi, Morgenshtern, and Bölcskei (2010).

As in Section 2.3, we consider, for simplicity of exposition, scattering functions that are compactly supported within the rectangle $[-\tau_0, \tau_0] \times [-\nu_0, \nu_0]$ and grid parameters satisfying the Nyquist condition $T \le 1/(2\nu_0)$ and $F \le 1/(2\tau_0)$. The resulting lower bound on capacity is presented in the following theorem, whose proof can be found in Durisi, Morgenshtern, and Bölcskei (2010).

Theorem 2.5 Let $(g(t), T, F)$ be a WH set and consider a Rayleigh-fading WSSUS channel with scattering function $C_{\mathbf{H}}(\tau,\nu)$. Then, for a given bandwidth B, the noncoherent capacity of the discretized channel (2.14) induced by $(g(t), T, F)$, with input subject to the average-power constraint (2.23) and the peak-power constraint (2.24), is lower-bounded as $C \ge L_1^{\text{int}}$, where

$$L_1^{\text{int}}(B) \triangleq \frac{B}{TF}\left[\mathsf{E}_h\left\{\ln\left(1+\frac{TF|h|^2\rho_s\sigma_h^2}{1+TF\rho_s\sigma_I^2}\right)\right\} - \left(\ln(\pi e \rho_s) - \mathsf{h}(s)\right)\right]$$
$$- \inf_{0<\alpha<1}\left\{\frac{1}{T}\int_{-1/2}^{1/2} \ln\det\left\{\mathbf{I}_L + \frac{TF\rho_s}{\alpha}\mathbf{P}_h(\theta)\right\}d\theta + \frac{B}{TF}\ln\left(1+\frac{TF\rho_s\sigma_I^2}{1-\alpha}\right)\right\}. \tag{2.47}$$

Here, $h \sim \mathcal{CN}(0,1)$, and s is a complex random variable that satisfies

$$\rho_s \triangleq \mathsf{E}\{|s|^2\} \le \frac{P}{B} \quad \text{and} \quad |s|^2 \le \beta\frac{P}{B}.$$

For WH sets that result in $\sigma_h^2 \approx 1$ and $\sigma_I^2 \approx 0$ (see Section 2.2.2.7), the lower bound L_1^{int} is close – for a large range of bandwidth values of practical interest – to the lower bound L_1 in (2.32). A qualitative justification of this statement is provided below, followed by numerical results in Section 2.4.2. When $\sigma_h^2 \approx 1$ and $\sigma_I^2 \approx 0$, we can approximate L_1^{int} as

$$L_1^{\text{int}}(B) \approx \frac{B}{TF}\left[\mathsf{E}_h\left\{\ln\left(1+TF|h|^2\rho_s\right)\right\} - \left(\ln(\pi e\rho_s) - \mathsf{h}(s)\right)\right.$$
$$\left. - \frac{1}{T}\int_{-1/2}^{1/2} \ln\det\{\mathbf{I}_L + TF\rho_s\mathbf{P}_h(\theta)\}d\theta\right]. \tag{2.48}$$

If, for simplicity, we neglect the peak constraint and take $s \sim \mathcal{CN}(0, \rho_s)$, we have that $\ln(\pi e \rho_s) - \mathsf{h}(s) = 0$. Furthermore, in the large-bandwidth regime of interest in this section, we can approximate the first term on the RHS of (2.48) by its first-order Taylor series expansion, and obtain (for $\rho_s = P/B$)

$$\mathsf{E}_h\left\{\ln\left(1 + TF|h|^2\rho_s\right)\right\} \approx P - \frac{P^2TF}{B}.$$

This expansion coincides with that of the mutual information $I(r; \sqrt{\gamma}u \,|\, h)$ in (2.32) (see also (2.34)). From this approximate analysis, it follows that, at least for Gaussian inputs, the difference between L_1^{int} and L_1 is small in the large-bandwidth regime (when $\sigma_h^2 \approx 1$ and $\sigma_I^2 \approx 0$). In Section 2.4.2, we present numerical results that support this statement. Specifically, these numerical results reveal that:

1. When the channel $\mathbf{H}$ is underspread, there exist WH sets for which (for Gaussian inputs) $L_1^{\text{int}} \approx L_1$ in the large-bandwidth regime; one such set is the root-raised-cosine WH set of Example 2.1, with T and F chosen so that the grid matching rule (2.19) is satisfied. Furthermore, both L_1^{int} and L_1 are close to the AWGN-capacity upper bound C_{AWGN} for a large range of bandwidth values of practical interest.
2. The difference between L_1^{int} and L_1 can be made small for a large range of bandwidth values of practical interest also when the Gaussian input distribution is replaced by an appropriate distribution that satisfies the peak constraint $|s|^2 \leq \beta P$.

Before providing the corresponding numerical results, we make three remarks on L_1^{int}, which simplify its numerical evaluation.

2.4.1.3 *Large-Bandwidth Approximation*

To ease the numerical evaluation of L_1^{int}, we proceed as in Section 2.3.5.3. Specifically, we use the asymptotic equivalence between Toeplitz and circulant matrices (Gray, 2005), to obtain the following large-bandwidth approximation of the first term inside the braces in (2.47):

$$\begin{aligned}\frac{1}{T}\int_{-1/2}^{1/2} \ln\det\left\{\mathbf{I}_L + \frac{TF\rho_s}{\alpha}\mathbf{P}_h(\theta)\right\} d\theta &\approx B\int_{-\infty}^{\infty}\int_{-\infty}^{\infty} \ln\left(1 + \frac{\rho_s}{\alpha}C_{\mathbf{H}}(\tau,\nu)|A_g(\tau,\nu)|^2\right) d\tau d\nu \\ &\leq B\int_{-\infty}^{\infty}\int_{-\infty}^{\infty} \ln\left(1 + \frac{\rho_s}{\alpha}C_{\mathbf{H}}(\tau,\nu)\right) d\tau d\nu. \end{aligned} \tag{2.49}$$

In the first step, we used the fact that $C_{\mathbf{H}}(\tau,\nu)$ is compactly supported and that T and F are chosen so as to satisfy the Nyquist condition $T \leq 1/(2\nu_0)$ and $F \leq 1/(2\tau_0)$. The last step follows because $|A_g(\tau,\nu)| \leq 1$. If we further use that $\sigma_h^2 \geq m_g$ and $\sigma_I^2 \leq M_g$ (see (2.20) and (2.21), respectively), we get the following large-bandwidth approximation of L_1^{int}, which is easier to evaluate numerically than L_1^{int}:

$$L_1^{\text{int}}(B) \gtrsim L_a^{\text{int}}(B) \triangleq \frac{B}{TF}\left[\mathsf{E}_h\left\{\ln\left(1+\frac{TF|h|^2\rho_s m_g}{1+TF\rho_s M_g}\right)\right\} - \left(\ln(\pi e \rho_s) - \mathsf{h}(s)\right)\right.$$
$$\left. - \inf_{0<\alpha<1}\left\{TF\int_{-\infty}^{\infty}\int_{-\infty}^{\infty}\ln\left(1+\frac{\rho_s}{\alpha}C_{\mathbb{H}}(\tau,\nu)\right)d\tau d\nu + \ln\left(1+\frac{TF\rho_s M_g}{1-\alpha}\right)\right\}\right]. \quad (2.50)$$

2.4.1.4 Reduction to a Square Setting

The evaluation of the lower bound is simplified if T and F are chosen according to the grid matching rule (2.19). In this case, a simple coordinate transformation yields the following result.

Lemma 2.6 Let $(g(t), T, F)$ be an orthonormal WH set, and assume that the channel scattering function $C_{\mathbb{H}}(\tau,\nu)$ is compactly supported within the rectangle $[-\tau_0,\tau_0]\times[-\nu_0,\nu_0]$. Then, for any $\zeta > 0$, we have

$$L_a^{\text{int}}(B, g(t), T, F, C_{\mathbb{H}}(\tau,\nu)) = L_a^{\text{int}}\left(B, \sqrt{\zeta}g(\zeta t), \frac{T}{\zeta}, \zeta F, C_{\mathbb{H}}\left(\frac{\tau_0}{\zeta}, \zeta\nu_0\right)\right).$$

In particular, assume that $\nu_0 T = \tau_0 F$ and let $\zeta = \sqrt{T/F} = \sqrt{\tau_0/\nu_0}$, $\tilde{g}(t) \triangleq \sqrt{\zeta}g(\zeta t)$, and $\widetilde{C}_{\mathbb{H}}(\tau,\nu) = C_{\mathbb{H}}(\tau_0/\zeta, \zeta\nu_0)$, where $\widetilde{C}_{\mathbb{H}}(\tau,\nu)$ is compactly supported within the square $[-\sqrt{\Delta_{\mathbb{H}}}/2, \sqrt{\Delta_{\mathbb{H}}}/2]\times[-\sqrt{\Delta_{\mathbb{H}}}/2, \sqrt{\Delta_{\mathbb{H}}}/2]$, with $\Delta_{\mathbb{H}} = 4\tau_0\nu_0$. Then,

$$L_a^{\text{int}}(B, g(t), T, F, C_{\mathbb{H}}(\tau,\nu)) = L_a^{\text{int}}(B, \tilde{g}(t), \sqrt{TF}, \sqrt{TF}, \widetilde{C}_{\mathbb{H}}(\tau,\nu)).$$

In what follows, for the sake of simplicity of exposition, we choose T and F such that the grid matching rule $\nu_0 T = \tau_0 F$ is satisfied. Then, as a consequence of Lemma 2.6, we will, without loss of generality, only consider WH sets of the form $(g(t), \sqrt{TF}, \sqrt{TF})$ and WSSUS channels with scattering function compactly supported within a square.

2.4.1.5 Maximization of the Lower Bound

The lower bound L_a^{int} in (2.50) can be tightened by maximizing it over the orthonormal WH set and the probability distribution on s. The maximization over all orthonormal WH sets is difficult to carry out because the dependency of m_g and M_g on the WH set $(g(t), T, F)$ is, in general, hard to characterize analytically. This problem can be partially overcome when $\Delta_{\mathbb{H}} \ll 1$. In this case, a first-order Taylor-series expansion of m_g and M_g around $\Delta_{\mathbb{H}} = 0$ yields an accurate picture, as discussed in Durisi et al. (2009). Here, we simplify the maximization problem by considering only WH sets that are based on a root-raised-cosine pulse (see Example 2.1).

For simplicity, we also avoid the maximization over the probability distribution of s. Instead, we consider a simple distribution, obtained by truncating a complex Gaussian distribution so that s satisfies the peak constraint $|s|^2 \le \beta P/B$. More specifically, we take $s = (\sqrt{P/B})\tilde{s}$, where the phase of $\tilde{s}$ is uniformly distributed on $[0, 2\pi)$; furthermore, $|\tilde{s}|^2 = z$ is distributed according to

$$f_z(z) = c_1 e^{-z/c_2} 1_{[0,\beta)}(z),$$

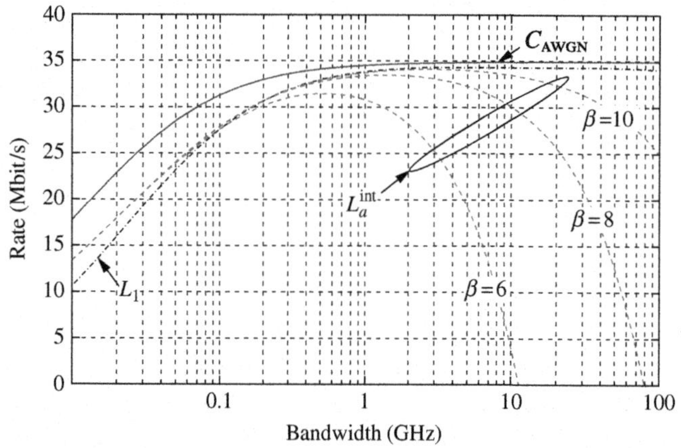

FIGURE 2.8

The AWGN-capacity upper bound C_{AWGN} in (2.35), the lower bound L_1 in (2.32) for $\beta = 10$ and the lower bound L_a^{int} in (2.50) for $\beta = 6$, 8, and 10. The channel scattering function is brick-shaped, with $\Delta_{\mathbb{H}} = 10^{-5}$.

where $c_1 = c_2/(1 - e^{-\beta/c_2})$, and c_2 is chosen so that $\mathbb{E}\{|\tilde{s}|^2\} = 1$. A straightforward calculation reveals that for this input distribution

$$\mathsf{h}(\tilde{s}) = \log c_1 + \frac{1}{c_2} + \ln \pi.$$

Hence, the penalty term $\ln(\pi e \rho_s) - \mathsf{h}(s)$ in (2.50) reduces to

$$\ln(\pi e \rho_s) - \mathsf{h}(s) \overset{(a)}{=} \ln(\pi e) - \mathsf{h}(\tilde{s})$$
$$= 1 - \log c_1 - \frac{1}{c_2}$$

where in (a) we used that $\rho_s = P/B$, by construction. Note that, for the input distribution under consideration, the penalty term $\ln(\pi e \rho_s) - \mathsf{h}(s)$ is independent of the SNR P/B, and depends only on the PAPR β through c_1 and c_2 (more precisely, the penalty term is monotonically decreasing in β, and vanishes in the limit $\beta \to \infty$). In the large-bandwidth regime, the lower bound L_a^{int} turns out to be highly sensitive to the value of this penalty term.

2.4.2 Numerical Examples

In this section, we evaluate the bound L_a^{int} for a set of parameters similar to the one considered in Section 2.3.6. In particular, we take $P/(1\,\text{W/Hz}) = 2.42 \cdot 10^7 \text{s}^{-1}$ and $TF = 1.25$, and consider a brick-shaped scattering function with $\Delta_{\mathbb{H}} = 10^{-5}$ (see Section 2.3.6 for a discussion on the practical relevance of this set of parameters). Furthermore, we focus on WH sets based on a root-raised-cosine pulse (see Example 2.1), and assume $\beta = 10$. Finally, we take as probability distribution of s the truncated complex Gaussian distribution discussed in Section 2.4.1.5.

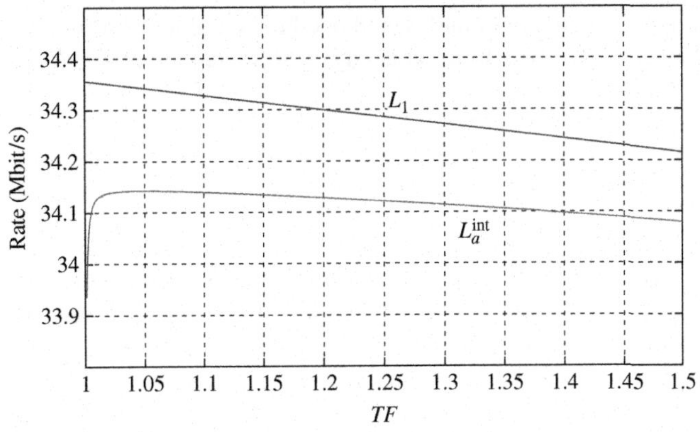

FIGURE 2.9

The lower bounds L_1 in (2.32) and L_a^{int} in (2.50), for $B = 3\text{GHz}$. The channel scattering function is brick-shaped with $\Delta_{\mathbb{H}} = 10^{-5}$. Furthermore, $\beta = 10$.

Figure 2.8 shows the AWGN-capacity upper bound C_{AWGN} in (2.35) (same curve as in Fig. 2.4), the lower bound L_1 in (2.32), which, differently from Fig. 2.4 is evaluated for $\beta = 10$, and the lower bound L_a^{int}, evaluated for $\beta = 6$, 8, and 10, to illustrate the impact of the penalty term $\ln(\pi e \rho_s) - \mathsf{h}(s)$ on L_a^{int} for different values of β. As mentioned earlier, L_a^{int} is extremely sensitive to β in the large-bandwidth regime. Fortunately, this sensitivity manifests itself only for bandwidth values that are significantly above those encountered in practical systems. We can observe that for $\beta = 10$ the difference between L_1 and L_a^{int} is negligible over a large range of bandwidth values of practical interest. Furthermore, also the difference between L_a^{int} and C_{AWGN} is small. We can, therefore, conclude that the presence of interference in (2.14) does not change the capacity behavior significantly for a large range of bandwidth values of practical interest.

We next analyze the dependency of L_1 and L_a^{int} on TF. Figure 2.9 shows the lower bounds L_1 and L_a^{int} as a function of TF for $B = 3\text{GHz}$, which roughly corresponds to the capacity-maximizing bandwidth for the setting considered in Fig. 2.8. Even though the lower bound L_1 decreases monotonically with TF, the rate loss experienced when TF is increased from 1 to 1.5 is extremely small (about 0.1%). This is in good agreement with our claim that, in the large-bandwidth regime, capacity is only mildly sensitive to a loss of degrees of freedom. The lower bound L_a^{int} is surprisingly close to L_1 over almost the entire range of TF values considered. Only for $TF \approx 1$, the gap widens significantly. As remarked in Example 2.1, when $TF = 1$, the pulse $g(t)$ reduces to a sinc function, which has poor time localization. As a consequence, the variance of the interference term increases (as illustrated in Fig. 2.3), and so does the gap between L_1 and L_a^{int}.

Compared with the characterization of the noncoherent capacity of the diagonalized channel (2.18), the characterization we provided here on the basis of L_a^{int} and C_{AWGN} is less accurate. At moderate bandwidth values, our analysis could be strengthened by replacing C_{AWGN} with a tighter upper bound based on perfect CSI at the receiver. Finding such a bound is an interesting open problem. In the very large bandwidth regime, the difference between C_{AWGN} and L_a^{int} is large. In fact C_{AWGN} approaches P as $B \to \infty$ (see (2.35)), while L_a^{int} approaches 0. To obtain a more accurate capacity characterization in the very large bandwidth regime, both upper and lower bound would need to be tightened. From an engineering point of view, however, our analysis is sufficient to (coarsely) determine the optimal (i.e., critical) bandwidth at which to operate.

2.5 THE HIGH-SNR REGIME

Next, we consider the high-SNR regime. As argued in Section 2.2.2.8, in this regime the choice of the WH set $(g(t), T, F)$ that leads to $\sigma_h^2 \approx 1$ and $\sigma_I^2 \approx 0$ might be suboptimal because of a loss of degrees of freedom. Hence, we work with the I/O relation (2.14) directly. In the previous section, we considered, for simplicity of exposition, only scattering functions that are compactly supported in the delay-Doppler plane. Here, we drop this assumption and consider the larger class of WSSUS channels that satisfy the more general underspread notion in Definition 2.1. We recall that the crucial parameters in Definition 2.1 are the channel spread $\Delta_{\mathbf{H}} = 4\tau_0\nu_0$, which is the area of the rectangle in the delay-Doppler plane that supports most of the volume of the scattering function, and ε, which is the maximal fraction of volume of $C_{\mathbf{H}}(\tau, \nu)$ that lies outside the rectangle of area $\Delta_{\mathbf{H}}$. The compact-support assumption on $C_{\mathbf{H}}(\tau, \nu)$ is dropped because the results summarized in Section 2.1.2 hint at a high sensitivity of capacity in the high-SNR regime to whether $C_{\mathbf{H}}(\tau, \nu)$ is compactly supported or not, a property of $C_{\mathbf{H}}(\tau, \nu)$ that, however, cannot be verified through measurements (see Section 2.2.1.2).

The goal of the high-SNR analysis presented in this section is twofold: (i) to shed light on the trade-off between interference reduction and maximization of the number of degrees of freedom (see also Section 2.2.2.7), and (ii) to assess the robustness of the Rayleigh-fading WSSUS underspread model by determining the SNR values at which capacity becomes sensitive to whether $C_{\mathbf{H}}(\tau, \nu)$ is compactly supported or not.

To this end, in the next section, we present a capacity lower bound that is explicit in the parameters $\Delta_{\mathbf{H}}$ and ε and in the WH set $(g(t), T, F)$. We then compare this lower bound to the AWGN-capacity upper bound C_{AWGN}. Our main result is the following: we show that for an appropriate choice of the grid-parameter product TF, the lower bound obtained using WH sets based on a root-raised-cosine pulse (see Example 2.1) is close to C_{AWGN} for all SNR values of interest in practical systems, and for all Rayleigh-fading WSSUS channels that are underspread according to Definition 2.1, independently of whether the scattering function $C_{\mathbf{H}}(\tau, \nu)$ is compactly supported or not.

2.5.1 A Lower Bound on Capacity

The lower bound we analyze in this section is obtained from the bound L_1^{int} in (2.47) through some additional bounding steps to make the dependency on $\Delta_{\mathbf{H}}$ and ε explicit.

More specifically, we use the fact that for any Rayleigh-fading WSSUS channel in the set $\mathcal{H}(\tau_0, \nu_0, \varepsilon)$ (see Definition 2.1), the following inequalities hold (Durisi, Morgenshtern & Bölcskei, 2010):

$$\begin{aligned} \sigma_h^2 &\geq m_g(1-\varepsilon) \\ \sigma_I^2 &\leq M_g + \varepsilon. \end{aligned} \tag{2.51}$$

The quantities m_g and M_g are defined in (2.20) and (2.21), respectively. Furthermore, whenever $\widetilde{\Delta}_{\mathsf{H}} \triangleq 2\nu_0 T < 1$, we also have (Durisi, Morgenshtern & Bölcskei, 2010)

$$\frac{1}{T}\int_{-1/2}^{1/2} \ln\det\left(\mathbf{I}_L + \frac{TF\rho_s}{\alpha}\mathbf{P}_h(\theta)\right)d\theta \le \widetilde{\Delta}_{\mathsf{H}}\ln\left(1+\frac{TF\rho_s}{\alpha\widetilde{\Delta}_{\mathsf{H}}}\right) + (1-\widetilde{\Delta}_{\mathsf{H}})\ln\left(1+\frac{TF\rho_s\varepsilon}{\alpha(1-\widetilde{\Delta}_{\mathsf{H}})}\right).$$

The condition $\widetilde{\Delta}_{\mathsf{H}} < 1$ is not restrictive for underspread channels if T and F are chosen according to the grid matching rule (2.19). In this case, $\widetilde{\Delta}_{\mathsf{H}} = \sqrt{\Delta_{\mathsf{H}} TF} \ll 1$ for all values of TF of practical interest. By using (2.51) in (2.47), we get the following result.

Theorem 2.7 Let $(g(t),T,F)$ be a WH set and consider any Rayleigh-fading WSSUS channel in the set $\mathscr{H}(\tau_0,\nu_0,\varepsilon)$ (see Definition 2.1). Then, for a given bandwidth B and a given SNR $\rho \triangleq P/B$, and under the technical condition $\widetilde{\Delta}_{\mathsf{H}} = 2\nu_0 T < 1$, the capacity of the discretized channel (2.14) induced by $(g(t),T,F)$, with input subject to the average-power constraint (2.23) and the peak-power constraint (2.24), is lower-bounded as $C \ge L_2^{\text{int}}$, where

$$\begin{aligned} L_2^{\text{int}}(\rho) \triangleq \frac{B}{TF}\Bigg[&\mathsf{E}\left\{\ln\left(1+\frac{TF\rho_s(1-\varepsilon)m_g|h|^2}{1+TF\rho_s(M_g+\varepsilon)}\right)\right\} - \left(\ln(\pi e\rho_s) - \mathsf{h}(s)\right)\\ &- \inf_{0<\alpha<1}\left\{\widetilde{\Delta}_{\mathsf{H}}\ln\left(1+\frac{TF\rho_s}{\alpha\widetilde{\Delta}_{\mathsf{H}}}\right) + (1-\widetilde{\Delta}_{\mathsf{H}})\ln\left(1+\frac{TF\rho_s\varepsilon}{\alpha(1-\widetilde{\Delta}_{\mathsf{H}})}\right)\right.\\ &\left.+\ln\left(1+\frac{TF\rho_s}{1-\alpha}(M_g+\varepsilon)\right)\right\}\Bigg]. \end{aligned} \tag{2.52}$$

Here, $h \sim \mathcal{CN}(0,1)$, and s is a complex random variable that satisfies

$$\rho_s \triangleq \mathsf{E}\{|s|^2\} \le \rho \qquad \text{and} \qquad |s|^2 \le \beta\rho.$$

2.5.1.1 Some Remarks on the Lower Bound

The remarks we listed in Sections 2.4.1.4 and 2.4.1.5 about the reduction to a square setting, and about the maximization of L_a^{int} with respect to the choice of the WH set and the probability distribution of s, apply to the bound L_2^{int} as well. As in Section 2.4, we consider only WH sets based on a root-raised-cosine pulse and use for s the truncated complex-Gaussian distribution described in Section 2.4.1.5. Differently from the behavior of L_a^{int} in the large-bandwidth regime, the bound L_2^{int} is hardly sensitive to the value of the penalty term $\ln(\pi e\rho_s) - \mathsf{h}(s)$.

For fixed $g(t)$ (in our case, a root-raised-cosine pulse), the lower bound L_2^{int} can be tightened by maximizing it over the product TF of the grid parameters. This provides an information-theoretic criterion for the design of the WH set $(g(t),T,F)$. Compared with the design criterion based on SIR maximization discussed in Section 2.2.2.7, the maximization of the lower bound L_2^{int} yields a more complete picture, because it reveals the influence of the number of degrees of freedom (reflected through the grid-parameter product TF), and of the SIR (reflected through the quantities m_g and M_g) on capacity.

The lower bound L_2^{int} is not useful in the asymptotic regime $\rho \to \infty$. In fact, the bound even turns negative when ρ is sufficiently large. Nevertheless, as shown in Section 2.5.2 below, when the channel is underspread, the bound L_2^{int}, evaluated for the TF value that maximizes it, is close to the AWGN-capacity upper bound for all SNR values of practical interest.

2.5.2 Numerical Examples

In this section, we evaluate the lower bound L_2^{int} in (2.52) for the WH set based on a root-raised-cosine pulse considered in Example 2.1, under the assumption that the underlying WSSUS channel is underspread according to Definition 2.1, i.e., $\Delta_{\mathsf{H}} \ll 1$ and $\varepsilon \ll 1$. More precisely, we assume $\Delta_{\mathsf{H}} \leq 10^{-4}$ and $\varepsilon \leq 10^{-4}$, respectively. We also assume that s follows the truncated complex-Gaussian distribution described in Section 2.4.1.5.

Trade-off between interference reduction and maximization of the number of degrees of freedom

In Fig. 2.10, we plot the ratio $L_2^{\text{int}}/C_{\text{AWGN}}$ for $\beta = 10$, $\varepsilon = 10^{-6}$, and $\Delta_{\mathsf{H}} = 10^{-4}$ and both $\Delta_{\mathsf{H}} = 10^{-6}$. The different curves correspond to different values of TF. We can observe that the choice $TF = 1$ is highly suboptimal. The reason for this suboptimality is the significant reduction in SIR this choice entails (see Fig. 2.3). In fact (as already mentioned), when $TF = 1$, the pulse $g(t)$ reduces to a sinc function, which has poor time localization. A value of TF slightly above 1 leads to a significant improvement in the time localization of $g(t)$ and to a corresponding increase in the lower bound L_2^{int} for all SNR values of practical interest, despite the (small) loss of degrees of freedom. A further increase of the product TF seems to be detrimental for all but very high SNR values, where the ratio $L_2^{\text{int}}/C_{\text{AWGN}}$ is much smaller than 1: the rate loss due to the reduction of the number of degrees of freedom is more significant than the rate increase due to the resulting SIR improvement.

Sensitivity to the PAPR β

Figure 2.11 shows the ratio $L_2^{\text{int}}/C_{\text{AWGN}}$ for $TF = 1.02$, $\Delta_{\mathsf{H}} = \varepsilon = 10^{-6}$, and different values of β. We can observe that in the high-SNR regime, L_2^{int} is only mildly sensitive to the value of the penalty term $\ln(\pi e \rho_s) - \mathsf{h}(s)$ (recall that the value of the penalty term decreases as β increases).

Sensitivity of capacity to the measure of the support of $C_{\mathsf{H}}(\tau, \nu)$

The results presented in Fig. 2.10 suggest that, for $TF = 1.02$, the lower bound L_2^{int} is close to the AWGN-capacity upper bound C_{AWGN} over a quite large range of SNR values. To make this statement precise, we compute the SNR interval $[\rho_{\min}, \rho_{\max}]$ over which

$$L_2^{\text{int}} \geq 0.75\, C_{\text{AWGN}}. \tag{2.53}$$

The interval end points $\rho_{\min}$ and $\rho_{\max}$ can easily be computed numerically; the corresponding values for $\rho_{\min}$ and $\rho_{\max}$ are illustrated in Figs 2.12 and 2.13, respectively, for different $(\Delta_{\mathsf{H}}, \varepsilon)$ pairs. For the WH set and WSSUS underspread channels considered in this section, we have $\rho_{\min} \in [-25\,\text{dB}, -7\,\text{dB}]$ and $\rho_{\max} \in [32\,\text{dB}, 68\,\text{dB}]$.

Insights into how $\rho_{\min}$ and $\rho_{\max}$ are related to the channel parameters Δ_{H} and ε can be obtained by replacing both sides of the inequality (2.53) by corresponding low-SNR approximations (to get $\rho_{\min}$) and high-SNR approximations (to get $\rho_{\max}$). Under the assumption that $\Delta_{\mathsf{H}} \leq 10^{-4}$ and $\varepsilon \leq 10^{-4}$, this analysis, detailed in Durisi, Morgenshtern, and Bölcskei (2010), yields $\rho_{\min} \approx 13\sqrt{\Delta_{\mathsf{H}}}$ and $\rho_{\max} \approx 0.22/(\Delta_{\mathsf{H}} + \varepsilon)$ for the WH set considered in this section. The following rule of thumb then holds: the capacity of all WSSUS underspread channels with scattering function $C_{\mathsf{H}}(\tau, \nu)$ having no

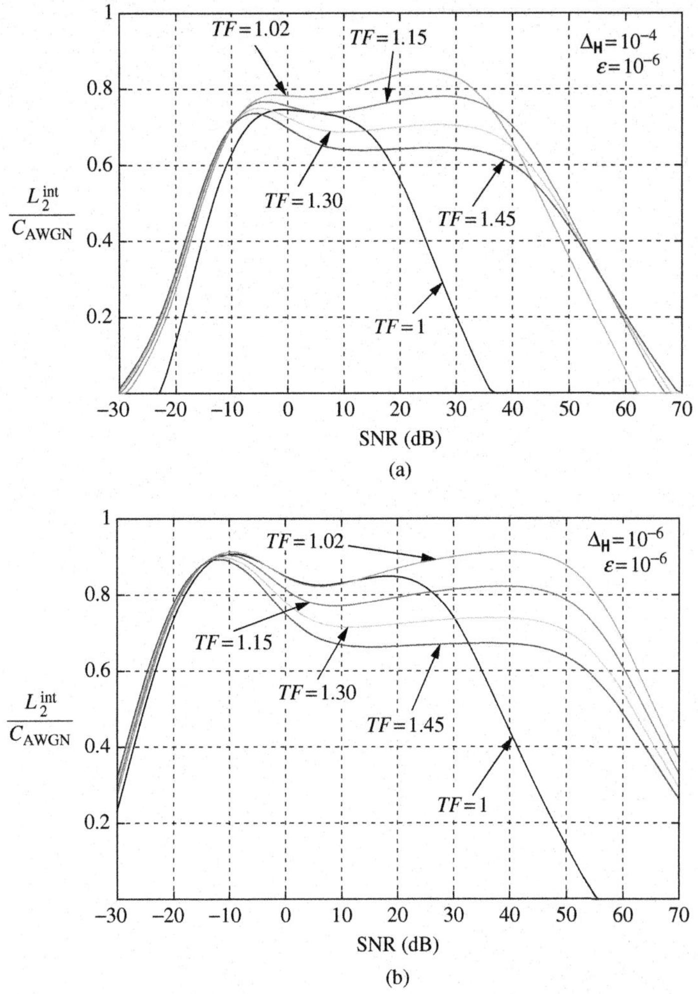

FIGURE 2.10

Lower bounds L_2^{int} on the capacity of the channel (2.14) normalized with respect to the AWGN capacity. The bounds are computed for WH sets based on a root-raised-cosine pulse (see Example 2.1), for different values of the grid-parameter product TF. The channel spread $\Delta_{\mathbf{H}}$ (see Definition 2.1) is equal to 10^{-4} in (a) and equal to 10^{-6} in (b); furthermore, in both cases $\varepsilon = 10^{-6}$ and $\beta = 10$.

more than ε of its volume outside a rectangle (in the delay-Doppler plane) of area $\Delta_{\mathbf{H}}$, is close to C_{AWGN} for all SNR values ρ that satisfy $\sqrt{\Delta_{\mathbf{H}}} \ll \rho \ll 1/(\Delta_{\mathbf{H}} + \varepsilon)$, independently of whether $C_{\mathbf{H}}(\tau, \nu)$ is compactly supported or not, and independently of the shape of $C_{\mathbf{H}}(\tau, \nu)$. To conclude, we stress that the condition $\sqrt{\Delta_{\mathbf{H}}} \ll \rho \ll 1/(\Delta_{\mathbf{H}} + \varepsilon)$ holds for all channels and SNR values of practical interest.

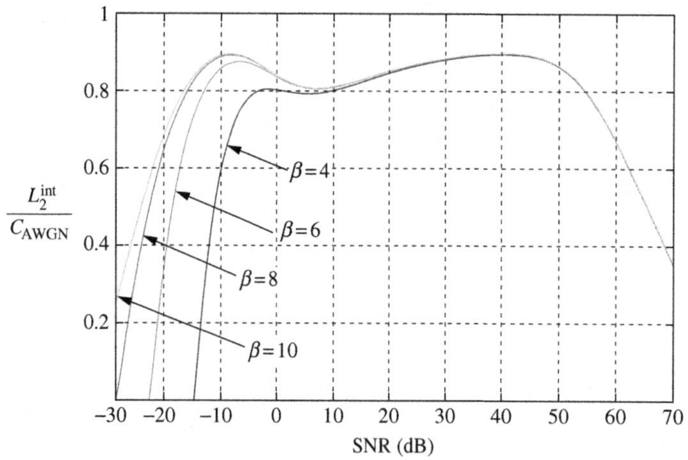

FIGURE 2.11

Lower bounds L_2^{int} on the capacity of the channel (2.14) normalized with respect to the AWGN capacity. The lower bounds are computed for WH sets based on a root-raised-cosine pulse (see Example 2.1), for different values of the PAPR β. The channel spread $\Delta_{\mathbb{H}}$ (see Definition 2.1) is equal to 10^{-6}; furthermore, $\varepsilon = 10^{-6}$ and $TF = 1.02$.

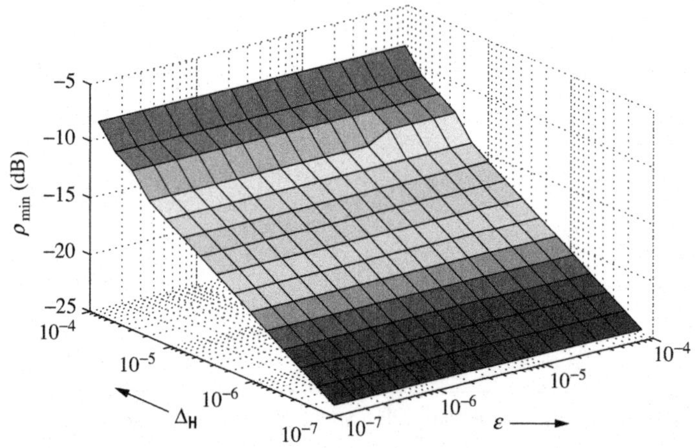

FIGURE 2.12

Minimum SNR value $\rho_{\min}$ for which (2.53) holds, as a function of $\Delta_{\mathbb{H}}$ and ε. The lower bound L_2^{int} is computed for a WH set based on a root-raised-cosine pulse (see Example 2.1); furthermore, $TF = 1.02$ and $\beta = 10$.

2.6 CONCLUSIONS

In this chapter, we provided an information-theoretic characterization of Rayleigh-fading channels that satisfy the WSSUS and the underspread assumptions. The information-theoretic analysis is built on

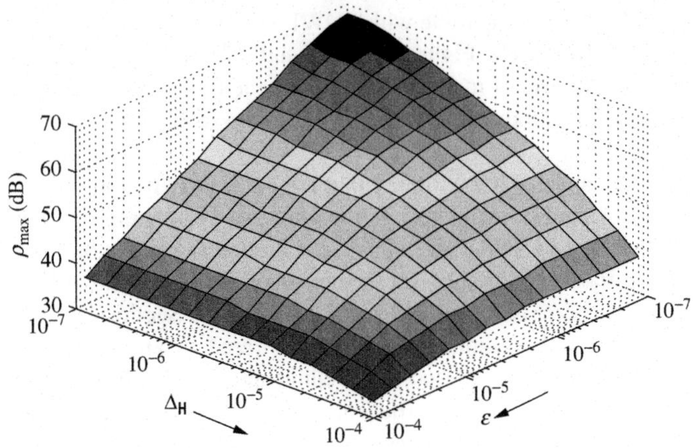

FIGURE 2.13

Maximum SNR value ρ_{max} for which (2.53) holds, as a function of Δ_H and ε. The lower bound L_2^{int} is evaluated for a WH set based on a root-raised-cosine pulse (see Example 2.1); furthermore, $TF = 1.02$ and $\beta = 10$.

a discretization of WSSUS underspread channels that takes the underspread property explicitly into account to minimize the ISI/ICI in the discretized I/O relation. The channel discretization is accomplished by transmitting and receiving on a WH set generated by a pulse that is well localized in time and frequency.

We obtained bounds on the noncoherent capacity under both an average and a peak-power constraint, and used these bounds to study engineering-relevant questions.

For the large-bandwidth regime, we provided upper and lower bounds on the noncoherent capacity on the basis of a diagonalized I/O relation that closely approximates the underlying continuous-time I/O relation. These upper and lower bounds provide important guidelines for the design of wireless communication systems operating over a Rayleigh-fading WSSUS underspread channel with known scattering function. Even if the scattering function is not known completely and the channel is only characterized coarsely by its maximum delay τ_0 and maximum Doppler shift ν_0, the bounds may serve as an efficient design tool. The maximum delay and the maximum Doppler shift are sufficient to characterize a brick-shaped scattering function, for which our upper and lower bounds are easy to evaluate. Furthermore, a brick-shaped scattering function was shown to minimize the capacity upper bound for a given τ_0 and ν_0. Therefore, the widely accepted practice to characterize wireless channels simply by means of these two parameters leads to robust design guidelines.

The numerical results in Section 2.3.6 indicate that the upper and lower bounds are surprisingly close over a large range of bandwidth values. In particular, our lower bound is close to the coherent-capacity upper bound in Fig. 2.4 for bandwidth values of up to 1 GHz. It is exactly this regime that is of most interest for current wideband and UWB communication systems; this is also the regime for which we can expect the WSSUS model to be most accurate.

The advent of UWB communication systems spurred the current interest in wireless communications over channels with very large bandwidth. Our bounds make it possible to assess whether multiple antennas at the transmitter are beneficial for UWB systems. The system parameters used to numerically evaluate the bounds in Section 2.3.8 are compatible with a UWB system that operates over a bandwidth of 7 GHz and transmits at −41.3 dBm/MHz. Figures 2.5, 2.6, and 2.7 show that the maximum rate increase that can be expected from the use of multiple antennas at the transmitter does not exceed 7%. For channels with smaller spreads than the one in Section 2.3.6 the possible rate increase is even smaller.

For the high-SNR regime, we provided a capacity lower bound that is obtained by treating interference as noise. This lower bound yields valuable insights into the capacity of Rayleigh-fading underspread WSSUS channels over a large range of SNR values of practical interest. On the basis of this lower bound, we derived an information-theoretic criterion for the design of capacity-optimal WH sets. This criterion is more fundamental than criteria based on SIR maximization (see Section 2.2.2.7) because it sheds light on the trade-off between the number of degrees of freedom and the time-frequency localization of the pulse $g(t)$. Unfortunately, the corresponding optimization problem is hard to solve. We simplified the problem by fixing $g(t)$ to be a root-raised-cosine pulse and performing an optimization over the grid parameters T and F. Our analysis shows that the optimal value of the grid-parameter product TF is close to 1 (but strictly larger than 1) for a large range of SNR values of practical interest. This result suggests that the maximization of the number of degrees of freedom should be privileged over the SIR maximization in the design of capacity-maximizing WH sets.

Even though our analysis was confined to a specific pulse shape (i.e., root-raised-cosine), we were able to show that for all Rayleigh-fading WSSUS channels that are underspread according to Definition 2.1, the corresponding capacity lower bound is close to the AWGN-capacity upper bound for all SNR values of practical interest, independently of whether the scattering function is compactly supported or not (a fine detail of the channel model). In other words, the capacity of Rayleigh-fading underspread WSSUS channels starts being sensitive to this fine detail of the channel model only for SNR values that lie outside the SNR range typically encountered in real-world systems. Hence, the Rayleigh-fading WSSUS underspread model is a robust model.

References

Abou-Faycal, I. C., Trott, M. D., & Shamai (Shitz), S. (2001). The capacity of discrete-time memoryless Rayleigh-fading channels. *IEEE Transactions on Information Theory*, *47*(4), 1290–1301.

Artés, H., Matz, G., & Hlawatsch, F. (2004). Unbiased scattering function estimators for underspread channels and extension to data-driven operation. *IEEE Transactions on Signal Processing*, *52*(5), 1387–1402.

Bello, P. (1963). Characterization of randomly time-variant linear channels. *IEEE Transactions on Communications*, *11*(4), 360–393.

Biglieri, E., Proakis, J., & Shamai (Shitz), S. (1998). Fading channels: Information-theoretic and communications aspects. *IEEE Transactions on Information Theory*, *44*(6), 2619–2692.

Bölcskei, H., & Janssen, A. J. E. M. (2000). Gabor frames, unimodularity, and window decay. *The Journal of Fourier Analysis and Applications*, *6*(3), 255–276.

Christensen, O. (2003). *An introduction to frames and Riesz bases*. Boston, MA: Birkhäuser.

Chuah, C. -N., Tse, D. N. C., Kahn, J. M., & Valenzuela, R. A. (2002). Capacity scaling in MIMO wireless systems under correlated fading. *IEEE Transactions on Information Theory*, *48*(3), 637–650.

Cover, T. M., & Thomas, J. A. (1991). *Elements of information theory*. New York, NY: Wiley.

Cox, D. C. (1973a). 910 MHz urban mobile radio propagation: Multipath characteristics in New York City. *IEEE Transactions on Communications, 21*(11), 1188–1194.

Cox, D. C. (1973b). A measured delay-Doppler scattering function for multipath propagation at 910 MHz in an urban mobile radio environment. *Proceedings of the IEEE, 61*(4), 479–480.

Daubechies, I., Landau, H., & Landau, Z. (1995). Gabor time-frequency lattices and the Wexler-Raz identity. *The Journal of Fourier Analysis and Applications, 1*(4), 437–478.

Dunford, N., & Schwarz, J. T. (1963). *Linear operators* (Vol. 2). New York, NY: Wiley.

Durisi, G., Morgenshtern, V. I., & Bölcskei, H. (2009). On the sensitivity of noncoherent capacity to the channel model. In *Proceedings of the IEEE International Symposium on Information Theory (ISIT)*, (pp. 2174–2178). Seoul, Korea.

Durisi, G., Schuster, U. G., Bölcskei, H., & Shamai (Shitz), S. (2010). Noncoherent capacity of underspread fading channels. *IEEE Transactions on Information Theory, 56*(1), 367–395.

Durisi, G., Morgenshtern, V. I., & Bölcskei, H. (2010). On the information-theoretic robustness of the WSSUS underspread channel model. Submitted.

Gallager, R. G. (1968). *Information theory and reliable communication*. New York, NY: Wiley.

Gray, R. M. (2005). Toeplitz and circulant matrices: A review. In *Foundations and trends in communications and information theory* (vol. 2). Delft: The Netherlands.

Gröchenig, K. (2001). *Foundations of time-frequency analysis*. Boston, MA: Birkhäuser.

Guo, D., Shamai (Shitz), S., & Verdú, S. (2005). Mutual information and minimum mean-square error in Gaussian channels. *IEEE Transactions on Information Theory, 51*(4), 1261–1282.

Hashemi, H. (1993). The indoor radio propagation channel. *Proceedings of the IEEE, 81*(7), 943–968.

Jakes, W. C. (Ed.) (1974). *Microwave mobile communications*. New York, NY: Wiley.

Janssen, A. J. E. M. (1995). Duality and biorthogonality for Weyl-Heisenberg frames. *The Journal of Fourier Analysis and Applications, 1*(4), 403–437.

Kennedy, R. S. (1969). *Fading dispersive communication channels*. New York, NY: Wiley.

Kermoal, J. P., Schumacher, L., Pedersen, K. I., Mogensen, P. E., & Frederiksen, F. (2002). A stochastic MIMO radio channel model with experimental validation. *IEEE Journal on Selected Areas in Communications, 20*(6), 1211–1226.

Kozek, W. (1997). *Matched Weyl-Heisenberg expansions of nonstationary environments*. (PhD thesis). Vienna, Austria: Vienna University of Technology, Department of Electrical Engineering.

Kozek, W., & Molisch, A. F. (1998). Nonorthogonal pulseshapes for multicarrier communications in doubly dispersive channels. *IEEE Journal on Selected Areas in Communications, 16*(8), 1579–1589.

Lapidoth, A. (2005). On the asymptotic capacity of stationary Gaussian fading channels. *IEEE Transactions on Information Theory, 51*(2), 437–446.

Lapidoth, A., & Moser, S. M. (2003). Capacity bounds via duality with applications to multiple-antenna systems on flat-fading channels. *IEEE Transactions on Information Theory, 49*(10), 2426–2467.

Liang, Y., & Veeravalli, V. V. (2004). Capacity of noncoherent time-selective Rayleigh-fading channels. *IEEE Transactions on Information Theory, 50*(12), 3096–3110.

Marzetta, T. L., & Hochwald, B. M. (1999). Capacity of a mobile multiple-antenna communication link in Rayleigh flat fading. *IEEE Transactions on Information Theory, 45*(1), 139–157.

Matz, G., & Hlawatsch, F. (1998). Time-frequency transfer function calculus (symbolic calculus) of linear time-varying systems (linear operators) based on a generalized underspread theory. *Journal of Mathematical Physics, 39*(8), 4041–4070.

Matz, G., & Hlawatsch, F. (2003). Time-frequency characterization of randomly time-varying channels. In Boashash, B., (Ed.), *Time-frequency signal analysis and processing: A comprehensive reference*, (Chap. 9.5, pp. 410–419). Oxford, UK: Elsevier.

Matz, G., Schafhuber, D., Gröchenig, K., Hartmann, M., & Hlawatsch, F. (2007). Analysis, optimization, and implementation of low-interference wireless multicarrier systems. *IEEE Transactions on Wireless Communications*, *6*(5), 1921–1931.

Médard, M. (2000). The effect upon channel capacity in wireless communications of perfect and imperfect knowledge of the channel. *IEEE Transactions on Information Theory*, *46*(3), 933–946.

Médard, M., & Gallager, R. G. (2002). Bandwidth scaling for fading multipath channels. *IEEE Transactions on Information Theory*, *48*(4), 840–852.

Naylor, A. W., & Sell, G. R. (1982). *Linear operator theory in engineering and science*. New York, NY: Springer.

Parsons, J. D. (2000). *The mobile radio propagation channel* (2nd ed.). Chichester, UK: Wiley.

Pierce, J. R. (1966). Ultimate performance of M-ary transmission on fading channels. *IEEE Transactions on Information Theory*, *12*(1), 2–5.

Poor, H. V. (1994). *An introduction to signal detection and estimation* (2nd ed.). New York, NY: Springer.

Porrat, D., Tse, D. N. C., & Nacu, Ş. (2007). Channel uncertainty in ultra-wideband communication systems. *IEEE Transactions on Information Theory*, *53*(1), 194–208.

Proakis, J. G. (2001). *Digital communications* (4th ed.). New York, NY: McGraw-Hill.

Raghavan, V., Hariharan, G., & Sayeed, A. M. (2007). Capacity of sparse multipath channels in the ultra-wideband regime. *IEEE Journal of Selected Topics in Signal Processing*, *1*(3), 357–371.

Rappaport, T. S. (2002). *Wireless communications: Principles and practice* (2nd ed.). Upper Saddle River, NJ: Prentice Hall.

Razavi, B. (1998). *RF microelectronics*. Upper Saddle River, NJ: Prentice Hall.

Ron, A., & Shen, Z. (1997). Weyl-Heisenberg frames and Riesz bases in $L_2(\mathbb{R}^d)$. *Duke Mathematical Journal*, *89*(2), 237–282.

Schuster, U. G. (2009). *Wireless communication over wideband channels*: vol. 4 of *Series in Communication Theory*. Konstanz, Germany: Hartung-Gorre Verlag.

Schuster, U. G., & Bölcskei, H. (2007). Ultrawideband channel modeling on the basis of information-theoretic criteria. *IEEE Transactions Wireless Communications*, *6*(7), 2464–2475.

Schuster, U. G., Durisi, G., Bölcskei, H., & Poor, H. V. (2009). Capacity bounds for peak-constrained multiantenna wideband channels. *IEEE Transactions Communications*, *57*(9), 2686–2696.

Sethuraman, V., & Hajek, B. (2005). Capacity per unit energy of fading channels with peak constraint. *IEEE Transactions on Information Theory*, *51*(9), 3102–3120.

Sethuraman, V., Hajek, B., & Narayanan, K. (2005). Capacity bounds for noncoherent fading channels with a peak constraint. In *Proceedings of the IEEE International Symposium on Information Theory (ISIT)*, (pp. 515–519). Adelaide: Australia.

Sethuraman, V., Wang, L., Hajek, B., & Lapidoth, A. (2009). Low SNR capacity of noncoherent fading channels. *IEEE Transactions on Information Theory*, *55*(4), 1555–1574.

Slepian, D. (1976). On bandwidth. *Proceedings of the IEEE*, *64*(3), 292–300.

Subramanian, V. G., & Hajek, B. (2002). Broad-band fading channels, Signal burstiness and capacity. *IEEE Transactions on Information Theory*, *48*(4), 809–827.

Taricco, G., & Elia, M. (1997). Capacity of fading channel with no side information. *Electronics Letters*, *33*(16), 1368–1370.

Telatar, İ. E. (1999). Capacity of multi-antenna Gaussian channels. *European Transactions on Telecommunication*, 10, 585–595.

Telatar, İ. E., & Tse, D. N. C. (2000). Capacity and mutual information of wideband multipath fading channels. *IEEE Transactions on Information Theory*, *46*(4), 1384–1400.

Tse, D. N. C., & Viswanath, P. (2005). *Fundamentals of wireless communication*. Cambridge, UK: Cambridge University Press.

Vaughan, R., & Bach Andersen, J. (2003). *Channels, propagation and antennas for mobile communications.* London, UK: The Institution of Electrical Engineers.

Verdú, S. (2002). Spectral efficiency in the wideband regime. *IEEE Transactions on Information Theory, 48*(6), 1319–1343.

Viterbi, A. J. (1967). Performance of an M-ary orthogonal communication system using stationary stochastic signals. *IEEE Transactions on Information Theory, 13*(3), 414–422.

Wyner, A. D. (1966). The capacity of the band-limited Gaussian channel. *Bell Systems Technical Journal, 45*(3), 359–395.

Zheng, L., & Tse, D. N. C. (2002). Communication on the Grassmann manifold: A geometric approach to the noncoherent multiple-antenna channel. *IEEE Transactions on Information Theory, 48*(2), 359–383.

CHAPTER

3 Algebraic Coding for Fast Fading Channels

Emanuele Viterbo and Yi Hong
Monash University, Clayton, Australia

3.1 INTRODUCTION

Elementary number theory has played a fundamental role in the development of error-correcting codes, in the early age of coding theory. Finite fields were the key tool in the design of powerful binary codes and gradually entered in the general mathematical background of communications engineers. Thanks to the technological developments and increased processing power available in digital receivers, the focus of coding theory moved to the design of signal space codes in the framework of coded modulation systems. In the 1980s, the theory of Euclidean lattices became of great interest for the design of dense signal constellations well suited for transmission over the additive white gaussian noise (AWGN) channel. More recently, the incredible boom of wireless communications forced coding theorists to deal with fading channels. New code design criteria had to be considered in order to improve the poor performance of wireless transmission systems. The need of bandwidth efficient coded modulation became even more important due to the scarce availability of radio bands.

Algebraic number theory was shown to be a very useful mathematical tool that enables the design of good coding schemes for fading channels (Sethuraman, Sundar Rajan, & Shashidhar, 2003). These codes are constructed as multidimensional *lattice* signal sets (or constellations) with particular geometric properties. Coding gain is obtained by introducing the so-called *modulation diversity* in the signal set, which results in a particular type of bandwidth efficient diversity technique.

Two approaches were proposed to construct high-modulation-diversity constellations. The first was based on the design of intrinsic high-diversity algebraic lattices, obtained by applying the *canonical embedding* of an *algebraic number field* into its *ring of integers*. Only later was it realized that high modulation diversity could also be achieved by applying a particular rotation to a multidimensional quadrature amplitude modulation (QAM) signal constellation in such a way that any two points achieve the maximum number of distinct components. Still, rotations giving diversity can be designed using algebraic number theory.

An attractive feature of modulation diversity technique is that a significant improvement in error performance is obtained, without requiring the use of conventional channel coding. In fact, we may think of the rotation as a precoder or a rate one code. This can always be added later if required.

Finally, dealing with lattice constellations has the major advantage that it is possible to use an efficient decoding algorithm known as the *sphere decoder*.

Research on coded modulation schemes obtained from lattice constellations with high diversity began more than 10 years ago, and extensive work has been done to improve the performance of these lattice codes. The goal of this chapter is to give both a unified point of view on the constructions obtained so far, and a brief tutorial on algebraic number theory methods useful for the design of algebraic lattice codes for the Rayleigh fading channel.

We start by detailing the channel and the transmission system model that we consider.

3.1.1 Fading Channel Model

We consider an independent Rayleigh fast-fading channel, where successively transmitted symbols are assumed to be sent over independent channel gains. Assuming that no inter-symbol interference is present at the receiver, the complex discrete-time model of the channel at time k is given by

$$r'[k] = h'[k]a_c[k] + w'[k],$$

where $a_c[k]$ is a symbol from a complex signal set, $w'[k]$ is the complex Gaussian noise, and $h'[k] = h[k]e^{j\varphi[k]}$ is the complex zero mean Gaussian fading coefficient. Under the independent fast fading assumption, the complex fading coefficients are independent from one symbol to the next. We note that the time index k may equivalently represent a subcarrier index in an orthogonal frequency division multiplexing (OFDM) scheme used in a rich multipath channel (see Chapter 7).

Assuming perfect channel state information (CSI) at the receiver but not at the transmitter, the phase $\varphi[k]$ of the fading coefficient can be removed, so that we get

$$r_c[k] = h[k]a_c[k] + w_c[k], \tag{3.1}$$

where $h[k] = |h'[k]|$ is now a real Rayleigh distributed fading coefficient, and $w_c[k] = w'[k]e^{-j\varphi[k]}$ remains complex Gaussian noise.

In this case, both in-phase and quadrature components of the transmitted symbol are subject to the same fading. By using an *in-phase/quadrature component interleaver*, we can consider the real fading channel model

$$r[k] = h[k]a[k] + w[k] \tag{3.2}$$

in which we assume that $a[k]$ is real and $w[k]$ is a real Gaussian random variable. The component interleaver ensures that the fading coefficients are independent from one real transmitted symbol to the next.

We now consider coded transmission, where the information is encoded into codewords that are transmitted over n successive real channel uses of (3.2). Let us denote the transmitted codewords as $\mathbf{x} = (x_1 \cdots x_n)$, i.e., n-dimensional real vectors taken from a finite signal constellation $\mathcal{S} \subseteq \mathbb{R}^n$. Each component x_i, $i = 1, \ldots, n$, is assumed to experience an independent real fading coefficient.

3.1.2 System Model

Given the above channel model, the communication system is shown in Fig. 3.1. Consider an n-dimensional signal constellation $\mathcal{S}$ carved from the set of lattice points $\Lambda = \{\mathbf{x} = \mathbf{uM}\}$, where $\mathbf{u}$ is an integer vector and $\mathbf{M}$ is the lattice generator matrix (see Section 3.3). At the transmitter, the information bits label the lattice points in $\mathcal{S}$ and the signal vector $\mathbf{x} = (x_1 \cdots x_n) \in \mathbb{R}^n$ is transmitted.

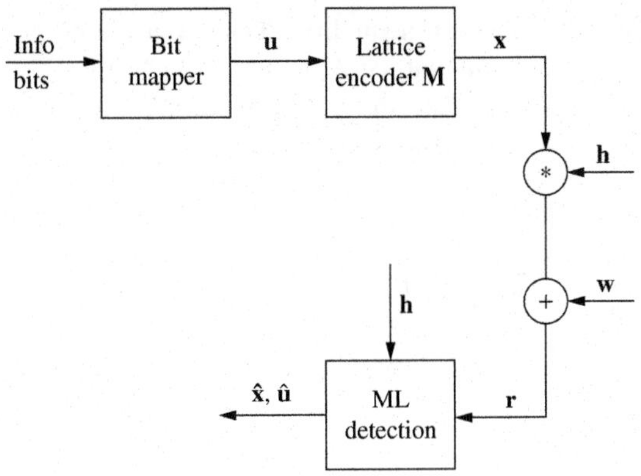

FIGURE 3.1

Transmission system model.

At the receiver, the received signal is given by $\mathbf{r} = (r_1 \cdots r_n)$ with $r_i = h_i x_i + w_i$ for $i = 1, 2, \cdots n$, where the h_i are independent real Rayleigh distributed random variables with unit second moment (i.e., $\mathsf{E}\{h_i^2\} = 1$), and w_i are real Gaussian random variables with zero mean and variance $N_0/2$ representing the additive noise. Assuming that the channel state information $\mathbf{h} = (h_1, \cdots h_n)$ is known at the receiver, *maximum likelihood* (ML) detection is used to minimize the metric

$$m(\mathbf{x}|\mathbf{r}, \mathbf{h}) \triangleq \sum_{i=1}^{n} |r_i - h_i x_i|^2 \tag{3.3}$$

with respect to $\mathbf{x} \in \mathcal{S}$. We obtain the decoded point $\hat{\mathbf{x}}$ and the corresponding integer component vector $\hat{\mathbf{u}}$, from which the decoded bits can be extracted. The minimization of (3.3) can be obtained more efficiently if a *sphere decoder* is applied, when compared with exhaustive search ML decoding, given lattice constellations (see Section 3.8).

3.2 PRODUCT DISTANCE BASED CODE DESIGN

In relation to the code design criteria, we first introduce two terms: *diversity* and *product distance*, then discuss lattice constellations.

3.2.1 Signal Space Diversity and Product Distance

In order to derive code design criteria, we estimate the codeword error probability $P_e(\mathcal{S})$ of the transmission system described in Section 3.1.2. Because the signal constellation $\mathcal{S}$ is carved from the infinite lattice Λ, it is convenient to evaluate the error probability of the infinite lattice $P_e(\Lambda)$.

Because a lattice is *geometrically uniform*, we may simply write $P_e(\Lambda) = P_e(\Lambda|\mathbf{0})$ for the point error probability (i.e., we assume the zero codeword $\mathbf{0}$ is transmitted). Applying the union bound, we get the following upper bound

$$P_e(\mathcal{S}) \leq P_e(\Lambda) \leq \sum_{\mathbf{y} \neq \mathbf{x}} \Pr\{\mathbf{x} \to \mathbf{y}\}, \tag{3.4}$$

where $\Pr\{\mathbf{x} \to \mathbf{y}\}$ is the pairwise error probability. Note that the first inequality takes into account the edge effects of the finite constellation $\mathscr{S}$ compared with the infinite lattice Λ.

Let us apply the standard Chernoff bound technique to estimate the pairwise error probability (Boutros, Viterbo, Rastello, & Belfiore, 1996). For large signal-to-noise ratios, we have

$$\Pr\{\mathbf{x} \to \mathbf{y}\} \leq \frac{1}{2} \prod_{x_i \neq y_i} \frac{4N_0}{(x_i - y_i)^2} = \frac{1}{2} \frac{(4N_0)^l}{\left(d_p^{(l)}(\mathbf{x},\mathbf{y})\right)^2}, \tag{3.5}$$

where $d_p^{(l)}(\mathbf{x},\mathbf{y})$ is the *l-product distance* of $\mathbf{x}$ from $\mathbf{y}$, when these two points differ in l components, i.e.,

$$d_p^{(l)}(\mathbf{x},\mathbf{y}) \triangleq \prod_{x_i \neq y_i} |x_i - y_i|. \tag{3.6}$$

We say that the pairwise error event has a diversity order of l, which defines the order of decay of the pairwise error probability for large signal-to-noise ratios (SNR) ($N_0 \to 0$). The dominant terms, for high SNR, in the sum in (3.4) are found for

$$L \triangleq \min_{\{\mathbf{x} \to \mathbf{y}\}} (l)$$

defined as the *modulation diversity* or *diversity order* of the signal constellation. In other words, L is the minimum number of distinct components between any two constellation points or the minimum Hamming distance between any two coordinate vectors of the constellation points. Among the terms with the same diversity order, the dominant term is found for $d_{p,\min} = \min d_p^{(L)}$.

We conclude that the error probability is determined asymptotically by the diversity order L and the minimum product distance $d_{p,\min}$. In particular, good signal sets have high L and $d_{p,\min}$.

If the diversity order L equals the dimension of the lattice n, we say that the constellation has *maximal diversity*.

As an example, take a 4-QAM constellation. In Fig. 3.2(a), the diversity is $L = 1$, whereas in Fig. 3.2(b), a rotated version of the constellation has diversity $L = 2$, thus maximal diversity. Suppose now a fading of $h = 0.5$ affects the second component. In case (a), the points will get closer to each other and eventually collapse together if the fading is deeper. In this case, a very small amount of noise will produce a decoding error. In case (b), the rotated version, where all coordinates are distinct, will be more resistant to noise, even in the presence of a deep fade.

It is clear that any small rotation would be enough to obtain maximal diversity, but in order to optimize the choice, we select the one that will give the lowest probability of error. This requires to consider the minimum product distance $d_{p,\min}$. In this particular case, the optimal rotation which maximizes $d_{p,\min}$ is by 13 degrees.

In Fig. 3.3, we show the diversity gain of the rotated constellation with respect to the nonrotated one, as well as the error probability of 4-QAM over the Gaussian channel. The gap between the curves represents the potential gain obtainable by increasing the diversity.

We will show that by increasing the diversity order of multidimensional constellations, it is possible to approach the performance of the transmission over an AWGN channel.

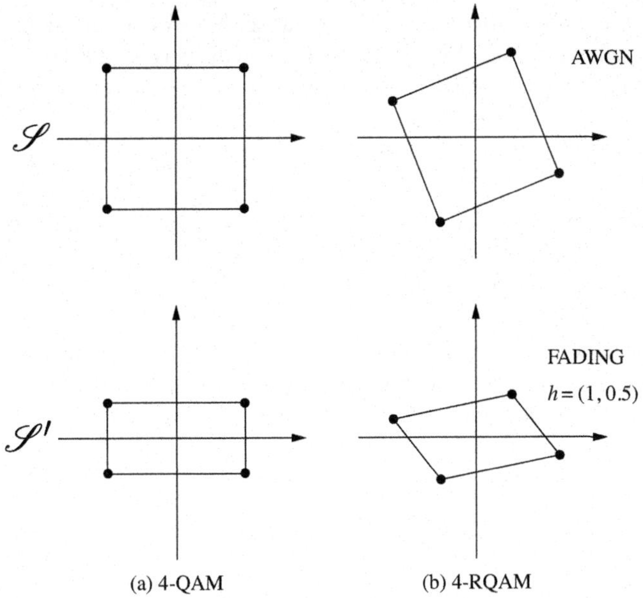

FIGURE 3.2

Example of modulation diversity with 4-QAM: (a) $L = 1$, (b) $L = 2$; top: AWGN channel, bottom: fading channel.

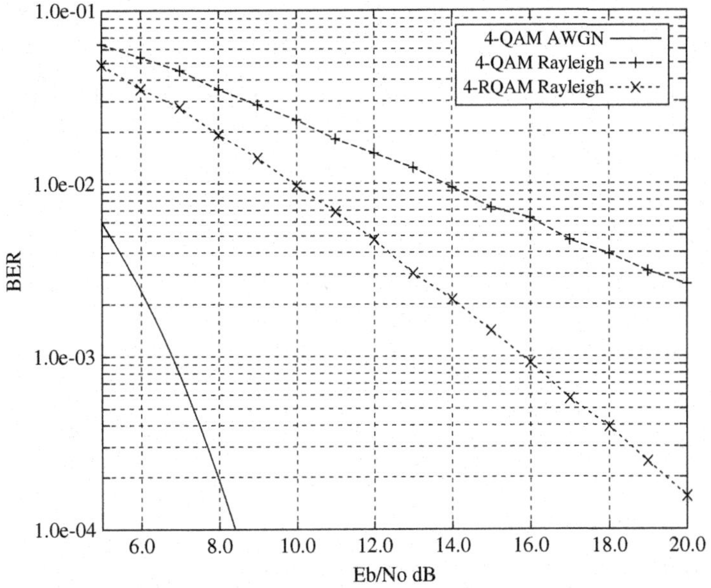

FIGURE 3.3

Bit error probability of 4-QAM and 4-RQAM over Gaussian and Rayleigh fading channels.

An interesting feature of the rotation operation is that the rotated signal set has exactly the same performance as the nonrotated one when used over a pure AWGN channel, while the performance over Rayleigh fading channels, for increasingly high-modulation diversity order, approaches that achievable over the Gaussian channel.

3.2.2 Lattice Constellations

In the design of signal constellations, two fundamental operations should always be kept in mind: *bit labeling* and *constellation shaping*. These may be very critical for the complexity of practical implementations and are strictly related to each other. If we want to avoid the use of a huge look-up table to perform bit labeling, we need to have a simple algorithm mapping bits to signal points and vice versa. However, it is well known that lattice constellations bounded by a sphere have the best shaping gain. Unfortunately, labeling a spherically shaped constellation without using a look-up table is not always an easy task. Cubic-shaped constellations offer a good trade-off: they are only slightly worse in terms of shaping gain but are usually easier to label.

The simplest labeling algorithm that we can use for a lattice constellation

$$\mathscr{S} = \{\mathbf{x} = \mathbf{uM} : \mathbf{u} = (u_1 \cdots u_n),\ u_i \in q\text{-PAM}\}$$

is obtained by performing the bit labeling on the integer components u_i of the vector $\mathbf{u}$. These are usually restricted to a q-PAM constellation $\{\pm 1, \pm 3, \ldots, \pm(2^{\eta/2} - 1)\}$, where η is the number of bits per 2 dimension (or bit/symbol). Gray bit labeling of each q-PAM one-dimensional component proved to be the most effective strategy to reduce the bit error rate.

If we restrict ourselves to the above very simple labeling algorithm, we observe that this induces a constellation shape similar to the fundamental parallelotope (see next section) of the underlying lattice. This means that the constellation shape will not be cubic and hence will produce an undesirable shaping loss for all lattices except for $\mathbb{Z}^n$–lattices.

We conclude that a good compromise is to work with $\mathbb{Z}^n$–lattices, which may be found in their fully diverse rotated versions by the use of algebraic constructions.

3.3 LATTICES

Lattices are periodic structures of points in an n-dimensional Euclidean space. Their rich mathematical structure has interested many mathematicians and has initially found applications to physical modeling (in two and three dimensions). Later, they were recognized also as a powerful tool for abstract algebra, providing a geometrical interpretation to many results in algebraic number theory. In this section, we review some basic definitions of lattice theory relevant to our problem.

3.3.1 First Definitions

We begin by recalling the definition of a group.

Definition 3.1 Let $\mathscr{G}$ be a set endowed with an internal operation (that we denote by the symbol +)

$$\begin{array}{ccc} \mathscr{G} \times \mathscr{G} & \rightarrow & \mathscr{G} \\ (a,b) & \mapsto & a+b. \end{array}$$

The pair $(\mathcal{G},+)$ is a *group* if

1. the operation is associative, i.e., $a+(b+c)=(a+b)+c$ for all $a,b,c \in \mathcal{G}$;
2. there exists a neutral element $0 \in \mathcal{G}$, such that $0+a=a+0=a$ for all $a \in \mathcal{G}$;
3. for all $a \in \mathcal{G}$, there exists an inverse $-a \in \mathcal{G}$ such that $a-a=-a+a=0$.

The group $\mathcal{G}$ is said to be *Abelian* if $a+b=b+a$ for all $a,b \in \mathcal{G}$, i.e., the internal operation is commutative.

Definition 3.2 Let $(\mathcal{G},+)$ be a group and $\mathcal{H}$ be a nonempty subset of $\mathcal{G}$. We say that $\mathcal{H}$ is a *subgroup* of $\mathcal{G}$ if $(\mathcal{H},+)$ is a group, where $+$ is the internal operation inherited from $\mathcal{G}$.

An interesting point in having a group structure is that one is sure that whenever two elements are in the group, then their sum is also in the group. We say the group $\mathcal{G}$ is *closed* under the group operation $+$.

Definition 3.3 Let $\mathbf{v}_1,\ldots,\mathbf{v}_m$ be a linearly independent set of vectors in $\mathbb{R}^n$ (so that $m \leq n$). The set of points

$$\Lambda = \left\{ \mathbf{x} = \sum_{i=1}^{m} u_i \mathbf{v}_i,\ u_i \in \mathbb{Z} \right\}$$

is called a *lattice* of dimension m, and $\{\mathbf{v}_1,\ldots,\mathbf{v}_m\}$ is called a *basis* of the lattice.

A lattice is a discrete set of points in $\mathbb{R}^n$. This is easily seen because we take integral linear combinations of $\mathbf{v}_1,\ldots,\mathbf{v}_m$. More precisely, it is a subgroup of $(\mathbb{R}^m,+)$, so that in particular the sum or difference of two vectors in the lattice is still in the lattice. We say that a lattice of dimension m *spans* $\mathbb{R}^m \subseteq \mathbb{R}^n$ (recall that $\mathbf{v}_1,\ldots,\mathbf{v}_m$ are linearly independent in $\mathbb{R}^n$). See Fig. 3.4.

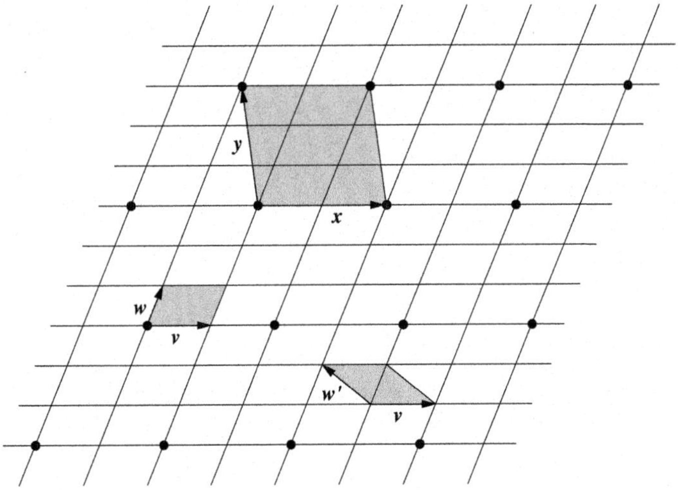

FIGURE 3.4

The points of the grid represent a lattice. The sets of vectors $\{\mathbf{v},\mathbf{w}\}$ and $\{\mathbf{v},\mathbf{w}'\}$ are two examples of a basis for this lattice. They both span a fundamental parallelotope for the lattice. Points • represent a sublattice. The set of vectors $\{\mathbf{x},\mathbf{y}\}$ forms a basis for this sublattice. It spans a fundamental parallelotope for the sublattice.

Definition 3.4 The parallelotope consisting of the points

$$\sum_{i=1}^{m} \theta_i \mathbf{v}_i \quad \text{with} \quad 0 \leq \theta_i < 1$$

is called a *fundamental parallelotope* of the lattice (see Fig. 3.4).

A fundamental parallelotope is an example of a *fundamental region* for the lattice, that is, a building block which when repeated many times fills the whole space with just one lattice point in each copy.

There are many different ways of choosing a basis for a given lattice, as shown in Fig. 3.4, where the lattice represented by the points of the grid can have $\{\mathbf{v}, \mathbf{w}\}$ or $\{\mathbf{v}, \mathbf{w}'\}$ as a basis.

Let the basis vectors be

$$\begin{aligned}
\mathbf{v}_1 &= (v_{11}\, v_{12} \cdots v_{1n}) \\
\mathbf{v}_2 &= (v_{21}\, v_{22} \cdots v_{2n}) \\
&\vdots \\
\mathbf{v}_m &= (v_{m1}\, v_{m2} \cdots v_{mn}),
\end{aligned}$$

where $n \geq m$.

Definition 3.5 The matrix

$$\mathbf{M} = \begin{pmatrix} v_{11} & v_{12} & \cdots & v_{1n} \\ v_{21} & v_{22} & \cdots & v_{2n} \\ \vdots & \vdots & & \vdots \\ v_{m1} & v_{m2} & \cdots & v_{mn} \end{pmatrix}$$

is called a *generator matrix* for the lattice. The matrix $\mathbf{G} = \mathbf{M}\mathbf{M}^T$ is called a *Gram matrix* for the lattice, where the superscript ^{T} denotes transposition.

More concisely, the lattice can be defined by its generator matrix as

$$\Lambda = \{\mathbf{x} = \mathbf{u}\mathbf{M} \mid \mathbf{u} = (u_1 \cdots u_m) \in \mathbb{Z}^m\}.$$

Definition 3.6 The *determinant of the lattice* Λ is defined to be the determinant of the matrix $\mathbf{G}$

$$\det\{\Lambda\} \triangleq \det\{\mathbf{G}\}.$$

This is an *invariant of the lattice* because it does not depend on the choice of the lattice basis.

Because the Gram matrix is given by $\mathbf{G} = \mathbf{M}\mathbf{M}^T$, where $\mathbf{M}$ contains the basis vectors $\{\mathbf{v}_i\}_{i=1}^m$ of the lattice, the (i,j)th entry of $\mathbf{G}$ is the inner product $\langle \mathbf{v}_i, \mathbf{v}_j \rangle = \mathbf{v}_i \mathbf{v}_j^T$.

Definition 3.7 A lattice Λ is called an *integer* lattice if its Gram matrix has coefficients in $\mathbb{Z}$.

A lattice Λ is integer if and only if $\langle \mathbf{x}, \mathbf{y} \rangle \in \mathbb{Z}$, for all $\mathbf{x}, \mathbf{y} \in \Lambda$. Indeed, take $\mathbf{x}, \mathbf{y} \in \Lambda$, $\mathbf{x} = \sum_{i=1}^m a_i \mathbf{v}_i$, $\mathbf{y} = \sum_{i=1}^m b_i \mathbf{v}_i$, with $a_i, b_i \in \mathbb{Z}$. Thus $\langle \mathbf{x}, \mathbf{y} \rangle = \sum_{i,j=1}^n a_i b_j \mathbf{v}_i \mathbf{v}_j^T = \sum_{i,j=1}^n a_i b_j g_{ij}$. If Λ is integer, $g_{ij} \in \mathbb{Z}$ for all i,j, and $\langle \mathbf{x}, \mathbf{y} \rangle \in \mathbb{Z}$.

In all the following, we will deal with *full-rank* lattices, i.e., $m = n$. In this case, $\mathbf{M}$ is a square matrix, and we have

$$\det\{\Lambda\} = [\det\{\mathbf{M}\}]^2.$$

Definition 3.8 For full-rank lattices, the volume of the fundamental parallelotope, is called *volume of the lattice*, and denoted by $\mathrm{vol}(\Lambda)$.

It can be easily shown that $\mathrm{vol}(\Lambda) = \sqrt{\det\{\Lambda\}}$.

3.3.2 Sublattices and Equivalent Lattices

Let Λ be a lattice of dimension n defined by its generator matrix $\mathbf{M}$.

Definition 3.9 Let $\mathbf{B}$ be an $n \times n$ integer matrix. A *sublattice* of Λ is given by

$$\Lambda' = \{\mathbf{x} = \mathbf{uBM} \mid \mathbf{u} \in \mathbb{Z}^n\}.$$

Because a lattice has a group structure, a sublattice Λ' is then a subgroup of Λ, and as such, we may consider the *quotient group* Λ/Λ'. For convenience, we recall how to define a quotient group.

Definition 3.10 Let $\mathcal{G}$ be a group (with operation +), and $\mathcal{H}$ be a subgroup of $\mathcal{G}$. Let $a \in \mathcal{G}$. The subset

$$a + \mathcal{H} = \{a + h,\ h \in \mathcal{H}\}\ (\text{resp. } \mathcal{H} + a = \{h + a,\ h \in \mathcal{H}\})$$

is called a left (resp. right) *coset* of $\mathcal{G}$ modulo $\mathcal{H}$.

If $\mathcal{G}$ is Abelian, then the distinction between left and right cosets modulo $\mathcal{H}$ is unnecessary. It can be shown (Lidl & Niederreiter, 1994, p. 6) that a group $\mathcal{G}$ can be partitioned into cosets modulo $\mathcal{H}$. For our purposes, we restrict the following definition to Abelian groups.

Definition 3.11 For a subgroup $\mathcal{H}$ of an Abelian group $\mathcal{G}$, the group formed by the cosets of $\mathcal{G}$ modulo $\mathcal{H}$ under the operation $(a + \mathcal{H}) + (b + \mathcal{H}) = (a + b)\mathcal{H}$ is called the *quotient group* of $\mathcal{G}$ modulo $\mathcal{H}$, and denoted by $\mathcal{G}/\mathcal{H}$.

We refer the reader to Lidl & Niederreiter (1994, p. 9) for more details and for the proof that the structure described in the definition is actually a group. Let us now return to the quotient of a lattice Λ modulo one of its sublattices Λ' (see Fig. 3.5).

Definition 3.12 The *index of the sublattice* $\Lambda' = \{\mathbf{x} = \mathbf{uBM} \mid \mathbf{u} \in \mathbb{Z}^n\}$ is the cardinality of the quotient group Λ/Λ' and is denoted by $|\Lambda/\Lambda'|$.

It can be easily shown that (Samuel, 1971):

$$|\Lambda/\Lambda'| = \frac{\mathrm{vol}(\Lambda')}{\mathrm{vol}(\Lambda)} = |\det\{\mathbf{B}\}|.$$

Example 3.1

Consider the lattice Λ and its sublattice Λ' given in Fig. 3.5, whose bases are $\{\mathbf{v}, \mathbf{w}\}$ resp. $\{\mathbf{x}, \mathbf{y}\}$. We have

$$\begin{pmatrix}\mathbf{x}\\ \mathbf{y}\end{pmatrix} = \mathbf{B}\begin{pmatrix}\mathbf{v}\\ \mathbf{w}\end{pmatrix} = \begin{pmatrix}2 & 0\\ 0 & 3\end{pmatrix}\begin{pmatrix}\mathbf{v}\\ \mathbf{w}\end{pmatrix}.$$

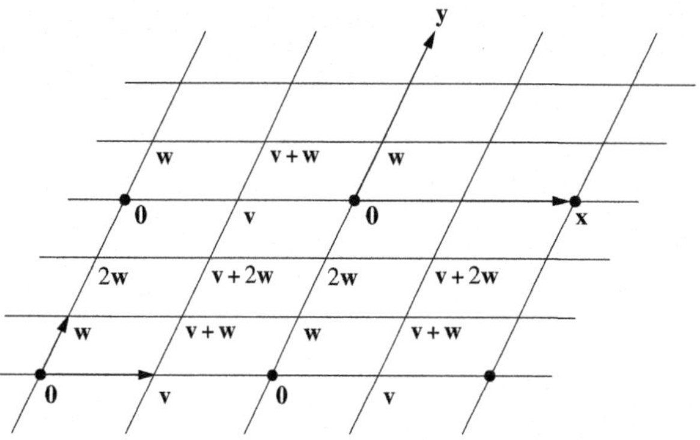

FIGURE 3.5

A way of visualizing the quotient group Λ/Λ': the grid represents a lattice Λ with basis $\{\mathbf{v},\mathbf{w}\}$, and the • represent a sublattice Λ' with basis $\{\mathbf{x},\mathbf{y}\}$. Points in Λ' are identified to zero in the quotient group Λ/Λ'.

The determinant of $\mathbf{B}$ is 6. It is the cardinality of the quotient group whose elements can be written as $\{(0,0),(0,1),(0,2),(1,0),(1,1),(1,2)\}$. The group operation is a component-wise addition modulo 2 and modulo 3, respectively.

■

It is always possible to find a sublattice of a given lattice considering its *scaled version* using an integer scaling factor.

Definition 3.13 Given a lattice Λ, a *scaled lattice* Λ' can be obtained by multiplying all the vectors of the lattice by a constant:

$$\Lambda' = c \cdot \Lambda,$$

where $c \in \mathbb{R}$. In particular, Λ' is a sublattice of Λ when $c \in \mathbb{Z}$.

More generally, we have the following definition.

Definition 3.14 If one lattice can be obtained from another by (possibly) a rotation, a reflection and a change of scale, we say that they are *equivalent*.

Consequently, two generator matrices $\mathbf{M}$ and $\mathbf{M}'$ define equivalent lattices if and only if they are related by $\mathbf{M}' = c\mathbf{UMB}$, where c is a nonzero constant, $\mathbf{U}$ is a matrix with integer entries and determinant ± 1 (*unimodular integer matrix*), and $\mathbf{B}$ is a real orthogonal matrix (with $\mathbf{BB}^T = \mathbf{I}_n$). The corresponding Gram matrices are related by $\mathbf{G}' = c^2\mathbf{UGU}^T$.

Thus one has to keep in mind that the equivalent lattice may be represented in several different ways. As a consequence, given a Gram (or generator) matrix, it is not easy to uniquely identify the corresponding lattice. Invariants such as the dimension and the determinant will help, but one has to be careful that having the same determinant is not a sufficient condition for two lattices to be equivalent. These considerations will be of importance later, when we will build algebraic lattice constellations where the particular orientation of the lattice within the Euclidean space becomes important.

3.4 ALGEBRAIC NUMBER THEORY

Algebraic number theory is roughly speaking the study of numbers. Typical questions that arise are related to the factorization of numbers, or to the solutions of algebraic equations. In this section, starting from the familiar sets $\mathbb{Z}$ and $\mathbb{Q}$, we define concepts such as

- a number field $\mathscr{K}$, its ring of integers $\mathscr{O}_{\mathscr{K}}$, and its integral basis;
- invariants of a number field: discriminant and signature;
- the embeddings of a number field into $\mathbb{C}$.

In Section 3.5, these concepts will be applied to the construction of *algebraic lattices*. In particular, we show how to build a lattice from a number field.

3.4.1 Algebraic Number Fields

Let $\mathbb{Z}$ be the set of rational integers $\{\ldots,-2,-1,0,1,2,\ldots\}$, and let $\mathbb{Q}$ be the set of rational numbers $\mathbb{Q} = \{\frac{a}{b} \mid a,b \in \mathbb{Z}\}$.[1] Starting from these two sets, the goal of this first subsection is to define algebraic structures so as to end up with the notion of *number field*.

Definition 3.15 Ring – Let $\mathscr{A}$ be a set endowed with two internal operations denoted by $+$ and $\cdot$

$$\begin{array}{ccc} \mathscr{A}\times\mathscr{A} & \to & \mathscr{A} \\ (a,b) & \mapsto & a+b \end{array} \quad \text{and} \quad \begin{array}{ccc} \mathscr{A}\times\mathscr{A} & \to & \mathscr{A} \\ (a,b) & \mapsto & a\cdot b \end{array}$$

The set $(\mathscr{A},+,\cdot)$ is a *ring* if

1. $(\mathscr{A},+)$ is an Abelian group (Definition 3.1);
2. the operation $\cdot$ is associative, i.e., $a\cdot(b\cdot c) = (a\cdot b)\cdot c$ for all $a,b,c \in \mathscr{A}$ and has a neutral element 1 such that $1\cdot a = a\cdot 1 = a$ for all $a \in \mathscr{A}$;
3. the operation $\cdot$ is distributive over $+$, i.e., $a\cdot(b+c) = a\cdot b + a\cdot c$ and $(a+b)\cdot c = a\cdot c + b\cdot c$ for all $a,b,c \in \mathscr{A}$.

The ring $\mathscr{A}$ is *commutative* if $a\cdot b = b\cdot a$ for all $a,b \in \mathscr{A}$. The set of elements of $\mathscr{A}$ that are invertible for the operation $\cdot$ is called the set of *units* of $\mathscr{A}$, and is denoted by $\mathscr{A}^*$.

The set $\mathbb{Z}$ is easily checked to be a ring. Its units are $\mathbb{Z}^* = \{1,-1\}$.

Definition 3.16 Field – Let $\mathscr{A}$ be a ring such that $\mathscr{A}^* = \mathscr{A}\backslash\{0\}$. Then $\mathscr{A}$ is said to be a *skew field*. If $\mathscr{A}$ is moreover commutative, it is said to be a *field*.

In other words, a field is a set where the four operations addition, subtraction, multiplication and division can be performed. The set $\mathbb{Q}$ is easily checked to be a field. Other examples of fields can be built starting from $\mathbb{Q}$. Take for example $\sqrt{2}$, which is not an element of $\mathbb{Q}$. One can build a new field by "adding" $\sqrt{2}$ to $\mathbb{Q}$. Note that in order to make this new set a field, we have to add all the multiples

[1] The term *rational integer* is used to describe the integers of the field $\mathbb{Q}$ in order to distinguish this concept from the more general concept of algebraic integers defined in the following section.

and all the powers of $\sqrt{2}$. We thus get a new field that contains both $\mathbb{Q}$ and $\sqrt{2}$, and that we denote by $\mathbb{Q}(\sqrt{2})$. We call it a field extension of $\mathbb{Q}$. Let us formalize this procedure.

Definition 3.17 Let $\mathcal{K}$ and $\mathcal{L}$ be two fields. If $\mathcal{K} \subseteq \mathcal{L}$, we say that $\mathcal{L}$ is a *field extension* of $\mathcal{K}$. We denote it by $\mathcal{L}/\mathcal{K}$.

It is useful to note that if $\mathcal{L}/\mathcal{K}$ is a field extension, then $\mathcal{L}$ has a natural structure as a vector space over $\mathcal{K}$, where vector addition is addition in $\mathcal{L}$ and scalar multiplication of $a \in \mathcal{K}$ by $v \in \mathcal{L}$ is just $av \in \mathcal{L}$. For example, an element $x \in \mathbb{Q}(\sqrt{2})$ can be written as $x = a + b\sqrt{2}$, where $\{1, \sqrt{2}\}$ are the basis "vectors" and $a, b \in \mathbb{Q}$ are the scalars. The dimension of $\mathbb{Q}(\sqrt{2})$ as a vector space over $\mathbb{Q}$ is 2.

Definition 3.18 Let $\mathcal{L}/\mathcal{K}$ be a field extension. The dimension of $\mathcal{L}$ as a vector space over $\mathcal{K}$ is called the *degree* of $\mathcal{L}$ over $\mathcal{K}$ and is denoted by $[\mathcal{L} : \mathcal{K}]$. If $[\mathcal{L} : \mathcal{K}]$ is finite, we say that $\mathcal{L}$ is a *finite extension* of $\mathcal{K}$.

A particular case of finite extension will be of great importance for us.

Definition 3.19 A finite extension of $\mathbb{Q}$ is called a *number field*.

Going on with our previous example, observe that a way to describe $\sqrt{2}$ is to say that this number is the solution of the equation $X^2 - 2 = 0$. Building $\mathbb{Q}(\sqrt{2})$, we thus add to $\mathbb{Q}$ the solution of a polynomial equation with integers coefficients. The number $\sqrt{2}$ is said to be algebraic.

Definition 3.20 Let $\mathcal{L}/\mathcal{K}$ be a field extension, and let $\alpha \in \mathcal{L}$. If there exists a non-zero irreducible monic (i.e., with highest coefficient 1) polynomial $p \in \mathcal{K}[X]$ (the ring of polynomials in the variable X with coefficients in $\mathcal{K}$) such that $p(\alpha) = 0$, we say that α is *algebraic* over $\mathcal{K}$. Such a polynomial is called the *minimal polynomial* of α over $\mathcal{K}$. We denote it by p_α.

In our example, the polynomial $X^2 - 2$ is the minimal polynomial of $\sqrt{2}$ over $\mathbb{Q}$.

Definition 3.21 Let $\mathcal{K}$ be a field extension of $\mathbb{Q}$. If all the elements of $\mathcal{K}$ are algebraic, we say that $\mathcal{K}$ is an *algebraic extension* of $\mathbb{Q}$.

Consider the field $\mathbb{Q}(\sqrt{2}) = \{a + b\sqrt{2},\ a, b \in \mathbb{Q}\}$. It is simple to see that any $x \in \mathbb{Q}(\sqrt{2})$ is a root of the polynomial $p_\alpha(X) = X^2 - 2aX + a^2 - 2b^2$ with rational coefficients. We conclude that $\mathbb{Q}(\sqrt{2})$ is an algebraic extension of $\mathbb{Q}$. Because it can be shown that a finite extension is an algebraic extension (see Stewart & Tall, 1979, p. 23), we also call equivalently a number field (Definition 3.19) an *algebraic number field*. Now that we have set up the framework, we will concentrate on the particular family of fields that are number fields, that is, field extensions $\mathcal{K}/\mathbb{Q}$, with $[\mathcal{K} : \mathbb{Q}]$ finite. Algebraic elements of $\mathcal{K}$ over $\mathbb{Q}$ are simply called algebraic numbers. In the following, $\mathcal{K}$ will denote a number field.

Theorem 3.1 (Stewart & Tall, 1979, p. 40) If $\mathcal{K}$ is a number field, then $\mathcal{K} = \mathbb{Q}(\theta)$ for some algebraic number $\theta \in \mathcal{K}$, called *primitive element*.

As a consequence, $\mathcal{K}$ is an n-dimensional $\mathbb{Q}$–vector space generated by the powers of θ, i.e., any element $\alpha \in \mathcal{K}$ can be written as

$$\alpha = \sum_{i=0}^{n-1} a_i \theta^i, \quad a_i \in \mathbb{Q}.$$

If $\mathcal{K}$ has degree n, then $\{1, \theta, \theta^2, \ldots, \theta^{n-1}\}$ is a *basis* of $\mathcal{K}$, and the degree of the minimal polynomial of θ is n. Computations involving the four field operations in $\mathcal{K} = \mathbb{Q}(\theta)$, are performed by using the polynomial representation of the field elements, as follows. Let $p_\theta(X) = \sum_{i=0}^{n} p_i X^i$, $p_i \in \mathbb{Q}$ for all i, $p_n = 1$, denote the minimal polynomial of θ. Because $p_\theta(\theta) = 0$, this yields an equation of degree n in θ:

$$\theta^n = -\sum_{i=0}^{n-1} p_i \theta^i.$$

Likewise, θ^{n+j} is given by

$$\theta^{n+j} = -\sum_{i=0}^{n-1} p_i \theta^{i+j},\ j \geq 1,$$

where each θ^{i+j} with $i+j \geq n$ can be reduced recursively so as to obtain an expression in the basis $\{1,\theta,\ldots,\theta^{n-1}\}$.

A similar way of looking at these computations is to represent an element $a = \sum_{i=0}^{n-1} a_i \theta^i \in \mathcal{K}$ as a polynomial $a(X) = \sum_{i=0}^{n-1} a_i X^i$. Operations between two elements $a, b \in \mathcal{K}$ are performed on the two corresponding polynomials $a(X)$ and $b(X)$, and the fact that $p_\theta(\theta) = 0$ translates into considering polynomial operations modulo $p_\theta(X)$.

One of the first goals of algebraic number theory was to study the solutions of polynomial equations with coefficients in $\mathbb{Z}$. Given the equation

$$\sum_{i=0}^{n-1} a_i X^i = 0, \quad a_i \in \mathbb{Z} \text{ for all } i,$$

what can we say about its solutions? It is first clear that there may be solutions not in $\mathbb{Q}$, as $\sqrt{2}$, which means that in order to find the solutions, we have to consider fields larger than $\mathbb{Q}$.

Definition 3.22 We say that $\alpha \in \mathcal{K}$ is an *algebraic integer* if it is a root of a monic polynomial with coefficients in $\mathbb{Z}$. The set of algebraic integers of $\mathcal{K}$ is a ring called the *ring of integers* of $\mathcal{K}$, denoted $\mathcal{O}_{\mathcal{K}}$.

The fact that the algebraic integers of $\mathcal{K}$ form a ring is a strong result (Stewart & Tall, 1979, p. 47), which is not so easy to see. The natural idea that comes to mind is to find the corresponding minimal polynomial. Take $\sqrt{2}$ and 2. Both are algebraic integers of $\mathbb{Q}(\sqrt{2})$. How easy is it to find the minimal polynomial of $\sqrt{2}+2$? How easy is it to find such a polynomial in general?

It can be shown (Stewart & Tall, 1979, p. 60) that the algebraic integers are the set $\mathbb{Z}[\sqrt{2}] \triangleq \{a + b\sqrt{2},\ a, b \in \mathbb{Z}\}$.[2] Care should be taken in generalizing this result. Note that $\mathbb{Z}[\sqrt{2}]$ is a ring because it is closed under all operations except for the inversion. For example, $(2+2\sqrt{2})^{-1} = (2-\sqrt{2})/6$ does not belong to $\mathbb{Z}[\sqrt{2}]$.

Theorem 3.2 **(Stewart & Tall, 1979, p. 49)** If $\mathcal{K}$ is a number field, then $\mathcal{K} = \mathbb{Q}(\theta)$ for an algebraic integer $\theta \in \mathcal{O}_{\mathcal{K}}$.

In other words, we can always find a primitive element, which is an algebraic integer. Consequently, the minimal polynomial $p_\theta(X)$ has coefficients in $\mathbb{Z}$.

[2]The notation $Z[\theta]$ is used to define a *module*, i.e., a vector space with integer scalars (see next section).

3.4.2 Integral Basis and Canonical Embeddings

In the following, we will first look at the structure of $\mathcal{O}_{\mathcal{K}}$, the ring of integers of a number field. We will also define two invariants of a number field: the *discriminant* and the *signature*. In the special case $\mathcal{K} = \mathbb{Q}(\sqrt{2})$, we have seen that $\mathcal{O}_{\mathcal{K}} = \mathbb{Z}[\sqrt{2}]$, which means that $\mathcal{O}_{\mathcal{K}}$ has a basis over $\mathbb{Z}$ given by $\{1, \sqrt{2}\}$. We call $\mathcal{O}_{\mathcal{K}}$ a $\mathbb{Z}$*–module*. The notion of $\mathcal{A}$–module, where $\mathcal{A}$ is a ring, is a generalization of a $\mathcal{K}$–vector space, where $\mathcal{K}$ is a field. In our case, we have that $\mathcal{K}$ has the structure of a vector space over the field $\mathbb{Q}$, while we only have the structure of a module for $\mathcal{O}_{\mathcal{K}}$ over the ring $\mathbb{Z}$. This is formalized as follows:

Theorem 3.3 (Stewart & Tall, 1979, p. 51) Let $\mathcal{K}$ be a number field of degree n. The ring of integers $\mathcal{O}_{\mathcal{K}}$ of $\mathcal{K}$ forms a free $\mathbb{Z}$–module of rank n (that is, there exists a basis of n elements over $\mathbb{Z}$).

Definition 3.23 Let $\{\omega_i\}_{i=1}^n$ be a basis of the $\mathbb{Z}$–module $\mathcal{O}_{\mathcal{K}}$, so that we can uniquely write any element of $\mathcal{O}_{\mathcal{K}}$ as $\sum_{i=1}^n a_i\omega_i$ with $a_i \in \mathbb{Z}$ for all i. We say that $\{\omega_i\}_{i=1}^n$ is an *integral basis* of $\mathcal{K}$.

We now give another example of a number field, where we summarize the different notions seen so far. Take $\mathcal{K} = \mathbb{Q}(\sqrt{5})$. We know that any algebraic integer β in $\mathcal{K}$ has the form $a + b\sqrt{5}$ with some $a, b \in \mathbb{Q}$, such that the polynomial $p_\beta(X) = X^2 - 2aX + a^2 - 5b^2$ has integer coefficients. By simple arguments, it can be shown that all the elements of $\mathcal{O}_{\mathcal{K}}$ take the form $\beta = (u + v\sqrt{5})/2$ with both u, v integers with the same parity. So we can write $\beta = h + k(1+\sqrt{5})/2$ with $h, k \in \mathbb{Z}$. This shows that $\{1, (1+\sqrt{5})/2\}$ is an integral basis. The basis $\{1, \sqrt{5}\}$ is not integral because $a + b\sqrt{5}$ with $a, b \in \mathbb{Z}$ is only a subset of $\mathcal{O}_{\mathcal{K}}$. Incidentally, $(1+\sqrt{5})/2$ is also a primitive element of $\mathcal{K}$ with minimal polynomial $X^2 - X - 1$. We will now see how a number field $\mathcal{K}$ can be represented, we say *embedded*, into $\mathbb{C}$.

Definition 3.24 Let $\mathcal{K}/\mathbb{Q}$ and $\mathcal{L}/\mathbb{Q}$ be two field extensions of $\mathbb{Q}$. We call $\varphi : \mathcal{K} \to \mathcal{L}$ a $\mathbb{Q}$*–homomorphism* if φ is a ring homomorphism that satisfies $\varphi(a) = a$ for all $a \in \mathbb{Q}$, i.e., that fixes $\mathbb{Q}$. If $\mathcal{A}$ and $\mathcal{B}$ are rings, a *ring homomorphism* is a map $\psi : \mathcal{A} \to \mathcal{B}$ that satisfies, for all $a, b \in \mathcal{A}$,

1. $\psi(a+b) = \psi(a) + \psi(b)$;
2. $\psi(a \cdot b) = \psi(a) \cdot \psi(b)$;
3. $\psi(1) = 1$.

Definition 3.25 A $\mathbb{Q}$–homomorphism $\varphi : \mathcal{K} \to \mathbb{C}$ is called an *embedding* of $\mathcal{K}$ into $\mathbb{C}$.

Note that the embedding is an injective map, so that we can really understand it as a way of representing elements of $\mathcal{K}$ as complex numbers.

Theorem 3.4 (Stewart & Tall, 1979, p. 41) Let $\mathcal{K} = \mathbb{Q}(\theta)$ be a number field of degree n over $\mathbb{Q}$. There are exactly n embeddings of $\mathcal{K}$ into $\mathbb{C}$: $\sigma_i : \mathcal{K} \to \mathbb{C}, i = 1, \ldots, n$, defined by $\sigma_i(\theta) = \theta_i$, where θ_i are the distinct zeros in $\mathbb{C}$ of the minimal polynomial of θ over $\mathbb{Q}$.

Notice that $\sigma_1(\theta) = \theta_1 = \theta$ and thus σ_1 is the identity mapping, $\sigma_1(\mathcal{K}) = \mathcal{K}$. When we apply the embedding σ_i to an arbitrary element x of $\mathcal{K}$, $x = \sum_{k=1}^n a_k\theta^k$ with $a_k \in \mathbb{Q}$, we get, using the properties of $\mathbb{Q}$-homomorphisms,

$$\begin{aligned}\sigma_i(x) &= \sigma_i\left(\sum_{k=1}^n a_k\theta^k\right)\\ &= \sum_{k=1}^n \sigma_i(a_k)\,(\sigma_i(\theta))^k = \sum_{k=1}^n a_k\theta_i^k \in \mathbb{C},\ a_k \in \mathbb{Q}\end{aligned}$$

and we see that the image of any x under σ_i is uniquely identified by θ_i. With the notion of embeddings, we define two quantities that will be very useful when considering algebraic lattices, namely the *norm* and the *trace* of an algebraic element.

Definition 3.26 Let $x \in \mathcal{K}$. The elements $\sigma_1(x), \sigma_2(x), \ldots \sigma_n(x)$ are called the *conjugates* of x and

$$N(x) = \prod_{i=1}^{n} \sigma_i(x), \quad \mathrm{Tr}\{x\} = \sum_{i=1}^{n} \sigma_i(x)$$

are called respectively the *norm* and the *trace* of x.

If the context is not clear, we write $N_{\mathcal{K}/\mathbb{Q}}$ $\mathrm{Tr}_{\mathcal{K}/\mathbb{Q}}$ resp. to avoid ambiguity.

Theorem 3.5 (Stewart & Tall, 1979, p. 54) For any $x \in \mathcal{K}$, we have $N(x) \in \mathbb{Q}$ and $\mathrm{Tr}\{x\} \in \mathbb{Q}$. If $x \in \mathcal{O}_{\mathcal{K}}$, we have $N(x) \in \mathbb{Z}$ and $\mathrm{Tr}\{x\} \in \mathbb{Z}$.

Let us come back to the example of $\mathbb{Q}(\sqrt{2})$, and illustrate these new definitions. The roots of the minimal polynomial $X^2 - 2$ are $\theta_1 = \sqrt{2}$ and $\theta_2 = -\sqrt{2}$. Thus

$$\sigma_1(\theta) = \sqrt{2} \quad \text{and} \quad \sigma_2(\theta) = -\sqrt{2}$$

and for $x \in \mathbb{Q}(\sqrt{2})$, $x = a + b\sqrt{2}$ with $a, b \in \mathbb{Q}$

$$\sigma_1(a + b\sqrt{2}) = a + b\sqrt{2} \quad \text{and} \quad \sigma_2(a + b\sqrt{2}) = a - b\sqrt{2}.$$

The norm of x is $N(x) = \sigma_1(x)\sigma_2(x) = a^2 - 2b^2$, while the trace of x is $\mathrm{Tr}\{x\} = \sigma_1(x) + \sigma_2(x) = 2a$.

These field embeddings enable to define a first *invariant* of a number field, that is, a property of the field that does not depend on the way it is represented.

Definition 3.27 Let $\{\omega_1, \omega_2, \ldots \omega_n\}$ be an integral basis of $\mathcal{K}$ and consider the matrix $\mathbf{M} = \{\sigma_j(\omega_i)\}$. The absolute *discriminant* of $\mathcal{K}$ is defined as $d_{\mathcal{K}} \triangleq (\det\{\mathbf{M}\})^2$.

It can be shown that the discriminant is independent of the choice of a basis (Samuel, 1971).

Theorem 3.6 (Stewart & Tall, 1979, p. 51) The absolute discriminant $d_{\mathcal{K}}$ of a number field belongs to $\mathbb{Z}$.

Let us compute the discriminant $d_{\mathcal{K}}$ of the field $\mathbb{Q}(\sqrt{5})$. Applying the two $\mathbb{Q}$-homomorphisms to the integral basis $\{\omega_1, \omega_2\} = \{1, (1+\sqrt{5})/2\}$, we obtain

$$d_{\mathcal{K}} = \det\begin{pmatrix} \sigma_1(1) & \sigma_2(1) \\ \sigma_1\left(\frac{1+\sqrt{5}}{2}\right) & \sigma_2\left(\frac{1+\sqrt{5}}{2}\right) \end{pmatrix}^2 = \det\begin{pmatrix} 1 & 1 \\ \frac{1+\sqrt{5}}{2} & \frac{1-\sqrt{5}}{2} \end{pmatrix}^2 = 5.$$

We have seen that the embeddings are uniquely identified by the n complex roots of the minimal polynomial defining the number field. We now define a second invariant of a number field related to the embeddings.

Definition 3.28 Let $\{\sigma_1, \sigma_2, \ldots \sigma_n\}$ be the n embeddings of $\mathcal{K}$ into $\mathbb{C}$. Let r_1 be the number of embeddings with image in $\mathbb{R}$, the field of real numbers, and $2r_2$ the number of embeddings with image in $\mathbb{C}$ so that

$$r_1 + 2r_2 = n.$$

The pair (r_1, r_2) is called the *signature* of $\mathcal{K}$. If $r_2 = 0$, we have a *totally real* algebraic number field. If $r_1 = 0$, we have a *totally complex* algebraic number field.

All the previous examples were totally real algebraic number fields with $r_1 = n$. Let us now consider $\mathcal{K} = \mathbb{Q}(\sqrt{-3})$. The minimal polynomial of $\sqrt{-3}$ is $X^2 + 3$ and has 2 complex roots so that the signature of $\mathcal{K}$ is (0, 1). Observe that $\{1, \sqrt{-3}\}$ is not an integral basis. If we take the third root of unity $\zeta_3 = e^{2\pi j/3} = (-1 + j\sqrt{3})/2$, where $j = \sqrt{-1}$, we have $\mathcal{K} = \mathbb{Q}(\zeta_3) = \mathbb{Q}(\sqrt{-3})$ and an integral basis is $\{1, \zeta_3\}$. The minimal polynomial of θ is $X^2 + X + 1$. The ring of integers of this field is also known as the *Eisenstein integer ring*.

We end this section with a key definition for the construction of algebraic lattices.

Definition 3.29 Let us order the σ_i's so that, for all $x \in \mathcal{K}$, $\sigma_i(x) \in \mathbb{R}$, $1 \le i \le r_1$, and $\sigma_{j+r_2}(x)$ is the complex conjugate of $\sigma_j(x)$ for $r_1 + 1 \le j \le r_1 + r_2$. We call *canonical embedding* $\sigma : \mathcal{K} \to \mathbb{R}^{r_1} \times \mathbb{C}^{r_2}$ the isomorphism defined by

$$\sigma(x) = \left(\sigma_1(x) \cdots \sigma_{r_1}(x)\ \sigma_{r_1+1}(x) \cdots \sigma_{r_1+r_2}(x)\right) \in \mathbb{R}^{r_1} \times \mathbb{C}^{r_2}.$$

If we identify $\mathbb{R}^{r_1} \times \mathbb{C}^{r_2}$ with $\mathbb{R}^n$, the canonical embedding can be rewritten as $\sigma : \mathcal{K} \to \mathbb{R}^n$

$$\sigma(x) = \left(\sigma_1(x) \cdots \sigma_{r_1}(x)\ \operatorname{Re}\{\sigma_{r_1+1}(x)\}\ \operatorname{Im}\{\sigma_{r_1+1}(x)\} \cdots \operatorname{Re}\{\sigma_{r_1+r_2}(x)\}\ \operatorname{Im}\{\sigma_{r_1+r_2}(x)\}\right) \in \mathbb{R}^n,$$

where Re is the real part and Im is the imaginary part.

The canonical embedding gives a geometrical representation of a number field, the one that will serve our purpose. In the following section, we will see how the canonical embedding applied to the ring of integers $\mathcal{O}_{\mathcal{K}}$ yields a so-called *algebraic lattice* $\Lambda = \sigma(\mathcal{O}_{\mathcal{K}})$.

3.5 ALGEBRAIC LATTICES

We are now ready to introduce algebraic lattices. The above definition of canonical embedding (Definition 3.29) establishes a one-to-one correspondence between the elements of an algebraic number field of degree n and the vectors of the n-dimensional Euclidean space. The final step for constructing an algebraic lattice is given by the following result.

Theorem 3.7 (Stewart & Tall, 1979, p. 155) Let $\{\omega_1, \omega_2, \ldots \omega_n\}$ be an integral basis of $\mathcal{K}$. The n vectors $\mathbf{v}_i = \sigma(\omega_i) \in \mathbb{R}^n$, $i = 1, \ldots, n$ are linearly independent, so they define a full rank algebraic lattice $\Lambda = \sigma(\mathcal{O}_{\mathcal{K}})$.

Recall (Definition 3.5) that the lattice $\Lambda = \sigma(\mathcal{O}_{\mathcal{K}})$ can be expressed by means of its generator matrix $\mathbf{M}$ as

$$\Lambda = \{\mathbf{x} = \mathbf{uM} \in \mathbb{R}^n \mid \mathbf{u} \in \mathbb{Z}^n\}$$

The lattice generator matrix $\mathbf{M}$ is given explicitly by

$$\begin{pmatrix} \sigma_1(\omega_1) & \cdots & \sigma_{r_1}(\omega_1) & \operatorname{Re}\{\sigma_{r_1+1}(\omega_1)\} & \operatorname{Im}\{\sigma_{r_1+1}(\omega_1)\} & \cdots & \operatorname{Re}\{\sigma_{r_1+r_2}(\omega_1)\} & \operatorname{Im}\{\sigma_{r_1+r_2}(\omega_1)\} \\ \sigma_1(\omega_2) & \cdots & \sigma_{r_1}(\omega_2) & \operatorname{Re}\{\sigma_{r_1+1}(\omega_2)\} & \operatorname{Im}\{\sigma_{r_1+1}(\omega_2)\} & \cdots & \operatorname{Re}\{\sigma_{r_1+r_2}(\omega_2)\} & \operatorname{Im}\{\sigma_{r_1+r_2}(\omega_2)\} \\ & & & & \vdots & & & \\ \sigma_1(\omega_n) & \cdots & \sigma_{r_1}(\omega_n) & \operatorname{Re}\{\sigma_{r_1+1}(\omega_n)\} & \operatorname{Im}\{\sigma_{r_1+1}(\omega_n)\} & \cdots & \operatorname{Re}\{\sigma_{r_1+r_2}(\omega_n)\} & \operatorname{Im}\{\sigma_{r_1+r_2}(\omega_n)\} \end{pmatrix}, \tag{3.7}$$

where the lattice generating vectors $\mathbf{v}_i, i = 1, \ldots n$ are the rows of $\mathbf{M}$.

Given the above lattice-generator matrix, it is easy to compute the determinant of the lattice.

Theorem 3.8 **(Samuel, 1971)** Let $d_{\mathcal{K}}$ be the discriminant of $\mathcal{K}$ and consider an algebraic lattice $\Lambda = \sigma(\mathcal{O}_{\mathcal{K}})$. The volume of the fundamental parallelotope of Λ is given by

$$\text{vol}(\Lambda) = |\det\{\mathbf{M}\}| = 2^{-r_2}\sqrt{|d_{\mathcal{K}}|} \tag{3.8}$$

Consequently,

$$\det\{\Lambda\} = 2^{-2r_2}|d_{\mathcal{K}}|.$$

Before going further, let us take some time to emphasize the correspondence between a lattice point $\mathbf{x} \in \Lambda \subset \mathbb{R}^n$ and an algebraic integer in $\mathcal{O}_{\mathcal{K}}$. A lattice point is of the form

$$\begin{aligned}\mathbf{x} &= (x_1, \ldots, x_{r_1}, x_{r_1+1}, \ldots, x_{r_1+2r_2}) \\ &= \left(\sum_{i=1}^{n}\lambda_i\sigma_1(\omega_i), \ldots, \sum_{i=1}^{n}\lambda_i\text{Re}\{\sigma_{r_1+1}(\omega_i)\}, \ldots, \sum_{i=1}^{n}\lambda_i\text{Im}\{\sigma_{r_1+r_2}(\omega_i)\}\right) \\ &= \left(\sigma_1\left(\sum_{i=1}^{n}\lambda_i\omega_i\right), \ldots, \text{Re}\left\{\sigma_{r_1+1}\left(\sum_{i=1}^{n}\lambda_i\omega_i\right)\right\}, \ldots, \text{Im}\left\{\sigma_{r_1+r_2}\left(\sum_{i=1}^{n}\lambda_i\omega_i\right)\right\}\right)\end{aligned}$$

for some $\lambda_i \in \mathbb{Z}$. Thus

$$\mathbf{x} = (\sigma_1(x), \ldots, \text{Re}\{\sigma_{r_1+1}(x)\}, \ldots, \text{Im}\{\sigma_{r_1+r_2}(x)\}) = \sigma(x) \tag{3.9}$$

for $x = \sum_{i=1}^{n}\lambda_i\omega_i$ an algebraic integer. This correspondence between a vector $\mathbf{x}$ in $\mathbb{R}^n$ and an algebraic integer x in $\mathcal{O}_{\mathcal{K}}$ makes it easy to compute the diversity of algebraic lattices.

Theorem 3.9 **(Boutros et al., 1996)** Algebraic lattices exhibit a diversity

$$L = r_1 + r_2.$$

Proof. Thanks to the fact that a lattice is geometrically uniform, it is sufficient to focus on the nonzero lattice points instead of all possible pairs of points. Let $\mathbf{x} \neq \mathbf{0}$ be an arbitrary point of Λ:

$$\mathbf{x} = \left(\sigma_1(x), \ldots, \sigma_{r_1}(x), \text{Re}\{\sigma_{r_1+1}(x)\}, \ldots, \text{Im}\{\sigma_{r_1+r_2}(x)\}\right)$$

with $x \in \mathcal{O}_{\mathcal{K}}$. Because $\mathbf{x} \neq \mathbf{0}$, we have $x \neq 0$, and the first r_1 coefficients are nonzero. The minimum number of nonzero coefficients among the remaining $2r_2$ coefficients is r_2 because the real and imaginary parts of any one of the complex embeddings may not both be zero. We, thus, have a diversity $L \geq r_1 + r_2$. Applying the canonical embedding to $x = 1$ gives exactly $r_1 + r_2$ nonzero coefficients ($\sigma_j(1) = 1$ for any j), which concludes the proof. ■

Corollary 3.1 Algebraic lattices built over totally real number fields (that is, with signature $(r_1, r_2) = (n, 0)$) have maximal diversity $L = n$.

Figure 3.6 shows an algebraic lattice from $\mathcal{K} = \mathbb{Q}(\sqrt{5})$. As seen before, the integral basis of $\mathcal{K}$ is $\left\{1, \frac{1+\sqrt{5}}{2}\right\}$. The two embeddings are $\sigma_1(\sqrt{5}) = \sqrt{5}$ and $\sigma_2(\sqrt{5}) = -\sqrt{5}$, and the lattice generator

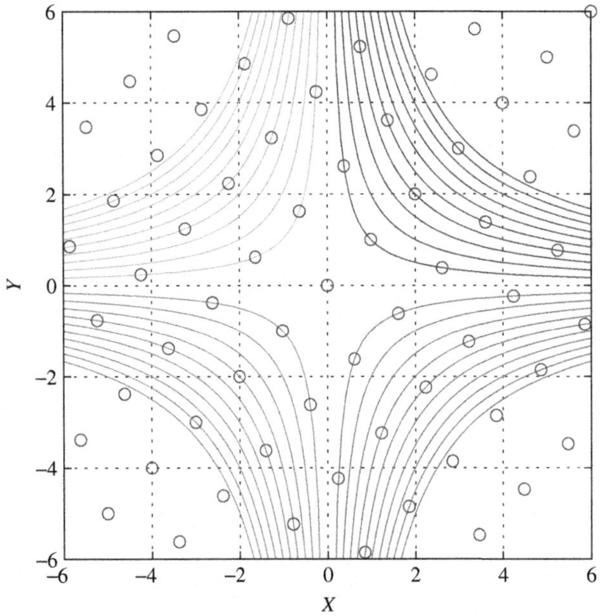

FIGURE 3.6

Algebraic lattice from $\mathbb{Q}(\sqrt{5})$.

matrix becomes

$$\mathbf{M} = \begin{pmatrix} \sigma_1(1) & \sigma_2(1) \\ \sigma_1\left(\frac{1+\sqrt{5}}{2}\right) & \sigma_2\left(\frac{1+\sqrt{5}}{2}\right) \end{pmatrix} = \begin{pmatrix} 1 & 1 \\ \frac{1+\sqrt{5}}{2} & \frac{1-\sqrt{5}}{2} \end{pmatrix}.$$

The fundamental volume is $\mathrm{vol}(\Lambda(\mathcal{O}_K)) = |\det\{\mathbf{M}\}| = \sqrt{5}$, furthermore $r_1 = 2, r_2 = 0$, and the diversity is $L = 2$. We note from Fig. 3.6 that all lattice points are on one of the hyperboles $XY = k$ for some integer $k \neq 0$ because we have that the corresponding algebraic integer has a norm equal to k.

So far, the key ingredient to build an algebraic lattice has been the existence of a $\mathbb{Z}$-basis in $\mathcal{K}$. Because it is known that $\mathcal{O}_{\mathcal{K}}$ has such a basis (more technically, that $\mathcal{O}_{\mathcal{K}}$ is a free $\mathbb{Z}$-module of rank n), we can embed it into $\mathbb{R}^n$ so as to obtain an algebraic lattice. However, there exist certain subsets of $\mathcal{O}_{\mathcal{K}}$ that also have this structure of free $\mathbb{Z}$-module of rank n. These are the *ideals* of $\mathcal{O}_{\mathcal{K}}$.

Definition 3.30 An *ideal* $\mathcal{I}$ of a commutative ring R is an additive subgroup of R which is stable under multiplication by R, i.e., $a\mathcal{I} \subseteq \mathcal{I}$ for all $a \in R$.

Among all the ideals of a ring, some of them have the special property of being generated by only one element. These will be of particular interest for us.

Definition 3.31 An ideal $\mathcal{I}$ is *principal* if it is of the form $\mathcal{I} = (x)R = \{xy,\ y \in R\}$, $x \in \mathcal{I}$ (also denoted as $\mathcal{I} = (x)$). The element x is called the generator of $\mathcal{I}$.

Example 3.2

If $R = \mathbb{Z}$, we have that $n\mathbb{Z}$ is a principal ideal of $\mathbb{Z}$ for all n.

We can define the *norm* of an ideal. If the ideal is principal, its norm is directly related to the norm of a generator of the ideal.

Definition 3.32 Let $\mathscr{I} = (x)\mathscr{O}_{\mathscr{K}}$ be a principal ideal of $\mathscr{O}_{\mathscr{K}}$. Its *norm* is defined by $N(\mathscr{I}) = |N(x)|$.

It can be shown that all ideals of $\mathscr{O}_{\mathscr{K}}$ have a $\mathbb{Z}$-basis of n elements.

Theorem 3.10 **(Stewart & Tall, 1979, p. 121)** Every ideal $\mathscr{I} \neq 0$ of $\mathscr{O}_{\mathscr{K}}$ has a $\mathbb{Z}$-basis $\{\nu_1, \ldots, \nu_n\}$ where n is the degree of $\mathscr{K}$.

Theorems 3.7 and 3.9 easily extend when replacing a basis of $\mathscr{O}_{\mathscr{K}}$ by a basis of $\mathscr{I} \subset \mathscr{O}_{\mathscr{K}}$. An algebraic lattice Λ' built from an ideal $\mathscr{I} \subset \mathscr{O}_{\mathscr{K}}$ gives a sublattice of the algebraic lattice Λ built from $\mathscr{O}_{\mathscr{K}}$.

Theorem 3.11 **(Samuel, 1971)** The volume of the fundamental parallelotope of Λ' is given by

$$\text{vol}(\Lambda') = |\det\{\mathbf{M}\}| = 2^{-r_2} N(\mathscr{I})\sqrt{|d_K|} \tag{3.10}$$

3.6 IDEAL LATTICES

In this section, we study a family of algebraic lattices endowed with a *trace form* called *ideal lattices*. Ideal lattices describe lattices with generator matrix of the type

$$\mathbf{M} = \{\sigma_i(\omega_j)\}_{i,j=1}^{n} \cdot \mathbf{A},$$

where $\mathbf{A}$ is a convenient diagonal matrix. We can think of the diagonal matrix $\mathbf{A}$ as a pre-fading, used to stretch an algebraic lattice into another, such as the $\mathbb{Z}^n$–lattice. We will restrict ourselves to totally real number fields in order to have maximum diversity. We will show how to derive an explicit formula for the minimum product distance. Furthermore, we will discuss the basic ideas for the construction of full-diversity rotated $\mathbb{Z}^n$–lattices from ideal lattices.

In the following, $\mathscr{K}$ denotes a totally real number field of degree n. Let $\{\sigma_i\}_{i=1}^{n}$ denote the n real embeddings of $\mathscr{K}$ into $\mathbb{C}$.

Definition 3.33 An *ideal lattice* is a lattice Λ defined by the pair $(\mathscr{I}, q_\alpha)$, where $\mathscr{I} \subseteq \mathscr{O}_{\mathscr{K}}$ is an ideal of $\mathscr{O}_{\mathscr{K}}$ and

$$q_\alpha : \mathscr{I} \times \mathscr{I} \to \mathbb{Z}$$
$$q_\alpha(x, y) = \text{Tr}\{\alpha xy\}, \ \forall x, y \in \mathscr{I}$$

where $\alpha \in \mathscr{K}$ is totally positive (i.e., $\sigma_i(\alpha) > 0$, $\forall i$). We denote such ideal lattice as $\Lambda = (\mathscr{I}, q_\alpha)$.

Let $\{\omega_1, \ldots, \omega_n\}$ be a $\mathbb{Z}$-basis of the above ideal $\mathscr{I} \subseteq \mathscr{O}_{\mathscr{K}}$. Using the above notations, we define a *twisted canonical embedding* $\sigma_\alpha : \mathscr{K} \to \mathbb{R}^n$ as

$$\sigma_\alpha(x) = (\sqrt{\alpha_1}\sigma_1(x) \ \cdots \ \sqrt{\alpha_n}\sigma_n(x)),$$

where $\alpha_i = \sigma_i(\alpha)$, $i = 1, \ldots, n$.

Using the twisted canonical embedding, the generator matrix $\mathbf{M}$ of the lattice $\Lambda = \sigma_\alpha(\mathscr{I})$ is given by

$$\mathbf{M} = \begin{pmatrix} \sqrt{\alpha_1}\sigma_1(\omega_1) & \sqrt{\alpha_2}\sigma_2(\omega_1) & \cdots & \sqrt{\alpha_n}\sigma_n(\omega_1) \\ \vdots & \vdots & \vdots & \vdots \\ \sqrt{\alpha_1}\sigma_1(\omega_n) & \sqrt{\alpha_2}\sigma_2(\omega_n) & \cdots & \sqrt{\alpha_n}\sigma_n(\omega_n) \end{pmatrix}$$

$$= (\sigma_j(\omega_i))_{i,j=1}^n \begin{pmatrix} \sqrt{\alpha_1} & & 0 \\ & \ddots & \\ 0 & & \sqrt{\alpha_n} \end{pmatrix}. \tag{3.11}$$

The corresponding Gram matrix $\mathbf{G}$ is given by $\mathbf{G} = \mathbf{M}\mathbf{M}^T = (g_{ij})_{i,j=1}^n$, where

$$\begin{aligned} g_{ij} &= \sum_{k=1}^n \sqrt{\alpha_k}\sigma_k(\omega_i)\sqrt{\alpha_k}\sigma_k(\omega_j) \\ &= \sum_{k=1}^n \alpha_k\sigma_k(\omega_i\omega_j) \\ &= \sum_{k=1}^n \sigma_k(\alpha_k)\sigma_k(\omega_i\omega_j) \\ &= \sum_{k=1}^n \sigma_k(\alpha_k\omega_i\omega_j) \\ &= \mathrm{Tr}\{\alpha\, \omega_i\omega_j\} \end{aligned}$$

Because the Gram matrix is a trace form, this shows that the generator matrix as given above indeed defines an ideal lattice. In the case of ideal lattices, the determinant of the lattice is related both to the discriminant $d_{\mathscr{K}}$ and to the norm of the ideal $\mathscr{I}$.

Theorem 3.12 **(Bayer-Fluckiger, 1999)** Let $\mathscr{I}$ be an ideal of $\mathscr{O}_{\mathscr{K}}$, and $\Lambda = (\mathscr{I}, q_\alpha)$ be an ideal lattice. Then

$$\det\{\Lambda\} = N(\alpha)\,[N(\mathscr{I})]^2\,|d_{\mathscr{K}}|.$$

The minimum product distance of an ideal lattice can be computed explicitly when the ideal is principal.

Theorem 3.13 If $\mathscr{I}$ is a principal ideal of $\mathscr{O}_{\mathscr{K}}$, then

$$\min_{x\in\mathscr{I}} N(x) = N(\mathscr{I}).$$

Proof. Since $\mathscr{I}$ is principal, $\mathscr{I} = (a)$ for $a \in \mathscr{I}$, and $N(\mathscr{I}) = |N(a)|$ (see Definition 3.32). Let $x \in \mathscr{I}$, so that $x = ay$ for some $y \in \mathscr{O}_{\mathscr{K}}$. Thus $|N(x)| = |N(a)||N(y)| \geq N(\mathscr{I})$ and equality holds if and only if $N(y) = \pm 1$. The minimum can be reached, e.g., by taking $y = 1$. ■

Exactly in the same way as for algebraic lattices (see Equation (3.9)), there is a correspondence between a point $\mathbf{x} \in \Lambda = (\mathscr{I}, q_\alpha) \subseteq \mathbb{R}^n$ and an algebraic integer:

$$\mathbf{x} = \boldsymbol{\sigma}_\alpha(x) = \left(\sum_{i=1}^{n} \lambda_i \sqrt{\alpha_1} \sigma_1(\omega_i) \cdots \sum_{i=1}^{n} \lambda_i \sqrt{\alpha_n} \sigma_n(\omega_i) \right), \quad \lambda_i \in \mathbb{Z}$$

for $x \in \mathscr{I} \subseteq \mathscr{O}_{\mathscr{K}}$.

Theorem 3.14 Let $\mathscr{I}$ be a principal ideal of $\mathscr{O}_{\mathscr{K}}$. The minimum product distance of an ideal lattice $\Lambda = (\mathscr{I}, q_\alpha)$ is

$$d_{p,\min}(\Lambda) = \sqrt{\frac{\det\{\Lambda\}}{|d_{\mathscr{K}}|}}.$$

Proof. Let $\mathbf{x}$ be a lattice point and $x \in \mathscr{I}$ be its corresponding algebraic integer, so that $\mathbf{x} = \boldsymbol{\sigma}_\alpha(x)$. We have:

$$\begin{aligned} d_{p,\min}(\Lambda) &= \min_{\mathbf{x} \in \Lambda} \prod_{j=1}^{n} |x_j| \\ &= \min_{x \in \mathscr{I}} \prod_{j=1}^{n} |\sqrt{\alpha_j} \sigma_j(x)| \\ &= \min_{x \in \mathscr{I}} \left\{ \prod_{j=1}^{n} \sqrt{\sigma_j(\alpha_j)} \cdot \prod_{j=1}^{n} |\sigma_j(x)| \right\} \\ &= \sqrt{N(\alpha)} \min_{x \in \mathscr{I}} |N(x)|. \end{aligned}$$

We conclude using Theorems 3.13 and 3.12 that

$$d_{p,\min}(\Lambda) = \sqrt{N(\alpha)} \min_{x \in \mathscr{I}} N(x) = \sqrt{N(\alpha) N(\mathscr{I})} = \sqrt{\frac{\det\{\Lambda\}}{|d_K|}}.$$

■

Less explicit results are available in the case of nonprincipal ideals (Oggier & Bayer-Fluckiger, 2003). The corresponding ideal lattices are conjectured to have a lower $d_{p,\min}$.

3.7 ALGEBRAIC ROTATIONS WITH HIGH PRODUCT DISTANCE

3.7.1 $\mathbb{Z}^n$ Ideal Lattices

We focus now on the construction of a particular lattice: the rotated $\mathbb{Z}^n$, $n \geq 2$. In terms of ideal lattices, this means that given n, we are looking for a number field $\mathscr{K}$ of degree n and an ideal $\mathscr{I} \subseteq \mathscr{O}_K$ such that $\Lambda' = (\mathscr{I}, q_\alpha)$ is equivalent to $\mathbb{Z}^n$, $n \geq 2$. The Gram matrix $\mathbf{G}$ of this lattice is the identity matrix, so that the lattice generator matrix $\mathbf{M}$ is an orthogonal matrix: $\mathbf{M}\mathbf{M}^T = \mathbf{I}_n$. From a geometrical point

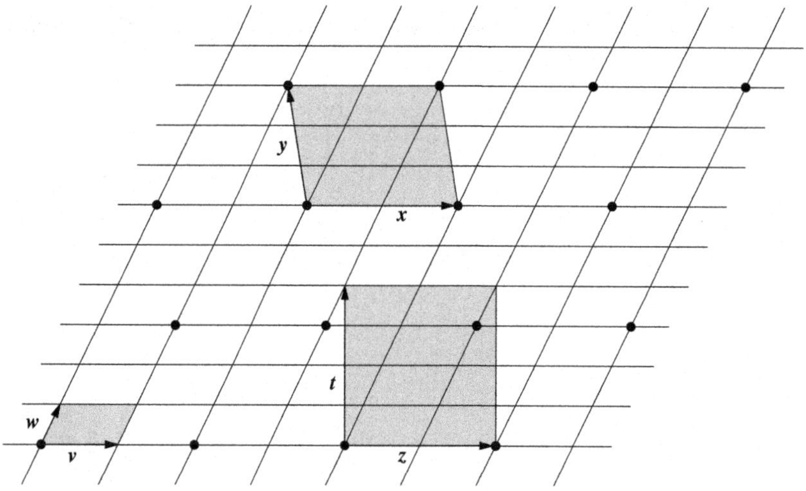

FIGURE 3.7

Looking for a square sublattice in a given lattice: the lattice Λ has basis $\{\mathbf{v},\mathbf{w}\}$, while its sublattices Λ' and Λ'' have bases $\{\mathbf{t},\mathbf{z}\}$, resp. $\{\mathbf{x},\mathbf{y}\}$. The lattice Λ' is a square lattice.

of view, a lattice $\Lambda' = (\mathscr{I}, q_\alpha)$ over $\mathscr{I} \subseteq \mathcal{O}_{\mathscr{K}}$ is a sublattice of $\Lambda = (\mathcal{O}_{\mathscr{K}}, q_\alpha)$. The idea is that in a given lattice Λ, we may find a sublattice which is $\mathbb{Z}^n$, as shown in Fig. 3.7 for $n = 2$.

The lattice determinant will be a useful criterion to help us find the $\mathbb{Z}^n$–lattice. Recall that the determinant of $\mathbb{Z}^n$ is 1, $n \geq 2$. A scaled version of $\mathbb{Z}^n$ is of the form $(\sqrt{c}\mathbb{Z})^n$ for some integer c, so that its determinant is $\det\{\mathbf{G}\} = [\det\{\mathbf{M}\}]^2 = c^n$. Using Theorem 3.12, we deduce the following necessary (but not sufficient!) condition for finding the scaled $\mathbb{Z}^n$ sublattice in Λ:

$$N(\alpha)[N(\mathscr{I})]^2|d_{\mathscr{K}}| = c^n, \tag{3.12}$$

where c is an integer. If we assume $\mathscr{I} = \mathcal{O}_{\mathscr{K}}$, this simplifies to

$$N(\alpha)|d_{\mathscr{K}}| = c^n. \tag{3.13}$$

If we take $\alpha = 1$, it simplifies to

$$[N(\mathscr{I})]^2|d_{\mathscr{K}}| = c^n. \tag{3.14}$$

This necessary condition will help us to choose α to build $\mathbb{Z}^n$–lattice codes as we will show in the following.

3.7.2 Rotated $\mathbb{Z}^n$–Lattice Codes

The question of finding algebraic lattices over totally real number fields with maximal minimum product distance has been extensively studied in the last decade.

The first examples using totally real algebraic number fields were given in Boullé and Belfiore (1992). Initially, no restriction on the shape of the lattice constellation was imposed, which resulted in either a complex bit labeling procedure or a loss in the average energy, as explained in Section 3.2.2. Further investigations were addressed to finding rotated $\mathbb{Z}^n$–lattices to avoid the above problems (Boutros & Viterbo, 1998; Giraud, Boutillon, & Belfiore, 1997). In Bayer-Fluckiger, Oggier, & Viterbo (2004), several families of full-diversity rotated $\mathbb{Z}^n$–lattices from totally real algebraic number fields were given and analyzed for all dimensions. We first show a simple example, then we explain how to construct one of these families.

3.7.3 A Simple Two-Dimensional Rotation Based on the Golden Number

Suppose we want to build the 2-dimensional lattice $\mathbb{Z}^2$ with full diversity. We take the field $\mathcal{K} = \mathbb{Q}(\sqrt{5})$, whose discriminant is $d_{\mathcal{K}} = 5$. We know that $\mathcal{K}$ is totally real, because its two embeddings are

$$\sigma_1(a+b\sqrt{5}) = a+b\sqrt{5} \quad \text{and} \quad \sigma_2(a+b\sqrt{5}) = a-b\sqrt{5}, \quad a,b \in \mathbb{Q}.$$

We have seen in (3.13) that a necessary condition for obtaining $\mathbb{Z}^2$ is to have an element α such that

$$N(\alpha)|d_{\mathcal{K}}| = N(\alpha)\cdot 5 = c^2, \quad c \in \mathbb{Z}.$$

It is natural to look for an element $\alpha \in \mathcal{K}$ whose norm is 5. It is a straightforward computation to check that the totally positive element

$$\alpha = 2 + \frac{1+\sqrt{5}}{2} \tag{3.15}$$

has the right norm:

$$N(\alpha) = \sigma_1(\alpha)\sigma_2(\alpha) = \left(2+\frac{1+\sqrt{5}}{2}\right)\left(2+\frac{1-\sqrt{5}}{2}\right) = 5.$$

A good choice for trying to build $\mathbb{Z}^2$ as an ideal lattice, thus, consists in taking $\mathcal{I} = \mathcal{O}_{\mathcal{K}} = \mathbb{Z}[\frac{1+\sqrt{5}}{2}]$ with α given by (3.15). The lattice generator matrix $\mathcal{M}$ is given by

$$\mathbf{M} = \begin{pmatrix} \sqrt{\sigma_1(\alpha)} & \sqrt{\sigma_2(\alpha)} \\ \sqrt{\sigma_1(\alpha)}\sigma_1\left(\frac{1+\sqrt{5}}{2}\right) & \sqrt{\sigma_2(\alpha)}\sigma_2\left(\frac{1+\sqrt{5}}{2}\right) \end{pmatrix}.$$

Let us now compute the Gram matrix $\mathbf{G} = \mathbf{M}\mathbf{M}^T$:

$$\mathbf{G} = \begin{pmatrix} \sigma_1(\alpha)+\sigma_2(\alpha) & \sigma_1\left(\alpha\frac{1+\sqrt{5}}{2}\right)+\sigma_2\left(\alpha\frac{1+\sqrt{5}}{2}\right) \\ \sigma_1\left(\alpha\frac{1+\sqrt{5}}{2}\right)+\sigma_2\left(\alpha\frac{1+\sqrt{5}}{2}\right) & \sigma_1\left(\alpha\left(\frac{1+\sqrt{5}}{2}\right)^2\right)+\sigma_2\left(\alpha\left(\frac{1+\sqrt{5}}{2}\right)^2\right) \end{pmatrix} = \begin{pmatrix} 5 & 0 \\ 0 & 5 \end{pmatrix}.$$

This shows that we get a scaled version of $\mathbb{Z}^2$. After normalization, we have that $\mathbb{Z}^2$ can be built over $\mathcal{O}_{\mathcal{K}} = \mathbb{Z}[\frac{1+\sqrt{5}}{2}]$, with generator matrix $\frac{1}{\sqrt{5}}\mathbf{M}$. By Theorem 3.14, the minimum product distance of this lattice code is

$$d_{p,\min} = \frac{1}{\sqrt{d_{\mathscr{K}}}} = \frac{1}{\sqrt{5}}.$$

3.7.4 The Cyclotomic Construction

We give a general construction that allows us to obtain $\mathbb{Z}^n$ for $n = (p-1)/2$, $p \geq 5$ a prime. The example of the previous section will appear to be a particular case. We start by introducing the so-called *cyclotomic fields* (Stewart & Tall, 1979; Washington, 1997).

Definition 3.34 A *cyclotomic field* is a number field $\mathbb{Q}(\zeta_m)$ generated by an m-th root of unity, $\zeta_m = e^{j2\pi/m}$.

We are interested in the particular case where $m = p \geq 5$ is a prime. The degree of $\mathbb{Q}(\zeta_p)$ over $\mathbb{Q}$ is $p-1$.

Definition 3.35 Let $\mathscr{K} = \mathbb{Q}(\zeta_p + \zeta_p^{-1})$ be a subfield of $\mathbb{Q}(\zeta_p)$ generated by $\zeta_p + \zeta_p^{-1} = 2\cos(2\pi/p)$, where ζ_p is a p-th root of unity. Since $[\mathbb{Q}(\zeta_p) : \mathscr{K}] = 2$ and $\mathscr{K}$ is totally real, then $\mathscr{K}$ is called the *maximal real subfield* of a cyclotomic field.

The degree of $\mathbb{Q}(\zeta_p + \zeta_p^{-1})$ over $\mathbb{Q}$ is $(p-1)/2$ (see Fig. 3.8), and its discriminant is

$$d_{\mathscr{K}} = p^{\frac{p-3}{2}}, \tag{3.16}$$

as it can be computed from Swinnerton-Dyer (2001, p. 46, Th. 21).

Theorem 3.15 **(Washington, 1997, p. 16)** Let $\mathscr{K} = \mathbb{Q}(\zeta_p + \zeta_p^{-1})$. Its ring of integers is $\mathcal{O}_{\mathscr{K}} = \mathbb{Z}[\zeta_p + \zeta_p^{-1}]$.

An integral basis (see Definition 3.23) of $\mathscr{K} = \mathbb{Q}(\zeta_p + \zeta_p^{-1})$ is given by $\left\{e_j = \zeta_p^j + \zeta_p^{-j}\right\}_{j=1}^n$. There are n embeddings of $\mathscr{K}$ into $\mathbb{C}$ given by

$$\sigma_k(e_j) = \zeta_p^{kj} + \zeta_p^{-kj}. \tag{3.17}$$

$\mathbb{Q}(\zeta_p)$

$|2$

$\mathbb{Q}\left(\zeta_p + \zeta_p^{-1}\right)$

$|\frac{p-1}{2}$

$\mathbb{Q}$

FIGURE 3.8

Cyclotomic field and its maximal real subfield. The labels on the branches indicate the degrees of the extensions.

We recall from (3.13) that a necessary condition for obtaining the $\mathbb{Z}^n$ ideal lattice is to find an element α such that

$$N(\alpha)p^{\frac{p-3}{2}} = c^n. \tag{3.18}$$

Taking the element $\alpha = (1-\zeta_p)(1-\zeta_p^{-1})$, which has $N(\alpha) = p$, we get

$$N(\alpha)p^{\frac{p-3}{2}} = p^{\frac{p-1}{2}}.$$

The following theorem shows that using this element, we are actually able to build the $\mathbb{Z}^n$–lattice.

Theorem 3.16 Let $\mathscr{K} = \mathbb{Q}(\zeta_p + \zeta_p^{-1})$ and $\alpha = (1-\zeta_p)(1-\zeta_p^{-1})$. Then

$$\Lambda = \left(\mathscr{O}_{\mathscr{K}}, \frac{1}{p}q_\alpha\right)$$

is equivalent to $\mathbb{Z}^n$, where we recall that $q_\alpha(x,y) = \mathrm{Tr}\{\alpha x y\}$.

Proof. The proof is a direct computation. To simplify notation, we write $\zeta = \zeta_p$.

We first compute $q_\alpha(e_i, e_j) = \mathrm{Tr}\{\alpha e_i e_j\}$ where $\{e_j\}_{j=1}^n$ is the usual integral basis of $\mathbb{Q}(\zeta + \zeta^{-1})$. From the matrix that we obtain, we find a new basis $\{e_j'\}_{j=1}^n$, where $\frac{1}{p}\mathrm{Tr}\{\alpha e_i' e_j'\}$ is exactly the identity matrix. Let

$$\alpha = (1-\zeta)(1-\zeta^{-1}) = 2 - (\zeta + \zeta^{-1}) \tag{3.19}$$

and denote by $\sigma_j(\zeta)$ and $\alpha_j = \sigma_j(\alpha)$ for $j = 1, \ldots, n$ the conjugates of ζ and α, respectively (see (3.17)). Note that

$$\mathrm{Tr}\{\zeta^k + \zeta^{-k}\} = \sum_{j=1}^{n} \sigma_j(\zeta^k + \zeta^{-k}) = -1,\ \forall\, k = 1, \ldots, n. \tag{3.20}$$

Using (3.19), we have

$$\begin{aligned}
\sum_{j=1}^{n} \alpha_j \sigma_j(\zeta^k + \zeta^{-k}) &= \sum_{j=1}^{n} \sigma_j(2 - (\zeta + \zeta^{-1}))\sigma_j(\zeta^k + \zeta^{-k}) \\
&= \sum_{j=1}^{n} 2\sigma_j(\zeta^k + \zeta^{-k}) - \sum_{j=1}^{n} \sigma_j(\zeta + \zeta^{-1})\sigma_j(\zeta^k + \zeta^{-k}) \\
&= -2 - \sum_{j=1}^{n} \sigma_j(\zeta^{k+1} + \zeta^{-k-1} + \zeta^{-k+1} + \zeta^{k-1}) \\
&= \begin{cases} -p & \text{if } k \equiv \pm 1 \pmod{p} \\ 0 & \text{otherwise.} \end{cases}
\end{aligned} \tag{3.21}$$

We now compute $q_\alpha(e_i, e_j)$ for $i = j$ using (3.19) and (3.20)

$$
\begin{aligned}
q_\alpha(e_i, e_i) &= \mathrm{Tr}\{\alpha e_i e_i\} = \mathrm{Tr}\{\alpha(\zeta^i + \zeta^{-i})^2\} \\
&= \sum_{j=1}^{n} \alpha_j \sigma_j(\zeta^{2i} + \zeta^{-2i} + 2) \\
&= \sum_{j=1}^{n} \alpha_j \sigma_j(\zeta^{2i} + \zeta^{-2i}) + 2\sum_{j=1}^{n}(2 - \sigma_j(\zeta + \zeta^{-1})) \\
&= \begin{cases} p & \text{if } i = n, \text{ i.e., } 2i \equiv -1 \pmod p \\ 2p & \text{otherwise.} \end{cases}
\end{aligned}
$$

Similarly for $i \neq j$,

$$
\begin{aligned}
q_\alpha(e_i, e_j) &= \sum_{k=1}^{n} \alpha_k \sigma_k(\zeta^{i+j} + \zeta^{-(i+j)}) + \sum_{k=1}^{n} \alpha_k \sigma_k(\zeta^{i-j} + \zeta^{-(i-j)}) \\
&= \begin{cases} -p & \text{if } |i-j| = 1 \\ 0 & \text{otherwise.} \end{cases}
\end{aligned}
$$

Thus, the matrix representing q_α in the basis $\{e_1, \ldots, e_n\}$ is given by

$$
\begin{pmatrix}
2p & -p & 0 & & \cdots & 0 \\
-p & 2p & -p & & & \\
0 & -p & 2p & & & \\
 & & & \ddots & -p & 0 \\
 & & & -p & 2 & -p \\
0 & \cdots & & 0 & -p & p
\end{pmatrix}.
$$

In the new basis $\{e'_1, \ldots, e'_n\}$, where $e'_n = e_n$ and $e'_j = e_j + e'_{j+1}, j = n-1, n-2, \ldots, 1$, the above matrix is the Gram matrix of the lattice $\mathbb{Z}^n$, i.e., p times the identity matrix. ■

The corresponding rotated $\mathbb{Z}^n$–lattice is obtained as follows. Recall from (3.17) that the n field embeddings of $\mathcal{K}$ are

$$
\sigma_k(e_j) = \zeta^{kj} + \zeta^{-kj} = 2\cos\left(\frac{2\pi kj}{p}\right).
$$

Then, the lattice generated by the ring of integers has the $n \times n$ generator matrix $\mathbf{M}$ with elements $[\mathbf{M}]_{k,j} = 2\cos\left(\frac{2\pi kj}{p}\right)$. The element α can be represented by the diagonal matrix

$$
\mathbf{A} = \mathrm{diag}\left\{\left\{\sqrt{\sigma_k(\alpha)}\right\}_{k=1}^{n}\right\}.
$$

The basis transformation matrix from $\{e_j\}$ to $\{e'_j\}$ is given by

$$\mathbf{T} = \begin{pmatrix} 1 & 1 & \cdots & 1 & 1 \\ 0 & 1 & 1 & \cdots & 1 \\ \vdots & & \ddots & & \vdots \\ 0 & \cdots & 0 & 1 & 1 \\ 0 & 0 & \cdots & 0 & 1 \end{pmatrix}.$$

Finally, the rotated $\mathbb{Z}^n$–lattice generator matrix is given by

$$\mathbf{R} = \frac{1}{\sqrt{p}}\mathbf{TMA}.$$

Following the above procedure, we obtain rotated $\mathbb{Z}^n$–lattices for $n = p = 2, 3, 5, 11, 23, 29, \ldots$. The missing prime dimensions are the ones where the rotated $\mathbb{Z}^n$–lattice is not found because the condition (3.12) is necessary but not sufficient.

Theorem 3.17 The minimum product distance of the ideal lattice Λ of dimension n as defined in Theorem 3.16 is

$$d_{p,\min}(\Lambda) = p^{-\frac{n-1}{2}}.$$

Proof. By Theorem 3.14, the minimum product distance is given by $d_{p,\min} = 1/\sqrt{|d_{\mathcal{K}}|}$ because we have normalized $\det\{\Lambda\} = 1$. We conclude recalling from (3.16) that the discriminant of $\mathcal{K}$ is $d_{\mathcal{K}} = p^{\frac{p-3}{2}} = p^{n-1}$. ■

It is also possible to construct rotated $\mathbb{Z}^n$–lattices for composite dimensions $n = p_1 p_2$ (where p_1 and p_2 are two primes) following the techniques presented in Bayer-Fluckiger, Oggier, & Viterbo (2004). Numerical values of $d_{p,\min}$ are given in Table 3.1.

Table 3.1 Minimum product distances for the cyclotomic construction.

n	$d_{p,\min}$	$\sqrt[n]{d_{p,\min}}$	n	$d_{p,\min}$	$\sqrt[n]{d_{p,\min}}$
2	$1/\sqrt{5}$	0.668740	15	$1/31^7$	0.201386
3	$1/7$	0.522757	18	$1/\sqrt{37^{17}}$	0.181744
5	$1/11^2$	0.383215	20	$1/\sqrt{41^{19}}$	0.171367
6	$1/\sqrt{13^5}$	0.343444	21	$1/43^{10}$	0.166785
8	$1/\sqrt{17^7}$	0.289520	23	$1/47^{11}$	0.158599
9	$1/19^4$	0.270187	26	$1/\sqrt{53^{25}}$	0.148259
11	$1/23^5$	0.240454	29	$1/59^{14}$	0.139670
14	$1/\sqrt{29^{13}}$	0.209425	30	$1/\sqrt{61^{29}}$	0.137116

3.8 SPHERE DECODING

The sphere decoder is an ML decoder for arbitrary lattice constellations. It solves the closest lattice point problem, i.e., it finds the closest lattice point to a given received point. At the basis of the sphere decoder is the Fincke–Pohst algorithm, which enumerates all lattice points within a sphere centered at the origin (Fincke & Pohst, 1985). With minor adaptations, it is possible to obtain an efficient lattice decoder. The key idea which makes the sphere decoder efficient is that the number of lattice points that are found inside a sphere is significantly smaller than the number of points within a hypercube containing the hypersphere, as the dimension of the space grows.

To avoid the exhaustive enumeration of all points of the constellation, the lattice decoding algorithm searches through the points of the lattice Λ which are found inside a sphere of given radius $\sqrt{C}$ centered at the received point as shown in Fig. 3.9. This guarantees that only the lattice points within the squared distance C from the received point are considered in the metric minimization.

The key steps of this algorithm are as follows:

1. Set the origin at the received point $\mathbf{r}$.
2. Consider the lattice $\Lambda = \{\mathbf{x} = \mathbf{uM} \mid \mathbf{u} \in \mathbb{Z}^n\}$.
3. Define the function $Q(\mathbf{u}) = \|\mathbf{x}\|^2 = \mathbf{xx}^T = \mathbf{uGu}^T$, where $\mathbf{G} = \mathbf{MM}^T$ is the Gram matrix.
4. Find all points in the sphere of square radius C by solving the inequality $Q(\mathbf{u}) \leq C$.
5. Choose $\mathbf{x}$ minimizing $\|\mathbf{r} - \mathbf{x}\|^2$.

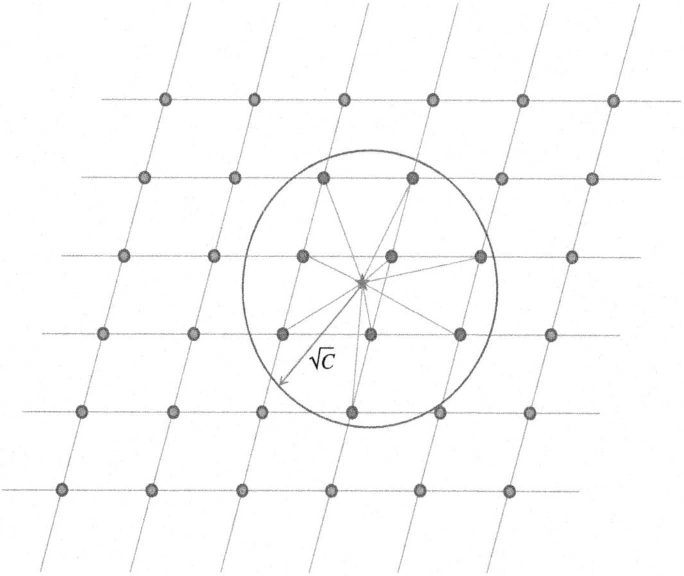

FIGURE 3.9

Sphere of radius $\sqrt{C}$ centered at the received point.

In order to perform ML decoding on high-diversity lattice constellations with fading, some further modifications are required.

3.8.1 The Sphere Decoder Algorithm

The closest lattice point algorithm was first presented in Pohst (1981) and further analyzed in Fincke & Pohst (1985). In Viterbo and Biglieri (1993) and Viterbo and Boutros (1999), the explicit geometric interpretation in terms of sphere decoder was shown.

In the following, it will be useful to think of the lattice Λ as the result of a linear transformation, defined by the matrix $\mathbf{M} : \mathbb{R}^n \to \mathbb{R}^n$, when applied to the $\mathbb{Z}^n$–lattice. So Λ can be seen as a skewed version of the $\mathbb{Z}^n$–lattice.

The problem to solve is the following:

$$\min_{\mathbf{x}\in\Lambda} \|\mathbf{r}-\mathbf{x}\|^2 = \min_{\mathbf{w}\in\mathbf{r}-\Lambda} \|\mathbf{w}\|^2 \tag{3.22}$$

that is, we search for the shortest vector $\mathbf{w}$ in the translated lattice $\mathbf{r}-\Lambda$ in the n-dimensional Euclidean space $\mathbb{R}^n$. We write $\mathbf{x} = \mathbf{u}\mathbf{M}$ with $\mathbf{u} \in \mathbb{Z}^n$, $\mathbf{r} = \boldsymbol{\rho}\mathbf{M}$ with $\boldsymbol{\rho} = (\rho_1 \cdots \rho_n) \in \mathbb{R}^n$, and $\mathbf{w} = \boldsymbol{\xi}\mathbf{M}$ with $\boldsymbol{\xi} = (\xi_1 \cdots \xi_n) \in \mathbb{R}^n$. Note that we have $\mathbf{w} = \sum_{i=1}^n \xi_i \mathbf{v}_i$, where the $\mathbf{v}_i$ are the lattice basis vectors, and the $\xi_i = \rho_i - u_i$, $i = 1,\ldots,n$ define the translated coordinate axes in the space of the integer component vectors $\mathbf{u}$ of the $\mathbb{Z}^n$–lattice.

The sphere of square radius C, centered at the received point, is transformed into an ellipsoid centered at the origin of the new coordinate system defined by $\boldsymbol{\xi}$:

$$\|\mathbf{w}\|^2 = Q(\boldsymbol{\xi}) = \boldsymbol{\xi}\mathbf{M}\mathbf{M}^T\boldsymbol{\xi}^T = \boldsymbol{\xi}\mathbf{G}\boldsymbol{\xi}^T = \sum_{i=1}^n \sum_{j=1}^n g_{ij}\xi_i\xi_j \le C. \tag{3.23}$$

Cholesky's factorization of the Gram matrix $\mathbf{G} = \mathbf{M}\mathbf{M}^T$ yields $\mathbf{G} = \mathbf{R}^T\mathbf{R}$, where $\mathbf{R}$ is an upper triangular matrix. Then,

$$Q(\boldsymbol{\xi}) = \boldsymbol{\xi}\mathbf{R}^T\mathbf{R}\boldsymbol{\xi}^T = \|\mathbf{R}\boldsymbol{\xi}^T\|^2 = \sum_{i=1}^n \left(r_{ii}\xi_i + \sum_{j=i+1}^n r_{ij}\xi_j \right)^2 \le C. \tag{3.24}$$

Substituting $q_{ii} = r_{ii}^2$ for $i = 1,\ldots,n$ and $q_{ij} = r_{ij}/r_{ii}$ for $i = 1,\ldots n, j = i+1,\ldots,n$, we can write

$$Q(\boldsymbol{\xi}) = \sum_{i=1}^n q_{ii} \left(\xi_i + \sum_{j=i+1}^n q_{ij}\xi_j \right)^2 = \sum_{i=1}^n q_{ii}U_i^2 \le C, \tag{3.25}$$

where the new coordinate system defined by the variables

$$U_i \triangleq \xi_i + \sum_{j=i+1}^n q_{ij}\xi_j, \; i = 1,\ldots n \tag{3.26}$$

defines an ellipsoid in its canonical form. Starting from U_n and working backwards, we find the equations of the border of the ellipsoid as

$$
\begin{aligned}
-\sqrt{\frac{C}{q_{nn}}} \leq\ & U_n && \leq \sqrt{\frac{C}{q_{nn}}} \\
-\sqrt{\frac{C-q_{nn}U_n}{q_{n-1,n-1}}} \leq\ & U_{n-1} && \leq \sqrt{\frac{C-q_{nn}U_n}{q_{n-1,n-1}}} \\
& \vdots &&
\end{aligned}
\tag{3.27}
$$

The corresponding ranges for the integer components u_n and u_{n-1} are found by replacing $\xi_i = \rho_i - u_i$ in (3.26) and (3.27):

$$
\begin{aligned}
\left\lceil -\sqrt{\frac{C}{q_{nn}}} + \rho_n \right\rceil \leq\ & u_n \leq \left\lfloor \sqrt{\frac{C}{q_{nn}}} + \rho_n \right\rfloor \\
\left\lceil -\sqrt{\frac{C-q_{nn}\xi_n^2}{q_{n-1,n-1}}} + \rho_{n-1} + q_{n-1,n}\xi_n \right\rceil \leq\ & u_{n-1} \leq \left\lfloor \sqrt{\frac{C-q_{nn}\xi_n^2}{q_{n-1,n-1}}} + \rho_{n-1} + q_{n-1,n}\xi_n \right\rfloor \\
& \vdots
\end{aligned}
$$

where $\lceil x \rceil$ is the smallest integer not smaller than x, and $\lfloor x \rfloor$ is the greatest integer not greater than x. For the ith integer component, we have

$$
\begin{aligned}
&\left\lceil -\sqrt{\frac{1}{q_{ii}}\left(C - \sum_{l=i+1}^{n} q_{ll}\left(\xi_l + \sum_{j=l+1}^{n} q_{lj}\xi_j\right)^2\right)} + \rho_i + \sum_{j=i+1}^{n} q_{ij}\xi_j \right\rceil \leq u_i \\
&\leq \left\lfloor \sqrt{\frac{1}{q_{ii}}\left(C - \sum_{l=i+1}^{n} q_{ll}\left(\xi_l + \sum_{j=l+1}^{n} q_{lj}\xi_j\right)^2\right)} + \rho_i + \sum_{j=i+1}^{n} q_{ij}\xi_j \right\rfloor .
\end{aligned}
\tag{3.28}
$$

To gain a simple geometric insight, we set the origin of the coordinate system in $\mathbf{r} = \mathbf{0}$ (i.e., $\rho_i = 0, i = 1, \ldots, n$), so that the sphere decoder reduces to the Fincke–Pohst enumeration algorithm. The three basic steps of the algorithm are illustrated in Figs. 3.10–3.12, which give the geometric interpretation of the operations involved in the sphere decoder.

1. The sphere is centered at the origin and includes the lattice points to be enumerated, see Fig. 3.10.
2. The sphere is transformed into an ellipsoid in the integer lattice domain, see Fig. 3.11.

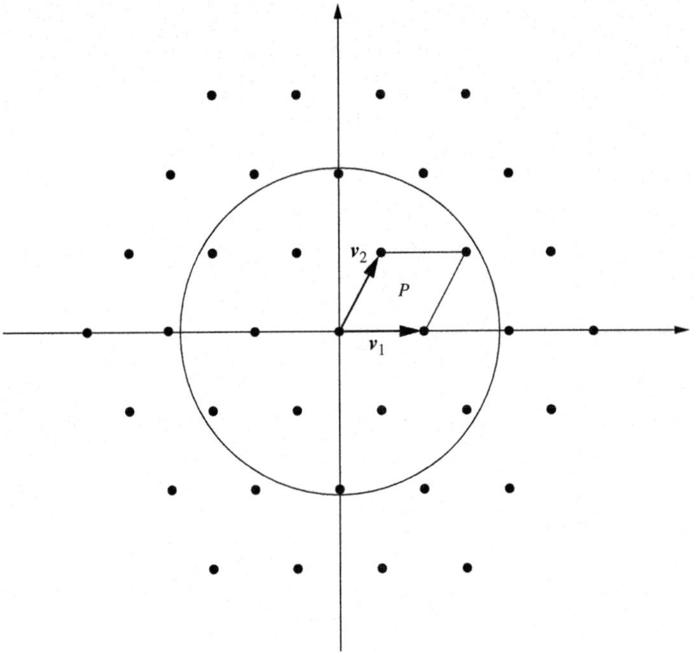

FIGURE 3.10

The sphere is centered at the origin and includes the lattice points to be enumerated.

3. The rotation into the new coordinate system defined by the U_i's enables an enumeration of the $\mathbb{Z}^n$–lattice points. The points inside the ellipse in Fig. 3.12 are visited from the bottom to the top and from left to right.

The search algorithm proceeds very much like a mixed radix counter on the digits u_i, with the addition that the bounds change whenever there is a carry operation from one digit to the next. In practice, the bounds can be updated recursively by using the following equations:

$$S_i = S_i(\xi_{i+1}, \ldots, \xi_n) = \rho_i + \sum_{l=i+1}^{n} q_{il}\xi_l$$

$$T_{i-1} = T_{i-1}(\xi_i, \ldots, \xi_n) = C - \sum_{l=i}^{n} q_{ll}\left(\xi_l + \sum_{j=l+1}^{n} q_{lj}\xi_j\right)^2$$

$$= T_i - q_{ii}(S_i - u_i)^2 .$$

When a vector inside the sphere is found, its square distance from the center (the received point) is given by

$$\hat{d}^2 = C - T_1 + q_{11}(S_1 - u_1)^2 .$$

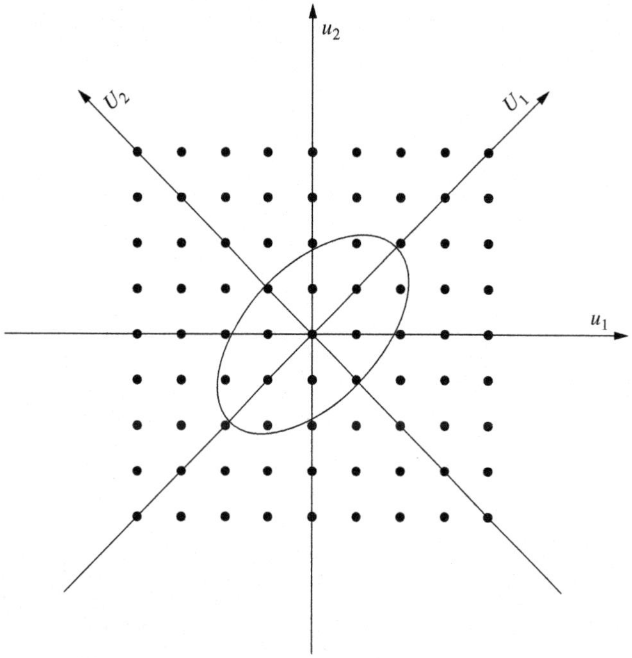

FIGURE 3.11

The sphere is transformed into an ellipsoid in the integer lattice domain.

This value is compared with the minimum square distance d^2 (initially set equal to C) found so far in the search. If it is smaller, then we have a new candidate closest point, and the search can go on using a new sphere with smaller radius $\hat{d}$.

The advantage of this method is that we never test vectors with a norm greater than the given radius. Every tested vector requires the computation of its norm, which entails n multiplications and $n-1$ additions. The increase in the number of operations needed to update the bounds (3.28) is largely compensated for by the enormous reduction in the number of vectors tested, especially when the dimension increases.

In order to be sure to always find a lattice point inside the sphere, we must select $\sqrt{C}$ equal to the *covering radius* of the lattice (Conway & Sloane, 1988). Otherwise, we do *bounded distance decoding*, and the decoder can signal an erasure whenever no point is found inside the sphere. A judicious choice of C can greatly speed up the decoder. In practice, the choice of C can be adjusted according to the noise variance N_0 so that the probability of a decoding failure is negligible. If a decoding failure is detected, the operation can either be repeated with a greater radius or an erasure can be declared.

The kernel of the sphere decoder (the enumeration of lattice points inside a sphere of radius $\sqrt{C}$) requires the greatest number of operations. The complexity is obviously independent of the constellation size, i.e., the number of operations does not depend on the spectral efficiency of the signal constellation.

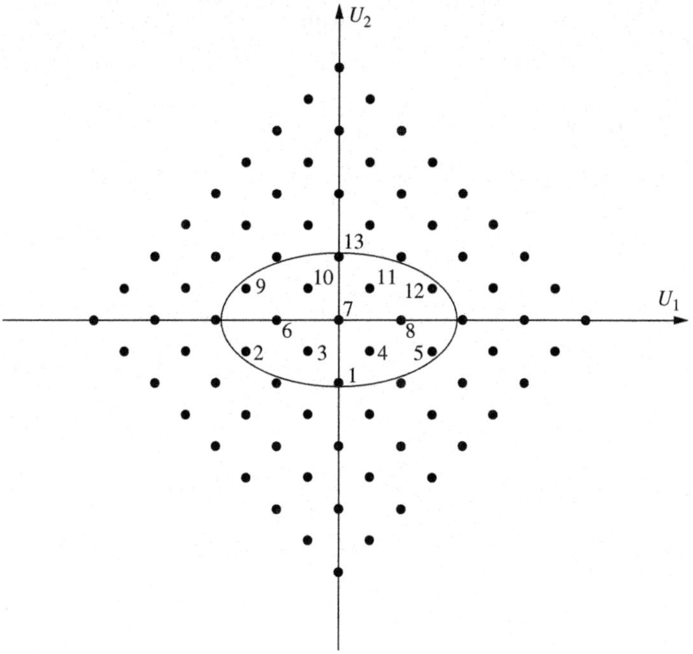

FIGURE 3.12

The coordinate rotation enables to enumerate the $\mathbb{Z}^n$–lattice points.

The complexity analysis presented in Fincke & Pohst (1985) shows that if λ^{-1} is a lower bound for the eigenvalues of the Gram matrix $\mathbf{G}$, then the number of arithmetical operations is

$$O\left(n^2\left(1+\frac{n-1}{4\lambda C}\right)^{4\lambda C}\right). \tag{3.29}$$

For a fixed radius and a given lattice (which fixes λ), the complexity of the decoding algorithm is polynomial. It should be noticed that this does not mean that the general lattice-decoding problem is not NP-hard. In fact, it is possible to construct a sequence of lattices of increasing dimension with an increasing value of the exponent λ.

The above complexity estimate is very pessimistic because it does not take into account the fact that we are dealing with an AWGN channel. In such a case, it was shown in Hassibi & Vikalo (2001) that for a wide range of signal-to-noise ratios and dimensions, the expected complexity is essentially polynomial as $O(n^3)$. For other results about the complexity of sphere decoding, the reader is addressed to Jaldén & Ottersten (2005).

When we deal with a lattice constellation, we must consider edge effects. During the search in the sphere, we discard the points, which do not belong to the lattice code; if no code vector is found, we declare an erasure. The complexity of this additional test depends on the shape of the constellation.

For cubic-shaped constellations, it only requires checking that the individual vector components lie within a given range. For a spherically shaped constellation, it is sufficient to compute the length of the lattice vector found in the search sphere in order to check if it is within the outermost shell of the constellation.

3.8.2 The Sphere Decoder with Fading

For ML decoding with perfect CSI at the receiver, the problem is to minimize the metric (3.3). Let **M** be the generator matrix of the lattice Λ, and let us consider the lattice $\Lambda^{(c)}$ with generator matrix

$$\mathbf{M}_c = \mathbf{M}\,\mathrm{diag}\{h_1,\ldots,h_n\}.$$

We can imagine this new lattice $\Lambda^{(c)}$ in a space where each component has been compressed or enlarged by a factor h_i, which represents the ith fading component of the channel **h**. A point of $\Lambda^{(c)}$ can be written as $\mathbf{x}^{(c)} = (x_1^{(c)} \cdots x_n^{(c)}) = (h_1x_1 \cdots h_nx_n)$. The metric to minimize is then

$$m(\mathbf{x}|\mathbf{r},\mathbf{h}) = \sum_{i=1}^{n} |r_i - x_i^{(c)}|^2.$$

This means that we can simply apply the lattice-decoding algorithm to the lattice $\Lambda^{(c)}$, when the received point is **r**. The decoded point $\hat{\mathbf{x}}^{(c)} \in \Lambda^{(c)}$ has the same integer components $(\hat{u}_1 \cdots \hat{u}_n)$ as $\hat{\mathbf{x}} \in \Lambda$.

The additional complexity required by this decoding algorithm comes from the fact that for each received point, we have a different compressed lattice $\Lambda^{(c)}$. So we need to compute a new Cholesky factorization of the Gram matrix for each $\Lambda^{(c)}$, which requires $O(n^3/3)$ operations. We also need $\mathbf{M}_c^{-1} = \mathrm{diag}\{1/h_1,\ldots,1/h_n\}\mathbf{M}^{-1}$ to find the ρ_i's, but this only requires a vector-matrix multiplication because $\mathbf{M}^{-1}$ is precomputed.

The choice of C in this case is more critical. In fact, whenever we are in the presence of deep fades, then many points fall inside the search sphere and the decoding can be very slow. This is also evident from the fact that the Gram matrix of $\Lambda^{(c)}$ may have a very small eigenvalue, which gives a large exponent λ in (3.29). This problem may be partially overcome by adapting C according to the values of the fading coefficients h_i. A good choice for C was found to be the smallest element of the diagonal of the Gram matrix of $\Lambda^{(c)}$. Note that the elements on the diagonal of the Gram matrix are the squared lengths of the basis vectors. A lattice basis reduction may be useful to reduce the search radius but requires additional overhead (see Agrell, Eriksson, Vardy, & Zeger, 2002).

3.9 PERFORMANCE OF ROTATED CONSTELLATIONS

A rotated $\mathbb{Z}^n$–lattice with diversity L is obtained by applying the rotation matrix **M** to the integer grid $\mathbb{Z}^n$, i.e.,

$$\Lambda = \{\mathbf{x} = \mathbf{uM},\ \mathbf{u} \in \mathbb{Z}^n\}.$$

The finite signal constellation is carved from this lattice by restricting the elements of **u** to a finite set of integers such as $\{\pm 1, \pm 3, \ldots, \pm(2^{\eta/2} - 1)\}$, where η is the spectral efficiency measured in bits per two dimensions.

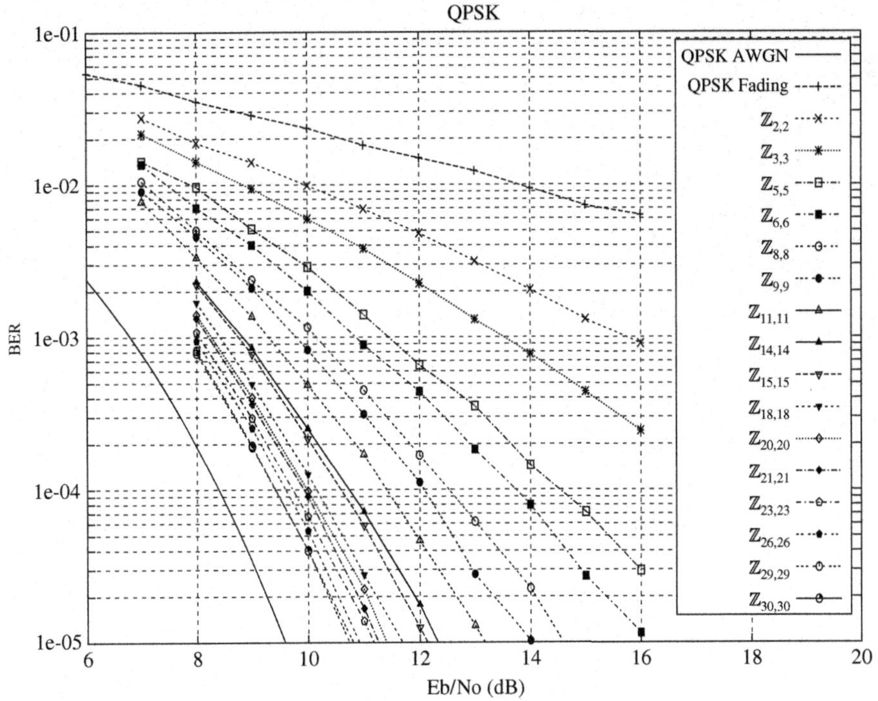

FIGURE 3.13

Performance (bit error rate versus channel SNR) for cyclotomic construction with QPSK ($\eta = 2$bit/symb).

The rotated $\mathbb{Z}^n$–lattice constellations based on the cyclotomic constructions have been simulated over an independent Rayleigh fading channel (i.e., the fading channel coefficients are iid Rayleigh distributed with $\mathsf{E}\{|h_i|^2\} = 1$). Figures 3.13 and 3.14 show the bit error rates versus E_b/N_0 of the rotated $\mathbb{Z}^n$ constellations for the cyclotomic constructions with QPSK and 16-QAM, respectively. The rotated $\mathbb{Z}^n$ constellations are denoted by $\mathbb{Z}_{n,L}$, where the two subscripts indicate dimension and diversity. For comparison, the performance of a standard component interleaved QPSK (resp. 16-QAM) over Gaussian and Rayleigh fading channels is also reported in the figures. We can observe how the bit error rate performance over the Rayleigh fading channel approaches that over the Gaussian channel as the diversity increases. Clearly, this gain is obtained at the expense of a higher decoding complexity because of the greater lattice dimension, but no extra bandwidth is used.

3.10 CONCLUSIONS

In this chapter, we have presented the mathematical theory of algebraic lattices which enables the construction of rotated $\mathbb{Z}^n$–lattice constellations with a desired modulation diversity and minimum product distance. These are the two key design parameters for the design of good codes over fast fading channels. The ultimate goal is to design codes for wireless fading channels whose performance

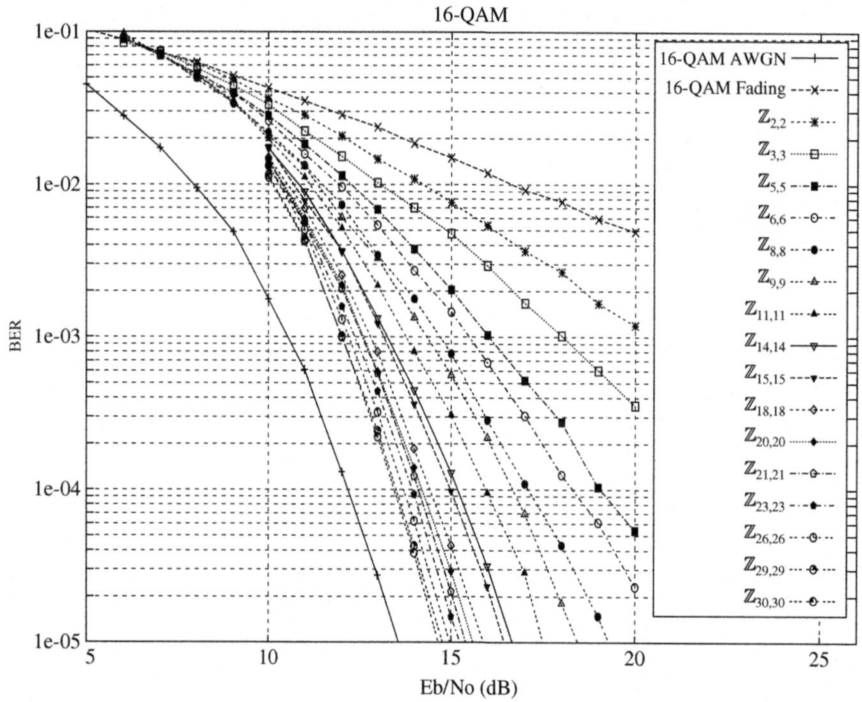

FIGURE 3.14

Performance (bit error rate versus channel SNR) for cyclotomic construction with 16-QAM ($\eta = 4$bit/symb).

approaches that of AWGN channels. This is indeed possible, given that the use of a sphere decoder at the receiver provides ML detection with the full diversity gain of the algebraically rotated lattices.

References

Agrell, E., Eriksson, T., Vardy, A., & Zeger, K. (2002). Closest point search in lattices. *IEEE Transactions on Information Theory*, *48*, 2201–2214.

Bayer-Fluckiger, E. (1999). Lattices and number fields. *Contemporary Mathematics*, *241*, 69–84.

Bayer-Fluckiger, E., Oggier, F., & Viterbo, E. (2004). New algebraic constructions of rotated $\mathbb{Z}^n$–lattice constellations for the Rayleigh fading channel. *IEEE Transactions on Information Theory*, *50*, 702–714.

Boullé, K., & Belfiore, J. C. (1992). Modulation schemes designed for the Rayleigh channel. *Proceedings of the CISS'92*, Princeton, NJ, *51*, 288–293.

Boutros, J., & Viterbo, E. (1998). Signal space diversity: A power and bandwidth efficient diversity technique for the Rayleigh fading channel. *IEEE Transactions on Information Theory*, *44*, 1453–1467.

Boutros, J., Viterbo, E., Rastello, C., & Belfiore, J. C. (1996). Good lattice constellations for both Rayleigh fading and Gaussian channels. *IEEE Transactions on Information Theory*, *42*, 502–518.

Conway, J. H., & Sloane, N. J. A. (1998). *Sphere packings, lattices and groups*. New York: Springer-Verlag.

Fincke, U., & Pohst, M. (1985). Improved methods for calculating vectors of short length in a lattice, including a complexity analysis. *Mathematics of Computation, 44*, 463–471.

Giraud, X., Boutillon, E., & Belfiore, J. C. (1997). Algebraic tools to build modulation schemes for fading channels. *IEEE Transactions on Information Theory, 43*, 938–952.

Hassibi, B., & Vikalo, H. (2001). On the expected complexity of sphere decoding. *Proceedings of the Thirty-Fifth Asilomar Conference on Signals, Systems and Computers*, Pacific Grove, CA, 1051–1055.

Jaldén, J., & Ottersten, B. (2005). On the complexity of sphere decoding in digital communications. *IEEE Transactions on Signal Processing*, 1474–1484.

Lidl, R., & Niederreiter, H. (1994). *Introduction to finite fields and their applications*. Cambridge, UK: Cambridge University Press.

Oggier, F., & Bayer-Fluckiger, E. (2003). Best rotated cubic lattice constellations for the Rayleigh fading channel. *Proceedings of the IEEE ISIT'03*, Yokohama, Japan, 37.

Pohst, M. (1981). On the computation of lattice vectors of minimal length, successive minima and reduced basis with applications. *ACM SIGSAM Bulletin, 15*, 37–44.

Samuel, P. (1971). *Théorie algébrique des nombres*. Paris, France: Hermann.

Sethuraman, B. A., Sundar Rajan, B., & Shashidhar, V. (2003). Full-diversity, high-rate space-time block codes from division algebras. *ACM SIGSAM Bulletin, 49*, 2596–2616.

Stewart, I. N., & Tall, D. O. (1979). *Algebraic number theory*. London, UK: Chapman and Hall.

Swinnerton-Dyer, H. P. F. (2001). *A brief guide to algebraic number theory*. Cambridge, UK: Cambridge University Press.

Viterbo, E., & Biglieri, E. (1993). A universal lattice decoder. *Proceedings of the 14eme Colloque GRETSI, Juan-les-Pins, France*, 611–614.

Viterbo, E., & Boutros, J. (1999). A universal lattice code decoder for fading channels. *IEEE Transactions on Information Theory, 45*, 1639–1642.

Washington, L. C. (1997). *Introduction to cyclotomic fields*. New York, NY: Springer.

CHAPTER

4 Estimation of Time-Varying Channels – A Block Approach

Geert Leus[1], Zijian Tang[2], Paolo Banelli[3]
[1]*Delft University of Technology, Delft, The Netherlands*
[2]*TNO Defence, Security and Safety, The Hague, The Netherlands*
[3]*University of Perugia, Perugia, Italy*

4.1 INTRODUCTION

For coherent detection in a wireless communication system, channel state information (CSI) is indispensable. Channel estimation has drawn tremendous attention in the literature (see Tong, Sadler, & Dong, 2004 and references therein), where the pilot-aided method is one of the most intensively studied approaches. This method is especially attractive for time-varying channels because of their short coherence time.

In this chapter, we will address pilot-aided channel estimation for both orthogonal frequency division multiplexing (OFDM) and single-carrier systems, where pilots are inserted in the frequency domain and time domain, respectively. We study these two systems under one framework because in the context of channel estimation, both systems can be characterized by data models of the same form. More specifically, the received samples can be expressed as the joint effect of the information part (due to the pilots), the interference part (due to the unknown data symbols), and the noise. Consequently, our task is to design a channel estimator that can combat both the interference and the noise. Such a data model is typical for OFDM over time-varying channels, where due to the Doppler effect, the orthogonality between the subcarriers is destroyed, and the channel matrix in the frequency domain becomes effectively a diagonally dominant yet full matrix instead of a diagonal matrix. As a result, the received frequency-domain samples depend on both the pilots and the unknown data symbols. For single-carrier systems, the channel matrix in the time domain is a strictly banded matrix if a finite impulse response (FIR) assumption for the channel is applied, and therefore, we can, in practice, find some received samples that solely depend on the pilots. However, it is sometimes beneficial to also consider received samples that depend on the unknown data symbols as well, to better suppress the interference and the noise. In any case, the resulting data model for single-carrier systems looks very similar to the data model for OFDM systems, and similar channel estimation techniques can be applied. Note that the considered data model can also account for superimposed pilot schemes (Ghogho & Swami, 2006; He & Tugnait, 2007), where the pilots and the data symbols coexist on the same subcarriers or time instants.

Whether we are dealing with OFDM or single-carrier systems, estimating a time-varying channel implies estimating a large number of unknowns, making the channel estimation problem much more

Wireless Communications Over Rapidly Time-Varying Channels

difficult than in the time-invariant case. As a remedy, we adopt in this chapter a parsimonious model, referred to as the basis expansion model (BEM), to approximate the time variation of the channel (see Section 1.6.1). If the BEM is accurate with negligible approximation error, channel estimation can be achieved by just estimating the BEM coefficients, which are much smaller in number than the actual unknowns, i.e., the channel tap values at different time instants.

In the remainder of the chapter, we first discuss the system and channel model in Section 4.2. In Section 4.3, we then present channel estimation algorithms within a single OFDM symbol/time block. We indicate how to position the pilots, where to select observation samples, and what is the best channel estimation strategy. In Section 4.4, we extend these methods to situations where multiple OFDM symbols/time blocks are utilized simultaneously. In this case, the position of the pilots plays an important role in the performance. Extensions to multiple-antenna systems are considered in Section 4.5 and adaptive channel estimation is briefly discussed in Section 4.6. We conclude this chapter in Section 4.7.

4.2 SYSTEM AND CHANNEL MODEL

4.2.1 System Model

Let us start with the channel input/output relationship in discrete form. The channel is assumed to be a doubly selective (doubly dispersive) channel that can be modeled by an FIR filter, which takes the effect of the transmitter filter, the propagation paths, and the receiver filter into account. Let us use $h[n,m]$ to denote the mth channel tap at the nth time instant and let us assume that the maximal channel order is $M-1$. For such a channel, if we use $s[n]$ and $r[n]$ to represent the transmitted and received signal at the nth time instant, respectively, they are related to each other through the channel as

$$r[n] = \sum_{m=0}^{M-1} h[n,m]s[n-m] + w[n], \tag{4.1}$$

where $w[n]$ stands for the additive noise.

Let us consider a block-wise transmission scheme, where we handle the transmitted and received sequences in blocks. Suppose the jth data block $\mathbf{a}[j]$ groups K data symbols and can be expressed as

$$\mathbf{a}[j] \triangleq (a[jK] \cdots a[(j+1)K-1])^T.$$

Before transmission, $\mathbf{a}[j]$ is first transformed by a linear precoding matrix $\mathbf{T}$ of size $(K+M-1)\times K$. The resulting $(K+M-1)\times 1$ vector is given by

$$\mathbf{s}[j] \triangleq (s[j(K+M-1)] \cdots s[(j+1)(K+M-1)-1])^T \triangleq \mathbf{T}\mathbf{a}[j].$$

Note that $\mathbf{T}$ introduces a redundancy equal to $M-1$ at the transmitter. Such a redundancy is useful in many applications, e.g., to exploit the channel diversity, or to better combat the interference coming from adjacent blocks due to the channel memory, as well as for many other purposes (Scaglione, Giannakis, & Barbarossa, 1999a,b). We will come back to this issue later on.

At the receiver, the received sample stream $r[n]$ is partitioned in blocks accordingly: $\mathbf{r}[j] \triangleq (r[j(K+M-1)] \cdots r[(j+1)(K+M-1)-1])^T$. Based on the FIR property of the channel, we can rewrite the channel I/O relationship of (4.1) in a block form as

$$\mathbf{r}[j] = \tilde{\mathbf{H}}[j](\mathbf{s}_{\mathrm{IBI}}^T[j]\ \mathbf{s}^T[j])^T + \mathbf{w}[j], \tag{4.2}$$

where $\mathbf{w}[j] \triangleq (w[j(K+M-1)]\ \cdots\ w[(j+1)(K+M-1)-1])^T$, and $\tilde{\mathbf{H}}[j]$ is the $(K+M-1)\times(K+2M-2)$ channel matrix representing the convolutive operation of the channel, which is given by $[\tilde{\mathbf{H}}[j]]_{nm} \triangleq h[j(K+M-1)+n-1, M-1+n-m]$. Because the channel has a memory of $M-1$, $\mathbf{r}[j]$ will not only depend on $\mathbf{s}[j]$, but also on the last $M-1$ symbols from the previous block $\mathbf{s}[j-1]$, which is denoted in (4.2) as $\mathbf{s}_{\mathrm{IBI}}[j] \triangleq (s[j(K+M-1)-M+1]\ \cdots\ s[j(K+M-1)-1])^T$. The effect of $\mathbf{s}_{\mathrm{IBI}}[j]$ is also known as the interblock interference (IBI) as indicated by its subscript. We can further rewrite (4.2) by splitting $\tilde{\mathbf{H}}[j]$ into its first $M-1$ columns and last $K+M-1$ columns, denoted by $\mathbf{H}_{\mathrm{IBI}}[j]$ and $\mathbf{H}[j]$, respectively, which leads to

$$\mathbf{r}[j] = \mathbf{H}_{\mathrm{IBI}}[j]\mathbf{s}_{\mathrm{IBI}}[j] + \mathbf{H}[j]\mathbf{s}[j] + \mathbf{w}[j].$$

Assuming perfect block and symbol synchronization, we first apply a linear decoder at the receiver, in the form of a $K\times(K+M-1)$ matrix $\mathbf{R}$. This results in

$$\mathbf{y}[j] \triangleq \mathbf{R}\mathbf{r}[j] = \mathbf{R}\mathbf{H}_{\mathrm{IBI}}[j]\mathbf{s}_{\mathrm{IBI}}[j] + \mathbf{R}\mathbf{H}[j]\mathbf{s}[j] + \mathbf{z}[j], \tag{4.3}$$

where $\mathbf{z}[j] \triangleq \mathbf{R}\mathbf{w}[j]$.

The decoder $\mathbf{R}$ as well as the precoder $\mathbf{T}$ depend on the specific (de)modulation scheme. Two particular cases are discussed below, namely OFDM and single-carrier modulation. For both systems, different forms of transmitter redundancy can be adopted, such as the use of a cyclic prefix (CP), zero padding (ZP), or nonzero padding (NZP). We will restrict ourselves to the CP case. Other forms of transmitter redundancy can be derived in a similar way.

4.2.1.1 OFDM System with CP

In an OFDM system where a CP is embedded at the transmitter, the precoder $\mathbf{T}$ and decoder $\mathbf{R}$ can be expressed as

$$\mathbf{T} = \mathbf{T}_{\mathrm{CP}}\mathbf{W}_K^H,$$
$$\mathbf{R} = \mathbf{W}_K\mathbf{R}_{\mathrm{CP}},$$

where $\mathbf{W}_K$ denotes the K-point unitary DFT matrix, i.e., $[\mathbf{W}_K]_{kl} \triangleq (1/\sqrt{K})e^{-\sqrt{-1}2\pi(k-1)(l-1)/K}$, the $(K+M-1)\times K$ matrix $\mathbf{T}_{\mathrm{CP}} \triangleq ((\mathbf{0}_{(M-1)\times(K-M+1)}\ \mathbf{I}_{M-1})^T\ \mathbf{I}_K)^T$ appends a CP, and the $K\times(K+M-1)$ matrix $\mathbf{R}_{\mathrm{CP}} \triangleq (\mathbf{0}_{K\times(M-1)}\ \mathbf{I}_K)$ discards the part of the received block corresponding to the CP. Because the transmitter redundancy equals the maximal channel order M, the IBI disappears and (4.3) becomes

$$\mathbf{y}[j] = \mathbf{W}_K\mathbf{R}_{\mathrm{CP}}\mathbf{H}[j]\mathbf{T}_{\mathrm{CP}}\mathbf{W}_K^H\mathbf{a}[j] + \mathbf{z}[j] = \mathbf{W}_K\mathbf{H}_{\mathrm{c}}[j]\mathbf{W}_K^H\mathbf{a}[j] + \mathbf{z}[j] = \mathbf{H}_{\mathrm{d}}[j]\mathbf{a}[j] + \mathbf{z}[j], \tag{4.4}$$

where $\mathbf{H}_{\mathrm{c}}[j] \triangleq \mathbf{R}_{\mathrm{CP}}\mathbf{H}[j]\mathbf{T}_{\mathrm{CP}}$ is the $K\times K$ time-domain channel matrix, and $\mathbf{H}_{\mathrm{d}}[j] \triangleq \mathbf{W}_K\mathbf{H}_{\mathrm{c}}[j]\mathbf{W}_K^H$ is the $K\times K$ frequency-domain channel matrix. If the channel is static or only slowly changing such that the time variation of the channel within an OFDM symbol can be neglected, $\mathbf{H}_{\mathrm{c}}[j]$ will be a circulant matrix (hence the subscript "c") with $(h[0]\ \cdots\ h[M-1]\ \mathbf{0}_{1\times(K-M+1)})^T$ on its first column and as a result $\mathbf{H}_{\mathrm{d}}[j]$ will be a diagonal matrix (hence the subscript "d") with

$\sqrt{K}\mathbf{W}_K(h[0] \cdots h[M-1]\ \mathbf{0}_{1\times(K-M+1)})^T$ on its diagonal. Note that we have dropped the time index n in the channel tap $h[n,m]$ because of the time-invariance assumption.

However, when the channel is varying faster, the circularity of $\mathbf{H}_c[j]$ is destroyed, and thus $\mathbf{H}_c[j]$ is not diagonalizable by (I)DFT operations. In principle, $\mathbf{H}_d[j]$ becomes a full matrix, where the nonzero off-diagonal entries induce intercarrier interference (ICI). The entries along the antidiagonal direction basically indicate how much the bandwidth will spread due to mobility-induced Doppler shifts. Because, in practice, this Doppler spread is limited, we can assume that most of the power in $\mathbf{H}_d[j]$ is concentrated on and close to the main diagonal, and it will gradually reduce in the antidiagonal direction. Hence, we may assume that $\mathbf{H}_d[j]$ is approximately circularly banded. This assumption has also been advocated in Stamoulis, Diggavi, & Al-Dhahir (2002) and Cai & Giannakis (2003).

4.2.1.2 *Single-Carrier System with CP*

In a single-carrier system with a CP, the (I)DFT operations are omitted at the transmitter and receiver, and the precoder **T** and decoder **R** simply become

$$\mathbf{T} = \mathbf{T}_{\mathrm{CP}},$$
$$\mathbf{R} = \mathbf{R}_{\mathrm{CP}}.$$

As in the OFDM case, the IBI is suppressed and (4.3) can be expressed as

$$\mathbf{y}[j] = \mathbf{R}_{\mathrm{CP}}\mathbf{H}[j]\mathbf{T}_{\mathrm{CP}}\mathbf{a}[j] + \mathbf{z}[j] = \mathbf{H}_c[j]\mathbf{a}[j] + \mathbf{z}[j]. \tag{4.5}$$

It is, in many applications, common to transform the single-carrier system model in (4.5) from the time domain to the frequency domain just like OFDM. For time-invariant channels, this operation will enable a simple one-tap channel equalizer (Falconer, Ariyavisitakul, Benyamin-Seeyar, & Eidson, 2002) because the channel matrix in the frequency domain $\mathbf{H}_d[j]$ will become diagonal, as explained earlier. Even if the channel is time-varying, it can still be attractive to consider a frequency-domain equalizer because of the circularly banded assumption on $\mathbf{H}_d[j]$, an idea that has been explored in Schniter & Liu (2003) and Tang & Leus (2008) (see also Chapter 6).

For channel estimation, however, it is not necessary to transform the single-carrier data model from the time domain to the frequency domain. The reason for this is that we will always work in the domain where the pilots will be embedded, which is in the frequency domain for the OFDM system and in the time domain for the single-carrier system.

4.2.2 BEM Channel Model

In the OFDM system as well as in the single-carrier system, estimating $h[n,m]$ requires the estimation of KM unknowns per OFDM symbol/time block, which is quite a lot. However, due to the fact that there is some correlation among these unknowns, the problem can be reduced. One approach to reduce the number of unknowns is to use a BEM to model the time variation of the channel (see also Section 1.6.1). To explain the BEM, let us assume that the time variation of the mth channel tap is smooth, and thus $h[n,m]$ is correlated in the time index n. Then, we can accurately model K consecutive samples of the mth channel tap, e.g., $h[j(K+M-1)+M-1,m],\ldots,h[(j+1)(K+M-1)-1,m]$, as

$$\begin{pmatrix} h[j(K+M-1)+M-1,m] \\ \vdots \\ h[(j+1)(K+M-1)-1,m] \end{pmatrix} \approx \underbrace{\begin{pmatrix} \mathbf{u}_0 & \cdots & \mathbf{u}_{I-1} \end{pmatrix}}_{\mathbf{U}} \begin{pmatrix} c_{0,j}[m] \\ \vdots \\ c_{I-1,j}[m] \end{pmatrix}, \tag{4.6}$$

where $I-1$ is the BEM order, the $K \times 1$ vector $\mathbf{u}_i$ is the ith BEM function, and $c_{i,j}[m]$ is the ith BEM coefficient of the mth channel tap within the jth block. Note that the BEM matrix $\mathbf{U}$ is a tall matrix because I is generally much smaller than K. It is furthermore predetermined and independent of the channel. The accuracy of (4.6) will depend on the design of the BEM matrix $\mathbf{U}$ and the choice of the BEM order $I-1$.

Different BEM designs are documented in various articles such as the discrete Karhunen-Loève BEM (DKL-BEM) (Visintin, 1996; Yip & Ng, 1997; Haykin, 1996; Teo & Ohno, 2005), the discrete prolate spheroidal BEM (DPS-BEM) (Zemen & Mecklenbräuker, 2005), the complex exponential BEM (CE-BEM) (Tsatsanis & Giannakis, 1996; Cirpan & Tsatsanis, 1999), and the polynomial BEM (P-BEM) (Borah & Hart, 1999a,b; Tomasin, Gorokhov, Yang, & Linnartz, 2005). Note that the CE-BEM can further be categorized into two types: the critically sampled CE-BEM (CCE-BEM) (Guillaud & Slock, 2003; Ma & Giannakis, 2003; Ma, Giannakis, & Ohno, 2003; Leus & Moonen, 2003; Kannu & Schniter, 2005) and the oversampled CE-BEM (OCE-BEM) (Thomas & Vook, 2000; Leus, 2004; Cui, Tellambura, & Wu, 2005). BEM designs other than the above are also reported, e.g., Zakharov, Tozer, & Adlard (2004) use a spline approach. Besides, it is also possible to combine the above BEMs for different purposes (Stamoulis, Diggavi, & Al-Dhahir, 2002; Nicoli, Simeone, & Spagnolini, 2003; Gorokhov & Linnartz, 2004). A comparison of the modeling performances of some BEMs is given in Zemen and Mecklenbräuker (2005) and in Tang (2007). It is noteworthy that all the above BEMs can, in principle, be utilized in this chapter with one minor adaptation: if $\tilde{\mathbf{U}}$ is one of the original BEM designs, we will construct a new BEM out of it by adopting

$$\mathbf{U} = \tilde{\mathbf{U}}\mathbf{Q}, \tag{4.7}$$

where $\mathbf{Q}$ is an $I \times I$ matrix that makes the columns of $\mathbf{U}$ orthonormal, i.e., $\mathbf{U}^H\mathbf{U} = \mathbf{I}_I$ (note that some original BEM designs already satisfy this property but not all of them). Such matrices will improve the numerical stability of the proposed algorithms.

Repeating the BEM approximation procedure of (4.6) for all channel taps, we obtain

$$\mathbf{h}[j] \approx (\mathbf{U} \otimes \mathbf{I}_M)\mathbf{c}[j], \tag{4.8}$$

where $\mathbf{h}[j] \triangleq (\mathbf{h}_0^T[j] \cdots \mathbf{h}_{K-1}^T[j])^T$, with $\mathbf{h}_k[j] \triangleq (h[j(K+M-1)+M-1+k,0] \cdots h[j(K+M-1)+M-1+k,M-1])^T$, stacks all channel taps in the jth block and where $\mathbf{c}[j] \triangleq (\mathbf{c}_0^T[j] \cdots \mathbf{c}_{I-1}^T[j])^T$, with $\mathbf{c}_i[j] \triangleq (c_{i,j}[0] \cdots c_{i,j}[M-1])^T$, stacks the BEM coefficients of all channel taps in the jth block. The notation $\otimes$ stands for the Kronecker product. By means of (4.8), we are able to reduce the total number of unknown channel parameters from KM to IM with $I \ll K$.

We can now also express the time-domain channel matrix $\mathbf{H}_c[j]$ and the frequency-domain channel matrix $\mathbf{H}_d[j]$ using the BEM. Observing that each diagonal in $\mathbf{H}_c[j]$ corresponds to one channel tap, we can derive after some algebra that

$$\mathbf{H}_c[j] = \sum_{i=0}^{I-1} \mathbf{U}_{d,i}\mathbf{C}_{c,i}[j], \tag{4.9}$$

where $\mathbf{U}_{d,i}$ is a diagonal matrix with $\mathbf{u}_i$ on its diagonal, and $\mathbf{C}_{c,i}[j]$ is a circulant matrix with $(\mathbf{c}_i^T[j]\ \mathbf{0}_{1\times(K-M)})^T$ on its first column. The frequency-domain channel matrix $\mathbf{H}_d[j]$ can then be expressed as

$$\mathbf{H}_d[j] = \mathbf{W}_K\mathbf{H}_c[j]\mathbf{W}_K^H = \sum_{i=0}^{I-1} \mathbf{W}_K\mathbf{U}_{d,i}\mathbf{W}_K^H\mathbf{W}_K\mathbf{C}_{c,i}[j]\mathbf{W}_K^H = \sum_{i=0}^{I-1} \mathbf{U}_{c,i}\mathbf{C}_{d,i}[j], \tag{4.10}$$

where $\mathbf{U}_{c,i} \triangleq \mathbf{W}_K\mathbf{U}_{d,i}\mathbf{W}_K^H$ is a circulant matrix with $(1/\sqrt{K})\mathbf{u}_i^T\mathbf{W}_K^H$ on its first row and $\mathbf{C}_{d,i}[j] \triangleq \mathbf{W}_K\mathbf{C}_{c,i}[j]\mathbf{W}_K^H$ is a diagonal matrix with $\sqrt{K}\mathbf{W}_K\left(\mathbf{c}_i^T[j]\ \mathbf{0}_{1\times(K-M)}\right)^T$ on its diagonal.

For the channel estimation schemes to be discussed in the ensuing sections, we will use the BEM channel matrices defined in (4.9) and (4.10) instead of the true channel matrices in the system model, and we will estimate the channel by estimating the BEM coefficients. The BEM modeling error will not be taken into account in the design of the channel estimators. This is motivated by the fact that if the BEM is accurate, the modeling error is usually on the order of 10^{-4} as reported in Zemen and Mecklenbräuker (2005) and Tang (2007), which is much smaller than the typical channel noise level. However, if the BEM is not accurate enough, then we will never obtain a reliable estimate, even if we design a channel estimator that takes the BEM modeling error into account, simply because the BEM itself is not accurate. Finally note that although the BEM modeling error is not taken into account in our channel estimator design, its effect will be considered in the simulations, where we will compare the estimated channel with the true channel instead of the BEM channel.

4.3 CHANNEL ESTIMATION BASED ON A SINGLE BLOCK

4.3.1 Introduction

In the previous section, we have shown how a communications scheme can be split into temporal blocks. In this section, we will discuss block-based channel estimation, or in other words, the channel is estimated every time a block of samples is received. This strategy is especially attractive for time-varying channels, where, due to a short coherence time, it is not possible to gather a large amount of received samples for which the channel is correlated. Because we only focus on a single received block in this section, we will drop the block index j for the sake of simplicity.

We will confine ourselves to pilot-aided channel estimators. For one transmitted symbol block $\mathbf{a}$, let $\mathbf{p}$ denote the pilot symbols and $\mathbf{d}$ denote the unknown data symbols. Defining $\mathcal{P}$ as the set of positions of the pilot symbols and $\mathcal{D}$ as the set of positions of the unknown data symbols, $\mathbf{p}$ has length $|\mathcal{P}|$ and $\mathbf{d}$ has length $|\mathcal{D}|$, where $|\mathcal{S}|$ denotes the cardinality of the set $\mathcal{S}$. Such a notation can account for various pilot insertion schemes. One such scheme is a multiplexed pilot scheme, where the pilot and unknown data symbols occupy different locations, i.e., $\mathcal{P} \cap \mathcal{D} = \emptyset$ and $\mathcal{P} \cup \mathcal{D} = \{0,\ldots,K-1\}$. But it also allows us to model a superimposed pilot scheme, where the pilot and unknown data symbols coexist on the same positions, i.e., $\mathcal{P} = \mathcal{D} = \{0,\ldots,K-1\}$. Let us at this point also introduce the notation $\mathbf{x}^{\{\mathscr{S}\}}$ to represent the subvector that collects the elements of the vector $\mathbf{x}$ with indices in the set $\mathscr{S}$, as well as $\mathbf{X}^{\{\mathscr{S}_1,\mathscr{S}_2\}}$ to represent the submatrix of $\mathbf{X}$ containing the rows with indices in the set $\mathscr{S}_1$ and the columns with indices in the set $\mathscr{S}_2$, where we use a colon instead of an index set if all elements are considered. This notation will be used throughout this chapter.

The multiplexed and superimposed pilot schemes will be unified under the same framework. Although no extra bandwidth is consumed with the superimposed pilot scheme, it is heavily influenced by the interference from the unknown data symbols. To some extent this also happens with the multiplexed pilot scheme because the delay (for the single-carrier system) or Doppler (for the OFDM system) spread of the channel will introduce some mixing between the pilot symbols and the unknown data symbols, but this mixing is much smaller than with the superimposed pilot scheme. For this reason, we will mainly focus on the multiplexed pilot scheme in this section, and we will write a remark on the superimposed pilot scheme from time to time. However, the expressions are basically the same for both pilot insertion schemes.

In the ensuing part, we will first rewrite the earlier data models as a function of the pilot symbols. Based on these data models, we will then discuss various channel estimation techniques and their characteristics. Some simulation results will be given at the end of this section.

4.3.2 Channel Estimation Data Model

As already indicated, we focus mainly on multiplexed pilot schemes in this chapter. This means that we consider time-multiplexed training for the single-carrier system and frequency-multiplexed training for the OFDM system. Let us assume that the pilot symbols are grouped in G clusters, each of length $P+1$. For the gth pilot cluster $\mathbf{p}_g$, suppose $\mathscr{P}_g \triangleq \{P_g,\ldots,P_g+P\}$ denotes the set of indices that contains all the pilot positions, with P_g standing for its starting position. Hence, $\mathbf{p}_g$ is related with the transmitted symbol block $\mathbf{a}$ as $\mathbf{p}_g \triangleq \mathbf{a}^{\{\mathscr{P}_g\}}$. Below, we will describe how to proceed for the two considered systems.

4.3.2.1 *Single-Carrier System with CP*

Because the time-domain channel matrix $\mathbf{H}_c$ is circularly banded with a band of length M (the discrete delay spread) on and below the main diagonal, we can assign to the gth pilot cluster $\mathbf{p}_g$ the observation samples $\mathbf{y}^{\{\mathscr{O}_g\}}$ whose indices are collected in the set

$$\mathscr{O}_g \triangleq \{P_g+\ell,\ldots,P_g+P+M-1-\ell\}, \tag{4.11}$$

where ℓ is an integer design parameter determining the number of observation samples that we want to take into account in our channel estimation. We will come back to this issue later on.

According to the input–output relationship in (4.5) and assuming that there is enough space in between two nonzero pilot symbols from different pilot clusters, i.e., the minimal number of samples in between two nonzero pilot symbols from different pilot clusters, denoted as Δ, satisfies $\Delta \geq M-\ell-1$, we can express $\mathbf{y}^{\{\mathscr{O}_g\}}$ as

$$\mathbf{y}^{\{\mathscr{O}_g\}} = \mathbf{H}_c^{\{\mathscr{O}_g,\mathscr{P}_g\}}\mathbf{p}_g + \mathbf{H}_c^{\{\mathscr{O}_g,\mathscr{D}_g\}}\mathbf{d}_g + \mathbf{z}^{\{\mathscr{O}_g\}}, \tag{4.12}$$

where $\mathscr{D}_g$ represents the set of indices of the unknown data symbols that are present in $\mathbf{y}^{\{\mathscr{O}_g\}}$, and $\mathbf{d}_g$ collects those unknown data symbols present in $\mathbf{y}^{\{\mathscr{O}_g\}}$, i.e., $\mathbf{d}_g \triangleq \mathbf{a}^{\{\mathscr{D}_g\}}$. The structure of the channel matrices in (4.12) is illustrated in Fig. 4.1.

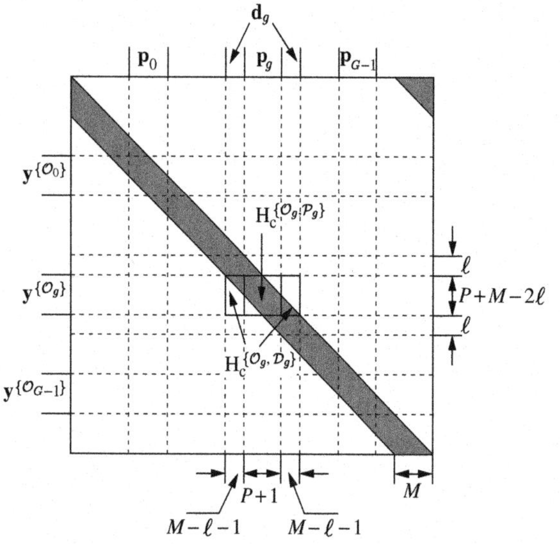

FIGURE 4.1

The partitioning of the time-domain channel matrix $\mathbf{H}_c$ into $\mathbf{H}_c^{\{\mathcal{O}_g,\mathcal{P}_g\}}$ and $\mathbf{H}_c^{\{\mathcal{O}_g,\mathcal{D}_g\}}$. The dark shaded areas represent large values, whereas the white areas represent zeros.

Using the BEM expression in (4.9), we can now rewrite (4.12) as

$$\mathbf{y}^{\{\mathcal{O}_g\}} = \sum_{i=0}^{I-1} \mathbf{U}_{d,i}^{\{\mathcal{O}_g,\mathcal{O}_g\}} \mathbf{C}_{c,i}^{\{\mathcal{O}_g,\mathcal{P}_g\}} \mathbf{p}_g + \underbrace{\sum_{i=0}^{I-1} \mathbf{U}_{d,i}^{\{\mathcal{O}_g,\mathcal{O}_g\}} \mathbf{C}_{c,i}^{\{\mathcal{O}_g,\mathcal{D}_g\}} \mathbf{d}_g}_{\mathbf{i}_g} + \mathbf{z}^{\{\mathcal{O}_g\}}, \tag{4.13}$$

where $\mathbf{U}_{d,i}^{\{\mathcal{O}_g,\mathcal{O}_g\}}$ is the submatrix of the diagonal matrix $\mathbf{U}_{d,i}$ consisting of the diagonal elements on the positions $\mathcal{O}_g$, $\mathbf{C}_{c,i}^{\{\mathcal{O}_g,\mathcal{P}_g\}}$ consists of the columns of $\mathbf{C}_{c,i}^{\{\mathcal{O}_g,:\}}$ corresponding to the positions of the pilot symbols in $\mathbf{p}_g$, and $\mathbf{C}_{c,i}^{\{\mathcal{O}_g,\mathcal{D}_g\}}$ consists of the columns of $\mathbf{C}_{c,i}^{\{\mathcal{O}_g,:\}}$ corresponding to the positions of the unknown data symbols in $\mathbf{d}_g$. Note that $\mathbf{C}_{c,i}^{\{\mathcal{O}_g,\mathcal{P}_g\}}$ and $\mathbf{C}_{c,i}^{\{\mathcal{O}_g,\mathcal{D}_g\}}$ have the same structure as $\mathbf{H}_c^{\{\mathcal{O}_g,\mathcal{P}_g\}}$ and $\mathbf{H}_c^{\{\mathcal{O}_g,\mathcal{D}_g\}}$ in Fig. 4.1.

The significance of ℓ becomes clearer from the definitions above. By adjusting the value of ℓ, the number of observation samples varies, and accordingly, the channel estimator can deal with different amounts of information regarding the pilot symbols and the unknown data symbols. For instance, in Ma, Giannakis, and Ohno (2003), ℓ is chosen as $\ell = M-1$, so that $\mathcal{D}_g = \emptyset$ and thus $\mathbf{H}_c^{\{\mathcal{O}_g,\mathcal{D}_g\}}$ vanishes. Moreover, under the assumption that $P \geq M-1$, we then have that $\mathcal{O}_g = \{P_g + M - 1, \ldots, P_g + P\}$ and thus $\mathbf{y}^{\{\mathcal{O}_k\}}$ contains the maximum number of received samples that depend on the pilot symbols without any interference from the unknown data symbols. Taking ℓ larger than $M-1$ degrades the performance but lowers the complexity. The largest value of ℓ is given by $(P+M-1)/2$ (assume $P+M-1$ is even), which is the case where only a single observation sample per pilot cluster is selected. Rousseaux, Leus, Stoica, and Moonen (2004) and Rousseaux, Leus, and Moonen (2006) chose $\ell=0$, which means that $\mathcal{O}_g=\{P_g,\ldots,P_g+P+M-1\}$ and thus $\mathbf{y}^{\{\mathcal{O}_k\}}$ contains the maximum number of received samples that depend on both the pilot and the interfering unknown data symbols. Taking ℓ smaller than 0 will not change the performance and only increases the complexity. Taking the earlier condition $\Delta \geq M-\ell-1$ into account, which is required for (4.12) to hold, the smallest value of ℓ is given by $\max\{0, M-\Delta-1\}$. In the sequel, we will follow an approach similar to the approach of Leus & van der Veen (2005), and let ℓ assume an arbitrary integer value, which we bound by $\max\{0, M-\Delta-1\} \leq \ell \leq (P+M-1)/2$.

Rewriting (4.13) as a function of the BEM coefficients $\mathbf{c}=(\mathbf{c}_0^T \cdots \mathbf{c}_{I-1}^T)^T$, we obtain

$$\mathbf{y}^{\{\mathcal{O}_g\}}=\mathbf{A}_g\mathbf{c}+\mathbf{i}_g+\mathbf{z}^{\{\mathcal{O}_g\}}, \tag{4.14}$$

with

$$\mathbf{A}_g \triangleq \left(\mathbf{U}_{\mathrm{d},0}^{\{\mathcal{O}_g,\mathcal{O}_g\}}\mathbf{P}_{\mathrm{c},g}^{\{\mathcal{O}_g,\mathcal{C}_g\}} \quad \cdots \quad \mathbf{U}_{\mathrm{d},I-1}^{\{\mathcal{O}_g,\mathcal{O}_g\}}\mathbf{P}_{\mathrm{c},g}^{\{\mathcal{O}_g,\mathcal{C}_g\}}\right), \tag{4.15}$$

where $\mathbf{P}_{\mathrm{c},g}$ is a circulant matrix with $(\mathbf{p}_g^T\ \mathbf{0}_{1\times(K-P-1)})^T$ on its first column, and $\mathcal{C}_g$ is the set of column indices of $\mathbf{P}_{\mathrm{c},g}$ that are hit by every $\mathbf{c}_i$, i.e., $\mathcal{C}_g \triangleq \{P_g,\ldots,P_g+M-1\}$. Note that $\mathcal{C}_g$ has the same starting point as $\mathcal{P}_g$, but it has length M instead of $P+1$. To derive (4.15), we have made use of the commutativity of the circular convolution, meaning that $\mathbf{C}_{\mathrm{c},i}^{\{:,\mathcal{P}_g\}}\mathbf{p}_g=\mathbf{P}_{\mathrm{c},g}^{\{:,\mathcal{C}_g\}}\mathbf{c}_i$.

Stacking the results obtained in (4.14) for all G pilot clusters, $\mathbf{y}^{\{\mathcal{O}\}} \triangleq (\mathbf{y}^{\{\mathcal{O}_0\}T} \ldots \mathbf{y}^{\{\mathcal{O}_{G-1}\}T})^T$, where the index set $\mathcal{O}$ is given by $\mathcal{O} \triangleq \{\mathcal{O}_0,\ldots,\mathcal{O}_{G-1}\}$, we obtain

$$\mathbf{y}^{\{\mathcal{O}\}}=\mathbf{A}\mathbf{c}+\mathbf{i}+\mathbf{z}^{\{\mathcal{O}\}}, \tag{4.16}$$

where $\mathbf{i}$ and $\mathbf{z}^{\{\mathcal{O}\}}$ are similarly defined as $\mathbf{y}^{\{\mathcal{O}\}}$, and

$$\mathbf{A} \triangleq (\mathbf{A}_0^T \cdots \mathbf{A}_{G-1}^T)^T = \begin{pmatrix} \mathbf{U}_{\mathrm{d},0}^{\{\mathcal{O}_0,\mathcal{O}_0\}}\mathbf{P}_{\mathrm{c},0}^{\{\mathcal{O}_0,\mathcal{C}_0\}} & \cdots & \mathbf{U}_{\mathrm{d},I-1}^{\{\mathcal{O}_0,\mathcal{O}_0\}}\mathbf{P}_{\mathrm{c},0}^{\{\mathcal{O}_0,\mathcal{C}_0\}} \\ \vdots & & \vdots \\ \mathbf{U}_{\mathrm{d},0}^{\{\mathcal{O}_{G-1},\mathcal{O}_{G-1}\}}\mathbf{P}_{\mathrm{c},G-1}^{\{\mathcal{O}_{G-1},\mathcal{C}_{G-1}\}} & \cdots & \mathbf{U}_{\mathrm{d},I-1}^{\{\mathcal{O}_{G-1},\mathcal{O}_{G-1}\}}\mathbf{P}_{\mathrm{c},G-1}^{\{\mathcal{O}_{G-1},\mathcal{C}_{G-1}\}} \end{pmatrix}. \tag{4.17}$$

4.3.2.2 OFDM System with CP

In contrast to the time-domain channel matrix $\mathbf{H}_\mathrm{c}$, the frequency-domain channel matrix $\mathbf{H}_\mathrm{d}$ basically is a full matrix, but with most of its power located on or close to the main diagonal, or in other words, the matrix $\mathbf{H}_\mathrm{d}$ is approximately circularly banded. There is only one case where the matrix $\mathbf{H}_\mathrm{d}$ is exactly circularly banded, and that is when the (C)CE-BEM is used to model the channel. In that case, $\mathbf{H}_\mathrm{d}$ will have I nonzero diagonals, where $I-1$ is the order of the (C)CE-BEM. However, the problem with this approach is that the (C)CE-BEM does not provide a good fit to the actual channel (Zemen & Mecklenbräuker, 2005; Tang, 2007).

Due to the fact that $\mathbf{H}_\mathrm{d}$ is a full matrix, it is less obvious to choose which received samples to use for channel estimation. However, similar to Section 4.3.2.1, if we assume that $\mathbf{H}_\mathrm{d}$ is approximately circularly banded with a band of length D (the significant discrete Doppler spread) centered around the main diagonal (assume $D-1$ is even), we can assign to the gth pilot cluster $\mathbf{p}_g$ the observation samples $\mathbf{y}^{\{\mathcal{O}_g\}}$ whose indices are collected in the set

$$\mathcal{O}_g \triangleq \{P_g - (D-1)/2 + \ell, \ldots, P_g + P + (D-1)/2 - \ell\}, \tag{4.18}$$

where ℓ is again an integer design parameter that controls the number of observation samples that will be used for channel estimation. Note that in contrast to other chapters, we use D here to indicate the double-sided discrete Doppler spread instead of the single-sided one, in order to better reflect the duality between M for the single-carrier system with CP and D for the OFDM system with CP.

According to the input–output relationship in (4.4), we can express $\mathbf{y}^{\{\mathcal{O}_g\}}$ as

$$\mathbf{y}^{\{\mathcal{O}_g\}} = \mathbf{H}_\mathrm{d}^{\{\mathcal{O}_g,\mathcal{P}\}}\mathbf{p} + \mathbf{H}_\mathrm{d}^{\{\mathcal{O}_g,\mathcal{D}\}}\mathbf{d} + \mathbf{z}^{\{\mathcal{O}_g\}}. \tag{4.19}$$

The difference with (4.12) for the time-domain case is that now all the pilot and unknown data symbols are present in $\mathbf{y}^{\{\mathcal{O}_g\}}$, due to the fact that $\mathbf{H}_\mathrm{d}$ is basically a full matrix. Moreover, there is no special constraint on Δ, which represents the minimal number of samples in between two nonzero pilot symbols from different pilot clusters. The structure of the channel matrices in (4.19) is illustrated in Fig. 4.2.

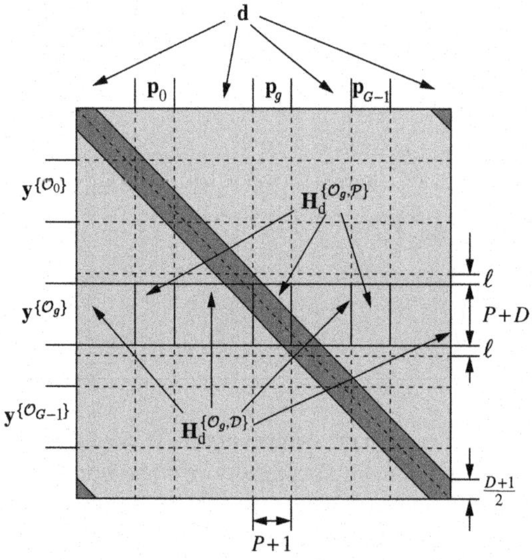

FIGURE 4.2

The partitioning of the frequency-domain channel matrix $\mathbf{H}_\mathrm{d}$ into $\mathbf{H}_\mathrm{d}^{\{\mathcal{O}_g,\mathcal{P}\}}$ and $\mathbf{H}_\mathrm{d}^{\{\mathcal{O}_g,\mathcal{D}\}}$. The dark shaded areas represent large values, whereas the light shaded areas represent small values.

Using the BEM expression in (4.10), we can now rewrite (4.19) as

$$\mathbf{y}^{\{\mathcal{O}_g\}} = \sum_{i=0}^{I-1} \mathbf{U}_{\mathrm{c},i}^{\{\mathcal{O}_g,\mathcal{P}\}} \mathbf{C}_{\mathrm{d},i}^{\{\mathcal{P},\mathcal{P}\}} \mathbf{p} + \underbrace{\sum_{i=0}^{I-1} \mathbf{U}_{\mathrm{c},i}^{\{\mathcal{O}_g,\mathcal{D}\}} \mathbf{C}_{\mathrm{d},i}^{\{\mathcal{D},\mathcal{D}\}} \mathbf{d}}_{\mathbf{i}_g} + \mathbf{z}^{\{\mathcal{O}_g\}}, \tag{4.20}$$

where $\mathbf{U}_{\mathrm{c},i}^{\{\mathcal{O}_g,\mathcal{P}\}}$ consists of the columns of $\mathbf{U}_{\mathrm{c},i}^{\{\mathcal{O}_g,:\}}$ corresponding to the positions of the pilot symbols in $\mathbf{p}$, $\mathbf{U}_{\mathrm{c},i}^{\{\mathcal{O}_g,\mathcal{D}\}}$ consists of the columns of $\mathbf{U}_{\mathrm{c},i}^{\{\mathcal{O}_g,:\}}$ corresponding to the positions of the unknown data symbols in $\mathbf{d}$, and finally $\mathbf{C}_{\mathrm{d},i}^{\{\mathcal{P},\mathcal{P}\}}$ and $\mathbf{C}_{\mathrm{d},i}^{\{\mathcal{D},\mathcal{D}\}}$ are the submatrices of the diagonal matrix $\mathbf{C}_{\mathrm{d},i}$ consisting of the diagonal elements on the positions $\mathcal{P}$ and $\mathcal{D}$, respectively. Note that $\mathbf{U}_{\mathrm{c},i}^{\{\mathcal{O}_g,\mathcal{P}\}}$ and $\mathbf{U}_{\mathrm{c},i}^{\{\mathcal{O}_g,\mathcal{D}\}}$ have the same structure as $\mathbf{H}_{\mathrm{d}}^{\{\mathcal{O}_g,\mathcal{P}\}}$ and $\mathbf{H}_{\mathrm{d}}^{\{\mathcal{O}_g,\mathcal{D}\}}$ in Fig. 4.2.

Similarly to the time-domain case, the relation of ℓ with the significant Doppler spread D will indicate how much significant interference from the unknown data symbols we will take into account. For instance, if we take $\ell = (D-1)/2$, we take no significant interference into account. Under the assumption that $P \geq D-1$, $\mathbf{y}^{\{\mathcal{O}_g\}}$ then contains the maximum number of received samples that depend on the pilot symbols without any significant interference from the unknown data symbols. However, this generally does not mean that the interference term, i.e., the second term in (4.20), vanishes. This will only be true when a (C)CE-BEM is used and $D = I$, as considered in Kannu & Schniter (2005), but as already mentioned earlier, the (C)CE-BEM is not a very accurate model. Taking ℓ larger than $(D-1)/2$ lowers the complexity as in the time-domain case but might not necessarily degrade the performance. Anyway, the largest value of ℓ is given by $(P+D-1)/2$ (assume $P+D-1$ is even), which is the case where only a single observation sample per pilot cluster is selected. If we take $\ell = 0$, $\mathbf{y}^{\{\mathcal{O}_g\}}$ contains the maximum number of received samples that depend on both the pilot symbols and the significantly interfering unknown data symbols. Note that in the time-invariant case, taking $\ell = (D-1)/2$ or $\ell = 0$ is the same because $D = 1$, which then leads to the traditional time-invariant OFDM channel estimation procedure presented for instance in Negi & Cioffi (1998). Taking ℓ smaller than zero increases the complexity as in the time-domain case, but might change the performance. Hence, we will not put any lower bound on ℓ yet. To conclude, we let ℓ assume an arbitrary integer value, which we bound by $\ell \leq (P+D-1)/2$.

Rewriting (4.20) as a function of the BEM coefficients $\mathbf{c} = (\mathbf{c}_0^T \cdots \mathbf{c}_{I-1}^T)^T$, we can obtain

$$\mathbf{y}^{\{\mathcal{O}_g\}} = \mathbf{A}_g \mathbf{c} + \mathbf{i}_g + \mathbf{z}^{\{\mathcal{O}_g\}}, \tag{4.21}$$

with

$$\mathbf{A}_g \triangleq \left(\mathbf{U}_{\mathrm{c},0}^{\{\mathcal{O}_g,\mathcal{P}\}} \cdots \mathbf{U}_{\mathrm{c},I-1}^{\{\mathcal{O}_g,\mathcal{P}\}}\right) \left(\mathbf{I}_I \otimes \mathrm{diag}\{\mathbf{p}\} \mathbf{V}_K^{\{\mathcal{P},:\}}\right), \tag{4.22}$$

where we have used that $\mathbf{C}_{\mathrm{d},i}^{\{\mathcal{P},\mathcal{P}\}} \mathbf{p} = \mathrm{diag}\{\mathbf{p}\} \sqrt{K} \mathbf{W}_K^{\{\mathcal{P},:\}} (\mathbf{c}_i^T \; \mathbf{0}_{1\times(K-M)})^T = \mathrm{diag}\{\mathbf{p}\} \mathbf{V}_K^{\{\mathcal{P},:\}} \mathbf{c}_i$, with $\mathbf{V}_K$ the first M columns of the DFT matrix $\sqrt{K}\mathbf{W}_K$.

Stacking the results obtained in (4.21) for all G pilot clusters, $\mathbf{y}^{\{\mathcal{O}\}} \triangleq (\mathbf{y}^{\{\mathcal{O}_0\}T} \ \cdots \ \mathbf{y}^{\{\mathcal{O}_{G-1}\}T})^T$, where the set $\mathcal{O}$ is given by $\mathcal{O} \triangleq \{\mathcal{O}_0, \ldots, \mathcal{O}_{G-1}\}$, we obtain

$$\mathbf{y}^{\{\mathcal{O}\}} = \mathbf{A}\mathbf{c} + \mathbf{i} + \mathbf{z}^{\{\mathcal{O}\}}, \tag{4.23}$$

where $\mathbf{i}$ and $\mathbf{z}^{\{\mathcal{O}\}}$ are similarly defined as $\mathbf{y}^{\{\mathcal{O}\}}$, and

$$\mathbf{A} \triangleq (\mathbf{A}_0^T \ \cdots \ \mathbf{A}_{G-1}^T)^T = \mathbf{A}_\mathrm{c}\mathbf{A}_\mathrm{d}, \tag{4.24}$$

with

$$\mathbf{A}_\mathrm{c} \triangleq \begin{pmatrix} \mathbf{U}_{\mathrm{c},0}^{\{\mathcal{O}_0,\mathcal{P}\}} & \cdots & \mathbf{U}_{\mathrm{c},I-1}^{\{\mathcal{O}_0,\mathcal{P}\}} \\ \vdots & & \vdots \\ \mathbf{U}_{\mathrm{c},0}^{\{\mathcal{O}_{G-1},\mathcal{P}\}} & \cdots & \mathbf{U}_{\mathrm{c},I-1}^{\{\mathcal{O}_{G-1},\mathcal{P}\}} \end{pmatrix}$$

$$= \mathbf{W}_K^{\{\mathcal{O},:\}}(\mathbf{U}_{\mathrm{d},0} \ \cdots \ \mathbf{U}_{\mathrm{d},I-1})\left(\mathbf{I}_I \otimes \mathbf{W}_K^{\{\mathcal{P},:\}H}\right), \tag{4.25}$$

and

$$\mathbf{A}_\mathrm{d} \triangleq \mathbf{I}_I \otimes \mathrm{diag}\{\mathbf{p}\}\mathbf{V}_K{}^{\{\mathcal{P},:\}}. \tag{4.26}$$

Remark 4.1 For the superimposed pilot scheme, the channel estimation data model can be derived in a similar way. However, since the pilot and unknown data symbols overlap in that case, i.e., $\mathbf{a} = \mathbf{p} + \mathbf{d}$, we basically have $G = 1$ and $\mathcal{P}_0 = \mathcal{D}_0 = \mathcal{O}_0 = \mathcal{P} = \mathcal{D} = \mathcal{O} = \{0, \ldots, K-1\}$. As a result, in (4.12) or (4.19) we then obtain $\mathbf{H}_\mathrm{c}^{\{\mathcal{O}_0,\mathcal{P}_0\}} = \mathbf{H}_\mathrm{c}^{\{\mathcal{O}_0,\mathcal{D}_0\}} = \mathbf{H}_\mathrm{c}$ or $\mathbf{H}_\mathrm{d}^{\{\mathcal{O}_0,\mathcal{P}\}} = \mathbf{H}_\mathrm{d}^{\{\mathcal{O}_0,\mathcal{D}\}} = \mathbf{H}_\mathrm{d}$, respectively. In spite of these differences, we can still show that the data model for the superimposed pilot scheme takes on a form that is similar to (4.16) or (4.23).

4.3.3 Channel Estimators

As is clear from the previous section, whether we work in the time or frequency domain, whether multiplexed or superimposed pilot symbols are used, the data model for channel estimation can always be expressed as a mixture of a useful part resulting from the pilot symbols, an interference part resulting from the interfering unknown data symbols, and a noise part:

$$\mathbf{y}^{\{\mathcal{O}\}} = \mathbf{A}\mathbf{c} + \mathbf{i}(\mathbf{c}) + \mathbf{z}^{\{\mathcal{O}\}}, \tag{4.27}$$

where we have made the dependence of the interference term $\mathbf{i}$ on the unknown BEM coefficients $\mathbf{c}$ explicit. Without loss of generality, we will present channel estimators based on the generic data model in (4.27). However, the inherent difference between the single-carrier channel estimators and the OFDM channel estimators will be pointed out from time to time.

We will focus on three channel estimators in this section: the linear minimum mean-squared error (LMMSE) estimator, the least squares (LS) estimator, and the best linear unbiased estimator (BLUE). All these estimators estimate the BEM coefficients $\mathbf{c}$ as

$$\hat{\mathbf{c}} = \mathbf{F}\mathbf{y}^{\{\mathcal{O}\}},$$

where $\mathbf{F}$ is the respective linear estimator. Whereas the LMMSE estimator relies on the statistics of $\mathbf{c}$, the LS estimator and the BLUE treat $\mathbf{c}$ as a deterministic variable. The unknown data symbols $\mathbf{d}$ and the noise $\mathbf{z}$ are always assumed to be stochastic and mutually uncorrelated, i.e., $\mathsf{E}\{\mathbf{dz}^H\} = \mathbf{0}$. For every estimator, we can then estimate the BEM channel vector $(\mathbf{U} \otimes \mathbf{I}_M)\mathbf{c}$ in (4.8) as $(\mathbf{U} \otimes \mathbf{I}_M)\hat{\mathbf{c}}$, and we can compute the related mean-squared error (MSE) as

$$\begin{aligned}\text{MSE} &\triangleq \mathsf{E}_{\mathbf{c},\mathbf{d},\mathbf{z}^{\{\mathcal{O}\}}}\left\{\|(\mathbf{U} \otimes \mathbf{I}_M)\hat{\mathbf{c}} - (\mathbf{U} \otimes \mathbf{I}_M)\mathbf{c}\|^2\right\} \\ &= \text{tr}\left\{(\mathbf{U} \otimes \mathbf{I}_M)\mathsf{E}_{\mathbf{c},\mathbf{d},\mathbf{z}^{\{\mathcal{O}\}}}\{(\hat{\mathbf{c}} - \mathbf{c})(\hat{\mathbf{c}} - \mathbf{c})^H\}(\mathbf{U} \otimes \mathbf{I}_M)^H\right\} \\ &= \text{tr}\left\{\mathsf{E}_{\mathbf{c},\mathbf{d},\mathbf{z}^{\{\mathcal{O}\}}}\left\{(\hat{\mathbf{c}} - \mathbf{c})(\hat{\mathbf{c}} - \mathbf{c})^H\right\}\right\},\end{aligned} \tag{4.28}$$

where $\text{tr}\{\cdot\}$ denotes the trace of a matrix and the last equality is due to the fact that we have designed $\mathbf{U}$ with orthonormal columns (cf. (4.7)). Note that this MSE does not include the modeling error of the BEM.

The performance of each channel estimator is sensitive to the design parameter ℓ. Taking a larger ℓ, we basically take less interference from unknown data symbols into account, which could improve our estimate. But on the other hand, increasing ℓ reduces the number of observation symbols and makes $\mathbf{A}$ a fatter matrix, which can be detrimental to the channel estimator.

4.3.3.1 The LMMSE Estimator

The LMMSE estimator treats $\mathbf{c}$ as a stochastic variable. We further assume it is uncorrelated with the unknown data symbols and the noise, i.e., $\mathsf{E}\{\mathbf{cz}^H\} = \mathbf{0}$ and $\mathsf{E}\{\mathbf{cd}^H\} = \mathbf{0}$. The LMMSE estimator is the linear filter $\mathbf{F}$ that minimizes the MSE between $\mathbf{c}$ and $\hat{\mathbf{c}}$:

$$\mathbf{F}_{\text{LMMSE}} \triangleq \arg\min_{\mathbf{F}} \mathsf{E}_{\mathbf{c},\mathbf{d},\mathbf{z}}\left\{\|\mathbf{F}\mathbf{y}^{\{\mathcal{O}\}} - \mathbf{c}\|^2\right\}.$$

Because the BEM coefficients $\mathbf{c}$, the unknown data symbols $\mathbf{d}$, and the noise $\mathbf{z}$ are all mutually uncorrelated, the LMMSE estimator can be derived as (Tang, Cannizzaro, Banelli, & Leus, 2007)

$$\mathbf{F}_{\text{LMMSE}} = \mathbf{R}_c\mathbf{A}^H(\mathbf{A}\mathbf{R}_c\mathbf{A}^H + \mathbf{R}_i + \mathbf{R}_z^{\{\mathcal{O}\}})^{-1}, \tag{4.29}$$

where the correlation matrices $\mathbf{R}_c$, $\mathbf{R}_i$, and $\mathbf{R}_z^{\{\mathcal{O}\}}$ are given by $\mathbf{R}_c \triangleq \mathsf{E}_{\mathbf{c}}\{\mathbf{cc}^H\}$, $\mathbf{R}_i \triangleq \mathsf{E}_{\mathbf{c},\mathbf{d}}\{\mathbf{ii}^H\}$, and $\mathbf{R}_z^{\{\mathcal{O}\}} \triangleq \mathsf{E}_{\mathbf{z}}\{\mathbf{z}^{\{\mathcal{O}\}}\mathbf{z}^{\{\mathcal{O}\}H}\}$. The computation of these correlation matrices in the time and frequency domains is given in Tang (2007).

Note that although (4.29) appears similar to the classical LMMSE estimator (Kay, 1993), it has the extra task to process the interference term $\mathbf{i}$, which depends on $\mathbf{c}$ itself. For this purpose, the proposed LMMSE estimator treats $\mathbf{i}$ as a random vector assuming the statistics of $\mathbf{c}$ and $\mathbf{d}$. A drawback of this approach is that the statistics of $\mathbf{c}$ are difficult to determine and not always reliable in practice. For instance, the Doppler spread could only be roughly known or the assumed Doppler spectrum could deviate from the actual shape. In such cases, the proposed LMMSE estimator will function suboptimally.

The MSE obtained with the LMMSE estimator (cf. (4.28)) can be expressed as

$$\text{MSE}_{\text{LMMSE}} = \text{tr}\left\{(\mathbf{A}^H(\mathbf{R}_i + \mathbf{R}_z^{\{\mathcal{O}\}})^{-1}\mathbf{A} + \mathbf{R}_c^{-1})^{-1}\right\}. \tag{4.30}$$

4.3.3.2 The Least-Squares Estimator

In contrast to the LMMSE estimator, the LS estimator treats $\mathbf{c}$ as a deterministic variable. It is the linear filter $\mathbf{F}$ that minimizes the squared error between $\mathbf{y}^{\{\mathcal{O}\}}$ and $\mathbf{A}\hat{\mathbf{c}}$:

$$\mathbf{F}_{\mathrm{LS}} \triangleq \arg\min_{\mathbf{F}} \|\mathbf{y}^{\{\mathcal{O}\}} - \mathbf{A}\mathbf{F}\mathbf{y}^{\{\mathcal{O}\}}\|^2.$$

The solution is well known (Kay, 1993) and given by the pseudo-inverse of $\mathbf{A}$, i.e.,

$$\mathbf{F}_{\mathrm{LS}} = \mathbf{A}^{\#}.$$

The LS estimator is the most robust estimator, in the sense that it requires no knowledge about the statistics of the BEM coefficients and the noise. This eliminates the risk of a mismatched knowledge. However, the LS estimator will perform poorly when the interference and/or the noise are prominent. More fundamentally, the conditioning of the matrix $\mathbf{A}$ plays a crucial role in the channel MSE of the LS estimator. Using (4.28), this channel MSE can be computed as

$$\begin{aligned}
\mathrm{MSE}_{\mathrm{LS}} &= \mathsf{E}_{\mathbf{c},\mathbf{d},\mathbf{z}}\left\{\mathrm{tr}\left\{\mathbf{A}^{\#}(\mathbf{i}+\mathbf{z}^{\{\mathcal{O}\}})(\mathbf{i}+\mathbf{z}^{\{\mathcal{O}\}})^H\mathbf{A}^{\#H}\right\}\right\} \\
&= \mathrm{tr}\left\{\mathbf{A}^{\#}\mathsf{E}_{\mathbf{c},\mathbf{d},\mathbf{z}}\left\{(\mathbf{i}+\mathbf{z}^{\{\mathcal{O}\}})(\mathbf{i}+\mathbf{z}^{\{\mathcal{O}\}})^H\right\}\mathbf{A}^{\#H}\right\} \\
&= \mathrm{tr}\left\{\mathbf{A}^{\#}(\mathbf{R}_i + \mathbf{R}_z^{\{\mathcal{O}\}})\mathbf{A}^{\#H}\right\}.
\end{aligned} \tag{4.31}$$

4.3.3.3 An Iterative BLUE

The BLUE yields a compromise between the LMMSE and LS estimators: it treats $\mathbf{c}$ as a deterministic variable, thus avoiding a possible error in calculating its statistics; at the same time, it leverages the statistics of the interference and the noise, such that they can be better suppressed than by the LS estimator. Simulation results show that the BLUE is able to yield a performance close to that of the LMMSE estimator, even if the latter uses perfect statistical knowledge.

The BLUE is the linear filter $\mathbf{F}$ that minimizes the MSE between $\mathbf{c}$ and $\hat{\mathbf{c}}$ subject to the condition that $\mathbf{c}$ is unbiased:

$$\mathbf{F}_{\mathrm{BLUE}} \triangleq \arg\min_{\mathbf{F}} \mathsf{E}_{\mathbf{d},\mathbf{z}}\left\{\|\mathbf{F}\mathbf{y}^{\{\mathcal{O}\}} - \mathbf{c}\|^2\right\}, \quad \text{s.t. } \mathsf{E}_{\mathbf{d},\mathbf{z}}\left\{\mathbf{F}\mathbf{y}^{\{\mathcal{O}\}}\right\} = \mathbf{c}.$$

The solution can be computed as in Kay (1993) by combining the interference $\mathbf{i}$ and the noise $\mathbf{z}^{\{\mathcal{O}\}}$ to a single disturbance term:

$$\mathbf{F}_{\mathrm{BLUE}} = (\mathbf{A}^H\tilde{\mathbf{R}}_\delta^{-1}(\mathbf{c})\mathbf{A})^{-1}\mathbf{A}^H\tilde{\mathbf{R}}_\delta^{-1}(\mathbf{c}), \tag{4.32}$$

where $\tilde{\mathbf{R}}_\delta(\mathbf{c})$ denotes the correlation matrix of the disturbance with $\mathbf{c}$ considered as a deterministic variable:

$$\tilde{\mathbf{R}}_\delta(\mathbf{c}) \triangleq \tilde{\mathbf{R}}_i(\mathbf{c}) + \mathbf{R}_z^{\{\mathcal{O}\}},$$

with $\tilde{\mathbf{R}}_i(\mathbf{c}) \triangleq \mathsf{E}_{\mathbf{d}}\{\mathbf{i}\mathbf{i}^H\}$. Note here that we do not take an average over $\mathbf{c}$ as in the LMMSE case, which is also the reason why we use a different notation. The calculation of $\tilde{\mathbf{R}}_\delta(\mathbf{c})$ in both the time and frequency domains can be found in Tang (2007).

The problem with (4.32) is that it cannot be calculated in closed form since due to the presence of $\tilde{\mathbf{R}}_\delta(\mathbf{c})$, its calculation requires the knowledge of $\mathbf{c}$ itself. As a remedy, we apply a recursive approach. Suppose that at the nth iteration, an estimate of $\mathbf{c}$ has been obtained, which is denoted as $\hat{\mathbf{c}}_{\mathrm{BLUE}}^{(n)}$. Then we utilize this temporary estimate to update the correlation matrix $\tilde{\mathbf{R}}_\delta(\mathbf{c})$, which in turn is used to produce the BLUE for the subsequent iteration, etc.:

$$\mathbf{F}_{\text{BLUE}}^{(n+1)} = (\mathbf{A}^H\tilde{\mathbf{R}}_\delta^{-1}(\hat{\mathbf{c}}_{\text{BLUE}}^{(n)})\mathbf{A})^{-1}\mathbf{A}^H\tilde{\mathbf{R}}_\delta^{-1}(\hat{\mathbf{c}}_{\text{BLUE}}^{(n)}),$$

$$\hat{\mathbf{c}}_{\text{BLUE}}^{(n+1)} = \mathbf{F}_{\text{BLUE}}^{(n+1)}\mathbf{y}^{\{\mathcal{O}\}}.$$

Note that a similar idea is adopted in Ghogho & Swami (2004) though in a different context. To initialize the iteration, we can set $\hat{\mathbf{c}}_{\text{BLUE}}^{(0)} = \mathbf{0}$, which results in the following expression for the first iteration:

$$\mathbf{F}_{\text{BLUE}}^{(1)} = (\mathbf{A}^H\mathbf{R}_z^{\{\mathcal{O}\}-1}\mathbf{A})^{-1}\mathbf{A}^H\mathbf{R}_z^{\{\mathcal{O}\}-1},$$

$$\hat{\mathbf{c}}_{\text{BLUE}}^{(1)} = \mathbf{F}_{\text{BLUE}}^{(1)}\mathbf{y}^{\{\mathcal{O}\}}.$$

The above expression actually corresponds to the LS estimator but weighted with the noise covariance matrix. Under the Gaussian noise assumption, this estimator is the maximum likelihood estimator (Kay, 1993) that is obtained by ignoring the interference $\mathbf{i}$.

Assuming that $\hat{\mathbf{c}}_{\text{BLUE}}^{(n)}$ will approach the theoretical BLUE $\hat{\mathbf{c}}_{\text{BLUE}}$ in (4.32), we can use (4.28) to compute an estimate of the channel MSE obtained by the iterative BLUE as

$$\begin{aligned}\text{MSE}_{\text{BLUE}} &= \mathsf{E}_{\mathbf{c},\mathbf{d},\mathbf{z}^{\{\mathcal{O}\}}}\left\{\text{tr}\left\{\mathbf{F}_{\text{BLUE}}\left(\mathbf{i}+\mathbf{z}^{\{\mathcal{O}\}}\right)\left(\mathbf{i}+\mathbf{z}^{\{\mathcal{O}\}}\right)^H\mathbf{F}_{\text{BLUE}}^H\right\}\right\} \\ &= \mathsf{E}_{\mathbf{c}}\left\{\text{tr}\left\{\mathbf{F}_{\text{BLUE}}\tilde{\mathbf{R}}_\delta(\mathbf{c})\mathbf{F}_{\text{BLUE}}^H\right\}\right\} \\ &= \mathsf{E}_{\mathbf{c}}\left\{\text{tr}\left\{(\mathbf{A}^H\tilde{\mathbf{R}}_\delta^{-1}(\mathbf{c})\mathbf{A})^{-1}\right\}\right\}.\end{aligned} \tag{4.33}$$

It must be remarked that the channel MSE of the BLUE is difficult to evaluate in closed form because the parameter $\mathbf{c}$ which is contained in $\tilde{\mathbf{R}}_\delta(\mathbf{c})^{-1}$ has to be averaged out. This forces us to resort to the Monte Carlo method to evaluate (4.33).

4.3.3.4 Optimization of ℓ

It is clear that the channel MSE of every estimator depends on ℓ (cf. (4.30), (4.31), and (4.33)). In this section, we will seek the optimal ℓ, or equivalently, the optimal number of observation samples used for channel estimation. With optimal ℓ we mean the ℓ that minimizes the channel MSE given in (4.30), (4.31), and (4.33) for the LMMSE estimator, the LS estimator, and the BLUE, respectively. A closed-form solution is, however, hard to obtain, especially for the BLUE. Alternatively, we can resort to an exhaustive search approach, which is feasible because the range of possible values for ℓ is limited. To be more specific, we have seen in Section 4.3.2 that the limits for ℓ are basically given by $\max\{0, M-\Delta-1\} \le \ell \le (P+M-1)/2$ and $\ell \le (P+D-1)/2$ for the single-carrier and OFDM system, respectively. Here, M stands for the discrete delay spread, D for the significant discrete Doppler spread, and Δ for the minimal number of samples in between two nonzero pilot symbols from different pilot clusters. In addition to these bounds, there also exist some other bounds for ℓ. First of all, we desire the matrix $\mathbf{A}$ to be of full column-rank, which is essential for the channel estimators to have a good performance in the absence of interference and noise. This means that $\ell \le (P+M)/2 - MI/(2G)$ and $\ell \le (P+D)/2 - MI/(2G)$ for the single-carrier and OFDM system, respectively. Second, we do not want the observation samples related to the nonzero pilot symbols from different pilot clusters to overlap. With δ denoting the minimal number of samples in between two pilot clusters (note that $\delta \leq \Delta$), this means that $\ell \geq (M - \delta - 1)/2$ and $\ell \geq (D - \delta - 1)/2$ for the single-carrier and OFDM system, respectively.

Fortunately, even the exhaustive search may be avoided as will become evident from the simulations, where the MSE-versus-ℓ curves for each channel estimator exhibit a monotonous track. In particular, the LS estimator yields the best performance when ℓ is maximized, whereas the LMMSE estimator and the BLUE perform best when ℓ is minimized. This can be explained as follows. As we already indicated before, there are two opposite effects. On one hand, increasing ℓ reduces the amount of interference from the unknown data symbols and hence improves the channel estimate. On the other hand, increasing ℓ makes $\mathbf{A}$ a fatter matrix and deteriorates the channel estimate. For the LS estimator, the first effect plays the most important role because the LS estimator is not good at suppressing this interference due to a lack of statistical knowledge. However, this interference poses no serious problems for the LMMSE estimator and the BLUE, since both of them can take the interference into account. Hence, for these estimators the second effect is stronger and thus ℓ should be as small as possible.

4.3.4 Channel Identifiability

In this chapter, we define channel identifiability in terms of the rank condition of the pilot-related matrix $\mathbf{A}$. It is obvious that for the LS estimator and the BLUE, $\mathbf{A}$ should be of full column-rank, i.e., $\text{rank}\{\mathbf{A}\} = MI$, where MI is the total number of BEM coefficients to be estimated. If $\mathbf{A}$ is not full column-rank, the channel cannot be correctly recovered even in an interference- and noise-free situation. Although not directly visible, the full column-rank condition for $\mathbf{A}$ is also significant for the performance of the LMMSE estimator. In this section, we will describe several sufficient conditions to obtain a full column-rank $\mathbf{A}$.

4.3.4.1 Channel Identifiability for Single-Carrier System

To discuss channel identifiability for the single-carrier system, we will adopt the following assumption on the pilot structure:

Assumption 4.1 The pilot structure satisfies the following conditions:

(C1) The length of each pilot cluster $P+1$ satisfies $P+1 \geq 2\ell + 1$.
(C2) Inside each pilot cluster, either the first or the last ℓ pilots are zero.
(C3) Every pilot cluster has at least one nonzero pilot in between the first and last ℓ pilots.
(C4) The nonzero pilot symbols from different pilot clusters should be separated by at least $M - 1$ symbols.

The structure of a pilot cluster satisfying conditions (C1), (C2), and (C3) is depicted in Fig. 4.3. Condition (C4) ensures that the observation samples related to a pilot cluster are not influenced by nonzero pilot symbols from another pilot cluster. Actually, condition (C4) corresponds to the earlier condition $\Delta \geq M - \ell - 1$ required for (4.12) to hold.

We can now state the following theorem:

Theorem 4.1 Under Assumption 4.1, the channel is identifiable if the number of pilot clusters G is greater than or equal to the BEM length, i.e.,

$$G \geq I.$$

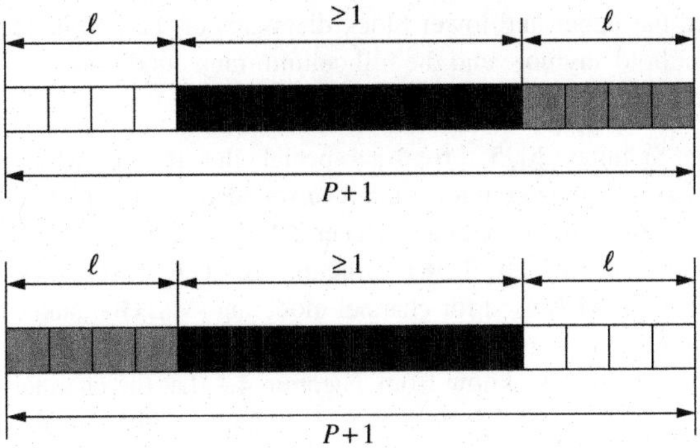

FIGURE 4.3

The two possible cases of the proposed pilot structure: the white boxes stand for the area where zero pilot symbols are located; the black boxes for the area where at least one nonzero pilot symbol is located; and the gray boxes for the area where arbitrary pilot symbols (zero or nonzero) are located.

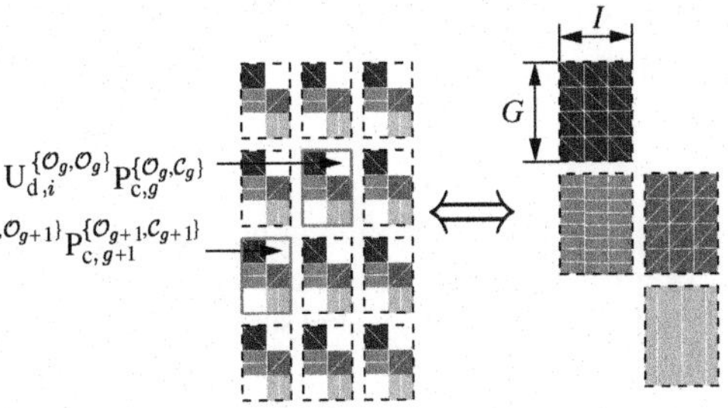

FIGURE 4.4

Changing a stack of banded matrices to a block banded matrix by row- and column-interleaving.

We only give a sketch of the proof of Theorem 4.1; a full detailed proof can be found in Tang (2007). Under (C1), (C2), (C3), and (C4), the $(P+M-2\ell)\times M$ matrix $\mathbf{P}_{c,g}^{\{\mathcal{O}_g,\mathcal{C}_g\}}$ used in (4.17) is a tall banded matrix with zeros above (below) the diagonal starting (ending) at the top left (bottom right) corner. From (4.17), the matrix $\mathbf{A}$ is a stack of G by I of such tall banded matrices, each one multiplied with a diagonal matrix, which does not change the tall banded structure. This is illustrated in the left part of Fig. 4.4. Let us now change the order of the rows and columns of $\mathbf{A}$ as illustrated in the right part of Fig. 4.4. Such a permutation does not affect the column-rank condition, and it enables us to obtain a new matrix that has a block-wise banded structure with blocks of size $G \times I$. It is then easy to prove that all the $G \times I$ blocks on the first (last) nonzero block diagonal of the permuted matrix have full column-rank, and this for all BEMs that have been considered in literature. This means that $\mathbf{A}$ has full column-rank.

Some remarks regarding Theorem 4.1 are now in order.

Remark 4.2 Assumption 4.1 is only sufficient for a full column-rank $\mathbf{A}$. However, it ensures that Theorem 4.1 holds for an arbitrary BEM and pilot choice. When $P+1<2\ell+1$, nonzero entries will have to emerge in the upper and lower diagonals of $\mathbf{P}_{c,g}^{\{\mathcal{O}_g,\mathcal{C}_g\}}$, and thus also in the upper and lower block diagonals of the permuted version of **A** (see Fig. 4.4). As a result, the proof does not hold anymore and the full column-rank condition of **A** cannot be guaranteed. It will depend on the choice of the BEM and the pilot symbols.

Remark 4.3 Ma, Giannakis, & Ohno (2003) and Kannu & Schniter (2005), propose a special pilot structure where the G pilot clusters are equidistant and every cluster consists of a nonzero pilot symbol surrounded by $M-1$ zeros to the left and right, i.e., $P+1=2M-1$. We borrow the term "time-domain Kronecker delta (TDKD) structure" from Kannu & Schniter (2005) to describe this special pilot structure. The TDKD structure has been proven to be optimal in the MSE and capacity sense provided the (C)CE-BEM is used for channel modeling (Ma, Giannakis, & Ohno, 2003; Kannu & Schniter, 2005). The TDKD structure is a special case of the pilot structure defined by Assumption 4.1 with $P+1=2M-1$ and $\ell=M-1$. In that case, we know from Theorem 4.1 that the channel is identifiable if $G \geq I$, as also argued in Ma, Giannakis, & Ohno (2003) and Kannu & Schniter (2005) for the (C)CE-BEM.

4.3.4.2 Channel Identifiability for OFDM System

For OFDM systems, the full column-rank condition of **A** is much harder to analyze, due to the fact that $\mathbf{H}_d$ is basically a full matrix. To simplify the analysis, we will adopt a specific pilot scheme here.

Assumption 4.2 The pilot structure satisfies the following conditions:

(C1) The length of each pilot cluster $P+1$ satisfies $P+1 \geq 1$.
(C2) Every pilot cluster has one nonzero pilot.
(C3) The nonzero pilots are equispaced.

Remark 4.4 In Kannu & Schniter (2005), the counterpart of the TDKD structure in the frequency domain is proposed, which is labeled the "frequency domain Kronecker delta (FDKD) structure." In the FDKD structure, the G pilot clusters are equidistant and every cluster consists of a nonzero pilot symbol surrounded by $D-1$ zeros to the left and right, i.e., $P+1=2D-1$, where D is the significant discrete Doppler spread. The FDKD structure is optimal in terms of the MSE if the (C)CE-BEM is used for channel modeling (Kannu & Schniter, 2005). The FDKD structure fits into Assumption 4.2 with $P+1=2D-1$.

Under Assumption 4.2, the full column-rank condition for the corresponding matrix **A** will be independent of the choice of the BEM and the pilot symbols, as we will show next. According to Assumption 4.2, the gth pilot cluster $\mathbf{p}_g$ can be expressed as

$$\mathbf{p}_g = (\mathbf{0}_{1\times L}\, p_g\, \mathbf{0}_{1\times R})^T,$$

where p_g stands for the nonzero pilot symbol, and L and R are the number of zeros left and right of this nonzero pilot. The total pilot vector $\mathbf{p}$ can therefore be written as

$$\mathbf{p} = \overline{\mathbf{p}} \otimes (\mathbf{0}_{1\times L}\, 1\, \mathbf{0}_{1\times R})^T,$$

where $\overline{\mathbf{p}} \triangleq (p_0 \ldots p_{G-1})^T$ contains all the G nonzero pilot symbols. The positions of these nonzero pilot symbols are collected in

$$\overline{\mathscr{P}} \triangleq \{\mu, \mu + K/G, \ldots, \mu + K(G-1)/G\}, \tag{4.34}$$

with μ denoting the position of the first nonzero pilot symbol.

With the help of the notations $\overline{\mathbf{p}}$ and $\overline{\mathscr{P}}$, and taking the zero pilot symbols into account, we can now rewrite the matrix **A** in (4.24) as

$$\mathbf{A} = \overline{\mathbf{A}}_c \overline{\mathbf{A}}_d,$$

with

$$\overline{\mathbf{A}}_c \triangleq \mathbf{W}_K^{\{\mathcal{O},:\}}(\mathbf{U}_{d,0} \ \dots \ \mathbf{U}_{d,I-1})\left(\mathbf{I}_I \otimes \mathbf{W}_K^{\{\overline{\mathcal{P}},:\}H}\right), \tag{4.35}$$

$$\overline{\mathbf{A}}_d \triangleq \mathbf{I}_I \otimes \text{diag}\{\overline{\mathbf{p}}\}\mathbf{V}_K^{\{\overline{\mathcal{P}},:\}}. \tag{4.36}$$

In the above equations, the columns of $\mathbf{A}_c$ and the rows of $\mathbf{A}_d$ corresponding to the positions of the zero pilot symbols have been discarded in $\overline{\mathbf{A}}_c$ and $\overline{\mathbf{A}}_d$, respectively, which obviously does not affect the result.

Let us now examine the matrix $\overline{\mathbf{A}}_d$. First of all, $\mathbf{V}_K^{\{\overline{\mathcal{P}},:\}}$ consists of the first M columns and G equidistant rows of the DFT matrix $\sqrt{K}\mathbf{W}_K$. Hence, $\mathbf{V}_K^{\{\overline{\mathcal{P}},:\}}$ is a $G \times M$ Vandermonde matrix with rank equal to $\min\{G,M\}$. Furthermore, $\text{diag}\{\overline{\mathbf{p}}\}$ is a diagonal matrix with the nonzero pilot symbols on its diagonal and as such will not change the rank of a matrix if multiplied with it. Hence, we can state the following lemma.

Lemma 4.1 Under Assumption 4.2, the $GI \times MI$ matrix $\overline{\mathbf{A}}_d$ depends on the pilot symbols but not on the BEM and its rank is given by

$$\text{rank}\{\overline{\mathbf{A}}_d\} = \min\{G,M\}I. \tag{4.37}$$

As a result, if $G \geq M$, the rank of $\overline{\mathbf{A}}_d$ is equal to MI.

To examine the matrix $\overline{\mathbf{A}}_c$, let us introduce an additional assumption.

Assumption 4.3 All the received samples will be used for channel estimation, i.e., $\mathcal{O} = \{0,1,\dots,K-1\}$.

This means that $\mathbf{W}_K^{\{\mathcal{O},:\}} = \mathbf{W}_K$, which can thus be omitted when analyzing the rank of $\overline{\mathbf{A}}_c$. Due to (4.34), the matrix $\mathbf{W}_K^{\{\overline{\mathcal{P}},:\}H}$ can be written as

$$\mathbf{W}_K^{\{\overline{\mathcal{P}},:\}H} = \sqrt{G/K}(1 \ e^{\sqrt{-1}2\pi\mu G/K} \ e^{\sqrt{-1}2\pi\mu 2G/K} \ \dots \ e^{\sqrt{-1}2\pi\mu(K/G-1)G/K})^T \tag{4.38}$$

$$\otimes \text{diag}\{(1 \ e^{\sqrt{-1}2\pi\mu/K} \ e^{\sqrt{-1}2\pi\mu 2/K} \ \dots \ e^{\sqrt{-1}2\pi\mu(G-1)/K})^T\}\mathbf{W}_G^H. \tag{4.39}$$

It is easy to show that this Kronecker structure of $\mathbf{W}_K^{\{\overline{\mathcal{P}},:\}H}$ relates the rank of $\overline{\mathbf{A}}_c$ to the rank of

$$\begin{pmatrix} \mathbf{U}_{d,0,0} & \cdots & \mathbf{U}_{d,I-1,0} \\ \vdots & & \vdots \\ \mathbf{U}_{d,0,K/G-1} & \cdots & \mathbf{U}_{d,I-1,K/G-1} \end{pmatrix}, \tag{4.40}$$

where the $G \times G$ matrix $\mathbf{U}_{d,i,k}$ is the diagonal submatrix of $\mathbf{U}_{d,i}$ with rows and columns going from kG to $(k+1)G-1$. Permutating the rows and columns, the above matrix can be transformed into

$$\begin{pmatrix} \mathbf{U}_0 & & \\ & \ddots & \\ & & \mathbf{U}_{G-1} \end{pmatrix}, \tag{4.41}$$

where the $K/G \times I$ matrix $\mathbf{U}_g$ selects a set of K/G equidistant rows from $\mathbf{U}$ starting at g. For all BEMs that have been considered in the literature, the matrix $\mathbf{U}_g$ has a rank equal to $\min\{K/G, I\}$. So we can formulate the following lemma.

Lemma 4.2 Under Assumptions 4.2 and 4.3, the $K \times GI$ matrix $\overline{\mathbf{A}}_c$ depends on the BEM but not on the pilot symbols and its rank is given by

$$\text{rank}\{\overline{\mathbf{A}}_c\} = \min\{K/G, I\}G. \tag{4.42}$$

Hence, if $K/G \geq I$, the rank of $\overline{\mathbf{A}}_c$ is equal to GI.

For an $m \times k$ matrix $\mathbf{A}$ and a $k \times n$ matrix $\mathbf{B}$, the rank inequality reads (Horn & Johnson, 1999)

$$\text{rank}\{\mathbf{A}\} + \text{rank}\{\mathbf{B}\} - k \leq \text{rank}\{\mathbf{AB}\} \leq \min\{\text{rank}\{\mathbf{A}\}, \text{rank}\{\mathbf{B}\}\}. \tag{4.43}$$

Applying this rank inequality to (4.37) and (4.42), we obtain the following theorem.

Theorem 4.2 Under Assumptions 4.2 and 4.3, the channel is identifiable if the number of pilot clusters G satisfies

$$M \leq G \leq K/I. \tag{4.44}$$

A few remarks are in order at this stage.

Remark 4.5 One can observe the link between Theorem 4.2 and the channel identifiability for a time-invariant channel. The latter can be viewed as a special case of the time-varying channel with $I = 1$. As per (4.44), in this case, we only require that the total number of pilot symbols should be larger than or equal to the channel length, which is consistent with the channel identifiability condition for time-invariant channels as given in Negi & Cioffi (1998).

Remark 4.6 Theorem 4.2 is a sufficient condition for an arbitrary choice of the nonzero pilot symbols. To understand this, it is interesting to study an extreme case where the whole block is occupied by nonzero pilot symbols. We can view such a pilot structure as a special case of Assumption 4.2 with $P = 0$ and $G = K$. In this case, the second inequality in (4.44) can only be satisfied if $I = 1$. Because $I > 1$ for time-varying channels, the column-rank of the matrix $\mathbf{A}$ will therefore not necessarily be full, and will depend on the specific values of the pilot symbols.

4.3.5 Simulation Results

In this section, we will show some simulation results illustrating the methods proposed in the previous sections. We only consider time-varying channels that follow Jakes' Doppler spectrum (Jakes, 1974). The methods given in Zheng & Xiao (2003) are used to generate these channels. Defining the normalized Doppler spread as $\xi_{\max} = \nu T_s$, where ν is the Doppler spread and T_s is the sampling period, we will concentrate on two normalized Doppler spreads: (1) $\xi_{\max} = 0.0008$ and (2) $\xi_{\max} = 0.004$.

4.3.5.1 *Results for the OFDM System*

For the OFDM system, we set the BEM length to $I = 3$. Further, we assume the channel to be an FIR filter with $M = 4$ taps, which are independent Gaussian random variables with zero mean and a uniform power delay profile, i.e., $\mathsf{E}\{|h[n,m]|^2\} = 1/M$. In summary, we can model the time-varying channel with $MI = 12$ BEM coefficients.

To examine the proposed channel estimation algorithms, we consider an OFDM system with $K = 64$ subcarriers. We adopt an FDKD structure with $G = 8$ equidistant pilot clusters, each containing

$P+1=3$ pilot symbols. One nonzero pilot symbol is positioned in the middle of the cluster with one zero pilot on both sides, i.e., the significant discrete Doppler spread D is assumed to be $D=2$. Note that roughly 37.5% of the subcarriers are devoted to training.

For the following simulations, we will use the DKL-BEM when constructing the LMMSE estimator because both of them rely on knowledge of the channel statistics. However, for both the DKL-BEM and the LMMSE estimator, we will allow for a mismatch of the statistics by assuming a fixed $\xi_{\max}=0.002$ for all the Doppler spreads under test. For the LS estimator and the BLUE, we will consider the (O)CE-BEM because both channel estimators and the BEM are independent of the channel statistics. In addition, we will also compare our results with the channel estimation method for the (C)CE-BEM presented in Kannu & Schniter (2005). Note that this method resembles our proposed LMMSE estimator (without mismatch) but uses a data model that is only applicable to the (C)CE-BEM, i.e., the frequency-domain channel matrix is viewed as strictly circularly banded.

We first study the effect of the parameter ℓ on the channel MSE of the various channel estimators. Note that we only consider the channel MSE, omitting the BEM modeling error at this point. We have mentioned in the previous section that ℓ is upper-bounded by the minimum of $(P+D-1)/2=1.5$ and $(P+D)/2-MI/(2G)=1.25$, and lower-bounded by $(D-\delta-1)/2=-2$, because $\delta=5$. For values of ℓ in this range, we evaluate the channel MSE of the LMMSE estimator (4.30) for SNR = 10, 20, 30, and 40 dB, and depict the results in Fig. 4.5. We observe that the effect of ℓ is very limited, but one can observe a slight decrease in channel MSE if ℓ decreases. The reason why we only see a limited effect is because of the mismatch in Doppler spread. If the exact Doppler spread was known, the influence of ℓ would be more pronounced. In any case, we choose $\ell=-2$ as the optimal value, which implies that the whole OFDM symbol will be invoked for channel estimation.

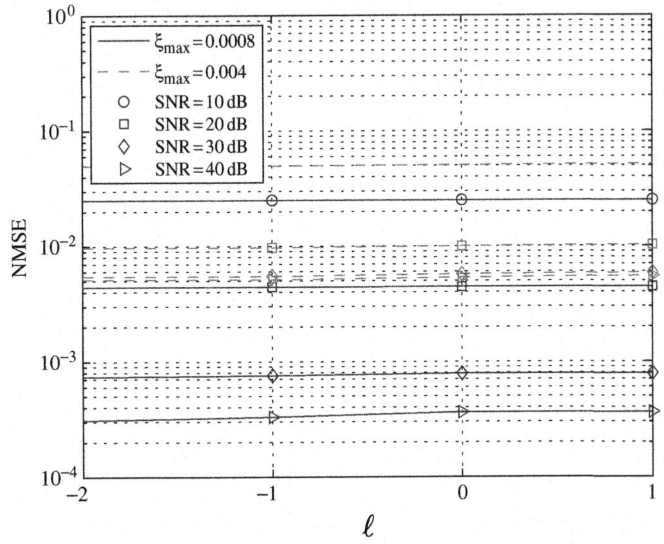

FIGURE 4.5

MSE versus ℓ for the LMMSE estimator in the OFDM case.

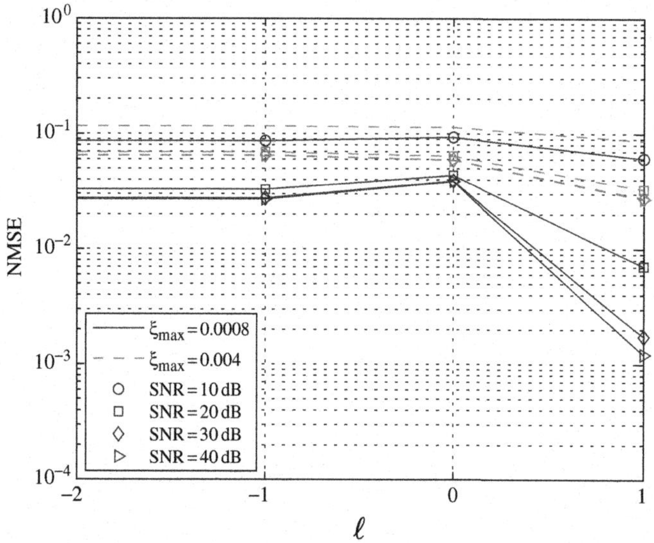

FIGURE 4.6

MSE versus ℓ for the LS estimator in the OFDM case.

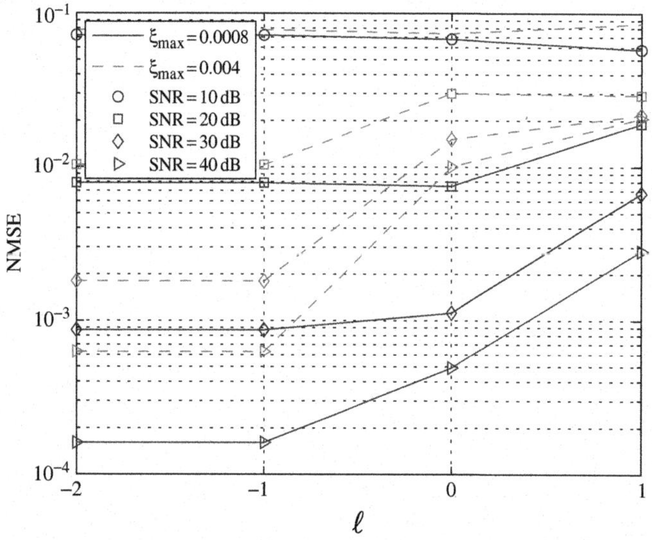

FIGURE 4.7

MSE versus ℓ for the BLUE in the OFDM case.

The results for the LS estimator are plotted in Fig. 4.6, where we observe that ℓ must be chosen as large as possible, i.e., $\ell = 1$.

For the BLUE in Fig. 4.7, a smaller ℓ always yields a lower channel MSE just as for the LMMSE estimator and we also take $\ell = -2$. Note though that for complexity reasons, one has to be careful

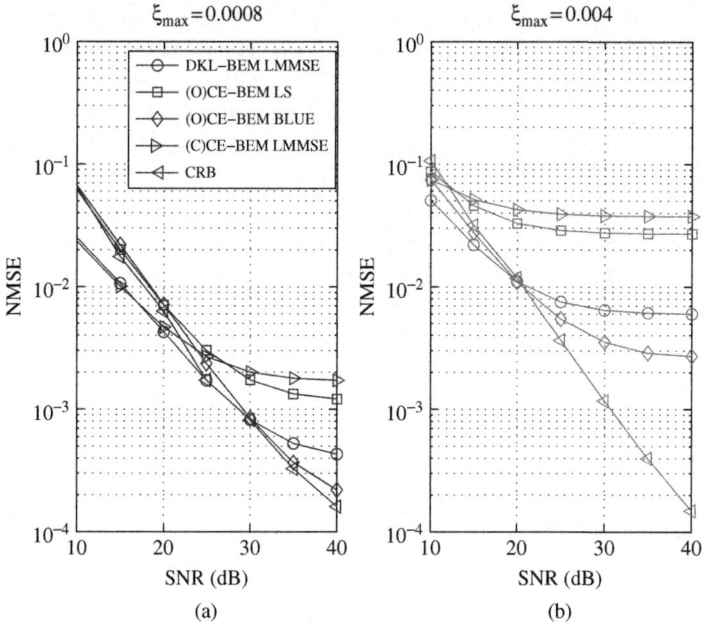

FIGURE 4.8

True channel NMSE versus SNR for different channel estimators. (a) $\xi_{\max} = 0.0008$, (b) $\xi_{\max} = 0.004$.

by taking ℓ too small because the BLUE has to be computed recursively and the procedure must be repeated for every OFDM symbol (the LMMSE estimator is in essence time-invariant and can thus be precomputed and stored off-line). A smaller ℓ often requires more iterations to reach convergence, and during each iteration, it requires a larger computational effort because more observation samples have to be processed.

Having set $\ell = -2$ for the LMMSE estimator, $\ell = 1$ for the LS estimator, and $\ell = -2$ for the BLUE, we now inspect their performance in terms of the true normalized mean squared error (NMSE), which can be defined as

$$\text{NMSE} \triangleq \frac{1}{K} \mathsf{E}_{\mathbf{h}} \left\{ \|\mathbf{h} - (\mathbf{U} \otimes \mathbf{I}_M)\hat{\mathbf{c}}\|^2 \right\}. \tag{4.45}$$

Note that the NMSE defined above explicitly takes the BEM modeling error into account. This is different from the channel MSE defined in (4.28), which merely indicates how close the estimated channel is to the best possible BEM fit. In Fig. 4.8, we depict the NMSE performance resulting from the LMMSE estimator based on a DKL-BEM, the LS estimator based on an (O)CE-BEM, and the BLUE based on an (O)CE-BEM. These results are compared with the LMMSE estimator that uses the (C)CE-BEM, which is proposed in Kannu & Schniter (2005). This estimator is different from the proposed LMMSE estimator because when adopting the (C)CE-BEM with $D = I = 3$, the frequency-domain channel matrix is strictly circularly banded and hence, the resulting LMMSE estimator cannot take the data-induced interference into account in the estimator design.

From Fig. 4.8, we can see that the NMSE performance of the LMMSE estimator based on the DKL-BEM is considerably worse if the channel varies faster. Recall that when constructing the DKL-BEM as well as the LMMSE estimator, we have assumed $\xi_{\max} = 0.002$ to emulate a mismatch of the channel statistics. The performance degradation suggests that underestimating the Doppler spread is more harmful than overestimating it. In contrast, the BLUE based on the (O)CE-BEM requires no knowledge of the channel statistics, and therefore exhibits a robust performance. We can see that its performance is quite close to the Cramer-Rao bound (CRB), whose derivation can be found in Tang, Cannizzaro, Banelli, & Leus (2007). In contrast, the LS estimator, although assuming no statistical knowledge at all, suffers from an inferior performance. However, it is still better than the LMMSE estimator based on the (C)CE-BEM (Kannu & Schniter, 2005) because the latter inherits a very large BEM modeling error.

4.3.5.2 *Results for the Single-Carrier System*

For the single-carrier system, we set the BEM length to $I = 5$. Furthermore, we assume the channel to be an FIR filter with $M = 6$ taps, which are independent Gaussian random variables with zero mean and an exponential power delay profile, i.e., $\mathsf{E}\{|h[n,m]|^2\} = c \cdot \mathrm{e}^{-m/10}$, for $m = 0, 1, \ldots, M-1$, where c denotes a normalization constant. So we can model the time-varying channel with $MI = 30$ BEM coefficients.

We consider a data block of length $K = 256$, out of which 55 symbols are reserved for pilots. The pilot symbols are grouped in $G = 5$ clusters, each cluster containing $P + 1 = 11$ pilots, so that the bandwidth efficiency is about 80%. In case Assumption 4.1 is applicable, we construct the pilot clusters accordingly, where in Fig. 4.3 the gray boxes are filled with zero pilots, and the black boxes are filled with nonzero pilots that are randomly picked constant-modulus symbols with the modulus chosen in such a way that the average pilot symbol power equals the data symbol power. We further take $G \geq I$ as indicated in Theorem 4.1. In case Assumption 4.1 is not applicable, we let all the pilot symbols take randomly picked constant-modulus symbols with the same modulus as the data symbols.

We first examine the effect of the parameter ℓ on the LMMSE estimator. As in the OFDM case, we use the DKL-BEM to approximate the time-varying channel. Both the DKL-BEM and the LMMSE estimator are based on the assumption of $\xi_{\max} = 0.002$ for both Doppler spreads under consideration. We know that ℓ is upper-bounded by the minimum of $(P + M - 1)/2 = 7.5$ and $(P + M)/2 - MI/(2G) = 5$ and lower-bounded by the maximum of $\max\{0, M - \Delta - 1\} = 0$ and $(M - \delta - 1)/2 = -17.5$ because $\Delta \geq 40$ and $\delta = 40$. Hence, we look at the performance for $\ell = 0, 3, 5$, and depict the results in Fig. 4.9. It can be seen that a smaller ℓ leads to an improved performance although the improvement is not very significant, as for the OFDM system. The reason is again the mismatch in Doppler spread as well as the fact that the exponential power delay profile limits the spread of the pilot power over multiple observation samples.

We carry out the same simulation for the LS estimator, which is based on an (O)CE-BEM. A much more pronounced performance discrepancy can be observed from Fig. 4.10. The LS estimator only functions with the largest ℓ. In that case, the observation samples only depend on the pilot symbols, which is possible in the time domain where the channel matrix is strictly circularly banded. For other values of ℓ, data-induced interference emerges, and the LS estimator suffers from a high noise floor.

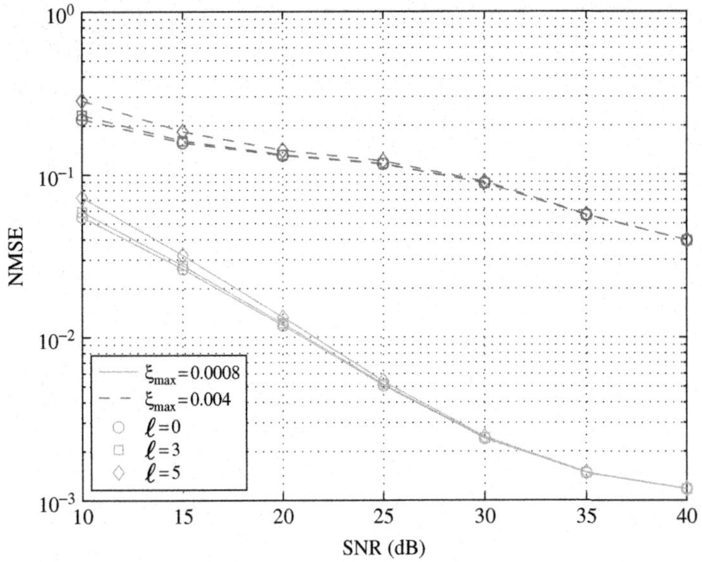

FIGURE 4.9

Performance of the LMMSE estimator in the single-carrier case.

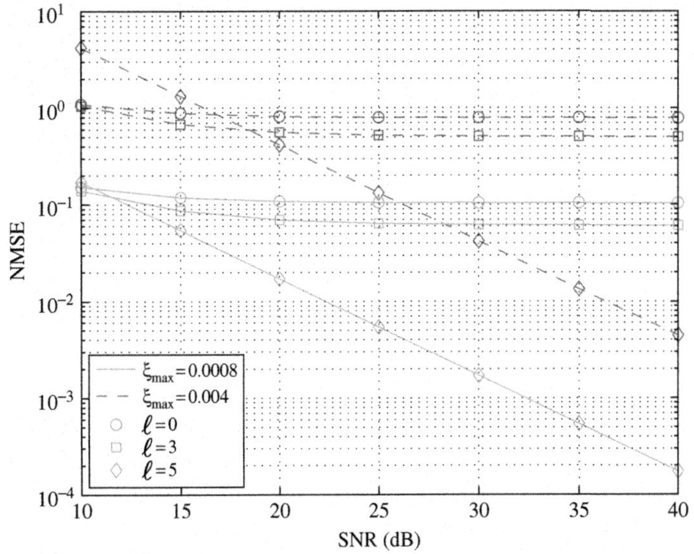

FIGURE 4.10

Performance of the LS estimator in the single-carrier case.

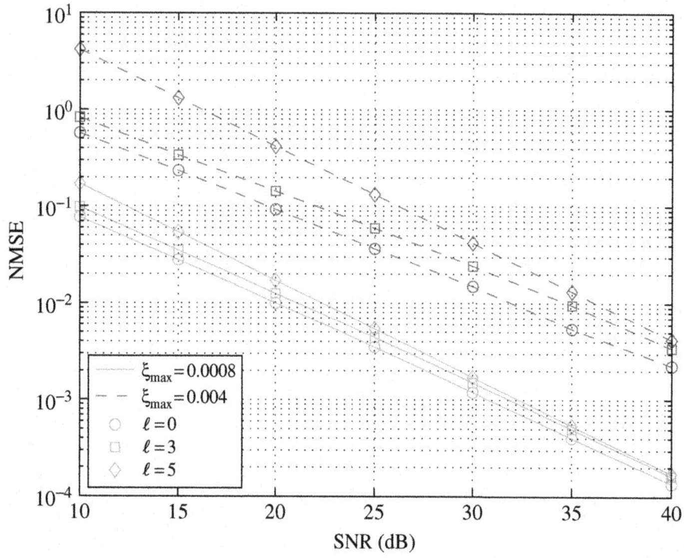

FIGURE 4.11

Performance of the BLUE in the single-carrier case.

The BLUE is also based on the (O)CE-BEM, and its behavior is analogous to that of the LMMSE estimator as can be seen in Fig. 4.11. However, the effect of ℓ can be more clearly observed than with the LMMSE estimator.

In Fig. 4.12, we compare the best performance of the different estimators, i.e., the LMMSE estimator with $\ell = 0$, the LS estimator with $\ell = 5$, and the BLUE with $\ell = 0$. The BLUE exhibits the best performance, especially if the channel varies fast. Note though that the complexity of the BLUE is the highest among the three methods.

4.4 CHANNEL ESTIMATION BASED ON MULTIPLE BLOCKS

4.4.1 Introduction

In the previous section, the channel is estimated for each block separately. To improve the performance, we will exploit the usage of multiple blocks in this section. It is, nonetheless, noteworthy that in the case of time-varying channels, the channel coherence time is rather short, which means that we cannot utilize an infinite number of blocks to enhance the estimation precision. Also note that because we are considering a time period that spans multiple blocks, the BEM could be less accurate, thereby inducing a larger modeling error.

For single-carrier systems, the channel estimation methods based on multiple blocks are basically similar to the ones based on a single block. We can simply consider a larger block size. As a result, we will not discuss this issue further.

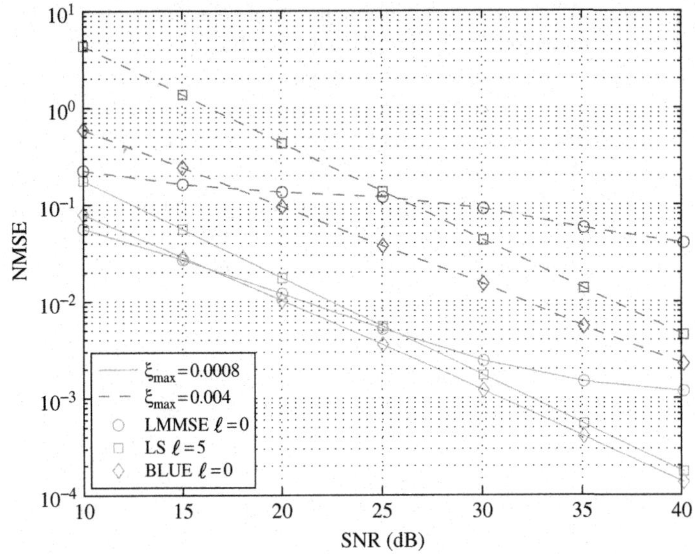

FIGURE 4.12

Comparison of the different channel estimators.

For OFDM systems, utilizing multiple blocks is, however, less straightforward. We have seen in the previous section that a pilot OFDM symbol[1] contains pilot symbols that are grouped in clusters along the frequency axis. Suppose that along the time axis, we consider J consecutive OFDM symbols, out of which V are pilot OFDM symbols. An interesting question can then be how to place the pilot symbols along both the frequency and time axes. To differentiate various pilot patterns, let us use the symbol $\mathscr{V}$ to denote the set that contains the indices of all the pilot OFDM symbols:

$$\mathscr{V} \triangleq \{j_0,\ldots,j_{V-1}\}, \tag{4.46}$$

where j_v stands for the position of the vth pilot OFDM symbol. The notation $\mathscr{P}[j_v]$ is comparable with the notation $\mathscr{P}$ introduced in the previous section, and it represents the set of pilot subcarriers within the vth pilot OFDM symbol.

Adopting the terms introduced in Coleri, Ergen, Puri, & Bahai (2002), we will basically focus on three pilot-placement scenarios.

1. The first scheme, referred to as the comb-type, is adopted in Li, Cimini, & Sollenberger (1998); Yang, Letaief, Cheng, & Cao (2001); Schafhuber, Matz, & Hlawatsch (2003); Athaudage & Jayalath (2004); Mostofi & Cox (2005); Schafhuber & Matz (2005); Tang, Leus, & Banelli (2006); Cannizzaro, Banelli, & Leus (2006). In this scheme, pilot symbols occupy only a fraction of the subcarriers, but such pilot symbols are carried by each OFDM symbol. In other words, we

[1]In this section, we refer to an OFDM symbol that carries some pilot symbols as a pilot OFDM symbol.

have $V = J$ and $|\mathscr{P}[j_v]| < K$, for $v = 0, \ldots, V-1$. This basically is the pilot scheme that we discussed in the previous section, but now extended to multiple OFDM symbols.

2. The second scheme, referred to as the block-type, is considered in Choi, Voltz, & Cassara (2001); Cui, Tellambura, & Wu (2005); Schniter (2006). In the block-type scheme, the pilot symbols occupy the entire OFDM symbol, and such pilot OFDM symbols are interleaved with data OFDM symbols. More specifically, this means that $V < J$ and $|\mathscr{P}[j_v]| = K$ for $v = 0, \ldots, V-1$.
3. The third scheme, considered in Choi & Lee (2005), is referred to as the mixed-type, which is a compromise between the comb-type and the block-type. To be more specific, the pilot symbols only occupy a fraction of the subcarriers, and such pilot OFDM symbols are interleaved with data OFDM symbols. So, we have $V < J$ and $|\mathscr{P}[j_v]| < K$ for $v = 0, \ldots, V-1$.

Some examples of these three pilot schemes are sketched in Fig. 4.13.

An intriguing question is which scheme is able to yield the most reliable channel estimate under the same bandwidth and power restrictions. Interestingly enough, conflicting results are reported; e.g., Negi & Cioffi (1998), Sanzi, Jelting, & Speidel (2003) and Choi & Lee (2005) advocate the comb-type or the mixed-type, while the block-type scheme is preferred in Cui, Tellambura, & Wu (2005) and Schniter (2006). Common to all pilot schemes is that the channel estimation method can be virtually decomposed into two steps: first, preliminary channel estimates are acquired for individual pilot OFDM symbols; second, these preliminary results are interpolated, with the aid of, e.g., channel statistics or basis functions, to obtain the final channel estimate. In terms of the intermediate channel estimation performance, the block-type scheme usually yields a better channel estimate because all the subcarriers are occupied by pilot symbols. The comb-type or mixed-type schemes work inferior because ICI leads inevitably to a noise floor.

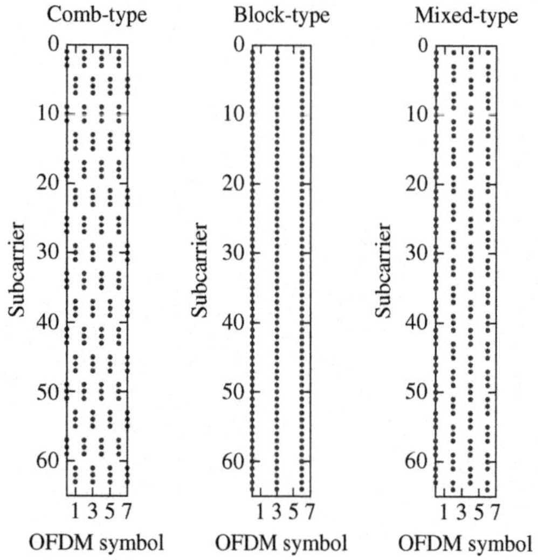

FIGURE 4.13

Illustration of different pilot schemes. The horizontal axis corresponds to the time; the vertical axis corresponds to the subcarrier positions; and the position where a pilot symbol is located is represented by a dot.

However, the block-type scheme is subject to a larger interpolation error with respect to the comb-type and mixed-type schemes because the pilot OFDM symbols are farther apart.

In the remainder of this section, we will provide a more detailed discussion of the impact of diverse pilot placement schemes on the channel estimation performance.

4.4.2 Data Model and BEM for Multiple OFDM Symbols

Compared with Section 4.3, we need to use a larger BEM to approximate the time variation of the channel over several OFDM symbols. More specifically, we will now model $J(K+M-1)$ consecutive samples of the mth channel tap, e.g., $h[0,m],\ldots,h[(J-1)(K+M-1)-1,m]$, as

$$\begin{pmatrix} h[0,m] \\ \vdots \\ h[(J-1)(K+M-1)-1,m] \end{pmatrix} \approx \underbrace{\begin{pmatrix} \mathbf{u}_0 & \cdots & \mathbf{u}_{I-1} \end{pmatrix}}_{\mathbf{U}} \begin{pmatrix} c_0[m] \\ \vdots \\ c_{I-1}[m] \end{pmatrix},$$

where $I-1$ is the BEM order, the $J(K+M-1)\times 1$ vector $\mathbf{u}_i$ is the ith BEM function, and $c_i[m]$ is the ith BEM coefficient of the mth channel tap over J OFDM symbols including the CP. Note that unlike in the single-symbol case, we also consider the part of the channel related to the CP in this model, although this is not very important.

As a result, for the jth OFDM symbol, we have

$$\begin{pmatrix} h[j(K+M-1)+M-1,m] \\ \vdots \\ h[(j+1)(K+M-1)-1,m] \end{pmatrix} \approx \underbrace{\begin{pmatrix} \mathbf{u}_0[j] & \cdots & \mathbf{u}_{I-1}[j] \end{pmatrix}}_{\mathbf{U}[j]} \begin{pmatrix} c_0[m] \\ \vdots \\ c_{I-1}[m] \end{pmatrix}, \tag{4.47}$$

where $\mathbf{U}[j]$ and $\mathbf{u}_i[j]$ are a selection of rows $j(K+M-1)+M-1$ through $(j+1)(K+M-1)-1$ from $\mathbf{U}$ and $\mathbf{u}_i$, respectively. So repeating (4.47) for all channel taps, we obtain

$$\mathbf{h}[j] \approx (\mathbf{U}[j] \otimes \mathbf{I}_M)\mathbf{c},$$

where $\mathbf{h}[j] \triangleq (\mathbf{h}_0^T[j] \cdots \mathbf{h}_{K-1}^T[j])^T$, with $\mathbf{h}_k[j] \triangleq (h[j(K+M-1)+M-1+k,0] \cdots h[j(K+M-1)+M-1+k,M-1])^T$, stacks all channel taps in the jth OFDM symbol and where $\mathbf{c} \triangleq (\mathbf{c}_0^T \cdots \mathbf{c}_{I-1}^T)^T$, with $\mathbf{c}_i \triangleq (c_i[0] \cdots c_i[M-1])^T$, stacks the BEM coefficients of all channel taps in all OFDM blocks.

By defining the BEM in this way, the resulting frequency-domain channel matrix of the jth OFDM symbol will admit an expression that is slightly different from (4.10):

$$\mathbf{H}_{\mathrm{d}}[j] = \sum_{i=0}^{I-1} \mathbf{W}_K^H \mathbf{U}_{\mathrm{d},i}[j] \mathbf{W}_K \mathbf{W}_K^H \mathbf{C}_{\mathrm{c},i} \mathbf{W}_K, \tag{4.48}$$

where $\mathbf{U}_{\mathrm{d},i}[j]$ is a diagonal matrix with $\mathbf{u}_i[j]$ on its diagonal, and $\mathbf{C}_{\mathrm{c},i}$ is a circulant matrix with $(\mathbf{c}_i^T\ \mathbf{0}_{1\times(K-M)})^T$ on its first column. It is clear that (4.48) is different from (4.10) in the sense that the BEM sequence $\mathbf{u}_i[j]$ is different for each OFDM symbol but the BEM coefficients $\mathbf{c}_i$ are not, while in (4.10), there is a common BEM for each OFDM symbol, but with different BEM coefficients.

For the same reason, we can state that for the νth pilot OFDM symbol, the observation samples collected in $\mathbf{y}^{\{\mathcal{O}[j_\nu]\}}[j_\nu]$ admit the following expression (cf. (4.23)):

$$\mathbf{y}^{\{\mathcal{O}[j_\nu]\}}[j_\nu] = \mathbf{A}[j_\nu]\mathbf{c} + \mathbf{i}[j_\nu] + \mathbf{z}^{\{\mathcal{O}[j_\nu]\}}[j_\nu],$$

with (cf. (4.24), (4.25), and (4.26))

$$\mathbf{A}[j_\nu] \triangleq \mathbf{W}_K^{\{\mathcal{O}[j_\nu],:\}}(\mathbf{U}_{\mathrm{d},0}[j_\nu] \cdots \mathbf{U}_{\mathrm{d},I-1}[j_\nu])\left(\mathbf{I}_I \otimes \mathbf{W}_K^{\{\mathcal{P}[j_\nu],:\}H}\right)$$
$$\times\left(\mathbf{I}_I \otimes \mathrm{diag}\{\mathbf{p}[j_\nu]\}\mathbf{V}_K{}^{\{\mathcal{P}[j_\nu],:\}}\right).$$

Note that we have now made the OFDM symbol index j_ν explicit to show which variables depend on the νth pilot OFDM symbol.

By repeating the above steps for all the pilot OFDM symbols and stacking the results, we obtain

$$\mathbf{y} = \mathbf{A}\mathbf{c} + \mathbf{i} + \mathbf{z}. \tag{4.49}$$

It is important to remark here that in the block-type scheme, the pilot symbols and unknown data symbols are in different OFDM symbols, and thus they are not interfering. In that case, (4.49) reduces to

$$\mathbf{y} = \mathbf{A}\mathbf{c} + \mathbf{z}.$$

Clearly, (4.49) bears an almost identical form as (4.27) for the single OFDM symbol case. Hence, all the channel estimation methods introduced in Section 4.3.3 can also be applied here with some trivial modifications. For the sake of space, we will not give all the details here, but refer the interested reader to Tang (2007).

4.4.3 Channel Identifiability Based on Multiple Blocks

In the previous section, while focusing on a single block, we said the channel is identifiable if the matrix $\mathbf{A}[j]$ has full column-rank. In practical situations, the conditions given in Theorem 4.2 may be hard to satisfy for a single OFDM symbol. We have mentioned one such example in Remark 4.6. Another quite common scenario is that the channel delay spread M could be too large to afford a sufficiently large number G of pilot clusters as per (4.44). A possible solution is to employ multiple pilot OFDM symbols. Now that the matrix $\mathbf{A}$ in (4.49) is a stack of several $\mathbf{A}[j_\nu]$ matrices, we will study in this section the impact of using multiple pilot OFDM symbols on the full column-rank condition for $\mathbf{A}$.

To understand this better, let us first rewrite $\mathbf{A}$ in (4.49) as a product of two matrices:

$$\mathbf{A} = \begin{pmatrix} \bar{\mathbf{A}}_\mathrm{c}[j_0] & & \\ & \ddots & \\ & & \bar{\mathbf{A}}_\mathrm{c}[j_{V-1}] \end{pmatrix} \begin{pmatrix} \bar{\mathbf{A}}_\mathrm{d}[j_0] \\ \vdots \\ \bar{\mathbf{A}}_\mathrm{d}[j_{V-1}] \end{pmatrix}, \tag{4.50}$$

where, by adopting the FDKD pilot structure of Assumption 4.2 in Section 4.3.4, we have (cf. (4.35) and (4.36))

$$\overline{\mathbf{A}}_{\mathrm{c}}[j_v] \triangleq \mathbf{W}_K^{\{\mathscr{O}[j_v],:\}}(\mathbf{U}_{\mathrm{d},0}[j_v] \cdots \mathbf{U}_{\mathrm{d},I-1}[j_v])\left(\mathbf{I}_I \otimes \mathbf{W}_K^{\{\mathscr{P}[j_v],:\}H}\right), \tag{4.51}$$

$$\overline{\mathbf{A}}_{\mathrm{d}}[j_v] \triangleq \mathbf{I}_I \otimes \mathrm{diag}\{\overline{\mathbf{p}}[j_v]\}\mathbf{V}_K^{\{\mathscr{P}[j_v],:\}}. \tag{4.52}$$

Using Assumption 4.3 in Section 4.3.4, we are now able to obtain two useful lemmas. The first lemma concerns the situation where every pilot OFDM symbol carries the same FDKD structure, i.e., the pilot values and positions are identical: $\overline{\mathbf{p}}[j_0] = \cdots = \overline{\mathbf{p}}[j_{V-1}]$ and $\mathscr{P}[j_0] = \cdots = \mathscr{P}[j_{V-1}]$. Here, $\overline{\mathbf{p}}[j_v]$ stands for the nonzero pilots, and $\mathscr{P}[j_v]$ stands for their positions in the vth pilot OFDM symbol.

Lemma 4.3 Under Assumptions 4.2 and 4.3, if each pilot OFDM symbol carries the same pilot structure, then the $VK \times MI$ matrix $\mathbf{A}$ will have full column-rank provided that

$$M \leq G \leq VK/I.$$

Proof. If each pilot OFDM symbol carries the same pilot structure, i.e., $\overline{\mathbf{A}}_{\mathrm{d}}[j_0] = \cdots = \overline{\mathbf{A}}_{\mathrm{d}}[j_{V-1}] = \overline{\mathbf{A}}_{\mathrm{d}}$, (4.50) becomes

$$\mathbf{A} = \begin{pmatrix} \overline{\mathbf{A}}_{\mathrm{c}}[j_0] \\ \vdots \\ \overline{\mathbf{A}}_{\mathrm{c}}[j_{V-1}] \end{pmatrix} \overline{\mathbf{A}}_{\mathrm{d}}.$$

As per Lemma 4.1, $\overline{\mathbf{A}}_{\mathrm{d}}$ has rank $\min\{G,M\}I$. Similar to the proof of Lemma 4.2, we can show that the rank of $(\overline{\mathbf{A}}_{\mathrm{c}}^T[j_0] \cdots \overline{\mathbf{A}}_{\mathrm{c}}^T[j_{V-1}])^T$ equals $\min\{VK/G,I\}G$. Then applying the rank inequality (4.43) concludes the proof. ■

Different from Lemma 4.3, let us consider a scenario where the nonzero pilot symbols are located on different subcarriers in each pilot OFDM symbol, i.e., $\mathscr{P}[j_0] \neq \cdots \neq \mathscr{P}[j_{V-1}]$. This leads to the following lemma.

Lemma 4.4 Under Assumptions 4.2 and 4.3, if the positions of the nonzero pilot symbols in each pilot OFDM symbol are different, the matrix $\mathbf{A}$ will have full column-rank provided that

$$M/V \leq G \leq K/I.$$

Proof. We can observe from (4.50) that the rank of the left matrix, which is a block diagonal matrix, is equal to the sum of the ranks of the diagonal blocks. As per Lemma 4.2, each of those diagonal blocks has rank $\min\{K/G,I\}G$, and the total rank of the left matrix is $\min\{K/G,I\}GV$. To study the rank of the right matrix in (4.50), let us permute its rows to obtain (cf. (4.52)):

$$\mathbf{I}_I \otimes \begin{pmatrix} \mathrm{diag}\{\overline{\mathbf{p}}[j_0]\} & & \\ & \ddots & \\ & & \mathrm{diag}\{\overline{\mathbf{p}}[j_{V-1}]\} \end{pmatrix} \begin{pmatrix} \mathbf{V}_K^{\{\mathscr{P}[j_0],:\}} \\ \vdots \\ \mathbf{V}_K^{\{\mathscr{P}[j_{V-1}],:\}} \end{pmatrix}. \tag{4.53}$$

Because the positions of the VG nonzero pilots are different in each pilot OFDM symbol, the $VG \times M$ matrix $\left(\mathbf{V}_K^{\{\mathcal{P}[j_0],:\}T} \cdots \mathbf{V}_K^{\{\mathcal{P}[j_{V-1}],:\}T}\right)^T$ consists of VG different rows of the Vandermonde matrix $\mathbf{V}_K$, and will have rank $\min\{VG, M\}$. Then again applying the rank inequality (4.43) concludes the proof. ■

Let us now combine the scenarios considered in Lemma 4.3 and Lemma 4.4 and introduce the following assumption.

Assumption 4.4 The total number of pilot OFDM symbols satisfies $V = V_a V_b$ with V_a and V_b being integers. Let us group the pilot OFDM symbols into V_a clusters, each containing V_b pilot OFDM symbols. Inside each cluster, we take exactly the same pilot structure, whereas among different clusters, we will give the nonzero pilot symbols different positions.

We then obtain the following theorem that guarantees the full column-rank condition for $\mathbf{A}$, or in other words channel identifiability.

Theorem 4.3 Under Assumptions 4.2, 4.3, and 4.4, the channel is identifiable over multiple OFDM symbols if the number of pilot clusters G inside each pilot OFDM symbol satisfies

$$M/V_a \leq G \leq V_b K/I. \tag{4.54}$$

The proof is essentially analogous to the proofs of Lemmas 4.3 and 4.4, and will be omitted here.

Compared with Theorem 4.2 in Section 4.3.4, we can easily observe that Theorem 4.3 relaxes the requirements on the number of pilot clusters G within a pilot OFDM symbol, which is beneficial to the bandwidth efficiency.

4.4.4 Simulation Results

In this section, we test different pilot schemes for an OFDM system, based on multiple blocks. The channel and OFDM parameters are the same as for the single OFDM symbol case.

We use the BLUE channel estimator and approximate the channel time-variation by means of the (O)CE-BEM, which spans in total $J = 8$ OFDM symbols including the CP, with each OFDM symbol containing $K = 64$ subcarriers. The order of the BEM is associated with the normalized Doppler spread $\xi_{\max}$. In the simulation, we set $I = 5$ if $\xi_{\max} \leq 0.002$; otherwise, we take $I = 9$. Note that for every pilot OFDM symbol, all observation samples are considered.

The pilot schemes to be compared are plotted in Fig. 4.13. In the comb-type scheme, all OFDM symbols are pilot OFDM symbols, thus $\mathcal{V} = \{0, 1, \ldots, 7\}$. Inside each pilot OFDM symbol, the pilots have the FDKD structure with $G = 8$ clusters of length $P + 1 = 3$. In the block-type scheme, the indices of the pilot OFDM symbols are given by $\mathcal{V} = \{0, 3, 6\}$. Inside each pilot OFDM symbol, we further set $(G, P+1) = (16, 3)$. Note that different from the traditional works, we let the block-type scheme also carry FDKD pilots to ensure the channel identifiability (see Remark 4.6). In the mixed-type scheme, we set $\mathcal{V} = \{0, 2, 4, 6\}$ and $(G, P+1) = (16, 3)$. In this way, all the pilot schemes result in an equal loss in bandwidth efficiency of about 37.5%.

As before, we use the true NMSE between the channel estimate and the actual time-varying channel as performance measure:

$$\text{NMSE}[j] \triangleq \frac{1}{K} \mathsf{E}_{\mathbf{h}}\{\|\mathbf{h}[j] - (\mathbf{U}[j] \otimes \mathbf{I}_M)\hat{\mathbf{c}}\|^2\}, \tag{4.55}$$

where $\mathbf{U}[j]$ is defined in (4.47) as the part of the BEM matrix $\mathbf{U}$ that corresponds to the jth OFDM symbol. To combat the BEM modeling error due to a large BEM window size, we adopt a sliding window approach. In other words, we will consider the NMSE only for the fourth and fifth OFDM symbol in the comb-type and mixed-type schemes, and the fifth and sixth OFDM symbols in the block-type scheme. The channels in the remaining OFDM symbols can be estimated when the window slides forward.

In the left and right plot of Fig. 4.14, we depict the performance of the three pilot schemes for short channels, $M = 4$, with normalized Doppler spreads $\xi_{\max} = 0.0008$ and $\xi_{\max} = 0.004$, respectively. In addition, we list the performance based on a single OFDM symbol, which is a special case of the comb-type scheme but considering $J = V = 1$. We can observe in the left plot of Fig. 4.14 that when the channel fading is slow, the three pilot schemes yield a similar performance, which is better than

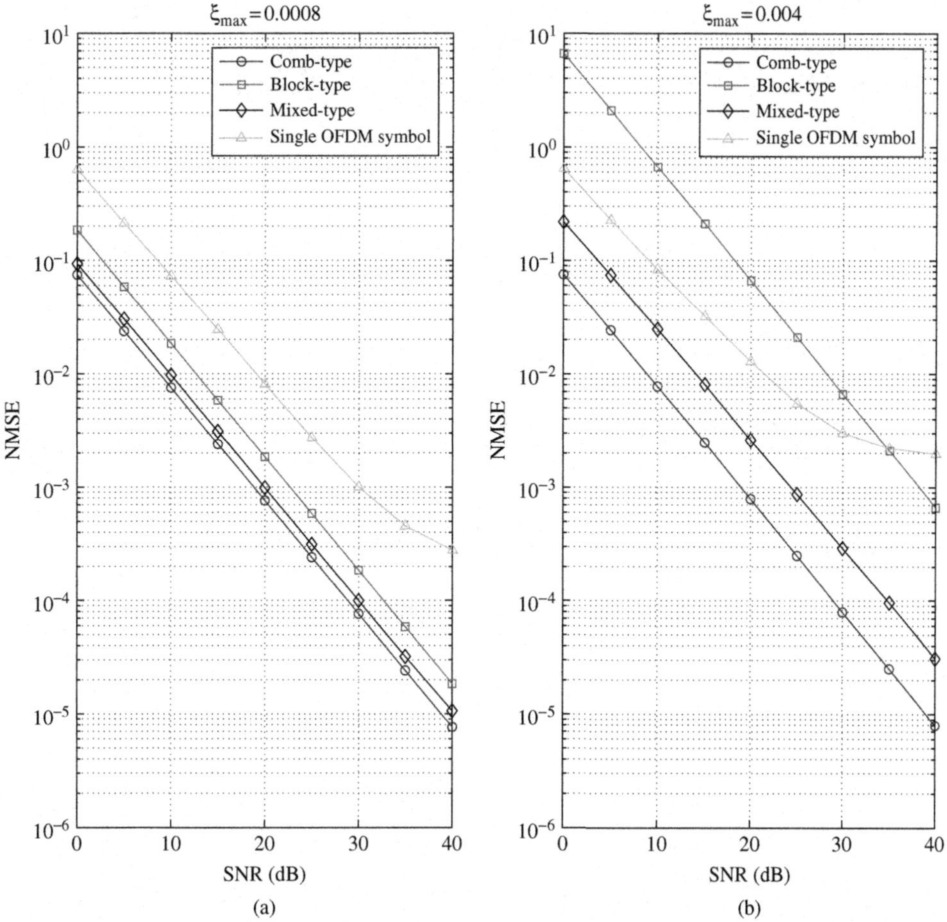

FIGURE 4.14

True channel NMSE versus SNR for $M = 4$. (a) $\xi_{\max} = 0.0008$, (b) $\xi_{\max} = 0.004$.

the performance of the single OFDM symbol case. When the channel fades faster, as illustrated in the right plot, the block-type scheme experiences more difficulty in tracking the channel compared with the other schemes. This is due to the fact that the pilot symbols are grouped in complete OFDM symbols and hence different pilot OFDM symbols are farther apart than in the other two schemes. Because the channel estimation methods implicitly carry out some interpolation over the considered time span, it is clear that the block-type scheme cannot promptly react to a time-varying channel.

The results for a much longer channel, $M = 16$, are depicted in Fig. 4.15. We observe again that when the channel varies slowly, the block-type scheme exhibits a similar performance as the other two schemes, but gets worse when the channel varies much faster as shown in the right plot. Interestingly, the performance of the comb-type scheme degrades in the right plot more severely than with $M = 4$, and suffers from a high noise floor at a moderate-to-high SNR. Compared with the other channel situations, this suggests that the comb-type scheme is inferior for channels that have a large spread

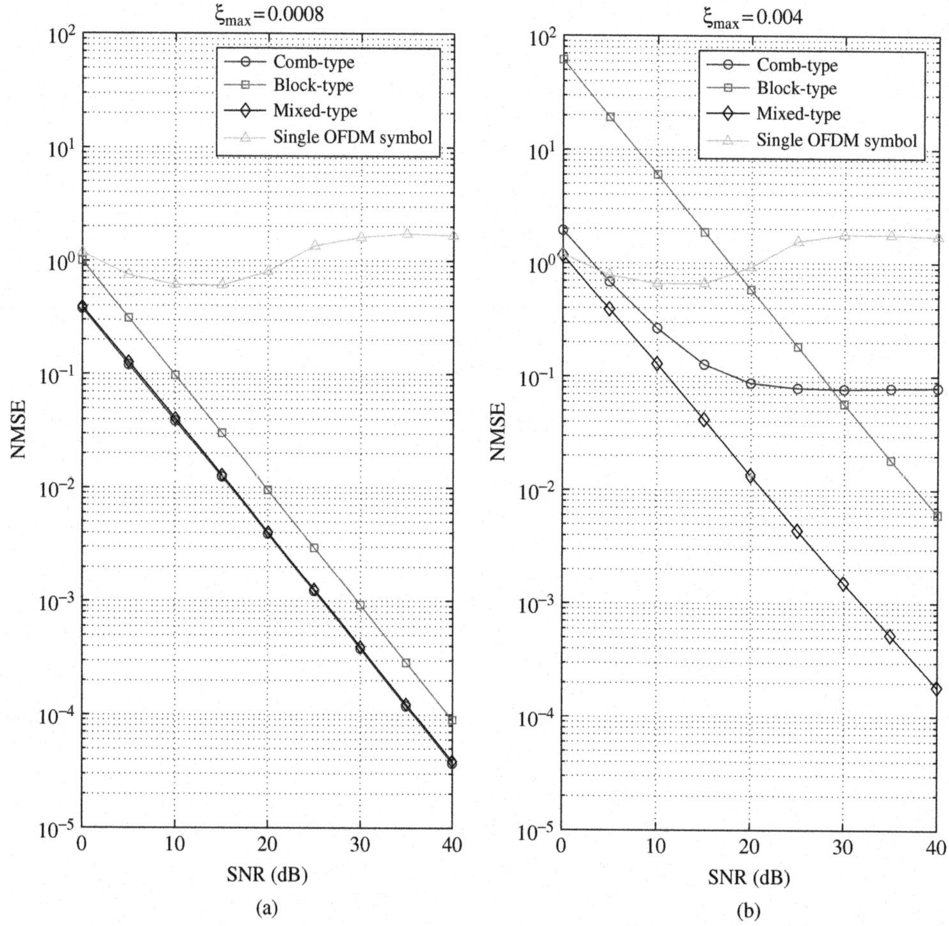

FIGURE 4.15

True channel NMSE versus SNR for $M = 16$. (a) $\xi_{max} = 0.0008$, (b) $\xi_{max} = 0.004$.

in both the Doppler and delay domains. It is noteworthy that the mixed-type scheme satisfies the conditions of both Theorem 4.2 and Theorem 4.3 for both channel lengths. In addition, it withstands the fast channel variation much better than the block-type scheme, thanks to the fact that the pilot OFDM symbols are closer to each other. These two factors endow the mixed-type scheme with a very robust channel estimation performance for all fading situations.

4.5 EXTENSION TO MIMO SYSTEMS

4.5.1 Introduction

In this section, we extend the methods discussed for a single transmit and receive antenna, or the single-input single-output (SISO) case, to multiple transmit and receive antennas, or the so-called multiple-input multiple-output (MIMO) case.

Consider a system with M_{T} transmit and M_{R} receive antennas. We assume again that every transmit antenna groups the pilot symbols in G clusters of length $P+1$, where the gth pilot cluster of the mth transmit antenna $\mathbf{p}_g^{(m)}$ has indices in $\mathscr{P}_g^{(m)} = \{P_g^{(m)}, \ldots, P_g^{(m)} + P\}$. To this pilot cluster, we again assign a set of observation samples with indices collected in the set $\mathscr{O}_g^{(m)}$, similarly defined as in (4.11) (for single-carrier systems) and (4.18) (for OFDM systems), with P_g replaced by $P_g^{(m)}$. To perform channel estimation, every receive antenna then selects the observation samples with indices in $\mathscr{O} = \{\mathscr{O}_0^{(1)}, \ldots, \mathscr{O}_{G-1}^{(1)}, \ldots, \mathscr{O}_{G-1}^{(M_{\mathrm{T}})}\}$. With this in mind, it is easy to extend the data models of (4.16) and (4.23) to multiple transmit and receive antennas. Also the channel estimators are straightforward extensions of the channel estimators derived for the SISO case.

For single-carrier systems, when $\ell \geq M-1$, there is no interference from unknown data symbols, and the channel estimation problem can actually be separated into the different receive antennas. However, when $\ell < M-1$, there is interference from unknown data symbols and the performance could benefit from using the observation samples from the other receive antennas.

For OFDM systems, however, we know that there is always interference from unknown data symbols and that all receive antennas should be treated together if we want to obtain the best possible performance.

Let us now briefly summarize some channel identifiability results for MIMO systems. Proofs can be derived in a similar manner as for the SISO case.

4.5.2 Single-Carrier System

Let us again start by adopting the following assumption on the pilot structure:

Assumption 4.5 The pilot structure satisfies the following conditions:

- **(C1)** The length $P+1$ of each pilot cluster satisfies $P+1 \geq 2\ell+1$.
- **(C2)** Inside each pilot cluster, either the first or the last ℓ pilots are zero.
- **(C3)** Every pilot cluster has at least one nonzero pilot in between the first and last ℓ pilots.
- **(C4)** The nonzero pilot symbols from different pilot clusters should be separated by at least $M-1$ symbols, viewed over all transmit antennas.

Note that these conditions are the same as in the SISO case (see Assumption 4.1 in Section 4.3.4) with the exception that condition (C4) should hold over all transmit antennas.

We can now state the following theorem:

Theorem 4.4 Under Assumption 4.5, the channel is identifiable if the number of pilot clusters per transmit antenna G is greater than or equal to the BEM length, i.e.,

$$G \geq I.$$

Note that this condition is the same as before (see Theorem 4.1), which is pleasing. However, in order to reduce the interantenna interference (IAI), it is better to overlay the nonzero pilot clusters on one transmit antenna with zero pilot clusters of the same length on the other transmit antennas. When $\ell \geq M-1$, this operation will actually separate the channel estimation problem into the different transmit antennas.

Remark 4.7 The TDKD structure has been extended to the MIMO case in Kannu & Schniter (2006) and Yang, Ma, & Giannakis (2004). It is a pilot structure where the pilot clusters on all transmit antennas overlap and where on the mth transmit antenna, the G pilot clusters are equidistant and every cluster consists of a nonzero pilot symbol surrounded by $M-1+(m-1)M$ zeros to the left and $M-1+(M_T-m)M$ zeros to the right, i.e., $P+1=(M_T+1)M-1$. As in the SISO case, the MIMO TDKD structure has been proven to be optimal in the MSE and capacity sense provided the (C)CE-BEM is used for channel modeling (Kannu & Schniter, 2006; Yang, Ma, & Giannakis, 2004). The MIMO TDKD structure is a special case of the pilot structure defined by Assumption 4.5 with $P+1=(M_T+1)M-1$ and $\ell = M-1$. In that case, we know from Theorem 4.4 that the channel is identifiable if $G \geq I$, as also argued in Kannu & Schniter (2006) and Yang, Ma, & Giannakis (2004) for the (C)CE-BEM.

4.5.3 OFDM System

For OFDM systems, we again adopt a more specific pilot structure.

Assumption 4.6 The pilot structure satisfies the following conditions:

(C1) The length $P+1$ of each pilot cluster satisfies $P+1 \geq 1$.
(C2) Every pilot cluster has one nonzero pilot.
(C3) The nonzero pilots are equispaced, viewed over all transmit antennas.

As in the single-carrier case, the conditions are the same as in the SISO case (see Assumption 4.2 in Section 4.3.4) with the exception that condition (C3) should hold over all transmit antennas. Note that this means that the nonzero pilots should not necessarily be equispaced on every transmit antenna.

Remark 4.8 In Kannu & Schniter (2006), also the FDKD structure has been extended to the MIMO case. It is a pilot structure where the pilot clusters on all transmit antennas overlap and where on the mth transmit antenna, the G pilot clusters are equidistant and every cluster consists of a nonzero pilot symbol surrounded by $D-1+(m-1)D$ zeros to the left and $D-1+(M_T-m)D$ zeros to the right, i.e., $P+1=(M_T+1)D-1$, where D is the significant discrete Doppler spread. As in the SISO case, the MIMO FDKD structure is optimal in the MSE sense if the (C)CE-BEM is used for channel modeling (Kannu & Schniter, 2006). However, this time, we cannot say that the MIMO FDKD structure is a special case of Assumption 4.6, since condition (C3) is not satisfied. The reason why we need condition (C3) is because we want to prove identifiability for a general BEM and not only for the (C)CE-BEM. Under the (C)CE-BEM, it would be enough to replace condition (C3) by the requirement that the nonzero pilots on the same transmit antenna are equispaced and the nonzero pilots from all transmit antennas are separated by at least $D-1$ symbols. In that case, the MIMO FDKD structure would fit into Assumption 4.6 with $P+1=(M_{\mathrm{T}}+1)D-1$.

Similarly to the SISO case (see Theorem 4.2), we can prove that for a single block, the following theorem holds:

Theorem 4.5 Under Assumptions 4.6 and 4.3, the channel is identifiable if the number of pilot clusters per transmit antenna G satisfies

$$M \leq G \leq K/(IM_T). \tag{4.56}$$

So the lower bound is the same as in the SISO case, but the upper bound is decreased with the number of transmit antennas M_T. As in the single-carrier case, we can reduce the IAI by overlaying the nonzero pilot clusters on one transmit antenna with zero pilot clusters of the same length on the other transmit antennas. In contrast to the single-carrier case though, we will generally not be able to completely separate the channel estimation problem into the different transmit antennas.

Remark 4.9 In Dai (2007), optimal pilot structures have been proposed for MIMO OFDM systems. However, it is assumed there that the discrete Doppler spread is exactly limited to D, the pilot clusters on all transmit antennas overlap and have length $P+1=D$, and one observation sample per pilot cluster is considered. The condition on the number of clusters G per transmit antenna then becomes $G \geq M_{\mathrm{T}}MI$, which is much worse than our condition in (4.56).

For multiple blocks and a mixed-mode pilot scheme, we can actually consider two cases, as illustrated in Fig. 4.16. When the pilot OFDM symbols from the different transmit antennas do not overlap (left side of Fig. 4.16), the different transmit antennas do not interfere with each other as far as channel identifiability is concerned, and we can actually use the results from the SISO case.

Theorem 4.6 Under Assumptions 4.2, 4.3, and 4.4, and assuming that the pilot OFDM symbols from the different transmit antennas do not overlap, the channel is identifiable over multiple OFDM symbols if the number of pilot clusters per transmit antenna G inside each pilot OFDM symbol satisfies

$$M/V_a \leq G \leq V_b K/I.$$

When the pilot OFDM symbols from the different transmit antennas overlap (right side of Fig. 4.16), however, we have a more complicated situation. In that case, the following theorem applies:

Theorem 4.7 Under Assumptions 4.6, 4.3, and 4.4, and assuming that the pilot OFDM symbols from the different transmit antennas overlap, the channel is identifiable over multiple OFDM symbols if the number of pilot clusters per transmit antenna G inside each pilot OFDM symbol satisfies

$$M/V_a \leq G \leq V_b K/(IM_T).$$

So as before, only the upper bound has been affected and is decreased by the factor M_T. The IAI can again be reduced by inserting additional zero pilot clusters overlaying the nonzero pilot clusters as indicated before, and this holds whether the pilot OFDM symbols from the different transmit antennas overlap or not.

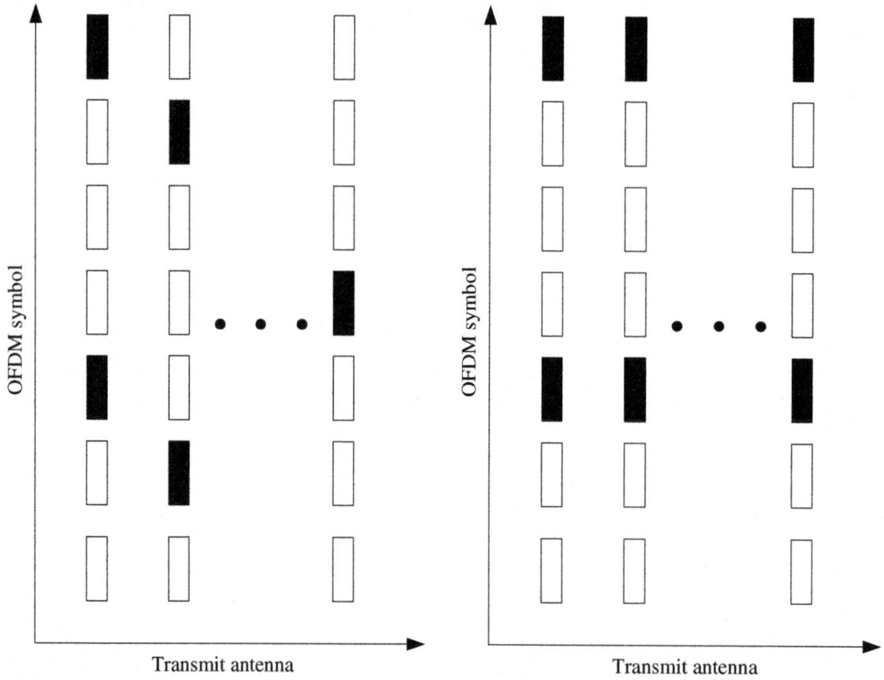

FIGURE 4.16

Two possible pilot structures for MIMO–OFDM, when multiple symbols are considered. The black boxes represent pilot OFDM symbols, whereas the white boxes represent data OFDM symbols.

4.6 ADAPTIVE CHANNEL ESTIMATION

In the earlier sections, we have shown batch approaches to estimate the time-varying channel BEM coefficients by dedicated pilot symbols in single-carrier and OFDM scenarios, with a single block or multiple blocks of observations. In this section, we will explain how all the proposed approaches can be cast in an adaptive framework, which could be useful to track not only the instantaneous channel variations but also the possible nonwide-sense stationary channel conditions and, potentially, to reduce the amount of redundant pilots. Whichever is the design of the pilots and of the observation samples, we have shown that the channel estimation relies on the observation equations (4.27). For notational simplicity, we will assume that each observation window is contiguous to the next one and coincident with a single-block duration, either in the single-carrier or the OFDM scenario. However, the time-domain formulation of the channel-estimation problem, especially in single-carrier scenarios, can naturally lead to observation windows that overlap in time. Thus, the results that follow can be easily extended to an overlapped observation approach. This overlapped observation approach may be particularly useful in very high Doppler scenarios in order to avoid abrupt changes in the evolution of the BEM coefficients, which may be hard to track otherwise.

To further simplify the problem, we also assume that the pilot design and the observation samples are chosen in such a way that the interference term in (4.27) can be neglected and, consequently, the observation equation reduces to

$$\mathbf{y}^{\{\mathcal{O}[n]\}}[n] = \mathbf{A}[n]\mathbf{c}[n] + \mathbf{z}^{\{\mathcal{O}[n]\}}[n], \tag{4.57}$$

where, as before, the index n has been used to identify the nth observation window. The observation matrix $\mathbf{A}[n]$, defined in (4.17), may depend on n or not, depending on how we select the observation index set $\mathcal{O}[n] = \{\mathcal{O}_0[n], \ldots, \mathcal{O}_G[n]\}$ and the pilot design $\mathbf{p}[n]$, which may change from one block to another. Note that $\mathbf{A}[n]$ may also depend on n via the underlaying BEM basis matrix $\mathbf{U}$, which, instead of being fixed for all n, can be smoothly linked from one block to another in order to facilitate tracking (see He & Tugnait, 2007). In the adaptive channel estimation literature, it is quite common to assume that a time-varying channel obeys an auto-regressive (AR) model evolution, which in the simplest case reduces to a Gauss–Markov process evolution. Several authors assume that also the BEM channel coefficients evolve in a similar fashion from one observation window to another, as expressed by

$$\mathbf{c}[n] = \mathbf{\Phi}[n]\mathbf{c}[n-1] + \boldsymbol{\beta}[n], \tag{4.58}$$

where $\boldsymbol{\beta}[n]$ represents the BEM coefficient innovation with respect to the $(n-1)$st observation window. Actually, (4.57) and (4.58) are the observation and the model evolution equations, respectively, that typically cast the channel-estimation problem in a vector-state vector-observation (VSVO) Kalman filter formulation. Thus, the Kalman filter estimate of the BEM coefficients can be recursively computed during each observation window according to the following set of equations:

$$\begin{aligned}
\mathbf{M}^F[n] &= \mathbf{\Phi}[n]\mathbf{M}^A[n-1]\mathbf{\Phi}^H[n] + \mathbf{Q}[n] \\
\mathbf{K}[n] &= \mathbf{M}^F[n]\mathbf{A}^H[n]\left(\mathbf{R}_{\tilde{n}} + \mathbf{A}[n]\mathbf{M}^F[n]\mathbf{A}^H[n]\right)^{-1} \\
\hat{\mathbf{c}}[n] &= \mathbf{\Phi}[n]\hat{\mathbf{c}}[n-1] + \mathbf{K}[n]\left(\mathbf{y}^{\{\mathcal{O}[n]\}}[n] - \mathbf{A}[n]\mathbf{\Phi}[n]\hat{\mathbf{c}}[n-1]\right) \\
\mathbf{M}^A[n] &= (\mathbf{I} - \mathbf{K}[n]\mathbf{A}[n])\,\mathbf{M}^F[n],
\end{aligned} \tag{4.59}$$

where $\mathbf{M}^F[n]$ and $\mathbf{M}^A[n]$ are the *forward* and the *a-posteriori* error covariance matrices, respectively, $\mathbf{K}[n]$ is the *Kalman gain* matrix, and $\mathbf{Q}[n]$ is the covariance matrix of the innovation of the BEM coefficients; their initial conditions for $n = 0$ can be set according to the classical Yule–Walker equations (Kay, 1993). Moreover, together with the evolution matrix $\mathbf{\Phi}[n]$, they can be matched to the channel statistics as specifically proposed in Cannizzaro, Banelli, & Leus (2006) for a VSVO Kalman formulation. A major complexity reduction associated with the matrix inversion in the Kalman gain update equation is obtained by an equivalent vector-state scalar-observation (VSSO) formulation that decouples the estimation problem into the different BEM coefficients, as proposed in Muralidhar & Kwok (2009). Another VSVO Kalman approach, which includes also a Kalman equalizer, has been proposed in He & Tugnait (2007) for single-carrier transmissions. This approach has also been extended to SIMO (He & Tugnait, 2008) and MIMO (Kim & Tugnait, 2008), where in addition a data-aided approach has been considered, as formerly explored in Banelli, Cannizzaro, & Rugini (2007) for OFDM. In order to reduce the complexity, He & Tugnait (2007); He & Tugnait (2008); Tugnait & He (2008); Kim & Tugnait (2008, 2009a,b); Song & Tugnait (2009) ignore the coupling among the evolutions of the BEM coefficients, by approximating the evolution matrix as a diagonal one (i.e., $\mathbf{\Phi}[n] = \phi[n]\mathbf{I}$), and they simply adapt $\phi[n]$ to the maximum Doppler frequency. Additionally, these single-carrier approaches take advantage of partially overlapped observation windows to improve tracking at the expense of some complexity increase. To further reduce the complexity, other adaptive techniques can be used, such as LMS algorithms, RLS algorithms (Cannizzaro, Banelli, & Leus, 2006; Kim & Tugnait, 2009a), or Wiener LMMSE algorithms (Lindbom, Sternad, & Ahlen, 2001), which usually trade off performance for complexity with respect to the optimum LMMSE Kalman solution.

4.7 CONCLUSIONS

In this chapter, we have discussed pilot-aided channel estimation for time-varying channels. We have focused on a block approach where the channel is modeled as a BEM within the block. Channel estimation has been studied for single-carrier as well as OFDM systems. The channel is always estimated in the time domain, but the pilot symbols are inserted in the time domain for single-carrier systems and in the frequency domain for OFDM systems. Different channel estimators have been studied and the number of observation samples taken into account for channel estimation has been optimized. We have first looked at the case where a single block of data is considered, and we have subsequently extended our discussion to the case where multiple blocks of data are used. The latter leads to less restrictive channel identifiability conditions, and can allow for a higher bandwidth efficiency. We have also considered extensions to multiple antenna systems and briefly discussed adaptive channel estimation.

References

Athaudage, C. R. N., & Jayalath, A. D. S. (2004). Enhanced MMSE channel estimation using timing error statistics for wireless OFDM systems. *IEEE Transactions on Broadcasting*, *50*(4), 369–376.

Banelli, P., Cannizzaro, R. C., & Rugini, L. (2007). Data-aided Kalman tracking for channel estimation in Doppler-affected OFDM systems. In *Proceedings of IEEE ICASSP-07* (pp. 133–136). Honolulu, HI.

Borah, D. K., & Hart, B. D. (1999a). A robust receiver structure for time-varying, frequency-flat Rayleigh fading channels. *IEEE Transactions on Communications*, *47*(3), 862–873.

Borah, D. K., & Hart, B. D. (1999b). Frequency-selective fading channel estimation with a polynomial time-varying channel model. *IEEE Transactions on Communications*, *47*(6), 862–873.

Cai, X., & Giannakis, G. B. (2003). Bounding performance and suppressing intercarrier interference in wireless mobile OFDM. *IEEE Transactions on Communications*, *51*(12), 2047–2056.

Cannizzaro, R. C., Banelli, P., & Leus, G. (2006, July). Adaptive channel estimation for OFDM systems with Doppler spread. In *Proceedings of IEEE SPAWC-06*, Cannes, France.

Choi, J.-W., & Lee, Y.-H. (2005). Optimum pilot pattern for channel estimation in OFDM systems. *IEEE Transactions on Wireless Communications*, *4*(5), 2083–2088.

Choi, Y.-S., Voltz, P. J., & Cassara, F. A. (2001). On channel estimation and detection for multicarrier signals in fast and selective Rayleigh fading channels. *IEEE Transactions on Communications*, *49*(8), 1375–1387.

Cirpan, H. A., & Tsatsanis, M. K. (1999). Maximum likelihood blind channel estimation in the presence of Doppler shifts. *IEEE Transactions on Signal Processing*, *47*(6), 1559–1569.

Coleri, S., Ergen, M., Puri, A., & Bahai, A. (2002). Channel estimation techniques based on pilot arrangement in OFDM systems. *IEEE Transactions on Broadcasting*, *48*(3), 223–229.

Cui, T., Tellambura, C., & Wu, Y. (2005, May). Low-complexity pilot-aided channel estimation for OFDM systems over doubly-selective channels. In *Proceedings of IEEE ICC-05*, Vol. 3 (pp. 1980–1984). Seoul, Korea.

Dai, X. (2007). Optimal training design for linearly time-varying MIMO/OFDM channels modelled by a complex exponential basis expansion. *IET Communications*, *1*(5), 945–953.

Falconer, D., Ariyavisitakul, S. L., Benyamin-Seeyar, A., & Eidson, B. (2002). Frequency domain equalization for single-carrier broadband wireless systems. *IEEE Communications Magazine*, *40*(9), 58–66.

Ghogho, M., & Swami, A. (2004, July). Improved channel estimation using superimposed training. In *Proceedings of IEEE SPAWC-04* (pp. 110–114). Lisbon, Portugal.

Ghogho, M., & Swami, A. (2006, September). Estimation of doubly-selective channels in block transmissions using data-dependent superimposed training. In *Proceedings of EUSIPCO-06*, Antalya, Turkey.

Gorokhov, A., & Linnartz, J.-P. (2004). Robust OFDM receivers for dispersive time-varying channels: equalization and channel acquisition. *IEEE Transactions on Communications*, *52*(4), 572–583.

Guillaud, M., & Slock, D. T. M. (2003, April). Channel modeling and associated intercarrier interference equalization for OFDM systems with high Doppler spread. In *Proceedings of IEEE ICASSP-03*, Vol. IV (pp. 237–240). Hong Kong.

Haykin, S. (1996). *Adaptive filter theory*. Englewood Cliffs: Prentice-Hall.

He, S., & Tugnait, J. (2007, April). Doubly-selective multiuser channel estimation using superimposed training and discrete prolate spheroidal basis expansion models. In *Proceedings of IEEE ICASSP-07*, Honolulu, HI.

He, S., & Tugnait, J. K. (2007). Doubly-selective channel estimation using exponential basis models and subblock tracking. In *Proceedings of IEEE GC-07* (pp. 2847–2851). Washington, DC.

He, S., & Tugnait, J. K. (2008). Decision-directed tracking of doubly-selective channels using exponential basis models. In *Proceedings of IEEE ICC-08* (pp. 5098–5102). Beijing, China.

Horn, R. A., & Johnson, C. R. (1999). *Matrix analysis*. Cambridge University Press, Cambridge, UK.

Jakes, W. C. (1974). *Microwave mobile channels*. New York: Wiley.

Kannu, A. P., & Schniter, P. (2005, March). MSE-optimal training for linear time-varying channels. In *Proceedings of IEEE ICASSP-05*, Vol. III (pp. 789–792). Philadelphia, PA.

Kannu, A. P., & Schniter, P. (2006, March). Minimum mean-squared error pilot-aided transmission for MIMO doubly selective channels. In *Proceedings of CISS-06* (pp. 134–139). Princeton, NJ.

Kay, S. M. (1993). *Fundamentals of statistical signal processing: Estimation theory*. Upper Saddle River, NJ: Prentice Hall.

Kim, H., & Tugnait, J. K. (2008). Doubly-selective MIMO channels estimation using exponential basis models and subblock tracking. In *Proceedings of IEEE SPAWC-08* (pp. 1258–1261). Recife, Brasil.

Kim, H., & Tugnait, J. K. (2009a). Recursive least-squares decision-directed tracking of doubly-selective channels using exponential basis models. In *Proceedings of IEEE ICASSP-09* (pp. 2689–2693). Taipei, Taiwan.

Kim, H., & Tugnait, J. K. (2009b). Turbo equalization of doubly-selective MIMO fading channels using exponential basis models. In *Proceedings of IEEE SPAWC-09* (pp. 21–24). Perugia, Italy.

Leus, G. (2004, September). On the estimation of rapidly time-varying channels. In *Proceedings of EUSIPCO-04* (pp. 2227–2230). Vienna, Austria.

Leus, G., & Moonen, M. (2003, June). Deterministic subspace based blind channel estimation for doubly-selective channels. In *Proceedings of IEEE SPAWC-2003* (pp. 210–214). Rome, Italy.

Leus, G., & van der Veen, A.-J. (2005). Channel estimation, in *Smart antennas – state of the art*. Ed. T. Kaiser, A. Bourdoux, H. Boche, J. R. Fonollosa, J. B. Andersen and W. Utschick, Hindawi, New York, USA.

Lindbom, L., Sternad, M., & Ahlen, A. (2001). Tracking of time-varying mobile radio channels–part I: The Wiener LMS algorithm. *IEEE Transactions on Communications*, *49*(12), 2207–2217.

Li, Y., Cimini, L. J., & Sollenberger, N. R. (1998). Robust channel estimation for OFDM systems with rapid dispersive fading channels. *IEEE Transactions on Communications*, *46*(7), 1146–1162.

Ma, X., & Giannakis, G. B. (2003). Maximum-diversity transmissions over doubly selective wireless channels. *IEEE Transactions on Information Theory*, *49*(7), 1832–1840.

Ma, X., Giannakis, G., & Ohno, S. (2003). Optimal training for block transmissions over doubly-selective fading channels. *IEEE Transactions on Signal Processing*, *51*(5), 1351–1366.

Mostofi, Y., & Cox, D. C. (2005). ICI mitigation for pilot-aided OFDM mobile systems. *IEEE Transactions on Wireless Communications*, *4*(2), 765–774.

Muralidhar, K., & Kwok, H. L. (2009). A low-complexity Kalman approach for channel estimation in doubly-selective OFDM systems. *IEEE Signal Processing Letters*, *16*(7), 632–635.

Negi, R., & Cioffi, J. (1998). Pilot tone selection for channel estimation in a mobile OFDM system. *IEEE Transactions on Consumer Electronics*, *44*(3), 1122–1128.

Nicoli, M., Simeone, O., & Spagnolini, U. (2003). Multislot estimation of frequency-selective fast-varying channels. *IEEE Transactions on Communications*, *51*(8), 1337–1347.

Rousseaux, O., Leus, G., Stoica, P., & Moonen, M. (2004). Gaussian maximum likelihood channel estimation with short training. *IEEE Transactions on Wireless Communications*, *4*(6), 2945–2955.

Rousseaux, O., Leus, G., & Moonen, M. (2006). Estimation and equalization of doubly-selective channels using known symbol padding. *IEEE Transactions on Signal Processing*, *54*(3), 979–990.

Sanzi, F., Jelting, S., & Speidel J. (2003). A comparative study of iterative channel estimation for mobile OFDM systems. *IEEE Transactions on Wireless Communications*, *2*(5), 849–859.

Scaglione, A., Giannakis, G. B., & Barbarossa, S. (1999a). Redundant filterbank precoders and equalizers – part I: Unification and optimal designs. *IEEE Transactions on Signal Processing*, *47*(7), 1988–2006.

Scaglione, A., Giannakis, G. B., & Barbarossa, S. (1999b). Redundant filterbank precoders and equalizers – part II: Blind channel estimation, synchronization and direct equalization. *IEEE Transactions on Signal Processing*, *47*(7), 2007–2022.

Schafhuber, D., & Matz, G. (2005). MMSE and adaptive prediction of time-varying channels for OFDM systems. *IEEE Transactions on Wireless Communications*, *4*(2), 593–602.

Schafhuber, D., Matz, G., & Hlawatsch, F. (2003, November). Kalman tracking of time-varying channels in wireless MIMO-OFDM systems. In *Proceedings of Asilomar-03* (pp. 1261–1265). Pacific Grove, CA.

Schniter, P. (2006, March). On doubly dispersive channel estimation for pilot-aided pulse-shaped multicarrier modulation. In *Proceedings of CISS-06* (pp. 1296–1301). Princeton, NJ.

Schniter, P., & Liu, H. (2003, November). Iterative equalization for single-carrier cyclic-prefix in doubly dispersive channels. In *Proceedings of Asilomar-03* (pp. 502–506). Pacific Grove, CA.

Song, L., & Tugnait, J. K. (2009). Doubly-selective fading channels equalization: A comparison of the Kalman filter approach with the exponential basis expansion model-based equalizers. *IEEE Transactions on Wireless Communications*, *1*(8), 60–64.

Stamoulis, A., Diggavi, S. N., & Al-Dhahir, N. (2002). Intercarrier interference in MIMO OFDM. *IEEE Transactions on Signal Processing*, *50*(10), 2451–2464.

Tang, Z. (2007). *OFDM transmission over rapidly changing channels*. Ph.D Disseration, Delft University of Technology, The Netherlands.

Tang, Z., & Leus, G. (2008). A novel receiver architecture for single-carrier transmission over time-varying channels. *IEEE Journal on Selected Areas in Communications*, *26*(2), 366–377.

Tang, Z., Leus, G., & Banelli, P. (2006, June). Pilot-assisted time-varying OFDM channel estimation based on multiple OFDM symbols. In *Proceedings of IEEE SPAWC-06*, Cannes, France.

Tang, Z., Cannizzaro, R. C., Banelli, P., & Leus, G. (2007). Pilot-assisted time-varying channel estimation for OFDM systems. *IEEE Transactions on Signal Processing*, *55*(5), 2226–2238.

Teo, K. D., & Ohno, S. (2005). Optimal MMSE finite parameter model for doubly-selective channels. In *Proceedings of IEEE GLOBECOM-05*, Vol. 6 (pp. 3503–3507). St. Louis, MO.

Thomas, T. A., & Vook, F. W. (2000, March). Multi-user frequency-domain channel identification, interference suppression, and equalization for time-varying broadband wireless communications. In *Proceedings of IEEE SAM-00* (pp. 444–448). Cambridge, MA.

Tomasin, S., Gorokhov, A., Yang, H., & Linnartz, J.-P. (2005). Iterative interference cancellation and channel estimation for mobile OFDM. *IEEE Transactions on Wireless Communications*, *4*(1), 238–245.

Tong, L., Sadler, B. M., & Dong, M. (2004). Pilot-assisted wireless transmissions: General model, design criteria, and signal processing. *IEEE Signal Processing Magazine*, *21*(6), 12–25.

Tsatsanis, M. K., & Giannakis, G. B. (1996). Modeling and equalization of rapidly fading channels. *International Journal of Adaptive Control and Signal Processing*, *10*(2/3), 159–176.

Tugnait, J. K., & He, S. (2008). Recursive least-squares doubly-selective channel estimation using exponential basis models and subblock-wise tracking. In *Proceedings of IEEE ICASSP-08* (pp. 2861–2864). Las Vegas, NV.

Visintin, M. (1996). Karhunen-Loève expansion of a fast Rayleigh fading process. *IET Electronics Letters*, *32*(18), 1712–1713.

Yang, L., Ma, X., & Giannakis, G. B. (2004, May). Optimal training for MIMO fading channels with time- and frequency-selectivity. In *Proceedings of IEEE ICASSP-04*, Vol. III (pp. 821–824). Montreal, Canada.

Yang, B., Letaief, K. B., Cheng, R. S., & Cao, Z. (2001). Channel estimation for OFDM transmission in multi-path fading channels based on parametric channel modeling. *IEEE Transactions on Communications*, *49*(3), 467–479.

Yip, K.-W., & Ng, T.-S. (1997). Karhunen-Loève expansion of the WSSUS channel output and its application to efficient simulation. *IEEE Journal on Selected Areas in Communications*, *15*(4), 640–646.

Zakharov, Y. V., Tozer, T. C., & Adlard, J. F. (2004). Polynomial spline-approximation of Clarke's model. *IEEE Transactions on Signal Processing*, *52*(5), 1198–1208.

Zemen, T., & Mecklenbräuker, C. F. (2005). Time-variant channel estimation using discrete prolate spheroidal sequences. *IEEE Transactions on Signal Processing*, *53*(9), 3597–3607.

Zheng, Y. R., & Xiao, C. (2003). Simulation models with correct statistical properties for Rayleigh fading channels. *IEEE Transactions on Communications*, *51*(6), 920–928.

CHAPTER

5

Pilot Design and Optimization for Transmission over Time-Varying Channels

Min Dong[1], Brian M. Sadler[2], Lang Tong[3]

[1]*University of Ontario Institute of Technology, Oshawa, Ontario, Canada*

[2]*Army Research Laboratory, Adelphi, MD, USA*

[3]*Cornell University, Ithaca, NY, USA*

5.1 INTRODUCTION

To facilitate data transmission over linear time-varying (LTV) channels, pilot symbols[1] are typically inserted in the information-bearing data stream. These symbols are known at the receiver and are exploited for channel estimation, receiver adaptation, and optimal decoding. Such design structure, also called pilot-assisted transmission (PAT) (Tong, Sadler, & Dong, 2004), is prevalent in modern communication systems, and a specific pilot design is included in almost any of the current standardized wireless systems, e.g., Global System for Mobile Communication (GSM), Wideband Code-Division Multiple Access (WCDMA), CDMA-2000, IEEE802.11, IEEE802.16, DVB-T, and 3GPP Long Term Evolution (LTE). See Fig. 5.1 for examples.

Pilot symbols carry no information about the data; the resource (time, frequency, power, and so on) used on sending pilot symbols is a resource taken away from transmitting information data. Therefore, pilot design requires optimization. This includes the amount of pilot symbols, the locations of these symbols in the data stream, and the power allocated for them. In this chapter, we will look at these design issues.

In Section 5.2, we present a general model that captures a number of pilot design schemes at the transmitter. Specifically, we view the problem of pilot design as one of allocating power in different design spaces. Receiver structures are discussed next, followed by possible design metrics.

Sections 5.3–5.5 consider the design issues in a traditional single antenna single carrier system. We first focus on the optimal pilot pattern from a channel estimation and symbol detection point of view. For any fixed percentage of pilot symbols and their power allocation, the optimal placement of pilot symbols in the data stream is derived for channel tracking and data decoding in Section 5.3. The LTV channel is assumed flat fading and modeled by a Gauss-Markov process. Time-division multiplexing (TDM) of pilot symbols and data are considered there, and a causal Kalman filter is used for channel tracking. The optimal pattern among all possible periodic placements is found to be single pilot periodic placement. In Section 5.4, we look at an alternative multiplexing scheme where pilot

[1]Pilot symbols are also called training symbols.

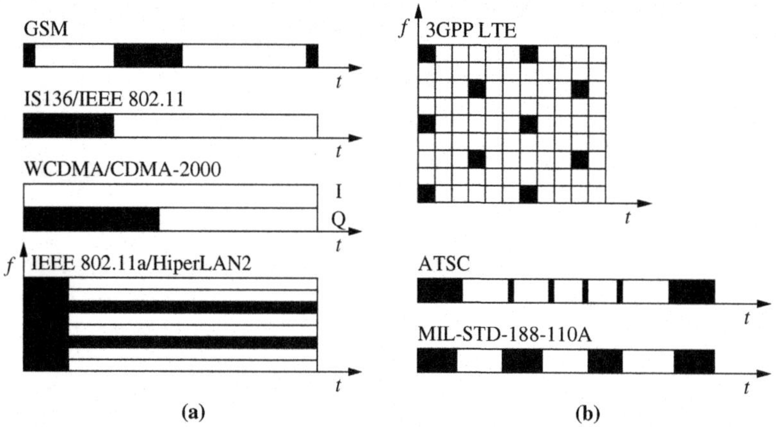

FIGURE 5.1

Pilot-placement patterns in existing wireless systems: (a) packet transmissions; (b) continuous transmissions. Black areas are pilot symbols.

symbols are superimposed onto data symbols. The advantage of such multiplexing is that it eliminates the bandwidth sacrifice for the pilot symbols as in the TDM case. Moreover, the constant presence of pilot symbols in the data stream brings the hope that it can better track the LTV channel in a high-mobility scenario. The other two factors for pilot design, i.e., the amount of pilot symbols and their power allocation, are discussed in Section 5.5, where the cutoff rate is considered as the optimization criterion. The cutoff rate is a lower bound on the channel capacity and provides an upper bound on the probability of block decoding error (through bounding the random coding exponent).

We next look at the pilot design in a MIMO system. For MIMO LTV channels, pilot design optimization is more difficult to obtain. Some recent results on MIMO pilot design are described in Section 5.6. In Section 5.7, we extend the pilot design to some newer generation wireless systems, such as OFDMA, CDMA, and ultra-wideband systems for broadband communications. Our goal in this section is to provide both a survey of methods and a description of issues associated with pilot optimization.

5.2 PILOT DESIGN: A FRAMEWORK

An optimal pilot design critically depends on a carefully chosen model and a well reasoned criterion. We first lay out the framework within which various optimal pilot-assisted transmission (PAT) schemes are to be developed.

5.2.1 Modeling of Pilot-Assisted Transmission

Pilot and data can be multiplexed in multiple dimensions. Pilot symbols are traditionally time-division multiplexed with data symbols. The use of an antenna array extends the multiplexing to the spatial dimension. Multicarrier transmissions and CDMA add frequency and code dimensions to the mix, respectively. Power allocation among data and pilots is another design factor. In addition to interleaving pilots and data by TDM, we may also consider superimposing pilots and data, an idea proposed as early as 1987 (Makrakis & Feher, 1987) which has also received more recent attention (Budianu & Tong, 2002; Dong, Tong, & Sadler, 2004; Hoeher & Tufvesson, 1999; Manton, Mareels, & Hua, 2000; Ohno & Giannakis, 2002; Tugnait, Meng, & He, 2006; Zhou, Viberg, & McKelvey, 2003). The combination of these factors multiplies the possible scenarios to be examined and motivates the model described next.

5.2.1.1 The Multidimensional PAT Model

The key to unifying various schemes is to view the problem of PAT design as one of power allocation. Specifically, in each design dimension – time, frequency, space, and so on – a pair of power allocation parameters are used to model the PAT. The simplest case is single-carrier transmission over a single input and (possibly) multiple output channel where each transmitted symbol $a[k]$ can be modeled as a linear combination of a known pilot $a_p[k]$ and an information-bearing data symbol $a_d[k]$. Specifically,

$$a[k] = \rho_{p,k} a_p[k] + \rho_{d,k} a_d[k], \quad k = 1, \ldots, L_B,$$

where $a_p[k]$ satisfying $|a_p[k]| = 1$ is known with allocated power $\rho^2_{p,k}$, and $a_d[k]$ is unknown data with zero mean, unit variance, and allocated power $\rho^2_{d,k}$. For the transmission of a packet of size L_B, a PAT scheme is defined by the L_B-dimensional pilot vector $\mathbf{a}_p = (a_p[k])$ and two power allocation vectors $\boldsymbol{\rho}_p = (\boldsymbol{\rho}_{p,k})$ and $\boldsymbol{\rho}_d = (\rho_{d,k})$. A graphical illustration of the one-dimensional scheme is shown in Fig. 5.2(a) where a partially shaded square indicates superimposed pilot and data symbols.

If the spatial domain is added, a two-dimensional description is necessary, as shown in Fig. 5.2(b). A block coded space-time transmission, for example, sends the data symbols in blocks, and each transmitted symbol $a[i,j]$ is indexed by the block number j and the position i within the block. If we assume that pilots may be superimposed in any position, we have

$$a[i,j] = \rho_{p,ij} a_p[i,j] + \rho_{d,ij} a_d[i,j], \quad i = 1, \ldots, N, \quad j = 1, \ldots, L_B,$$

and the PAT scheme is parameterized by the $N \times L_B$ pilot matrix $\mathbf{A}_p = (a_p[i,j])$, and non-negative power allocation matrices $\mathbf{P}_p = (\rho_{p,ij})$ and $\mathbf{P}_d = (\rho_{d,ij})$. The same formulation naturally applies to an OFDM system by treating i as the frequency index. The generalization to higher dimensions is straightforward; the idea is illustrated in Fig. 5.2(c).

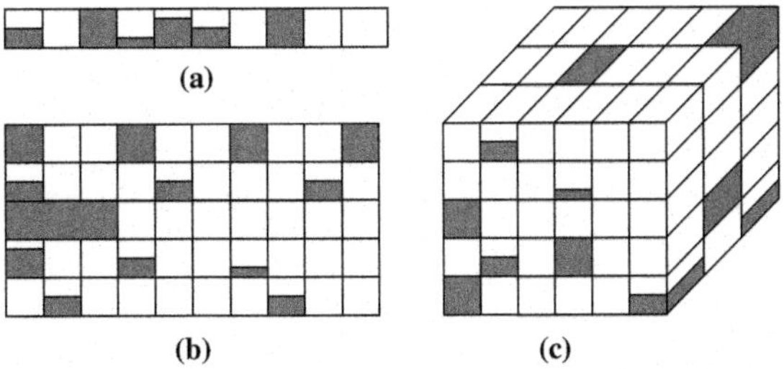

FIGURE 5.2

Illustration of the multidimensional PAT Model (with a per-symbol average power constraint). Each square is a symbol boundary. The pilot symbol at each square has power proportional to the area of the shaded region.

5.2.1.2 Power Constraints

All transmissions are subject to power constraints, and there are many ways in which such constraints can be imposed on PAT. It is sufficient to consider the two-dimensional case. Given $N \times L_B$ matrices $\mathbf{P}_p$ and $\mathbf{P}_d$, the average power constraint is given by

$$\frac{1}{NL_B}\sum_{j=1}^{L_B}\sum_{i=1}^{N}\mathrm{E}\left\{|a[i,j]|^2\right\} = \frac{1}{NL_B}\sum_{j=1}^{L_B}\sum_{i=1}^{N}\left(\rho_{p,ij}^2 + \rho_{d,ij}^2\right) = P.$$

As a special case, the per-symbol average power constraint imposes a more stringent condition:

$$\mathrm{E}\left\{|a[i,j]|^2\right\} = \rho_{p,ij}^2 + \rho_{d,ij}^2 = P.$$

In this case, the power allocation matrices $\mathbf{P}_p$ and $\mathbf{P}_d$ are complementary, and one power allocation matrix is sufficient.

5.2.2 Transceiver Architectures

The presence of pilots naturally implies that they will be used at the PAT receiver explicitly or implicitly. Parametric approaches, as illustrated in Fig. 5.3, estimate channel parameters and use the estimated channel for demodulation and decoding. The channel estimator takes the pilot vector $\mathbf{a}_p$ (and possibly the entire observation $\mathbf{y}$), produces a channel estimate $\hat{\boldsymbol{\theta}}$, and feeds the estimate to the decoder. A practical decoder may assume that the estimated channel parameters are perfect. Such an assumption is, of course, not valid, and the corresponding scheme is referred to as a mismatched decoder (Lapidoth & Narayan, 1998; Merhav, Kaplan, Lapidoth, & Shamai, 1994). An alternative is to treat the estimated channel parameters as part of the observation. The optimal decoder then exploits the joint statistics of $(\hat{\boldsymbol{\theta}},\mathbf{y})$.

Nonparametric approaches, in contrast, treat the pilot symbols as side information. The channel estimator in Fig. 5.3 is bypassed, and pilots are used to tune the receiver directly. In some voice-band modems, for example, explicit channel estimates are not obtained, rather, the training is used for adaptively updating an equalizer.

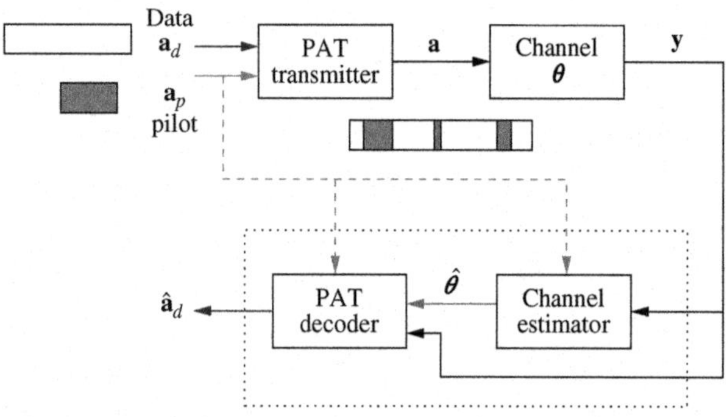

FIGURE 5.3

The structure of PAT transceivers.

5.2.3 Performance Criteria

Pilot design is primarily a transmitter technique, although receiver characteristics must also be taken into account. The PAT scheme used for a specific application often needs to be standardized. It is therefore important that the chosen PAT scheme is optimal or near-optimal for a wide range of channel conditions. Furthermore, because designers may have different design constraints and objectives, it is preferable that the PAT scheme is optimal for different design criteria. We outline next a few commonly used design criteria.

5.2.3.1 Information-Theoretic Metrics

The information-theoretic metrics for PAT apply to the class of systems constrained to using pilots in specific ways. In other words, we are interested in the PAT design with some fixed transceiver structure (such as that shown in Fig. 5.3) while allowing the design of optimal signaling and coding schemes.

Shannon capacity measures the maximum rate of reliable transmission among all transceiver designs. Reliable transmission at rate R requires the existence of encoding and decoding schemes that make the detection error probability arbitrarily small, when the code length is sufficiently long. Shannon's characterization of the reliable rate of transmission (the achievable rate) is through the use of mutual information (Cover & Thomas, 1991). The optimal PAT design that maximizes channel capacity requires expressing mutual information as a function of PAT parameters, and maximizing the mutual information with respect to these parameters and the channel input distribution. Unfortunately, the required mutual information expressions are often difficult to obtain. In some cases, however, bounds (Lapidoth & Shamai, 2002; Medard, 2000) on the achievable rate can be obtained and optimized with respect to PAT parameters (Adireddy, Tong, & Viswanathan, 2002; Baltersee, Fock, & Meyr, 2001; Hassibi & Hochwald, 2003; Ohno & Giannakis, 2002).

Codes that ensure reliable transmission do not always exist. For slow fading channels, practical constraints on decoding delay make the Shannon capacity zero. In such cases, the outage probability may be used. Specifically, given a transmission rate R, an optimal PAT scheme maximizes the probability that rate R can be reliably achieved.

The channel reliability function, random coding exponent, and cutoff rate (Gallager, 1968; Viterbi & Omura, 1979) are explicit measures that relate detection error probability with data rate and codeword length. For PAT systems, they are all functions of PAT parameters. These information-theoretic metrics allow us to quantify the decay rate of error probability with respect to the code length.

5.2.3.2 Channel Estimation: Mean Square Error and Cramér-Rao Bound

The information-theoretic metrics are global measures of a communication system. Often, we are interested in the optimality of specific configurations or system components. For the receiver structure shown in Fig. 5.3, one may be interested in the PAT scheme that minimizes the channel estimator error. A natural criterion is the mean square error (MSE) of the estimator. Because it is desirable that the optimal PAT design does not depend on the specific algorithm used at the receiver, the Cramér-Rao bound (CRB) (Kay, 1993) is a natural choice as a figure of merit. Specifically, the MSE of an unbiased estimator $\hat{\boldsymbol{\theta}}$, under regularity conditions, is lower bounded by

$$\mathrm{E}\{||\hat{\boldsymbol{\theta}} - \boldsymbol{\theta}||^2\} \geq \mathrm{tr}\{\mathbf{F}^{-1}(\mathbf{P}_p, \mathbf{P}_d, \mathbf{V})\},$$

where $\mathbf{F}(\mathbf{P}_p, \mathbf{P}_d, \mathbf{V})$ is the Fisher information matrix[2] (Kay, 1993). Note, however, that the CRB might not be achievable with finite data, although the existence of asymptotically efficient algorithms (e.g., maximum likelihood estimation in many cases) justifies the use of CRB as a design criterion, and the CRB is achievable with finite data in many instances (Kay, 1993).[3] The CRB may be formulated with both random and deterministic parameter models (Van Trees, 1968); random models lead to useful insights into ensemble behavior, whereas deterministic models provide means for assessing specific realizations of channels and sources. The channel CRB may also be a function of the unknown data transmitted simultaneously with the pilots. When the unknown data are treated as random parameters, they need to be marginalized to obtain the likelihood function. If, however, these unknown data are treated as deterministic unknown parameters, then the data may be viewed as nuisance parameters that affect the CRB of the channel estimator. Incorporation of known pilots into the bounds yields performance limits for semiblind estimators (Dong & Tong, 2002).

5.2.3.3 Source Estimation: BER, Error-Exponent Function, and MSE

For detection, bit (or symbol) error rate (BER) is the most appropriate performance metric; it is also one of the most difficult to precisely characterize (Proakis & Salehi, 2007). A more tractable approach is to use BER bounds as the figure of merit. To that end, Bhattacharyya and random coding bounds (Blahut, 1987; Viterbi & Omura, 1979) can be considered. Also relevant is the error-exponent function that measures the decay rate of the error probability (Gallager, 1968). One can also treat symbol detection as a problem of parameter estimation and use the MSE as the metric for optimization. For example, MSE is widely used in the design of equalizers (Proakis & Salehi, 2007).

5.3 OPTIMAL TDM PILOT INSERTION PATTERN IN SINGLE CARRIER SYSTEMS

The pilot design in a single-carrier system boils down to the optimization of three main system parameters: the ratio of power allocated to pilot and data symbols, the fraction of time allocated to pilot symbols, and the placement of pilot symbols in the data stream. We will discuss these optimization issues in the rest of this section.

The way that pilot symbols are multiplexed into the data stream affects the receiver performance for LTV channels. Under TDM training, the presence of pilot symbols makes channel estimates accurate at some periods of time and coarse at others. If the percentage of pilot symbols is fixed, we then have to choose between obtaining accurate estimates infrequently, or frequent but less accurate estimates. Our starting point here is to fix the percentage (in power or in the number of channel uses) of pilot symbols, and optimize the pilot symbol placement.

Periodic placement is a class of placement schemes where pilot symbols are inserted in the data stream with a periodic pattern. In the following, we derive the optimal periodic pattern among all periodic placements. The restriction to periodic placements is mild; a system with aperiodic training will not reach a steady state, and is seldom considered in practice.

[2]For deterministic parameters, $\mathbf{F}(\mathbf{P}_p, \mathbf{P}_d, \mathbf{V})$ is also a function of the unknown parameters. For random parameters, the Fisher information is a function of the (prior) parameter distribution.

[3]This contrasts with information-theoretic metrics where there is a coding theorem that ensures the achievability of capacity, although the code length may be very long.

5.3.1 Channel Model

Consider a flat Rayleigh fading channel modeled as

$$y[k] = h[k]a[k] + z[k], \quad k = 1,2,\ldots, \tag{5.1}$$

where $y[k]$ is the received observation after demodulation, $a[k]$ is the transmitted symbol, $h[k]$ is the complex Gaussian channel state with zero mean and variance σ_h^2, and $z[k]$ is the circular complex additive white Gaussian noise (AWGN) at time k with zero mean and variance σ_z^2. We assume that data $a[k]$, channel $h[k]$, and noise $z[k]$ are mutually independent.

Modeling the channel statistics as a Gauss–Markov process is frequently used as a simple and effective model to characterize the fading process (Gaston, Chriss, & Walker, 1973; Iltis, 1990; Medard, 2000; Stojanovic, Proakis, & Catipovic, 1995). The Gauss–Markov model provides a reasonable fit to the widely used Jakes' model that characterizes the power spectral density of the channel process $h[k]$. Here, we model the dynamics of the channel state $h[k]$ by a first-order Gauss–Markov process

$$h[k] = \alpha h[k-1] + u[k], \tag{5.2}$$

where the fading correlation coefficient $\alpha \in [0,1]$ characterizes the degree of time variation, and $u[k]$ is the white complex Gaussian driving noise with distribution

$$u[k] \overset{\text{i.i.d.}}{\sim} \mathcal{CN}\left(0, \left(1-\alpha^2\right)\sigma_h^2\right).$$

Small α models correspond to fast fading, and large α corresponds to slow fading. The value of α is determined by the channel Doppler spread and the transmission bandwidth, where the relation among the three is found in Eqn (79) of Medard (2000). It can be accurately obtained at the receiver for a variety of channels (Abou-Faycal, Medard, & Madhow, 2005; Iltis, 1990; Stojanovic et al., 1995). Here, we assume α is known.

5.3.2 Periodic TDM Pilot Placement

We consider the class of periodic placements where the placement pattern of pilot symbols repeats periodically in the data stream, as shown in Fig. 5.4. We define the placement period, denoted by T_p, to be the length of the smallest symbol block over which the placement pattern repeats. Note that the starting point of a period can be chosen arbitrarily. Without loss of generality, we assume that each period starts with a pilot symbol and ends with a data symbol.

In general, any periodic placement with n clusters of pilot symbols in a period of length T_p can be specified by a 2-tuple $\mathscr{P} = (\boldsymbol{\gamma}, \boldsymbol{\nu})$, where $\boldsymbol{\gamma} = (\gamma_1, \ldots, \gamma_n)^T$ is the pilot cluster length vector and $\boldsymbol{\nu} = (\nu_1, \ldots, \nu_n)^T$ is the data block length vector, as illustrated in Fig. 5.5. Note that the placement period T_p satisfies $T_p = \sum_{i=1}^{n}(\gamma_i + \nu_i)$. We further denote by $\mathscr{I}_p(\mathscr{P})$ the index set that contains the relative positions of the pilot symbols within one period.

For different placement schemes, we make the following two assumptions:

A1: All pilot symbols have equal power, denoted by σ_p^2; the power for data symbols is denoted by σ_d^2.

A2: The percentage of pilot symbols in a data stream is fixed, and is given by $\eta = \sum_{i=1}^{n} \gamma_i / T_p$.

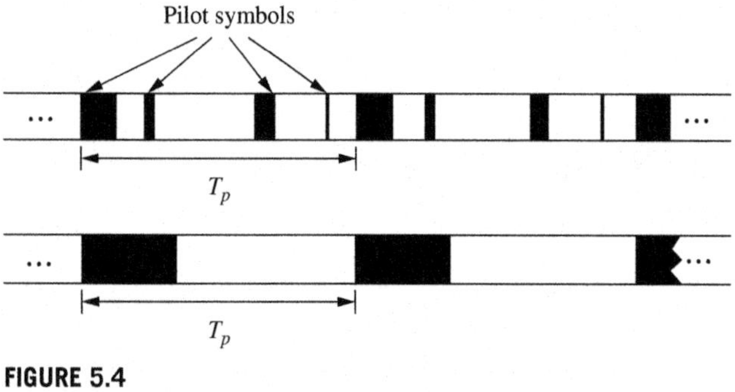

FIGURE 5.4

Data streams with periodic placements.

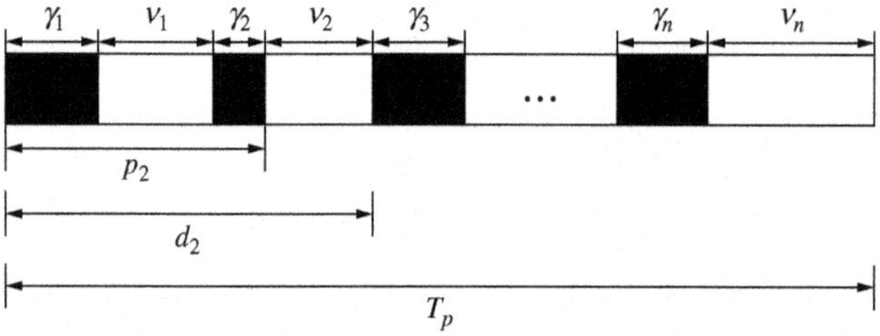

FIGURE 5.5

Representation of pilot placement within one placement period.

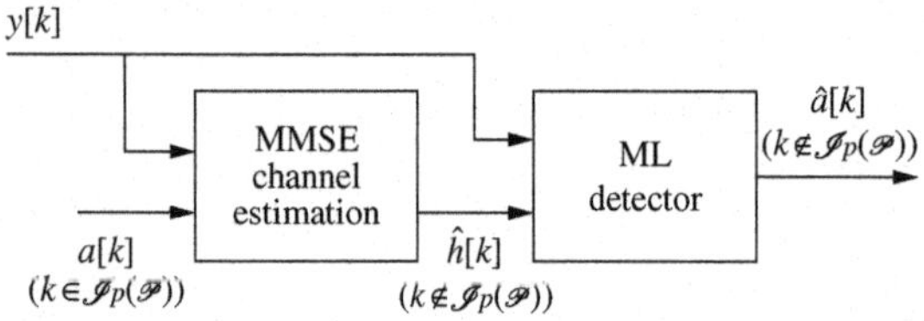

FIGURE 5.6

Receiver structure.

5.3.3 Receiver Structure

We consider a typical receiver structure shown in Fig. 5.6: the channel estimator provides the channel estimate $\hat{h}[k]$ to the demodulator, and the data symbol $a[k]$ $(k \notin \mathscr{I}_p(\mathscr{P}))$ is detected based on $\hat{h}[k]$ and the received sample $y[k]$ using the symbol-by-symbol maximum likelihood (ML) detector.[4]

For a given placement $\mathscr{P}$, the observations received from pilot symbols are given by $\{y[lT_p+k] : k \in \mathscr{I}_p(\mathscr{P}), l=0,1,\ldots\}$. We consider the minimum mean square error (MMSE) channel estimator

[4]Note that the symbol-by-symbol ML detector is optimal only when the channel is memoryless. When a channel has memory, the optimal detector detects the transmitted symbol or sequence based on the sequence of received symbols. The globally optimum sequence detector is the ML sequence detector.

based on the current and all past pilot symbols and their corresponding observations. The MMSE channel estimator at time $(lT_p + k)$, denoted by $\hat{h}[lT_p + k]$, is given by

$$\hat{h}[lT_p + k; \mathscr{P}] = \mathrm{E}\big\{h[lT_p + k] \,|\, \{y[lT_p + j] : j \leq k, j \in \mathscr{I}_p(\mathscr{P})\}, \\ \{y[(l-m)T_p + j] : j \in \mathscr{I}_p(\mathscr{P}), m = 1,2,\ldots\}\big\}.$$

For a channel with Gaussian statistics, it is equivalent to the linear MMSE (LMMSE) estimator. Implementation-wise, the direct use of the MMSE channel estimator requires storage of all the past pilot symbol observations at the receiver, as well as computing a matrix inversion at each symbol. As the number of observations grows, this requires excessive storage and computation at the receiver. In practice, an adaptive filter is desired, especially for LTV channels, and the Kalman filter (KF) has been widely used because of its optimality and adaptivity. Based on the Gauss–Markov model (5.2), the above MMSE channel estimator can be implemented recursively by a KF (Kailath, 1980; Kay, 1993). At each sample time, we only need to store the most recent channel estimate, and the KF produces a channel estimate with simple scalar operations. For the transmission with TDM pilot placements, as in Fig. 5.4, the KF switches between two modes: during the training, it updates the channel estimate using the pilot symbol period; and during data transmission, it predicts the channel state.

Given the estimated channel state $\hat{h}[k]$ and assuming equally probable data symbols, the optimal detection is given by the ML detector. For a fixed placement $\mathscr{P}$, let

$$M[k;\mathscr{P}] \triangleq \mathrm{E}\left\{\left|\hat{h}[k;\mathscr{P}] - h[k]\right|^2\right\}$$

be the MSE of the channel estimate at time k produced by the KF. Conditioned on any data symbol $a[k]$ $(k \notin \mathscr{I}_p(\mathscr{P}))$, $y[k]$ and $\hat{h}[k;\mathscr{P}]$ are jointly Gaussian with zero mean and covariance

$$\boldsymbol{\Sigma}(a[k]) = \begin{pmatrix} \sigma_z^2 + |a[k]|^2\sigma_h^2 & a[k](\sigma_h^2 - M[k;\mathscr{P}]) \\ a^*[k](\sigma_h^2 - M[k;\mathscr{P}]) & \sigma_h^2 - M[k;\mathscr{P}] \end{pmatrix}.$$

For any phase-shift keying (PSK) constellation, we have $|a[k]|^2 = \sigma_d^2$ $(k \notin \mathscr{I}_p(\mathscr{P}))$. Let $\mathcal{A}$ be the symbol alphabet set for the given constellation. Given $\hat{h}[k;\mathscr{P}]$ and $y[k]$ at the receiver, the ML decision rule for detecting $a[k]$ $(k \notin \mathscr{I}_p(\mathscr{P}))$ is given by (here, $f(\cdot\,|\,\cdot)$ denotes the conditional probability density function)

$$\begin{aligned} \hat{a}[k] &= \arg\max_{a[k]\in\mathcal{A}} f(y[k], \hat{h}[k;\mathscr{P}] | a[k]) \\ &= \arg\min_{a[k]\in\mathcal{A}} (y^*[k]\ \hat{h}^*[k;\mathscr{P}])(\boldsymbol{\Sigma}(a[k]))^{-1}(y[k]\ \hat{h}[k;\mathscr{P}])^T \\ &= \arg\max_{a[k]\in\mathcal{A}} \mathrm{Re}\{a^*[k]\hat{h}^*[k;\mathscr{P}]y[k]\} \\ &= \arg\min_{a[k]\in\mathcal{A}} \left|y[k] - \hat{h}[k;\mathscr{P}]a[k]\right|^2. \end{aligned} \tag{5.3}$$

The last equation is similar to the ML detector with a known channel, but now substituting the channel estimate.

5.3.4 Optimization Criteria

For TDM schemes, the MSE and BER performance are not stationary. During a training block, the KF uses pilot symbols to produce increasingly more accurate channel estimates until the pilot cluster is exhausted and data transmission resumes, during which time the KF can only predict the channel state based on the Gauss–Markov model, and therefore produces increasingly inaccurate estimates.

The KF updating algorithm can be obtained from standard KF theory (Kailath, Sayed, & Hassibi, 2000; Kay, 1993), and is detailed in Appendix 5.A. Here we only give the expressions for channel MMSE updates needed for our analysis.

In a pilot cluster, we obtain recursive expressions for the channel MMSE as

$$M[lT_p+k;\mathscr{P}] = \frac{\sigma_z^2\left[\alpha^2 M[lT_p+k-1;\mathscr{P}]+(1-\alpha^2)\sigma_h^2\right]}{\sigma_z^2+\left[\alpha^2 M[lT_p+k-1;\mathscr{P}]+(1-\alpha^2)\sigma_h^2\right]\sigma_p^2} \tag{5.4}$$

for $k \in \mathscr{I}_p(\mathscr{P})$ and all integer l. Once the ith pilot cluster in a placement period ends, for which the relative index within the period is denoted by p_i as shown in Fig. 5.5, the KF predicts the channel state during data transmissions of duration ν_i. The resulting MMSE is given by

$$M[lT_p+p_i+k;\mathscr{P}] = \alpha^{2k} M[lT_p+p_i;\mathscr{P}]+\sigma_h^2(1-\alpha^{2k}) \tag{5.5}$$

for $k = 1,\ldots,\nu_i$.

As $l \to \infty$, the system converges to a periodic steady state, and we are naturally interested in the steady-state performance

$$M_k(\mathscr{P}) \triangleq \lim_{l\to\infty} M[lT_p+k;\mathscr{P}], \quad k=1,\ldots,T_p. \tag{5.6}$$

Furthermore, we are only interested in the MSE of the channel estimate during data transmission. Thus, from (5.5)–(5.6),

$$M_{p_i+k}(\mathscr{P}) = \alpha^{2k} M_{p_i}(\mathscr{P}) + \left(1-\alpha^{2k}\right)\sigma_h^2, \quad k=1,\ldots,\nu_i, \tag{5.7}$$

which monotonically increases with increasing symbol index k.

Within a placement period, define $d_i \triangleq p_i+\nu_i$ as the index of the position of the last symbol in the ith data block, as shown in Fig. 5.5. Then, the maximum steady-state MMSE in this block is reached at d_i. The maximum steady-state channel MMSE over data symbols is then given by[5]

$$\mathscr{E}(\mathscr{P}) \triangleq \max_{k\notin \mathscr{I}_p(\mathscr{P})} M_k(\mathscr{P}) = \max_{1\le i\le n} M_{d_i}(\mathscr{P}).$$

The optimal placement $\mathscr{P}^*_{\text{MMSE}}$ that minimizes the maximum steady-state channel MMSE is then obtained as

$$\mathscr{P}^*_{\text{MMSE}} = \arg\min_{\mathscr{P}} \mathscr{E}(\mathscr{P}) = \arg\min_{\mathscr{P}} \max_{1\le i\le n} M_{d_i}(\mathscr{P}). \tag{5.8}$$

[5] We assume there are n pilot clusters for a given placement $\mathscr{P}$.

Our goal is to find the optimal placement that minimizes the maximum steady-state BER. Specifically, let $P_e[k;\mathscr{P}]$ be the steady-state BER at the kth position relative to the beginning of the placement period. We are interested in the following optimization

$$\mathscr{P}^*_{\text{BER}} = \arg\min_{\mathscr{P}} \max_{k \notin \mathscr{I}_p(\mathscr{P})} P_e[k;\mathscr{P}]. \tag{5.9}$$

The BER performance is directly affected by the MMSE of the channel estimate. We show next that $\mathscr{P}^*_{\text{MMSE}} = \mathscr{P}^*_{\text{BER}}$ for BPSK and QPSK constellations.

Proposition 5.1 *Under the Gauss–Markov channel model with BPSK or QPSK input symbols, if the MMSE channel estimator is used along with the ML detector, then*

$$\mathscr{P}^*_{\text{MMSE}} = \mathscr{P}^*_{\text{BER}}.$$

Proof. See Appendix 5.B. ■

5.3.5 Optimal TDM Pilot Placement

We first find the optimal placement for a special class of placements called *regular periodic placements*. The extension to the general class follows.

The regular periodic placement RPP-γ has only one pilot cluster of size γ and one data cluster of size ν. In Fig. 5.4, the second example belongs to this class.

From (5.8), it follows that for RPP-γ, the optimal placement is obtained as

$$\gamma_* = \arg\min_{\gamma} \mathscr{E}(\gamma) = \arg\min_{\gamma} M_{T_p}(\gamma). \tag{5.10}$$

Our problem now is to find the explicit expression of the steady-state MMSE $M_{T_p}(\gamma)$, and analyze its behavior as a function of pilot cluster size γ.

A useful quantity in the sequel is the steady-state MMSE when all symbols are pilots. Define M_∞ as the steady-state MMSE in this case by

$$M_\infty \triangleq \lim_{\gamma\to\infty} M_\gamma(\gamma).$$

At the steady-state, M_∞ satisfies the Riccati equation

$$M_\infty = \frac{\sigma_z^2\left[\alpha^2 M_\infty + (1-\alpha^2)\sigma_h^2\right]}{\sigma_z^2 + \left[\alpha^2 M_\infty + (1-\alpha^2)\sigma_h^2\right]\sigma_p^2},$$

and the solution to the above steady-state Riccati equation is given by

$$M_\infty = \frac{\sigma_h^2}{\frac{1}{2}(1+\lambda_p) + \sqrt{\left[\frac{1}{2}(1+\lambda_p)\right]^2 + \frac{\alpha^2}{1-\alpha^2}\lambda_p}},$$

where λ_p is the *received* pilot signal-to-noise ratio, i.e., $\lambda_p \triangleq \sigma_h^2\sigma_p^2/\sigma_z^2$. Obviously, M_∞ is a lower bound on the MMSE for any placement.

The following lemma provides the closed-form MMSE expression for the RPP-γ scheme.

Lemma 5.1 For any RPP-γ scheme, the steady-state channel MMSE at relative time k is given by

$$M_k(\gamma) = \frac{M_{T_p}(\gamma) - \sigma_h^2\left(1-\alpha^{2(T_p-k)}\right)}{\alpha^{2(T_p-k)}}, \quad \text{for } k = \gamma+1, \ldots, T_p - 1, \tag{5.11}$$

and

$$M_{T_p}(\gamma) = \delta_{T_p}(\gamma) + M_\infty, \tag{5.12}$$

where $\delta_{T_p}(\gamma)$ is computed as follows

$$\delta_{T_p}(\gamma) = -b_\gamma + \sqrt{b_\gamma^2 + c_\gamma}, \tag{5.13}$$

with

$$b_\gamma \triangleq \frac{1-\left(\frac{\alpha^{2\frac{1-\eta}{\eta}}}{\beta_1}\right)^\gamma}{1-\frac{1}{\beta_1^\gamma}} \frac{\beta_1 - 1}{2\beta_2} - \frac{1-\alpha^{2\frac{1-\eta}{\eta}\gamma}}{2}\rho_\infty\sigma_h^2,$$

$$c_\gamma \triangleq \frac{1-\alpha^{2\frac{1-\eta}{\eta}\gamma}}{1-\frac{1}{\beta_1^\gamma}} \frac{\beta_1-1}{\beta_2}\rho_\infty\sigma_h^2,$$

$$\beta_1 \triangleq \frac{[1+(1-\alpha^2\rho_\infty)\lambda_p]^2}{\alpha^2}, \tag{5.14}$$

$$\beta_2 \triangleq \frac{\lambda_p}{\sigma_h^2}[1+(1-\alpha^2\rho_\infty)\lambda_p], \tag{5.15}$$

$$\rho_\infty \triangleq 1 - \frac{M_\infty}{\sigma_h^2},$$

$$\eta = \frac{\gamma}{T_p}.$$

Proof. See Appendix 5.C. ■

From (5.12), because M_∞ is not a function of γ, the optimization in (5.10) can now be rewritten as

$$\gamma_* = \arg\min_\gamma \delta_{T_p}(\gamma)$$

where the explicit expression of $\delta_{T_p}(\gamma)$ as a function of γ is obtained in (5.13). By analyzing the behavior of $\delta_{T_p}(\gamma)$ as a function of γ, we obtain the optimal placement for RPP schemes in the following theorem.

Theorem 5.1 For the class of RPP schemes, under assumptions A1–A2 in Section 5.3.2, the maximum MMSE $\mathscr{E}(\gamma)$ of the channel estimates during data transmission is a monotone increasing function of γ. Thus RPP-1 is optimal among all RPP placements. Define $\mathscr{E}^{\text{TDM}} \triangleq \min_\gamma \mathscr{E}(\gamma)$, then

$$\mathscr{E}^{\text{TDM}} = \mathscr{E}(1),$$

and $\mathscr{E}(1)$ is given by

$$\mathscr{E}(1) = \sigma_h^2 - \alpha^{2\frac{1-\eta}{\eta}}\left(\sigma_h^2 - M_1\right) \tag{5.16}$$

where

$$M_1 = \frac{\sigma_h^2}{\frac{1}{2}\left(1+\lambda_p\right) + \sqrt{\left[\frac{1}{2}\left(1+\lambda_p\right)\right]^2 + \frac{\alpha^{2/\eta}}{1-\alpha^{2/\eta}}\lambda_p}}. \tag{5.17}$$

Proof. See Appendix 5.D. ■

Theorem 5.1 shows that decreasing the pilot cluster length and estimating the channel (through pilots) more frequently results in decreased steady-state maximum channel MMSE and thus lower BER. This immediately implies that if there is a constraint on the minimum pilot cluster size γ_o, RPP-γ_o is optimal.

Extending the above result, we show in the following theorem that RPP-1 is, in fact, optimal among all periodic placements.

Theorem 5.2 Given a fixed percentage of pilot symbols η, the optimal placement for periodic TDM training that minimizes the maximum steady-state MMSE and BER for any first-order Gauss–Markov channel is RPP-γ_o, with γ_o the minimum pilot cluster size allowed.

Proof. We outline the main steps required to prove the optimality of RPP-1, leaving the details to Dong et al. (2004). Consider first the case with two pilot clusters of lengths γ_1 and γ_2, and two data blocks of lengths ν_1 and ν_2, present in each period. Let $d_1 = \gamma_1 + \nu_1$ and $d_2 = d_1 + \gamma_2 + \nu_2$ be the end positions of the two data blocks, where the MMSE $M_k(\mathscr{P})$ reaches the maximum. Intuition suggests that moving the second pilot cluster away from the first, i.e., increasing ν_1 and decreasing ν_2, increases $M_{d_1}(\mathscr{P})$ and decreases $M_{d_2}(\mathscr{P})$. (However, this is not obvious because moving the second pilot cluster will also affect the initial MMSE $M_1(\mathscr{P})$.) It follows that to minimize the maximum MMSE for the entire period suggests the equalization rule that forces $M_{d_1}(\mathscr{P}) = M_{d_2}(\mathscr{P})$, which leads to making pilot clusters equal, and eventually, results in the reduction to the RPP-γ scheme. Extending to any placement with n pilot clusters in a period, using a similar equalization rule and applying the above result to each two consecutive pilot clusters repeatedly, leads to the same reduction to RPP-γ. Combining Theorem 5.1 and Proposition 5.1, we then have the optimality of RPP-1. ■

Note that the optimality of RPP-1 in Theorem 5.2 holds regardless of the values of the received pilot SNR λ_p and channel correlation α. Fig. 5.7 provides a comparison of the performance of different TDM RPP-γ placement schemes. The MMSE and BER are calculated using the MMSE expressions in Lemma 5.1 and the BER expression in (5.B.3), respectively. The powers of data and pilot symbols are set to be equal, i.e., $\sigma_d^2 = \sigma_p^2 = P$, and the received signal-to-noise ratio is $\text{SNR} \triangleq \sigma_h^2 P/\sigma_z^2$. Figures 5.7(a) and (b), respectively, show the maximum steady-state MMSE and BER performance for BPSK signals as a function of α for SNR = 20 dB. The percentage of pilot symbols in the stream was $\eta = 20\%$. We observe that the largest gain obtained by placing pilot symbols optimally occurs when α is in the range from 0.9 to 1, which is a common range of channel time variation.[6]

[6] For example, for bandwidths in the 10 kHz range and Doppler spreads of order 100 Hz, the parameter α typically ranges between 0.9 and 0.99 (Abou-Faycal et al., 2005).

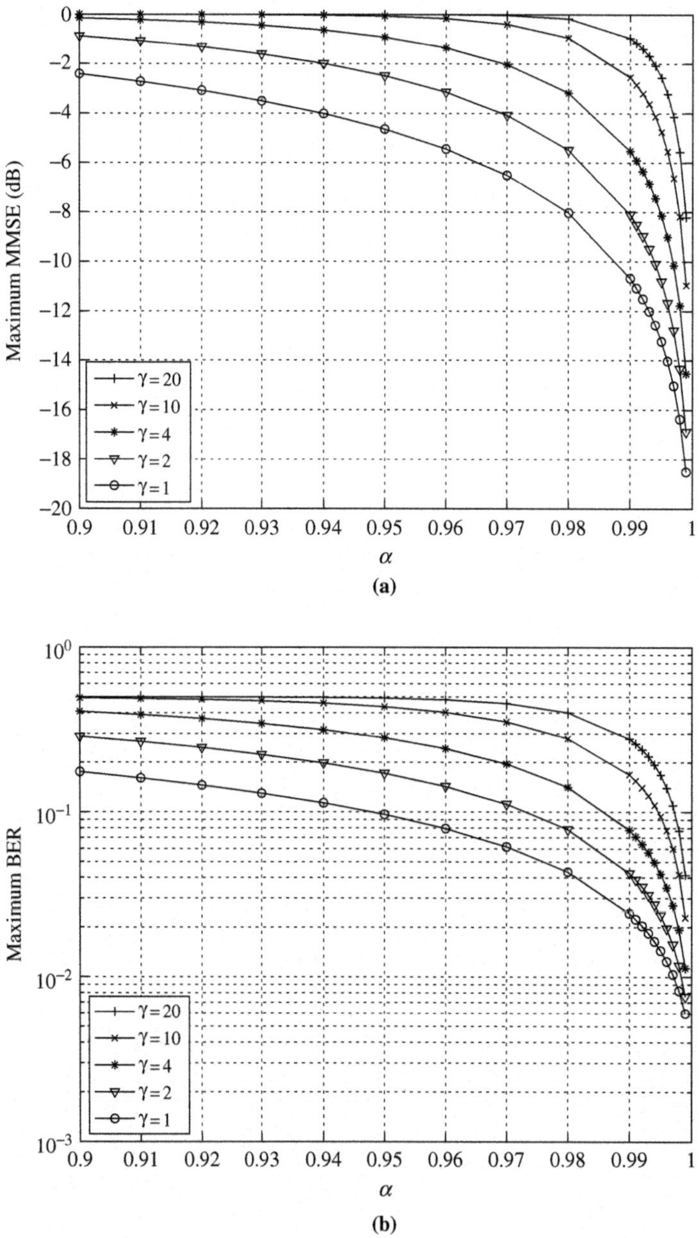

FIGURE 5.7

(a) Maximum steady-state channel MMSE $\mathscr{E}(\gamma)$ versus α; (b) maximum steady-state BER versus α. (SNR $= 20$ dB, $\eta = 20\%$.)

5.3.6 Bibliographical Notes

5.3.6.1 Training Design for Block Fading Channels

In a slow fading environment, the classical treatment is to use block fading to model the channel (Tse & Viswanath, 2005). Block fading channels are time-invariant for the coherence time, and then change to a different fading state. This, of course, is an approximation, but a reasonable one for many applications. Channel estimation and detection under such models are often performed within each block with the help of pilot symbols.

When training-based estimators are used for single-carrier systems over intersymbol interference channels, the placement design is simple and is the same as that for time-invariant channels. That is, all pilots should be clustered into a single block. The only design choice left is the selection of the pilot sequence. Designing the optimal sequence is an old problem dating back to the early 1960s (Frank, 1962), and the work continues for different settings and objectives.

For semiblind channel estimation, however, the positions of pilot symbols do make a difference as pilots and data are both part of the estimation model. Using the Cramér-Rao Bound (CRB) as the design metric, and under a random channel model, optimal placement of pilot symbols and optimal power allocations are derived for SISO and MIMO intersymbol interference channels in Dong & Tong (2002). It is shown that the optimal placement scheme that minimizes the CRB is independent of the prior channel distribution and the SNR. The placement does depend on the power allocation and the number of pilot symbols in a block. The CRB has also been used to study the impact of side information, including training, with MIMO channels in Sadler & Kozick (2000, 2001); see also the discussion in Section 5.6.

5.3.6.2 Training Design for Fast Fading Channels

In a fast fading environment, channels experience fast symbol-by-symbol variation. Using the RPP-1 scheme for pilot-assisted transmission was first proposed in the late 1980s (Lodge & Moher, 1987; Sampei & Sunaga, 1989) for tracking LTV channels, and first analyzed in Cavers (1991). Although no optimality among all other placement schemes was shown, it has been widely applied for its practical simplicity and studied in various settings (Abou-Faycal et al., 2005; Gansman, Fitz, & Krogmeier, 1997; Ohno & Giannakis, 2002; Torrance & Hanzo, 1995; Tsatsanis & Xu, 2000). The optimality of RPP-1 was established in Dong et al. (2004) over flat fading channels for minimizing the MMSE of channel tracking as well as decoding BER, and some details are also discussed in this section.

5.3.6.3 Training Design for Doubly Selective Channels

In some applications, one must model the channel as doubly selective, where both frequency selectivity and fading time correlation introduce memory to the channel. The pilot placement is optimized for packetized transmission in Dong, Adireddy, & Tong (2003) for semiblind LMMSE channel tracking, where the dynamics of each tap of the frequency selective channel are modeled as a Gauss–Markov process. An alternative model for the doubly selective channel is the basis expansion model (BEM) discussed in Chapter 1. The pilot placement in this scenario is analyzed in Ma et al. (2003) for block transmissions, where periodic data sub-blocks and pilot sub-blocks are inserted in each transmission block. The training is optimized for LMMSE channel estimation by maximizing a lower bound on the capacity. The optimal training strategy for periodic placement pattern is given by equispaced and equipowered pilot symbols surrounded by a number of zeros dictated by the channel delay spread, and the period is dictated by the channel Doppler spread. For this BEM, the number of basis coefficients grows with the packet size, so the method is generally most appropriate for short blocks. The BEM used in Ma et al. (2003) is termed the critically sampled (CS) complex exponential (CE) BEM. In Tang & Leus (2007), a noncritically sampled (NCS)-CE model is used in simulations and is shown to more accurately model the doubly selective channel. For this NCS-CE BEM, the periodic placement and the power of pilot symbols are optimized numerically in Whitworth, Ghogho, & McLernon (2009) for zero-padded block transmission to minimize the total MSE, which includes both noise and channel modeling inaccuracies. A more detailed description of pilot-aided channel estimation using a block approach with BEM as the channel model is given in Chapter 4.

5.4 ALTERNATIVE PILOT INSERTION STRATEGY – SUPERIMPOSED TRAINING

The pilot design for superimposed training takes the form of allocating power to pilot and data symbols at each time index. For the stationary Gauss–Markov channel considered here, it is reasonable to consider a time-invariant power allocation where the transmitted symbol, $a[k] = \rho_p a_p[k] + \rho_d a_d[k]$, is the superposition of pilot and data symbols. The observation is given by

$$y[k] = (\rho_p a_p[k] + \rho_d a_d[k])h[k] + z[k] \tag{5.18}$$

where $\{a_p[k]\}$ is the pilot sequence, and $\{a_d[k]\}$ is the data sequence, which is drawn from an i.i.d. zero-mean sequence. We assume that $a_p[k]$ and $a_d[k]$ have unit powers, i.e., $\mathrm{E}\{|a_p[k]|^2\} = \mathrm{E}\{|a_d[k]|^2\} = 1$, and we denote by ρ_p and ρ_d the pilot and data power allocation coefficients, respectively.

5.4.1 Kalman Tracking with Superimposed Training

A complication of superimposed training is that $h[k]$ and $(y[k], y[k-1], \ldots)$ are not jointly Gaussian. Therefore, the MMSE channel estimator based on the conditional expectation is difficult to compute and implement (Kay, 1993). Instead, we consider the LMMSE channel estimator implemented through the KF. Rewrite (5.18) as

$$y[k] = \rho_p a_p[k]h[k] + v[k]$$

where $v[k] \triangleq \rho_d a_d[k]h[k] + z[k]$. Because $\{a_d[k]\}$ is i.i.d. zero mean and independent of $h[k]$, we have

$$\mathrm{E}\{v[i]v^*[j]\} = \left(\rho_d^2\sigma_h^2 + \sigma_z^2\right)\delta_{ij}, \quad \mathrm{E}\{h[i]v^*[j]\} = 0 \quad \forall i,j.$$

Let $\hat{h}[k]$ be the LMMSE estimator of $h[k]$ based on the current and all past observations $\{y[k], y[k-1], \ldots\}$. Then the KF can again be used to implement this LMMSE estimator, and thus, to track the channel. Let $M[k] \triangleq \mathrm{E}\{|h[k] - \hat{h}[k]|^2\}$. Following the KF (Kailath et al., 2000; Kay, 1993), we then obtain the KF algorithm for channel tracking under superimposed training:

$$K[k] = \frac{\mathrm{E}\left\{h[k]\left(y[k] - \alpha\rho_P a_P[k]\hat{h}[k-1]\right)^*\right\}}{\mathrm{E}\{|y[k] - \alpha\rho_P a_P[k]\hat{h}[k-1]|^2\}}$$

$$= \frac{[\alpha^2 M[k-1] + (1-\alpha^2)\sigma_h^2]\rho_P a_P[k]}{\rho_t^2[\alpha^2 M[k-1] + (1-\alpha^2)\sigma_h^2] + \rho_d^2\sigma_h^2 + \sigma_z^2}, \tag{5.19}$$

$$\hat{h}[k] = \alpha\hat{h}[k-1] + K[k]\left(y[k] - \alpha\rho_P a_P[k]\hat{h}[k-1]\right),$$

$$M[k] = (1 - K[k]\rho_P a_P[k])\left[\alpha^2 M[k-1] + (1-\alpha^2)\sigma_h^2\right]. \tag{5.20}$$

Combining (5.19) and (5.20), we obtain

$$M[k] = \frac{\left(\frac{\sigma_z^2}{\sigma_h^2\rho_P^2} + \frac{\rho_d^2}{\rho_P^2}\right)\left(\frac{\alpha^2}{1-\alpha^2}M[k-1] + \sigma_h^2\right)}{\frac{1}{1-\alpha^2}\left(\frac{\sigma_z^2}{\sigma_h^2\rho_P^2} + \frac{\rho_d^2}{\rho_P^2}\right) + \frac{\alpha^2}{(1-\alpha^2)\sigma_h^2}M[k-1] + 1}.$$

The steady-state MSE, defined as $\mathscr{E}^{\,\mathrm{sup}} \triangleq \lim_{k\to\infty} M[k]$, is then given by

$$\mathscr{E}^{\,\mathrm{sup}} = \frac{\frac{\alpha^2}{1-\alpha^2}\mathscr{E}^{\,\mathrm{sup}} + \sigma_h^2}{\frac{1}{1-\alpha^2} + \left(\frac{\alpha^2}{(1-\alpha^2)\sigma_h^2}\mathscr{E}^{\,\mathrm{sup}} + 1\right)\kappa}$$

where

$$\kappa \triangleq \frac{\sigma_h^2\rho_P^2}{\sigma_h^2\rho_d^2 + \sigma_z^2} \tag{5.21}$$

is the received signal-to-interference-and-noise ratio (SINR). The solution of the above equation is

$$\mathscr{E}^{\,\mathrm{sup}} = \frac{\sigma_h^2}{\frac{1}{2}(1+\kappa) + \sqrt{\left[\frac{1}{2}(1+\kappa)\right]^2 + \frac{\alpha^2}{1-\alpha^2}\kappa}}. \tag{5.22}$$

Note that in the steady-state, in contrast to the periodic placement scheme, the channel MSE in this case is time-invariant.

For data symbol decoding, we again consider BPSK signaling. The detector estimates $d[k]$ based on $\hat{h}[k]$ and $y[k]$ by

$$\hat{d}[k] = \mathrm{sign}\left(\mathrm{Re}\left\{\hat{h}^*[k](y[k] - \rho_P a_d[k]\hat{h}[k])\right\}\right).$$

Notice that this detector is not the true ML detector based on $y[k]$ and $\hat{h}[k]$; it is a pseudo ML detector that assumes the estimated $\hat{h}[k]$ has no error.

5.4.2 Superimposed versus TDM Schemes: Performance Comparison

We now compare the two types of pilot design schemes so far discussed, i.e., superimposed training and the optimal TDM training RPP-1. One apparent benefit of superimposed training is that there is no bandwidth reduction (rate loss) due to pilot symbols as in the TDM case. In addition to this, we are interested to see how the channel tracking and decoding performance with superimposed training compares with that of TDM training.

For the performance comparison of the two schemes, we need to impose the following power constraints:

$$\rho_p^2 + \rho_d^2 = \eta\sigma_p^2 + (1-\eta)\sigma_d^2 = P, \tag{5.23}$$

$$\frac{\rho_d^2}{\rho_p^2} = \frac{\sigma_d^2}{\sigma_p^2}\frac{1-\eta}{\eta}. \tag{5.24}$$

The first constraint keeps the transmission power P used in each scheme the same, and the second keeps the ratio of power allocated to pilots and data in each scheme the same. Then, for a TDM scheme with percentage of pilot symbols η, pilot power σ_p^2, and data power σ_d^2, the corresponding power allocation coefficients ρ_d and ρ_p for the superimposed scheme are given as

$$\rho_p^2 = \eta\sigma_p^2, \quad \rho_d^2 = (1-\eta)\sigma_d^2$$

and κ in (5.21) can be rewritten as

$$\kappa = \frac{1}{\frac{\sigma_d^2}{\sigma_p^2}\frac{1-\eta}{\eta} + \frac{1}{\eta\lambda_p}},$$

where again $\lambda_p = \sigma_h^2\sigma_p^2/\sigma_z^2$. The normalized MMSEs (NMMSEs) corresponding to the MMSEs in (5.16) and (5.22) are given by

$$\begin{aligned}\bar{\mathscr{E}}^{\text{TDM}}(\eta,\alpha,\lambda_p) &\triangleq \frac{\mathscr{E}^{\text{TDM}}}{\sigma_h^2}\\ &= 1-\alpha^{2\frac{1-\eta}{\eta}}\left(1-\frac{1}{\frac{1}{2}(1+\lambda_p)+\sqrt{\left[\frac{1}{2}(1+\lambda_p)\right]^2+\frac{\alpha^{2/\eta}}{1-\alpha^{2/\eta}}\lambda_p}}\right)\end{aligned} \tag{5.25}$$

$$\begin{aligned}\bar{\mathscr{E}}^{\text{sup}}\left(\eta,\alpha,\lambda_p,\frac{\sigma_d^2}{\sigma_p^2}\right) &\triangleq \frac{\mathscr{E}^{\text{sup}}}{\sigma_h^2}\\ &= \frac{1}{\frac{1}{2}(1+\kappa)+\sqrt{\left[\frac{1}{2}(1+\kappa)\right]^2+\frac{\alpha^2}{1-\alpha^2}\kappa}}.\end{aligned} \tag{5.26}$$

We note that, for the superimposed scheme, channel tracking benefits from the constant presence of pilot symbols. However, it is affected by both noise and interference from the data. This effect is evident from (5.26), where $\bar{\mathscr{E}}^{\,\sup}\left(\eta,\alpha,\lambda_p,\frac{\sigma_d^2}{\sigma_p^2}\right)$ is a function of κ, which indicates the SINR level. The higher κ is, the smaller is $\bar{\mathscr{E}}^{\,\sup}\left(\eta,\alpha,\lambda_p,\frac{\sigma_d^2}{\sigma_p^2}\right)$. In contrast, the RPP-1 scheme has the advantage of updating the channel state during training with no data interference, but there is no information sent to facilitate tracking during data transmission. For a given η, differing from the RPP-1 scheme where the NMMSE in (5.25) is only affected by λ_p, the NMMSE for the superimposed scheme is also a function of the data versus pilot power ratio σ_d^2/σ_p^2. Thus, the performance under superimposed training varies with the data power, whereas that under RPP-1 does not.

To gain insights into the fundamental differences, we analyze the performance difference of the two schemes in a few limiting cases.

- Fast fading ($\alpha \to 0$): For this case, we have almost i.i.d. fading. For fixed λ_p,

$$\bar{\mathscr{E}}^{\rm TDM}(\eta,\alpha,\lambda_p) = 1 - O(\alpha^{2\frac{1-\eta}{\eta}})$$

$$\bar{\mathscr{E}}^{\,\sup}\left(\eta,\alpha,\lambda_p,\frac{\sigma_d^2}{\sigma_p^2}\right) = \frac{1}{1+\kappa} - O(\alpha^2).$$

  As expected, due to the constant presence of pilot symbols in the data stream, the superimposed scheme provides a better tracking ability than that of the RPP-1 scheme. Note that this is in addition to the benefit of no bandwidth loss using superimposed training.
- Slow fading ($\alpha \to 1$): As $\alpha \to 1$, the channel becomes constant. In the limit, channel estimation becomes perfect for both schemes, and we have

$$\bar{\mathscr{E}}^{\rm TDM}(\eta,\alpha,\lambda_p) = O\left(\sqrt{1-\alpha^{\frac{1}{\eta}}}\right) \tag{5.27}$$

$$\bar{\mathscr{E}}^{\,\sup}\left(\eta,\alpha,\lambda_p,\frac{\sigma_d^2}{\sigma_p^2}\right) = O(\sqrt{1-\alpha}). \tag{5.28}$$

  For constant channels, it is intuitive that both schemes should give perfect estimation in the steady state. However, it is evident from (5.27) and (5.28) that, as α becomes less than 1, the increase rate of $\bar{\mathscr{E}}^{\rm TDM}(\eta,\alpha,\lambda_p)$ is faster than that of $\bar{\mathscr{E}}^{\,\sup}\left(\eta,\alpha,\lambda_p,\frac{\sigma_d^2}{\sigma_p^2}\right)$. In other words, as the channel varies faster, the NMMSE of TDM training deteriorates at a much more rapid rate than that of superimposed training. This again demonstrates that the superimposed scheme provides better tracking performance because of the constant presence of pilot symbols.

  We should note that because some power is taken away from the data for training in the superimposed scheme, the decoding performance under superimposed training may still be worse than that under TDM training as a result of reduced data SNR. We will later rely on numerical results to compare the decoding performance.
- High SNR ($\lambda_p \to \infty$): This corresponds to the noiseless case. For RPP-1, the channel state over each pilot symbol can be perfectly estimated. Estimation errors during data transmission are due to

the imperfect tracking of the KF, and we have

$$\bar{\mathscr{E}}^{\text{TDM}}\left(\eta,\alpha,\lambda_p\right) = 1-\alpha^{2\frac{1-\eta}{\eta}} + O\left(\frac{1}{\lambda_p}\right). \tag{5.29}$$

For the superimposed scheme, although there is no noise, interference from the data is always present. Therefore, the tracking error is resulting from the data symbol interference, and we have

$$\bar{\mathscr{E}}^{\text{sup}}\left(\eta,\alpha,\lambda_p,\frac{\sigma_d^2}{\sigma_p^2}\right) = \frac{1}{\frac{1}{2}(1+\kappa)+\sqrt{\left[\frac{1}{2}(1+\kappa)\right]^2+\frac{\alpha^2}{1-\alpha^2}\kappa}} + O\left(\frac{1}{\lambda_p}\right)$$

where κ in this case is given by

$$\kappa = \frac{\sigma_p^2}{\sigma_d^2}\frac{\eta}{1-\eta}.$$

The NMMSEs for both schemes vary with λ_p on the same order. The limiting NMMSEs depend on the channel fading rate, characterized by α.

Under the power constraints in (5.23) and (5.24), we calculate the channel tracking MMSE under superimposed training using the expression in (5.22) and compare it with that under RPP-1. Figures 5.8(a) and (b) provide a numerical comparison of channel tracking MMSE and decoding BER for BPSK inputs at various values of α under the two training schemes, respectively. We allocate half of the total transmission power P to the pilot symbols, i.e., $\rho_d^2 = \rho_p^2$. Furthermore, $\eta = 10\%$, and SNR $= 20$ dB. As mentioned earlier, we see that the tracking performance for superimposed training is better than that for TDM training over the entire range of α. However, the final decoding performance comparison depends on α. For sufficiently slow LTV channels, TDM training has a better BER performance than superimposed training does. In this case, depending on the amount of gain, TDM training may be preferred even though it requires a rate reduction. As the channel varies more rapidly, the performance of TDM training deteriorates faster. When α drops under a threshold, superimposed training provides a better BER performance than TDM training does, in addition to the benefit of no rate loss. The performance of superimposed training can be further improved by an iterative method for joint channel estimation and symbol detection, in which the interfering data symbols are estimated and subsequently removed from the observation for further improved channel estimation and detection (Tugnait et al., 2006).

5.4.3 Bibliographical Notes

For time-invariant channels, channel estimation using a superimposed periodic pilot sequence was initially considered in Farhang-Boroujeny (1995) and Mazzenga (2000). This idea of exploiting the underlying cyclostationary first-order statistics induced by the periodic training sequence was further explored and analyzed in Orozco-Lugo, Lara, & McLernon (2004); Tugnait & Luo (2003) and Zhou et al. (2003). Although superimposed training may not be the optimal approach for block fading frequency-selective channels, periodic superimposed training leads to a low-complexity channel estimation algorithm (Zhou et al., 2003). Data-dependent superimposed training is proposed for further channel tracking improvement and is analyzed in Alameda-Hemandex, McLernon, Orozco-Lugo,

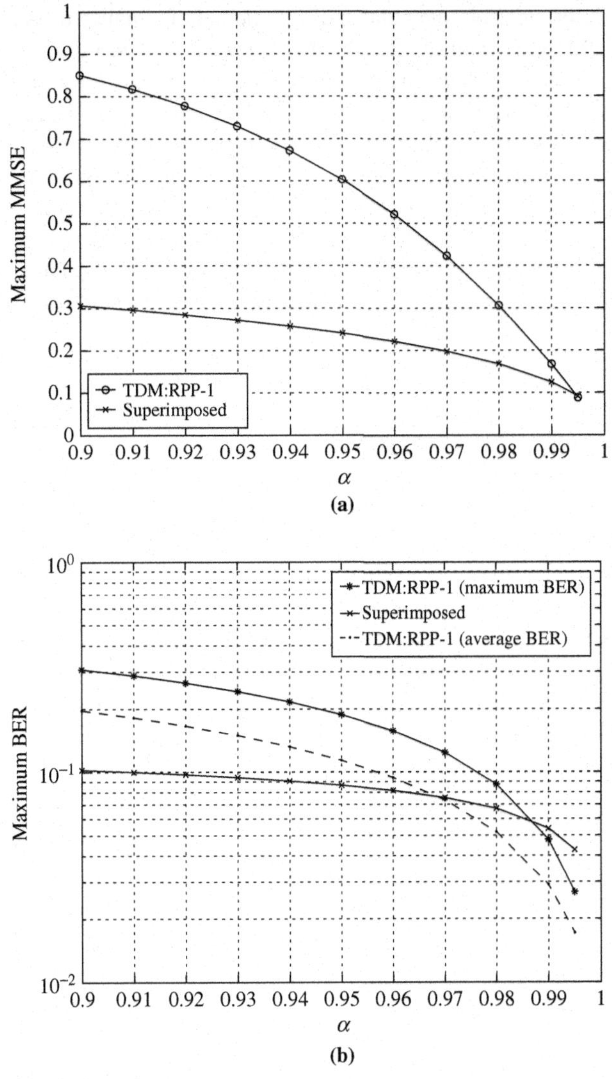

FIGURE 5.8

(a) Maximal steady-state channel tracking MMSE versus α; (b) maximal steady-state BER versus α. (SNR = 20 dB, $\eta = 10\%$.)

Lara, & Ghogho (2001) and Ghogho, McLernon, Alameda-Hernandez, & Swami (2005), where both a deterministic periodic pilot sequence and a data-dependent sequence are superimposed onto the data sequence. The latter is used to eliminate the interference due to the unknown data during channel estimation. It distorts the data by removing the components at equally spaced frequencies, leaving these positions for pilot symbols for channel estimation.

For LTV channels, using superimposed pilot symbols over flat fading channels is analyzed and its performance is compared with that of TDM pilot symbols for channel tracking and decoding in Dong et al. (2004). Some details have been described earlier. The advantage of superimposed training for fast fading environments or low-SNR scenarios is observed. However, the weakness of superimposed training is apparent at high SNR or slow fading resulting from interference from data symbols. The idea of exploring the first-order statistics of the periodic superimposed pilot sequence for channel estimation for time-invariant channels is further extended to doubly selective channels in Tugnait et al. (2006), where the basis expansion channel model is used. In addition, an iterative approach to joint channel estimation and symbol detection with Viterbi detectors is proposed, in which the interfering data symbols are detected and subsequently removed from the observation for further improved channel estimation and data detection.

5.5 RESOURCE ALLOCATION: AMOUNT OF TRAINING AND POWER OPTIMIZATION

As mentioned earlier, the resources allocated to pilot symbols are taken away from the data. If the channel is perfectly known at the receiver, the insertion of pilot symbols reduces the information rate. However, the channel needs to be estimated at the receiver, and it has been shown that the estimation error will reduce the achievable rate (Medard, 2000). Our earlier analysis shows that increasing the number of pilot symbols and their power reduces the channel estimation error, and thus improves the achievable information rate. Therefore, there is a trade-off on the number of pilot symbols and the power allocated to them to maximize the achievable rate.

5.5.1 Cutoff Rate Analysis

Let us look at the optimal RPP-1 TDM pilot pattern derived in Section 5.3.5. Assume the channel model given in Section 5.3.1 and no feedback to the transmitter. For an LTV channel with channel estimation at the receiver, a closed-form expression of the channel capacity is not available because the capacity depends on the channel estimate, which may be non-Gaussian. As an alternative, a lower bound on the channel capacity which can be expressed in closed form is often evaluated and maximized. We consider pilot resource optimization for maximizing the cutoff rate. The cutoff rate is a lower bound on the channel capacity and provides an upper bound on the probability of a block decoding error. It has been used to establish practical limits on coded performance under complexity constraints (Arikan, 1998).

The system transmits codewords, each of length $N(T_p - 1)$, for $N > 0$ a positive integer. Without loss of generality, consider transmitting a codeword starting at time $k = 0$ denoted by

$$\mathbf{a}_d = (a[1],\ldots,a[T_p-1],a[T_p+1],\ldots,a[2T_p-1],\ldots,a[(N-1)T_p+1],\ldots,a[NT_p-1])^T,$$

and correspondingly define

$$\begin{aligned}
\mathbf{h} &= (h[1],\ldots,h[T_p-1],h[T_p+1],\ldots,h[NT_p-1])^T,\\
\hat{\mathbf{h}} &= (\hat{h}[1],\ldots,\hat{h}[T_p-1],\hat{h}[T_p+1],\ldots,\hat{h}[NT_p-1])^T,\\
\mathbf{y} &= (y[1],\ldots,y[T_p-1],y[T_p+1],\ldots,y[NT_p-1])^T.
\end{aligned}$$

We have the received SNR on data and pilot symbols as

$$\lambda_d = \frac{\sigma_d^2 \sigma_h^2}{\sigma_z^2},$$

$$\lambda_p = \frac{\sigma_p^2 \sigma_h^2}{\sigma_z^2}.$$

The cutoff rate, measured in bits per channel use, is defined by (Wozencraft & Jacobs, 1965)

$$R_o = -\lim_{N\to\infty} \min_{Q(\cdot)} \frac{1}{NT_p} \log_2 \int_{\mathbf{y}} \int_{\hat{\mathbf{h}}} \left[\sum_{\mathbf{a}_d \in \mathscr{S}} Q(\mathbf{a}_d) \sqrt{f(\mathbf{y}, \hat{\mathbf{h}} | \mathbf{a}_d)} \right]^2 d\hat{\mathbf{h}}\, d\mathbf{y} \tag{5.30}$$

where $Q(\mathbf{a}_d)$ is the probability of transmitting codeword $\mathbf{a}_d$, and $\mathscr{S}$ is the set consisting of all possible input sequences for $\mathbf{a}_d$. The normalization factor $1/(NT_p)$ is used instead of $1/[N(T_p-1)]$ to account for the information rate loss because of the insertion of the pilot symbols.

Because the observations at $lT_p + k$, for $l = 0, 1, \ldots$, are all k symbols away from that of the last pilot symbol, the channels at these locations have the same estimation error statistics $M[lT_p + k]$. If we define this kth location relative to the last pilot symbol as the kth subchannel, we have a total of $T_p - 1$ subchannels for a codeword transmission. We assume that interleaving is done only among symbols within the same subchannel, thus preserving the channel estimation error statistics. We further assume that the interleaver uses a different interleaving scheme on each subchannel, so that the correlation between any two codeword symbols is zero. For K-PSK modulation, the communication channel is symmetric in its input; therefore, the cutoff rate is maximized by the equiprobable distribution $Q_k(\cdot) = 1/K$. Evaluating (5.30) with the above assumptions, and using the constant modulus property of K-PSK, we have

$$R_o = -\frac{1}{T_p} \sum_{l=1}^{T_p-1} \log_2 \left(\frac{1}{K} \sum_{k=0}^{K-1} \frac{1 + \lambda_d \frac{M_l(1)}{\sigma_h^2}}{1 + \lambda_d \left[1 - \left(1 - \frac{M_l(1)}{\sigma_h^2} \right) \cos^2\left(\frac{\pi k}{K}\right) \right]} \right), \tag{5.31}$$

where $M_l(1)$, for $l = 1, \ldots, T_p - 1$, can be obtained by (5.12) with $\gamma = 1$. A detailed derivation of (5.31) is given in Misra, Swami, & Tong (2006).

5.5.1.1 *Energy Allocation*

The energy constraint for the total energy allocated in one transmission period is given by

$$\lambda_p + (T_p - 1)\lambda_d \le \lambda_{av} T_p, \tag{5.32}$$

where $\lambda_{av} > 0$ is the allowable average energy per symbol. Note that the inequality in the constraint will be met with equality for R_o maximization because R_o is increasing with both λ_d and λ_p.

For BPSK signaling, the cutoff rate optimizing pilot energy λ_p^* is given by the following one-dimensional optimization problem, which only involves the channel estimation error at the pilot symbol location:

$$\lambda_p^* = \underset{0 \le \lambda_p \le \lambda_{av} T_p}{\arg\max} \left[\frac{\lambda_{av} T_p - \lambda_p}{\lambda_{av} T_p - \lambda_p + (T_p - 1)} \left(1 - \frac{M_{T_p}(1; \lambda_p)}{\sigma_h^2} \right) \right], \tag{5.33}$$

where we include λ_p in $M_{T_p}(1;\lambda_p)$ to emphasize its dependency on λ_p. This is obtained by substituting for λ_d in (5.31) based on energy constraint (5.32), and using the relation of $M_{T_p}(1;\lambda_p)$ and $M_k(1;\lambda_p)$ in (5.11) to obtain the following:

$$1-\frac{M_k(1;\lambda_p)}{\sigma_h^2}=\alpha^{2k}\left(1-\frac{M_{T_p}(1;\lambda_p)}{\sigma_h^2}\right),\quad k=1,\ldots,T_p-1.$$

Note that the optimal pilot energy λ_p^* in (5.33) depends on the channel estimator used. We next derive the optimal pilot energy at low and high SNR when the KF is used for causal channel estimation.

- Low SNR ($\lambda_{av}\to 0$): From (5.17), based on a second-order approximation, we obtain $M_{T_p}(1;\lambda_p)$ at low SNR as

$$M_{T_p}(1;\lambda_p)\simeq\left(1-\frac{\lambda_p}{1-\alpha^{2T_p}}\right)\sigma_h^2, \tag{5.34}$$

  where the approximation becomes tighter as $\lambda_p\to 0$. Substituting (5.34) into (5.33), and considering the fraction of energy per pilot symbol, $\lambda_p/(\lambda_{av}T_p)$, instead of λ_p, we have

$$\frac{\lambda_p^*}{\lambda_{av}T_p}=\underset{0\le\frac{\lambda_p}{\lambda_{av}T_p}\le 1}{\arg\max}\left[\frac{(1-\frac{\lambda_p}{\lambda_{av}T_p})\frac{\lambda_p}{\lambda_{av}T_p}}{1-\frac{\lambda_p}{\lambda_{av}T_p}+\frac{T_p-1}{\lambda_{av}T_p}}\cdot\frac{1}{1-\alpha^{2T_p}}\right].$$

  As $\lambda_{av}\to 0$, the maximization yields

$$\lim_{\lambda_{av}\to 0}\frac{\lambda_p^*}{\lambda_{av}T_p}=\frac{1}{2}.$$

  This shows that at low SNR, half of the total energy per period should be allocated to the pilot symbol.
- High SNR ($\lambda_{av}\to\infty$): At high SNR, as $\lambda_p\to\infty$, the channel MMSE on the pilot symbol approaches zero, i.e., the channel state in the most recent pilot symbol transmission $k=lT$ is learnt perfectly at high SNR. From (5.17), for $\lambda_p\to\infty$, we have $M_{T_p}(1;\lambda_p)\simeq\sigma_h^2/(1+\lambda_p)$. Effectively, the channel estimator based on all past pilot observations converges to the one based on the most recent pilot observation. Again, consider the fraction of pilot energy $\lambda_p/(\lambda_{av}T_p)$ in (5.33). At high SNR, it is given by

$$\lim_{\lambda_{av}\to\infty}\frac{\lambda_p^*}{\lambda_{av}T_p}=\frac{1}{1+\sqrt{T_p-1}}.$$

- Slow fading ($\alpha\to 1$): From (5.12) with $\gamma=1$, we have $M_{T_p}(1;\lambda_p)\to 0$ as $\alpha\to 1$, therefore we have $\lambda_p^*\to 0$. This is because the channel becomes time-invariant; and as $l\to\infty$, we have an infinite number of noisy pilot observations.
- Fast fading ($\alpha\to 0$): As $\alpha\to 0$, the channel becomes memoryless, and only the most recent pilot symbol provides useful information. The channel estimator essentially converges to the one based

on the most recent pilot observation. Thus, λ_p^* is obtained as

$$\lambda_p^* = \begin{cases} \dfrac{\sqrt{(T_p-1)(\lambda_{av}T_p+1)[(\lambda_{av}+1)T_p-1]}-(\lambda_{av}+1)T_p+1}{T_p-2}, & T_p > 2, \\ \dfrac{\lambda_{av}T_p}{2}, & T_p = 2. \end{cases}$$

For general K-PSK signaling, a similar optimal pilot energy allocation in the low/high SNR regime is also found in Misra et al. (2006). Furthermore, for a given channel correlation α, the closed-form solution for the optimal energy can also be found in certain cases (see Misra et al. (2006)).

5.5.1.2 *Amount of Training*

We consider the optimal value of T_p (the number of pilot symbols to be inserted into the transmission). In general, this depends on the channel correlation α, the cardinality of the input K, the total energy constraint λ_{av} and allocation to pilot symbols, and the channel estimator used at the receiver. However, the analysis is greatly simplified at high SNR.

At high SNR, the optimal value of T_p can be found from (5.30). As mentioned in the high-SNR case, as $\lambda_{av} \to \infty$, the channel estimator converges to the one based on the most recent pilot observation. By maximizing R_o in (5.31), we have

$$T_c = \underset{2\le x<\infty, x\in\mathbb{N}}{\arg\max} \left[\prod_{l=1}^{x-1} \frac{1}{K} \sum_{k=0}^{K-1} \frac{1-\alpha^{2l}}{1-\alpha^{2l}\cos^2\left(\frac{\pi k}{K}\right)} \right]^{-1/x}, \tag{5.35}$$

where T_c denotes the optimal period T_p as $\lambda_{av} \to \infty$. We see that (5.35) depends only on K and α; it is independent of the energy allocation strategy and channel estimator used.

For a wide range of SNR, examples provided in Misra et al. (2006) show that (5.35) yields a good approximation to the optimal training period. Table 5.1 lists a few examples from Misra et al. (2006) for QPSK ($K = 4$), where T_p^* denotes the optimal period T_p under the optimal energy allocation.

Table 5.1 The Optimal Training Period for QPSK.

α	λ_{av}	T_p^*	T_c
0.95	0 dB	5	4
	10 dB	4	4
	20 dB	4	4
0.99	0 dB	10	7
	10 dB	8	7
	20 dB	7	7

5.5.2 Bibliographical Notes

A direct attack on the problem of training design using Shannon capacity as the metric was made by Marzetta (1999), and later by Hassibi & Hochwald (2003), where the class of ergodic block fading channels was considered. Assuming TDM training, Hassibi and Hochwald maximized a lower bound on channel capacity with respect to the number of pilots used in the block and the power allocated to the pilots. For a single-antenna system, they showed that when the pilot power is optimized, only one pilot symbol per block should be sent to maximize the capacity lower bound. Hassibi and Hochwald showed that, in the low-SNR regime and when the coherence time (block length) is short, the optimal PAT scheme incurred a substantial penalty; training may lead to bad channel estimates, and no training may be preferable. However, PAT was close to being optimal in the high SNR and long coherence time regimes. This is consistent with the intuition that, when a negligible price is required to obtain high-quality estimates, we can assume that the channel is approximately known at the receiver.

The optimization of training amount and power allocation for LTV channels was considered in Abou-Faycal et al. (2005), Baltersee et al. (2001) and Ohno & Giannakis (2002). Periodic pilot symbol placements with a cluster size of one were used. Mutual information using a binary input was discussed in Abou-Faycal et al. (2005), where a channel capacity lower bound was maximized numerically with respect to the percentage of pilot symbols. An optimization to maximize a lower bound on channel capacity was carried out for a band-limited LTV channel model in Baltersee et al. (2001) and Ohno & Giannakis (2002), where a noncausal LMMSE channel estimator based on all the past and future pilot symbols was considered. A closed-form expression for the optimal power allocation and training period was obtained for LTV channels with ideal low-pass Doppler spectrum. Unlike the mutual information measure, the use of cutoff rate leads to an analytically more tractable framework that gives, for some cases, closed-form expressions for the optimal power allocation (Ling, 1999; Misra et al., 2006; Phoel & Honig, 2002).

5.6 PILOT DESIGN FOR MIMO CHANNELS

With the additional spatial dimension in the multiple-antenna system, the pilot design optimization for LTV MIMO channels is a much more challenging task. For the case of fast LTV channels, no clear guidance on the optimal design has been found. For slow fading scenarios, to make the problem more tractable, a simplified block fading MIMO channel model is commonly used. With this model, the amount of training needed in a MIMO system is investigated in Hassibi & Hochwald (2003), where power allocation and training amount are obtained to maximize a lower bound on the channel capacity. The optimal power allocation turns out to depend on the number of pilot symbols in the block. At low SNR, the optimal strategy is to allocate half of the total energy to pilot symbols; and at high SNR, the ratio of overall pilot energy to total energy in a block is given by

$$\frac{\sqrt{N_t}}{\sqrt{T_p - T_{tr}} + \sqrt{N_t}},$$

where T_p is the block length, T_{tr} is the number of pilot symbols in the block, and N_t is the number of transmit antennas. Interestingly, these results at low and high SNR, when $N_t = 1$, coincide with the optimal power allocation results in Section 5.5.1 where the cutoff rate is maximized for the Gauss–Markov channel.

In terms of training amount, Hassibi & Hochwald (2003) show that if power allocation to data and pilot symbols is allowed to be optimized, the optimal number of pilot symbols in a block is always equal to the number of transmit antennas. If equal power allocation is required, the optimal number of pilot symbols can be larger than the number of antennas. In particular, at low SNR, half of the block should be devoted to pilot symbols.

Given the number of pilot symbols and the power assigned to them, the pilot placement for MIMO channels is optimized in Dong & Tong (2002) for semiblind channel estimation by minimizing the CRB, and in Ma, Yang, & Giannakis (2005) by maximizing a channel capacity lower bound which further is shown to be equivalent to minimizing the LMMSE channel estimation error.

Pilot sequence optimization in training-based preamble design for MIMO channels and carrier frequency offset (CFO) estimation is investigated in Ghogho & Swami (2006). The CRB is used as the performance metric and is minimized through the pilot sequence optimization. The optimal pilot sequence associated with different transmit antennas is shown to satisfy a certain orthogonality condition. Furthermore, a subclass of the optimal sequence designs is shown to render the CRB for the CFO independent of the channel zeros.

5.7 PILOT DESIGN FOR WIDEBAND SYSTEMS

Increasing demands for high data rate ubiquitous wireless service pushed the technology evolution of third-generation (3G) and fourth-generation (4G) wideband wireless systems in both static and mobile environments. Pilot design in such systems becomes especially important to ensure high data rate and system efficiency. Our goal in this section is to provide a brief survey of the literature and a few pointers with respect to the three basic questions discussed previously, i.e., pilot placement, amount of training, and power allocation.

5.7.1 OFDMA Systems

For OFDM transmission, pilot symbols are placed in frequency, and so are referred to as pilot subcarriers. The first attempts to design an optimal pilot placement was made by Rinne & Renfors (1996) and Negi & Cioffi (1998) for OFDM in SISO systems over time-invariant channels. In Negi & Cioffi (1998), the authors optimized the pilot subcarrier spacing for training-based MMSE estimation. For a channel with order[7] L, and for which $L+1$ subcarriers are selected for training, they showed that selecting pilot subcarriers periodically (in frequency) results in the minimum MSE.

From an information-theoretic point of view, the optimal pilot placement for an ergodic block frequency-selective fading channel of order L with OFDM transmissions was obtained in Adireddy et al. (2002). It was shown that the periodic placement in frequency maximizes a capacity lower bound.

Pilot placement for a fast fading LTV channel in an OFDM system is considered in Dong, Tong, & Sadler (2002), Fernandez-Getino, Garcia, Paez Borrallo, & Zazo Bello (1999), and Stamoulis,

[7] A channel with $L+1$ taps is said to have order L.

Diggavi, & Al-Dhahir (2002). It is shown that uniformly spreading the pilots over frequency and time minimizes the channel estimation error. When the channel exhibits symbol-by-symbol variation, the orthogonality of OFDM is destroyed, causing inter-carrier interference (ICI), which complicates channel estimation. In Stamoulis et al. (2002), the authors analyzed the effect of ICI on MIMO-OFDM. They proposed an ICI-mitigating linear filter, as well as a channel estimation and tracking scheme. They showed that grouping pilot subcarriers into equally spaced clusters is more effective for time-varying channels than using equally spaced pilot subcarriers, whereas the latter is shown to be optimal for time-invariant channels. Hexagonal placement is considered in Fernandez-Getino et al. (1999), with the intuition that hexagonal placement has the best coverage in the time-frequency plane. Such placement is also adopted in the design specification of the 3GPP Long Term Evolution (LTE) standard for the next-generation OFDMA system (see Fig. 5.9 for an example). Some common ICI mitigation techniques are described in Chapter 7. More detailed discussions on pilot-aided channel estimation for OFDM transmission are given in Chapters 4 and 7.

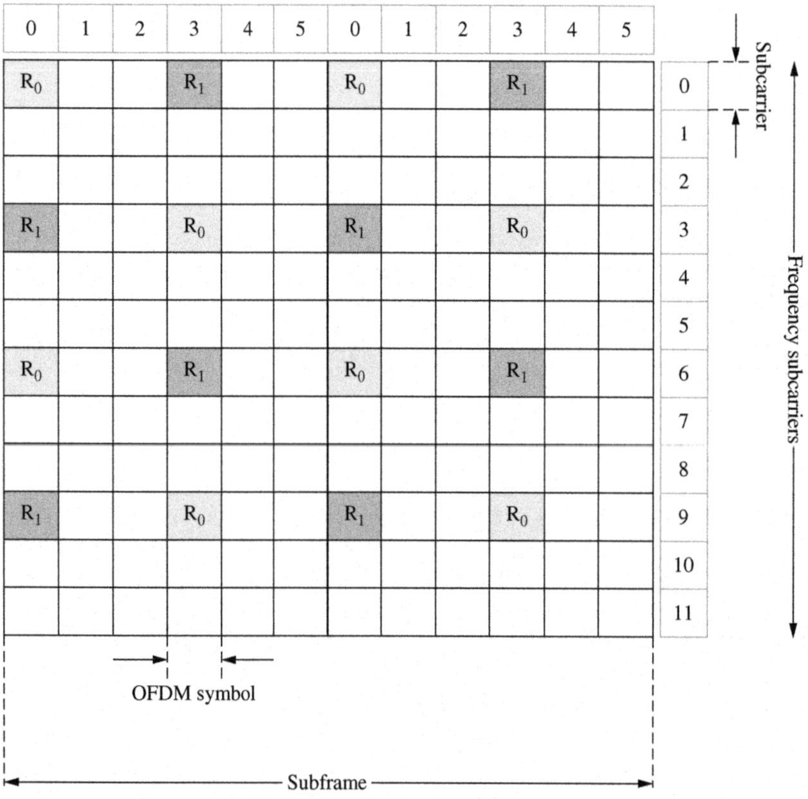

FIGURE 5.9

Pilot placement in a subframe in the current LTE working standard for the case of two transmit antennas (3GPP TS36.201, 2009). R_o: pilot symbol over transmit antenna 1; R_1: pilot symbol over transmit antenna 2.

5.7.2 CDMA Systems

Pilot channel design for CDMA systems has been studied by many authors. The optimal power ratio of pilot and data channels was obtained in Schramm (1998). Under the optimal power ratio, the loss resulting from imperfect channel estimation was calculated. In Schramm & Muller (1998), optimal pilot symbol spacing for pilot symbol–assisted BPSK over Rayleigh fading channels with L_d diversity paths was obtained. The loss resulting from imperfect channel estimation was also calculated. In Ling (1999), the author provided a comprehensive analysis and performance comparison, in terms of channel estimation, detection, and cut-off rate, under two training schemes, i.e., pilot channel-assisted and pilot symbol-assisted schemes, in CDMA systems. Training parameters were optimized based on Bhattacharyya bounds and cutoff rates. The optimal spacing of periodically inserted pilot symbols for an LMMSE channel and data estimator over flat Rayleigh fading channels for multiuser CDMA systems was discussed in Evans (2002). The estimation of both time- and frequency-selective channels for short-code CDMA systems with single-user receiver was addressed in Baissas & Sayeed (2002), where a pilot channel was used in parallel to the data channel.

5.7.3 Ultra Wideband

Pilot design is highly warranted in ultra wideband (UWB) communications because of the significant challenges of acquisition, synchronization, and equalization in this high-bandwidth regime. An interesting way to incorporate training is through a transmitted reference (TR) approach, an old idea that has received attention recently in the UWB context (Choi & Stark, 2002; Hoctor & Tomlinson, 2002; Hoyos, Sadler, & Arce, 2005). TR schemes significantly reduce receiver complexity but come at a high performance penalty (Tang, Xu, & Sadler, 2007; Xu & Sadler, 2006). More generally, optimal training in the UWB context is considered in Yang & Giannakis (2004), where the optimal pilot waveform is reminiscent of the TDM RPP-1 pattern but tailored to the UWB system. The impact of imperfect channel estimates with training in a direct-sequence-type UWB approach is analyzed in Sadler & Swami (2004).

5.8 CONCLUSION

In this chapter, we addressed some main issues in pilot design optimization for transmission over LTV channels. A general PAT model was given, and common design criteria were reviewed. Focusing on a single-antenna single-carrier system, we first derived the TDM pilot pattern for optimal channel tracking and symbol decoding performance, assuming a fixed percentage of pilots, as well as a fixed amount of power allocated to them. Superimposed training was then discussed for channel tracking as an alternative to TDM training. We compared its performance with that of TDM training and discussed its advantages and disadvantages with respect to TDM training. The optimal number of pilots and their power allocation were then derived for TDM training based on a cutoff rate analysis. Finally, pilot designs over LTV channels in MIMO systems and other wideband systems were reviewed with a brief literature survey.

Acknowledgment

This work was supported, in part, by the U.S. Army Research Office under Grant ARO-W911NF-06-1-0346 and NSERC Discovery Grant RGPIN 372059.

Appendix

5.A THE KALMAN FILTER FOR TDM TRAINING

The Kalman filter for channel tracking during each pilot cluster transmission or data block transmission consists of computing the Kalman gain, a channel estimate update, and a channel MMSE update. The channel MMSE update in (5.4) and (5.5) is obtained from the following:

- During the pilot cluster transmissions ($k \in \mathcal{I}_p(\mathcal{P})$):
  Kalman gain:

$$K[lT_p+k] = \frac{\left[\alpha^2 M\left[lT_p+k-1;\mathcal{P}\right]+(1-\alpha^2)\sigma_h^2\right]a\left[lT_p+k\right]}{\sigma_z^2+\left[\alpha^2 M\left[lT_p+k-1;\mathcal{P}\right]+(1-\alpha^2)\sigma_h^2\right]\sigma_p^2}.$$

  Channel estimate update:

$$\hat{h}\left[lT_p+k\right]=\alpha\hat{h}\left[lT_p+k-1\right]+K\left[lT_p+k\right]\left(y\left[lT_p+k\right]-\alpha\hat{h}\left[lT_p+k-1\right]a\left[lT_p+k\right]\right).$$

  Channel MMSE update:

$$M\left[lT_p+k;\mathcal{P}\right]=\left(1-K\left[lT_p+k\right]a\left[lT_p+k\right]\right)\left[\alpha^2 M\left[lT_p+k-1;\mathcal{P}\right]+\left(1-\alpha^2\right)\sigma_h^2\right].$$

- During the data block transmissions ($k \notin \mathcal{I}_p(\mathcal{P})$):
  Channel estimate update:

$$\hat{h}\left[lT_p+k\right]=\alpha\hat{h}\left[lT_p+k-1\right].$$

  Channel MMSE update:

$$M\left[lT_p+k;\mathcal{P}\right]=\alpha^2 M\left[lT_p+k-1;\mathcal{P}\right]+\left(1-\alpha^2\right)\sigma_h^2.$$

5.B PROOF OF PROPOSITION 5.1

The ML detection rule is given in (5.3). For a system using BPSK, it can be simplified as

$$\hat{a}[k]=\operatorname{sign}\left(\operatorname{Re}\left\{\hat{h}^*[k]y[k]\right\}\right)\sigma_d.$$

Under the system equation in (5.1), the decision statistic is

$$\begin{aligned} d[k] &= \operatorname{Re}\left\{\hat{h}^*[k]y[k]\right\}=\operatorname{Re}\left\{a[k]\hat{h}^*[k](\hat{h}[k]+\tilde{h}[k])+\hat{h}^*[k]z[k]\right\} \\ &= a[k]|\hat{h}[k]|^2+a[k]\operatorname{Re}\left\{\hat{h}^*[k]\tilde{h}[k]\right\}+\operatorname{Re}\left\{\hat{h}^*[k]z[k]\right\} \end{aligned} \tag{5.B.1}$$

where $\tilde{h}[k] \triangleq h[k] - \hat{h}[k]$. Conditioned on $\hat{h}[k]$ and $a[k]$, the second and third terms in (5.B.1) are independent zero-mean complex Gaussian random variables. At the steady state,

$$d[k] \sim \mathcal{CN}\left(a[k]|\hat{h}[k]|^2, \frac{1}{2}\left[\sigma_d^2 M_k(\mathcal{P}) + \sigma_z^2\right]|\hat{h}[k]|^2\right).$$

Therefore, for a system using BPSK, the bit error probability conditioned on $\hat{h}[k]$ and $a[k]$ is

$$\Pr\left\{\hat{a}[k] \neq a[k] \middle| \hat{h}[k], a[k]\right\} = Q\left(\sqrt{\frac{2|\hat{h}[k]|^2\sigma_d^2}{\sigma_d^2 M_k(\mathcal{P}) + \sigma_z^2}}\right),$$

where $Q(\cdot)$ is the Q-function. The BER for data symbols at the kth position of a placement period is thus given by

$$P_e[k;\mathcal{P}] = \mathrm{E}\left\{Q\left(\sqrt{\frac{2|\hat{h}[k]|^2\sigma_d^2}{\sigma_d^2 M_k(\mathcal{P}) + \sigma_z^2}}\right)\right\}, \tag{5.B.2}$$

where the expectation is taken with respect to $|\hat{h}[k]|^2$. Note that $\hat{h}[k]$ is a zero-mean Gaussian random variable with variance $\sigma_h^2 - M_k(\mathcal{P})$; therefore $|\hat{h}[k]|^2$ is exponentially distributed. For an exponentially distributed random variable X with probability density function $\frac{1}{c}\mathrm{e}^{-x/c}$ there is $\mathrm{E}\{Q(\sqrt{x})\} = \int_0^\infty Q(\sqrt{x})\frac{1}{c}\mathrm{e}^{-x/c}\mathrm{d}x = \frac{1}{2}\left[1 - \sqrt{c/(2+c)}\right]$. Using this result, we obtain the following closed-form expression for (5.B.2):

$$\begin{aligned} P_e[k;\mathcal{P}] &= \int_0^\infty Q\left(\sqrt{\frac{2x\sigma_d^2}{\sigma_d^2 M_k(\mathcal{P}) + \sigma_z^2}}\right)\frac{1}{\sigma_h^2 - M_k(\mathcal{P})}\mathrm{e}^{-\frac{x}{\sigma_h^2 - M_k(\mathcal{P})}}dx \\ &= \frac{1}{2}\left[1 - \sqrt{\frac{1 - \frac{M_k(\mathcal{P})}{\sigma_h^2}}{1 + \frac{1}{\sigma_h^2 \mathrm{SNR}_d}}}\right], \end{aligned} \tag{5.B.3}$$

where $\mathrm{SNR}_d \triangleq \sigma_d^2/\sigma_z^2$.

For QPSK signaling ($a[k] = \frac{\sigma_d}{\sqrt{2}}(\pm 1 \pm j)$), the decision rule is

$$\mathrm{Re}\{\hat{a}[k]\} = \mathrm{sign}\left(\mathrm{Re}\{\hat{h}^*[k]y[k]\}\right)\frac{\sigma_d}{\sqrt{2}}, \quad \mathrm{Im}\{\hat{a}[k]\} = \mathrm{sign}\left(\mathrm{Im}\{\hat{h}^*[k]y[k]\}\right)\frac{\sigma_d}{\sqrt{2}}.$$

The bit error probability conditioned on $\hat{h}[k]$ and $a[k]$ can be derived similarly as in the BPSK case,

$$\Pr\left\{\text{bit error}\middle| \hat{h}[k], a[k]\right\} = Q\left(\sqrt{\frac{|\hat{h}[k]|^2\sigma_d^2}{\sigma_d^2 M_k(\mathcal{P}) + \sigma_z^2}}\right)$$

and the BER at the kth position of a period is given by

$$P_e[k;\mathscr{P}] = \mathrm{E}\left\{Q\left(\sqrt{\frac{|\hat{h}[k]|^2\sigma_d^2}{\sigma_d^2 M_k(\mathscr{P})+\sigma_z^2}}\right)\right\}$$

$$= \frac{1}{2}\left[1-\sqrt{\frac{1-\frac{M_k(\mathscr{P})}{\sigma_h^2}}{1+\frac{M_k(\mathscr{P})}{\sigma_h^2}+\frac{2}{\sigma_h^2 \mathrm{SNR}_d}}}\right]. \tag{5.B.4}$$

The BER expressions for BPSK and QPSK signaling are now obtained as functions of the steady-state channel MMSE with placement $\mathscr{P}$ in (5.B.3) and (5.B.4), respectively. In both cases, it is clear that increasing $M_k(\mathscr{P})$ results in increased $P_e[k;\mathscr{P}]$. It immediately follows that, in either case, the optimization in (5.9) is equivalent to that in (5.8). ■

5.C PROOF OF LEMMA 5.1

At the steady state of an RPP-γ scheme, the channel MMSE attains a periodic steady state. During a training period ($1 \le k \le \gamma$), M_k obeys the same update recursion as in (5.4):

$$M_k(\gamma) = \frac{\sigma_z^2[\alpha^2 M_{k-1}(\gamma)+\sigma_u^2]}{\sigma_z^2+[\alpha^2 M_{k-1}(\gamma)+\sigma_u^2]\sigma_p^2}$$

where $\sigma_u^2 \triangleq \sigma_h^2(1-\alpha^2)$. Define $\delta_k(\gamma)$ as the difference of M_k and M_∞, then

$$\begin{aligned}\delta_k(\gamma) &\triangleq M_k(\gamma) - M_\infty \\ &= \frac{\sigma_z^2\left[\alpha^2 M_{k-1}(\gamma)+\sigma_u^2\right]}{\sigma_z^2+\left[\alpha^2\delta_{k-1}(\gamma)+\alpha^2 M_\infty+\sigma_u^2\right]\sigma_p^2} - \frac{\sigma_z^2(\alpha^2 M_\infty+\sigma_u^2)}{\sigma_z^2+(\alpha^2 M_\infty+\sigma_u^2)\sigma_p^2} \\ &= \frac{\alpha^2\sigma_z^4\delta_{k-1}(\gamma)}{\left[\alpha^2\sigma_p^2\delta_{k-1}(\gamma)+\sigma_z^2+(\alpha^2 M_\infty+\sigma_u^2)\sigma_p^2\right]\left[\sigma_z^2+(\alpha^2 M_\infty+\sigma_u^2)\sigma_p^2\right]}.\end{aligned}$$

We then have the following first-order difference equation for $\delta_k(\gamma)$:

$$\frac{1}{\delta_k(\gamma)} = \beta_1\frac{1}{\delta_{k-1}(\gamma)}+\beta_2, \quad k=1,\ldots,\gamma$$

where β_1 and β_2 are defined in (5.14) and (5.15), respectively. Note that, in the above recursion, when $k=1$, $\delta_0(\gamma)$ corresponds to the value of $\delta_k(\gamma)$ at the end position of the previous placement period,

$\delta_0(\gamma) = \delta_{T_p}(\gamma)$. Therefore, we can express $\delta_k(\gamma)$ in terms of $\delta_{T_p}(\gamma)$ from the above recursion as

$$\begin{aligned}\frac{1}{\delta_k(\gamma)} &= \beta_1^2 \frac{1}{\delta_{k-2}(\gamma)} + \beta_2 + \beta_1\beta_2 \\ &= \beta_1^k \frac{1}{\delta_0(\gamma)} + \beta_2 \frac{1-\beta_1^k}{1-\beta_1} \\ &= \beta_1^k \frac{1}{\delta_{T_p}(\gamma)} + \beta_2 \frac{1-\beta_1^k}{1-\beta_1}, \quad 1 \le k \le \gamma. \end{aligned} \tag{5.C.1}$$

During data transmission ($\gamma + 1 \le k \le T_p$), the updating recursion for $M_k(\gamma)$ is given by (5.7). Therefore,

$$\begin{aligned}\delta_k(\gamma) &= \alpha^{2(k-\gamma)} M_\gamma(\gamma) + (1-\alpha^{2(k-\gamma)})\sigma_h^2 - M_\infty \\ &= \alpha^{2(k-\gamma)} \delta_\gamma(\gamma) + (1-\alpha^{2(k-\gamma)})(\sigma_h^2 - M_\infty) \\ &= \alpha^{2(k-\gamma)} \delta_\gamma(\gamma) + (1-\alpha^{2(k-\gamma)})\rho_\infty \sigma_h^2, \quad k = \gamma+1, \ldots, T_p. \end{aligned} \tag{5.C.2}$$

From (5.C.1)–(5.C.2), $\delta_\gamma(\gamma)$ and $\delta_{T_p}(\gamma)$ satisfy the following relations:

$$\begin{cases} \delta_\gamma(\gamma) = \dfrac{\delta_{T_p}(\gamma)}{1 + \delta_{T_p}(\gamma) \dfrac{\beta_2\left(1-\frac{1}{\beta_1^\gamma}\right)}{\beta_1 - 1}} \left(\dfrac{1}{\beta_1}\right)^\gamma \\ \\ \delta_{T_p}(\gamma) = \alpha^{2\gamma\frac{1-\eta}{\eta}} \delta_\gamma(\gamma) + \left(1 - \alpha^{2\gamma\frac{1-\eta}{\eta}}\right)\rho_\infty \sigma_h^2, \end{cases} \tag{5.C.3}$$

where we have used the relation $T_p = \gamma/\eta$. From (5.C.3), we obtain a quadratic equation for $\delta_{T_p}(\gamma)$ as

$$\begin{aligned}&\frac{\beta_2\left(1-\frac{1}{\beta_1^\gamma}\right)}{\beta_1 - 1} \delta_{T_p}^2(\gamma) + \left[1 - \left(\frac{\alpha^{2\frac{1-\eta}{\eta}}}{\beta_1}\right)^\gamma - \frac{\beta_2\left(1-\frac{1}{\beta_1^\gamma}\right)}{\beta_1 - 1}(1-\alpha^{2\gamma\frac{1-\eta}{\eta}})\rho_\infty \sigma_h^2\right] \delta_{T_p}(\gamma) \\ &\quad - (1-\alpha^{2\gamma\frac{1-\eta}{\eta}})\rho_\infty \sigma_h^2 = 0. \end{aligned} \tag{5.C.4}$$

Solving the above equation, we have the expression of $\delta_{T_p}(\gamma)$, as a function of γ, given in (5.13).

From (5.7), for any RPP-γ scheme, the expression of the steady-state channel MMSE $M_k(\gamma)$ over each data symbol is then obtained in (5.12) and (5.11). ■

5.D PROOF OF THEOREM 5.1

The algebraic proof of Theorem 5.1, based on the expression for $\delta_{T_p}(\gamma)$ in (5.13) as a function of γ, can be found in the technical report (Dong, Tong, & Sadler, 2003). Here we give a more intuitive graphic-aided proof using Fig. 5.D.1.

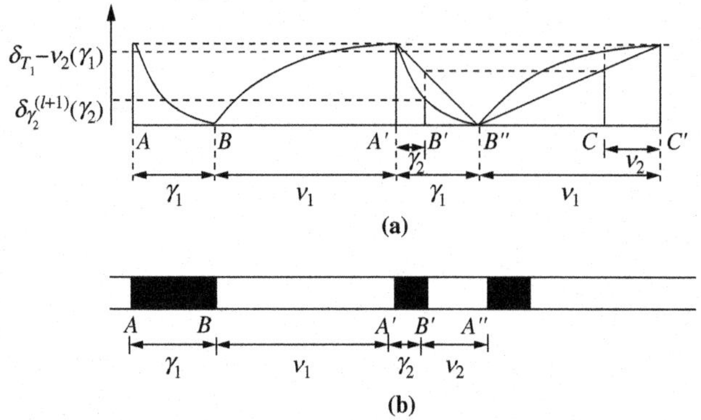

FIGURE 5.D.1

Proof of Theorem 5.1.

The basic idea in this proof is the following. We first let the data stream contain larger pilot clusters (thus longer period T_p). After the process reaches its steady state, we change the pilot placement to the one with smaller pilot clusters (thus shorter period T_p). After this rearrangement, we show that the channel MMSE at the last position of a period ($M_{T_p}(\gamma)$) is smaller in the new placement than that in the previous placement. This eventually results in a decreased MMSE when the new process goes to its steady state.

For a RPP-γ scheme, following similar derivations in (5.C.1)–(5.C.4), we have $\delta_k(\gamma)$ during training as

$$\delta_k(\gamma) = \frac{\delta_{T_p}(\gamma)}{1 + \delta_{T_p}(\gamma)\frac{\beta_2\left(1-\frac{1}{\beta_1^k}\right)}{\beta_1 - 1}}\left(\frac{1}{\beta_1}\right)^k, \quad k = 1,\ldots,\gamma.$$

Thus, $\delta_k(\gamma)$ decreases exponentially at rate $(1/\beta_1)^k$. During a data block, it follows from (5.C.2) that $\rho_\infty\sigma_h^2 - \delta_k(\gamma)$ decreases exponentially at rate $\alpha^{2(k-\gamma)}$.

Figure 5.D.1(b) and (a) describe the placement pattern $\mathscr{P}$ and the corresponding steady-state trajectory of $\delta_k(\gamma)(= M_k(\gamma) - M_\infty)$, respectively. For fixed pilot percentage η, let us consider two schemes: RPP-γ_1 with $\mathscr{P}_1 = (\gamma_1, \nu_1)$, and RPP-$\gamma_2$ with $\mathscr{P}_2 = (\gamma_2, \nu_2)$, where $\gamma_1 > \gamma_2$. Shown in the left part of Fig. 5.D.1(b) is the RPP-γ_1 scheme, where the placement period is T_1. Indexes A and A' denote the end positions of the data blocks in the $(l-1)$th and lth placement periods under RPP-γ_1, respectively. Index B denotes the end position of the pilot cluster in the lth period. Assume that in the $(l-1)$th and lth placement periods, the channel MMSE is in its steady state. The corresponding $\delta_k(\gamma_1)$ at A and A' is $\delta_{T_1}(\gamma_1)$. The trajectory curve $C_{A\to B\to A'}$ of $\delta_k(\gamma_1)$ is shown in Fig. 5.D.1(a). Because the changing rates of $\delta_k(\gamma_1)$ during pilot and data cluster are exponential, $C_{A\to B}$ and $C_{B\to A'}$ are both exponential. If RPP-γ_1 is still used in the $(l+1)$th placement period, then the trajectory of $\delta_k(\gamma_1)$ is the curve $C_{A'\to B''\to C'}$ in Fig. 5.D.1(a). It is equivalent to $C_{A\to B\to A'}$. Now, after the lth period, we change the placement to RPP-γ_2, shown in the right part of Fig. 5.D.1(b), where indices B' and A'' denote the new end positions of pilot and data cluster in the period, respectively. The change of placement results in a new

MMSE value. Denote $\delta_{\gamma_2}^{(l+1)}(\gamma_2) = M[lT_1 + \gamma_2] - M_\infty$ over B', and similarly $\delta_{T_2}^{(l+1)}(\gamma_2)$ over A''. Note that $\delta_{\gamma_2}^{(l+1)}(\gamma_2)$ is still on the trajectory curve $C_{A'\to B'}$ in Fig. 5.D.1(a). Therefore, $\delta_{\gamma_2}^{(l+1)}(\gamma_2) = \delta_{\gamma_2}(\gamma_1)$. Now let C in Fig. 5.D.1(a) be the point such that the length $CC' = \nu_2$. Because for fixed pilot percentage η, $\gamma_1/\gamma_2 = \nu_1/\nu_2$, we have $A'B'/A'B'' = CC'/B''C'$, shown in Fig. 5.D.1(a). Then, because $C_{A'\to B''}$ and $C_{B''\to C'}$ are exponential, where the former is convex and the latter is concave, from the geometry, we have

$$\delta_{\gamma_2}^{(l+1)}(\gamma_2) = \delta_{\gamma_2}(\gamma_1) \le \delta_{T_1-\nu_2}(\gamma_1) \tag{5.D.1}$$

shown in Fig. 5.D.1(a). From (5.C.2), since

$$\delta_{T_2}^{(l+1)}(\gamma_2) = \alpha^{2\nu_2}\delta_{\gamma_2}^{(l+1)}(\gamma_2) + (1-\alpha^{2\nu_2})\rho_\infty\sigma_h^2$$
$$\delta_{T_1}(\gamma_1) = \alpha^{2\nu_2}\delta_{T_1-\nu_2}(\gamma_1) + (1-\alpha^{2\nu_2})\rho_\infty\sigma_h^2,$$

it follows from (5.D.1) that $\delta_{T_2}^{(l+1)}(\gamma_2) \le \delta_{T_1}(\gamma_1)$. Consequently, $\delta_{T_2}^{(j)}(\gamma_2) \le \delta_{T_1}(\gamma_1)$, for $j > l$. At the steady state of RPP-γ_2, we have

$$\delta_{T_2}(\gamma_2) = \lim_{l\to\infty} \delta_{T_2}^{(l+1)}(\gamma_2) \le \delta_{T_1}(\gamma_1).$$

Because $\mathscr{E}(\gamma_i) = \delta_{T_i}(\gamma_i) + M_\infty$ and M_∞ is not a function of γ, we have $\mathscr{E}(\gamma_1) \ge \mathscr{E}(\gamma_2)$, for $\gamma_1 > \gamma_2$. The minimum $\mathscr{E}(\gamma)$ can be obtained using Lemma 5.1. ■

References

3GPP TS36.201. Evolved Universal Terrestrial Radio Access (E-UTRA); *LTE physical layer general description.* http://www.3gpp.org/ftp/specs/html-info/36-series.htm.

Abou-Faycal, I., Medard, M., & Madhow, U. (2005). Binary adaptive coded pilot assisted modulation over Rayleigh fading channels without feedback. *IEEE Transactions on Communications, 53*(6), 1036–1046.

Adireddy, S., Tong, L., & Viswanathan, H. (2002). Optimal placement of known symbols for frequency-selective block-fading channels. *IEEE Transactions on Information Theory, 48*, 2338–2353.

Alameda-Hemandex, E., McLernon, D. C., Orozco-Lugo, A. G., Lara, M. M., & Ghogho, M. (2001). Frame/training sequence synchronization and DC-offset removal (data-dependent) superimposed training based channel estimation. *IEEE Transactions on Signal Processing, 55*(6), 2557–2569.

Arikan, E. (1998). Upper bound on the cutoff rate of sequential decoding. *IEEE Transactions on Information Theory, 34*(1), 55–63.

Baissas M. A., & Sayeed, A. (2002). Pilot-based estimation of time-varying multipath channels for coherent CDMA receivers. *IEEE Transactions on Signal Processing, 50*(8), 2037–2049.

Baltersee, J., Fock, G., & Meyr, H. (2001). An information theoretic foundation of synchronized detection. *IEEE Transactions Communication, 49*(12), 2115–2123.

Blahut, R. (1987). *Principles and practice of information theory.* Reading, MA: Addison-Wesley.

Budianu C., & Tong, L. (2002). Channel estimation for space-time block coding systems. *IEEE Transactions on Signal Processing, 50*, 2515–2528.

Cavers, J. K. (1991). An analysis of pilot symbol assisted modulation for Rayleigh fading channels. *IEEE Transactions on Vehicular Technology, 40*(11), 686–693.

Choi J. D., & Stark, W. E. (2002). Performance of ultra-wideband communications with suboptimal receivers in multipath channels. *IEEE Journal of Selected Areas Communication, 20*(9), 1754–1766.

Cover T., & Thomas, J. (1991). *Elements of information theory*. New York, NY: John Wiley & Sons Inc.

Dong, M., & Tong, L. (2002). Optimal design and placement of pilot symbols for channel estimation. *IEEE Transactions on Signal Processing, 50*, 3055–3069.

Dong, M., Tong, L., & Sadler, B. M. (2002). Optimal pilot placement for channel tracking in OFDM. In *Proceedings of IEEE Military Communications Conference* (pp. 602–606), Anaheim, CA.

Dong, M., Adireddy, S., & Tong, L. (2003). Optimal pilot placement for semi-blind channel tracking of packetized transmission over time-varying channels. *IEICE Transactions on Fundamentals of Electronics, Communications and Computer Sciences*, E86-A, 550–563.

Dong, M., Tong, L., & Sadler, B. M. (2003). Optimal placement of training over time-varying channel. *ACSP TR-01-03-02* (http://acsp.ece.cornell.edu/papers/TR-01-03-02.pdf).

Dong, M., Tong, L., & Sadler, B. M. (2004). Optimal insertion of pilot symbols for transmissions over time-varying flat fading channels. *IEEE Transactions on Signal Processing, 52*(5), 1403–1418.

Evans, J. (2002, August). Optimal resource allocation for pilot symbol aided multiuser receivers in Rayleigh faded CDMA channels. *IEEE Transactions Communication, 50*(8), 1316–1325.

Farhang-Boroujeny, B. (1995). Pilot-based channel identification: Proposal for semi-blind identification of communication channels. *Electronics Letters, 31*(13), 1044–1046.

Fernandez-Getino, M. J. Garcia, M. J., Paez Borrallo, J. M., & Zazo Bello, S. (1999). Novel pilot patterns for channel estimation in OFDM mobile systems over frequency selective fading channels. In *Proceedings of the 10th International Symposium on Personal, Indoor and Mobile Radio Communications* (pp. 363–367), Osaka, Japan.

Frank, R. L. (1962). Phase shift pulse codes with good periodic correlation properties. *IRE Transactions on Information Theory, 8*, 381–382.

Gallager, P.G. (1968). *Information theory and reliable communication*. New York, NY: John Wiley & Sons, Inc.

Gansman, J. A., Fitz, M. P., & Krogmeier, J. V. (1997). Optimum and suboptimum frame synchronization for pilot-symbol-assisted modulation. *IEEE Transactions Communication, 45*, 1327–1337.

Gaston, A., Chriss, W., & Walker, E. (1973). A multipath fading simulator for radio. *IEEE Transactions on Vehicular Technology, 22*, 241–244.

Ghogho M., & Swami, A. (2006). Training design for multipath channel and frequency-offset estimation in MIMO systems. *IEEE Transactions on Signal Processing, 54*(10), 3957–3965.

Ghogho, M., McLernon, D. C., Alameda-Hernandez, E., & Swami, A. (2005). Channel estimation and symbol detection for block transmission using data-dependent superimposed training. *IEEE Signal Processing Letters, 12*(3), 226–229.

Hassibi B., & Hochwald, B. (2003). How much training is needed in multiple-antenna wireless links? *IEEE Transactions on Information Theory, 49*(4), 951–963.

Hoctor, R., & Tomlinson, H. (2002). Delay-hopped transmitted-reference RF communications. In *Proceedings of IEEE Conference on Ultra Wideband Systems and Technologies* (pp. 265–269), Baltimore, MA.

Hoeher, P., & Tufvesson, F. (1999). Channel estimation with superimposed pilot sequence. In *Proceedings of IEEE Global Communications Conference*, Vol. 4. (pp. 2162–2166), Rio de Janeiro, Brazil.

Hoyos, S., Sadler, B. M., & Arce, G. (2005). Monobit digital receivers for ultra-wideband communications. *IEEE Transactions Wireless Communication, 4*(4), 1337–1344.

Iltis, R. (1990). Joint estimation of PN code delay and multipath using the extended Kalman filter. *IEEE Transactions Communication, 38*, 1677–1685.

Kailath, T. (1980). *Linear systems*. Englewood Cliffs, NJ: Prentice Hall.

Kailath, T., Sayed, A. H., & Hassibi, B. (2000). *Linear estimation*. Englewood Cliffs, NJ: Prentice Hall.

Kay, S. (1993). *Fundamentals of statistical signal processing: Estimation theory*. Englewood Cliffs, NJ: Prentice Hall.

Lapidoth, A., & Narayan, P. (1998). Reliable communication under channel uncertainty. *IEEE Transactions on Information Theory*, *44*(6), 2148–2177.

Lapidoth, A., & Shamai, S. (2002). Fading channels: How perfect need "perfect side information" be? *IEEE Transactions on Information Theory*, *48*(5), 1118–1134.

Ling, F. (1999). Optimal reception, performance bound, and cutoff rate analysis of reference-assisted coherent CDMA communications with applications. *IEEE Transactions Communication*, COM-47, 1583–1592.

Lodge J. H., & Moher, M. L. (1987). Time diversity for mobile satellite channels using trellis coded modulations. In *Proceedings of IEEE Global Communications Conference* (pp. 303–307), Tokyo, Japan.

Ma, X., Giannakis, G., & Ohno, S. (2003). Optimal training for block transmission of doubly selective wireless fading channels. *IEEE Transactions on Signal Processing*, *51*(5), 1351–1366.

Ma, X., Yang, L., & Giannakis, G. (2005). Optimal training for MIMO frequency-selective fading channels. *IEEE Transactions Wireless Communication*, *4*(2), 453–466.

Makrakis D., & Feher, K. (1987). A novel pilot insertion-extraction technique based on spread spectrum affine precoders for reliable communications. In *Proceedings Miami technicon* (pp. 128–132), Miami, FL.

Manton, J., Mareels, I., & Hua, Y. (2000). Affine precoders for reliable communications. In *Proceedings of IEEE International Conference on Acoustics, Speech, and Signal Processing* (Vol. 5, pp. 249–2752), Istanbul, Turkey.

Marzetta, T. (1999). BLAST training: Estimating channel characteristics for high capacity space-time wireless. In *Proceedings of 37th Annual Allerton Conference on Communication, Control and Computing* (pp. 958–966), Monticello, IL.

Mazzenga, F. (2000). Channel estimation and equalization for M-QAM transmission with a hidden pilot sequence. *IEEE Transactions on Broadcasting*, *46*(6), 170–176.

Medard, M. (2000). The effect upon channel capacity in wireless communication of perfect and imperfect knowledge of the channel. *IEEE Transactions on Information Theory*, *46*(5), 933–946.

Merhav, N., Kaplan, G., Lapidoth, A., & Shamai, S. (1994). On information rates for mismatched decoders. *IEEE Transactions on Information Theory*, *40*(11), 1953–1967.

Misra, S., Swami, A., & Tong, L. (2006). Optimal training for time-selective wireless fading channels using cutoff rate. *EURASIP Journal on Applied Signal Processing*, 2006, 1–15.

Negi, R., & Cioffi, J. (1998). Pilot tone selection for channel estimation in a mobile OFDM system. *IEEE Transactions on Consumer Electronics*, *44*, 1122–1128.

Ohno, S., & Giannakis, G. B. (2002). Average-rate optimal PSAM transmissions over time-selective fading channels. *IEEE Transactions Wireless Communication*, *1*, 712–720.

Ohno, S., & Giannakis, G. B. (2002). Optimal training and redundant precoding for block transmissions with application to wireless OFDM. *IEEE Transactions Communication*, *50*(12), 2113–2123.

Orozco-Lugo, A. G., Lara, M. M., & McLernon, D. C. (2004). Channel estimation using implicit training. *IEEE Transactions on Signal Processing*, *52*(1), 240–254.

Phoel W., & Honig, M. (2002). Performance of coded DS-CDMA with pilot-assisted channel estimation and linear interference suppression. *IEEE Transactions Communication*, *50*(5), 822–832.

Proakis, J. G., & Salehi, M. (2007). *Digital communications* (5th ed.). New York, NY: McGraw-Hill.

Rinne, J., & Renfors, M. (1996). Pilot spacing in orthogonal frequency division multiplexing system. *IEEE Transactions on Consumer Electronics*, *42*(11), 959–962.

Sadler, B. M., & Kozick, R. J. (2000). Bounds on uncalibrated array signal processing. In *Proceedings of the 10th IEEE Workshop on Statistical Signal and Array Processing* (pp. 73–77), Pacono Manor, PA.

Sadler, B. M., & Kozick, R. J. (2001). Bounds on MIMO channel estimation and equalization with side information. In *Proceedings IEEE International Conference on Acoustics, Speech, and Signal Processing* (Vol. 4. pp. 2145–2148), Salt Lake City, UT.

Sadler, B. M., & Swami, A. (2004). On the performance of episodic UWB and direct-sequence communication systems. *IEEE Transactions Wireless Communication*, *3*(6), 2246– 2255.

Sampei, S., & Sunaga, T. (1989). Rayleigh fading compensation method for 16QAM in digital land mobile radio channels. In *Proceedings of IEEE Vehicular Technology Conference* (pp. 640–646), San Francisco, CA.

Schramm, P. (1998, September). Analysis and optimization of pilot-channel-assisted BPSK for DS-CDMA systems. *IEEE Transactions Communication*, *46*(9), 1122–1124.

Schramm, P., & Muller, R. (1998). Pilot symbol assisted BPSK and Rayleigh fading channels with diversity: Performance analysis and parameter optimization. *IEEE Transactions Communication*, *46*(12), 1560–1563.

Stamoulis, A. Diggavi, S., & Al-Dhahir, N. (2002). Intercarrier interference in MIMO OFDM. *IEEE Transactions on Signal Processing*, *50*(10), 2451–2464.

Stojanovic, M., Proakis, J., & Catipovic, J. (1995). Analysis of the impact of channel estimation errors on the performance of a decision-feedback equalizer in fading multipath channels. *IEEE Transactions Communication*, *43*, 877–886.

Tang, Z., & Leus, G. (2007). Time-multiplexed training for time-selective channels. *IEEE Signal Processing Letters*, *14*(9), 585–588.

Torrance, J. M., & Hanzo, L. (1995). Comparative study of pilot symbol assisted modem schemes. In *Proceedings of 6th International Conference on Radio Receivers and Associated Systems* (pp. 36–41). Bath, UK.

Tugnait, J. K., & Luo, W. (2003). On channel estimation using superimposed training and first-order statistics. *IEEE Communications Letters*, *7*(9), 413–415.

Tugnait, J., Meng, X., & He, S. (2006). Doubly selective channel estimation using superimposed training and exponential bases models. *EURASIP Journal on Applied Signal Processing*, *2006*, 22, 1–11.

Tang, J., Xu, Z., & Sadler, B. M. (2007). Performance analysis of *b*-bit digital receivers for TR-UWB systems with inter-pulse interference. *IEEE Transactions Wireless Communication*, *6*(2), 494–505.

Tong, L., Sadler, B. M., & Dong, M. (2004). Pilot-assisted wireless transmissions general model, design criteria, and signal processing. *IEEE Signal Processing Magazine*, *21*(6), 12–25.

Tsatsanis, M. K., & Xu, Z. (2000). Pilot symbol assisted modulation in frequency selective fading wireless channels. *IEEE Transactions on Signal Processing*, *7*, 2353–2365.

Tse, D., & Viswanath, P. (2005). *Fundamentals of wireless communication.* New York: Cambridge University Press.

Van Trees, H. L. (1968). *Detection, estimation and modulation theory* (Vol. 1). New York: Wiley.

Viterbi, A., & Omura, J. (1979). *Principles of digital communication and coding*. New York, NY: McGraw-Hill.

Whitworth, T., Ghogho, M., & McLernon, D. (2009). Optimized training and basis expansion model parameters for doubly-selective channel estimation. *IEEE Transactions Wireless Communication*, *8*(3), 1490–1498.

Wozencraft, J. M., & Jacobs, I. M. (1965). *Principles of communication engineering*. New York, NY: John Wiley & Sons.

Xu, Z., & Sadler, B. M. (2006). Multiuser transmitted reference ultra-wideband communication systems. *IEEE Journal of Selected Areas Communication*, *24*(4), 766–772.

Yang, L., & Giannakis, G. (2004). Optimal pilot waveform assisted modulation for ultra-wideband communications. *IEEE Transactions Wireless Communication*, 3(4), 1236–1249.

Zhou, G. T., Viberg, M., & McKelvey, T. (2003). A first-order statistical method for channel estimation. *IEEE Signal Processing Letters*, *10*(3), 57–60.

CHAPTER

Equalization of Time-Varying Channels

6

Philip Schniter[1], Sung-Jun Hwang[2], Sibasish Das[3], Arun P. Kannu[4]

[1]*The Ohio State University, Columbus, OH, USA*

[2]*Qualcomm, Inc., Santa Clara, CA, USA*

[3]*Qualcomm, Inc., San Diego, CA, USA*

[4]*Indian Institute of Technology, Madras, Chennai, India*

6.1 INTRODUCTION

As discussed in Chapter 1, the wireless communication channel can be modeled as a time-varying (TV) linear[1] system whose output is corrupted by additive noise. To reliably recover the transmitted information from the channel output, the receiver must address the effects of both linear distortion and additive noise. Although, in theory, the mitigation of linear distortion and additive noise should be done jointly, in practice the task is often partitioned into two tasks, equalization and decoding, in order to reduce implementation complexity.

Roughly speaking, *equalization* leverages knowledge of channel structure to mitigate the effects of the linear distortion, whereas *decoding* leverages knowledge of code structure to mitigate the channel's additive noise component. The equalizer might be well informed about the channel (e.g., knowing the complete channel impulse response) or relatively uninformed (e.g., knowing only the maximum channel length). In some cases, knowledge of symbol structure (e.g., the symbol alphabet or, if applicable, the fact that the symbols have a constant modulus) is assumed to be in the domain of the equalizer, whereas in other cases, it is assumed to be in the domain of the decoder; because the equalizer and decoder work together to infer the transmitted information from the channel output, the role of equalization versus decoding is somewhat a matter of definition. For this chapter, however, we assume that exploitation of code structure is *not* in the domain of the equalizer.

Generally speaking, the output of the equalizer is a sequence of symbol (or bit) estimates which have been, to the best of the equalizer's ability, freed of channel corruption. These estimates are then passed to the decoder for further refinement and final decision making. In so-called *turbo equalization* schemes (Douillard et al., 1995; Koetter, Singer, & Tüchler, 2004), the decoder passes refined soft bit estimates back to the equalizer for further refinement, and the equalizer passes further refined soft bit estimates to the decoder. The process is then iterated until the equalizer and decoder "agree" on the soft bit estimates. Note that the use of soft bit estimates implies that the equalizer treats the bits as (a priori) independent. Turbo equalization is illustrated in Fig. 6.1 and will be discussed in more detail later.

[1]Some channels are better modeled as nonlinear, but such channels are not the focus of this book.

Wireless Communications Over Rapidly Time-Varying Channels

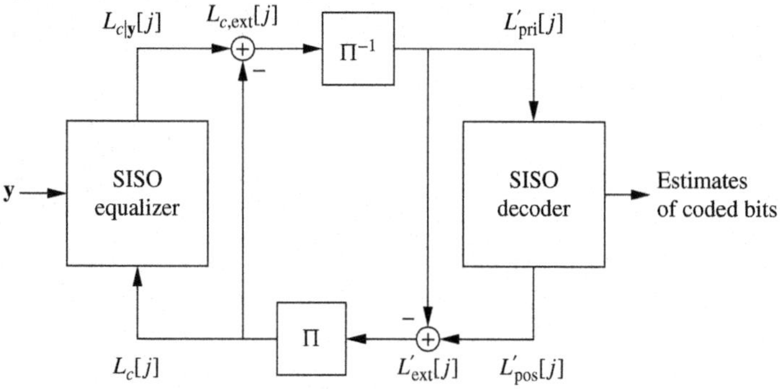

FIGURE 6.1

Soft-input soft-output (SISO) equalizer connected with a SISO decoder in a turbo configuration. Note the presence of deinterleaver Π^{-1} fed by the extrinsic equalizer LLRs $L_{c,\text{ext}}[j] = L_{c|\mathbf{y}}[j] - L_c[j]$, and interleaver Π fed by the extrinsic decoder LLRs $L'_{\text{ext}}[j] = L'_{\text{pos}}[j] - L'_{\text{pri}}[j]$.

The inputs to an equalizer depend on its design. So-called *coherent equalizers* are assumed to know the parameters describing the *state* of the TV linear system that they are trying to mitigate, or an estimate thereof. Typical examples of channel state parameters include impulse response coefficients or intercarrier interference coefficients. Coherent equalization requires the simultaneous operation of a *channel estimator*, whose main purpose is to provide accurate and up-to-date estimates of the TV channel state to the equalizer. Channel estimation is discussed in Chapter 4. The idea to separate channel estimation from equalization can be traced back to early work by Kailath (1960).

So-called *noncoherent equalizers* operate without explicit knowledge of the channel state, and therefore are not dependent on the implementation of a channel estimator. Noncoherent equalizers, however, are sometimes assumed to know the *channel statistics* (e.g., the scattering function) or an estimate thereof. In the case of a nonstationary channel, the statistics themselves would need to be tracked. In their most general form, noncoherent equalizers treat the channel parameters as "nuisance parameters" that complicate data estimation. In some cases, they explicitly estimate the channel state parameters in conjunction with the data (i.e., *joint channel/symbol estimation*), whereas in other cases they compute data estimates without ever computing a channel estimate.

The equalization of rapidly TV communication channels is much more challenging than the equalization of their time-invariant (or slowly TV) counterparts. This can be understood intuitively as follows. From the perspective of coherent equalization, a rapidly TV channel implies that the channel state is constantly changing, which implies that the equalizer must be constantly redesigned in order to stay well matched to the channel. From the perspective of noncoherent equalization, a rapidly TV channel has more degrees of freedom (over a given bandwidth and signaling epoch) than a slowly varying channel, and thus more nuisance parameters to contend with.

Beyond these intuitive considerations, there is another important reason why rapidly TV channels are more difficult to equalize than slowly TV ones. For time-invariant linear channels, information can be split up and transmitted in parallel on noninterfering subcarriers. In this case, equalization becomes a simple matter of adjusting the gain and phase on each received subcarrier. This is, in fact, the main idea behind multicarrier modulation schemes like *orthogonal frequency division multiplexing* (OFDM)

(Cimini, 1985). For slowly TV channels, the same approach can be easily extended: to mimic a time-invariant channel, the OFDM symbol duration can be chosen shorter than the channel's coherence time. But, as now explained, such an approach turns out to be impractical for rapidly TV channels. To prevent interference between adjacent OFDM symbols, guard intervals are typically inserted. For time-invariant or slowly TV channels, the loss in spectral efficiency due to the inclusion of these guards can be made small because the channel delay spread (and hence the guard interval) is much smaller than the channel coherence time (and hence the OFDM symbol length). For rapidly TV channels, the OFDM symbol length would need to be made extremely short, at which point the loss of spectral efficiency resulting from guard insertion would be severe. If one tried to optimize the modulation strategy, one would find that it is, in fact, impossible to prevent interference among the subcarriers without significant compromise in spectral efficiency (Strohmer & Beaver, 2003) – a consequence of the Balian-Low theorem (Daubechies, 1992). To summarize: while the equalization of slowly TV channels can be trivialized through suitable choice of the transmission scheme, the equalization of rapidly TV channels cannot.

The remainder of this chapter will be organized as follows. In Section 6.2, we outline the system model assumed throughout the chapter and detail the essential features that result from rapid channel time-variation. In Section 6.3, we describe coherent approaches to equalization of rapidly TV channels and, in Section 6.4, we describe noncoherent approaches. In Section 6.5, we conclude.

6.2 SYSTEM MODEL

We now outline the system model used in the remainder of the chapter. In this chapter, we focus on systems which use a single transmitter antenna and a single receiver antenna; multiantenna systems will be discussed in Chapter 8.

6.2.1 Basic Assumptions

As discussed in Chapter 1, the time-domain received sample $r[n]$ can be written in terms of the transmitted sequence $(s[n])_{n\in\mathbb{Z}}$, the TV time-n length-M impulse response $(h[n,m])_{m=0}^{M-1}$, and additive white Gaussian noise process $(w[n])_{n\in\mathbb{Z}}$ of variance σ_w^2 as follows:

$$r[n] = \sum_{m=0}^{M-1} h[n,m]s[n-m] + w[n]. \tag{6.1}$$

In this chapter, we assume that the transmitted sequence $(s[n])_{n\in\mathbb{Z}}$ is generated from the finite-alphabet symbol sequence $(a[k])_{k\in\mathbb{Z}}$ using a generic *finite-memory linear modulation* scheme, and that the demodulated sequence $(y[k])_{k\in\mathbb{Z}}$ is generated from the received sequence $(r[n])_{n\in\mathbb{Z}}$ using a corresponding *finite-memory linear demodulation* scheme. Prior to modulation, the symbol sequence $(a[k])_{k\in\mathbb{Z}}$ is mapped from a coded-bit sequence $(c[j])_{j\in\mathbb{Z}}$, which is generated from an information-bit sequence $(b[i])_{i\in\mathbb{Z}}$ through rate-R_c coding and interleaving. We denote the symbol alphabet by $\mathscr{A}$, its cardinality by $|\mathscr{A}|$, and the set of admissible symbol sequences (as allowed by coding/interleaving) by $\mathscr{A}$.

For ease of notation, we find it convenient to assume *block transmission* with block length K, where the symbols

$$\mathbf{a} \triangleq (a[0]\, a[1] \cdots a[K-1])^T \in \mathscr{A}^K$$

can be related to the demodulated channel outputs

$$\mathbf{y} \triangleq (y[0]\, y[1] \cdots y[K-1])^T \in \mathbb{C}^K$$

through the matrix/vector equation

$$\mathbf{y} = \underbrace{\boldsymbol{\Gamma}\mathbf{H}\mathbf{G}}_{\triangleq \mathbf{Q}}\mathbf{a} + \mathbf{z}. \tag{6.2}$$

We note, however, that the block length K can be arbitrarily large and that the receiver might not be able to store/process the entire vector $\mathbf{y}$. In (6.2), $\boldsymbol{\Gamma}$, $\mathbf{H}$, and $\mathbf{G}$ are matrix representations of the linear demodulation operator, the linear TV channel, and the linear modulation operator, respectively, and

$$\mathbf{z} \triangleq (z[0]\, z[1] \cdots z[K-1])^T \in \mathbb{C}^K$$

represents the noise after demodulation. We note that the *effective channel matrix* $\mathbf{Q} = \boldsymbol{\Gamma}\mathbf{H}\mathbf{G} \in \mathbb{C}^{K\times K}$ represents the combined effects of modulation, channel propagation, and demodulation, and will be used extensively throughout the chapter. Finally, we collect, in the vector $\mathbf{c}$, the $K\log_2|\mathscr{A}|$ coded bits that determine the K symbols in $\mathbf{a}$. Note that, with a block length of K, we have $\boldsymbol{\mathscr{A}} \subset \mathscr{A}^K$.

In writing (6.2), we have assumed that the K demodulated samples in $\mathbf{y}$ are sufficient for equalization/decoding of the K symbols in $\mathbf{a}$ (i.e., that $(y[k])_{k<0}$ and $(y[k])_{k\geq K}$ can be ignored), and that interblock interference (IBI) is negligible. These assumptions will be satisfied for any well-designed block transmission scheme. Furthermore, we will assume that the noise $\mathbf{z}$, the symbols $\mathbf{a}$, and the effective channel $\mathbf{Q}$ are mutually independent, and that (unless otherwise noted) the symbols $\mathbf{a}$ are zero-mean (i.e., $\boldsymbol{\mu}_a = \mathbf{0}$) and white (i.e.,[2] $\mathbf{C}_\mathbf{a} = \sigma_a^2\mathbf{I}$). Finally, it should be noted that the demodulated noise $\mathbf{z}$ is *not* assumed to be white unless otherwise noted; although $(w[n])_{n\in\mathbb{Z}}$ is white, the demodulation process does not necessarily guarantee white $(z[k])_{k\in\mathbb{Z}}$.

Throughout the chapter, we assume that the equalizer knows the symbol alphabet $\mathscr{A}$ but not the code structure, i.e., $\boldsymbol{\mathscr{A}}$. Thus, the topic of *joint* equalization/decoding lies outside the scope of this chapter. Turbo equalization, where separate equalization and decoding steps are iterated (as illustrated in Fig. 6.1) will, however, be discussed.

6.2.2 The Structure of the Effective Channel Matrix Q

In block equalization, if it can be assumed that certain coefficients of $\mathbf{Q}$ will be negligible for nearly all realizations of $\mathbf{Q}$, then it is reasonable to conclude that an equalizer that ignores these coefficients will perform nearly as good as an equalizer that incorporates these coefficients. However, the equalizer which ignores these coefficients may be significantly cheaper to implement, especially if the proportion of negligible coefficients is large. This is, in fact, the guiding principle behind the design of practical equalization algorithms for rapidly TV channels.

[2]Throughout the chapter, we use subscripted versions of $\mathbf{C}$ to denote covariance matrices.

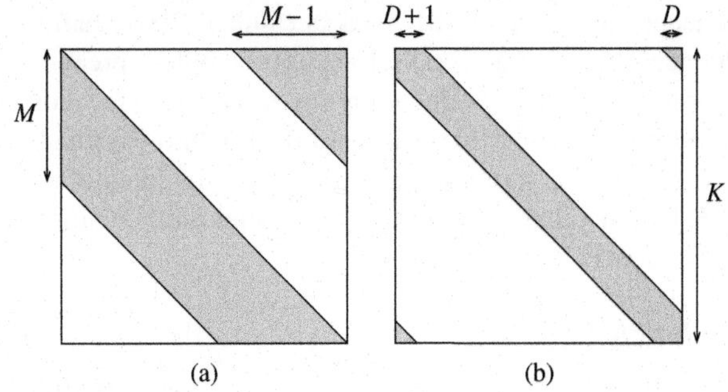

FIGURE 6.2

Support region of (a) "widely quasibanded" and (b) "narrowly quasibanded" matrices. While M is often large (e.g., in the hundreds), D is usually very small (e.g., 1 or 2).

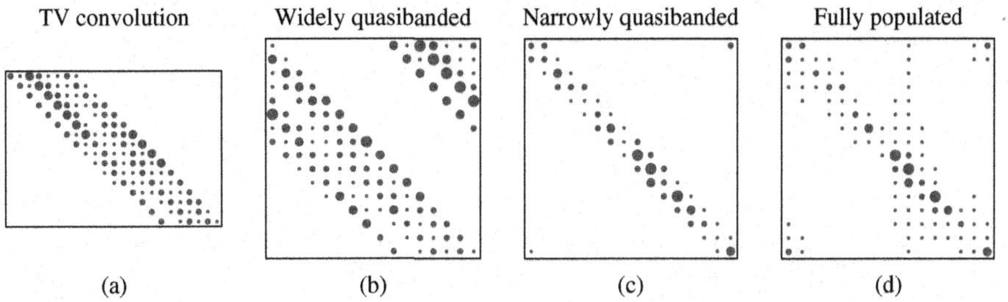

FIGURE 6.3

Example of (a) a TV channel's propagation matrix and the corresponding effective channel matrices that result from (b) CP-SCM, (c) CP-OFDM with max-SINR receiver windowing (Schniter, 2004), and (d) CP-OFDM with rectangular receiver windowing. The dot size is proportional to the coefficient magnitude.

Based on the characteristics of rapidly TV channels and commonly used modulation/demodulation schemes, we partition effective channel matrices $\mathbf{Q}$ into three classes based on the support region of non-negligible coefficients within the matrix: (1) *widely quasibanded*, (2) *narrowly quasibanded*, and (3) *fully populated* matrices. The support regions of widely quasibanded and narrowly quasibanded matrices are defined in Fig. 6.2, and illustrative examples of $\mathbf{Q}$ based on a randomly generated channel impulse response and several modulation/demodulation schemes are given in Fig. 6.3 (the construction of which will be detailed later). Note that we use the term "quasibanded" as opposed to "banded" due to the corner[3] support regions in Fig. 6.2. Banded matrices, like that illustrated in Fig. 6.5(b) on page 248, will also be discussed in the sequel.

To understand how these patterns manifest in $\mathbf{Q} = \mathbf{\Gamma H G}$, we must consider the composite effect of linear modulation $\mathbf{G}$, propagation through the TV linear channel $\mathbf{H}$, and demodulation $\mathbf{\Gamma}$. As implied by (6.1), the channel propagation matrix $\mathbf{H}$ is a TV convolution matrix whose n^{th} row contains the impulse response coefficients $(h[n,m])_{m=0}^{M-1}$. For example, Fig. 6.3(a) shows a TV channel propagation

[3]Note that the one-corner support of the widely quasibanded matrix in Fig. 6.2(a) can be transformed into the two-corner support of the narrowly quasibanded matrix in Fig. 6.2(b) by simply rotating the columns of the former matrix right by $M/2$ places. Thus, the essential difference between these matrices is really the width of the support region (i.e., M versus $2D+1$).

matrix for $M = 8$ that was randomly generated according to the WSSUS Jakes (Stüber, 2001) fading assumption with $\nu_{\max} T_s = 0.03$, where $\nu_{\max}$ denotes the maximum (single-sided) Doppler spread in Hz and T_s the channel-use interval (i.e., the symbol period in a single-carrier system) in seconds. If the channel was time-invariant, the propagation matrix would have a Toeplitz structure. But here, since the channel is rapidly TV, each coefficient's magnitude varies smoothly along its diagonal of the propagation matrix. Given the construction of $\mathbf{H}$, the characteristics of $\mathbf{Q}$ will depend on the choices of $\mathbf{G}$ and $\boldsymbol{\Gamma}$ and their interaction with $\mathbf{H}$, as discussed next.

6.2.2.1 *Single-Carrier Modulation/Demodulation*

For single-carrier modulation/demodulation schemes, $\mathbf{G}$ and $\boldsymbol{\Gamma}$ accomplish little more than insertion and removal of a guard interval (of length $N_g \geq M - 1$). In this case, $\mathbf{Q}$ is created from the propagation matrix $\mathbf{H}$ by simply cutting the first N_g columns of $\mathbf{H}$ out and superimposing them onto the last N_g columns of $\mathbf{H}$. This operation was used, e.g., to create the widely quasibanded matrix in Fig. 6.3(b) from the TV convolution matrix in Fig. 6.3(a). More precisely, when $\mathbf{H}$ has dimensions $K \times (K + M - 1)$, cyclic-prefixed single-carrier modulation (CP-SCM) (Falconer, Ariyavisitakul, Benyamin-Seeyar, & Eidson, 2002) uses

$$\mathbf{G} = \begin{pmatrix} \mathbf{0} & \mathbf{I}_{M-1} \\ \mathbf{I}_{K-M+1} & \mathbf{0} \\ \mathbf{0} & \mathbf{I}_{M-1} \end{pmatrix} \quad \text{and} \quad \boldsymbol{\Gamma} = \mathbf{I}_K, \tag{6.3}$$

whereas zero-padded single-carrier modulation (ZP-SCM) (Wang, Ma, & Giannakis, 2004) uses a slightly different construction of $\mathbf{G}$, $\mathbf{H}$, and $\boldsymbol{\Gamma}$ that results in an equivalent $\mathbf{Q}$ matrix. We consider the effective channel matrix generated from SCM to be *widely quasibanded* because M, the width of the non-negligible band in $\mathbf{Q}$, is typically large: because $M \triangleq \lceil \tau_{\max}/T_s \rceil$ is the discrete delay spread of the channel, it is not unusual for M to be in the hundreds (e.g., delay spread $\tau_{\max} = 20\ \mu$s and bandwidth $1/T_s = 10$ MHz yield $M = 200$). Although small-M applications do exist, they yield equalization problems that are not very challenging, and hence not very interesting, especially in the coherent setting. Hence, we focus on the case of large M.

6.2.2.2 *Time-Frequency Concentrated Modulation/Demodulation*

The effect, on the transmitted signal $\{s[n]\}$, of propagation through the linear TV channel $\{h[n,m]\}$ can be understood as *simultaneous delay and Doppler spreading*. Thus, if each symbol $a[k]$ is modulated on a time-frequency concentrated waveform $\mathbf{g}_k \triangleq (g_k[0] \cdots g_k[N-1])^T$ for suitable[4] N, so that

$$s[n] = \sum_{k=0}^{K-1} a[k] g_k[n] \quad \text{for} \quad n = 0, \ldots, N - M, \tag{6.4}$$

where $\mathbf{g}_k$ is sufficiently "isolated" from the other waveforms $\{\mathbf{g}_{k'}\}_{k' \neq k}$ in the time-frequency domain, then propagation through the delay/Doppler spreading channel should cause only mild interference

[4] If N exceeds the time period between consecutive block transmissions, then interblock interference (IBI) can result. In this case, the model (6.2) can be generalized to $\mathbf{y} = \mathbf{Q}\mathbf{a} + \mathbf{Q}_{\text{pre}}\mathbf{a}_{\text{pre}} + \mathbf{Q}_{\text{pst}}\mathbf{a}_{\text{pst}} + \mathbf{z}$, where $\mathbf{Q}_{\text{pre}}\mathbf{a}_{\text{pre}}$ accounts for precursor IBI and $\mathbf{Q}_{\text{pst}}\mathbf{a}_{\text{pst}}$ accounts for postcursor IBI. The IBI can be made negligible, however, with suitable design of modulation/demodulation pulses $\{\mathbf{g}_k\}$ and $\{\boldsymbol{\gamma}_k\}$.

between these $\{a[k]\}$. Extraction of the kth symbol's contribution from the received signal $\{r[n]\}$ would then be accomplished through the linear demodulation operation

$$y[k] = \sum_{n=0}^{N-1} r[n]\gamma_k^*[n] \quad \text{for} \quad k = 0, \ldots, K-1, \tag{6.5}$$

for $\boldsymbol{\gamma}_k = (\gamma_k[0] \cdots \gamma_k[N-1])^T$ concentrated at the same time and frequency as $\mathbf{g}_k$. This is the main idea behind pulse-shaped multicarrier schemes (Bölcskei, 2002; Das & Schniter, 2007; Haas & Belfiore, 1997; Kozek & Molisch, 1998; Le Floch, Alard, & Berrou, 1995; Matheus & Kammeyer, 1997; Matz, Schafhuber, Gröchenig, Hartmann, & Hlawatsch, 2007; Rugini, Banelli, & Leus, 2006; Schniter, 2004; Strohmer & Beaver, 2003) as well as Slepian schemes (Sigloch, Andrews, Mitra, & Thomson, 2005).

With suitably designed modulation/demodulation waveforms $\{\mathbf{g}_k\}$ and $\{\boldsymbol{\gamma}_k\}$, the combined channel matrix $\mathbf{Q}$ under (6.4)–(6.5) can be ensured to have the *narrowly quasibanded structure* illustrated in Fig. 6.2(b). There, D can be interpreted as the (single-sided) discrete Doppler spread of the effective channel and $2D+1$ can be recognized as the width of the non-negligible interference band. Typically D is chosen as

$$D = \lceil \nu_{\max} T_s K + D_0 \rceil, \tag{6.6}$$

where D_0 is a small non-negative constant (e.g., $0 \le D_0 \le 2$ for a well-designed modulation/demodulation scheme), as discussed in the sequel. We can see that $\mathbf{Q}$ will be *narrowly quasibanded*, so that $2D+1 \ll M$, by plugging the typical block-length choice of $K = 4M$ into (6.6) and then using the definition $M = \tau_{\max}/T_s$ to see that (Hwang & Schniter, 2006)

$$D \le \lceil 4\nu_{\max}\tau_{\max} \rceil + \lceil D_0 \rceil \tag{6.7}$$

$$= 1 + \lceil D_0 \rceil \quad \text{when} \quad 0 < 2\nu_{\max}\tau_{\max} \le 0.5. \tag{6.8}$$

The quantity $2\nu_{\max}\tau_{\max}$, sometimes referred to as the "spreading index," describes the total severity of delay-Doppler spreading. The boundary between *underspread* and *overspread* channels occurs at $2\nu_{\max}\tau_{\max} = 1$, and it can be safely assumed that $2\nu_{\max}\tau_{\max} \ll 1$ for practical applications. Thus, from (6.8), we conclude that the width of the non-negligible coefficient band is $2D+1 \le 3 + 2\lceil D_0 \rceil$ when suitable modulation/demodulation waveforms are used. In summary, $2D+1 \ll M$ is a reasonable claim for the values of M that are of interest in this chapter.

As an example, Fig. 6.3(c) shows $\mathbf{Q}$ constructed through cyclic-prefixed orthogonal frequency division multiplexing (CP-OFDM) (Cimini, 1985) with max-SINR receiver pulse-shaping (Schniter, 2004) using the TV convolution matrix shown in Fig. 6.3(a). Although the channel has an extremely high spreading index of $2\nu_{\max}\tau_{\max} = 0.8$, all coefficients in $\mathbf{Q}$ outside of the 3-wide band are negligible. As another example, Fig. 6.4 shows $\mathsf{E}\{|[\mathbf{Q}]_{k,k+d}|^2\}$ (in dB) versus d for several modulation/demodulation schemes and a channel with a spreading index of 0.1. ($\mathsf{E}\{|[\mathbf{Q}]_{k,k+d}|^2\}$ is invariant to k.) As can be seen in Fig. 6.4, the JOMS scheme from Das & Schniter (2007) suppresses coefficients outside the band of radius $D = \lceil \nu_{\max} T_s K \rceil = 1$ by at least 44 dB.

At this point, we make one final observation about a narrowly quasibanded matrix $\mathbf{Q}$. If we upper-triangularize $\mathbf{Q}$, e.g., through the QR decomposition $\mathbf{Q} = \mathbf{V}\bar{\mathbf{Q}}$ where $\mathbf{V}$ is unitary and $\bar{\mathbf{Q}}$ is upper triangular, then $\bar{\mathbf{Q}}$ will have the "V-shaped" structure shown in Fig. 6.5(c) on page 248. Such upper-triangularization of $\mathbf{Q}$ occurs before decision feedback and tree-search based equalization, as discussed in Section 6.3.2.

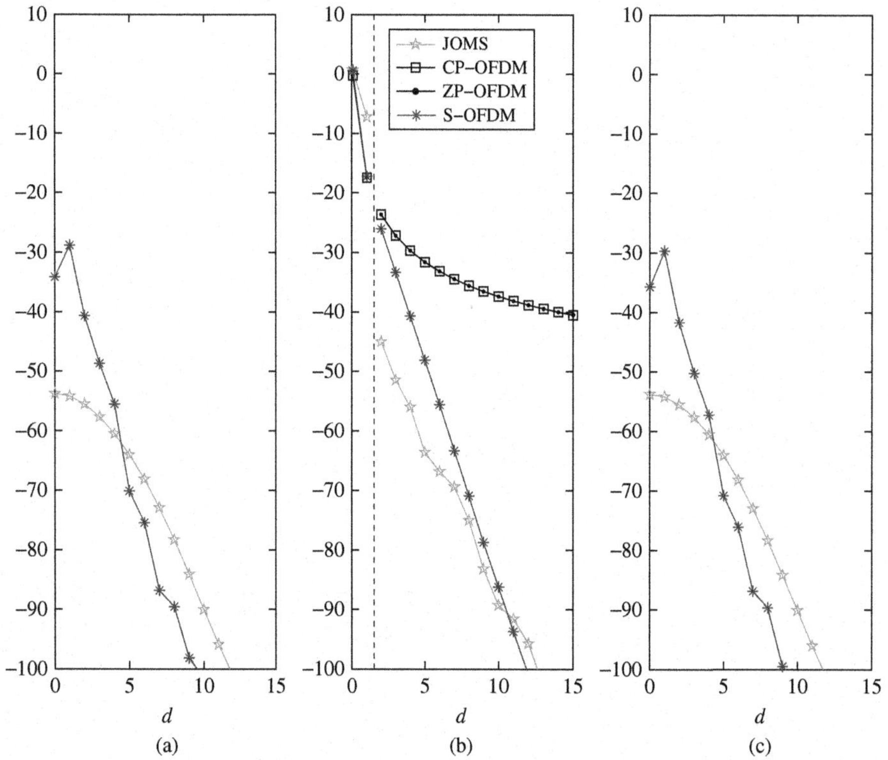

FIGURE 6.4

For the IBI model $\mathbf{y} = \mathbf{Q}\mathbf{a} + \mathbf{Q}_{\text{pre}}\mathbf{a}_{\text{pre}} + \mathbf{Q}_{\text{pst}}\mathbf{a}_{\text{pst}} + \mathbf{z}$, with $K = 64, N_g = 15, M = 16, \nu_{\max}T_s = 0.003$, and WSSUS Jakes fading (see Chapter 1), subplot (b) shows the mean-square value (in decibel) of a coefficient in $\mathbf{Q}$ versus its distance "d" from the main diagonal of $\mathbf{Q}$, while subplots (a) and (c) show the same for the coefficients in $\mathbf{Q}_{\text{pre}}$ and $\mathbf{Q}_{\text{pst}}$, respectively. The dashed vertical line indicates $D = \lceil \nu_{\max} T_s K \rceil$. JOMS refers to Das & Schniter's joint transmitter/receiver optimization max-SINR scheme (Das & Schniter, 2007) while S-OFDM refers to Strohmer and Beaver's orthogonal scheme (Strohmer & Beaver, 2003).

6.2.2.3 *Other Modulation/Demodulation Schemes*

When the modulation and demodulation pulses $\{\mathbf{g}_k\}_{k=0}^{K-1}$ and $\{\boldsymbol{\gamma}_k\}_{k=0}^{K-1}$ are *not* designed to curb the effects of delay/Doppler spreading, the support of non-negligible coefficients within $\mathbf{Q}$ can be widespread, to the point where $\mathbf{Q}$ must be considered as *fully populated.* Examples of such modulation/demodulation schemes include the wavelet-based schemes (Martone, 2000; Wornell, 1996), the chirp-based schemes (Barbarossa & Torti, 2001; Kannu & Schniter, 2008; Martone, 2001), a scheme designed to maximize a lower bound on capacity (Yun, Chung, & Lee, 2007), and the diversity maximizing schemes (Hwang & Schniter, 2007; Ma & Giannakis, 2003).

Even popular multicarrier schemes like CP-OFDM, when used with a rectangular receiver pulse, yield a near-fully populated $\mathbf{Q}$ when the channel is TV rapidly enough. Figure 6.3 shows this by example: the effective channel matrix in Fig. 6.3(d) was constructed from the TV convolution matrix in Fig. 6.3(a) through standard CP-OFDM. Notice that the non-negligible coefficients in $\mathbf{Q}$ are not all located in the central band of the matrix. Figure 6.4 shows a similar phenomenon: for CP-OFDM, $\mathsf{E}\{|[\mathbf{Q}]_{k,k+d}|^2\}$ decays very slowly with d, the distance from the main diagonal of $\mathbf{Q}$.

6.3 COHERENT EQUALIZATION

In this section, we focus on *coherent equalization*, i.e., equalization under the assumption that the channel matrix $\mathbf{H}$, and thus the effective channel matrix $\mathbf{Q}$, is known. The noncoherent case will be discussed in Section 6.4.

In Section 6.3.1, we discuss several criteria (i.e., notions of optimality) under which coherent equalizers are designed, and, in Section 6.3.2, we describe classical equalization algorithms for generic $\mathbf{Q}$. Then, in Sections 6.3.3–6.3.4, we focus on coherent equalization techniques for the specific types of $\mathbf{Q}$ anticipated for rapidly TV channels in Section 6.2.2.

6.3.1 Coherent Equalization Criteria

Referring to (6.2), the goal of coherent block equalization is estimation of the symbol vector $\mathbf{a}$, or the corresponding coded-bit vector $\mathbf{c}$, from the linearly distorted and noisy demodulator output vector $\mathbf{y}$, assuming knowledge of the channel matrix $\mathbf{Q}$ and the noise statistics. Note that this may or may not include a hard-decision or quantization step, as explained later. In any case, we are fundamentally interested in identifying the "optimal" method to generate these symbol estimates. The answer, however, depends on how optimality is defined, i.e., which *equalization criterion* is used.

In organizing the criteria that are most often used for equalizer design, it helps to consider how the equalizer outputs will be used by the receiver (e.g., by the decoder).

6.3.1.1 *Hard Symbol or Bit Estimates*

If there is no decoder or if the decoder wants *hard estimates* of the symbols or bits, then the goal is to produce a finite-alphabet estimate $\hat{\mathbf{a}} \in \mathcal{A}^K$. (Recall that the equalizer is assumed to know the symbol alphabet $\mathcal{A}$ but not the set of coded symbol sequences $\boldsymbol{\mathcal{A}}$.)

Maximum a posteriori (MAP) *sequence detection* (SD) (Poor, 1994) minimizes the probability of sequence error. By definition, the MAPSD estimate is

$$\hat{\mathbf{a}}_{\mathsf{cMAPSD}} \triangleq \arg \max_{\mathbf{a}' \in \mathcal{A}^K} \Pr\{\mathbf{a} = \mathbf{a}' \mid \mathbf{y}, \mathbf{Q}\}. \tag{6.9}$$

In (6.9), we use the notation "cMAPSD" to emphasize that this is the *coherent* version of the MAP criterion applied to *sequence* detection. In contrast, coherent MAP *symbol* and *bit* detection takes the form

$$\hat{a}_{\mathsf{cMAP}}[k] \triangleq \arg\max_{a \in \mathcal{A}} \Pr\{a[k] = a \mid \mathbf{y}, \mathbf{Q}\} \quad \text{for} \quad k \in 0, \ldots, K-1 \tag{6.10}$$

$$\hat{c}_{\mathsf{cMAP}}[j] \triangleq \arg \max_{c \in \{0,1\}} \Pr\{c[j] = c \mid \mathbf{y}, \mathbf{Q}\} \quad \text{for} \quad j \in 0, \ldots, K \log_2 |\mathcal{A}| - 1. \tag{6.11}$$

In writing (6.9)–(6.11), we have treated the channel $\mathbf{Q}$ as a random quantity.

If we assume that each of the symbol sequences in $\mathscr{A}^K$ has equal prior probability, i.e., $\Pr\{\mathbf{a} = \mathbf{a}'\} = 1/|\mathscr{A}|^K \ \forall \mathbf{a}' \in \mathscr{A}^K$, then coherent MAPSD reduces to coherent *maximum likelihood* (ML) SD (Poor, 1994):

$$\hat{\mathbf{a}}_{\mathsf{cMLSD}} \triangleq \arg\max_{\mathbf{a}\in\mathscr{A}^K} f(\mathbf{y} \mid \mathbf{a}, \mathbf{Q}), \tag{6.12}$$

where $f(\mathbf{y} \mid \mathbf{a}, \mathbf{Q})$ denotes the probability density function of $\mathbf{y}$ conditioned on $\mathbf{a}$ and $\mathbf{Q}$, also known as the *likelihood function*. To see this, notice from Bayes rule that

$$\Pr\{\mathbf{a} = \mathbf{a}' \mid \mathbf{y}, \mathbf{Q}\} = \frac{f(\mathbf{y} \mid \mathbf{a}', \mathbf{Q}) \Pr\{\mathbf{a} = \mathbf{a}' \mid \mathbf{Q}\}}{f(\mathbf{y} \mid \mathbf{Q})} = \frac{1}{|\mathscr{A}|^K} \frac{f(\mathbf{y} \mid \mathbf{a}', \mathbf{Q})}{f(\mathbf{y} \mid \mathbf{Q})}, \tag{6.13}$$

from which it becomes clear that maximizing $\Pr\{\mathbf{a} = \mathbf{a}' \mid \mathbf{y}, \mathbf{Q}\}$ over $\mathbf{a}'$ is equivalent to maximizing $f(\mathbf{y} \mid \mathbf{a}', \mathbf{Q})$ over $\mathbf{a}'$. Due to our assumption of zero-mean Gaussian noise with covariance $\mathbf{C}_\mathbf{z}$, we have $f(\mathbf{y} \mid \mathbf{a}', \mathbf{Q}) = \frac{1}{\pi^K \det\{\mathbf{C}_\mathbf{z}\}} \exp(-\|\mathbf{y} - \mathbf{Q}\mathbf{a}'\|^2_{\mathbf{C}_\mathbf{z}^{-1}})$, so that coherent MLSD reduces to

$$\hat{\mathbf{a}}_{\mathsf{cMLSD}} = \arg\min_{\mathbf{a}\in\mathscr{A}^K} \|\mathbf{y} - \mathbf{Q}\mathbf{a}\|^2_{\mathbf{C}_\mathbf{z}^{-1}}. \tag{6.14}$$

Above, we used the quadratic-form notation $\|\mathbf{z}\|^2_\mathbf{A} \triangleq \mathbf{z}^H \mathbf{A} \mathbf{z}$, where $\mathbf{A}$ is any positive semidefinite Hermitian matrix.

6.3.1.2 *Complex-Field Symbol Estimates*

If the decoder prefers or tolerates complex-valued symbol estimates, rather than finite-alphabet symbol estimates, then one can consider equalization schemes that yield $\hat{\mathbf{a}} \in \mathbb{C}^K$. Note, however, that we still assume $\mathbf{a} \in \mathscr{A}^K$.

A popular criterion for this case is *minimum mean-squared error* (MMSE) (Poor, 1994). The coherent unconstrained MMSE sequence estimate is defined as

$$\hat{\mathbf{a}}_{\mathsf{cMMSE}} \triangleq \arg\min_{\mathbf{a}'\in\mathbb{C}^K} \mathsf{E}\{\|\mathbf{a} - \mathbf{a}'\|^2 \mid \mathbf{y}, \mathbf{Q}\}. \tag{6.15}$$

Because the MMSE estimate equals the conditional mean (Poor, 1994), we have

$$\hat{\mathbf{a}}_{\mathsf{cMMSE}} = \mathsf{E}\{\mathbf{a} \mid \mathbf{y}, \mathbf{Q}\} = \sum_{\mathbf{a}\in\mathscr{A}^K} \mathbf{a}\, p(\mathbf{a} \mid \mathbf{y}, \mathbf{Q}) \tag{6.16}$$

$$= \sum_{\mathbf{a}\in\mathscr{A}^K} \mathbf{a} \frac{f(\mathbf{y} \mid \mathbf{a}, \mathbf{Q}) p(\mathbf{a})}{\sum_{\mathbf{a}'\in\mathscr{A}^K} f(\mathbf{y} \mid \mathbf{a}', \mathbf{Q}) p(\mathbf{a}')}. \tag{6.17}$$

If we assume that $p(\mathbf{a})$ is uniformly distributed over $\mathscr{A}^K$, then

$$\hat{\mathbf{a}}_{\mathsf{cMMSE}} = \frac{\sum_{\mathbf{a}\in\mathscr{A}^K} \mathbf{a} \exp\left(-\|\mathbf{y} - \mathbf{Q}\mathbf{a}\|^2_{\mathbf{C}_\mathbf{z}^{-1}}\right)}{\sum_{\mathbf{a}'\in\mathscr{A}^K} \exp\left(-\|\mathbf{y} - \mathbf{Q}\mathbf{a}'\|^2_{\mathbf{C}_\mathbf{z}^{-1}}\right)}. \tag{6.18}$$

Notice from (6.18) that the finite-alphabet nature of $\mathbf{a}$ makes the conditional mean difficult to evaluate because it requires the evaluation of $|\mathscr{A}|^K$ terms.

To reduce complexity, the MMSE criterion is often used in conjunction with particular constraints on how the symbol estimates are generated from $\mathbf{y}$. The most common examples are MMSE *linear equalization* (6.22) and MMSE *decision feedback equalization* (6.27). Note that, if one assumes that $\mathbf{a}|\mathbf{y},\mathbf{Q}$ is Gaussian distributed, then the *un*constrained MMSE estimator (6.15) itself becomes a linear function of $\mathbf{y}$ (Poor, 1994).

As the signal-to-noise ratio (SNR) increases, the effect of linear channel distortion overwhelms that of additive noise, motivating the so-called *zero-forcing* (ZF) criterion. Effectively, ZF equalizers "invert" the effect of the linear channel distortion while ignoring the presence of additive channel noise. The most common examples are ZF linear equalization and ZF decision feedback equalization, both described in Section 6.3.2. In the absence of additive noise, ZF equalizers are equivalent to their MMSE counterparts.

6.3.1.3 Soft Bit Estimates

If the decoder prefers *soft bit estimates*, then the goal is to produce reliability information on each of the coded bits in $\mathbf{c}$. Typically, bit reliabilities are expressed in the form of a *log likelihood ratio* (LLR) for each bit. The goal of coherent equalization, thus, becomes the computation of the coherent posterior LLRs[5]

$$L_{c|\mathbf{y},\mathbf{Q}}[j] \triangleq \ln \frac{\Pr\{c[j]=1 \mid \mathbf{y},\mathbf{Q}\}}{\Pr\{c[j]=0 \mid \mathbf{y},\mathbf{Q}\}} \quad \text{for} \quad j=0,\ldots,K\log_2|\mathscr{A}|-1, \tag{6.19}$$

given the a priori LLRs

$$L_c[j] \triangleq \ln \frac{\Pr\{c[j]=1\}}{\Pr\{c[j]=0\}} \quad \text{for} \quad j=0,\ldots,K\log_2|\mathscr{A}|-1. \tag{6.20}$$

When nothing is a priori known about the bit $c[j]$, the value $L_c[j]=0$ is used. Nonzero a priori LLRs are used, e.g., when the equalizer is fed by the outputs of a soft decoder, as in turbo equalization (see Fig. 6.1) or when certain bits are known pilots. If $c[j]$ is a pilot (or otherwise known with complete confidence), then $L_c[j]=\pm\infty$. Recall that the use of a priori LLRs implies that the equalizer treats the coded bits as independent.

Hard MAP bit estimates can be generated by quantizing the posterior LLRs as follows:

$$\hat{c}_{\mathsf{cMAP}}[j] = \frac{1}{2}\big(1+\operatorname{sign}(L_{c|\mathbf{y},\mathbf{Q}}[j])\big). \tag{6.21}$$

6.3.2 Coherent Equalization Tools

The coherent equalization criteria discussed in Section 6.3.1 each describe a particular goal for equalization, but not how equalization would be practically implemented. For example, the MAPSD, MLSD, and (unconstrained) MMSE estimates described in Section 6.3.1 require the evaluation of $\mathscr{O}(|\mathscr{A}|^K)$ metrics if computed through brute force, which is not practical for typical values of K. In this section, we review classical equalization implementations whose designs are guided by the various criteria in Section 6.3.1.

[5]Sometimes, it is more practical to calculate posteriors using only a limited number of (say J) future observations (Li, Vucetic, & Sato, 1995). In this so-called "fixed lag" case, the conditioning in (6.19) is performed on $\big(y[0],\ldots,y[\lceil \frac{j}{\log_2|\mathscr{A}|}\rceil + J]\big)^T$ instead of $\mathbf{y}$.

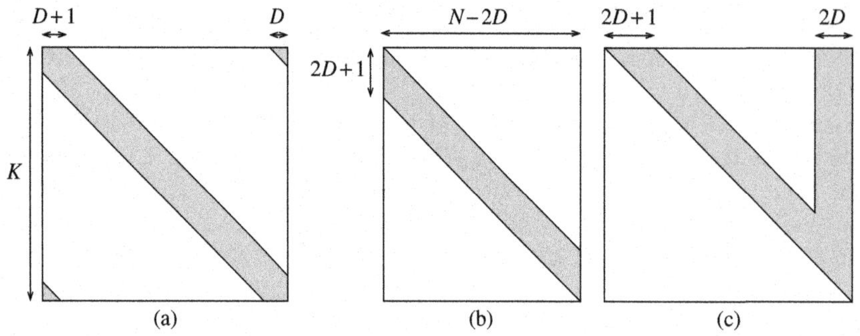

FIGURE 6.5

Support region of (a) "quasibanded," (b) "banded," and (c) "V-shaped" matrices. From quasibanded (a), banded (b) is obtained by deleting the first and last D columns, while V-shaped (c) is obtained by upper triangularization.

6.3.2.1 *Trellis-Based Equalization*

Trellis methods can be used to implement MLSD and MAP equalization when $\mathbf{Q}$ is a *banded* matrix. As illustrated in Fig. 6.5, a banded matrix differs from its quasibanded counterpart because of the lack of corner elements. A banded matrix manifests when, e.g., the first and last few elements of $\mathbf{a}$ are known or zero-valued.[6] If $\mathbf{Q}$ is a banded matrix with a $2D+1$ wide band, then the *Viterbi algorithm* (Forney, 1972) can perform MAPSD/MLSD equalization using $\mathcal{O}(KD|\mathscr{A}|^{2D+1})$ operations. Similarly, the *forward–backward* (or BCJR) algorithm (Bahl, Cocke, Jelinek, & Raviv, 1974) can be used to accomplish MAP symbol/bit equalization with a complexity of $\mathcal{O}(KD|\mathscr{A}|^{2D+1})$ operations (Forney, 1973, Appendix). Lower-complexity trellis-based approximate MAP equalizers include fixed-lag approaches (Li et al., 1995) and the soft-output Viterbi algorithm (SOVA) (Hagenauer & Hoeher, 1989). In all cases, the complexity is linear in the block length K and exponential in the effective channel length $2D+1$. Thus, these techniques will be practical if and only if $2D+1$ is very small.

Trellis methods can be modified to work on *quasi*banded $\mathbf{Q}$ using, e.g., a "tail-biting" approach. Here, from an arbitrary location within the block, the Viterbi algorithm is initialized from each of the possible $|\mathscr{A}|^{2D+1}$ states and forced to terminate in the same state; the initialization leading to the optimum sequence metric is then chosen. This approach requires running the Viterbi algorithm $|\mathscr{A}|^{2D+1}$ times, for a total cost of $\mathcal{O}(KD|\mathscr{A}|^{4D+2})$ operations.

6.3.2.2 *Linear Equalization*

In *linear equalization*, the symbol estimates are a linear function of the observation $\mathbf{y}$, i.e.,

$$\hat{\mathbf{a}}_{\mathsf{LIN}} = \mathbf{E}\mathbf{y}, \tag{6.22}$$

[6]More precisely, consider the system model (6.2). If $\mathbf{Q}$ is as illustrated in Fig. 6.5(a) and the last $M-1$ elements of $\mathbf{a}$ are zero-valued, then we can write $\mathbf{y} = \check{\mathbf{Q}}\check{\mathbf{a}} + \mathbf{z}$ where $\check{\mathbf{a}} = (a[0] \cdots a[K-M])^T$ and where $\check{\mathbf{Q}}$ is a banded matrix (as illustrated in Fig. 6.5(b)) with an M-wide band. Or, if $\mathbf{Q}$ is as illustrated in Fig. 6.5(b) and the first and last D elements of $\mathbf{a}$ are zero-valued, then we can write $\mathbf{y} = \check{\mathbf{Q}}\check{\mathbf{a}} + \mathbf{z}$ where $\check{\mathbf{a}} = (a[D] \cdots a[K-D-1])^T$ and where $\check{\mathbf{Q}}$ is a banded matrix (as illustrated in Fig. 6.5(b)) with a $2D+1$ wide band.

for a suitably chosen matrix $\mathbf{E} \in \mathbb{C}^{K\times K}$. In some cases, such as when it is impractical to process the entire observation $\mathbf{y}$ at once, additional constraints are placed on $\mathbf{E}$. Because linear equalization ignores the finite-alphabet property of $\mathbf{a}$, its performance is generally much worse than that of techniques that leverage the finite-alphabet property.

The coherent *linear MMSE* (LMMSE)[7] equalizer uses, for $\mathbf{E}$ in (6.22),

$$\mathbf{E}_{\mathsf{LMMSE}} \triangleq \arg \min_{\mathbf{E}\in\mathbb{C}^{K\times K}} \mathsf{E}\{\|\mathbf{a} - \hat{\mathbf{a}}_{\mathsf{LIN}}\|^2 \mid \mathbf{Q}\}. \tag{6.23}$$

Given the symbol and noise statistics assumed in Section 6.2, it can be shown (Verdú, 1998) that

$$\mathbf{E}_{\mathsf{LMMSE}} = \mathbf{Q}^H(\mathbf{Q}\mathbf{Q}^H + \sigma_a^{-2}\mathbf{C}_\mathbf{z})^{-1} \tag{6.24}$$

$$= (\mathbf{Q}^H\mathbf{C}_\mathbf{z}^{-1}\mathbf{Q} + \sigma_a^{-2}\mathbf{I}_K)^{-1}\mathbf{Q}^H\mathbf{C}_\mathbf{z}^{-1}. \tag{6.25}$$

The matrix inversion lemma[8] can be used to relate (6.24) and (6.25). The *linear ZF* (LZF) estimator uses (6.22) with $\mathbf{E}$ set to

$$\mathbf{E}_{\mathsf{LZF}} = \mathbf{Q}^{-1}, \tag{6.26}$$

assuming that $\mathbf{Q}$ is invertible. When $\mathbf{Q}$ is not invertible, the LZF equalizer is said not to exist.

Due to the matrix inversions in (6.24)–(6.26), the complexity of LMMSE and LZF equalization is $\mathcal{O}(K^3)$, which is much less than the $\mathcal{O}(|\mathcal{A}|^K)$ complexity of unconstrained MMSE estimation in (6.18). Still, $\mathcal{O}(K^3)$ may be impractical when K is large.

6.3.2.3 Decision Feedback Equalization

Decision feedback equalization (DFE) exploits the finite-alphabet symbol property while keeping complexity close to that of linear equalization. Essentially, it makes hard symbol decisions sequentially and leverages past decisions for future symbol estimates.

The DFE generates complex-valued symbol estimates as follows:[9]

$$\hat{\mathbf{a}}_{\mathsf{DFE}} = \mathbf{E}\mathbf{y} - (\mathbf{U} - \mathbf{I}_K)\mathcal{D}_{\mathcal{A}}(\hat{\mathbf{a}}_{\mathsf{DFE}}). \tag{6.27}$$

In (6.27), $\mathcal{D}_{\mathcal{A}}(\cdot) : \mathbb{C}^K \to \mathcal{A}^K$ denotes element-wise quantization w.r.t. the symbol alphabet $\mathcal{A}$, $\mathbf{U} \in \mathbb{C}^{K\times K}$ is monic upper triangular (to ensure that decision feedback is strictly causal), and $\mathbf{E} \in \mathbb{C}^{K\times K}$. Keeping the monic upper-triangular property of $\mathbf{U}$ in mind, (6.27) can be understood as follows: the estimate $\hat{a}_{\mathsf{DFE}}[K-1]$ is linearly computed from $\mathbf{y}$ using the last row in $\mathbf{E}$; then, the estimate $\hat{a}_{\mathsf{DFE}}[K-2]$ is linearly computed from $\mathbf{y}$ and quantized $\hat{a}_{\mathsf{DFE}}[K-1]$ using the second-to-last rows in $\mathbf{E}$ and $\mathbf{U}$, respectively; then, the estimate $\hat{a}_{\mathsf{DFE}}[K-3]$ is linearly computed from $\mathbf{y}$ and quantized $\{\hat{a}_{\mathsf{DFE}}[K-2], \hat{a}_{\mathsf{DFE}}[K-1]\}$ using the third-to-last rows in $\mathbf{E}$ and $\mathbf{U}$, respectively; and so on.

[7] Note that the LMMSE equalizer described here is a generalization of the classical tapped delay-line LMMSE equalizer (Proakis, 2001).

[8] The matrix inversion lemma can be stated as $(\mathbf{A}^{-1} + \mathbf{B}\mathbf{C}^{-1}\mathbf{B}^H)^{-1} = \mathbf{A} - \mathbf{A}\mathbf{B}(\mathbf{C} + \mathbf{B}^H\mathbf{A}\mathbf{B})^{-1}\mathbf{B}^H\mathbf{A}$, assuming the inverses exist.

[9] The DFE described here is sometimes referred to as a "generalized" DFE to distinguish it from the classical DFE implemented using tapped delay-line forward and feedback filters (Al-Dhahir & Cioffi, 1995).

The DFE matrices $\mathbf{E}$ and $\mathbf{U}$ are typically designed according to the MMSE or ZF criteria. As with linear equalization, additional constraints may be placed on $\mathbf{E}$ and/or $\mathbf{U}$. The coherent *MMSE-DFE* (Cioffi & Forney, 1997) uses (6.27) with $\{\mathbf{E},\mathbf{U}\}$ set to

$$\{\mathbf{E}_{\text{MMSE-DFE}},\mathbf{U}_{\text{MMSE-DFE}}\} = \arg\min_{\mathbf{E},\mathbf{U}} \mathsf{E}\{\|\mathbf{a}-\hat{\mathbf{a}}_{\text{DFE}}\|^2 \mid \mathbf{Q}\}$$

$$\text{assuming } \mathscr{D}_{\mathscr{A}}(\hat{\mathbf{a}}_{\text{DFE}}) = \mathbf{a}, \tag{6.28}$$

i.e., set to minimize the MSE of $\hat{\mathbf{a}}_{\text{DFE}}$ *under the assumption of perfect decision feedback*. It can be shown that $\mathbf{U}_{\text{MMSE-DFE}}$ and $\mathbf{E}_{\text{MMSE-DFE}}$ can be computed with the aid of an LDU decomposition (Al-Dhahir & Sayed, 2000):

$$\mathbf{U}^H_{\text{MMSE-DFE}}\boldsymbol{\Delta}_{\text{MMSE-DFE}}\mathbf{U}_{\text{MMSE-DFE}} = \mathbf{Q}^H\mathbf{C}_{\mathbf{z}}^{-1}\mathbf{Q}+\sigma_a^{-2}\mathbf{I} \tag{6.29}$$

$$\mathbf{E}_{\text{MMSE-DFE}} = \mathbf{U}_{\text{MMSE-DFE}}\mathbf{E}_{\text{LMMSE}}, \tag{6.30}$$

with $\mathbf{E}_{\text{LMMSE}}$ given by (6.24)–(6.25). The *ZF-DFE* takes the form of (6.27) with $\mathbf{U}_{\text{ZF-DFE}}$ computed through the LDU decomposition $\mathbf{U}^H_{\text{ZF-DFE}}\boldsymbol{\Delta}_{\text{ZF-DFE}}\mathbf{U}_{\text{ZF-DFE}} = \mathbf{Q}^H\mathbf{Q}$, and with $\mathbf{E}_{\text{ZF-DFE}} = \mathbf{U}_{\text{ZF-DFE}}\mathbf{Q}^{-1}$.

In practice, the hard decisions in (6.27) are not always perfect, which leads to the phenomenon known as *error propagation* (O'Reilly & de Oliviera Duarte, 1985). There, a decision error on $a[k]$ has the effect of amplifying, rather than canceling, the interference that $a[k]$ causes to the not-yet-estimated symbols $\{a[k']\}_{k'=0}^{k-1}$. Although error propagation can be somewhat alleviated by detecting symbols with higher signal-to-interference noise ratio (SINR) first (e.g., by V-BLAST detection ordering (Wolniansky, Foschini, Golden, & Valenzuela, 1998)), error propagation is better avoided through tree search or iterative soft equalization, as discussed in Sections 6.3.2.4 and 6.3.2.5.

Finally, we note that hybrid trellis/DFE techniques have been proposed with complexities and performances that lie between trellis and DFE methods. Two of the more well-known techniques are *reduced state sequence estimation* (Eyuboglu & Qureshi, 1988) and *delayed decision feedback estimation* (Duel-Hallen & Heegard, 1989).

6.3.2.4 Equalization Based on Tree Search

In DFE, a single hypothesis of the sequence $(a[k+1],\ldots,a[K-1])$ is used to aid the estimation of $a[k]$. Tree-search[10] methods improve on this idea by keeping and using several hypotheses of the sequence $(a[k+1],\ldots,a[K-1])$ until it is clear which is the single best hypothesis.

Tree-search algorithms can be partitioned into optimal and suboptimal approaches. *Optimal tree search* methods are capable of implementing MLSD with a complexity that is *on average* much less than that of brute-force search (Mow, 1994). Although this average complexity has been claimed to grow as roughly $\mathcal{O}(K^3)$ at sufficiently high SNR (Hassibi & Vikalo, 2005), a careful analysis shows that, in fact, the average complexity of optimal tree search is exponential in K (Jalden & Ottersten, 2005). To circumvent the potentially high complexity of exact tree search (especially at low SNR), *suboptimal tree search* may be considered because a very small performance sacrifice can

[10]What we call "tree search" is sometimes referred to as closest lattice point search, lattice decoding, sequential decoding, or sphere decoding.

often lead to a huge reduction in complexity. In fact, a well-designed suboptimal tree search can achieve near-ML performance with near-DFE complexity (Murugan, El Gamal, Damen, & Caire, 2006). When assessing a suboptimal tree search algorithm, it is most appropriate to think in terms of its *performance/complexity tradeoff*.

Before conducting a tree search, the observations $\mathbf{y}$ in (6.2) must be preprocessed to yield a *causal observation model* of the form

$$\bar{\mathbf{y}} = \bar{\mathbf{Q}}\bar{\mathbf{a}} + \bar{\mathbf{z}}, \tag{6.31}$$

where $\bar{\mathbf{Q}}$ is upper triangular and $\bar{\mathbf{a}}$ is some permutation of $\mathbf{a}$. Ignoring permutation for the moment (so that $\bar{\mathbf{a}} = \mathbf{a}$), the standard approach to upper-triangularization of (6.2) is *QR decomposition*: if $\mathbf{Q} = \mathbf{V}_{\mathsf{QR}}\bar{\mathbf{Q}}_{\mathsf{QR}}$, where $\mathbf{V}_{\mathsf{QR}}$ is unitary and $\bar{\mathbf{Q}}_{\mathsf{QR}}$ is upper triangular, then preprocessing according to $\bar{\mathbf{y}} = \mathbf{V}_{\mathsf{QR}}^H\mathbf{y} \triangleq \bar{\mathbf{y}}_{\mathsf{QR}}$ yields (6.31) with $\bar{\mathbf{Q}} = \bar{\mathbf{Q}}_{\mathsf{QR}}$ and $\bar{\mathbf{z}} = \mathbf{V}_{\mathsf{QR}}^H\mathbf{z} \triangleq \bar{\mathbf{z}}_{\mathsf{QR}}$. Notice that $\bar{\mathbf{z}}_{\mathsf{QR}}$ is statistically equivalent to $\mathbf{z}$. As we show below, QR preprocessing is closely related to the feedforward filtering operation in ZF-DFE. Because the MMSE-DFE is known to outperform the ZF-DFE in noisy environments, it has been suggested (Damen, El Gamal, & Caire, 2003) to replace the QR preprocessing step with its MMSE-DFE equivalent, at least for suboptimal tree search. To see this from another perspective, imagine for the moment that suboptimal tree search is conducted according to the most greedy method possible, i.e., with a single surviving hypothesis per stage. Then, if QR preprocessing is used, this suboptimal tree search is exactly the ZF-DFE, whereas, if MMSE-DFE preprocessing is used, this suboptimal tree search is exactly the MMSE-DFE.

We will now provide the technical link between the QR decomposition and the ZF-DFE, as well as the details of the MMSE-DFE preprocessor. Comparing the LDU decomposition $\mathbf{Q}^H\mathbf{Q} = \mathbf{U}_{\mathsf{ZF\text{-}DFE}}^H\boldsymbol{\Delta}_{\mathsf{ZF\text{-}DFE}}\mathbf{U}_{\mathsf{ZF\text{-}DFE}}$ to the QR decomposition $\mathbf{Q} = \mathbf{V}_{\mathsf{QR}}\bar{\mathbf{Q}}_{\mathsf{QR}}$, it becomes evident that $\bar{\mathbf{Q}}_{\mathsf{QR}} = \boldsymbol{\Delta}_{\mathsf{ZF\text{-}DFE}}^{1/2}\mathbf{U}_{\mathsf{ZF\text{-}DFE}}$, from which it follows that $\mathbf{V}_{\mathsf{QR}}^H = \bar{\mathbf{Q}}_{\mathsf{QR}}\mathbf{Q}^{-1} = \boldsymbol{\Delta}_{\mathsf{ZF\text{-}DFE}}^{1/2}\mathbf{U}_{\mathsf{ZF\text{-}DFE}}\mathbf{Q}^{-1} = \boldsymbol{\Delta}_{\mathsf{ZF\text{-}DFE}}^{1/2}\mathbf{E}_{\mathsf{ZF\text{-}DFE}}$. Thus, $\bar{\mathbf{y}}_{\mathsf{QR}} = \boldsymbol{\Delta}_{\mathsf{ZF\text{-}DFE}}^{1/2}\mathbf{E}_{\mathsf{ZF\text{-}DFE}}\mathbf{y}$ can be recognized as a scaled version of the ZF-DFE feedforward filter output $\mathbf{E}_{\mathsf{ZF\text{-}DFE}}\mathbf{y}$. If we repeat the same steps with MMSE-DFE quantities in place of ZF-DFE quantities, we obtain the *MMSE-DFE preprocessed observation* (Damen et al., 2003)

$$\bar{\mathbf{y}}_{\mathsf{MMSE\text{-}DFE}} \triangleq \boldsymbol{\Delta}_{\mathsf{MMSE\text{-}DFE}}^{1/2}\mathbf{E}_{\mathsf{MMSE\text{-}DFE}}\mathbf{y}, \tag{6.32}$$

and the corresponding causal model

$$\bar{\mathbf{y}}_{\mathsf{MMSE\text{-}DFE}} = \boldsymbol{\Delta}_{\mathsf{MMSE\text{-}DFE}}^{1/2}\mathbf{U}_{\mathsf{MMSE\text{-}DFE}}\mathbf{a} + \bar{\mathbf{z}}_{\mathsf{MMSE\text{-}DFE}}, \tag{6.33}$$

where $\bar{\mathbf{z}}_{\mathsf{MMSE\text{-}DFE}} \triangleq \bar{\mathbf{y}}_{\mathsf{MMSE\text{-}DFE}} - \boldsymbol{\Delta}_{\mathsf{MMSE\text{-}DFE}}^{1/2}\mathbf{U}_{\mathsf{MMSE\text{-}DFE}}\mathbf{a}$.

MMSE-DFE preprocessed tree search proceeds from the causal model (6.33), where the interference $\bar{\mathbf{z}}_{\mathsf{MMSE\text{-}DFE}}$ is treated as (signal-independent) additive white Gaussian noise (AWGN). Although it can be shown that $\bar{\mathbf{z}}_{\mathsf{MMSE\text{-}DFE}}$ is white (in fact, $\mathbf{C}_{\bar{\mathbf{z}}_{\mathsf{MMSE\text{-}DFE}}} = \mathbf{I}$ for any $\mathbf{C}_{\mathbf{z}}$), it can readily be seen that $\bar{\mathbf{z}}_{\mathsf{MMSE\text{-}DFE}}$ is signal-dependent (and hence non-Gaussian) (Damen et al., 2003). Thus, treating $\bar{\mathbf{z}}_{\mathsf{MMSE\text{-}DFE}}$ as if it were AWGN will produce suboptimal[11] sequence estimates. However, it turns out that the increase in prequantization SINR (from the use of MMSE-DFE in place of ZF-DFE) more

[11] Interestingly, it has been shown that $\bar{\mathbf{z}}_{\mathsf{MMSE\text{-}DFE}}$ can be treated as AWGN when $\mathscr{A}$ is constant modulus. In other words, $\hat{\mathbf{a}}_{\mathsf{cMLSD}} = \arg\min_{\mathbf{a}\in\mathscr{A}^K} \|\bar{\mathbf{y}}_{\mathsf{MMSE\text{-}DFE}} - \boldsymbol{\Delta}_{\mathsf{MMSE\text{-}DFE}}^{1/2}\mathbf{U}_{\mathsf{MMSE\text{-}DFE}}\mathbf{a}\|^2$ for constant modulus $\mathscr{A}$ (Hwang & Schniter, 2005).

than compensates for the loss in optimality (due to non-AWGN $\bar{\mathbf{z}}_{\text{MMSE-DFE}}$). Thus, relative to QR preprocessing, MMSE-DFE preprocessing has been observed to yield significant improvements in the performance/complexity tradeoff of suboptimal tree search (Murugan et al., 2006).

Other types of preprocessing include lattice reduction (e.g., the method of Lenstra, Lenstra, & Lovász (1982)) and column permutation (e.g., reordering of **a** so that stronger symbols are decided first, as in V-BLAST ordering (Wolniansky et al., 1998)). Because these techniques would destroy the quasibanded structure of **Q**, however, we will not elaborate on them further.

Tree search algorithms (whether optimal or suboptimal) can be categorized as breadth-first, depth-first, or best-first (Anderson & Mohan, 1984; Murugan et al., 2006). Breadth-first search algorithms include, e.g., the M-algorithm (Anderson & Mohan, 1984), the T-algorithm (Simmons, 1990), statistical pruning algorithms (Gowaikar & Hassibi, 2003), the Wozencraft sequential decoder (Wozencraft & Reiffen, 1961), and the Pohst sphere decoder (Fincke & Pohst, 1985). Depth-first search algorithms include, e.g., the Schnorr–Euchner sphere decoder and its variants (Agrell, Eriksson, Vardy, & Zeger, 2002; Damen et al., 2003; Viterbo & Boutros, 1999). Best-first search algorithms include, e.g., the stack and Fano algorithms (Fano, 1963; Murugan et al., 2006; Viterbi & Omura, 1979). Because a thorough description and comparison of these approaches are outside the scope of this chapter, we make only a few remarks. The Fano algorithm was recently found to yield a superior complexity/performance tradeoff when **Q** was either a convolution matrix or fully populated (Murugan et al., 2006). The superiority of the Fano algorithm does not appear to hold when **Q** is quasibanded, though (Hwang & Schniter, 2006). The M-algorithm is popular for two reasons: simplicity and fixed complexity (i.e., complexity invariant to channel/noise realizations and SNR).

Although so far we have focused on tree-search implementations of MLSD, we now describe how tree search can be used to find (approximate) posterior LLRs, and thus MAP symbol and bit estimates, using the method of Hochwald & ten Brink (2003). First, we define the *coherent MAP sequence metric*

$$\zeta_{\text{coh}}(\mathbf{c}) \triangleq \ln f(\mathbf{y} \mid \mathbf{c}, \mathbf{Q}) + \mathbf{l}_c^T \mathbf{c} \tag{6.34}$$

$$= -\|\mathbf{y} - \mathbf{Q}\mathbf{a}\|^2_{\mathbf{C}_\mathbf{z}^{-1}} - \ln\left(\pi^K \det\{\mathbf{C}_\mathbf{z}\}\right) + \mathbf{l}_c^T \mathbf{c}, \tag{6.35}$$

where $\mathbf{l}_c \triangleq (L_c[0] \cdots L_c[K-1])^T$, with $L_c[k]$ being a priori LLRs from (6.20), and where the symbols **a** are determined by the hypothesized bit vector **c**. As previously remarked, the use of a priori LLRs implies that the coded bits $\{c[j]\}$ are treated as independent. It is straightforward to show (see Appendix 6.A) that the posterior LLR defined in (6.19) can be written as

$$L_{c|\mathbf{y},\mathbf{Q}}[j] = \ln \frac{\sum_{\mathbf{c}:c[j]=1} e^{\zeta_{\text{coh}}(\mathbf{c})}}{\sum_{\mathbf{c}:c[j]=0} e^{\zeta_{\text{coh}}(\mathbf{c})}}. \tag{6.36}$$

Note that, in the summations of (6.36), all possibilities of $\mathbf{c} \in \{0,1\}^{K \log_2 |\mathcal{A}|}$ are considered, not only those in the codebook. (The same holds true in related equations throughout the chapter.) This reflects our assumption that the equalizer does *not* use knowledge of the code structure to generate posterior LLRs; code structure is exploited only by the decoder.

Computing $L_{c|\mathbf{y},\mathbf{Q}}[j]$ through (6.36) would require $2^{K \log_2 |\mathcal{A}|}$ evaluations of the MAP metric $\zeta_{\text{coh}}(\mathbf{c})$, and hence would be impractical. However, as suggested in Hochwald & ten Brink (2003), the "max-log" approximation $\ln \sum_{\mathbf{c}} e^{\zeta(\mathbf{c})} \approx \max_{\mathbf{c}} \zeta(\mathbf{c})$ can be applied to yield

$$L_{c|\mathbf{y},\mathbf{Q}}[j] \approx \max_{\mathbf{c}:c[j]=1} \zeta_{\mathsf{coh}}(\mathbf{c}) - \max_{\mathbf{c}:c[j]=0} \zeta_{\mathsf{coh}}(\mathbf{c}). \tag{6.37}$$

Suboptimal tree search can then be used to find the set of all bit vectors $\mathbf{c} \in \{0,1\}^{K \log_2 |\mathscr{A}|}$ which yield non-negligible coherent MAP metrics $\zeta_{\mathsf{coh}}(\mathbf{c})$, as detailed in de Jong & Willink (2005). Once the posterior LLRs have been calculated, it is possible to generate hard bit estimates through (6.21), if needed. In a turbo configuration, though, the equalizer passes the posterior LLRs to a soft-input/soft-output decoder. After decoding, the refined LLRs are passed back to the equalizer to be used as priors, i.e., $\mathbf{l}_c$. (Recall Fig. 6.1.)

6.3.2.5 *Iterative Soft Equalization*

For approximate symbol/bit MAP equalization, one can consider using *iterative soft equalization* techniques (Tüchler, Koetter, & Singer, 2002; Wang & Poor, 1999) as an alternative to the trellis and tree-search approaches described earlier. The iterative soft equalization techniques described here use *linear estimation* strategies in conjunction with *evolving beliefs* of the interfering bits. After estimating a given bit, the equalizer updates its belief about that bit to better estimate the other bits. Once all bit beliefs (e.g., LLRs) have been updated, the process repeats. The equalizer may itself iterate several times and/or it may trade soft bit information with a soft-input/soft-output decoder in a turbo configuration.

Below we detail the main concepts behind iterative soft equalization for the simple case of BPSK.[12] This simplification allows us to make a direct mapping between each bit and a corresponding symbol, e.g., $a[j] = 2b[j] - 1$ for $j = 0, \ldots, K-1$, where $b[j] \in \{0,1\}$ and $a[j] \in \mathscr{A} = \{-1, +1\}$. In this case, the a priori LLR from (6.20) can be rewritten as

$$L_c[j] = \ln \frac{\Pr\{a[j] = +1\}}{\Pr\{a[j] = -1\}}. \tag{6.38}$$

Suppose that we are interested in estimating the jth bit, $c[j]$, or equivalently the jth symbol, $a[j]$. And say that, when doing so, we have prior information on the *other* bits, and thus the other symbols $\bar{\mathbf{a}}_j \triangleq (a[0] \cdots a[j-1] \;\; 0 \;\; a[j+1] \cdots a[K-1])^T$, that come in the form of a priori LLRs. To facilitate the use of linear operations, the symbol estimation stage treats the elements in $\bar{\mathbf{a}}_j$ as independent Gaussian with means and variances that are calculated from the respective LLRs. In particular, the calculated mean of $a[k]$ (for $k \neq j$) is computed through $\mu_a[k] \triangleq \sum_{a \in \{-1,+1\}} a \Pr\{a[k] = a\}$ using the identity

$$\Pr\{a[k] = a\} = \frac{\exp\big((a-1)L_c[k]/2\big)}{1 + \exp(-L_c[k])} \quad \text{for} \quad a \in \{-1, +1\}, \tag{6.39}$$

from which it can be shown that

$$\mu_a[k] = \frac{1 - \exp(-L_c[k])}{1 + \exp(-L_c[k])} = \tanh(L_c[k]/2). \tag{6.40}$$

[12] The case of nonbinary alphabets follows similar principles but is more tedious to describe.

Similarly, the calculated variance of $a[k]$ (for $k \neq j$) is computed through

$$v_a[k] \triangleq -\mu_a[k]^2 + \sum_{a \in \{-1,+1\}} a^2 \Pr\{a[k]=a\} \;=\; 1-\mu_a[k]^2. \tag{6.41}$$

The estimation of $a[j]$ proceeds by writing the observation as

$$\mathbf{y} = \mathbf{q}_j a[j] + \mathbf{Q}\bar{\mathbf{a}}_j + \mathbf{z}, \tag{6.42}$$

where $\mathbf{q}_j$ denotes the jth column of $\mathbf{Q}$. For convenience, we collect the calculated means into $\bar{\boldsymbol{\mu}}_j \triangleq (\mu_a[0]\cdots\mu_a[j-1]\;\;0\;\;\mu_a[j+1]\cdots\mu_a[K-1])^T$ and the calculated variances into $\bar{\mathbf{v}}_j \triangleq (v_a[0]\cdots v_a[j-1]\;\;0\;\;v_a[j+1]\cdots v_a[K-1])^T$.

In the classical iterative soft equalization approach proposed by Wang & Poor (1999), soft interference cancelation:

$$\mathbf{x}_j = \mathbf{y} - \mathbf{Q}\bar{\boldsymbol{\mu}}_j \tag{6.43}$$

is followed by LMMSE combining:

$$\hat{a}_{\mathsf{LMMSE}}[j] = \mathbf{e}_j^H\mathbf{x}_j \quad \text{with} \quad \mathbf{e}_j = \arg\min_{\mathbf{e}\in\mathbb{C}^K} \mathsf{E}\{|a[j]-\mathbf{e}^H\mathbf{x}_j|^2\}. \tag{6.44}$$

Writing the interference-canceled vector as

$$\mathbf{x}_j = \mathbf{q}_j a[j] + \mathbf{r}_j, \tag{6.45}$$

with residual interference vector

$$\mathbf{r}_j = \mathbf{Q}(\bar{\mathbf{a}}_j - \bar{\boldsymbol{\mu}}_j) + \mathbf{z}, \tag{6.46}$$

it can be seen that $\mathbf{C}_{\mathbf{r}_j} = \mathbf{Q}\mathrm{diag}\{\bar{\mathbf{v}}_j\}\mathbf{Q}^H + \mathbf{C}_{\mathbf{z}}$. Withholding prior belief on $a[j]$, so that $\mathsf{E}\{a[j]\}=0$ and $\mathrm{var}\{a[j]\}=1$, the LMMSE combiner in (6.44) becomes

$$\mathbf{e}_j = \mathbf{C}_{\mathbf{x}_j}^{-1}\mathbf{C}_{\mathbf{x}_j,a[j]} = \left(\mathbf{q}_j\mathbf{q}_j^H + \mathbf{C}_{\mathbf{r}_j}\right)^{-1}\mathbf{q}_j = \frac{1}{1+\mathbf{q}_j^H\mathbf{C}_{\mathbf{r}_j}^{-1}\mathbf{q}_j}\mathbf{C}_{\mathbf{r}_j}^{-1}\mathbf{q}_j, \tag{6.47}$$

where the matrix inversion lemma was used to obtain the right side of (6.47). Thus, $\hat{a}_{\mathsf{LMMSE}}[j]$ becomes

$$\hat{a}_{\mathsf{LMMSE}}[j] = \frac{\mathbf{q}_j^H\mathbf{C}_{\mathbf{r}_j}^{-1}\mathbf{x}_j}{1+\mathbf{q}_j^H\mathbf{C}_{\mathbf{r}_j}^{-1}\mathbf{q}_j}. \tag{6.48}$$

From straightforward arguments,[13] one can conclude that any scaled version of the statistic

$$g[j] \triangleq \mathbf{q}_j^H\mathbf{C}_{\mathbf{r}_j}^{-1}\mathbf{x}_j, \tag{6.49}$$

[13]Sufficiency can be understood as follows. After constructing the interference-canceled/whitened observation $\mathbf{C}_{\mathbf{r}_j}^{-1/2}\mathbf{x}_j = \mathbf{C}_{\mathbf{r}_j}^{-1/2}\mathbf{q}_j a[j] + \mathbf{C}_{\mathbf{r}_j}^{-1/2}\mathbf{r}_j$, the application of the matched filter $\mathbf{C}_{\mathbf{r}_j}^{-1/2}\mathbf{q}_j$, or any scaling thereof, yields a sufficient statistic for the detection of $a[j]$ (Poor, 1994). These two steps are combined in writing (6.49).

including the LMMSE estimate $\hat{a}_{\mathsf{LMMSE}}[j]$, is sufficient (Poor, 1994) for ML[14] detection of $a[j]$ (and thus of $b[j]$) from $\mathbf{x}_j$. In fact, the ML symbol decision is simply the sign of

$$L_g[j] \triangleq \ln\frac{f(g[j]\mid a[j]=+1)}{f(g[j]\mid a[j]=-1)} = \ln\frac{f(g[j]\mid c[j]=1)}{f(g[j]\mid c[j]=0)}. \tag{6.50}$$

Expanding $g[j]$ as

$$g[j] = \mathbf{q}_j^H \mathbf{C}_{\mathbf{r}_j}^{-1} \mathbf{q}_j a[j] + \mathbf{q}_j^H \mathbf{C}_{\mathbf{r}_j}^{-1} \mathbf{r}_j, \tag{6.51}$$

it can be seen that $g[j]|a[j]$ is circular Gaussian with mean $a[j]\mu_{g[j]}$ and variance $\sigma_{g[j]}^2$, where $\mu_{g[j]} = \mathbf{q}_j^H \mathbf{C}_{\mathbf{r}_j}^{-1} \mathbf{q}_j = \sigma_{g[j]}^2$. Hence,

$$L_g[j] = \ln\frac{\exp(-|g[j]-\mu_{g[j]}|^2/\sigma_{g[j]}^2)}{\exp(-|g[j]+\mu_{g[j]}|^2/\sigma_{g[j]}^2)} \tag{6.52}$$

$$= -|g[j]-\mu_{g[j]}|^2/\sigma_{g[j]}^2 + |g[j]+\mu_{g[j]}|^2/\sigma_{g[j]}^2 \tag{6.53}$$

$$= 4\mathrm{Re}\{g[j]\}. \tag{6.54}$$

Finally, a posterior LLR on $a[j]$ (and hence on $c[j]$) can be generated through

$$\ln\frac{\Pr\{a[j]=+1\mid g[j]\}}{\Pr\{a[j]=-1\mid g[j]\}} = \ln\frac{\Pr\{c[j]=1\mid g[j]\}}{\Pr\{c[j]=0\mid g[j]\}} = L_g[j] + L_c[j], \tag{6.55}$$

where (6.50) and Bayes rule were used to obtain the right side of (6.55). The posterior LLR (6.55) can then be used in place of $L_c[j]$ in (6.40)–(6.41) to calculate the mean $\mu_a[j]$ and variance $v_a[j]$ for subsequent estimation of $\{c[k]\}_{k\neq j}$.

6.3.2.6 **Remarks on Complexity**

The coherent equalization tools described in this section are quite general; they apply to any $\mathbf{Q}$ and, thus, to any type of linear modulation/demodulation combined with any type of linear channel propagation (whether the channel is rapidly TV or not). In fact, when structure in $\mathbf{Q}$ is lacking or ignored, equalization can be viewed as a form of *CDMA multiuser detection* (Moshavi, 1996; Verdú, 1998) where the code matrix (in this case $\mathbf{Q}$) changes from one bit to the next, or as a form of *MIMO decoding* (Tse & Viswanath, 2005) for communication over a flat-fading channel with K transmit and K receive antennas. For the case of generic $\mathbf{Q}$, however, the cost of implementing the equalization criteria rises rapidly with K, the block size. For example, we saw that linear and DFE schemes consume $\mathcal{O}(K^3)$ operations per block, and that more sophisticated schemes can be significantly more expensive. Because, for the applications we envision, typical values of K can be in the hundreds or thousands, equalization is made practical only by leveraging the structural properties of $\mathbf{Q}$ discussed in Section 6.2.2. Using these properties, Sections 6.3.3–6.3.4 describe equalization algorithms specifically tailored to rapidly TV channels.

[14]Because we assume a uniform prior on $a[j]$, ML detection is equivalent to MAP detection.

6.3.3 Coherent Equalization for Time-Frequency Concentrated Modulation/Demodulation

Recall from Section 6.2.2 that, when sufficiently time-frequency concentrated modulation/demodulation pulses are used, the effective channel matrix $\mathbf{Q}$ falls into the "narrowly quasibanded" class. Here, $\mathbf{Q}$ contains only negligible coefficients outside of the shaded region in Fig. 6.2(b), for some $D \ll K$. The main idea behind the equalization algorithms discussed in this section is that, by ignoring these negligible coefficients, the complexity of equalization can be significantly reduced without a significant loss in performance.

In this section, we will treat the interference caused by the negligible coefficients in $\mathbf{Q}$ as if it were part of the additive noise $\mathbf{z}$, allowing us to regard the negligible coefficients in $\mathbf{Q}$ as if they were zero-valued. In doing so, we will assume that the interference radius D has been chosen large enough so that these additional contributions to $\mathbf{z}$ are relatively small (for the SNRs of interest). In particular, we will assume that the value of D allows us to continue treating $\mathbf{z}$ as statistically independent of $\mathbf{a}$, as assumed in Section 6.2. With suitably designed modulation/demodulation schemes like the max-SINR schemes in Das & Schniter (2007), these assumptions have been shown (Hwang & Schniter, 2006)[15] to be satisfied with $D = \lceil \nu_{\max} T_s K \rceil + 1$ at SNRs up to at least 10 dB and with $D = \lceil \nu_{\max} T_s K \rceil + 2$ at SNRs up to at least 30 dB. Less time-frequency-concentrated schemes require larger values of D, making equalization more expensive to implement for the same level of residual interference. For example, the interference profiles in Fig. 6.4 suggest that Strohmer & Beaver's scheme (Strohmer & Beaver, 2003) requires a radius D at least 2 higher than the max-SINR scheme of Das & Schniter (2007) for the same level of residual interference.

In the remainder of this section, we provide some insight into how the narrowly quasibanded structure of $\mathbf{Q}$ can be leveraged to lower the complexity of the equalization strategies described in Section 6.3.2. In particular, we identify two principal approaches to this problem: *fast serial equalization* and *fast joint equalization*. We keep our description brief because equalization for narrowly quasibanded $\mathbf{Q}$ is related to particular forms of equalization for OFDM, which is the topic of Chapter 7.

6.3.3.1 Fast Serial Equalization

Many techniques that leverage the narrowly quasibanded structure of $\mathbf{Q}$ can be classified as *fast serial equalization* techniques. These techniques avoid the $K \times K$ matrix operations (e.g., inversion and LDU decomposition) specified in Section 6.3.2 for, e.g., linear equalization (6.24)–(6.26), decision feedback equalization (6.29)–(6.30), tree-search-based equalization (6.35), and iterative soft equalization (6.49). Instead, the fast serial techniques work on the *local observation model*

$$\mathbf{y}_k = \mathbf{Q}_k \mathbf{a}_k + \mathbf{z}_k \tag{6.56}$$

[15] For the specified D and SNR range, MLSD performance was found to be identical whether the out-of-band $\mathbf{Q}$ coefficients were treated as part of the channel or as part of the noise. We note that the variable "D" in Hwang & Schniter (2006) is defined to have twice the value of D in this chapter because in Hwang & Schniter (2006) the effective channel matrix is real-valued.

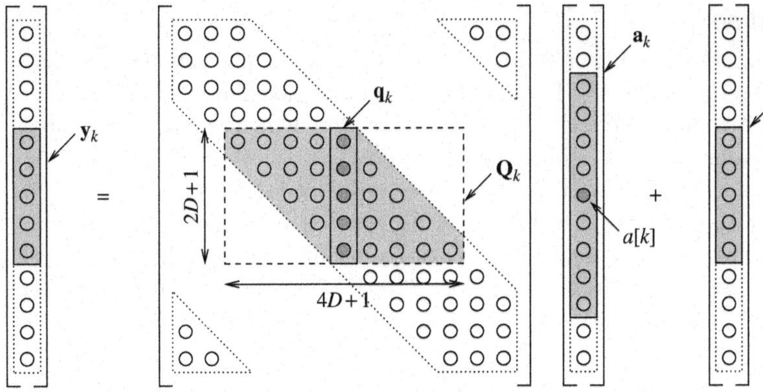

FIGURE 6.6

The local observation model used for fast serial equalization.

when estimating the symbol $a[k]$ (or any coded bits represented by $a[k]$) for $k = 0,\ldots,K-1$. Here, $\mathbf{y}_k \triangleq (y[k-D]\cdots y[k+D])^T$ and $\mathbf{a}_k \triangleq (a[k-2D]\cdots a[k+2D])^T$ are illustrated in Fig. 6.6, along with $\mathbf{Q}_k$ and $\mathbf{z}_k$. The principal idea behind the local model is the following. *Because $a[k]$ affects only the local observations $\mathbf{y}_k \in \mathbb{C}^{2D+1}$ within $\mathbf{y} \in \mathbb{C}^K$, use only these local observations to estimate $a[k]$.* We can, thus, think of $\mathbf{Q}_k \in \mathbb{C}^{(2D+1)\times(4D+1)}$ as the "*local* effective channel matrix." It is usually convenient to increment the index k in steps of 1, so that estimation is performed *serially*, i.e., one symbol at a time. And sometimes it helps to start over at $k = 0$ after $k = K-1$ has been reached. Notice that, due to the corner support regions of the quasibanded $\mathbf{Q}$ in Fig. 6.6, the local observation window shifts cyclically within $\mathbf{y}$.

To our knowledge, Jeon, Chang, & Cho (1999) were the first to apply this fast serial approach to the equalization of rapidly TV channels. In particular, they proposed an LMMSE approximation that required only $\mathcal{O}(KD^3)$ operations per block. Note that, when $D \ll K$, their approach is much cheaper than standard $\mathcal{O}(K^3)$ LMMSE, i.e., (6.24)–(6.25). Cai & Giannakis (2003) proposed LMMSE and MMSE-DFE extensions of Jeon et al. (1999), where the inversion of the $(2D+1)\times(2D+1)$ covariance matrix $\mathbf{C}_{\mathbf{y}_k}$ was accomplished using a rank-one update. However, their schemes require $\mathcal{O}(K^2D)$ operations per block, where the quadratic dependence on K remained as a result of not fully exploiting the quasibanded property of $\mathbf{Q}$. Barhumi, Leus, & Moonen (2004, 2006) proposed generalized $\mathcal{O}(K^2D^2)$ per-tone linear equalization schemes that allowed oversampling in the frequency domain.

Hunziker & Dahlhaus (2003) proposed an iterative approximation to ML symbol detection in which the likelihoods of individual symbols were serially maximized assuming tentative hard decisions on the other symbols. To reduce error propagation, they initialized using an approximation of LZF that was implemented serially using Gauss–Seidel iterations. Das & Schniter (2007) and Schniter (2004) proposed iterative soft equalization based on the local observation model (6.56), requiring only $\mathcal{O}(KD^3)$ operations per block iteration. As discussed in Section 6.3.2, a well-designed iterative soft equalizer is effective at preventing error propagation and can be used in a turbo configuration, as in Das & Schniter (2007). A similar iterative soft equalization scheme was proposed later by Peng & Ryan (2006).

6.3.3.2 Fast Joint Equalization

Fast joint equalization techniques have also been proposed for the coherent equalization of narrowly quasibanded versions of $\mathbf{Q}$ that result when time-frequency concentrated modulation/demodulation is used with rapidly TV channels. As opposed to serial equalization schemes, which estimate the symbols in $\mathbf{a}$ one-at-a-time, joint equalization schemes estimate the K symbols in $\mathbf{a}$ jointly.

Early joint techniques assumed not only that $\mathbf{Q}$ is narrowly quasibanded, but also that the off-diagonal coefficients within the support region of $\mathbf{Q}$ are themselves relatively small. For example, iterative LZF approximation techniques that require $\mathcal{O}(KD)$ operations per block iteration were proposed by Toeltsch & Molisch (2001) and Guillaud & Slock (2003). Gorokhov & Linnartz (2004) proposed $\mathcal{O}(KD)$ approximate LMMSE and DFE-like schemes using a first-order Taylor series approximation of the $K \times K$ LMMSE matrix inverse. Tomasin, Gorokhov, Yang, & Linnartz (2005) extended the techniques in Gorokhov & Linnartz (2004) to incorporate iterative hard interference cancelation. Hou & Chen (2005) proposed a $\mathcal{O}(KD^2)$ nonlinear estimator of the form $\hat{\mathbf{a}} = \mathbf{E}_{\mathsf{FF}}\mathbf{y} - \mathbf{E}_{\mathsf{FB}}\mathcal{D}_{\mathcal{A}}(\mathbf{E}_{\mathsf{FF}}\mathbf{y})$, where $\mathbf{E}_{\mathsf{FF}}$ and $\mathbf{E}_{\mathsf{FB}}$ are both narrowly banded. Note that, in Hou & Chen (2005), quantization is performed on the linear estimates $\mathbf{E}_{\mathsf{FF}}\mathbf{y}$ rather than (causally) on the final estimates $\hat{\mathbf{a}}$, as in DFE (6.27).

More recently, Rugini, Banelli, & Leus proposed $\mathcal{O}(KD^2)$ exact LMMSE (Rugini, Banelli, & Leus, 2005) and MMSE-DFE (Rugini et al., 2006) schemes for narrowly banded[16] $\mathbf{Q}$ based on fast LDU decomposition. Furthermore, they showed how to design a receiver window to ensure that the noise covariance $\mathbf{C}_{\mathbf{z}}$ is quasibanded, making the observation covariance $\mathbf{C}_{\mathbf{y}} = \sigma_a^2\mathbf{Q}^H\mathbf{Q} + \mathbf{C}_{\mathbf{z}}$ banded as well. These nonapproximate LMMSE and MMSE-DFE equalizers are expected to outperform their approximate counterparts. (See Chapter 7 for more details.)

Joint MLSD-based schemes exploiting the quasibanded structure of $\mathbf{Q}$ have also been proposed. For example, Matheus & Kammeyer (1997) applied the Viterbi algorithm to implement exact MLSD on a $(2D+1)$-banded $\mathbf{Q}$ with complexity $\mathcal{O}(KD|\mathcal{A}|^{2D+1})$. The same idea was reinvented in later works, e.g., (Sigloch et al., 2005). For typical values of $|\mathcal{A}|$ and D, however, the complexity of Viterbi equalization can be orders-of-magnitude higher than that of MMSE-DFE. Thus, Hwang & Schniter (2006) investigated tree-search approaches to approximate MLSD. Their techniques use fast MMSE-DFE preprocessing, costing $\mathcal{O}(KD^2)$ operations, followed by a tree-search that uses a fast metric update and is tuned to the V-shaped structure of the upper triangular matrix $\mathbf{\Delta}^{1/2}_{\mathsf{MMSE\text{-}DFE}}\mathbf{U}_{\mathsf{MMSE\text{-}DFE}}$ in the causal model (6.33). (Recall the V-shaped illustration in Fig. 6.5(c).) The resulting scheme has approximately the same complexity as the fast MMSE-DFE from Rugini et al. (2006), yet results in performance that is almost indistinguishable from MLSD.

MAP schemes exploiting the quasibanded structure of $\mathbf{Q}$ have also been proposed. For $(2D+1)$-banded $\mathbf{Q}$, e.g., Liu & Fitz (2007) used reduced-state sequence estimation (Eyuboglu & Qureshi, 1988) to compute approximate soft bit estimates while Hwang & Schniter (2009) applied tree-search (with a fast metric update). Finally, using the technique of Tüchler, Koetter, & Singer (2002), it is straightforward to translate any set of LMMSE estimates into soft bit estimates. Leveraging this idea, Fang, Rugini, & Leus (2008) turned the fast joint LMMSE estimation scheme of Rugini et al. (2005) into a soft bit estimation scheme.

[16]With minor modifications, the fast LDU decomposition that Rugini, Banelli, & Leus developed for banded matrices (i.e., those matching Fig. 6.5(b)) can be extended to quasibanded matrices (i.e., those matching Fig. 6.5(a)) (Hwang & Schniter, 2006).

6.3.3.3 Other Approaches to Equalization for Time-Frequency Concentrated Schemes

For completeness, we mention two other schemes proposed for the equalization of channels yielding a narrowly quasibanded $\mathbf{Q}$. The article by Choi, Voltz, & Cassara (2001) was among the first to consider equalization for multicarrier modulation over doubly selective (i.e., time- and frequency-selective) channels, and it proposed ZF, LMMSE, and ZF-DFE schemes for doing so. However, these schemes consumed $\mathcal{O}(K^3)$ operations per block because the narrowly quasibanded structure of $\mathbf{Q}$ was not leveraged. For the same application, Stamoulis, Diggavi, & Al-Dhahir (2002) proposed an $\mathcal{O}(K^2)$ LMMSE approximation where the matrix to be inverted during each block interval is replaced by its time average. As we have seen, however, near-optimal schemes can be designed with complexities that are *linear* in K.

6.3.4 Coherent Equalization for Single-Carrier Modulation/Demodulation

Recall from Section 6.2.2 that, when single-carrier modulation/demodulation is used, the effective channel matrix $\mathbf{Q}$ falls into the "widely quasibanded" class. Here, $\mathbf{Q}$ has only negligible coefficients outside the shaded region in Fig. 6.2(a), where M denotes the discrete channel delay spread. Because, in this case, $\mathbf{Q}$ is quasibanded, the equalization techniques described in Section 6.3.3 can *in principle* be applied here as well (e.g., Ahmed, Sellathurai, Lambotharan, & Chambers, 2006). However, this approach will only be practical when M is small. Because M is often large (e.g., in the hundreds), there is good reason to study equalization schemes whose complexities are robust to large M.

6.3.4.1 Frequency-Domain Equalization

Frequency-domain equalization (FDE) (Falconer et al., 2002) is one approach to make the equalization complexity of single-carrier schemes reasonable when M is large. To describe FDE, we focus on the case of CP-SCM modulation/demodulation, assuming adequate guard length (i.e., $N_g \geq M-1$) and white noise (i.e., $\mathbf{C_z} = \sigma_z^2 \mathbf{I}_K$). The first step of FDE is transformation of the observations $\mathbf{y}$ to the frequency domain. Denoting the $K \times K$ unitary discrete Fourier transform (DFT) matrix by $\mathbf{W}$ and using under-bars to identify frequency-domain vectors (e.g., $\underline{\mathbf{y}} \triangleq \mathbf{Wy}$, $\underline{\mathbf{a}} \triangleq \mathbf{Wa}$, and $\underline{\mathbf{z}} \triangleq \mathbf{Wz}$), it follows from (6.2) that

$$\underline{\mathbf{y}} = \underline{\mathbf{Q}}\underline{\mathbf{a}} + \underline{\mathbf{z}} \tag{6.57}$$

where $\underline{\mathbf{Q}} = \mathbf{WQW}^H$ and where $\underline{\mathbf{z}}$ also has covariance $\sigma_z^2 \mathbf{I}_K$. The second step of FDE is estimation of $\underline{\mathbf{a}}$ from $\underline{\mathbf{y}}$. Although the elements of $\mathbf{a}$ belong to a finite alphabet, the elements of $\underline{\mathbf{a}}$ do not, and hence the estimator must be linear. One option is LMMSE estimation, i.e., $\hat{\underline{\mathbf{a}}}_{\mathsf{LMMSE}} = (\underline{\mathbf{Q}}^H \underline{\mathbf{Q}} + \frac{\sigma_z^2}{\sigma_a^2} \mathbf{I}_K)^{-1} \underline{\mathbf{Q}}^H \underline{\mathbf{y}}$. The third and final step of FDE is transformation of $\hat{\underline{\mathbf{a}}}$ back to the time domain, yielding the symbol estimates $\hat{\mathbf{a}} \triangleq \mathbf{W}^H \hat{\underline{\mathbf{a}}}$. If the DFTs are implemented using radix-2 FFTs, they will consume only $\mathcal{O}(K \log_2 K)$ operations.

With a time-invariant channel, the use of CP-SCM makes $\mathbf{Q}$ circulant (recalling Section 6.2.2) and hence $\underline{\mathbf{Q}}$ diagonal. In this case, the LMMSE estimation step consumes only $\mathcal{O}(K)$ operations (because the matrix to invert is diagonal), and FDE consumes $\mathcal{O}(K \log_2 K)$ operations in total. Note that, for large M, FDE would be significantly cheaper than LMMSE estimation of $\mathbf{a}$ from $\mathbf{y}$ through fast LDU (Rugini et al., 2005) (as discussed in Section 6.3.3), which consumes $\mathcal{O}(KM^2)$ operations, where typically $M \approx K/4$.

With a TV channel, $\mathbf{Q}$ will not be circulant, and thus $\underline{\mathbf{Q}}$ will not be diagonal. In this case, the off-diagonal terms of $\underline{\mathbf{Q}}$ will be nonzero, complicating the estimation of $\underline{\mathbf{a}}$ from $\underline{\mathbf{y}}$. In fact, the interference power profile of $\underline{\mathbf{Q}}$ with CP-SCM is identical to that of $\mathbf{Q}$ with CP-OFDM, which (as shown in Fig. 6.4) decays quite slowly with distance from the diagonal. However, through the application of time-domain windowing[17] at the demodulator (Schniter & Liu, 2003), it is possible to give $\underline{\mathbf{Q}}$ the narrowly quasi-banded support of Fig. 6.2(b), in which case any of the fast *linear* equalization techniques described in Section 6.3.3 can be used to estimate $\underline{\mathbf{a}}$ from $\underline{\mathbf{y}}$. For example, Tang & Leus (2008) proposed a method to equalize a single-carrier system using the OFDM fast LMMSE technique (Rugini et al., 2005).

Because $\underline{\mathbf{a}}$ does not have a finite-alphabet structure, the trellis, DFE, and tree-search-based techniques discussed in Section 6.3.3 are not directly applicable to the estimation of $\underline{\mathbf{a}}$. Iterative soft equalization, however, is applicable. We now summarize the approach proposed by Schniter & Liu (2003). First, the fast serial iterative soft equalization technique of Schniter (2004) is used to compute the LMMSE interference-canceled estimate $\hat{\underline{\mathbf{a}}}$ from the frequency-domain windowed observations $\underline{\mathbf{y}}$ (given current estimates of the time-domain symbol means and variances). Next, the estimates $\hat{\underline{\mathbf{a}}}$ are transformed to the time domain through $\hat{\mathbf{a}} = \mathbf{W}^H\hat{\underline{\mathbf{a}}}$, from which posterior LLRs are calculated for each of the bits $c[j]$. The posterior LLR computation is more complicated than (6.54)–(6.55), though, due to the correlation that results from the time-frequency transformation. Finally, the posterior LLRs are used as priors in the next iteration, which begins by recalculating the time-domain symbol means and variances. In Schniter & Liu (2003), a fast algorithm for the entire procedure was derived that consumes only $\mathcal{O}(D^2 K \log K)$ operations per block iteration. Ng & Falconer (2004) later extended the technique of Schniter & Liu (2003) to include widely linear estimation (although they neglected the receiver windowing step).

Although the windowed FDE method above focuses on CP-SCM, similar techniques can be applied to ZP-SCM under appropriate processing of the received guard samples. For single-carrier modulation *without* a prefix, the use of *IBI-cancellation and cyclic-prefix reconstruction* (Kim & Stüber, 1998) enables the application of the CP-SCM-based windowed FDE methods discussed earlier, as demonstrated by Schniter & Liu (2004).

6.3.4.2 Other Approaches to Equalization for Single-Carrier Schemes

Barhumi, Leus, & Moonen (2005) proposed a CP-SCM equalization technique based on linear TV filters whose time-variations were constrained to obey a (possibly oversampled) complex-exponential basis expansion model of order $I-1$. Under these constraints, LZF and LMMSE equalizers, requiring $\mathcal{O}(KI^3M^3)$ operations per block, were designed. However, because of the cubic complexity in M, these schemes are much more expensive than frequency-domain equalization when M is large.

6.4 NONCOHERENT EQUALIZATION

In Section 6.3, we discussed the coherent approach to equalization, i.e., estimation of $\mathbf{a}$ from $\mathbf{y}$ in (6.2), where the channel $\mathbf{H}$, and hence the effective channel $\mathbf{Q}$, was assumed to be known. Here, we discuss *noncoherent equalization*, where the channel realization $\mathbf{H}$ is unknown but its statistics may be known.

[17] With time-domain windowing, $\underline{\mathbf{y}} = \mathbf{W}\boldsymbol{\Delta}\mathbf{y}$ and $\underline{\mathbf{Q}} = \mathbf{W}\boldsymbol{\Delta}\mathbf{Q}\mathbf{W}^H$ for suitably chosen diagonal $\boldsymbol{\Delta}$.

In Section 6.4.1, we rewrite the system model in a form that is more convenient for noncoherent equalization. Then, in Sections 6.4.2 and 6.4.3, we describe criteria and algorithms for noncoherent equalization, respectively. Finally, in Sections 6.4.4–6.4.5, we describe specific strategies suitable for the noncoherent equalization of rapidly TV channels.

6.4.1 Noncoherent System Model

Because the effective channel **Q** is now unknown, it helps to reformulate the system model developed in Section 6.2 into a more convenient form. In particular, we rewrite (6.2) using an efficient parameterization for the entries of the matrix[18] **Q**. To do this, we build a *basis expansion model* (BEM) for the trajectories of the effective channel coefficients that make up **Q**, and then we write the observation in terms of these BEM coefficients.

From (6.2), we can see that

$$y[l] = \sum_{k=0}^{K-1} [\mathbf{Q}]_{l,k} a[k] + z[l] \;=\; \sum_{d=l}^{l-K+1} [\mathbf{Q}]_{l,l-d} a[l-d] + z[l] \tag{6.58}$$

$$= \sum_{d=0}^{K-1} q[l,d] a[l-d] + z[l] \tag{6.59}$$

for $q[l,d] \triangleq [\mathbf{Q}]_{l,\langle l-d \rangle_K}$ and where the index of $a[\cdot]$ in (6.59), and henceforth, is taken modulo-K. Here, $\langle j \rangle_K$ denotes "j modulo K." Notice that (6.59) expresses the relationship between $\{a[l]\}$ and $\{y[l]\}$ in exactly the same way as (6.1) expressed the relationship between $\{s[n]\}$ and $\{r[n]\}$: using a TV convolution. In fact, when the modulation and demodulation operations are trivial, as in single-carrier modulation, we have $q[l,d] = h[l,d]$. Because of the support of **Q**, as described in Section 6.2.2, the summation range in (6.59) can be truncated to $d \in \{-D,\ldots,D\}$ for narrowly quasibanded **Q** and to $d \in \{0,\ldots,M-1\}$ for widely quasibanded **Q**. In this section, we will assume the general case that $d \in \{0,\ldots,N_q-1\}$, so that the widely quasibanded case follows directly from $N_q = M$ and the narrowly quasibanded case follows from $N_q = 2D+1$ after cyclically left-shifting the columns of **Q** by D places. (Recall Fig. 6.2.)

Although BEMs are usually applied to the channel impulse response trajectories $\{h[n,m]\}_{n=0}^{N-1}$ (e.g., Tsatsanis & Giannakis, 1996), here we apply a BEM to the *effective* channel impulse response trajectory, which includes the effects of modulation/demodulation. In particular, we model the dth trajectory $\{q[l,d]\}_{l=0}^{K-1}$ using the BEM coefficients $\{\theta[i,d]\}_{i=0}^{I-1}$ and basis waveforms constructed from $\{\beta[l,i]\}$, as follows:

$$q[l,d] = \sum_{i=0}^{I-1} \beta[l,i]\theta[i,d] \quad \text{for } l = 0,\ldots,K-1. \tag{6.60}$$

If one prefers not to use a BEM, then the *trivial BEM*, specified by $I = K$ and $\beta[l,i] = \delta[l-i]$, where $\delta[\cdot]$ denotes the Kronecker delta, guarantees $\theta[l,d] = q[l,d]$ $\forall l,d$. Using $\boldsymbol{\theta}_d \triangleq$

[18] We choose to parameterize the entries of **Q** rather than those of **H** to avoid explicitly defining the modulation and demodulation operations.

$(\theta[0,d] \;\cdots\; \theta[I-1,d])^T \in \mathbb{C}^I$ and $\boldsymbol{\beta}_l \triangleq (\beta[l,0] \;\cdots\; \beta[l,I-1])^H \in \mathbb{C}^I$, we have $q[l,d] = \boldsymbol{\beta}_l^H \boldsymbol{\theta}_d$ and hence equation (6.59) can be rewritten in terms of BEM quantities as

$$y[l] = \boldsymbol{\beta}_l^H \sum_{d=0}^{N_q-1} a[l-d]\boldsymbol{\theta}_d + z[l]. \tag{6.61}$$

Collecting the demodulator outputs $\{y[l]\}_{l=0}^{k}$ in a vector, (6.61) implies

$$\underbrace{\begin{pmatrix} y[0] \\ \vdots \\ y[k] \end{pmatrix}}_{\triangleq\, \mathbf{y}_k} = \underbrace{\begin{pmatrix} a[0]\boldsymbol{\beta}_0^H & \cdots & a[-N_q+1]\boldsymbol{\beta}_0^H \\ \vdots & & \vdots \\ a[k]\boldsymbol{\beta}_k^H & \cdots & a[k-N_q+1]\boldsymbol{\beta}_k^H \end{pmatrix}}_{\triangleq\, \boldsymbol{\Lambda}_{\mathbf{a}_k}} \underbrace{\begin{pmatrix} \boldsymbol{\theta}_0 \\ \vdots \\ \boldsymbol{\theta}_{N_q-1} \end{pmatrix}}_{\triangleq\, \boldsymbol{\theta}} + \underbrace{\begin{pmatrix} z[0] \\ \vdots \\ z[k] \end{pmatrix}}_{\triangleq\, \mathbf{z}_k} \tag{6.62}$$

summarized by

$$\mathbf{y}_k = \boldsymbol{\Lambda}_{\mathbf{a}_k}\boldsymbol{\theta} + \mathbf{z}_k. \tag{6.63}$$

The matrix $\boldsymbol{\Lambda}_{\mathbf{a}_k}$ is constructed from the partial symbol vector $\mathbf{a}_k \triangleq (a[0]\cdots a[k])^T$ and the BEM waveforms $\{\boldsymbol{\beta}_l\}_{l=0}^{k}$. Notice that $\mathbf{a}_{K-1} = \mathbf{a}$, $\mathbf{y}_{K-1} = \mathbf{y}$, and $\mathbf{z}_{K-1} = \mathbf{z}$ for the previously defined vectors $\mathbf{a}$, $\mathbf{y}$, and $\mathbf{z}$. Notice also that, in (6.63), the channel realization is represented by the BEM coefficient vector $\boldsymbol{\theta}$.

Throughout this section, we assume (for simplicity) that the symbol block is zero-prefixed, i.e., $a[l] = 0$ for $l \in \{-N_q+1,\ldots,0\}$. Note that, due to the cyclic indexing assumption on $a[\cdot]$, this implies that $a[l] = 0$ for $l \in \{K-N_q+1,\ldots,K-1\}$. With this assumption, (6.61) yields a causal relationship between the symbols and the demodulator outputs, i.e., $\{y[l]\}_{l\le k}$ depends only on $\{a[l]\}_{l\le k}$.

We also assume that the channel is *Rayleigh fading*, i.e., that the impulse response coefficients are zero-mean[19] Gaussian distributed. Because of the linearity of modulation/demodulation and basis expansion modeling, this implies that the BEM coefficients $\boldsymbol{\theta}$ will also be zero-mean Gaussian. All other model assumptions stated in Section 6.2 apply here as well.

6.4.2 Noncoherent Equalization Criteria

In this section, we review several well-known noncoherent equalization criteria. As in our previous discussion of coherent criteria, we partition the discussion into criteria that apply to hard symbol estimates, complex-field symbol estimates, and soft bit estimates.

Before continuing, though, we discuss the important *ambiguity* phenomenon that can arise in noncoherent equalization. For example, if $C\mathbf{a} \in \mathscr{A}^K$ for some $\mathbf{a} \in \mathscr{A}^K$ and some $C \neq 1$, then it is impossible to distinguish, from the output $\mathbf{y}$, between the hypotheses $(\mathbf{a},\boldsymbol{\theta})$ and $(C\mathbf{a}, C^{-1}\boldsymbol{\theta})$. Notice that C accounts for both phase and/or gain ambiguity. To prevent ambiguity, one could, e.g., use an asymmetric scalar alphabet $\mathscr{A}$ or treat a single symbol (e.g., $a[0]$) as a known pilot, so that the set of candidate symbol vectors becomes asymmetric (Hart, 2000). In stating the criteria below, we assume that the ambiguity issue has been taken care of.

[19] Some details of the nonzero mean case can be found in Reader & Cowley (1996).

6.4.2.1 Hard Symbol Estimates

Similar to coherent equalization, the minimal probability of sequence error is guaranteed by *noncoherent maximum a posteriori sequence detection* (MAPSD):

$$\hat{\mathbf{a}}_{\mathsf{ncMAPSD}} \triangleq \arg \max_{\mathbf{a}' \in \mathscr{A}^K} \Pr\{\mathbf{a} = \mathbf{a}' \mid \mathbf{y}\}. \tag{6.64}$$

Note that the noncoherent posterior in (6.64) is not conditioned on the effective channel matrix $\mathbf{Q}$, as was the coherent posterior in (6.9). Similarly, noncoherent MAP symbol and bit detection are defined as

$$\hat{a}_{\mathsf{ncMAP}}[k] \triangleq \arg \max_{a \in \mathscr{A}} \Pr\{a[k] = a \mid \mathbf{y}\} \text{ for } k \in 0,\ldots,K-1 \tag{6.65}$$

$$\hat{c}_{\mathsf{ncMAP}}[j] \triangleq \arg \max_{c \in \{0,1\}} \Pr\{c[j] = c \mid \mathbf{y}\} \text{ for } j \in 0,\ldots,K\log_2|\mathscr{A}|-1. \tag{6.66}$$

If the equalizer assumes that $\mathbf{a}$ is uniformly distributed over $\mathscr{A}^K$, then noncoherent MAPSD reduces to *noncoherent maximum likelihood sequence detection* (MLSD):

$$\hat{\mathbf{a}}_{\mathsf{ncMLSD}} \triangleq \arg \max_{\mathbf{a} \in \mathscr{A}^K} f(\mathbf{y} \mid \mathbf{a}). \tag{6.67}$$

(The justification is similar to (6.13).) Due to the Rayleigh fading assumption, $f(\mathbf{y} \mid \mathbf{a})$ is a Gaussian distribution with zero mean and covariance

$$\mathbf{C}_{\mathbf{y}|\mathbf{a}} = \mathbf{C}_{\mathbf{z}} + \mathbf{\Lambda}_{\mathbf{a}} \mathbf{C}_{\theta} \mathbf{\Lambda}_{\mathbf{a}}^H. \tag{6.68}$$

Given this conditional distribution for $\mathbf{y}$, (6.67) reduces to

$$\hat{\mathbf{a}}_{\mathsf{ncMLSD}} = \arg \min_{\mathbf{a} \in \mathscr{A}^K} \left\{ \mathbf{y}^H \mathbf{C}_{\mathbf{y}|\mathbf{a}}^{-1} \mathbf{y} + \ln(\pi^K \det\{\mathbf{C}_{\mathbf{y}|\mathbf{a}}\}) \right\}. \tag{6.69}$$

There is an interesting connection between the noncoherent MAPSD/MLSD criteria and MMSE channel estimation (Haeb & Meyr, 1989; Kailath, 1969). To see this, we first write the MMSE estimate of $\boldsymbol{\theta}$ from $\mathbf{y}$ under the sequence hypothesis $\mathbf{a}$ as

$$\hat{\boldsymbol{\theta}}_{\mathsf{MMSE}|\mathbf{a}} \triangleq \mathsf{E}\{\boldsymbol{\theta} \mid \mathbf{y}, \mathbf{a}\} = \mathbf{C}_{\theta,\mathbf{y}|\mathbf{a}} \mathbf{C}_{\mathbf{y}|\mathbf{a}}^{-1} \mathbf{y} \tag{6.70}$$

$$= \mathbf{C}_{\theta} \mathbf{\Lambda}_{\mathbf{a}}^H (\mathbf{C}_{\mathbf{z}} + \mathbf{\Lambda}_{\mathbf{a}} \mathbf{C}_{\theta} \mathbf{\Lambda}_{\mathbf{a}}^H)^{-1} \mathbf{y} \tag{6.71}$$

$$= (\mathbf{C}_{\theta}^{-1} + \mathbf{\Lambda}_{\mathbf{a}}^H \mathbf{C}_{\mathbf{z}}^{-1} \mathbf{\Lambda}_{\mathbf{a}})^{-1} \mathbf{\Lambda}_{\mathbf{a}}^H \mathbf{C}_{\mathbf{z}}^{-1} \mathbf{y}, \tag{6.72}$$

where the matrix inversion lemma was used to obtain (6.72). In Appendix 6.B, we use (6.72) to show that (6.69) can be rewritten as

$$\hat{\mathbf{a}}_{\mathsf{ncMLSD}} = \arg \min_{\mathbf{a} \in \mathscr{A}^K} \left\{ \|\mathbf{y} - \mathbf{\Lambda}_{\mathbf{a}} \hat{\boldsymbol{\theta}}_{\mathsf{MMSE}|\mathbf{a}}\|_{\mathbf{C}_{\mathbf{z}}^{-1}}^2 + \|\hat{\boldsymbol{\theta}}_{\mathsf{MMSE}|\mathbf{a}}\|_{\mathbf{C}_{\theta}^{-1}}^2 + \ln(\pi^K \det\{\mathbf{C}_{\mathbf{y}|\mathbf{a}}\}) \right\}. \tag{6.73}$$

Equation (6.73) states that the noncoherent MLSD metric can be written as the coherent MLSD metric based on the *implicit channel estimate* $\hat{\boldsymbol{\theta}}_{\mathsf{MMSE}|\mathbf{a}}$, plus a term that penalizes the deviation in $\hat{\boldsymbol{\theta}}_{\mathsf{MMSE}|\mathbf{a}}$ from the prior statistics on $\boldsymbol{\theta}$, plus what is sometimes referred to as a "bias" term. Thus, while the noncoherent MLSD/MAPSD estimates can be found without computing a channel estimate (as in (6.69)), they can also be found via joint channel/symbol estimation (as in (6.73)).

If the channel statistics (i.e., $\mathbf{C}_{\boldsymbol{\theta}}$) are unknown, then the noncoherent ML and MAP criteria do not apply. In this case, it may be more appropriate to use the *generalized likelihood ratio test* (GLRT) criterion (Warrier & Madhow, 2002):

$$\hat{\mathbf{a}}_{\mathsf{GLRT}} \triangleq \arg\max_{\mathbf{a}\in\mathscr{A}^K} \max_{\boldsymbol{\theta}\in\mathbb{C}^{N_q I}} f(\mathbf{y} \mid \mathbf{a}, \boldsymbol{\theta}). \tag{6.74}$$

Because $\ln f(\mathbf{y} \mid \mathbf{a}, \boldsymbol{\theta}) = -\|\mathbf{y} - \mathbf{\Lambda}_{\mathbf{a}}\boldsymbol{\theta}\|^2_{\mathbf{C}_{\mathbf{z}}^{-1}} - \ln(\pi^K \det\{\mathbf{C}_{\mathbf{z}}\})$, the GLRT sequence estimate can be expressed as

$$\hat{\mathbf{a}}_{\mathsf{GLRT}} = \arg\min_{\mathbf{a}\in\mathscr{A}^K} \left\|\mathbf{y} - \mathbf{\Lambda}_{\mathbf{a}}\hat{\boldsymbol{\theta}}_{\mathsf{ML}|\mathbf{a}}\right\|^2_{\mathbf{C}_{\mathbf{z}}^{-1}}, \tag{6.75}$$

where $\hat{\boldsymbol{\theta}}_{\mathsf{ML}|\mathbf{a}} \triangleq \arg\max_{\boldsymbol{\theta}\in\mathbb{C}^{N_q I}} f(\mathbf{y} \mid \mathbf{a}, \boldsymbol{\theta})$ denotes the $\mathbf{a}$-conditional maximum likelihood (ML) channel estimate, i.e.,

$$\hat{\boldsymbol{\theta}}_{\mathsf{ML}|\mathbf{a}} = \arg\min_{\boldsymbol{\theta}\in\mathbb{C}^{N_q I}} \|\mathbf{y} - \mathbf{\Lambda}_{\mathbf{a}}\boldsymbol{\theta}\|^2_{\mathbf{C}_{\mathbf{z}}^{-1}} \tag{6.76}$$

$$= \left(\mathbf{\Lambda}_{\mathbf{a}}^H \mathbf{C}_{\mathbf{z}}^{-1} \mathbf{\Lambda}_{\mathbf{a}}\right)^{-1} \mathbf{\Lambda}_{\mathbf{a}}^H \mathbf{C}_{\mathbf{z}}^{-1} \mathbf{y}. \tag{6.77}$$

Combining (6.77) and (6.75), the GLRT sequence estimate becomes

$$\hat{\mathbf{a}}_{\mathsf{GLRT}} = \arg\max_{\mathbf{a}\in\mathscr{A}^K} \mathbf{y}^H \mathbf{C}_{\mathbf{z}}^{-1} \mathbf{\Lambda}_{\mathbf{a}} \left(\mathbf{\Lambda}_{\mathbf{a}}^H \mathbf{C}_{\mathbf{z}}^{-1} \mathbf{\Lambda}_{\mathbf{a}}\right)^{-1} \mathbf{\Lambda}_{\mathbf{a}}^H \mathbf{C}_{\mathbf{z}}^{-1} \mathbf{y}. \tag{6.78}$$

Notice that, when the noise is white, $\hat{\boldsymbol{\theta}}_{\mathsf{ML}|\mathbf{a}}$ in (6.77) reduces to the conditional *least-squares* (LS) channel estimate: $\hat{\boldsymbol{\theta}}_{\mathsf{LS}|\mathbf{a}} = \left(\mathbf{\Lambda}_{\mathbf{a}}^H \mathbf{\Lambda}_{\mathbf{a}}\right)^{-1} \mathbf{\Lambda}_{\mathbf{a}}^H \mathbf{y}$.

Finally, we note that the GLRT metric in (6.78) equals the limiting case of the $\mathbf{y}^H \mathbf{C}_{\mathbf{y}|\mathbf{a}}^{-1} \mathbf{y}$ component of the noncoherent MLSD metric in (6.69) when $\mathbf{C}_{\boldsymbol{\theta}} = \sigma_\theta^2 \mathbf{I}$ and $\sigma_\theta^2 \to \infty$, i.e., when the signal is white and the noise power is negligible relative to the signal power.

6.4.2.2 Complex-Field Symbol Estimates

The *noncoherent minimum mean-squared error* (MMSE) criterion specifies the complex-valued sequence estimate

$$\hat{\mathbf{a}}_{\mathsf{ncMMSE}} \triangleq \arg\min_{\mathbf{a}'\in\mathbb{C}^K} \mathsf{E}\left\{\|\mathbf{a} - \mathbf{a}'\|^2 \mid \mathbf{y}\right\}. \tag{6.79}$$

Note that, unlike the coherent case (6.15), the expectation in (6.79) is not conditioned on the channel $\mathbf{Q}$. Writing the MMSE estimate as the conditional mean (Poor, 1994), we find

$$\hat{\mathbf{a}}_{\mathsf{ncMMSE}} = \mathsf{E}\{\mathbf{a} \mid \mathbf{y}\} = \sum_{\mathbf{a}\in\mathscr{A}^K} \mathbf{a}\, p(\mathbf{a} \mid \mathbf{y}) \tag{6.80}$$

$$= \sum_{\mathbf{a}\in\mathscr{A}^K} \mathbf{a} \frac{f(\mathbf{y} \mid \mathbf{a}) p(\mathbf{a})}{\sum_{\mathbf{a}'\in\mathscr{A}^K} f(\mathbf{y} \mid \mathbf{a}') p(\mathbf{a}')} \tag{6.81}$$

$$= \sum_{\mathbf{a}\in\mathscr{A}^K} \mathbf{a} \frac{p(\mathbf{a}) \int f(\mathbf{y} \mid \mathbf{a}, \boldsymbol{\theta}) f(\boldsymbol{\theta}) d\boldsymbol{\theta}}{\sum_{\mathbf{a}'\in\mathscr{A}^K} p(\mathbf{a}') \int f(\mathbf{y} \mid \mathbf{a}', \boldsymbol{\theta}) f(\boldsymbol{\theta}) d\boldsymbol{\theta}}. \tag{6.82}$$

If we assume that $p(\mathbf{a})$ is uniformly distributed over $\mathscr{A}^K$, then

$$\hat{\mathbf{a}}_{\mathsf{ncMMSE}} = \frac{\sum_{\mathbf{a}\in\mathscr{A}^K} \mathbf{a} \int \exp\left(-\|\mathbf{y}-\mathbf{\Lambda}_{\mathbf{a}}\boldsymbol{\theta}\|^2_{\mathbf{C}_{\mathbf{z}}^{-1}} - \|\boldsymbol{\theta}\|^2_{\mathbf{C}_{\theta}^{-1}}\right)\mathrm{d}\boldsymbol{\theta}}{\sum_{\mathbf{a}'\in\mathscr{A}^K} \int \exp\left(-\|\mathbf{y}-\mathbf{\Lambda}_{\mathbf{a}'}\boldsymbol{\theta}\|^2_{\mathbf{C}_{\mathbf{z}}^{-1}} - \|\boldsymbol{\theta}\|^2_{\mathbf{C}_{\theta}^{-1}}\right)\mathrm{d}\boldsymbol{\theta}}. \tag{6.83}$$

Note that the finite-alphabet nature of $\mathbf{a}$ makes the conditional mean difficult to compute because it requires the evaluation of $|\mathscr{A}|^K$-term summations. Unlike the coherent case, imposing constraints (e.g., linear) on the noncoherent MMSE estimator does not significantly simplify its design, and so this approach is not very popular.

6.4.2.3 Soft Bit Estimates

In soft noncoherent equalization (as used in, e.g., turbo equalization), the equalizer computes posterior LLRs[20] for the coded bits $c[j]$

$$L_{c|\mathbf{y}}[j] \triangleq \ln \frac{\Pr\{c[j]=1 \mid \mathbf{y}\}}{\Pr\{c[j]=0 \mid \mathbf{y}\}} \quad \text{for} \quad j=0,\ldots,K\log_2|\mathscr{A}|-1, \tag{6.84}$$

given the a priori LLRs defined in (6.20). Note that the posterior probabilities in (6.84) are not conditioned on the channel, unlike the coherent case (6.19). If needed, the noncoherent MAP bit estimates can be generated from the noncoherent posterior LLRs using

$$\hat{c}_{\mathsf{ncMAP}}[j] = \frac{1}{2}\left(1+\operatorname{sign}(L_{c|\mathbf{y}}[j])\right). \tag{6.85}$$

6.4.3 Noncoherent Equalization Tools

The noncoherent MLSD, MAPSD, MAP, and MMSE estimates, as outlined in Section 6.4.2, require the evaluation of $\mathscr{O}(|\mathscr{A}|^K)$ metrics if computed through brute force, which is not practical for the anticipated values of K. In this section, we review algorithms that are designed for practical noncoherent equalization.

6.4.3.1 Suboptimality of Trellis-Based Noncoherent Equalization

Among the optimal coherent MAPSD, MLSD, and symbol/bit MAP algorithms in Section 6.3.2 were trellis-based methods. We now investigate whether similar approaches exist for optimal noncoherent equalization. In doing so, we use the causal[21] model summarized by (6.63). To simplify the notation, we assume BPSK, allowing us to make a direct mapping between each bit and symbol, e.g., $a[j] = 2b[j]-1$ for $j=0,\ldots,K-1$, where $a[j]\in\mathscr{A}=\{-1,+1\}$. Finally, we allow prior beliefs on the bits (and thus symbols) in the form of a priori LLRs $\{L_c[k]\}_{k=0}^{K-1}$, defined in (6.20) and simplified for BPSK symbols in (6.38).

---

[20]Under the "fixed lag" constraint, the conditioning in (6.84) is performed on $\left(y[0]\cdots y\left[\lceil\frac{j}{\log_2|\mathscr{A}|}\rceil+J\right]\right)^T$ instead of $\mathbf{y}$ (Zhang, Fitz, & Gelfand, 1997), where the look-ahead interval $J\geq 0$ trades off between performance and complexity.

[21]Note that this causal model is different from the one used for DFE in Section 6.3.2. For DFE, $\mathbf{a}$ was estimated *backward* from the last symbol, whereas here $\mathbf{a}$ is estimated *forward* from the first symbol.

Analogous to the coherent MAP sequence metric (6.34), we now define a *noncoherent MAP sequence metric*. In particular, we define a *partial* noncoherent MAP sequence metric that depends on the partial observation $\mathbf{y}_k$ from (6.63) and the partial bit vector $\mathbf{c}_k \triangleq (c[0] \;\cdots\; c[k])^T$:

$$\zeta_{\mathsf{nc}}(\mathbf{c}_k) \triangleq \ln f(\mathbf{y}_k \mid \mathbf{c}_k) + \mathbf{l}_k^T \mathbf{c}_k. \tag{6.86}$$

Here, $\mathbf{l}_k \triangleq (L_c[0] \;\cdots\; L_c[k])^T$ is a partial version of the a priori LLR vector $\mathbf{l}_c$ defined just after (6.35). Notice that the complete noncoherent MAP sequence metric $\zeta_{\mathsf{nc}}(\mathbf{c})$ is obtained when $k = K-1$. If we are interested in noncoherent MLSD rather than noncoherent MAPSD, then we would use $\zeta_{\mathsf{nc}}(\mathbf{c})$ with $\mathbf{l}_c = \mathbf{0}$, and if we are interested in the GLRT criterion, then we would additionally assume that $\mathbf{C}_\theta = \sigma_\theta^2 \mathbf{I}$ with $\sigma_\theta^2 \to \infty$.

Under the Rayleigh fading assumption, $f(\mathbf{y}_k \mid \mathbf{c}_k)$ is Gaussian with zero mean and covariance $\mathbf{C}_{\mathbf{y}_k|\mathbf{a}_k} = \mathbf{C}_{\mathbf{z}_k} + \mathbf{\Lambda}_{\mathbf{a}_k} \mathbf{C}_\theta \mathbf{\Lambda}_{\mathbf{a}_k}^H$, where $\mathbf{a}_k$ denotes the BPSK symbol vector corresponding to the bit vector $\mathbf{c}_k$. Thus, we have

$$\zeta_{\mathsf{nc}}(\mathbf{c}_k) = -\mathbf{y}_k^H \mathbf{C}_{\mathbf{y}_k|\mathbf{a}_k}^{-1} \mathbf{y}_k - \ln(\pi^{k+1} \det\{\mathbf{C}_{\mathbf{y}_k|\mathbf{a}_k}\}) + \mathbf{l}_k^T \mathbf{c}_k, \tag{6.87}$$

which reduces to the noncoherent MLSD metric in (6.69) when $\mathbf{l}_k = \mathbf{0}$. Using the matrix inversion lemma, (6.87) can be rewritten as

$$\zeta_{\mathsf{nc}}(\mathbf{c}_k) = -\mathbf{y}_k^H \mathbf{C}_{\mathbf{z}_k}^{-1} \mathbf{\Lambda}_{\mathbf{a}_k} \mathbf{\Sigma}_{\mathbf{a}_k}^{-1} \mathbf{\Lambda}_{\mathbf{a}_k}^H \mathbf{C}_{\mathbf{z}_k}^{-1} \mathbf{y}_k - \ln\left(\pi^{k+1} \det\{\mathbf{C}_{\mathbf{y}_k|\mathbf{a}_k}\}\right) + \mathbf{l}_k^T \mathbf{c}_k \tag{6.88}$$

with $\mathbf{\Sigma}_{\mathbf{a}_k} \triangleq \mathbf{C}_\theta^{-1} + \mathbf{\Lambda}_{\mathbf{a}_k}^H \mathbf{C}_{\mathbf{z}_k}^{-1} \mathbf{\Lambda}_{\mathbf{a}_k}$. From (6.63), it follows that the partial observation can be decomposed as

$$\mathbf{y}_k = \begin{pmatrix} \mathbf{y}_{k-1} \\ y[k] \end{pmatrix} = \begin{pmatrix} \mathbf{\Lambda}_{\mathbf{a}_{k-1}} \\ \boldsymbol{\lambda}_{\mathbf{a}_k}^H \end{pmatrix} \boldsymbol{\theta} + \begin{pmatrix} \mathbf{z}_{k-1} \\ z[k] \end{pmatrix}, \tag{6.89}$$

where $\boldsymbol{\lambda}_{\mathbf{a}_k}^H$ is the last row of $\mathbf{\Lambda}_{\mathbf{a}_k}$. When the noise is white (i.e., $\mathbf{C}_{\mathbf{z}_k} = \sigma_z^2 \mathbf{I}_{k+1}$), it can then be shown that (6.88) and (6.89) can be combined to write the noncoherent metric *recursively*:

$$\begin{aligned} \zeta_{\mathsf{nc}}(\mathbf{c}_k) &= \zeta_{\mathsf{nc}}(\mathbf{c}_{k-1}) + L_c[k]c[k] + \ln \eta_{\mathbf{a}_k} \\ &\quad - \begin{pmatrix} \boldsymbol{\lambda}_{\mathbf{a}_k} y[k] \\ \hat{\boldsymbol{\theta}}_{\mathsf{MMSE}|\mathbf{a}_{k-1}} \end{pmatrix}^H \begin{pmatrix} \sigma_z^{-2} \mathbf{\Sigma}_{\mathbf{a}_k}^{-1} & \mathbf{I} - \eta_{\mathbf{a}_k} \mathbf{\Sigma}_{\mathbf{a}_{k-1}}^{-1} \boldsymbol{\lambda}_{\mathbf{a}_k} \boldsymbol{\lambda}_{\mathbf{a}_k}^H \\ \mathbf{I} - \eta_{\mathbf{a}_k} \boldsymbol{\lambda}_{\mathbf{a}_k} \boldsymbol{\lambda}_{\mathbf{a}_k}^H \mathbf{\Sigma}_{\mathbf{a}_{k-1}}^{-1} & -\eta_{\mathbf{a}_k} \sigma_z^2 \boldsymbol{\lambda}_{\mathbf{a}_k} \boldsymbol{\lambda}_{\mathbf{a}_k}^H \end{pmatrix} \begin{pmatrix} \boldsymbol{\lambda}_{\mathbf{a}_k} y[k] \\ \hat{\boldsymbol{\theta}}_{\mathsf{MMSE}|\mathbf{a}_{k-1}} \end{pmatrix}. \end{aligned} \tag{6.90}$$

In (6.90), $\eta_{\mathbf{a}_k} \triangleq (\sigma_z^2 + \boldsymbol{\lambda}_{\mathbf{a}_k}^H \mathbf{\Sigma}_{\mathbf{a}_{k-1}}^{-1} \boldsymbol{\lambda}_{\mathbf{a}_k})^{-1}$ and $\hat{\boldsymbol{\theta}}_{\mathsf{MMSE}|\mathbf{a}_{k-1}}$ denote the MMSE estimate of $\boldsymbol{\theta}$ from $\mathbf{y}_{k-1}$ under the sequence hypothesis $\mathbf{a}_{k-1}$.

From (6.90), we can make two important observations. First, we know that a trellis-based implementation exists only if the metric update depends on a *fixed* number of past symbols. Although $\boldsymbol{\lambda}_{\mathbf{a}_k}$ depends only on the past N_q symbols $\{a[k], \ldots, a[k-N_q+1]\}$, the terms $\mathbf{\Sigma}_{\mathbf{a}_k}^{-1}$ and $\hat{\boldsymbol{\theta}}_{\mathsf{MMSE}|\mathbf{a}_{k-1}}$ depend, in general, on the full sequence $\mathbf{a}_k$, implying that optimal noncoherent MAPSD/MLSD/GLRT cannot be implemented by a trellis-based technique. Second, when the BEM coefficient trajectories (i.e., $\{\theta[i,d]\}_{i=0}^{I-1}$ for each d) satisfy an order-N_{AR} Gauss–Markov model and the trellis has $|\mathscr{A}|^{N_q+N_{\mathsf{AR}}}$ states, the Kalman filter can be used to recursively compute the MMSE channel estimate $\hat{\boldsymbol{\theta}}_{\mathsf{MMSE}|\mathbf{a}_{k-1}}$ and its error covariance, $\boldsymbol{\Sigma}_{\mathbf{a}_{k-1}}^{-1}$, conditioned on the symbols $\mathbf{a}_{k-1}$ that define each surviving path. Thus, while a trellis can facilitate the computation of the exact partial sequence metric, it cannot guarantee optimal pruning. The literature is not always clear about these points, however. For example, Chugg (1998) points out that some seminal and often-cited works (e.g., Dai & Shwedyk, 1994; Morley & Snyder, 1979) seem to claim that noncoherent MLSD can be implemented with a trellis, and shows precisely why this cannot be the case.

Not surprisingly, trellis-based implementations of the noncoherent MAP symbol and bit criteria (6.65)–(6.66) are also suboptimal. With the forward–backward algorithm,[22] there is no concept of surviving paths, and channel state information is required for each state of the trellis. When the channel is unknown, the trellis can be expanded so that a channel estimate can be calculated at each state, after which the forward–backward algorithm can again be applied, although not optimally: the performance (and complexity) depends on the amount of trellis expansion (Anastasopoulos & Chugg, 2000; Davis, Collings, & Hoeher, 2001; Gertsman & Lodge, 1997; Hart & Pasupathy, 2000).

6.4.3.2 *Noncoherent Equalization through Per-Survivor Processing*

Although trellis-based noncoherent equalization is suboptimal, a trellis can be used for *approximate* noncoherent MAPSD/MLSD/GLRT and MAP symbol and bit estimation.

Of the many practical suboptimal trellis-based schemes that have been proposed, a good number can be classified as *per-survivor processing* (PSP) (Raheli, Polydoros, & Tzou, 1995). There, the idea is to compute (at stage k of the trellis) a separate channel estimate $\hat{\boldsymbol{\theta}}_{\mathbf{a}_k}$ for each surviving path extension $\mathbf{a}_k$, and then evaluate a partial metric corresponding to the pair $(\mathbf{a}_k, \hat{\boldsymbol{\theta}}_{\mathbf{a}_k})$. Only the surviving path extensions leading to the best metrics are retained as survivors, after which the process repeats at the next stage of the trellis. As discussed earlier, an $|\mathscr{A}|^{N_q+N_{\mathsf{AR}}}$-state trellis facilitates recursive MMSE estimation of order-N_{AR} Gauss–Markov BEM trajectories through Kalman filtering, and thus recursive computation of MAPSD and MLSD partial metrics (recalling (6.73)). Thus, for that channel class, the Viterbi algorithm with per-survivor Kalman filtering provides near-optimal noncoherent MAPSD/MLSD. This idea seems to have been first proposed in Morley & Snyder (1979) using continuous-time filtering. Lodge & Moher (1990) considered discrete-time filters and realized that, if the observations are first whitened (which requires only a bank of LTI filters), then the metric calculation simplifies in a way that eliminates the need for Kalman filtering. This latter approach is known as the "innovations" approach. Similar ideas can be applied to fixed-lag MAP symbol/bit estimation processing, as proposed by Iltis, Shynk, & Giridhar (1994).

Because the complexity of these trellis-based PSP methods grows exponentially in $N_q + N_{\mathsf{AR}}$, however, PSP methods based on more general tree-searches may be more practical for near-optimal noncoherent detection, especially at high SNR, where sphere decoders can find the optimal solution without visiting many nodes. In fact, the proposal of noncoherent tree-search can already be found in early works, e.g., Dai & Shwedyk (1994). Notice that, with appropriate definition of the metric, the

[22]For fixed-lag MAP symbol estimates, the situation is a bit different because the forward–backward algorithm does not apply. There, the posterior symbol probabilities can be calculated recursively (assuming Gauss–Markov BEM trajectories), but they require averaging over all possible past-symbol sequences and thus cannot be folded into a trellis (Zhang et al. 1997).

tree-search methods discussed in Section 6.3.2 – in the context of coherent equalization – apply here too, except that the preprocessing used there now becomes unnecessary because the model (6.63) is already causal. A further advantage of tree-search is that it does not require BEM coefficients to satisfy a Gauss–Markov property, which can be useful in, e.g., multicarrier applications. In any case, the key to a computationally efficient tree-search is minimizing both the number of nodes visited and the complexity per visited node. In regards to the latter, Hwang & Schniter (2007, November) have shown that the quantities $\hat{\boldsymbol{\theta}}_{\mathsf{MMSE}|\mathbf{a}_{k-1}}$ and $\boldsymbol{\Sigma}_{\mathbf{a}_k}^{-1}$ can be updated recursively (for the generic modulation/demodulation and BEM setup of (6.63)), yielding an $\mathcal{O}(N_q^2 I^2)$ update to the noncoherent metric (6.90).

6.4.3.3 *Iterative Noncoherent Equalization through the EM Algorithm*

The *expectation-maximization* (EM) algorithm (Dempster, Laird, & Rubin, 1977) is a well-known approach for ML estimation in the presence of "missing data." If $\mathbf{y}$ is the observation, $\mathbf{x}$ is the vector to be estimated, and $\mathbf{u}$ is the "missing data," then the EM algorithm attempts to find $\hat{\mathbf{x}}_{\mathsf{ML}} = \arg\max_{\mathbf{x}} f(\mathbf{y}|\mathbf{x}) = \arg\max_{\mathbf{x}} \ln f(\mathbf{y}|\mathbf{x})$ iteratively using the following recursion[23] (where i denotes the iteration index):

$$\hat{\mathbf{x}}^{(i+1)} = \arg\max_{\mathbf{x}} \int f(\mathbf{u} \mid \mathbf{y}, \hat{\mathbf{x}}^{(i)}) \ln f(\mathbf{y}, \mathbf{u} \mid \mathbf{x}) d\mathbf{u}. \tag{6.91}$$

In Appendix 6.C, we show that (6.91) arises from the goal of maximally increasing the likelihood at each iteration. So called *generalized EM* algorithms, which increase but do not necessarily maximally increase the likelihood at each iteration, have also been proposed (e.g., Fessler & Hero, 1994). The EM recursion is sometimes regarded as having two separate steps: an "E step," which computes the conditional expectation (i.e., the integral) in (6.91), and an "M step," which performs the maximization in (6.91). A *Bayesian EM* (EMB) algorithm, with the goal to find $\hat{\mathbf{x}}_{\mathsf{MAP}} = \max_{\mathbf{x}} f(\mathbf{x} \mid \mathbf{y})$, follows by direct extension. Using Bayes rule and disregarding irrelevant terms, we can write $\hat{\mathbf{x}}_{\mathsf{MAP}} = \arg\max_{\mathbf{x}} \{\ln f(\mathbf{y} \mid \mathbf{x}) + \ln f(\mathbf{x})\}$ and attempt to find $\hat{\mathbf{x}}_{\mathsf{MAP}}$ using the recursion

$$\hat{\mathbf{x}}^{(i+1)} = \arg\max_{\mathbf{x}} \left\{ \int f(\mathbf{u} \mid \mathbf{y}, \hat{\mathbf{x}}^{(i)}) \ln f(\mathbf{y}, \mathbf{u} \mid \mathbf{x}) d\mathbf{u} + \ln f(\mathbf{x}) \right\}. \tag{6.92}$$

One can immediately think of two ways that the EM(B) algorithms could be applied to noncoherent equalization: (1) the coded bits could be estimated while treating the channel as missing (i.e., "info EM(B)") (Georghiades & Han, 1997) or (2) the channel could be estimated, while treating the data as missing, and later used for coherent sequence detection (i.e., "channel EM(B)") (Antón-Haro, Fonollosa, & Fonollosa, 1997; Chiavaccini & Vitetta, 2001; Cirpan & Tsatsanis, 2001; Kaleh & Vallet, 1994; Nguyen & Levy, 2005; Nissilä & Pasupathy, 2003; Yan & Rao, 2003). EM(B) algorithms that treat both channel and data values as parameters to be estimated have also been proposed (e.g., Zamiri-Jafarian & Pasupathy, 1999).

We now describe the *channel EM(B)* algorithm for noncoherent equalization. (See Appendix 6.D for a discussion of the less practical *info EM(B)* algorithm.) For this, we use the model $\mathbf{y} = \boldsymbol{\Lambda}_{\mathbf{a}}\boldsymbol{\theta} + \mathbf{z}$ from (6.63), where we once again find it convenient to assume BPSK in order to ensure a one-to-one

[23] If the missing data $\mathbf{u}$ was discrete, integration in (6.91) would be replaced by summation.

correspondence between symbols **a** and bits **c**. For noncoherent equalization, $\boldsymbol{\theta}$ is the vector to estimate and **c** is the missing data, so that (from (6.92)) channel EMB[24] performs the recursion

$$\hat{\boldsymbol{\theta}}^{(i+1)} = \arg\max_{\hat{\boldsymbol{\theta}}} \left\{ \sum_{\mathbf{c}\in\{0,1\}^K} p(\mathbf{c} \mid \mathbf{y}, \hat{\boldsymbol{\theta}}^{(i)}) \ln f(\mathbf{y}, \mathbf{c} \mid \hat{\boldsymbol{\theta}}) + \ln f(\hat{\boldsymbol{\theta}}) \right\}. \tag{6.93}$$

Using the property $\ln f(\mathbf{y}, \mathbf{c} \mid \boldsymbol{\theta}) = \ln f(\mathbf{y} \mid \mathbf{c}, \boldsymbol{\theta}) + \ln f(\mathbf{c})$ in conjunction with the Rayleigh fading and Gaussian noise assumptions, (6.93) reduces to

$$\hat{\boldsymbol{\theta}}^{(i+1)} = \arg\min_{\hat{\boldsymbol{\theta}}} \left\{ \sum_{\mathbf{c}\in\{0,1\}^K} p(\mathbf{c} \mid \mathbf{y}, \hat{\boldsymbol{\theta}}^{(i)}) \|\mathbf{y} - \mathbf{A_a}\hat{\boldsymbol{\theta}}\|^2_{\mathbf{C_z}^{-1}} + \hat{\boldsymbol{\theta}}^H \mathbf{C}_{\theta}^{-1} \hat{\boldsymbol{\theta}} \right\} \tag{6.94}$$

$$= \left(\mathbf{C}_{\theta}^{-1} + \sum_{\mathbf{c}\in\{0,1\}^K} p(\mathbf{c} \mid \mathbf{y}, \hat{\boldsymbol{\theta}}^{(i)}) \mathbf{A_a}^H \mathbf{C_z}^{-1} \mathbf{A_a} \right)^{-1} \sum_{\mathbf{c}\in\{0,1\}^K} p(\mathbf{c} \mid \mathbf{y}, \hat{\boldsymbol{\theta}}^{(i)}) \mathbf{A_a}^H \mathbf{C_z}^{-1} \mathbf{y}. \tag{6.95}$$

Above, **a** denotes the symbol vector corresponding to the bit vector **c**. From (6.95), channel EM(B) can be interpreted as performing *iterative soft decision-directed channel estimation*, using soft decisions computed from the previous channel estimate. In fact, with constant modulus (CM) $\mathscr{A}$ and white noise, the summed terms in (6.95) can be rewritten using the tanh(·) operator, reminiscent of (6.40). Soft decision-directed channel estimation can be contrasted with per-survivor channel estimation, as used in noncoherent MAPSD/MLSD and GLRT, in that soft decision-directed channel estimation generates a *single* channel estimate after "averaging" the soft bit estimates, whereas per-survivor channel estimation generates *multiple* channel estimates, one for each hypothesized bit sequence. Iterative soft decision-directed channel estimation has also been considered outside of the EM(B) context in, e.g., Baccarelli & Cusani (1998), Fang et al. (2008), Liu & Fitz (2008), Otnes & Tüchler (2004), Song, Singer, & Sung (2004).

The posterior bit probabilities $\{p(\mathbf{c} \mid \mathbf{y}, \hat{\boldsymbol{\theta}}^{(i)})\}_{\mathbf{c}\in\{0,1\}^K}$ required for (6.95) can be obtained in various ways. For the case of BEM trajectories that satisfy a Gauss–Markov model, the forward–backward algorithm can be used (Kaleh & Vallet, 1994). A different approach was proposed in Hwang & Schniter (2009) that allows the use of a priori LLRs and more general BEM statistics. We now briefly describe this approach. Writing

$$p(\mathbf{c} \mid \mathbf{y}, \hat{\boldsymbol{\theta}}^{(i)}) = \frac{f(\mathbf{y} \mid \mathbf{c}, \hat{\boldsymbol{\theta}}^{(i)}) p(\mathbf{c})}{\sum_{\mathbf{c}'\in\{0,1\}^K} f(\mathbf{y} \mid \mathbf{c}', \hat{\boldsymbol{\theta}}^{(i)}) p(\mathbf{c}')} = \frac{e^{\zeta_{\mathsf{coh}}(\mathbf{c}; \hat{\boldsymbol{\theta}}^{(i)})}}{\sum_{\mathbf{c}'\in\{0,1\}^K} e^{\zeta_{\mathsf{coh}}(\mathbf{c}'; \hat{\boldsymbol{\theta}}^{(i)})}}, \tag{6.96}$$

where we have used the coherent MAP sequence metric $\zeta_{\mathsf{coh}}(\mathbf{c}; \hat{\boldsymbol{\theta}}^{(i)}) = \ln f(\mathbf{y} \mid \mathbf{c}, \hat{\boldsymbol{\theta}}^{(i)}) + \mathbf{l}_c^T \mathbf{c}$ from (6.34) with the $\hat{\boldsymbol{\theta}}^{(i)}$-dependence explicitly noted, the EM recursion (6.95) can be restated as

$$\hat{\boldsymbol{\theta}}^{(i+1)} = \left(\sum_{\mathbf{c}\in\{0,1\}^K} e^{\zeta_{\mathsf{coh}}(\mathbf{c}; \hat{\boldsymbol{\theta}}^{(i)})} (\mathbf{C}_{\theta}^{-1} + \mathbf{A_a}^H \mathbf{C_z}^{-1} \mathbf{A_a}) \right)^{-1} \sum_{\mathbf{c}\in\{0,1\}^K} e^{\zeta_{\mathsf{coh}}(\mathbf{c}; \hat{\boldsymbol{\theta}}^{(i)})} \mathbf{A_a}^H \mathbf{C_z}^{-1} \mathbf{y}. \tag{6.97}$$

[24]Channel EM would yield (6.93) without the $\ln f(\hat{\boldsymbol{\theta}})$ term and hence (6.94)–(6.97) without the $\mathbf{C}_{\theta}^{-1}$ term. This relationship is reminiscent of that between GLRT and noncoherent MLSD.

Equation (6.97) suggests iterating a soft decision-directed channel estimator, with input $\{\zeta_{\text{coh}}(\mathbf{c};\hat{\boldsymbol{\theta}}^{(i)})\}_{\mathbf{c}\in\{0,1\}^K}$ and output $\hat{\boldsymbol{\theta}}^{(i+1)}$, and a soft-input/soft-output coherent equalizer, with input $\hat{\boldsymbol{\theta}}^{(i)}$ and output $\{\zeta_{\text{coh}}(\mathbf{c};\hat{\boldsymbol{\theta}}^{(i)})\}_{\mathbf{c}\in\{0,1\}^K}$. Together, the pair forms a soft-input/soft-output *non*coherent equalizer, which could be iterated with a soft-input/soft-output decoder for turbo reception.

When the channel is frequency-nonselective, the noise is white, and the BEM is trivial, $\ln f(\mathbf{y} \mid \mathbf{c},\hat{\boldsymbol{\theta}}^{(i)}) = C - \frac{1}{\sigma_w^2}\sum_{k=0}^{K-1}\left|y[k]-c[k]\hat{\theta}[k,0]\right|^2$, so that the 2^K term summations in (6.97) decouple into K binary summations (Chiavaccini & Vitetta, 2001), greatly simplifying the evaluation of (6.97). In the general case, the 2^K-term summations do not decouple, but not all 2^K metrics $\{\zeta_{\text{coh}}(\mathbf{c};\hat{\boldsymbol{\theta}}^{(i)})\}_{\mathbf{c}\in\{0,1\}^K}$ need to be calculated because very few of them yield non-negligible $e^{\zeta_{\text{coh}}(\mathbf{c};\hat{\boldsymbol{\theta}}^{(i)})}$. The dominant posterior probabilities can be found without too much effort using, e.g., M-algorithm tree-search (Hwang & Schniter, 2009), as discussed for the coherent case in Section 6.3.2.

6.4.3.4 Other Noncoherent Equalization Schemes

Other approaches to noncoherent equalization exist as well. For example, Anastasopoulos, Chugg, Colavolpe, Ferrari, & Raheli (2007) and Motedayen-Aval, Krishnamoorthy, & Anastasopoulos (2007) applied message-passing algorithms to MLSD and MAP symbol detection over time-selective flat-fading channels. They have shown that, under certain conditions, MLSD complexity scales as $\mathcal{O}(K^{2\text{rank}\{\mathbf{C}_{\boldsymbol{\theta}}\}})$. Because of space limitations, these techniques will not be discussed here. One can also imagine ad hoc combination of decoupled (coherent) equalization and channel estimation. Examples will be provided in the sequel.

6.4.4 Noncoherent Equalization for Single-Carrier Modulation/Demodulation

When single-carrier modulation/demodulation is used, the effective channel coincides with the propagation channel (i.e., $q[l,d]=h[l,d]\ \forall l,d$ and $N_q=M$), whose trajectories $\{h[n,m]\}_{n=0}^{N-1}$ (for each $m\in\{0,\ldots,M-1\}$) are well described by a Gauss–Markov model of suitable order N_{AR}. Thus, when the trivial BEM is used to write (6.2), so that $\theta[i,d]=q[i,d]\ \forall i,d$, the BEM trajectories themselves are well described by a Gauss–Markov model of order N_{AR}. In this case, and assuming the noise $\mathbf{z}$ is white, an $|\mathscr{A}|^{M+N_{\text{AR}}}$-state trellis can be used to implement near-optimal noncoherent MAPSD/MLSD, as well as near-optimal MAP symbol and bit detection, as described in Section 6.4.3. In fact, these ideas dominated much of the early literature on noncoherent equalization of rapidly TV channels.

6.4.4.1 Near-Optimal Trellis-PSP Equalization for Single-Carrier Schemes

As mentioned earlier, Lodge & Moher (1990) were some of the first authors to propose a near-optimal trellis-based implementation of noncoherent MLSD. In particular, they proposed to use the Viterbi algorithm with per-branch linear prediction for MLSD of CM signals with ARMA time-selective channels. Soon after, Iltis (1992) proposed to use the Viterbi algorithm in conjunction with an extended Kalman filter for per-survivor joint estimation of symbol timing offset and an AR doubly selective channel. Dai & Shwedyk (1994) proposed similar near-optimal trellis-based implementations of noncoherent MLSD for general signal alphabets and ARMA doubly selective channels, using per-branch Kalman filtering. Yu & Pasupathy (1995) then extended Lodge & Moher (1990) to general signal alphabets and ARMA doubly selective channels. The latter technique was extended further to carrier-frequency-offset Rician channels by Hart & Taylor (1998).

In related work, Gertsman & Lodge (1997) showed that the forward–backward algorithm, with per-branch linear prediction, can be used for near-MAP symbol detection under AR time-selective channels and CM alphabets. Independently, these ideas were generalized to doubly selective channels and general signal alphabets by Hart & Pasupathy (2000) and Davis et al. (2001). For fixed-lag MAP symbol estimation, Zhang et al. (1997) proposed to use per-survivor Kalman filtering, echoing earlier work by Iltis et al. (1994). Anastasopoulos & Chugg (2000) then presented two general families of trellis algorithms, one based on parameter-first combining and the other based on sequence-first combining, that yield both forward–backward and fixed-lag algorithms.

6.4.4.2 Reduced-Complexity Trellis-PSP Equalization for Single-Carrier Schemes

Because of the complexity of near-optimal PSP methods, which are typically based on per-sequence Kalman filtering, simpler PSP techniques have also been proposed based on simpler forms of adaptive filtering, such as the RLS[25] and LMS algorithms. For example, Kubo, Murakami, & Fujino (1994) proposed to use the Viterbi algorithm in conjunction with the LMS algorithm (Haykin, 2001) for per-survivor channel estimation, whereas Raheli et al. (1995) proposed to use the Viterbi algorithm in conjunction with the RLS algorithm (Haykin, 2001). Other LMS and RLS approaches were discussed in Anastasopoulos & Chugg (2000).

While the previously described PSP algorithms assumed a trivial BEM, PSP for more general BEMs has also been considered. For example, trellis-based PSP algorithms for joint estimation of symbols and polynomial BEM (Borah & Hart, 1999a) coefficients were proposed, for time-selective channels, by Borah & Hart (1999c) and Leon & Taylor (2003). DFE and trellis methods for both polynomial and Karhunen-Loève BEMs (Borah & Hart, 1999b) were studied by Borah & Hart (1999b) for doubly selective channels. Trellis-based PSP using a complex-exponential BEM (Tsatsanis & Giannakis, 1996) was discussed by El-Mahdy (2004).

6.4.4.3 Near-Optimal Tree-PSP Equalization for Single-Carrier Schemes

Because the trellis-based approaches to noncoherent equalization typically use an $(M+N_{\mathsf{AR}})$-state trellis, with $\mathcal{O}(K|\mathscr{A}|^{M+N_{\mathsf{AR}}})$ complexity, they are impractical for all but very short delay spreads. Tree-search-based PSP is one way to circumvent this complexity. In one of the earliest proposals, Dai & Shwedyk (1994) suggested to use a Fano-like tree-search with per-survivor Kalman estimation to noncoherently equalize a doubly selective ARMA channel (assuming a trivial BEM). The method in Zhang et al. (1997) can be considered as using the T-algorithm to obtain symbol-MAP fixed-lag metrics for the same channel. For a doubly selective channel modeled by a generic BEM, Hwang & Schniter proposed PSP-based noncoherent M-algorithm tree-searches that accomplish approximate MLSD, in Hwang & Schniter (2007, September), and approximate MAP bit detection, in Hwang & Schniter

[25]Recall that the ML channel estimate (6.77) reduces to an LS estimate in the case of white noise, which can be computed recursively using RLS.

(2007, November). The latter, with complexity $\mathcal{O}(KM^2I^2)$, was combined with soft decoding in a turbo receiver.

6.4.4.4 *Iterative Noncoherent Equalization for Single-Carrier Schemes*

For EM-based iterative noncoherent equalization for single-carrier systems, the channel EM(B) algorithm described in Section 6.4.3 is the most popular approach; the info EM(B) algorithm, proposed by Georghiades & Han (1997) for AR time-selective fading and CM signaling and described in Appendix 6.D, was found, in the more recent studies (Chen, Perry, & Buckley, 2003; Yan & Rao, 2003), to have convergence problems.

One of the first applications of the channel EMB algorithm to noncoherent equalization of frequency-selective channels was given by Kaleh & Vallet (1994). Antón-Haro, Fonollosa, & Fonollosa (1997) proposed a channel EM algorithm for doubly selective channels that used a polynomial BEM, assuming CM signaling. More recently, Yan & Rao (2003) proposed a channel EMB method for AR-1 time-selective channels and CM signaling that uses a Kalman filter. Nissilä & Pasupathy (2003) generalized these ideas to AR doubly selective channels and arbitrary constellations through the use of Kalman smoothing. All of these approaches used the forward–backward algorithm to evaluate the posterior bit probabilities in (6.95). Because the number of trellis states in the forward–backward algorithm is $|\mathcal{A}|^{M+N_{\mathsf{AR}}}$, these approaches are practical for only very short delay spread M.

To circumvent the complexity of trellis processing, Hwang & Schniter (2009) proposed to use suboptimal tree-search to compute the dominant posterior bit probabilities, using (6.96)–(6.97), leading to a complexity of only $\mathcal{O}(KM^2I^2)$. This approach can be considered an extension of the technique originally proposed by Chiavaccini & Vitetta (2001) for a time-selective channel and trivial BEM, to doubly selective channels modeled by generic BEMs.

6.4.5 Noncoherent Equalization for Time-Frequency Concentrated Modulation/Demodulation

We saw, in the previous section, that the use of single-carrier modulation/demodulation facilitated Gauss–Markov modeling of the effective channel trajectory $\{q[l,d]\}_{l=0}^{K-1}$. Time-frequency concentrated modulation/demodulation schemes[26] generally do not facilitate the use of a Gauss–Markov model with order $N_{\mathsf{AR}} \ll K$ because $q[l,d]$ can change very quickly in l. For example, with multicarrier schemes, $\{q[l,0]\}_{l=0}^{K-1}$ represents the channel frequency response, which may exhibit deep and sudden nulls. Thus, the trellis-based approaches to noncoherent equalization (whether optimal, PSP approximate, or EM-based) do not apply here. For this reason, the literature on noncoherent equalization for time-frequency concentrated modulation/demodulation schemes is somewhat sparse.

For time-frequency concentrated modulation/demodulation, the dimensionality of the effective channel response $\{q[l,d]\}$ is more efficiently reduced by a BEM, e.g., the complex-exponential BEM (as proposed in the classical OFDM work (Edfors, Sandell, van de Beek, Wilson, & Börjesson, 1998)).

[26] In this section, we will include OFDM in the "time-frequency concentrated" class under the assumption that the Doppler spread is mild enough to guarantee a short intercarrier interference spread.

Cui & Tellambura (2007) applied the complex exponential (CE)-BEM, made several approximations to the noncoherent MLSD metric in (6.69) to reduce it to a simple quadratic form $\mathbf{a}^H\mathbf{R}\mathbf{a}$ (with **a**-independent **R**), and then used tree-search to find the optimal $\mathbf{a} \in \mathscr{A}^K$, all under the assumption that $\mathscr{A}$ was constant modulus. Hwang & Schniter took a more direct approach, leveraging the CE-BEM to design PSP-based and channel EMB-based noncoherent MAP bit equalization algorithms in Hwang & Schniter (2008) and Hwang & Schniter (2009), respectively, whose complexities scale as $\mathcal{O}\big(K(2D+1)^2I^2\big)$. The key to these low complexities is the use of a fast metric update. Here, the BEM dimension I refers to the number of *active* channel taps; it is typical that $I \ll M$ when the delay power profile is *sparse*. These latter algorithms achieve near-MAP performance with a complexity that is quite reasonable, even for large simultaneous channel delay and Doppler spreads.

6.5 CONCLUSION

In this chapter, we have given a broad overview of coherent and noncoherent equalization for rapidly TV channels, focusing on the case of significant delay spread. To better understand the problem, we described the combined effect of modulation, channel propagation, and demodulation using an effective channel matrix **Q**, and then examined the key features of **Q**. We found that the support of the significant coefficients within **Q** can be described as widely quasibanded when single-carrier modulation/demodulation is used, and narrowly quasibanded when time-frequency concentrated modulation/demodulation is used. This structure of **Q** was later used to explain the design of low-complexity equalization algorithms.

We then discussed coherent equalization, where **Q** is assumed to be known. Various equalization criteria were described, including those based on ML, MAP, MMSE, and the computation of posterior LLRs. Equalization tools were described next, including trellis-based, linear, decision feedback, tree-search based, and iterative methods. We then described how these criteria and tools have been applied to the design of coherent equalizers for time-frequency concentrated modulation/demodulation over rapidly TV channels, highlighting fast serial and fast joint equalization schemes. For equalization for single-carrier modulation/demodulation over rapidly TV channels, we focused on frequency-domain equalization approaches that yield high performance with low complexity.

Finally, we discussed noncoherent equalization, where **Q** is assumed unknown (although sometimes its statistics are known). For this, the system model was reformulated to accommodate an efficient BEM-based parameterization of the effective channel **Q**. Various equalization criteria were described, including those based on ML, MAP, GLRT, MMSE, and posterior LLRs. Equalization tools were described next, including those based on trellis, tree search, per-survivor processing, and the EM algorithm. We then described how these criteria and tools have been applied to the design of noncoherent equalizers for single-carrier modulation/demodulation over rapidly TV channels. While the traditional approach was to leverage a Gauss–Markov fading model for the channel trajectory, general BEM approaches have been developed more recently. Finally, we described noncoherent equalization for time-frequency concentrated modulation/demodulation over rapidly TV channels, a problem which has received attention only recently.

Appendices

6.A DERIVATION OF POSTERIOR LLR EXPRESSION (6.36)

Applying Bayes rule to the numerator of the posterior LLR in (6.19), we find

$$\Pr\{c[j]=1 \mid \mathbf{y},\mathbf{Q}\} = \sum_{\mathbf{c}:c[j]=1} p(\mathbf{c} \mid \mathbf{y},\mathbf{Q}) = \sum_{\mathbf{c}:c[j]=1} \frac{f(\mathbf{y} \mid \mathbf{c},\mathbf{Q})p(\mathbf{c} \mid \mathbf{Q})}{f(\mathbf{y} \mid \mathbf{Q})}. \tag{6.98}$$

As in the text, all possibilities of bit vectors $\mathbf{c} \in \{0,1\}^{K\log_2|\mathscr{A}|}$ are considered in the summations, not only those in the codebook. Doing the same to the denominator of (6.19) and then taking the log of their ratio yields

$$L_{c|\mathbf{y},\mathbf{Q}}[j] = \ln \frac{\sum_{\mathbf{c}:c[j]=1} f(\mathbf{y} \mid \mathbf{c},\mathbf{Q})p(\mathbf{c} \mid \mathbf{Q})}{\sum_{\mathbf{c}:c[j]=0} f(\mathbf{y} \mid \mathbf{c},\mathbf{Q})p(\mathbf{c} \mid \mathbf{Q})}. \tag{6.99}$$

Assuming independent coded bits, so that $p(\mathbf{c} \mid \mathbf{Q}) = p(\mathbf{c}) = \prod_{j'=0}^{K\log_2|\mathscr{A}|-1} p(c[j'])$, and using the identity

$$p(c[j']) = \frac{\exp\left((c[j']-1)L_c[j']\right)}{1+\exp(-L_c[j'])} \quad \text{for} \quad c[j'] \in \{0,1\}, \tag{6.100}$$

we can rewrite (6.99) as

$$L_{c|\mathbf{y},\mathbf{Q}}[j] = \ln \frac{\sum_{\mathbf{c}:c[j]=1} f(\mathbf{y} \mid \mathbf{c},\mathbf{Q}) \exp(\mathbf{l}_c^T \mathbf{c})}{\sum_{\mathbf{c}:c[j]=0} f(\mathbf{y} \mid \mathbf{c},\mathbf{Q}) \exp(\mathbf{l}_c^T \mathbf{c})} \tag{6.101}$$

for $\mathbf{l}_c \triangleq (L_c[0] \cdots L_c[K-1])^T$. Finally, writing the LLR expression (6.101) in terms of the MAP metric (6.34) yields (6.36).

6.B DERIVATION OF THE NONCOHERENT MLSD EXPRESSION (6.73)

Using (6.68), we can write the first term in (6.69) as

$$\mathbf{y}^H \mathbf{C}_{\mathbf{y}|\mathbf{a}}^{-1} \mathbf{y} = \mathbf{y}^H (\mathbf{C}_\mathbf{z} + \mathbf{\Lambda}_\mathbf{a} \mathbf{C}_\theta \mathbf{\Lambda}_\mathbf{a}^H)^{-1} \mathbf{y}. \tag{6.102}$$

Applying the matrix inversion lemma, and introducing a pair of terms that sum to zero,

$$\begin{aligned}
\mathbf{y}^H \mathbf{C}_{\mathbf{y}|\mathbf{a}}^{-1} \mathbf{y} &= \mathbf{y}^H \mathbf{C}_\mathbf{z}^{-1} \mathbf{y} - \mathbf{y}^H \mathbf{C}_\mathbf{z}^{-1} \mathbf{\Lambda}_\mathbf{a} (\mathbf{C}_\theta^{-1} + \mathbf{\Lambda}_\mathbf{a}^H \mathbf{C}_\mathbf{z}^{-1} \mathbf{\Lambda}_\mathbf{a})^{-1} \mathbf{\Lambda}_\mathbf{a}^H \mathbf{C}_\mathbf{z}^{-1} \mathbf{y} \\
&\quad - \mathbf{y}^H \mathbf{C}_\mathbf{z}^{-1} \mathbf{\Lambda}_\mathbf{a} (\mathbf{C}_\theta^{-1} + \mathbf{\Lambda}_\mathbf{a}^H \mathbf{C}_\mathbf{z}^{-1} \mathbf{\Lambda}_\mathbf{a})^{-1} \mathbf{\Lambda}_\mathbf{a}^H \mathbf{C}_\mathbf{z}^{-1} \mathbf{y} \\
&\quad + \mathbf{y}^H \mathbf{C}_\mathbf{z}^{-1} \mathbf{\Lambda}_\mathbf{a} (\mathbf{C}_\theta^{-1} + \mathbf{\Lambda}_\mathbf{a}^H \mathbf{C}_\mathbf{z}^{-1} \mathbf{\Lambda}_\mathbf{a})^{-1} \mathbf{\Lambda}_\mathbf{a}^H \mathbf{C}_\mathbf{z}^{-1} \mathbf{y}.
\end{aligned} \tag{6.103}$$

Plugging in the expression for $\hat{\theta}_{\text{MMSE}|\mathbf{a}}$ in (6.72),

$$\mathbf{y}^H\mathbf{C}_{\mathbf{y}|\mathbf{a}}^{-1}\mathbf{y} = \mathbf{y}^H\mathbf{C}_{\mathbf{z}}^{-1}\mathbf{y} - \hat{\theta}_{\text{MMSE}|\mathbf{a}}^H\mathbf{\Lambda}_{\mathbf{a}}^H\mathbf{C}_{\mathbf{z}}^{-1}\mathbf{y} - \mathbf{y}^H\mathbf{C}_{\mathbf{z}}^{-1}\mathbf{\Lambda}_{\mathbf{a}}\hat{\theta}_{\text{MMSE}|\mathbf{a}}$$

$$+\hat{\theta}_{\text{MMSE}|\mathbf{a}}^H(\mathbf{C}_{\theta}^{-1} + \mathbf{\Lambda}_{\mathbf{a}}^H\mathbf{C}_{\mathbf{z}}^{-1}\mathbf{\Lambda}_{\mathbf{a}})\hat{\theta}_{\text{MMSE}|\mathbf{a}} \tag{6.104}$$

$$= \|\mathbf{y} - \mathbf{\Lambda}_{\mathbf{a}}\hat{\theta}_{\text{MMSE}|\mathbf{a}}\|_{\mathbf{C}_{\mathbf{z}}^{-1}}^2 + \|\hat{\theta}_{\text{MMSE}|\mathbf{a}}\|_{\mathbf{C}_{\theta}^{-1}}^2. \tag{6.105}$$

6.C EXPLANATION OF EM RECURSION (6.91)

Given an estimate $\hat{\mathbf{x}}^{(i)}$ at the ith iteration, the EM algorithm (Dempster et al., 1977) attempts to find $\mathbf{x}$, which maximizes the *increase* in log-likelihood, i.e.,

$$\ln f(\mathbf{y} \mid \mathbf{x}) - \ln f(\mathbf{y} \mid \hat{\mathbf{x}}^{(i)})$$

$$= \ln \int f(\mathbf{y},\mathbf{u} \mid \mathbf{x})d\mathbf{u} - \ln f(\mathbf{y} \mid \hat{\mathbf{x}}^{(i)}) \tag{6.106}$$

$$= \ln \int f(\mathbf{u} \mid \mathbf{y},\hat{\mathbf{x}}^{(i)})\frac{f(\mathbf{y},\mathbf{u} \mid \mathbf{x})}{f(\mathbf{u} \mid \mathbf{y},\hat{\mathbf{x}}^{(i)})}d\mathbf{u} - \ln f(\mathbf{y} \mid \hat{\mathbf{x}}^{(i)}) \tag{6.107}$$

$$\geq \int f(\mathbf{u} \mid \mathbf{y},\hat{\mathbf{x}}^{(i)})\ln\frac{f(\mathbf{y},\mathbf{u} \mid \mathbf{x})}{f(\mathbf{u} \mid \mathbf{y},\hat{\mathbf{x}}^{(i)})}d\mathbf{u} - \ln f(\mathbf{y} \mid \hat{\mathbf{x}}^{(i)}) \triangleq \Delta(\mathbf{x} \mid \hat{\mathbf{x}}^{(i)}), \tag{6.108}$$

where Jensen's inequality was used in (6.108). From (6.108), it can be seen that $\hat{\mathbf{x}}^{(i+1)} \triangleq \arg\max_{\mathbf{x}} \Delta(\mathbf{x} \mid \hat{\mathbf{x}}^{(i)})$ can be written as (6.91) after dropping nonessential terms. Because

$$\Delta(\hat{\mathbf{x}}^{(i)} \mid \hat{\mathbf{x}}^{(i)}) = \int f(\mathbf{u} \mid \mathbf{y},\hat{\mathbf{x}}^{(i)})\ln\frac{f(\mathbf{y},\mathbf{u} \mid \hat{\mathbf{x}}^{(i)})}{f(\mathbf{u} \mid \mathbf{y},\hat{\mathbf{x}}^{(i)})}d\mathbf{u} - \ln f(\mathbf{y} \mid \hat{\mathbf{x}}^{(i)}) \tag{6.109}$$

$$= \int f(\mathbf{u} \mid \mathbf{y},\hat{\mathbf{x}}^{(i)})\ln\frac{f(\mathbf{y},\mathbf{u} \mid \hat{\mathbf{x}}^{(i)})}{f(\mathbf{u} \mid \mathbf{y},\hat{\mathbf{x}}^{(i)})}d\mathbf{u}$$

$$+\ln\frac{1}{f(\mathbf{y} \mid \hat{\mathbf{x}}^{(i)})}\int f(\mathbf{u} \mid \mathbf{y},\hat{\mathbf{x}}^{(i)})d\mathbf{u} \tag{6.110}$$

$$= \int f(\mathbf{u} \mid \mathbf{y},\hat{\mathbf{x}}^{(i)})\ln\underbrace{\frac{f(\mathbf{y},\mathbf{u} \mid \hat{\mathbf{x}}^{(i)})}{f(\mathbf{u} \mid \mathbf{y},\hat{\mathbf{x}}^{(i)})f(\mathbf{y} \mid \hat{\mathbf{x}}^{(i)})}}_{=1}d\mathbf{u} \tag{6.111}$$

$$= 0, \tag{6.112}$$

if follows that the increase in log-likelihood associated with the EM estimate $\hat{\mathbf{x}}^{(i+1)}$ equals $\Delta(\hat{\mathbf{x}}^{(i+1)} \mid \hat{\mathbf{x}}^{(i)}) = \max_{\mathbf{x}} \Delta(\mathbf{x} \mid \hat{\mathbf{x}}^{(i)}) \geq \Delta(\hat{\mathbf{x}}^{(i)} \mid \hat{\mathbf{x}}^{(i)}) = 0$, and hence the EM recursion never decreases the log likelihood. Thus, when the likelihood $f(\mathbf{y} \mid \mathbf{x})$ is unimodal in $\mathbf{x}$, the EM recursions will converge to $\hat{\mathbf{x}}_{\text{ML}}$.

6.D INFO EM(B) ALGORITHMS FOR NONCOHERENT EQUALIZATION

In info EMB, the bits $\mathbf{c}$ are estimated while treating the channel parameters $\boldsymbol{\theta}$ as missing data, so that (6.92) becomes (with $\mathbf{x}=\mathbf{c}$ and $\mathbf{u}=\boldsymbol{\theta}$)

$$\hat{\mathbf{c}}^{(i+1)} = \arg \max_{\mathbf{c}\in\{0,1\}^K} \left\{ \int f(\boldsymbol{\theta} \mid \mathbf{y}, \hat{\mathbf{c}}^{(i)}) \ln f(\mathbf{y},\boldsymbol{\theta} \mid \mathbf{c}) d\boldsymbol{\theta} + \ln p(\mathbf{c}) \right\}. \tag{6.113}$$

The identity $\ln f(\mathbf{y},\boldsymbol{\theta} \mid \mathbf{c}) = \ln f(\mathbf{y} \mid \boldsymbol{\theta},\mathbf{c}) + \ln f(\boldsymbol{\theta})$, in conjunction with the Gaussian noise assumption, yields

$$\hat{\mathbf{c}}^{(i+1)} = \arg \min_{\mathbf{c}\in\{0,1\}^K} \left\{ \int f(\boldsymbol{\theta} \mid \mathbf{y}, \hat{\mathbf{c}}^{(i)}) \, \|\mathbf{y} - \mathbf{\Lambda}_{\mathbf{a}'}\boldsymbol{\theta}\|^2_{\mathbf{C}_\mathbf{z}^{-1}} d\boldsymbol{\theta} - \ln p(\mathbf{c}) \right\}, \tag{6.114}$$

where $\mathbf{a}$ is a one-to-one function of $\mathbf{c}$. Then using Bayes rule for

$$f(\boldsymbol{\theta} \mid \mathbf{y}, \hat{\mathbf{c}}^{(i)}) = \frac{f(\mathbf{y} \mid \boldsymbol{\theta}, \hat{\mathbf{c}}^{(i)}) f(\boldsymbol{\theta})}{\int f(\mathbf{y} \mid \boldsymbol{\theta}', \hat{\mathbf{c}}^{(i)}) f(\boldsymbol{\theta}') d\boldsymbol{\theta}'},$$

in conjunction with the Rayleigh fading assumption, (6.113) reduces to

$$\hat{\mathbf{c}}^{(i+1)} = \arg \min_{\mathbf{c}\in\{0,1\}^K} \left\{ \frac{\int \|\mathbf{y} - \mathbf{\Lambda}_{\mathbf{a}}\boldsymbol{\theta}\|^2_{\mathbf{C}_\mathbf{z}^{-1}} f(\mathbf{y} \mid \boldsymbol{\theta}, \hat{\mathbf{c}}^{(i)}) f(\boldsymbol{\theta}) d\boldsymbol{\theta}}{\int f(\mathbf{y} \mid \boldsymbol{\theta}', \hat{\mathbf{c}}^{(i)}) f(\boldsymbol{\theta}') d\boldsymbol{\theta}'} - \ln p(\mathbf{c}) \right\} \tag{6.115}$$

$$= \arg \min_{\mathbf{c}\in\{0,1\}^K} \left\{ \frac{\int \exp\left(-\|\mathbf{y} - \mathbf{\Lambda}_{\hat{\mathbf{a}}^{(i)}}\boldsymbol{\theta}\|^2_{\mathbf{C}_\mathbf{z}^{-1}} - \|\boldsymbol{\theta}\|^2_{\mathbf{C}_{\boldsymbol{\theta}}^{-1}}\right) \|\mathbf{y} - \mathbf{\Lambda}_{\mathbf{a}}\boldsymbol{\theta}\|^2_{\mathbf{C}_\mathbf{z}^{-1}} d\boldsymbol{\theta}}{\int \exp\left(-\|\mathbf{y} - \mathbf{\Lambda}_{\hat{\mathbf{a}}^{(i)}}\boldsymbol{\theta}'\|^2_{\mathbf{C}_\mathbf{z}^{-1}} - \|\boldsymbol{\theta}'\|^2_{\mathbf{C}_{\boldsymbol{\theta}}^{-1}}\right) d\boldsymbol{\theta}'} - \ln p(\mathbf{c}) \right\}. \tag{6.116}$$

The optimization problem (6.116) is, in general, difficult to solve. In the simplified case of frequency-nonselective fading, white noise, CM alphabet, and the trivial BEM, though, $\|\mathbf{y} - \mathbf{\Lambda}_{\mathbf{a}}\boldsymbol{\theta}\|^2_{\mathbf{C}_\mathbf{z}^{-1}} = C_1 + C_2 \prod_{k=0}^{K-1} \operatorname{Re}\{a[k]\hat{\theta}[k,0]\}$ for C_1 and C_2 that do not depend on $\mathbf{a}$, making the optimization problem (6.116) tractable (Georghiades & Han, 1997). Even then, the hard-decision nature of info EMB makes it subject to error propagation. Thus, it is not surprising that channel EMB has been shown to outperform info EMB for this simplified setup (Yan & Rao, 2003).

For info EM, or without an informative prior distribution for $\mathbf{c}$, the $\ln p(\mathbf{c})$ term can be neglected, so that

$$\hat{\mathbf{c}}^{(i+1)} = \arg \min_{\mathbf{c}\in\{0,1\}^K} \int \exp\left(-\|\mathbf{y} - \mathbf{\Lambda}_{\hat{\mathbf{a}}^{(i)}}\boldsymbol{\theta}\|^2_{\mathbf{C}_\mathbf{z}^{-1}} - \|\boldsymbol{\theta}\|^2_{\mathbf{C}_{\boldsymbol{\theta}}^{-1}}\right) \|\mathbf{y} - \mathbf{\Lambda}_{\mathbf{a}}\boldsymbol{\theta}\|^2_{\mathbf{C}_\mathbf{z}^{-1}} d\boldsymbol{\theta}. \tag{6.117}$$

Note, however, that the minimization is not significantly simplified relative to (6.116).

References

Agrell, E., Eriksson, T., Vardy, A., & Zeger, K. (2002). Closest point search in lattices. *IEEE Transactions on Information Theory*, *48*(8), 2201–2214.

Ahmed, S., Sellathurai, M., Lambotharan, S., & Chambers, J. A. (2006). Low-complexity iterative method of equalization for single carrier with cyclic prefix in doubly selective channels. *IEEE Signal Processing Letters*, *13*(1), 5–8.

Al-Dhahir, N., & Cioffi, J. M. (1995). MMSE decision feedback equalizers: Finite-length results. *IEEE Transactions on Information Theory*, *41*(4), 961–976.

Al-Dhahir, N., & Sayed, A. H. (2000). The finite-length multi-input multi-output MMSE-DFE. *IEEE Transactions on Signal Processing*, *48*(10), 2921–2936.

Anastasopoulos, A., & Chugg, K. M. (2000). Iterative detection for channels with memory. *IEEE Transactions on Communications*, *48*(10), 1638–1649.

Anastasopoulos, A., Chugg, K. M., Colavolpe, G., Ferrari, G., & Raheli, R. (2007). Iterative detection for channels with memory. *Proceedings of the IEEE*, *95*(6), 1272–1294.

Anderson, J. B., & Mohan, S. (1984). Sequential decoding algorithms: A survey and cost analysis. *IEEE Transactions on Communications*, *32*, 169–172.

Antón-Haro, C., Fonollosa, J. A. R., & Fonollosa, J. R. (1997). Blind channel estimation and data detection using hidden Markov models. *IEEE Transactions on Signal Processing*, *45*(1), 241–247.

Baccarelli, E., & Cusani, R. (1998). Combined channel estimation and data detection using soft statistics for frequency-selective fast-fading digital links. *IEEE Transactions on Communications*, *46*(4), 424–427.

Bahl, L. R., Cocke, J., Jelinek, F., & Raviv, J. (1974). Optimal decoding of linear codes for minimizing symbol error rate. *IEEE Transactions on Information Theory*, *20*, 284–287.

Barbarossa, S., & Torti, R. (2001, May). Chirped-OFDM for transmissions over time-varying channels with linear delay/Doppler spreading. In *Proceedings of the IEEE International Conference on Acoustics, Speech, and Signal Processing*, Vol. 4, (pp. 2377–2380). Salt Lake City, UT.

Barhumi, I., Leus, G., & Moonen, M. (2004). Time-domain and frequency-domain per-tone equalization for OFDM in doubly selective channels. *Signal Processing*, *84*, 2055–2066.

Barhumi, I., Leus, G., & Moonen, M. (2005). Time-varying FIR equalization for doubly selective channels. *IEEE Transactions on Wireless Communications*, *4*(1), 202–214.

Barhumi, I., Leus, G., & Moonen, M. (2006). Equalization for OFDM over doubly selective channels. *IEEE Transactions on Signal Processing*, *54*(4), 1445–1458.

Bölcskei, H. (2002). Orthogonal frequency division multiplexing based on offset QAM. In H. G. Feichtinger & T. Strohmer (Eds.), *Advances in Gabor Analysis*, (pp. 321–352). Boston, MA: Birkhäuser.

Borah, D. K., & Hart, B. D. (1999a). Frequency-selective fading channel estimation with a polynomial time-varying channel model. *IEEE Transactions on Communications*, *47*(6), 862–873.

Borah, D. K., & Hart, B. D. (1999b). Receiver structures for time-varying frequency-selective fading channels. *IEEE Journal on Selected Areas in Communications*, *17*(11), 1863–1875.

Borah, D. K., & Hart, B. D. (1999c). A robust receiver structure for time-varying, frequency-flat, Rayleigh fading channels. *IEEE Transactions on Communications*, *47*(3), 360–364.

Cai, X., & Giannakis, G. B. (2003). Bounding performance and suppressing inter-carrier interference in wireless mobile OFDM. *IEEE Transactions on Communications*, *51*(12), 2047–2056.

Chen, H., Perry, R., & Buckley, K. (2003). On MLSE algorithms for unknown fast time-varying channels. *IEEE Transactions on Communications*, *51*(5), 730–734.

Chiavaccini, E., & Vitetta, G. M. (2001). MAP symbol estimation on frequency-flat Rayleigh fading channels via a Bayesian EM algorithm. *IEEE Transactions on Communications*, *49*(11), 1869–1872.

Choi, Y.-S., Voltz, P. J., & Cassara, F. A. (2001). On channel estimation and detection for multicarrier signals in fast and selective Rayleigh fading channels. *IEEE Transactions on Communications*, *49*(8), 1375–1387.

Chugg, K. (1998). The condition for the applicability of the Viterbi algorithm with implications for fading channel MLSD. *IEEE Transactions on Communications*, *46*(9), 1112–1116.

Cimini, L. J., Jr. (1985). Analysis and simulation of a digital mobile radio channel using orthogonal frequency division multiplexing. *IEEE Transactions on Communications*, *33*, 665–765.

Cioffi, J. M., & Forney, G. D. (1997). Generalized decision-feedback equalization for packet transmission with ISI and Gaussian noise. In A. Paulraj, V. Roychowdhury, & C. Schaper (Eds.), *Communication, Computation, Control and Signal Processing*, chapter 4, (pp. 79–127). Boston, MA: Kluwer.

Cirpan, H. A., & Tsatsanis, M. K. (2001). Maximum-likelihood estimation of FIR channels excited by convolutionally encoded inputs. *IEEE Transactions on Communications*, *49*, 1125–1128.

Cui, T., & Tellambura, C. (2007). Blind receiver design for OFDM systems over doubly selective channels. *IEEE Transactions on Communications*, *55*(5), 906–917.

Damen, M. O., El Gamal, H., & Caire, G. (2003). On maximum-likelihood detection and the search for the closest lattice point. *IEEE Transactions on Information Theory*, *49*(10), 2389–2402.

Dai, Q., & Shwedyk, E. (1994). Detection of bandlimited signals over frequency selective Rayleigh fading channels. *IEEE Transactions on Communications*, *42*(2/3/4), 941–950.

Das, S., & Schniter, P. (2007). Max-SINR ISI/ICI-shaping multi-carrier communication over the doubly dispersive channel. *IEEE Transactions on Signal Processing*, *55*(12), 5782–5795.

Daubechies, I. (1992). *Ten lectures on wavelets*. Philadelphia, PA: SIAM.

Davis, L. M., Collings, I. B., & Hoeher, P. (2001). Joint MAP equalization and channel estimation for frequency-selective and frequency-flat fast-fading channels. *IEEE Transactions on Communications*, *49*(12), 2106–2114.

de Jong, Y. L. C., & Willink, T. J. (2005). Iterative tree search detection for MIMO wireless systems. *IEEE Transactions on Communications*, *53*, 930–935.

Dempster, A., Laird, N. M., & Rubin, D. B. (1977). Maximum-likelihood from incomplete data via the EM algorithm. *Journal of the Royal Statistical Society*, *39*, 1–17.

Douillard, C., Jezequel, M., Berrou, C., Picart, A., Didier, P., & Glavieux, A. (1995). Iterative correction of intersymbol interference: Turbo equalization. *European Transactions on Telecommunications*, *6*, 507–511.

Duel-Hallen, A., & Heegard, C. (1989). Delayed decision feedback sequence equalization. *IEEE Transactions on Communications*, *37*, 428–436.

Edfors, O., Sandell, M., van de Beek, J.-J., Wilson, S. K., & Börjesson, P. O. (1998). OFDM channel estimation by singular value decomposition. *IEEE Transactions on Communications*, *46*, 931–939.

El-Mahdy, A. E.-S. (2004). Adaptive channel estimation and equalization for rapidly mobile communication channels. *IEEE Transactions on Communications*, *52*(7), 1126–1135.

Eyuboglu, M. V., & Qureshi, S. U. (1988). Reduced-state sequence estimation with set partitioning and decision feedback. *IEEE Transactions on Communications*, *36*(1), 12–20.

Falconer, D., Ariyavisitakul, S. L., Benyamin-Seeyar, A., & Eidson, B. (2002). Frequency domain equalization for single-carrier broadband wireless systems. *IEEE Communications Magazine*, *40*(4), 58–66.

Fang, K., Rugini, L., & Leus, G. (2008). Iterative channel estimation and turbo equalization for time-varying OFDM systems. In *Proceedings of the IEEE International Conference on Acoustics, Speech, and Signal Processing*, (pp. 2909–2912). Las Vegas, NV.

Fano, R. M. (1963). A heuristic discussion of probabilistic decoding. *IEEE Transactions on Information Theory*, *9*(2), 64–74.

Fessler, J. A., & Hero, A. O. (1994). Space-alternating generalized expectation-maximization algorithm. *IEEE Transactions on Signal Processing*, *42*, 2664–2677.

Fincke, U., & Pohst, M. (1985). Improved methods for calculating vectors of short length in a lattice. *Mathematics of Computation, 44*, 463–471.

Forney, G. (1972). Maximum-likelihood sequence estimation of digital sequences in the presence of inter symbol interference. *IEEE Transactions on Information Theory, 18*(3), 363–378.

Forney, G. D., Jr. (1973). The Viterbi algorithm. *IEEE Transactions on Information Theory, 61*, 262–278.

Georghiades, C. N., & Han, J. C. (1997, March). Sequence estimation in the presence of random parameters via the EM algorithm. *IEEE Transactions on Communications, 45*, 300–308.

Gertsman, M. J., & Lodge, J. H. (1997). Symbol-by-symbol MAP demodulation of CPM and PSK signals on Rayleigh flat-fading channels. *IEEE Transactions on Communications, 45*(7), 788–799.

Gorokhov, A., & Linnartz, J.-P. (2004). Robust OFDM receivers for dispersive time-varying channels: Equalization and channel acquisition. *IEEE Transactions on Communications, 52*(4), 572–583.

Gowaikar, R., & Hassibi, B. (2003, April). Efficient statistical pruning algorithms for maximum likelihood decoding. In *Proceedings of the IEEE International Conference on Acoustics, Speech, and Signal Processing*, Vol. 5, (pp. 49–52). Hong Kong.

Guillaud, M., & Slock, D. T. M. (2003, April). Channel modeling and associated inter-carrier interference equalization for OFDM systems with high Doppler spread. In *Proceedings of the IEEE International Conference on Acoustics, Speech, and Signal Processing*, Vol. 4, (pp. 237–240). Hong Kong.

Haeb, R., & Meyr, H. (1989). A systematic approach to carrier recovery and detection of digitally phase modulated signals on fading channels. *IEEE Transactions on Communications, 37*(7), 748–754.

Hagenauer, J., & Hoeher, P. (1989, November). A Viterbi algorithm with soft-decision outputs and its applications. In *Proceedings of the IEEE Global Telecommunications Conference* (pp. 1680–1686). Dallas, TX.

Haas, R., & Belfiore, J.-C. (1997). A time-frequency well-localized pulse for multiple carrier transmission. *Wireless Personal Communications, 5*, 1–18.

Hart, B. D. (2000). Maximum likelihood sequence detection using a pilot tone. *IEEE Transactions on Vehicular Technology, 49*(2), 550–560.

Hart, B. D., & Pasupathy, S. (2000). Innovations-based MAP detection for time-varying frequency-selective channels. *IEEE Transactions on Vehicular Technology, 48*(9), 1507–1519.

Hart, B. D., & Taylor, D. P. (1998). Maximum-likelihood synchronization, equalization, and sequence estimation for unknown time-varying frequency-selective Rician channels. *IEEE Transactions on Communications, 46*(2), 211–221.

Hassibi, B., & Vikalo, H. (2005). On the sphere-decoding algorithm I. Expected complexity. *IEEE Transactions on Signal Processing, 53*(8), 2806–2818.

Haykin, S. (2001). *Adaptive filter theory.* (4th ed.). Englewood Cliffs, NJ: Prentice-Hall.

Hochwald, B. M., & ten Brink, S. (2003). Achieving near-capacity on a multiple-antenna channel. *IEEE Transactions on Communications, 51*, 389–399.

Hou, W.-S., & Chen, B.-S. (2005). ICI cancellation for OFDM communication systems in time-varying multipath fading channels. *IEEE Transactions on Wireless Communications, 4*(5), 2100–2110.

Hunziker, T., & Dahlhaus, D. (2003). Iterative detection for multicarrier transmission employing time-frequency concentrated pulses. *IEEE Transactions on Communications, 51*(4), 641–651.

Hwang, S.-J., & Schniter, P. (2005, October). On the optimality of MMSE-GDFE pre-processed sphere decoding. In *Proceedings of Allerton Conference on Communication, Control, and Computing*, Monticello, IL.

Hwang, S.-J., & Schniter, P. (2006). Efficient sequence detection of multi-carrier transmissions over doubly dispersive channels. *EURASIP Journal on Applied Signal Processing*, Vol. 2006, Article ID 93638, pages 17.

Hwang, S.-J., & Schniter, P. (2007, June). Maximum-diversity affine precoding for the noncoherent doubly dispersive channel. In *Proceedings of the IEEE Workshop on Signal Processing Advances in Wireless Communications*, (pp. 1–5). Helsinki, Finland.

Hwang, S.-J., & Schniter, P. (2007, September). Efficient communication over highly spread underwater acoustic channels. In *Proceedings of the 2nd ACM International Workshop on Under Water Networks (WUWNet)*. Montreal, QC.

Hwang, S.-J., & Schniter, P. (2007, November). Fast noncoherent decoding of block transmissions over doubly dispersive channels. In *Proceedings of the Asilomar Conference on Signals, Systems, and Computers*. Pacific Grove, CA.

Hwang, S.-J., & Schniter, P. (2008). Efficient multicarrier communication for highly spread underwater acoustic channels. *IEEE Journal on Selected Areas in Communications*, *26*(9), 1674–1683.

Hwang, S.-J., & Schniter, P. (2009, June). EM-based soft noncoherent equalization of doubly selective channels using tree search and basis expansion. In *Proceedings of the IEEE Workshop on Signal Processing Advances in Wireless Communications*. Perugia, Italy.

Iltis, R. A. (1992). A Bayesian maximum-likelihood sequence estimation algorithm for a priori unknown channels and symbol timing. *IEEE Journal on Selected Areas in Communications*, *10*(3), 579–588.

Iltis, R. A., Shynk, J. J., & Giridhar, K. (1994). Bayesian algorithms for blind equalization using parallel adaptive filtering. *IEEE Transactions on Communications*, *42*(2/3/4), 1017–1032.

Jalden, J., & Ottersten, B. (2005). On the complexity of sphere decoding in digital communications. *IEEE Transactions on Signal Processing*, *53*(4), 1474–1484.

Jeon, W. G., Chang, K. H., & Cho, Y. S. (1999). An equalization technique for orthogonal frequency-division multiplexing systems in time-variant multipath channels. *IEEE Transactions on Communications*, *47*(1), 27–32.

Kailath, T. (1960). Correlation detection of signals perturbed by a random channel. *IRE Transactions on Information Theory*, *6*, 361–366.

Kailath, T. (1969). A general likelihood formula for random signals in Gaussian noise. *IEEE Transactions on Information Theory*, *15*(3), 350–361.

Kaleh, G. K., & Vallet, R. (1994). Joint parameter estimation and symbol detection for linear or nonlinear unknown channels. *IEEE Transactions on Communications*, *42*, 2406–2413.

Kannu, A. P., & Schniter, P. (2008). Design and analysis of MMSE pilot-aided cyclic-prefixed block transmissions for doubly selective channels. *IEEE Transactions on Signal Processing*, *56*(3), 1148–1160.

Kim, D., & Stüber, G. (1998). Residual ISI cancellation for OFDM with application to HDTV broadcasting. *IEEE Journal on Selected Areas in Communications*, 16(8), 1590–1599.

Koetter, R., Singer, A. C., & Tüchler, M. (2004). Turbo equalization. *IEEE Signal Processing Magazine*, *21*(1), 67–80.

Kozek, W., & Molisch, A. F. (1998). Nonorthogonal pulseshapes for multicarrier communications in doubly dispersive channels. *IEEE Journal on Selected Areas in Communications*, *16*(8), 1579–1589.

Kubo, H., Murakami, K., & Fujino, T. (1994). An adaptive maximum-likelihood sequence estimator for fast time-varying intersymbol interference channels. *IEEE Transactions on Communications*, *42*(2/3/4), 1872–1880.

Le Floch, B., Alard, M., & Berrou, C. (1995, June). Coded orthogonal frequency division multiplex. *Proceedings of the IEEE*, *83*(6), 982–996.

Lenstra, A. K., Lenstra, H. W., & Lovász, L. (1982). Factoring polynomials with rational coefficients. *Mathematische Annalen*, *261*(4), 515–534.

Leon, W. S., & Taylor, D. P. (2003). Generalized polynomial-based receiver for the flat fading channel. *IEEE Transactions on Communications*, *51*(6), 896–899.

Li, Y., Vucetic, B., & Sato, Y. (1995). Optimum soft-output detection for channels with intersymbol interference. *IEEE Transactions on Information Theory*, *41*, 704–713.

Liu, D., & Fitz, M. P. (2007, September). Reduced state iterative MAP equalization and decoding in wireless mobile coded OFDM. In *Proceedings of the Allerton Conference on Communication, Control, and Computing*. Monticello, IL.

Liu, D., & Fitz, M. P. (2008, September). Joint turbo channel estimation and data recovery in fast fading mobile coded OFDM. In *Proceedings of the IEEE International Symposium on Personal, Indoor and Mobile Radio Communications*, (pp. 1–6). Cannes, France.

Lodge, J. H., & Moher, M. L. (1990). Maximum likelihood sequence estimation of CPM signals transmitted over Rayleigh flat-fading channels. *IEEE Transactions on Communications*, *38*(6), 787–794.

Ma, X., & Giannakis, G. B. (2003). Maximum-diversity transmissions over doubly selective wireless channels. *IEEE Transactions on Information Theory*, *49*(7), 1832–1840.

Martone, M. (2000). Wavelet-based separating kernels for array processing of cellular DS/CDMA signals in fast fading. *IEEE Transactions on Communications*, *48*(6), 979–995.

Martone, M. (2001). A multicarrier system based on the fractional Fourier transform for time-frequency selective channels. *IEEE Transactions on Communications*, *49*(6), 1011–1020.

Matheus, K., & Kammeyer, K.-D. (1997, November). Optimal design of a multicarrier system with soft impulse shaping including equalization in time or frequency direction. In *Proceedings of the IEEE Global Telecommunications Conference*, Vol. 1, (pp. 310–314). Phoenix, AZ.

Matz, G., Schafhuber, D., Gröchenig, K., Hartmann, M., & Hlawatsch, F. (2007). Analysis, optimization, and implementation of low-interference wireless multicarrier systems. *IEEE Transactions on Wireless Communications*, *6*(5), 1921–1931.

Morley, R. E., Jr. & Snyder, D. L. (1979). Maximum likelihood sequence estimation for randomly dispersive channels. *IEEE Transactions on Communications*, *27*(6), 833–839.

Moshavi, S. (1996). Multi-user detection for DS-CDMA systems. *IEEE Communications Magazine*, *34*(10), 124–136.

Motedayen-Aval, I., Krishnamoorthy, A., & Anastasopoulos, A. (2007). Optimal joint detection/estimation in fading channels with polynomial complexity. *IEEE Transactions on Information Theory*, *53*(1), 209–223.

Mow, W. H. (1994). Maximum likelihood sequence estimation from the lattice viewpoint. *IEEE Transactions on Information Theory*, *40*(5), 1591–1600.

Murugan, A., El Gamal, H., Damen, M. O., & Caire, G. (2006). A unified framework for tree search decoding: Rediscovering the sequential decoder. *IEEE Transactions on Information Theory*, *52*(3), 933–953.

Ng, B. K., & Falconer, D. (2004, December). A novel frequency domain equalization method for single-carrier wireless transmissions over doubly-selective fading channels. In *Proceedings of the IEEE Global Telecommunications Conference*, Vol. 1, (pp. 237–241). Dallas, TX.

Nguyen, H., & Levy, B. C. (2005). Blind equalization of dispersive fast fading Ricean channels via the EMV algorithm. *IEEE Transactions on Vehicular Technology*, *54*(5), 1793–1801.

Nissilä, M., & Pasupathy, S. (2003). Adaptive Bayesian and EM-based detectors for frequency-selective fading channels. *IEEE Transactions on Communications*, *51*(8), 1325–1336.

O'Reilly, J. J., & de Oliviera Duarte, A. M. (1985). Error propagation in decision feedback receivers. *IEE Proceedings F: Image Radar and Signal Processing*, *132*(7), 561–566.

Otnes, R., & Tüchler, M. (2004). Iterative channel estimation for turbo equalization of time-varying frequency-selective channels. *IEEE Transactions on Wireless Communications*, *3*(6), 1918–1923.

Peng, F., & Ryan, W. E. (2006, April). A low-complexity soft demapper for OFDM fading channels with ICI. In *Proceedings of the IEEE Wireless Communications and Networking Conference*, (pp. 1549–1554). Las Vegas, NV.

Poor, H. V. (1994). *An Introduction to signal detection and estimation*. (2nd ed.). New York: Springer.

Proakis, J. G. (2001). *Digital communications*. (4th ed.). New York: McGraw-Hill.

Raheli, R., Polydoros, A., & Tzou, C. K. (1995). Per-survivor processing: A general approach to MLSE in uncertain environments. *IEEE Transactions on Communications*, *43*(2/3/4), 354–364.

Reader, D. J., & Cowley, W. G. (1996, September). Blind maximum likelihood sequence detection over fast fading channels. In *Proceedings of the European Signal Processing Conference*. Trieste, Italy.

Rugini, L., Banelli, P., & Leus, G. (2005). Simple equalization of time-varying channels for OFDM. *IEEE Communications Letters*, *9*(7), 619–621.

Rugini, L., Banelli, P., & Leus, G. (2006). Low-complexity banded equalizers for OFDM systems in Doppler spread channels. *EURASIP Journal on Applied Signal Processing*, Vol. 2006, Article ID 67404, pp. 1–13.

Schniter, P. (2004). Low-complexity equalization of OFDM in doubly selective channels. *IEEE Transactions on Signal Processing*, *52*(4), 1002–1011.

Schniter, P., & Liu, H. (2003, October). Iterative equalization for single carrier cyclic-prefix in doubly-dispersive channels. In *Proceedings of the Asilomar Conference on Signals Systems, and Computers*, Vol. 1, (pp. 502–506). Pacific Grove, CA.

Schniter, P., & Liu, H. (2004, November). Iterative frequency-domain equalization for single-carrier systems in doubly-dispersive channels. In *Proceedings of the Asilomar Conference on Signals Systems, and Computers*, Vol. 1, (pp. 667–671). Pacific Grove, CA.

Sigloch, K., Andrews, M. R., Mitra, P. P., & Thomson, D. J. (2005). Communicating over nonstationary nonflat wireless channels. *IEEE Transactions on Signal Processing*, *53*(6), 2216–2227.

Simmons, S. J. (1990). Breadth-first trellis decoding with adaptive effort. *IEEE Transactions on Communications*, *38*(1), 3–12.

Song, S., Singer, A. C., & Sung, K.-M. (2004). Soft input channel estimation for turbo equalization. *IEEE Transactions on Signal Processing*, *52*(10), 2885–2894.

Stamoulis, A., Diggavi, S. N., & Al-Dhahir, N. (2002). Intercarrier interference in MIMO OFDM. *IEEE Transactions on Signal Processing*, *50*(10), 2451–2464.

Strohmer, T., & Beaver, S. (2003). Optimal OFDM design for time-frequency dispersive channels. *IEEE Transactions on Communications*, *51*(7), 1111–1122.

Stüber, G. L. (2001). *Principles of mobile communication*. (2nd ed.). New York: Springer.

Tang, Z., & Leus, G. (2008). A novel receiver architecture for single-carrier transmission over time-varying channels. *IEEE Journal on Selected Areas in Communications*, *26*(2), 366–377.

Toeltsch, M., & Molisch, A. F. (2001, June). Equalization of OFDM-systems by interference cancellation techniques. In *Proceedings of the IEEE International Conference on Communications*, Vol. 6, (pp. 1950–1954). Helsinki, Finland.

Tomasin, S., Gorokhov, A., Yang, H., & Linnartz, J.-P. (2005). Iterative interference cancellation and channel estimation for mobile OFDM. *IEEE Transactions on Wireless Communications*, *4*(1), 238–245.

Tsatsanis, M. K., & Giannakis, G. B. (1996). Modeling and equalization of rapidly fading channels. *International Journal of Adaptive Control and Signal Processing*, *10*(2/3), 159–176.

Tse, D., & Viswanath, P. (2005). *Fundamentals of wireless communication*. New York, NY: Cambridge University Press.

Tüchler, M., Koetter, R., & Singer, A. C. (2002). Turbo equalization: Principles and new results. *IEEE Transactions on Communications*, *50*(5), 754–767.

Verdú, S. (1998). *Multiuser detection*. New York: Cambridge University Press.

Viterbi, A. J., & Omura, J. (1979). *Principles of digital communication and decoding*. New York: McGraw-Hill.

Viterbo, E., & Boutros, J. (1999). A universal lattice code decoder for fading channels. *IEEE Transactions on Information Theory*, *45*(5), 1639–1642.

Wang, X., & Poor, H. V. (1999). Iterative (turbo) soft interference cancellation and decoding for coded CDMA. *IEEE Transactions on Communications*, *47*, 1046–1061.

Wang, Z., Ma, X., & Giannakis, G. B. (2004). OFDM or single-carrier block transmissions? *IEEE Transactions on Communications*, *52*(3), 380–394.

Warrier, D., & Madhow, U. (2002). Spectrally efficient noncoherent communication. *IEEE Transactions on Information Theory*, *48*(3), 651–668.

Wolniansky, P. W., Foschini, G. J., Golden, G. D., & Valenzuela, R. A. (1998, September). V-BLAST: An architecture for realizing very high data rates over the rich-scattering wireless channel. In *Proceedings of URSI ISSSE-98*, (pp. 295–300). Pisa, Italy.

Wornell, G. W. (1996). Emerging applications of multirate signal processing and wavelets in digital communications. *Proceedings of the IEEE*, *84*(4), 586–603.

Wozencraft, J. M., & Reiffen, B. (1961). *Sequential decoding*. New York: MIT Press and Wiley.

Yan, M., & Rao, B. D. (2003). Soft decision-directed MAP estimate of fast Rayleigh flat fading channels. *IEEE Transactions on Communications*, *51*(12), 1965–1969.

Yu, X., & Pasupathy, S. (1995). Innovations-based MLSE for Rayleigh fading channels. *IEEE Transactions on Communications*, *43*(2/3/4), 1534–1544.

Yun, J. Y., Chung, S.-Y., & Lee, Y. H. (2007). Design of ICI canceling codes for OFDM systems based on capacity maximization. *IEEE Signal Processing Letters*, *14*(3), 169–172.

Zamiri-Jafarian, H., & Pasupathy, S. (1999). Adaptive MLSDE using the EM algorithm. *IEEE Transactions on Communications*, *47*, 1181–1193.

Zhang, Y., Fitz, M. P., & Gelfand, S. B. (1997, October). Soft output demodulation on frequency-selective Rayleigh fading channels using AR channel models. In *Proceedings of the IEEE Global Telecommunications Conference*, (pp. 720–724). Phoenix, AZ.

CHAPTER

OFDM Communications over Time-Varying Channels

7

Luca Rugini[1], Paolo Banelli[1], Geert Leus[2]

[1]*University of Perugia, Perugia, Italy*

[2]*Delft University of Technology, Delft, The Netherlands*

7.1 OFDM SYSTEMS

Orthogonal frequency-division multiplexing (OFDM), also known as multicarrier modulation (Bingham, 1990; Cimini Jr, 1985; Keller & Hanzo, 2000; Le Floch, Alard, & Berrou, 1995; Sari, Karam, & Jeanclaude, 1995; Wang & Giannakis, 2000; Zou & Wu, 1995), relies on the concept of parallel data transmission in the frequency domain and mainly owes its success to the easy equalization for linear time-invariant (LTI) frequency-selective channels. In OFDM systems, the data symbol stream is split into L parallel flows, which are transmitted on equispaced frequencies called subcarriers, each one characterized by a transmission rate that is $1/L$ times lower than the original data rate. This is obtained by splitting the original data stream into multiple blocks, which are transmitted in consecutive time intervals, where each symbol of a block is associated to a specific subcarrier. This frequency-domain multiplexing can be efficiently performed by means of fast Fourier transform algorithms.

Due to the use of orthogonal (equispaced) subcarriers, OFDM systems with LTI frequency-selective channels avoid the so-called intercarrier interference (ICI) among the data symbols of the same OFDM block. Differently from conventional frequency-division multiplexing, a frequency overlapping among the spectra associated to different substreams is permitted, resulting in a significant reduction of the bandwidth requirements. Moreover, for LTI frequency-selective channels, the absence of ICI allows an easy channel equalization, which can be performed on a per-subcarrier basis by means of scalar divisions. The *intersymbol interference* (ISI)[1] among data symbols of different OFDM blocks, induced by multipath propagation, is avoided by a suitable cyclic extension of each OFDM block, usually referred to as *cyclic prefix* (CP) (Sari et al., 1995; Wang & Giannakis, 2000; Zou & Wu, 1995).

However, when the channel experiences a nonnegligible time variation, each subcarrier undergoes a Doppler spreading effect that destroys the subcarrier orthogonality, producing significant ICI (Robertson & Kaiser, 1999; Russell & Stüber, 1995; Stantchev & Fettweis, 2000). Dually to the ISI in single-carrier systems, the ICI power reduces the *signal-to-interference-plus-noise ratio* (SINR) and, when left uncompensated, impairs the performance of OFDM systems. A simple method that reduces the ICI is the shortening of the OFDM block duration. This way the channel becomes (almost) constant

[1]The ISI is also known as *interblock interference*, while the OFDM blocks are also known as *OFDM symbols*.

over each block. However, the block-length shortening is capacity inefficient, because the CP has to be inserted more frequently. Therefore, other ICI mitigation techniques are necessary. These techniques are reviewed in Section 7.2. In addition, the rapid time variation of the channel makes its estimation more complicated. This issue is discussed in Section 7.3.

In this section, we first set up the system model and review the behavior of OFDM systems with LTI channels, focusing on the most popular OFDM wireless standards. Subsequently, we show the effects of rapidly time-varying channels on conventional OFDM systems by analyzing the ICI power, the SINR degradation, and the bit-error rate (BER) performance loss. Finally, we extend the system model to multiantenna OFDM systems.

7.1.1 System Model

We consider an OFDM system with L equispaced subcarriers, where F is the subcarrier separation and L_{CP} is the size of the CP that is prepended to each OFDM block. The whole OFDM system we are going to describe is depicted in Fig. 7.1.

After serial-to-parallel conversion, the stream of symbols is split into data blocks. Each OFDM block, of size L, can contain either data symbols or pilot symbols, or both data and pilots, depending on the training pattern. The pilot symbols may be used at the receiver side for time and frequency synchronization, channel estimation, phase offset correction, and so on. Virtual carriers, which are included in every OFDM system as guard bands to prevent adjacent-channel interference, are considered as pilot symbols. The generic symbol transmitted on the lth subcarrier of the kth OFDM block is denoted by $x[l,k]$. Defining $\mathbf{x}[k] \triangleq (x[0,k]\cdots x[L-1,k])^T$ as the vector that collects the data $\mathbf{a}[k]$ and

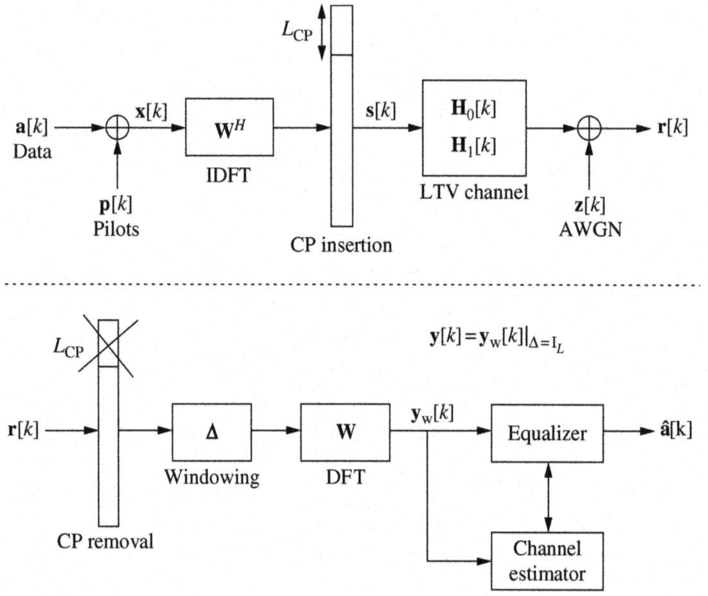

FIGURE 7.1

OFDM system model. Top: transmitter and channel. Bottom: receiver.

pilots $\mathbf{p}[k]$ of the kth block, i.e., $\mathbf{x}[k] = \mathbf{a}[k] + \mathbf{p}[k]$, the kth transmitted block $\mathbf{s}[k]$, of size $N = L + L_{\text{CP}}$, can be expressed as (Wang & Giannakis, 2000)

$$\mathbf{s}[k] = \mathbf{T}_{\text{CP}}\mathbf{W}^H\mathbf{x}[k]. \tag{7.1}$$

Here, $\mathbf{W}$ is the $L \times L$ unitary discrete Fourier transform (DFT) matrix, defined by $[\mathbf{W}]_{i,l} \triangleq \frac{1}{\sqrt{L}} e^{-j2\pi il/L}$, $0 \le i, l \le L-1$, and $\mathbf{T}_{\text{CP}} \triangleq \left(\mathbf{I}_{\text{CP}}^T \, \mathbf{I}_L\right)^T$ is the $N \times L$ matrix that inserts the CP, where $\mathbf{I}_{\text{CP}}$ contains the last L_{CP} rows of the identity matrix $\mathbf{I}_L$. Thus, OFDM can be seen as a particular linearly precoded block transmission, with precoding matrix $\mathbf{T}_{\text{CP}}\mathbf{W}^H$.

After parallel-to-serial conversion, the signal stream $s[kN+n] \triangleq [\mathbf{s}[k]]_n$ is transmitted through a linear time-varying (LTV) multipath channel with discrete-time impulse response $h[n,m]$, where n is the time index and m is the time-delay (lag) index. We assume a finite impulse response LTV channel, i.e., $h[n,m]$ has zero entries outside $0 \le m \le M-1$. Assuming time and frequency synchronization at the receiver side, the received samples can be expressed as

$$r[n] = \sum_{m=0}^{M-1} h[n,m]s[n-m] + w[n],$$

where $w[n]$ represents additive white Gaussian noise (AWGN). The N received samples relative to the kth OFDM block are grouped in the vector $\mathbf{r}[k]$, with $[\mathbf{r}[k]]_n = r[kN+n]$, thus obtaining

$$\mathbf{r}[k] = \mathbf{H}_0[k]\mathbf{s}[k] + \mathbf{H}_1[k]\mathbf{s}[k-1] + \mathbf{w}[k]. \tag{7.2}$$

Here, $\mathbf{H}_0[k]$ and $\mathbf{H}_1[k]$ are $N \times N$ matrices with elements $[\mathbf{H}_0[k]]_{n,m} = h[kN+n, n-m]$ and $[\mathbf{H}_1[k]]_{n,m} = h[kN+n, N+n-m]$:

$$\mathbf{H}_0[k] \triangleq \begin{pmatrix} h[kN,0] & 0 & \cdots & \cdots & 0 \\ \vdots & \ddots & \ddots & & \vdots \\ h[kN+M-1,M-1] & & \ddots & \ddots & \vdots \\ \vdots & \ddots & & \ddots & 0 \\ 0 & \cdots & h[kN+N-1,M-1] & \cdots & h[kN+N-1,0] \end{pmatrix},$$

$$\mathbf{H}_1[k] \triangleq \begin{pmatrix} 0 & \cdots & h[kN,M-1] & \cdots & h[kN,1] \\ \vdots & \ddots & & \ddots & \vdots \\ 0 & & \ddots & & h[kN+M-2,M-1] \\ \vdots & \ddots & & \ddots & \vdots \\ 0 & \cdots & 0 & \cdots & 0 \end{pmatrix}.$$

In obtaining (7.2), we have implicitly assumed that the block length N is greater than the channel order $M-1$ so that ISI is possible only from the previous data block $\mathbf{s}[k-1]$. At the receiver, $\mathbf{r}[k]$ in (7.2) is left-multiplied by the matrix $\mathbf{R}_{\text{CP}} \triangleq \left(\mathbf{0}_{L \times L_{\text{CP}}} \, \mathbf{I}_L\right)$ that removes the CP. In what follows, we assume $L_{CP} \geq M-1$. Then, the ISI is completely eliminated, since $\mathbf{R}_{CP}\mathbf{H}_1[k]\mathbf{s}[k-1] = \mathbf{0}_{L\times 1}$ (Wang & Giannakis, 2000).

Next, the received signal is converted to the frequency domain by applying a DFT, as expressed by $\mathbf{y}[k] \triangleq \mathbf{W}\mathbf{R}_{CP}\mathbf{r}[k]$, which by (7.1) and (7.2) can be rewritten as

$$\mathbf{y}[k] = \mathbf{W}\mathbf{H}_T[k]\mathbf{W}^H\mathbf{x}[k] + \mathbf{W}\mathbf{R}_{CP}\mathbf{w}[k] = \mathbf{H}_F[k]\mathbf{x}[k] + \mathbf{z}[k]. \tag{7.3}$$

Here, $\mathbf{H}_T[k] \triangleq \mathbf{R}_{CP}\mathbf{H}_0[k]\mathbf{T}_{CP}$ is the $L \times L$ matrix that summarizes the LTV channel in the time domain, including CP insertion and removal, with elements expressed by

$$[\mathbf{H}_T[k]]_{n,m} = h[kN + L_{CP} + n, (n-m) \bmod L], \tag{7.4}$$

while $\mathbf{H}_F[k] \triangleq \mathbf{W}\mathbf{H}_T[k]\mathbf{W}^H$ is the $L \times L$ frequency-domain channel matrix, with elements expressed by

$$\begin{aligned}[\mathbf{H}_F[k]]_{l+d,l} &= \frac{1}{L}\sum_{n=0}^{L-1}\sum_{m=0}^{L-1}[\mathbf{H}_T[k]]_{n,m}\mathrm{e}^{-j2\pi(l(n-m)+dn)/L} \\ &= \frac{1}{L}\sum_{n=0}^{L-1}\sum_{m=0}^{M-1}h[kN + L_{CP} + n, m]\mathrm{e}^{-j2\pi(lm+dn)/L},\end{aligned} \tag{7.5}$$

where l represents the subcarrier index and d is the discrete Doppler index. Specifically, the off-diagonal elements of the lth column of $\mathbf{H}_F[k]$ represent the discrete Doppler spread associated with the lth subcarrier, which is responsible for the ICI induced by the lth symbol of the OFDM block on the other symbols.

Summarizing, clearly $\mathbf{H}_F[k]$ plays a crucial role, since it describes how the transmitted frequency-domain block $\mathbf{x}[k]$ is modified by the LTV channel. In addition, in (7.3), $\mathbf{z}[k] \triangleq \mathbf{W}\mathbf{R}_{CP}\mathbf{w}[k]$ is the frequency-domain noise, which is AWGN because $\mathbf{W}$ is unitary.

7.1.1.1 *LTI Channels and One-Tap Equalizers*

When either the channel time variation is absent, i.e., for LTI multipath channels, or it can be neglected, the channel impulse response (CIR) is constant over time. Hence, (7.4) becomes $[\mathbf{H}_T[k]]_{n,m} = h[0, (n-m) \bmod L]$, i.e., $\mathbf{H}_T[k] = \mathbf{H}_T$ is circulant and constant over the OFDM blocks. In this scenario, the CP not only eliminates the ISI, which could be removed by any kind of sufficiently long guard interval, e.g., by trailing zeros (Wang & Giannakis, 2000). In addition, the CP induces a time-domain circular convolution of the transmitted signal with the CIR, which corresponds to a scalar multiplication in the discrete frequency domain. Because the columns of the DFT matrix, which linearly precodes the OFDM data, are eigenvectors of circulant matrices, the eigenvalue decomposition of $\mathbf{H}_T$ is given by $\mathbf{H}_T = \mathbf{W}^H\mathbf{\Lambda}\mathbf{W}$. Consequently, $\mathbf{H}_F[k] = \mathbf{H}_F = \mathbf{\Lambda}$ is diagonal, which shows that in LTI channels there is no ICI. A continuous-time interpretation of OFDM systems is that, for every OFDM block, the lth symbol is transmitted in the frequency domain by a *sinc* function centered on the lth subcarrier. The zeros of this *sinc* function are located on the other equispaced subcarriers, which guarantees ICI-free reception by DFT spectrum sampling. From (7.5), it is easy to derive

$$\lambda_{ll} \triangleq [\mathbf{\Lambda}]_{l,l} = \sum_{m=0}^{M-1} h[0,m]e^{-j2\pi lm/L},$$

i.e., $\mathbf{H}_{\mathrm{F}}$ contains on its diagonal the DFT of the CIR. Due to the diagonal frequency-domain channel matrix, the input–output relation can be expressed as

$$y[k,l] = \lambda_{ll} x[k,l] + z[k,l].$$

Hence, in OFDM systems, the equalization of LTI channels is rather simple and may be computed as $\hat{x}[k,l] = y[k,l]/\lambda_{ll}$ (Sari et al., 1995). This is usually referred to as *one-tap equalization*.

In general, the channel transfer function is estimated, for pilot locations, by $\hat{\lambda}_{ll} = y[k,l]/p[k,l]$ or by $\hat{\lambda}_{ll} = \frac{1}{K}\sum_{k=0}^{K-1} \frac{y[k,l]}{p[k,l]}$ when the pilot positions are constant for K OFDM blocks. Estimates of λ_{ll} for the data subcarriers are usually obtained by interpolating the channel values estimated for the pilot subcarriers (Ozdemir & Arslan, 2007).

7.1.1.2 OFDM Standards

In this section, we compare some popular wireless OFDM standards to appreciate their sensitivity to the Doppler spread induced by LTV channels. The OFDM standards under investigation are DVB-T/H (ETSI, 2004), DAB (ETSI, 2005), IEEE 802.11a (IEEE, 1999), and IEEE 802.16e (WiMAX) (IEEE, 2006). Clearly, the time variability of the channel, summarized by the channel coherence time T_{c}, should be compared with the duration of the OFDM block: standards with longer OFDM block duration are more sensitive to the Doppler effect; they feature bigger channel matrices $\mathbf{H}_{\mathrm{T}}[k]$, whose diagonals display a larger time variability of the channel.

Dually, we can compare the maximum Doppler frequency $\nu_{\max}$ with the subcarrier separation F: indeed, as it will be clarified in the next section, the ICI power is roughly a quadratic function of the normalized maximum Doppler shift $\vartheta_{\max} \triangleq \nu_{\max}/F$. As a consequence, robust standards have a small $\vartheta_{\max}$. This quantity can be calculated as

$$\vartheta_{\max} = \frac{f_{\mathrm{c}}}{F}\frac{\upsilon}{c_0},$$

where f_{c} is the carrier frequency, v is the relative speed between transmitter and receiver, and c_0 is the speed of light.

For the IEEE 802.11a WLAN standard (IEEE, 1999), $F = 312.5\,\mathrm{kHz}$, while the frequency band is around 5 GHz. Specifically, for the maximum carrier frequency $f_{\mathrm{c}} = 5.825$ GHz and speed $\upsilon = 100$ km/h, we obtain $\vartheta_{\max} \approx 0.0017$.

For the IEEE 802.16e WiMAX standard (IEEE, 2006), the subcarrier separation depends on the ratio between the allocated bandwidth B and the number L of subcarriers as $F \approx \tilde{n}B/L$, where $\tilde{n}$ is a rational scaling factor between 1.12 and 1.152. Typical low values of B and L give $F \approx 9.77\tilde{n} \approx 10$ kHz, which equals the value obtained for typical high values of B and L. Since the maximum carrier frequency is $f_{\mathrm{c}} = 10.68$ GHz, for $\upsilon = 100$ km/h, we obtain $\vartheta_{\max} \approx 0.089$. This value, which is roughly 50 times higher than for WLAN, explains why the channel time variation could be a problem for WiMAX, while it can be ignored in WLAN systems. Indeed, the emerging IEEE 802.11p standard amendment for vehicular communications further increases the number of subcarriers with respect to IEEE 802.11a, without significantly degrading the Doppler resistance. The CP length is increased to guarantee ISI-free transmission in outdoor environments. The resulting loss in spectral efficiency is kept down by an increased duration of the IEEE 802.11p OFDM block.

For DVB-T/H (ETSI, 2004), the value of ϑ_{max} is highly dependent on the channel bandwidth B, which ranges from 5 to 8 MHz for different countries, on the carrier frequency f_c, and on the transmission mode, which determines the number of subcarriers L for a given bandwidth B. The available modes are Mode 2k ($L = 2048$), Mode 4k ($L = 4096$), and Mode 8k ($L = 8192$). In the following, we will focus on the 8k mode, which is the most sensitive to the Doppler spread. Assuming again $v = 100$ km/h, for $f_c = 230$ MHz, the normalized maximum Doppler shift is between $\vartheta_{max} \approx 0.019$ (for $B = 8$ MHz) and $\vartheta_{max} \approx 0.031$ (for $B = 5$ MHz). Hence, in this case, the Doppler effect for DVB-T/H 8k is less pronounced than for WiMAX. For $f_c = 862$ MHz, we have $\vartheta_{max} \approx 0.071$ for $B = 8$ MHz and $\vartheta_{max} \approx 0.11$ for $B = 5$ MHz. Therefore, in this second case, the sensitivity of DVB-T/H 8k to the channel time variation is similar to WiMAX. In addition, for $f_c = 1.492$ GHz, the performance degradation of DVB-T/H 8k is even higher than for WiMAX, since $\vartheta_{max} \approx 0.12$ for $B = 8$ MHz and $\vartheta_{max} \approx 0.20$ for $B = 5$ MHz. The results for $B = 6$ MHz and $B = 7$ MHz can be found in Table 7.1. With respect to Mode 8k, the subcarrier separation F of Modes 4k and 2k is the double and quadruple, respectively, and consequently, the values of ϑ_{max} are one-half and one-quarter of those for the 8k mode listed in Table 7.1.

Also for DAB (ETSI, 2005), and for its evolution known as T-DMB, we have to distinguish among different cases, depending on the carrier frequency f_c and the transmission mode. However, the transmission bandwidth is fixed to $B = 1.536$ MHz. The values of ϑ_{max}, listed in Table 7.2, show that the sensitivity of DAB Mode I to Doppler is similar to that of DVB-T/H Mode 8k with $B = 7$ MHz. The sensitivity of DAB Mode IV is similar to that of DVB-T/H Mode 4k, and the sensitivity of DAB Mode II is similar to that of DVB-T/H Mode 2k. Indeed, in all cases, the ratio of the number

Table 7.1 Normalized Maximum Doppler Shift ϑ_{max} for DVB-T/H (Mode 8k), Assuming $v = 100$ km/h.

	$B = 5$ MHz	$B = 6$ MHz	$B = 7$ MHz	$B = 8$ MHz
$f_c = 230$ MHz	$\vartheta_{max} \approx 0.031$	$\vartheta_{max} \approx 0.025$	$\vartheta_{max} \approx 0.022$	$\vartheta_{max} \approx 0.019$
$f_c = 862$ MHz	$\vartheta_{max} \approx 0.11$	$\vartheta_{max} \approx 0.095$	$\vartheta_{max} \approx 0.082$	$\vartheta_{max} \approx 0.071$
$f_c = 1492$ MHz	$\vartheta_{max} \approx 0.20$	$\vartheta_{max} \approx 0.16$	$\vartheta_{max} \approx 0.14$	$\vartheta_{max} \approx 0.12$

Table 7.2 Normalized Maximum Doppler Shift ϑ_{max} for DAB, Assuming $v = 100$ km/h

	Mode I, $L = 2048$	Mode IV, $L = 1024$	Mode II, $L = 512$	Mode III, $L = 256$
$f_c = 230$ MHz	$\vartheta_{max} \approx 0.021$	$\vartheta_{max} \approx 0.011$	$\vartheta_{max} \approx 0.0053$	$\vartheta_{max} \approx 0.0027$
$f_c = 862$ MHz	$\vartheta_{max} \approx 0.080$	$\vartheta_{max} \approx 0.040$	$\vartheta_{max} \approx 0.020$	$\vartheta_{max} \approx 0.010$
$f_c = 1492$ MHz	$\vartheta_{max} \approx 0.14$	$\vartheta_{max} \approx 0.069$	$\vartheta_{max} \approx 0.035$	$\vartheta_{max} \approx 0.017$

of subcarriers of DVB-T/H versus DAB is constant (equal to 4), and approximately equal to the bandwidth ratio.

7.1.2 Effects of Rapidly Time-Varying Channels

When the channel is LTV, the frequency-domain channel matrix $\mathbf{H}_\mathrm{F}[k]$ in (7.5) is neither diagonal nor constant over successive OFDM blocks. Therefore, both the useful channel part and the ICI change from block to block. Let us split the frequency-domain channel matrix into two parts, as expressed by

$$\mathbf{H}_\mathrm{F}[k] = \mathbf{Q}[k] + \mathbf{\Phi}[k],$$

where $\mathbf{Q}[k]$ is the diagonal part of $\mathbf{H}_\mathrm{F}[k]$ and $\mathbf{\Phi}[k] \triangleq \mathbf{H}_\mathrm{F}[k] - \mathbf{Q}[k]$ is the corresponding off-diagonal matrix. Then, (7.3) can be rewritten as

$$\mathbf{y}[k] = \mathbf{Q}[k]\mathbf{x}[k] + \mathbf{\Phi}[k]\mathbf{x}[k] + \mathbf{z}[k], \tag{7.6}$$

where the three terms on the right-hand side of (7.6) represent the useful signal, the ICI, and the AWGN, respectively. Since $q_{ll}[k] \triangleq [\mathbf{Q}[k]]_{l,l} = [\mathbf{H}_\mathrm{F}[k]]_{l,l}$, from (7.5), the useful channel can be written as

$$q_{ll}[k] = \sum_{m=0}^{M-1} \left[\frac{1}{L} \sum_{n=0}^{L-1} h[kN + L_\mathrm{CP} + n, m] \right] \mathrm{e}^{-j2\pi lm/L}. \tag{7.7}$$

This is obtained as the DFT of the *time-averaged CIR*, which is the expression within the square brackets in (7.7). When the CIR varies rapidly with time, the time-averaged CIR in (7.7) decreases, because the elements $\{h[kN + L_\mathrm{CP} + n, m]\}$ add incoherently. As a result, the average power of the frequency-domain useful channel (i.e., the power of the elements of $\mathbf{Q}[k]$) decreases. In addition, a rapid time variation of the CIR also leads to an increased ICI power in $\mathbf{\Phi}[k]$, as detailed in the next subsection.

7.1.2.1 *ICI and SINR Analysis*

Since conventional one-tap equalizers do not take the ICI into account, it is important to quantify the effect of the ICI on the decision variable. Herein, we present a statistical analysis of both the ICI and the SINR. For simplicity, in this subsection, we assume that only data symbols are transmitted, i.e., $x[k,l] = a[k,l], \forall l, \forall k$. From (7.6), we obtain

$$y[k,l] = q_{ll}[k]a[k,l] + \sum_{d=0, d\neq l}^{L-1} \phi_{dl}[k]a[k,d] + z[k,l]. \tag{7.8}$$

Assuming that

1. data and noise terms have zero mean;
2. all data symbols on different subcarriers and in different OFDM blocks are uncorrelated and have equal mean power σ_a^2;
3. the LTV channel is wide-sense stationary with uncorrelated scattering (WSSUS) with normalized path loss, i.e., $\rho_\mathbf{H}^2$ defined in Chapter 1 is equal to one;
4. the noise is independent of the data and the channel;

then, the mean power received on the lth subcarrier of the kth OFDM block is expressed as

$$\sigma_y^2 \triangleq \mathsf{E}\{|y[k,l]|^2\} = \mathsf{E}\{|q_{ll}[k]|^2\}\sigma_a^2 + P_{\text{ICI}}\sigma_a^2 + \sigma_z^2, \tag{7.9}$$

where $P_{\text{ICI}} \triangleq \sum_{d=0,d\neq l}^{L-1} \mathsf{E}\{|\phi_{dl}[k]|^2\} = 1 - \mathsf{E}\{|q_{ll}[k]|^2\}$ is the ICI power normalized by the data power σ_a^2.

The value of P_{ICI} can be well approximated by assuming an infinite number of subcarriers. When the time-frequency correlation function $R_{\mathbf{H}}(\Delta t, \Delta f)$, defined in Chapter 1, is separable, i.e., $R_{\mathbf{H}}(\Delta t, \Delta f) = r_{\mathbf{H}}^{(2)}(\Delta t) r_{\mathbf{H}}^{(1)}(\Delta f)$, or equivalently when the scattering function $C_{\mathbf{H}}(\tau, \nu)$, defined in Chapter 1, is separable, i.e., $C_{\mathbf{H}}(\tau,\nu) = c_{\mathbf{H}}^{(1)}(\tau) c_{\mathbf{H}}^{(2)}(\nu)$, P_{ICI} can be expressed by (Li & Cimini Jr, 2001)

$$\begin{aligned} P_{\text{ICI}} &= 1 - 2\int_0^1 (1-x)\, r_{\mathbf{H}}^{(2)}\left(\frac{x}{F}\right) \mathrm{d}x \\ &= 1 - \int_{-\nu_{\max}}^{\nu_{\max}} c_{\mathbf{H}}^{(2)}(\nu)\,\mathrm{sinc}^2\left(\frac{\pi\nu}{F}\right) \mathrm{d}\nu. \end{aligned} \tag{7.10}$$

In this case, the ICI power does not depend on the delay power profile of the channel, whereas it depends on the Doppler power profile. For instance, in case of Jakes' Doppler power profile with $r_{\mathbf{H}}^{(2)}(\Delta t) = J_0(2\pi \vartheta_{\max} F \Delta t)$, where $\vartheta_{\max}$ is the normalized maximum Doppler shift, (7.10) becomes (Robertson & Kaiser, 1999)

$$\begin{aligned} P_{\text{ICI}} &= 1 - {}_1F_2\left(\frac{1}{2}; \frac{3}{2}, 2; -(\pi\vartheta_{\max})^2\right) \\ &= 1 - 2\sum_{i=0}^{\infty} (-1)^i \frac{(\pi\vartheta_{\max})^{2i}}{(i!)^2\,(2i+1)\,(2i+2)} \\ &\approx \frac{\pi^2}{6}\vartheta_{\max}^2 - \frac{\pi^4}{60}\vartheta_{\max}^4 + \frac{\pi^6}{1008}\vartheta_{\max}^6, \end{aligned} \tag{7.11}$$

where ${}_pF_q$ stands for the generalized hypergeometric function (Gradshteyn & Ryzhik, 1994). The ICI power can be calculated also for different Doppler power profiles, such as uniform, Gaussian, and the two-path model (Robertson & Kaiser, 1999). In particular, the two-path model characterizes the Doppler shift caused by a carrier frequency offset (CFO). In this case, we have

$$P_{\text{ICI}} = 1 - \mathrm{sinc}^2(\pi\vartheta_{\max}) \approx \frac{\pi^2}{3}\vartheta_{\max}^2 - \frac{2\pi^4}{45}\vartheta_{\max}^4 + \frac{\pi^6}{315}\vartheta_{\max}^6. \tag{7.12}$$

It is noteworthy that the ICI power produced by a CFO is roughly twice the ICI power related to a classical Jakes' Doppler power profile and is quite close to the universal upper bound $P_{\text{ICI}} \leq (\pi\vartheta_{\max})^2/3$ (Li & Cimini Jr, 2001).

When guard bands are present, the ICI power $\tilde{P}_{\text{ICI}}[l] \triangleq \sum_{d\,\text{active}, d\neq l} \mathsf{E}\{|\phi_{dl}[k]|^2\}$ depends on the subcarrier index l. For subcarriers far away from the guard subcarriers, $\tilde{P}_{\text{ICI}}[l] \approx P_{\text{ICI}}$, expressed for instance by (7.11) or (7.12), while $\tilde{P}_{\text{ICI}}[l] \approx P_{\text{ICI}}/2$ for the edge subcarriers, since they receive most interference from a single side only. The exact value of $\tilde{P}_{\text{ICI}}[l]$ can be determined by summing up the elements $\mathsf{E}\{|\phi_{dl}[k]|^2\}$ for all the indices $d \neq l$ corresponding to the active subcarriers, where (Schniter, 2004)

$$\begin{aligned}\mathsf{E}\{|\phi_{dl}[k]|^2\} &= \frac{1}{L^2}\sum_{m=-L+1}^{L-1}(L-|m|)r_{t,\mathbf{H}}[m]\mathrm{e}^{-j2\pi dm/L} \\ &= \left(\frac{\sin^2(\omega L/2)}{L^2\sin^2(\omega/2)} * s_{\nu,\mathbf{H}}(\omega)\right)\Bigg|_{\omega=2\pi d/L}.\end{aligned} \tag{7.13}$$

Observe that (7.13) is the DFT of the product of the triangular function $L-|m|$ and the discrete-time correlation $r_{t,\mathbf{H}}[m] \triangleq r_{\mathbf{H}}^{(2)}(m/(LF))$ of the channel, or, dually, the sampled version of the frequency-domain convolution of a squared digital-sinc function with the Doppler power profile of the discrete-time channel $s_{\nu,\mathbf{H}}(\omega) \triangleq \sum_{m=-\infty}^{\infty} r_{t,\mathbf{H}}[m]\mathrm{e}^{-j\omega m}$. This convolution destroys the zeros of the squared digital-sinc function and hence generates ICI. From (7.13), an important result is that $\mathsf{E}\{|\phi_{dl}[k]|^2\}$ rapidly decreases with increasing Doppler index d, because the squared-sinc function tends to zero quadratically. Hence, most of the ICI is due to only a few subcarriers, especially for small values of $\vartheta_{\max}$. Therefore, when the number L of subcarriers is large, $\tilde{P}_{\text{ICI}}[l] \approx P_{\text{ICI}}$ for almost all the subcarriers. It should be noted that (7.13) does not depend on the subcarrier index l, but only on the Doppler index d.

From (7.9), the SINR is expressed by

$$\rho \triangleq \frac{\mathsf{E}\{|q_{ll}[k]|^2\}\sigma_a^2}{P_{\text{ICI}}\sigma_a^2+\sigma_z^2} = \frac{1-P_{\text{ICI}}}{P_{\text{ICI}}+\sigma_z^2/\sigma_a^2}.$$

Hence, when the ICI is left uncompensated, the SINR cannot exceed the maximum value $\rho_{\max} = P_{\text{ICI}}^{-1}-1$. When there are virtual subcarriers, the SINR on the lth subcarrier is expressed by $\rho_l = \frac{1-P_{\text{ICI}}}{\tilde{P}_{\text{ICI}}[l]+\sigma_z^2/\sigma_a^2}$, and the maximum SINR is $\rho_{\max} \approx 2(P_{\text{ICI}}^{-1}-1)$ for the edge subcarriers.

7.1.2.2 BER Performance with One-Tap Equalizers

While the analysis of the ICI power is relatively straightforward, a theoretical BER analysis is quite difficult, apart from some specific cases. As a consequence, we assume that

1. a linear modulation scheme (e.g., PSK or QAM) is used;
2. the channel $h[n,m]$ is WSSUS with Rayleigh fading statistics;
3. a receiver with perfect time and frequency synchronization is used;
4. the one-tap equalizer for the lth subcarrier has perfect knowledge of the useful channel coefficient $q_{ll}[k]$.

First, we review some theoretical models for the uncoded BER, and then, we extend the discussion to the coded BER, which is usually investigated by simulations.

For theoretical purposes, the *power series model* of an LTV channel is often used (Bello, 1963). With this model, the time variation of the channel is represented by a Taylor series expansion, usually truncated to the first term, as expressed by

$$h(t,\tau) \approx h(t_0,\tau) + h'(t_0,\tau)(t-t_0), \tag{7.14}$$

where t_0 is the time instant in the center of the OFDM block, and $h'(t_0,\tau) \triangleq \frac{\partial}{\partial t}h(t,\tau)\big|_{t=t_0}$. In the linear model (7.14), $h(t_0,\tau)$ stands for the useful component, and $h'(t_0,\tau)$ represents the slope of the channel time variability, assumed linear during the block interval. The approximation (7.14) is very accurate for relatively small time variability, e.g., when $\vartheta_{\max} \leq 0.1$ (Chiavaccini & Vitetta, 2000), but can also be used when the Doppler spread is larger (Wang, Proakis, Masry, & Zeidler, 2006).

Since for Rayleigh fading $h(t_0,\tau)$ and $h'(t_0,\tau)$ are complex Gaussian and independent, the useful signal and the ICI will be independent, too. By dropping the block index k for simplicity, (7.8) becomes

$$y_l = q_{ll}a_l + i_l + z_l,$$

where the useful coefficient q_{ll} induced by $h(t_0,\tau)$ is Gaussian, and the ICI $i_l \triangleq \sum_{d=0,d\neq l}^{L-1} \phi_{dl}a_d$, related to $h'(t_0,\tau)$, is a Gaussian mixture.

When the number L of subcarriers is sufficiently high, due to the central limit theorem, the probability density function (pdf) of the ICI i_l can be well approximated as Gaussian, and hence, also $i_l + z_l$ is Gaussian. By means of this *Gaussian ICI approximation*, the BER can be obtained with standard approaches. For instance, for QPSK with Gray coding, the conditional bit error probability $\Pr\{\mathrm{Re}\{\hat{a}_l\} \neq \mathrm{Re}\{a_l\}|q_{ll}\}$ can be expressed as

$$\Pr\{\mathrm{Re}\{\hat{a}_l\} \neq \mathrm{Re}\{a_l\}|q_{ll}\} = Q\left(\sqrt{\frac{|q_{ll}|^2\sigma_a^2}{\sigma_i^2[l]+\sigma_z^2}}\right), \quad (7.15)$$

where $Q(x) \triangleq \frac{1}{\sqrt{2\pi}}\int_x^{\infty} \mathrm{e}^{-u^2/2}du$. The average of (7.15) over the Rayleigh pdf of $|q_{ll}|$ leads to

$$\Pr\{\mathrm{Re}\{\hat{a}_l\} \neq \mathrm{Re}\{a_l\}\} = \frac{1}{2}\left(1 - \sqrt{\frac{\rho_l}{\rho_l+2}}\right), \quad (7.16)$$

where ρ_l is the SINR per symbol on the lth subcarrier. The same expression is also valid for the imaginary part. According to (7.16), the BER only depends on the SINR and does not depend on the delay power profile of the channel. Chiavaccini & Vitetta (2000) have shown that this approach is very accurate for QPSK when $L = 1024$. A similar approach has been used by Russell & Stüber (1995) to evaluate the symbol-error rate for 16-QAM. However, the numerical approximation of the symbol-error rate, expressed by $\Pr\{\hat{a}_l \neq a_l\} \approx 6.48/\rho_l$, is valid only for large SINR. Al-Gharabally & Das (2006) have used a Gaussian ICI approximation that also incorporates the effect of channel estimation errors.

An improved BER approximation can be obtained by avoiding the Gaussian ICI approximation. By denoting with $v_l \triangleq i_l/q_{ll}$ the ICI after equalization, the Gaussian mixture conditional pdf $f_{v_l|q_{ll}}(\mathrm{Re}\{v_l\},\mathrm{Im}\{v_l\}|q_{ll})$ can be expressed as a two-dimensional Gram–Charlier series, whose coefficients depend on the joint moments of $\mathrm{Re}\{v_l\}$ and $\mathrm{Im}\{v_l\}$ (Wang, Proakis, Masry, & Zeidler, 2006); then, the conditional BER is obtained after series truncation, and the average over the statistics of q_{ll} can be done by means of semianalytical computation. Wang et al. (2006) have shown that a series truncation order equal to 4 produces a good accuracy for 16-QAM when $L = 128$. Interestingly, the Gram–Charlier series approach highlights that the uncoded BER is moderately dependent on the frequency selectivity of the channel. When truncated up to the second order, the Gram–Charlier series expansion reduces to the Gaussian ICI approximation.

For the coded BER, a theoretical characterization is rather difficult even for LTI channels, apart from some specific cases. Consequently, we only discuss some results obtained by simulations by Poggioni, Rugini, & Banelli (2008). We assume that the information bit sequence $b[i]$ is convolutionally encoded to obtain the coded bit sequence $c[j]$, whose length is $KA\log_2(N_a)$, where N_a is the constellation size, A is the number of data subcarriers, and K is the number of OFDM blocks within the interleaver time span. After interleaving and mapping, $P = L - A$ pilot symbols per block are added, and the KL resulting symbols are transmitted within K OFDM blocks. While for the uncoded BER the delay power profile of the channel has little importance, its effect on the coded BER is relevant, since channel coding is able to exploit the frequency selectivity of the channel. Moreover, when the interleaver time-span $T_{\text{int}} \triangleq K(1 + L_{\text{CP}}/L)/F$ is greater than the channel coherence time T_c, the OFDM system can benefit from the time selectivity of the channel.

While the coded BER performance highly depends on the specific channel encoder and interleaver, only a few channel parameters have a significant impact on the coded BER. To explain this point, we introduce the equivalent frequency-domain OFDM model (EFDOM) of Poggioni et al. (2008), which, combined with the specific channel encoder and interleaver, produces the same BER as the original OFDM model with LTV channels. Basically, the EFDOM is a simple approximate model obtained using only a reduced number of parameters, which are the most important for the coded BER. First, (7.8) is rewritten as

$$\underline{\mathbf{y}} = \underline{\mathbf{Q}}\underline{\mathbf{a}} + \underline{\mathbf{i}} + \underline{\mathbf{z}}, \tag{7.17}$$

where the underlined vectors, of size KL, are obtained by collecting the elements on the L subcarriers of the K OFDM blocks. In (7.17), the diagonal matrix $\underline{\mathbf{Q}}$ contains the useful part of the LTV channel, $\underline{\mathbf{a}}$ is the data vector, $\underline{\mathbf{i}}$ is the ICI vector, and $\underline{\mathbf{z}}$ stands for the AWGN vector. In order to speed up simulations for the coded BER, the EFDOM replaces (7.17) with

$$\underline{\mathbf{y}}^{(\mathrm{E})} = \underline{\mathbf{Q}}^{(\mathrm{E})}\underline{\mathbf{a}} + \sqrt{\varphi^{(\mathrm{E})}}\underline{\mathbf{i}}^{(\mathrm{E})} + \underline{\mathbf{z}}, \tag{7.18}$$

where $\underline{\mathbf{Q}}^{(\mathrm{E})}$ has the same statistical properties as $\underline{\mathbf{Q}}$, dictated by the delay power profile and by the Doppler power profile, $\underline{\mathbf{i}}^{(\mathrm{E})}$ is a Gaussian random vector with the same mean and the same covariance as $\underline{\mathbf{i}}$, and $\varphi^{(\mathrm{E})}$ is a real positive random variable that models the energy variability of the ICI with respect to its mean value. Specifically, $\varphi^{(\mathrm{E})}$ is a computer-generated random variable that has approximately the same pdf of the random variable

$$\varphi \triangleq \frac{\|\underline{\mathbf{i}}\|^2}{\mathsf{E}\{\|\underline{\mathbf{i}}\|^2\}} \approx \frac{\|\underline{\mathbf{i}}\|^2}{KLP_{\text{ICI}}\sigma_a^2},$$

whose pdf is well approximated by the pdf of the sum of exponential random variables (Poggioni et al., 2008). In the coded case, since K OFDM blocks are processed together, the time variability of the channel has a greater impact than in the uncoded case, where single blocks are separately considered. Therefore, in the coded case, the linear approximation (7.14) is not valid in general, and hence, the useful part of the channel $\underline{\mathbf{Q}}$ can be correlated with the ICI $\underline{\mathbf{i}}$. The EFDOM generates $\underline{\mathbf{Q}}^{(\mathrm{E})}$ and $\underline{\mathbf{i}}^{(\mathrm{E})}$ in (7.18) in such a way that $\rho_{\mathrm{P}}^{(\mathrm{E})}$, defined as the correlation coefficient between $\|\underline{\mathbf{Q}}^{(\mathrm{E})}\underline{\mathbf{a}}\|^2$ and $\|\underline{\mathbf{i}}^{(\mathrm{E})}\|^2$, is equal to ρ_{P}, which is the correlation coefficient between $\|\underline{\mathbf{Q}}\underline{\mathbf{a}}\|^2$ and $\|\underline{\mathbf{i}}\|^2$. Indeed, simulation results

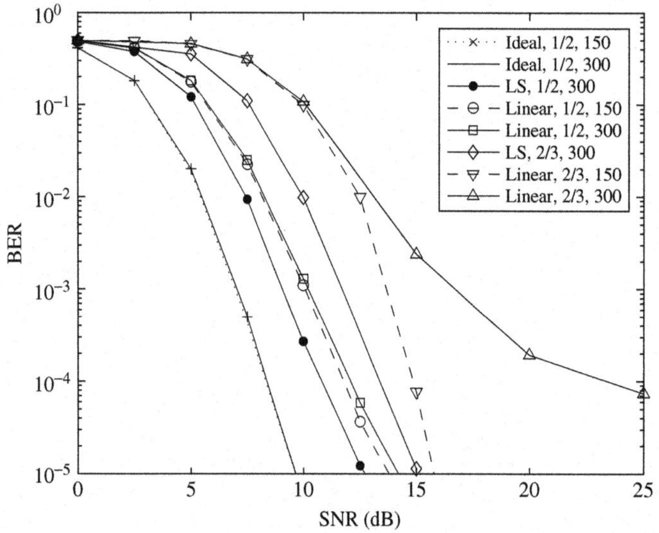

FIGURE 7.2

BER performance of DVB-H. In the legend, the first term indicates the type of channel estimation, the second term represents the code rate of the convolutional code, and the third term is the speed of the mobile receiver expressed in kilometer/hour.

have shown that the single coefficient ρ_P is able to summarize the whole correlation effect over the K blocks (Poggioni et al., 2008). For $K = 1$ (frequency-domain-only interleaver), ρ_P is practically zero for $\vartheta_{\max} \leq 0.5$. Due to the EFDOM, fast simulation of coded OFDM standards is enabled.

Figure 7.2 shows the BER performance of DVB-H at the output of the Viterbi decoder. We consider Mode 2k ($L = 2048$) with carrier frequency $f_c = 800\,\text{MHz}$ and channel bandwidth $B = 8\,\text{MHz}$. For a mobile receiver with speed $\upsilon = 150\,\text{km/h}$, this corresponds to a normalized maximum Doppler shift $\vartheta_{\max} \approx 0.025$. We assume QPSK modulation, a Rayleigh fading multipath channel with Jakes' Doppler power profile (Poggioni, Rugini, & Banelli, 2009), soft Viterbi decoding, and perfect time and frequency synchronization. The receiver assumes a time-invariant channel within the OFDM block. The CIR estimation is performed by interpolation or fitting of the frequency-domain channel estimates obtained on equispaced pilot subcarriers: in Fig. 7.2, *Linear* stands for linear interpolation, *LS* stands for least-squares fitting, and *Ideal* stands for perfect knowledge of the average CIR. Figure 7.2 shows that when the code rate of the convolutional encoder is $1/2$, increasing the mobile speed from $\upsilon = 150\,\text{km/h}$ to $\upsilon = 300\,\text{km/h}$ produces a small performance degradation. On the contrary, when the code rate is $2/3$, the performance degradation due to the increased mobile speed is significant, especially if the channel estimator employs linear interpolation. Using least-squares fitting instead of linear interpolation, a big performance improvement can be obtained, at the price of increased complexity.

Figure 7.3 illustrates the BER performance of DAB at the output of the Viterbi decoder. We consider Mode III ($L = 256$) with carrier frequency $f_c = 800\,\text{MHz}$ and channel bandwidth $B = 1.536\,\text{MHz}$. For a mobile receiver with speed $\upsilon = 150\,\text{km/h}$, this corresponds to a normalized maximum Doppler shift $\vartheta_{\max} \approx 0.014$. We assume $\pi/4$-DQPSK modulation, a Rayleigh fading multipath channel with Jakes'

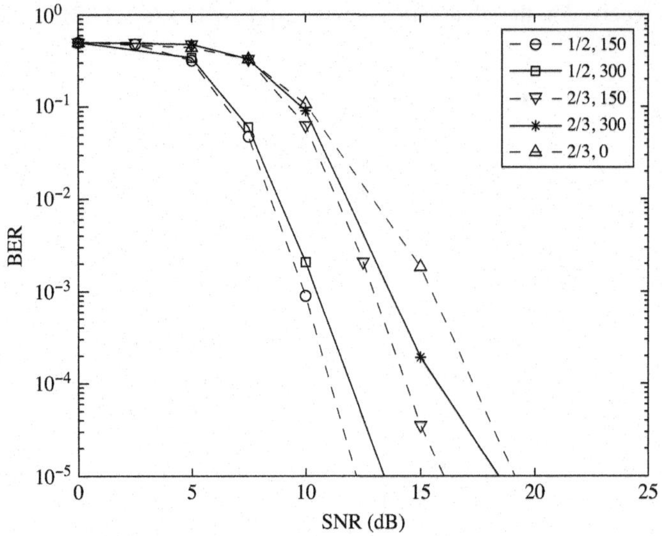

FIGURE 7.3

BER performance of DAB. In the legend, the first term represents the code rate of the convolutional code, and the second term is the speed of the mobile receiver expressed in kilometer/hour.

Doppler power profile, and soft Viterbi decoding (Poggioni et al., 2009). Differential demodulation is used. When increasing the mobile speed from $\upsilon = 0$ km/h to $\upsilon = 150$ km/h, the performance improves due to the time diversity gathered by the interleaver. However, when the mobile speed increases from $\upsilon = 150$ km/h to $\upsilon = 300$ km/h, the ICI causes a performance loss.

Additionally, Reed–Solomon encoding is incorporated in DVB-T/H and in T-DMB as outer code. A detailed performance comparison of DVB-T/H and T-DMB have been presented by Poggioni et al. (2009). For DVB-T/H, differently from the uncoded BER, the coded BER highly depends on the delay power profile of the channel (Poggioni et al., 2009). On the other hand, for T-DMB, the delay power profile of the channel has only a slight impact, because the effect of the time-domain interleaver is dominant (Poggioni et al., 2009).

7.1.3 MIMO-OFDM

We now extend the OFDM model with LTV channels to multiple-input multiple-output (MIMO) OFDM systems with M_T transmit antennas and M_R receive antennas. Denoting by $\mathbf{x}^{(j)}[k]$ the frequency-domain vector containing data and pilots of the jth transmit antenna, the vector transmitted from the jth antenna can be expressed as (see (7.1))

$$\mathbf{s}^{(j)}[k] \triangleq \mathbf{T}_{CP}\mathbf{W}^H\mathbf{x}^{(j)}[k].$$

The signal transmitted from the jth antenna arrives at the ith receive antenna after passing through an LTV channel with impulse response $h^{(i,j)}[n,m]$. We denote by M the maximum of the $M_T M_R$ maximum discrete-time delay spreads. The vector received at the ith antenna can be expressed by (see (7.2))

$$\mathbf{r}^{(i)}[k] = \sum_{j=1}^{M_\mathrm{T}} (\mathbf{H}_0^{(i,j)}[k]\mathbf{s}^{(j)}[k] + \mathbf{H}_1^{(i,j)}[k]\mathbf{s}^{(j)}[k-1]) + \mathbf{w}^{(i)}[k].$$

After CP removal and DFT, this becomes $\mathbf{y}^{(i)}[k] \triangleq \mathbf{W}\mathbf{R}_{\mathrm{CP}}\mathbf{r}^{(i)}[k]$, which is expressed by (see (7.3))

$$\mathbf{y}^{(i)}[k] = \sum_{j=1}^{M_\mathrm{T}} \mathbf{H}_\mathrm{F}^{(i,j)}[k]\mathbf{x}^{(j)}[k] + \mathbf{z}^{(i)}[k].$$

We now stack the vectors related to all the receive antennas in a single vector, denoted as $\underline{\mathbf{y}}[k] \triangleq \left(\mathbf{y}^{(1)T}[k] \cdots \mathbf{y}^{(M_\mathrm{R})T}[k]\right)^T$, and similarly for the transmit antennas, i.e., $\underline{\mathbf{x}}[k] \triangleq \left(\mathbf{x}^{(1)T}[k] \cdots \mathbf{x}^{(M_\mathrm{T})T}[k]\right)^T$, and we define the $LM_\mathrm{R} \times LM_\mathrm{T}$ matrix

$$\underline{\mathbf{H}}_\mathrm{F}[k] \triangleq \begin{pmatrix} \mathbf{H}_\mathrm{F}^{(1,1)}[k] & \dots & \mathbf{H}_\mathrm{F}^{(1,M_\mathrm{T})}[k] \\ \vdots & & \vdots \\ \mathbf{H}_\mathrm{F}^{(M_\mathrm{R},1)}[k] & \dots & \mathbf{H}_\mathrm{F}^{(M_\mathrm{R},M_\mathrm{T})}[k] \end{pmatrix}.$$

The MIMO-OFDM system can then be described as

$$\underline{\mathbf{y}}[k] = \underline{\mathbf{H}}_\mathrm{F}[k]\underline{\mathbf{x}}[k] + \underline{\mathbf{z}}[k]. \tag{7.19}$$

Expression (7.19) shows that in MIMO-OFDM systems, the ICI increases due to the presence of multiple transmit antennas. The ICI power, whose analysis has been presented by Stamoulis, Diggavi, & Al-Dhahir (2002), can be roughly estimated as M_T times the ICI for the single antenna case. In addition, as usual in MIMO schemes, there exists some inter-antenna interference (IAI). Despite the increased interference, multiple receive antennas provide additional degrees of freedom in order to mitigate both ICI and IAI.

In addition, we can stack the vectors related to K successive OFDM blocks, resulting in $\underline{\underline{\mathbf{y}}} \triangleq \left(\underline{\mathbf{y}}^T[0] \cdots \underline{\mathbf{y}}^T[K-1]\right)^T$ and $\underline{\underline{\mathbf{x}}} \triangleq \left(\underline{\mathbf{x}}^T[0] \cdots \underline{\mathbf{x}}^T[K-1]\right)^T$, and define the block diagonal matrix

$$\underline{\underline{\mathbf{H}}}_\mathrm{F} \triangleq \begin{pmatrix} \underline{\mathbf{H}}_\mathrm{F}[0] & & \mathbf{0} \\ & \ddots & \\ \mathbf{0} & & \underline{\mathbf{H}}_\mathrm{F}[K-1] \end{pmatrix}.$$

We then obtain

$$\underline{\underline{\mathbf{y}}} = \underline{\underline{\mathbf{H}}}_\mathrm{F}\underline{\underline{\mathbf{x}}} + \underline{\underline{\mathbf{z}}}. \tag{7.20}$$

In (7.20), the elements are ordered in such a way that first there is a change in the subcarrier index, then in the antenna index, and finally in the OFDM block index. However, this order can be changed by using suitable permutation matrices.

7.2 ICI MITIGATION TECHNIQUES

In this section, we present some common techniques for reducing the ICI produced by LTV channels. Some of these techniques have also been discussed in Chapter 6. Throughout this section, we assume that the LTV channel is unknown at the transmitter and perfectly known at the receiver. We first present techniques that make use of receiver processing only. These receiver-only techniques, which will be further divided into linear and nonlinear, are similar to those used for multiuser detection for code-division multiple-access (CDMA) systems. However, the specific structure of the ICI allows for some specific methods. Subsequently, we describe ICI mitigation techniques that employ transmitter preprocessing. These transmitter methods can be powerful, but in general are not compliant with the current OFDM standards. Finally, we extend the ICI mitigation techniques to MIMO-OFDM systems.

7.2.1 Linear Equalization

Among the receiver equalization methods, linear algorithms construct a soft data estimate by a linear combination of the received samples. For convenience, we rewrite (7.3) by dropping the OFDM block index as

$$\mathbf{y} = \mathbf{H}_{\mathrm{F}}\mathbf{x} + \mathbf{z}, \tag{7.21}$$

where $\mathbf{y}$ represents the L-dimensional frequency-domain received vector, $\mathbf{H}_{\mathrm{F}}$ is the $L \times L$ frequency-domain nondiagonal matrix that induces ICI, $\mathbf{x}$ is the frequency-domain transmitted vector, and $\mathbf{z}$ stands for the AWGN. Since the channel matrix $\mathbf{H}_{\mathrm{F}}$ is assumed known at the receiver, to simplify the explanation, we assume that no pilot symbols are transmitted except for P guard subcarriers, which are consecutive and typically present in any OFDM standard. Here, P is assumed even. These guard bands correspond to the edge positions of the analog bandpass frequency-domain transmitted signal and hence to the central positions of the corresponding discrete-time baseband signal. For convenience, we reorder the subcarriers by a cyclic shift in such a way that the $A = L - P$ data positions are in the center. Denoting by $\mathbf{T}_{\mathrm{GB}} \triangleq \left(\mathbf{0}_{A\times P/2}\ \mathbf{I}_A\ \mathbf{0}_{A\times P/2}\right)^T$ the $L \times A$ matrix that inserts the guard subcarriers, and by $\mathbf{a}$ the A-dimensional subvector of $\mathbf{x}$ containing the data symbols, we obtain $\mathbf{x} = \mathbf{T}_{\mathrm{GB}}\mathbf{a}$. At the receiver, we can exclude the P virtual subcarriers by applying $\mathbf{R}_{\mathrm{GB}} \triangleq \mathbf{T}_{\mathrm{GB}}^T$, as expressed by $\mathbf{y}_{\mathrm{A}} \triangleq \mathbf{R}_{\mathrm{GB}}\mathbf{y}$. This becomes

$$\mathbf{y}_{\mathrm{A}} = \mathbf{H}_{\mathrm{A}}\mathbf{a} + \mathbf{z}_{\mathrm{A}}, \tag{7.22}$$

where $\mathbf{H}_{\mathrm{A}} \triangleq \mathbf{R}_{\mathrm{GB}}\mathbf{H}_{\mathrm{F}}\mathbf{T}_{\mathrm{GB}}$ is the $A \times A$ ICI matrix relative to the data subcarriers and $\mathbf{y}_{\mathrm{A}}$ ($\mathbf{z}_{\mathrm{A}}$) is the A-dimensional received (AWGN) vector. Equalizers designed using the model (7.22) will be referred to as *block equalizers*, since the data subcarriers of the whole OFDM block are jointly equalized.

As explained by (7.13), due to the structure of the Doppler spreading, the ICI on the lth subcarrier mainly comes from a few subcarriers. This means that the matrix $\mathbf{H}_{\mathrm{A}}$ can be well approximated by a banded matrix $\mathbf{B}_{\mathrm{A}}^{(D_{\mathrm{b}})}$, where D_{b} denotes the number of retained subdiagonals and, at the same time, superdiagonals of $\mathbf{H}_{\mathrm{A}}$. An intuitive example of $\mathbf{B}_{\mathrm{A}}^{(D_{\mathrm{b}})}$ is given in Fig. 7.4. Therefore, in the banded case, the block model (7.22) becomes

$$\mathbf{y}_{\mathrm{A}} = \mathbf{B}_{\mathrm{A}}^{(D_{\mathrm{b}})}\mathbf{a} + \mathbf{z}_{\mathrm{A}}. \tag{7.23}$$

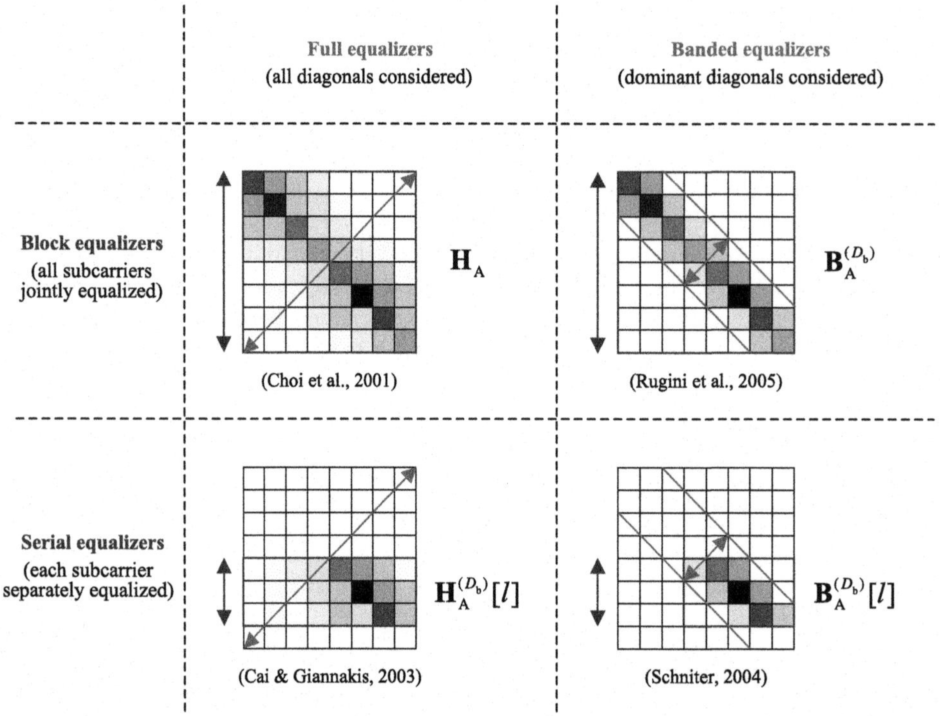

FIGURE 7.4

Possible approximations of the frequency-domain channel matrix. The gray intensity is proportional to the magnitude of the corresponding element.

The integer parameter D_{b} represents the (single-sided) discrete Doppler support that is used for equalization. Since the ICI coefficients $\mathsf{E}\{|\phi_{dl}[k]|^2\}$ in (7.13) have a rapid decay, the significant discrete Doppler support D is usually quite low. Anyway, we can select $D_{\mathrm{b}} < D$ to reduce complexity. The value of D_{b} is usually chosen according to some empirical rules, such as proportionally to $\vartheta_{\max}$, or as the value that reduces the Frobenius norm $\|\mathbf{H}_{\mathrm{A}} - \mathbf{B}_{\mathrm{A}}^{(D_{\mathrm{b}})}\|_{\mathrm{F}}^2$ below a given threshold. A common choice is $D_{\mathrm{b}} = \lceil \vartheta_{\max} + D_0 \rceil$, where D_0 is a small nonnegative number (Schniter, 2004; Hwang & Schniter, 2006) (see also (6.6) in Chapter 6). This rule usually leads to $2D_{\mathrm{b}} + 1 \ll L$, which allows for low-complexity equalization algorithms.[2]

It is noteworthy that the relations (7.22) and (7.23) only consider the A active subcarriers. When the equalizer considers all the L subcarriers, the frequency-domain channel matrix $\mathbf{H}_{\mathrm{F}}$ can be approximated by a matrix with cyclically banded structure,[3] since the upper-right and the lower-left corners are significant (see Fig. 6.2(b) in Chapter 6). This effect, due to frequency-domain aliasing, disappears

[2]When the ICI mitigation support $2D_{\mathrm{b}} + 1$ exceeds the channel length M, time-domain equalizers are less complex than frequency-domain equalizers (Hrycak & Matz, 2006).
[3]Cyclically banded matrices are also known as quasi-banded matrices.

in the presence of guard subcarriers, which cancel the first and the last columns of the channel matrix.

7.2.1.1 Serial Equalizers

Alternatively to the block models expressed by (7.22) and (7.23), a reduced model for the lth subcarrier can be exploited. Indeed, due to the banded structure of the channel matrix, the energy of the lth data symbol a_l mostly falls onto a subvector of $\mathbf{y}_\mathrm{A}$ with size $2D_\mathrm{b}+1$, denoted by $\mathbf{y}_\mathrm{A}^{(D_\mathrm{b})}[l] \triangleq \left(y_{l-D_\mathrm{b}} \cdots y_l \cdots y_{l+D_\mathrm{b}}\right)^T$, which can be expressed by

$$\mathbf{y}_\mathrm{A}^{(D_\mathrm{b})}[l] = \mathbf{H}_\mathrm{A}^{(D_\mathrm{b})}[l]\mathbf{a} + \mathbf{z}_\mathrm{A}^{(D_\mathrm{b})}[l]. \tag{7.24}$$

Here, $\mathbf{H}_\mathrm{A}^{(D_\mathrm{b})}[l]$ is the $(2D_\mathrm{b}+1)\times A$ submatrix of $\mathbf{H}_\mathrm{A}$ that contains the rows with index from $l-D_\mathrm{b}$ to $l+D_\mathrm{b}$, as shown in Fig. 7.4, and $\mathbf{z}_\mathrm{A}^{(D_\mathrm{b})}[l]$ is the AWGN subvector, defined similarly to $\mathbf{y}_\mathrm{A}^{(D_\mathrm{b})}[l]$. The equalizers designed using (7.24) will be refered to as *serial equalizers*. Indeed, since (7.24) is valid for the lth subcarrier only, the data have to be equalized serially (sequentially). Including the band approximation, the serial model (7.24) becomes

$$\mathbf{y}_\mathrm{A}^{(D_\mathrm{b})}[l] = \mathbf{B}_\mathrm{A}^{(D_\mathrm{b})}[l]\mathbf{a}^{(D_\mathrm{b})}[l] + \mathbf{z}_\mathrm{A}^{(D_\mathrm{b})}[l], \tag{7.25}$$

where $\mathbf{B}_\mathrm{A}^{(D_\mathrm{b})}[l]$ is the $(2D_\mathrm{b}+1)\times(4D_\mathrm{b}+1)$ submatrix of $\mathbf{H}_\mathrm{A}$ with row index from $l-D_\mathrm{b}$ to $l+D_\mathrm{b}$ and column index from $l-2D_\mathrm{b}$ to $l+2D_\mathrm{b}$ (see Fig. 7.4), and $\mathbf{a}^{(D_\mathrm{b})}[l] \triangleq \left(a_{l-2D_\mathrm{b}} \cdots a_l \cdots a_{l+2D_\mathrm{b}}\right)^T$.

In the context of LTV channel equalization for OFDM, different linear serial equalizers have been proposed so far. Indeed, although the reduced model (7.24) is suboptimal with respect to the full one (7.22), serial equalization deals with matrices and vectors with smaller dimension and hence reduces the memory requirements of the equalizer. One of the most popular serial equalizers is the zero-forcing (ZF) or least-squares (LS) banded approach of Jeon, Chang, & Cho (1999), which estimates the soft data as

$$\hat{a}_l = \mathbf{e}_{2D_\mathrm{b}+1,D_\mathrm{b}+1}^T \tilde{\mathbf{B}}_\mathrm{A}^{(D_\mathrm{b})-1}[l]\mathbf{y}_\mathrm{A}^{(D_\mathrm{b})}[l]. \tag{7.26}$$

Here, $\mathbf{e}_{m,n}$ is the nth column of $\mathbf{I}_m$, and $\tilde{\mathbf{B}}_\mathrm{A}^{(D_\mathrm{b})}[l]$ is the $(2D_\mathrm{b}+1)\times(2D_\mathrm{b}+1)$ central block of $\mathbf{B}_\mathrm{A}^{(D_\mathrm{b})}[l]$. Therefore, the ICI is completely eliminated, at the price of some noise enhancement, quantitatively summarized by the condition number of $\tilde{\mathbf{B}}_\mathrm{A}^{(D_\mathrm{b})}[l]$. The computational complexity of banded linear serial equalizers can be reduced from $\mathcal{O}(D_\mathrm{b}^3 A)$ to $\mathcal{O}(D_\mathrm{b}^2 A)$ per block, using recursive inversion algorithms that compute $\tilde{\mathbf{B}}_\mathrm{A}^{(D_\mathrm{b})-1}[l+1]$ by updating the already calculated $\tilde{\mathbf{B}}_\mathrm{A}^{(D_\mathrm{b})-1}[l]$ (Cai & Giannakis, 2003).

To reduce the noise enhancement, serial equalizers based on the linear minimum mean-squared error (MMSE) criterion have been proposed. For instance, the nonbanded approach of Cai & Giannakis (2003) is expressed by

$$\hat{a}_l = \mathbf{e}_{A,l}^T \mathbf{H}_\mathrm{A}^{(D_\mathrm{b})H}[l]\mathbf{R}_\mathrm{A}^{(D_\mathrm{b})-1}[l]\mathbf{y}_\mathrm{A}^{(D_\mathrm{b})}[l], \tag{7.27}$$

where $\mathbf{R}_\mathrm{A}^{(D_\mathrm{b})}[l] = \mathbf{H}_\mathrm{A}^{(D_\mathrm{b})}[l]\mathbf{H}_\mathrm{A}^{(D_\mathrm{b})H}[l] + \gamma\mathbf{I}_{2D_\mathrm{b}+1}$ and $\gamma = \sigma_z^2/\sigma_a^2$ is the noise-to-signal ratio. With respect to (7.26), the approach in (7.27) produces an improved performance for two reasons: (1) differently from a ZF equalizer, an MMSE equalizer balances ICI reduction and noise enhancement; (2) there is no band approximation error. Since nonbanded approaches model the out-of-band (OOB) elements of the ICI matrix, they have a larger computational complexity, which is $\mathcal{O}(D_b A^2)$ per block when recursive inversion is employed to obtain $\mathbf{R}_A^{(D_b)-1}[l+1]$ from $\mathbf{R}_A^{(D_b)-1}[l]$ (Cai & Giannakis, 2003).

A different linear serial equalizer has been proposed by Barhumi, Leus, & Moonen (2004) exploiting a basis expansion model (BEM) for both the LTV channel and the equalizer.[4] Using complex exponential basis functions, the linear equalizer of Barhumi et al. (2004) is modeled as banded with bandwidth parameter $\tilde{D}_b > D$, i.e., greater than the bandwidth of the channel matrix. The resulting complexity is $\mathcal{O}(\tilde{D}_b^2 A^2)$ per block.

7.2.1.2 Block Equalizers

In the literature, many linear block equalizers have been proposed, relying on either the LS or the MMSE criterion, sometimes exploiting the band approximation. LS and linear MMSE equalizers based on the full (nonbanded) model (7.22) have been proposed by Choi, Voltz, & Cassara (2001). However, due to the high complexity ($\mathcal{O}(A^3)$ per block), nonbanded block equalizers have limited applicability in OFDM systems with many subcarriers, such as DVB-T/H.

Indeed, in block equalization, a structured model of the frequency-domain channel matrix is essential to reduce the computational complexity of the equalizer and is instrumental for LTV channel estimation, too. For instance, by exploiting the band approximation, a linear block MMSE equalizer based on (7.23) can be expressed by (Rugini, Banelli, & Leus, 2005)

$$\hat{\mathbf{a}} = \mathbf{B}_A^{(D_b)H}(\mathbf{B}_A^{(D_b)}\mathbf{B}_A^{(D_b)H} + \gamma \mathbf{I}_A)^{-1}\mathbf{y}_A. \tag{7.28}$$

Since in (7.28) the matrix to be inverted is banded, the estimated data can be obtained by exploiting banded linear system solving techniques (such as band LDL^H factorization), whose complexity is $\mathcal{O}(D_b^2 A)$ like in the corresponding serial case. Figure 7.4 summarizes the four possible combinations that can be obtained by selecting a block or a serial equalizer and a banded or a full (nonbanded) equalizer. For each of the four models, different equalization criteria and structures are possible, including ZF, MMSE, and nonlinear equalizers.

Alternatively to direct equalization, block (and serial) equalization can be performed relying on *iterative linear equalization*. In contrast to iterative nonlinear equalization, which will be discussed in Section 7.2.2, iterative linear equalizers do not use hard decisions or nonlinearly modified (e.g., hyperbolic tangent) soft decisions of the data. Specifically, the matrix inversion in ZF or MMSE equalizers is avoided by performing an iterative procedure that produces an increasingly improved approximation of the exact result. For example, Li, Yang, Cai, & Gui (2003) have presented an iterative banded block ZF equalizer based on Jacobi iterations. Denoting by κ the iteration index, the received data vector is estimated as

$$\begin{aligned}\hat{\mathbf{a}}^{(\kappa)} &= \mathbf{Q}_A^{-1}\left[\mathbf{y}_A - (\mathbf{B}_A^{(D_b)} - \mathbf{Q}_A)\hat{\mathbf{a}}^{(\kappa-1)}\right] \\ &= \hat{\mathbf{a}}^{(\kappa-1)} + \mathbf{Q}_A^{-1}(\mathbf{y}_A - \mathbf{B}_A^{(D_b)}\hat{\mathbf{a}}^{(\kappa-1)}),\end{aligned}$$

where $\mathbf{Q}_A$ is a diagonal matrix that contains the main diagonal of $\mathbf{B}_A^{(D_b)}$. The term $(\mathbf{B}_A^{(D_b)} - \mathbf{Q}_A)\hat{\mathbf{a}}^{(\kappa-1)}$ represents the soft ICI reconstructed from the previous iteration. Therefore, the ZF equalizer of

[4] We will discuss BEM techniques in the context of channel estimation in Section 7.3.1.

Li et al. (2003) implements a linear parallel ICI cancelation scheme. Since the matrix to be inverted is diagonal, each iteration requires very few computations. Moreover, the convergence of this algorithm to the exact solution $\hat{\mathbf{a}}^{(\infty)} = \mathbf{B}_{\mathrm{A}}^{(D_\mathrm{b})-1}\mathbf{y}_{\mathrm{A}}$ is always guaranteed. The speed of convergence can be slow, especially for some bad channel realizations. However, an acceleration of convergence can be achieved (Molisch, Toeltsch, & Vermani, 2007). The number of iterations and the choice of the initial estimate $\hat{\mathbf{a}}^{(0)}$ highly affect the approximation error of the final data estimate.

Undoubtedly, in block equalization, the system matrix is bigger than for serial approaches. As a consequence, a high condition number can be a significant issue. This problem can be reduced by using Tikhonov regularization, which adds a small term to γ in (7.28) to improve conditioning. Alternatively, γ can be replaced by the inverse modified SINR $\tilde{\rho}^{-1}$, where the modified SINR $\tilde{\rho} \triangleq \frac{1-P_{\mathrm{OOB}}^{(D_\mathrm{b})}}{P_{\mathrm{OOB}}^{(D_\mathrm{b})}+\gamma}$ is obtained by considering the elements of $\mathbf{H}_{\mathrm{A}}$ within the main band as useful terms, and the OOB elements as effective ICI, as expressed by $P_{\mathrm{OOB}}^{(D_\mathrm{b})} \triangleq \|\mathbf{H}_{\mathrm{A}} - \mathbf{B}_{\mathrm{A}}^{(D_\mathrm{b})}\|_{\mathrm{F}}^2/A$. A third option is to employ an iterative equalization with implicit regularization: Tauböck, Hampejs, Matz, Hlawatsch, & Gröchenig (2007) have used an LSQR algorithm to perform iterative banded block ZF equalization constrained to the Krylov subspace generated by $\mathbf{B}_{\mathrm{A}}^{(D_\mathrm{b})H}\mathbf{B}_{\mathrm{A}}^{(D_\mathrm{b})}$ and $\mathbf{B}_{\mathrm{A}}^{(D_\mathrm{b})H}\mathbf{y}_{\mathrm{A}}$. In the LSQR algorithm, the conditioning improvement is obtained by early termination of the iterative algorithm, which also helps in saving complexity.

7.2.1.3 Receiver Windowing

Banded equalizers sometimes employ time-domain receiver windowing techniques to concentrate the ICI into the main band of $\mathbf{H}_{\mathrm{A}}$ so that the band approximation is more accurate. This ICI shortening technique can be viewed as the dual of ISI channel shortening for single-carrier systems with LTI multipath channels. Receiver windowing is compatible with both serial (Schniter, 2004) and block (Rugini, Banelli, & Leus, 2006) approaches and can be used also in conjunction with nonlinear equalization. To examine receiver windowing, we define $\mathbf{y}_{\mathrm{W}} \triangleq \mathbf{W}\boldsymbol{\Delta}\mathbf{R}_{\mathrm{CP}}\mathbf{r}$, where $\boldsymbol{\Delta}$ is the $L \times L$ diagonal matrix representing the time-domain windowing operation, performed before the DFT at the receiver. The OFDM signal model of (7.3) can then be replaced by

$$\mathbf{y}_{\mathrm{W}} = \mathbf{W}\boldsymbol{\Delta}\mathbf{H}_{\mathrm{T}}\mathbf{W}^H\mathbf{x} + \mathbf{W}\boldsymbol{\Delta}\mathbf{R}_{\mathrm{CP}}\mathbf{w} = \mathbf{H}_{\mathrm{W}}\mathbf{x} + \mathbf{z}_{\mathrm{W}}, \tag{7.29}$$

where $\mathbf{H}_{\mathrm{W}} \triangleq \mathbf{W}\boldsymbol{\Delta}\mathbf{H}_{\mathrm{T}}\mathbf{W}^H$ is the frequency-domain windowed channel matrix and $\mathbf{z}_{\mathrm{W}} \triangleq \mathbf{W}\boldsymbol{\Delta}\mathbf{R}_{\mathrm{CP}}\mathbf{w}$ is the noise after windowing. It is interesting to note that $\mathbf{H}_{\mathrm{W}} = \boldsymbol{\Gamma}\mathbf{H}_{\mathrm{F}}$ and $\mathbf{z}_{\mathrm{W}} = \boldsymbol{\Gamma}\mathbf{z}$, where $\boldsymbol{\Gamma} \triangleq \mathbf{W}\boldsymbol{\Delta}\mathbf{W}^H$ is a circulant filtering matrix that models receiver windowing (i.e., ICI shortening) in the frequency-Doppler domain. As a result of $\boldsymbol{\Gamma}$, the noise $\mathbf{z}_{\mathrm{W}}$, though Gaussian, is no longer white. Obviously, by selecting $\boldsymbol{\Delta} = \mathbf{I}_L$, (7.29) reduces to classical OFDM and coincides with (7.3).

Receiver windowing does not affect the performance of nonbanded linear block equalizers, since it only performs a linear operation on the received signal. Nevertheless, when coupled with the band approximation, the OOB ICI energy, which is neglected by banded equalizers, can be greatly reduced, thereby improving performance considerably. From a performance viewpoint, a good window design criterion could be the minimization of the mean-squared error (MSE) on the decision variable. However, a closed-form solution to this minimization problem is hard to find. Therefore, common design criteria target the windowed matrix $\mathbf{H}_{\mathrm{W}}$ rather than the MSE on the data. For instance, the *Max-Average*

SINR criterion of Schniter (2004) maximizes the average input SINR, expressed by

$$\rho_{\mathrm{IN}}^{(D_b)}(\boldsymbol{\Delta}) = \frac{\mathsf{E}\{\|\mathbf{B}_{\mathrm{W}}^{(D_b)}\|_F^2\}}{\mathsf{E}\{\|\mathbf{H}_{\mathrm{W}} - \mathbf{B}_{\mathrm{W}}^{(D_b)}\|_F^2\} + \sigma_z^2 \mathrm{tr}\{\boldsymbol{\Gamma}\boldsymbol{\Gamma}^H\}}, \tag{7.30}$$

where $\mathbf{B}_{\mathrm{W}}^{(D_b)}$ is the cyclically banded matrix that contains the $2D_b + 1$ central diagonals of $\mathbf{H}_{\mathrm{W}}$. Of course, the maximization of (7.30) is subject to the window energy constraint $\mathrm{tr}\{\boldsymbol{\Delta}\boldsymbol{\Delta}^H\} = L$. Similarly, the *minimum band approximation error (MBAE)* criterion of Rugini et al. (2006) looks for the window that minimizes the OOB ICI energy $\mathsf{E}\{\|\mathbf{H}_{\mathrm{W}} - \mathbf{B}_{\mathrm{W}}^{(D_b)}\|_F^2\}$, with the additional constraint that the window is the sum of $2D_b + 1$ exponential (SOE) functions, as expressed by

$$\boldsymbol{\delta} \triangleq \mathrm{diag}\{\boldsymbol{\Delta}\} = \tilde{\mathbf{W}}^{(D_b)} \boldsymbol{\eta}^{(D_b)},$$

where $\tilde{\mathbf{W}}^{(D_b)}$ is an $L \times (2D_b + 1)$ matrix that contains the first $D_b + 1$ and the last D_b columns of $\mathbf{W}$, and $\boldsymbol{\eta}^{(D_b)}$ is a vector of size $2D_b + 1$ containing the window coefficients. The MBAE solution $\boldsymbol{\eta}_{\mathrm{MBAE}}^{(D_b)}$ with the SOE constraint is the eigenvector that corresponds to the maximum eigenvalue of $\tilde{\mathbf{W}}^{(D_b)H} \mathbf{A} \tilde{\mathbf{W}}^{(D_b)}$, where $\mathbf{A}$ is an $L \times L$ Toeplitz matrix defined by

$$[\mathbf{A}]_{m,n} \triangleq r_{t,\mathbf{H}}[n-m] \frac{\sin(\pi(2D_b + 1)(n-m)/L)}{L \sin(\pi(n-m)/L)}. \tag{7.31}$$

Hence, the window depends on the selected parameter D_b, and on the Doppler power profile (of the discrete-time channel) through the time-domain autocorrelation $r_{t,\mathbf{H}}[m]$ (see comments after (7.13)).

Other criteria than Max-Average SINR and MBAE-SOE are possible. For instance, different types of input SINR could be defined. The Max-SINR criterion of Schniter (2004) considers the instantaneous input SINR rather than the average input SINR. This translates into a window that depends on the LTV channel realization rather than on the LTV channel statistics. In this case, the window design must be repeated for each OFDM block. Das & Schniter (2007) have proposed a window design that considers: the elements on the main diagonal as useful signal; the other elements on the dominant diagonals as *don't-care* values; and the elements on the other diagonals as interference. The interference power also includes other disturbances, such as the ISI coming from the previous OFDM block when the CP is short or absent.

A nice feature of the window design with the SOE constraint is that the circulant matrix $\boldsymbol{\Gamma}$, which represents the frequency-domain noise after windowing, is cyclically banded (with bandwidth $2D_b + 1$). This can be exploited for low-complexity equalization. The banded linear block MMSE equalizer (Rugini et al., 2006) can be expressed by (see (7.28))

$$\hat{\mathbf{a}} = \mathbf{B}_{\mathrm{WA}}^{(D_b)H} \left(\mathbf{B}_{\mathrm{WA}}^{(D_b)} \mathbf{B}_{\mathrm{WA}}^{(D_b)H} + \gamma \boldsymbol{\Gamma}_{\mathrm{A}} \boldsymbol{\Gamma}_{\mathrm{A}}^H\right)^{-1} \mathbf{y}_{\mathrm{WA}}, \tag{7.32}$$

where $\mathbf{B}_{\mathrm{WA}}^{(D_b)} \triangleq \mathbf{R}_{\mathrm{GB}} \mathbf{B}_{\mathrm{W}}^{(D_b)} \mathbf{T}_{\mathrm{GB}}$, $\boldsymbol{\Gamma}_{\mathrm{A}} \triangleq \mathbf{R}_{\mathrm{GB}} \boldsymbol{\Gamma}$, and $\mathbf{y}_{\mathrm{WA}} \triangleq \mathbf{R}_{\mathrm{GB}} \mathbf{y}_{\mathrm{W}}$ are obtained by excluding the guard bands. Since $\mathbf{B}_{\mathrm{W}}^{(D_b)}$ and $\boldsymbol{\Gamma}$ are cyclically banded, when the guard band on each side has size $P/2 \geq D_b$, $\mathbf{B}_{\mathrm{WA}}^{(D_b)}$ is banded with bandwidth $2D_b + 1$, and the matrix to be inverted in (7.32) is banded with bandwidth $4D_b + 1$. Therefore, as in the absence of windowing, simple equalizers can be employed, with linear complexity in the number of subcarriers (Rugini et al., 2006).

The main advantage of receiver windowing lies in its extremely low additional complexity, despite the significant performance improvement. We note that good window designs require the knowledge of the channel statistics, such as the normalized maximum Doppler shift $\vartheta_{\max}$ and the shape of the Doppler power profile. In the absence of channel statistics, suboptimal windows can be employed, such as those used for spectral estimation (e.g., Hamming, Bartlett, Gaussian) (Harris, 1978), at the price of a reduced performance improvement. A performance comparison of different windows has been presented by Peiker, Dominicus, Teich, & Lindner (2008), assuming one-tap equalization and an additional cyclic extension (postfix).

7.2.1.4 *Performance-Complexity Trade-Off*

We now compare some representative linear equalizers in terms of simulated BER performance and computational complexity. We consider an OFDM system with $L = 128$ subcarriers, of which $A = 96$ are active, and QPSK modulated data. We assume a WSSUS Rayleigh fading channel with truncated exponential delay power profile $\mathsf{E}\{|h[n,m]|^2\} = \alpha e^{-0.6m}$, where α is a normalization constant. The channel length is chosen as $M = 9$, and consequently, the CP length is set to $L_{\mathrm{CP}} = 8$. Regarding the time variation of the channel, we assume a Jakes' Doppler power profile with $\vartheta_{\max} = 0.12$, i.e., the maximum Doppler frequency $\nu_{\max}$ is 12% of the subcarrier spacing F.

Figure 7.5 shows the BER performance of the following linear equalizers:

- Conventional one-tap equalizer;
- Full block ZF and MMSE equalizers (Choi et al., 2001);
- Banded serial ZF equalizer (Jeon et al., 1999);

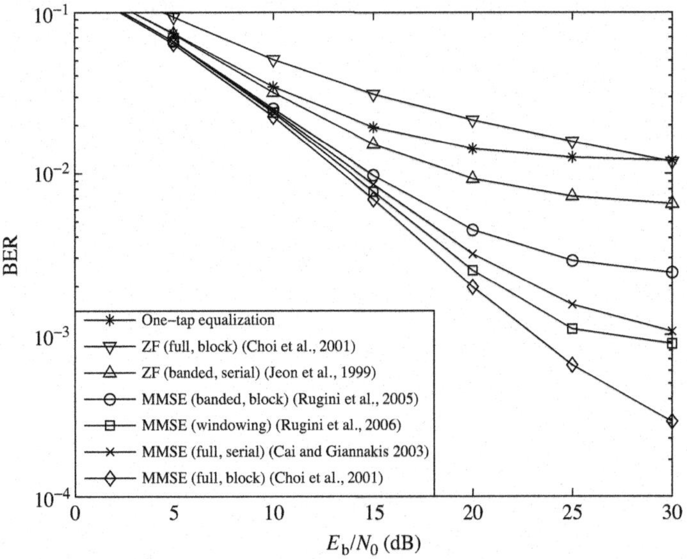

FIGURE 7.5

BER performance comparison of linear equalizers.

- Full serial MMSE equalizer (Cai & Giannakis, 2003);
- Banded block MMSE equalizer (Rugini et al., 2005) and its window-aided version (Rugini et al., 2006).

The matrix bandwidth parameter of banded equalizers is $D_b = 2$, i.e., only $2D_b + 1 = 5$ diagonals are considered. Similarly, serial equalizers only consider $D_b = 2$ subcarriers for each side, and hence, the observation vector length is $2D_b + 1 = 5$. The receiver window is designed using the MBAE-SOE criterion (Rugini et al., 2006), assuming perfect knowledge of the Doppler power profile. To avoid ill-conditioning problems at high SNR, in the absence of windowing, the banded block MMSE equalizer (Rugini et al., 2005) exploits a Tikhonov regularization, i.e., when the SNR $E_s/N_0 = \log_2(N_a)E_b/N_0$ exceeds 20 dB, the equalizer assumes a virtual SNR of 20 dB. All the equalizers exploit perfect channel-state information (CSI) at the receiver.

From the results of Fig. 7.5, it is clear that there exists a big performance gap between the ZF and MMSE equalizers. This confirms that doubly selective channels are ill conditioned, since an MMSE equalizer can be interpreted as a regularized ZF equalizer. Among the MMSE equalizers, the best performance is obtained by the full block approach of Choi et al. (2001), whose complexity per OFDM block is however cubic in the number of subcarriers. Therefore, the complexity for the full block MMSE equalizer of Choi et al. (2001) is $\mathcal{O}(A^2)$ per symbol, where $A^2 \approx 10^4$. The full serial MMSE equalizer of Cai & Giannakis (2003) is able to reduce the computational complexity to about $\mathcal{O}(D_b A)$ per symbol, with $(2D_b + 1)A \approx 500$, at a price of a modest performance loss. The banded block MMSE equalizer is able to significantly reduce complexity, since the number of complex operations per equalized symbol is $C = 8D_b^2 + 22D_b + 4 = 80$ (Rugini et al., 2005), plus $2D_b + 1 = 5$ additional complex operations per symbol when windowing is included (Rugini et al., 2006).

Despite the lower complexity, the banded block MMSE equalizers maintain a good BER performance: specifically, due to the statistical CSI knowledge (summarized by the Doppler power profile), the window-aided banded block MMSE equalizer (Rugini et al., 2006) is able to outperform the full serial MMSE equalizer (Cai & Giannakis, 2003) with respect to both performance and complexity.

Figure 7.6 presents a BER performance comparison of banded block MMSE equalizers as a function of the normalized maximum Doppler shift ϑ_{max}, for the same scenario previously described, when $E_b/N_0 = 20$ dB. For comparison purposes, also the conventional one-tap equalizer and the full block MMSE equalizer (Choi et al., 2001) are considered. Clearly, to maintain a fixed performance, the matrix bandwidth size D_b should be increased as ϑ_{max} grows, especially when receiver windowing is not used. However, the computational complexity increases quadratically with D_b, ranging from $C = 8D_b^2 + 22D_b + 4 = 34$ complex operations per symbol when $D_b = 1$ to $C = 220$ complex operations per symbol when $D_b = 4$. Moreover, when D_b increases, more matrix parameters have to be estimated, and hence, a more powerful channel estimator is required.

7.2.2 Nonlinear Equalization

A nonlinear equalizer estimates the data symbols by applying a nonlinear operation on the received vector. A typical configuration of a nonlinear equalizer consists of a first linear stage that produces some tentative data decisions and a second nonlinear stage that cancels the ICI using the tentative decisions. This configuration includes decision-feedback equalization, parallel ICI cancelation, successive ICI cancelation, and many other types of interference cancelation techniques. In addition, similarly to

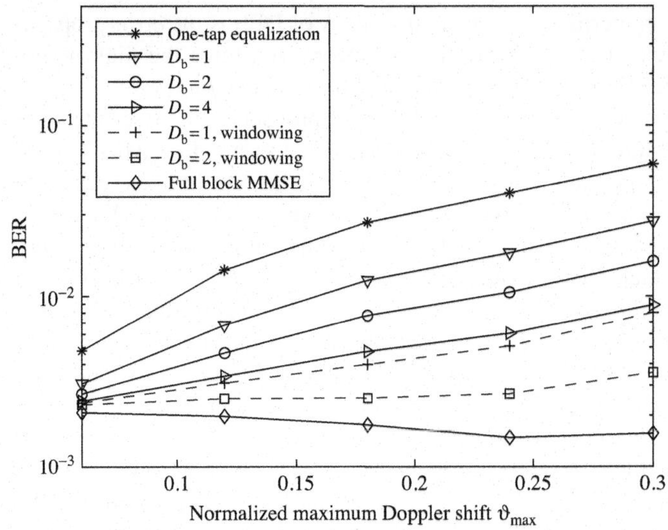

FIGURE 7.6

BER performance comparison of banded block MMSE equalizers.

multipath channel equalization for single-carrier systems, there exists a large variety of other nonlinear equalizer structures, including maximum-likelihood (ML) methods and turbo approaches. In all cases, as for linear equalizers, nonlinear equalizers can be classified as serial or block methods, banded or nonbanded approaches, window-aided or nonwindow-aided techniques.

Generally, a nonlinear equalizer performs better than a linear equalizer, although in many cases the computational complexity increases, especially for ML approaches. In the following, we review the most common techniques for OFDM systems with LTV channels.

7.2.2.1 Decision-Feedback Equalizers

Decision-feedback equalization (DFE) is characterized by a feedforward filter that reduces the ICI produced by the not-yet-detected symbols and a feedback filter that cancels the ICI produced by the already-detected symbols. For block equalizers, the soft-detected data can be expressed by

$$\hat{\mathbf{a}} = \mathbf{F}_{\mathrm{F}}\mathbf{y}_{\mathrm{A}} - \mathbf{F}_{\mathrm{B}}\check{\mathbf{a}}, \tag{7.33}$$

where $\mathbf{F}_{\mathrm{F}}$ is the $A \times A$ feedforward filter matrix, $\mathbf{F}_{\mathrm{B}}$ is the $A \times A$ feedback filter matrix, and $\check{\mathbf{a}}$ contains the already-estimated hard-detected data symbols. Usually, the data symbols are detected sequentially, starting from the first (last) subcarrier; in this case, $\mathbf{F}_{\mathrm{B}}$ should be strictly lower (upper) triangular, which guarantees that the not-yet-detected symbols are not fed back.

To design the DFE filters, the ZF or MMSE criterion can be used. Since a linear (ZF or MMSE) equalizer can be obtained as a degenerate case of DFE with $\mathbf{F}_{\mathrm{B}} = \mathbf{0}_{A\times A}$, DFE approaches generally outperform their corresponding linear counterparts. The main drawback of DFE is the error propagation due to the bad cancelation of an incorrectly detected symbol. Moreover, often the filter design optimistically assumes perfect (error-free) feedback, neglecting the error propagation.

For DFE, both serial and block approaches are possible. In block approaches, the two filters are jointly designed for all the subcarriers. Rugini et al. (2006) have presented some block MMSE DFE receivers that also incorporate a band approximation and receiver windowing. As in linear equalization, the band approximation is used to obtain $\mathbf{F}_\mathrm{F}\mathbf{y}_\mathrm{A}$ with reduced complexity. Moreover, $\mathbf{F}_\mathrm{B}$ is also banded so that only $2D_\mathrm{b}$ symbols are fed back, thereby reducing the error propagation. As a result, the computational complexity of the block DFE of Rugini et al. (2006) is $\mathcal{O}(D_\mathrm{b}^2 A)$ per block. This complexity, which is lower compared to nonbanded approaches, is balanced by a performance loss that increases the error floor. However, also in the banded case, DFE outperforms linear equalization with basically the same complexity. The banded block DFE can also be coupled with receiver windowing, leading to a significant reduction of the error floor. However, the complexity is approximately doubled with respect to the nonwindowed DFE (Rugini et al., 2006).

In the serial case, the feedforward filter (different from subcarrier to subcarrier) acts on a few elements of the received vector, e.g., on $\mathbf{y}_\mathrm{A}^{(D_\mathrm{b})}[l]$ in (7.25). One example is the serial MMSE DFE proposed by Cai & Giannakis (2003), where the filters are computed recursively from the filters used for the previously detected subcarrier. This recursive procedure reduces the complexity to $\mathcal{O}(D_\mathrm{b}A^2)$ per block. A specific feature of the DFE of Cai & Giannakis (2003) relies on its cyclic ordering for successive cancelation. Consequently, the "best" subcarrier can be chosen as starting point instead of one of the edge subcarriers. This produces a clear connection with SIC equalizers, discussed in the following subsection.

7.2.2.2 ICI Cancellers

The concept of ICI cancelation, introduced for DFE above, is exploited also by other equalization structures, such as *successive interference cancelation (SIC)* equalizers. Also SIC equalizers have two stages, with the first producing tentative decisions and the second subtracting the regenerated ICI. However, differently from DFE, SIC equalizers perform an ordered ICI cancelation, in such a way that reliably detected subcarriers are detected first. Due to this subcarrier ordering, the probability of error propagation is reduced, especially for the first subtractions. However, a sorting procedure is necessary to establish the subcarrier ordering. This can be problematic for banded equalizers, since subcarrier sorting destroys the banded structure of the frequency-domain channel matrix.

In the technical literature, different options have been considered for the first tentative data detection: conventional one-tap equalization (Leung & Ho, 1998), nonbanded linear block MMSE equalization (Choi et al., 2001), and banded linear serial MMSE equalization (Kim & Park, 2006; Lu, Kalbasi, & Al-Dhahir, 2006). Also the detection order can be chosen using different criteria: postdetection SINR (Choi et al., 2001; Lu et al., 2006), magnitude of the diagonal elements of the channel matrix $\mathbf{H}_\mathrm{F}$ (Kim & Park, 2006), and distance between soft and hard estimates produced by the first stage (Leung & Ho, 1998). The subcarrier order can be updated during the ICI subtraction, as in the nulling-canceling approach of Choi et al. (2001). This implies an increased complexity due to multiple sorting. In addition, many cancelation stages can be employed, as proposed by Leung & Ho (1998), who basically used an iterative nonlinear equalizer.

A closely related technique is *parallel interference cancelation (PIC)*, where the ICI of all the symbols is jointly canceled in a block fashion. The first estimate is typically obtained by one-tap MMSE equalization (Chen & Yao, 2004; Gorokhov & Linnartz, 2004) or by serial approaches (Chang, Han, Ha, & Kim, 2006; How & Chen, 2005). Banded cancelation is used to save complexity with small performance loss, since only the relevant ICI produced by a few subcarriers is subtracted. An improved PIC approach can be obtained by replacing the hard cancelation by reliability-based nonlinear soft cancelation (Molisch et al., 2007), where the hyperbolic tangent function is used to control the amount of ICI cancelation. Molisch et al. (2007) have also included a performance comparison with a SIC scheme.

Huang, Letaief & Lu (2005) have applied bit-interleaved coded modulation over multiple OFDM blocks. This scheme employs a reduced ML decoder obtained by approximating the LTV channel as constant over a single OFDM block. Since the reduced ML decoder neglects the ICI, its effectiveness is limited to low $\vartheta_{\max}$. For high Doppler spreads, the uncoded BER floor is too high, and the channel decoder even increases the BER. Therefore, Huang et al. (2005) have also included a PIC equalizer driven by a linear MMSE equalizer that works over multiple OFDM symbols.

Hybrid approaches that combine PIC with SIC are also possible, such as in *groupwise interference cancelation*. Basically, the set of subcarriers is split into a certain number of groups of subcarriers. Then, ICI cancelation within the group is performed in a parallel way, whereas the ICI among different groups is subtracted in a successive way. This reduces the ordering problem, because the number of groups is smaller than the number of subcarriers. Commonly, the groups contain consecutive subcarriers, but reliability-based subcarrier grouping criteria are sometimes used. Some examples of these techniques have been presented by Vogeler, Brötje, Klenner, Kühn, & Kammeyer (2004); Tran & Fujino (2005); Song, Kim, Nam, Yu, & Hong (2008); and Hampejs et al. (2009).

7.2.2.3 Near-ML Equalizers

Assuming the block model (7.22), the ML equalizer is expressed by

$$\hat{\mathbf{a}} = \arg\min_{\mathbf{s}} \|\mathbf{y}_{\mathrm{A}} - \mathbf{H}_{\mathrm{A}}\mathbf{s}\|, \qquad (7.34)$$

where $\mathbf{s}$ is a generic possible data vector. Among the various approaches, the ML equalizer gives the best performance, since (7.34) minimizes the conditional probability of block error $\Pr\{\hat{\mathbf{a}} \neq \mathbf{a}|\mathbf{H}_{\mathrm{A}}\}$. On the other hand, the ML approach is characterized by the worst complexity $\mathcal{O}(N_a^A)$, where N_a is the constellation size, i.e., the complexity is exponential in the number of active subcarriers A. Hence, a major goal is to find a low-complexity yet good approximation of the exact ML equalizer. In theory, most of the methods already developed for multiuser detection of CDMA signals could be employed, but the specific structure of the OFDM channel matrix and the potentially large number of subcarriers should be taken into account to avoid prohibitive complexity.

Using the band approximation $\mathbf{H}_{\mathrm{A}} \approx \mathbf{B}_{\mathrm{A}}^{(D_{\mathrm{b}})}$, Ohno (2005) proposed a banded block ML equalizer that reduces the equalization complexity up to $\mathcal{O}(D_{\mathrm{b}}N_a^{2D_{\mathrm{b}}+1}A)$. This is achieved by employing a Viterbi algorithm with a reduced number of surviving paths. However, in the case of a large constellation size N_a, complexity is still an issue. In addition, the Viterbi equalizer proposed by Ohno (2005) assumed white noise and therefore is not compatible with receiver windowing.

A second type of ML approximation consists in performing, for a specific subcarrier, a *local ML search* that only considers the neighboring subcarriers, with a philosophy that is similar to the blind time-domain equalizer of Cui & Tellambura (2007). This approach can be regarded as the serial version of the banded ML equalizer.

Similarly, a third type of quasi-ML equalizers can be established by employing *sphere decoding* (see Section 3.8, Section 8.3.4 and references therein) or other *tree-search* techniques (see Section 6.3.2.4). For instance, Hwang & Schniter (2006) have applied a breadth-first search based on the T-algorithm, in a multicarrier system with transmitter and receiver windowing. This specific tree-search algorithm is coupled with a banded block MMSE DFE preprocessing with cyclic ordering, practically leading to ML performance with reduced complexity (below $\mathcal{O}(L^{2.4})$ per block) (Hwang & Schniter, 2006). Another tree-search algorithm has been investigated by Chow & Jeremic (2006).

As a fourth option, the optimization problem (7.34) can be relaxed to an equality-constrained *quadratic programming* problem, which is solved iteratively (Kou, Lu, & Antoniou, 2005). This approach can be extended to QAM as proposed by Zhang, Lu, & Gulliver (2007), which also reduces the equalization complexity by using a subspace constraint.

Mixed approaches are also possible. For instance, the groupwise approach of Feng, Minn, Yan, & Jinhui (2010) employs semidefinite relaxation to mitigate the ICI within a group of adjacent subcarriers and a PIC technique to reduce the ICI coming from the other groups of subcarriers.

7.2.2.4 Iterative and Turbo Approaches

Differently from the linear ones, iterative nonlinear equalizers perform nonlinear operations to iteratively update the data estimate. From this viewpoint, many ICI cancelation schemes that we described previously are also iterative. As a consequence, in this section, we describe those iterative nonlinear equalizers that have also other specific features.

As an example, equalization can be combined with channel estimation and with forward error correction decoding. Tomasin, Gorokhov, Yang, & Linnartz (2005) have presented an iterative channel estimator and a PIC equalizer that reuses the output of the convolutional decoder. The frequency-diversity gain provided by channel coding allows for a reliable ICI estimate, which produces improved performance (at least at medium-to-high SINR) but also increases the decoding delay. Joint equalization and channel estimation is also performed by Mostofi & Cox (2005).

A remarkable iterative structure is the turbo equalizer proposed by Schniter (2004), which is based on a window-assisted serial linear MMSE equalizer. In the equalizer of Schniter (2004), the symbol a_l is iteratively estimated using a linear MMSE (LMMSE) criterion, as expressed by

$$\hat{a}_l = \mu_l + \mathbf{e}^T_{4D_b+1,2D_b+1}\mathbf{B}_W^{(D_b)H}[l]\mathbf{R}_W^{(D_b)-1}[l](\mathbf{y}_W^{(D_b)}[l] - \mathbf{B}_W^{(D_b)}[l]\boldsymbol{\mu}^{(D_b)}[l]). \tag{7.35}$$

Here, μ_l is the a priori mean of a_l, $\boldsymbol{\mu}^{(b)}[l] \triangleq \left(\mu_{l-2D_b} \cdots \mu_{l-1}\ 0\ \mu_{l+1} \cdots \mu_{l+2D_b}\right)^T$ is the a priori mean of the data vector $\mathbf{a}^{(D_b)}[l]$ (except for the middle symbol, which is set to 0 instead of μ_l), $\mathbf{R}_W^{(D_b)}[l] \triangleq \mathbf{B}_W^{(D_b)}[l]\mathbf{V}^{(D_b)}[l]\mathbf{B}_W^{(D_b)H}[l] + \gamma\boldsymbol{\Gamma}^{(D_b)}[l]\boldsymbol{\Gamma}^{(D_b)H}[l]$ is the matrix to be inverted, with $\mathbf{V}^{(D_b)}[l] \triangleq \text{diag}\left(\nu_{l-2D_b} \cdots \nu_{l-1}\ 1\ \nu_{l+1} \cdots \nu_{l+2D_b}\right)$ the diagonal matrix that contains the a priori variances (except for a_l) and $\boldsymbol{\Gamma}^{(D_b)}[l]$ the $(2D_b+1) \times L$ matrix obtained by selecting the rows of $\boldsymbol{\Gamma}$ from index $l - D_b$ to $l + D_b$. After LMMSE symbol estimation, the iterative procedure of Schniter (2004) updates the log-likelihood ratio (LLR) $\text{L}(a_l|\hat{a}_l)$, which is used to update the a priori means and variances (for QPSK, $\mu_l = \tanh(\text{L}(a_l|\hat{a}_l)/2)$ and $\nu_l = 1 - |\mu_l|^2$), to be used for symbol estimation in the next iteration. Different algorithms are possible depending on how the a priori quantities are updated: for instance, in order to obtain $\hat{a}_l^{(\kappa)}$ in the κth iteration, (7.35) can use the a priori quantities calculated in the previous iteration $\left\{\mu_{l-2D_b}^{(\kappa-1)}, \nu_{l-2D_b}^{(\kappa-1)}, \ldots, \mu_{l-1}^{(\kappa-1)}, \nu_{l-1}^{(\kappa-1)}, \mu_{l+1}^{(\kappa-1)}, \nu_{l+1}^{(\kappa-1)}, \ldots, \mu_{l+2D_b}^{(\kappa-1)}, \nu_{l+2D_b}^{(\kappa-1)}\right\}$, or it can employ also some a priori values already calculated in the current iteration $\left\{\mu_{l-2D_b}^{(\kappa)}, \nu_{l-2D_b}^{(\kappa)}, \ldots, \mu_{l-1}^{(\kappa)}, \nu_{l-1}^{(\kappa)}, \mu_{l+1}^{(\kappa-1)}, \nu_{l+1}^{(\kappa-1)}, \ldots, \mu_{l+2D_b}^{(\kappa-1)}, \nu_{l+2D_b}^{(\kappa-1)}\right\}$. These two updating strategies correspond to a block-wise update and a serial-wise update, respectively. The simulation results of Schniter (2004) show that the serial-wise update produces a better performance, since the newly acquired a priori information is used as soon as it is available, thereby improving the convergence of the iterative algorithm. In both cases, the computational complexity is linear in the number of subcarriers and in the number of iterations.

Alternatively, the serial MMSE equalizer (7.35) can be replaced by a block MMSE equalizer that jointly calculates all the a priori values (Fang, Rugini, & Leus, 2008). In this case, only the block-wise a priori update is possible. Turbo block MMSE equalization can be related to probabilistic data association, which is commonly considered a quasi-ML technique. Indeed, in the presence of receiver windowing, the turbo block MMSE equalizer (Fang et al., 2008) outperforms the corresponding serial version, improving the BER performance at medium SNR. The computational complexity is similar to that of Schniter (2004).

Summarizing, many iterative ICI mitigation techniques have been presented in the literature, exploiting serial or block MMSE equalization, receiver windowing, and serial or block a priori LLR updating, sometimes incorporating channel estimation or channel decoding into the turbo loop. Obviously, performance and complexity highly depend on the type of iterative scheme and on the number of iterations. Therefore, a thorough comparison is difficult, except for some specific cases. A comparison between some selected schemes has been performed by Schniter (2004) and Fang et al. (2008). A general drawback of iterative nonlinear equalizers is the difficulty of a theoretical convergence analysis. Usually, the number of iterations is selected heuristically or by means of EXIT charts (ten Brink, 2001).

7.2.2.5 *Performance-Complexity Trade-Off*

First, we compare a linear block equalizer with a nonlinear block DFE approach. Both equalizers are designed using the MMSE criterion. The simulation scenario is that of Fig. 7.5, i.e., $L = 128$ subcarriers ($A = 96$ active), QPSK modulation, WSSUS Rayleigh channel with Jakes' Doppler power profile with $\vartheta_{\max} = 0.12$, and with truncated exponential delay power profile $\mathsf{E}\{|h[n,m]|^2\} = \alpha e^{-0.6m}$ of length $M = 9$. Banded equalizers use perfect CSI, $D_b = 2$, and Tikhonov regularization. MBAE-SOE windowing is employed.

Figure 7.7 shows the BER performance of three linear block MMSE equalizers (banded, window-aided, and full) and of three block MMSE DFE receivers (banded, window-aided, and full). The use of DFE produces a noticeable performance improvement. This improvement is more evident in the presence of windowing, e.g., when the computational complexity of DFE is roughly doubled with respect to linear approaches (Rugini et al., 2006).

We next compare two nonlinear approaches, focusing on iterative (turbo) MMSE equalizers. We again assume the simulation parameters of Fig. 7.7, except the channel length, which is $M = 17$ in this case, and the CP length $L_{CP} = 16$. Figure 7.8 displays the BER performance of two window-aided turbo banded MMSE equalizers using either a block approach (Fang et al., 2008) or a serial approach (Schniter, 2004). For both cases, $D_b = 2$ and MBAE-SOE windowing is adopted. The block equalizer employs Tikhonov regularization when the SNR E_s/N_0 exceeds 25 dB. For the serial equalizer, the

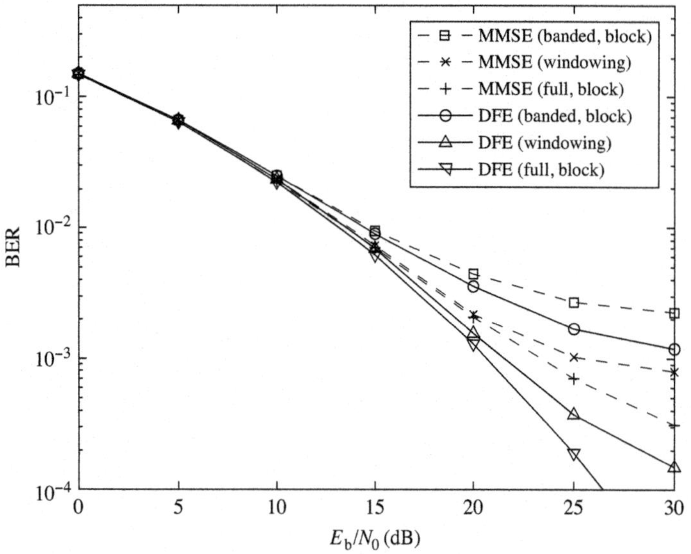

FIGURE 7.7

Block MMSE equalizers: BER performance comparison between linear and DFE versions.

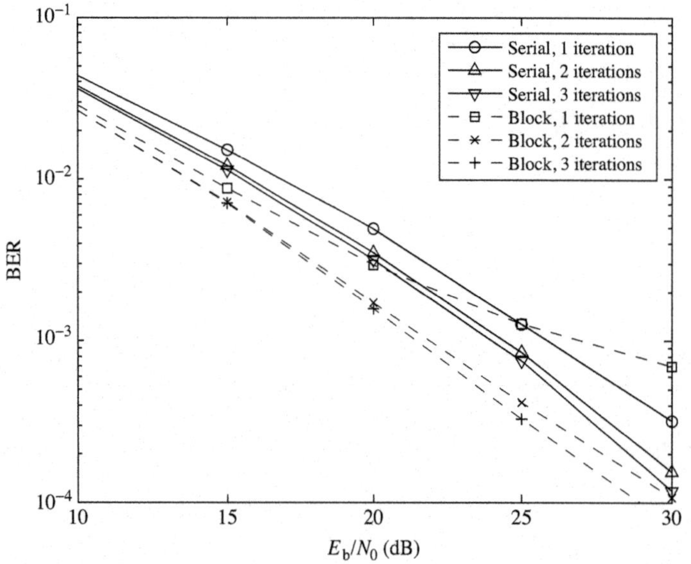

FIGURE 7.8

Iterative (turbo) banded MMSE equalizers: BER performance comparison between serial and block versions.

serial-wise LLR updating is considered, since it outperforms the block-wise LLR updating (Schniter, 2004). Figure 7.8 shows that the serial equalizer exhibits a slight performance loss at low SNR. This is mainly caused by the presence of windowing: since the window produces noise correlation across the subcarriers, considering only few subcarriers simultaneously is suboptimal. Block approaches jointly equalize all the subcarriers and therefore do not suffer from this loss. By focusing on the first iteration, obtained by linear equalization, it can be observed that the serial equalizer outperforms the block equalizer at very high SNR. This is mainly due to the ill-conditioning of the frequency-domain channel matrix, which is more problematic in the block case because of the larger size of the channel matrix. However, after the second iteration, the block equalizer is able to recover the gap and to outperform the corresponding serial version. More iterations improve the performance only slightly, in both cases. As far as complexity is concerned, both versions have linear complexity with respect to the number of subcarriers and the number of iterations. However, serial equalizers are more complex, with $\mathcal{C}_{\text{serial}} \approx 1.75\,\mathcal{C}_{\text{block}}$ when $D_{\text{b}} = 2$ and $\mathcal{C}_{\text{serial}} \approx 2.50\,\mathcal{C}_{\text{block}}$ when $D_{\text{b}} = 4$. On the other hand, serial equalizers deal with matrices of smaller size and hence in general are characterized by reduced memory requirements.

7.2.3 Transmitter Preprocessing

In Sections 7.2.1 and 7.2.2, we presented a broad overview of receiver processing techniques that are able to reduce the ICI power in CP-based OFDM systems with LTV channels. Current OFDM systems, which are designed for LTI channels, allow for the use of alternative receiver techniques when the channel is LTV. More refined approaches modify the OFDM transmission scheme in order to counteract the Doppler effect *before* the signal is received. As a result of transmitter processing, the need for equalization is highly reduced.

When the time variation of the channel is significant, it is not reasonable to assume full knowledge of the CSI at the transmitter. Therefore, CSI transmitter processing techniques used for LTI channels, such as ZF pre-equalization or Tomlinson–Harashima precoding, are not appropriate for LTV channels. However, statistical CSI knowledge of the WSSUS channel can be helpful.

In this section, we describe some common techniques to cope with the channel time variation at the transmitter. Some of these techniques have been originally proposed to counteract the ICI produced by a CFO, but have been successfully applied also to other types of Doppler distortions. We split these transmitter techniques into two categories: (1) data precoding, which only acts on the data to be transmitted, and (2) pulse shaping, which instead operates on the transmitted signal waveform. In general, data precoding requires a minor modification of current OFDM standards, since the precoded data can be transmitted by standard OFDM systems. On the other hand, pulse-shaping techniques require additional filtering at the transmitter and at the receiver.

7.2.3.1 Data Precoding

In OFDM, linear precoding of the frequency-domain data vector can result in a multipath diversity gain even in the absence of Galois-field channel coding (Wang & Giannakis, 2003). Similarly, data precoding across tones can be exploited aiming at ICI reduction (Zhang & Li, 2003; Zhao & Häggman, 2001). Two types of data precoding methods exist: redundant and nonredundant. Redundant linear precoding is performed by means of a tall $L \times A$ precoding matrix $\mathbf{P}$ applied to the data vector $\mathbf{a}$.

By using (7.21), we can express the received data as

$$\mathbf{y} = \mathbf{H}_{\mathrm{F}}\mathbf{P}\mathbf{a} + \mathbf{z}.$$

In other words, A data symbols share $L > A$ subcarriers. The most popular redundant precoding approach, known as *self-cancelation* or *polynomial cancelation coding* (Armstrong, 1999; Zhao & Häggman, 2001), exploits the ICI correlation of nearby subcarriers. For instance, the rate-$1/2$ scheme uses $\mathbf{P} = \mathbf{I}_{L/2} \otimes (1\ -1)^T$, where $\otimes$ denotes the Kronecker product. Hence, the same symbol is transmitted on two consecutive subcarriers. At the receiver, the data symbols are estimated as

$$\hat{\mathbf{a}} = \mathbf{P}^H\mathbf{Q}^{-1}\mathbf{y} = \mathbf{P}^H\mathbf{Q}^{-1}\mathbf{H}_{\mathrm{F}}\mathbf{P}\mathbf{a} + \mathbf{P}^H\mathbf{Q}^{-1}\mathbf{z},$$

where $\mathbf{Q} = \mathrm{diag}\{\mathbf{H}_{\mathrm{F}}\}$, i.e., one-tap equalization is used. We note that the data received on two consecutive subcarriers are subtracted to recover the original data. Since there exists a strong correlation among nearby ICI elements of $\mathbf{H}_{\mathrm{F}}$, significant ICI cancelation is achieved. In other words, the off-diagonal elements of the $L/2 \times L/2$ system matrix $\mathbf{P}^H\mathbf{Q}^{-1}\mathbf{H}_{\mathrm{F}}\mathbf{P}$ are by far smaller than those of $\mathbf{Q}^{-1}\mathbf{H}_{\mathrm{F}}$, which results in a reduced ICI. Consequently, significant performance gains are possible compared to convolutional coding (Zhao & Häggman, 2001), especially at high Doppler spread and at low SNR. As explained by Armstrong (1999) for CFO, this technique effectively eliminates the constant and linear components of the ICI variation over the rows of the frequency-domain channel matrix.

Even higher performance gains can be obtained by self-canceling more than two subcarriers. In general, $\mathbf{P} = \mathbf{I}_A \otimes \tilde{\mathbf{p}}$ is used, where $\tilde{\mathbf{p}}$ is a size-G vector obtained using the coefficients of the polynomial $p(x) = (1-x)^{G-1}$ and $G \triangleq L/A$ (assumed integer) denotes the number of subcarriers per symbol (Zhao & Häggman, 2001). For instance, in the CFO case, $G = 3$ permits the elimination of the ICI up to the cubic component (Armstrong, 1999). The main drawback is the information rate reduction, which, however can be eliminated using higher-order constellations.

An extension of the self-cancelation method consists in precoding each data symbol over G consecutive carriers, which are used to transmit several frequency-shifted replicas of the same data symbol (Seyedi & Saulnier, 2005). The frequency-shift values can be integer (Zhao & Häggman, 2001) or noninteger numbers. Noninteger frequency-shifted data precoding can be implemented using frequency-domain upsampling and time-domain windowing, producing an additional computational complexity of only $\mathcal{O}(L)$ (Seyedi & Saulnier, 2005). The main drawback of this approach lies in the complexity of the window design, which requires a numerical maximization (Seyedi & Saulnier, 2005). Other ICI cancelation codes, based on capacity maximization, are investigated by Yun, Chung, & Lee (2007).

Nonredundant precoding techniques are commonly based on *correlative coding* (Zhao, Leclercq, & Häggman, 1998) and *partial response coding* (Zhang & Li, 2003), applied in the frequency domain. This corresponds to a square precoder $\mathbf{P}$ with triangular and banded structure applied to (modulo-precoded) data. Precoder designs that approximately minimize the ICI power have been presented by Zhang & Li (2003). Similarly to the time-domain case, the data detection can be performed using a per-subcarrier detector (Zhao et al., 1998) or a joint (ML-based) detector (Zhang & Li, 2003). This last case is more complex but provides significant ICI reduction (more than 4 dB).

7.2.3.2 Pulse Shaping and Transmitter Windowing

Pulse-shaping techniques use transmitter (and receiver) windows in order to reduce the sensitivity to ICI (Bölcskei, 2003; Haas & Belfiore, 1997; Hunziker & Dahlhaus, 2003; Kozek & Molisch, 1998; Matz, Schafhuber, Gröchenig, Hartmann, & Hlawatsch, 2007; Strohmer & Beaver, 2003). In this case, the signal waveform is modified without changing the transmitted data symbols. In addition to the reduced ICI, pulse-shaping techniques also provide additional robustness to frequency synchronization errors and reduce the spurious emissions into adjacent channels. To describe this set of techniques, it is useful to adopt a continuous-time model of the OFDM transmission. Similarly to (2.13) in Chapter 2, the transmitted signal can be expressed by

$$s(t) = \sum_{k=0}^{K-1}\sum_{l=0}^{L-1} a[k,l]g(t-kT)e^{j2\pi lFt},$$

where $a[k,l]$ is the data symbol transmitted on the lth subcarrier of the kth OFDM block, F is the subcarrier separation, T is the OFDM block duration, including a possible cyclic extension, and $g(t)$ is the transmitted pulse, which is rectangular in conventional OFDM systems. For simplicity, we assume that only data are transmitted. After passing through an LTV channel with continuous-time impulse response $h(t,\tau)$, the received signal is obtained as

$$r(t) = \int_{-\infty}^{\infty} h(t,\tau)s(t-\tau)d\tau + w(t),$$

which is demodulated by computing the inner products with the receiver pulse-shaped waveforms $\{\gamma(t-kT)e^{j2\pi lFt}\}$, as expressed by

$$y[k,l] = \int_{-\infty}^{\infty} r(t)\gamma^*(t-kT)e^{-j2\pi lFt}dt.$$

Conventional OFDM systems, which employ rectangular pulse shapes for both the transmitter pulse $g(t)$ and the receiver pulse $\gamma(t)$, are orthogonal for both ideal and LTI channels. Orthogonality means that in the absence of noise, $y[k,l]$ is a scaled version of $a[k,l]$ and hence does not contain any unwanted contribution from symbols $a[\tilde{k},\tilde{l}]$ with $(\tilde{k},\tilde{l}) \neq (k,l)$. In other words, ISI is absent, since $y[k,l]$ does not depend on $a[\tilde{k},\tilde{l}]$ with $\tilde{k} \neq k$, and also intra-block ICI is avoided,[5] because $y[k,l]$ does not depend on $a[k,\tilde{l}]$ with $\tilde{l} \neq l$. A necessary condition for orthogonality is $TF \geq 1$. For noiseless ideal channels, where the CP is not necessary, OFDM systems avoid the ISI by rectangular windowing, i.e., $g(t)$ is constant when $t \in [0,T]$ and $g(t) = 0$ elsewhere, and avoid the ICI by choosing $F = 1/T$. Indeed, the time-domain truncated complex exponentials $\{e^{j2\pi lFt}g(t), l = 0,...,L-1\}$ with $F = 1/T$ lead to a frequency-domain sinc-shaped waveform centered at $lF = l/T$ with zeros on a regular grid, at $l'F = l'/T$, $l' \neq l$, so that only a single *sinc* waveform contributes to the signal components at the frequency $lF = l/T$. In this case, $TF = 1$, i.e., the spectral efficiency is maximum. OFDM signals maintain their orthogonality even for LTI multipath channels, provided that a cyclic extension (or a guard interval) is inserted

[5]Note that we have included the inter-block ICI in the ISI component.

to guarantee ISI-free reception and $F = 1/T_U$, where T_U is the useful (CP-free) part of the signal. Note that in CP-based OFDM, orthogonality is maintained at the price of reduced spectral efficiency, due to the insertion of a guard period of duration $T_{CP} = T - T_U$, which makes $TF > 1$. However, the orthogonality of OFDM systems is lost in the case of LTV channels: the Doppler spread modifies the *sinc* waveforms by a frequency-domain convolution. Therefore, as explained after (7.13), the zeros of the resulting function do not fall on the regular frequency grid anymore, and consequently ICI arises. This ICI is significant for time-domain rectangular windowing in the case of LTV channels, since the *sinc* function decays only as $1/f$. On the contrary, using pulse-shaping (windowing) functions with better frequency-domain decay properties, less ICI would be introduced by Doppler spread, thereby decreasing the need for complex LTV equalization. Note that for LTV channels, OFDM systems with rectangular pulses avoid ISI, provided that a sufficiently long guard interval is inserted.

Let us first assume that the pulses $g(t)$ and $\gamma(t)$ have the same shape, as suggested by optimal (matched) filtering in AWGN. A first pulse-shaping approach to reduce the ICI for LTV channels, while maintaining orthogonality in the ideal case, is to use the Nyquist criterion (Muschallik, 1996) to design the pulses $g(t)$ and $\gamma(t)$. For instance, dually to the ISI mitigation principle for single-carrier transmissions, a time-domain raised cosine window, which decays as $1/f^3$, can be evenly split between transmitter and receiver. However, the decay rate is not the only factor that induces ICI mitigation (Tan & Beaulieu, 2004). Different kinds of orthogonal pulses have been proposed. The use of time-domain *sinc* pulse shapes makes the subcarrier spectrum rectangular and therefore is dual to conventional OFDM with rectangular pulses. This is the idealized version of the *filtered multitone* approach (Amini & Farhang-Boroujeny, 2009; Tonello & Pecile, 2008; Wang, Proakis, & Zeidler, 2007), which can use guard bands between subcarriers to completely eliminate the ICI in the case of LTV channels. Other designs adopt pulses that are well localized in both time and frequency. Some examples include the quasi-orthogonal pulses of Haas & Belfiore (1997), which are based on Hermite functions, and the scale-adapted pulses of Liu, Kadous, & Sayeed (2004), which are matched to the spread factor of the channel. Although the orthogonal approach is optimum for (nondispersive) AWGN channels, there is a price to be paid for the obtained ICI reduction. Indeed, any window with good spectral properties has a larger duration than the rectangular window, and consequently, an additional guard period is necessary to avoid ISI, thus reducing the spectral efficiency. Otherwise, inter-block orthogonality is lost, and ISI equalization is required even for nondispersive channels.

Another orthogonal scheme is the *lattice OFDM* approach of Strohmer & Beaver (2003), where the temporal locations of the OFDM blocks at a given subcarrier are staggered with respect to those at the two adjacent subcarriers. This gives rise to hexagonal lattices in the time-frequency plane, which is consistent with the sphere-packing principle. Well-localized pulses are designed by orthogonalization of Gaussian pulses, taking into account the shape of the channel scattering function $C_{\mathbf{H}}(\tau, \nu)$. Since this scheme uses well-localized orthogonal pulses, $TF > 1$.

Let us now assume that the pulses $g(t)$ and $\gamma(t)$ have different shapes. *Biorthogonal* approaches rely on transmit and receive pulses $g(t)$ and $\gamma(t)$ that are characterized by the ambiguity function according to

$$A_{g\gamma}(kT, lF) \triangleq \int_{-\infty}^{\infty} g(t)\gamma^*(t - kT)e^{-j2\pi lFt}dt = \delta[k]\delta[l]. \tag{7.36}$$

The key idea is that different transmit and receive pulses provide more degrees of freedom for ICI and ISI mitigation, at the price of a slightly reduced performance for idealized AWGN channels due to mismatched receive filtering. Kozek & Molisch (1998) have proposed a transmit pulse design based on the maximization of the useful signal, leading to a large $|A_{g\gamma}(\tau,\nu)|^2$ wherever $C_{\mathbf{H}}(\tau,\nu)$ is large, while the receive pulse is chosen to fulfill the biorthogonality constraint. Matz et al. (2007) have proposed a receive pulse design that minimizes the ICI power (for a fixed transmit pulse).

Alternatively, *nonbiorthogonal* approaches are possible. In this case, different pulses $g(t)$ and $\gamma(t)$ that do not satisfy (7.36) are chosen. For instance, the joint transmit-receive pulse design of Matz et al. (2007) is obtained by disregarding the biorthogonality constraint. On the other hand, the pulse designs of Das & Schniter (2007) adopt an input SINR criterion that neglects the ICI due to nearby subcarriers, which is subsequently mitigated by an iterative banded nonlinear equalizer.

A different class of techniques is based on *offset QAM*, with real-valued pulses. The main advantage of pulse-shaped OFDM with offset QAM is the existence of well-localized functions even for $TF = 1$, which gives the maximum spectral efficiency. Some examples have been presented by Bölcskei (2003), Jung & Wunder (2007), Le Floch et al. (1995), and Vahlin & Holte (1995).

A totally different approach relies on using *chirp waveforms* for the "pulses" $g(t)$ and $\gamma(t)$. In this case, perfect ICI elimination is possible when the delay-Doppler spreading function of the channel is a straight line in the time-frequency plane (Barbarossa & Torti, 2001). Chirp approaches based on the fractional Fourier transform and on the affine Fourier transform have been presented by Martone (2001) and by Erseghe, Laurenti, & Cellini (2005), respectively.

7.2.4 Extension to MIMO-OFDM

The transmitter-based and receiver-based ICI mitigation methods discussed in the previous sections can be extended to MIMO-OFDM systems, using convenient methods to deal with the IAI arising from multiple receive antennas. The research on ICI mitigation in MIMO-OFDM systems is quite recent, and relatively few techniques have been proposed so far. Therefore, differently from the single-antenna case, a meticulous categorization of the proposed techniques is not opportune. In the following sections, we distinguish between ICI mitigation techniques proposed for spatial multiplexing systems, which aim at increasing the data rate, and techniques proposed for space-time-frequency coding systems, which seek to improve the performance.

7.2.4.1 *OFDM with Spatial Multiplexing*

In spatial multiplexing systems, different data symbols are transmitted from different antennas using shared frequency-time slots. Since (7.19) is formally similar to its single-antenna version (7.3), the equalization methods previously described can be used for MIMO-OFDM with minor modifications. For instance, a block linear MMSE equalizer has been investigated by Stamoulis et al. (2002), and a banded version that includes receiver windowing in the MBAE sense has been considered by Rugini & Banelli (2006). We note that linear equalization is appropriate only when $M_{\mathrm{R}} \geq M_{\mathrm{T}}$; otherwise there are not enough degrees of freedom for data recovery. Therefore, when $M_{\mathrm{R}} < M_{\mathrm{T}}$, nonlinear equalization or ICI cancelation is necessary.

Among the ICI cancelation techniques proposed in the literature, a simple approach consists in applying an iterative interference canceller with two stages: the first counteracts the ICI due to the Doppler spread, while the second reduces the IAI due to multiple transmit antennas. In the first stage, a banded PIC is usually adopted (Li, Li, & Vucetic, 2008; Song & Lim, 2006), while the second stage could employ a VBLAST-like nulling-canceling method (Song & Lim, 2006) or a PIC with linear combining of the outputs of different iterations (Li et al., 2008). Alternatively, joint cancelation of both ICI and IAI can be performed by turbo banded approaches, as in the window-assisted equalizer of Rugini, Banelli, Fang, & Leus (2009) and in the turbo decoder of Liu & Fitz (2008).

Correlative coding has been extended to MIMO-OFDM by Zhang & Liu (2006). However, the data recovery by means of ML sequence estimation can be quite expensive when the number of subcarriers (and the number of transmit antennas) is large.

7.2.4.2 *OFDM with Space-Time-Frequency Coding*

It is well known that the BER performance for MIMO channels can be improved by space-time coding (STC). In MIMO-OFDM, there is an additional domain, and hence, space-frequency coding (SFC) or space-time-frequency coding (STFC) is also possible. The choice between STC and SFC depends on the channel selectivity: intuitively, STC is able to collect the time diversity, while SFC can gather the frequency diversity. STFC can collect both gains, but it requires additional processing.

OFDM is a type of block transmission, and consequently, most of the techniques proposed for MIMO-OFDM in LTV channels are based on block codes. Usually, the employed block codes are orthogonal, such as the well-known Alamouti code (Alamouti, 1998). However, the ICI induced by the channel Doppler spread destroys the code orthogonality, and consequently, Alamouti combining at the receiver is no longer equal to the ML detector. Lin, Chiang, & Li (2005) have compared different receiver combining methods (namely Alamouti, ZF, decision-feedback, and ML) for space-time block coding (STBC) and space-frequency block coding (SFBC).

Kim, Heath Jr, & Powers (2005) have proposed an STBC receiver that switches between decision-feedback and Alamouti combining. Fang, Leus, & Rugini (2006) have investigated a banded MMSE approach for STBC. For CFO distortions, a Tomlinson–Harashima precoder with partial CSI at the transmitter has been proposed by Fu, Tellambura, & Krzymien (2007). SFBC ICI self-cancelation codes are proposed in Dao & Tellambura (2005). The SFBC of Park & Cho (2005) is an Alamouti technique applied to a group of consecutive subcarriers, which contain redundant frequency-domain precoded data. Zhu, Wen, & Du (2008) have shown that space-time-frequency block coding (STFBC) outperforms STBC and SFBC, as expected by intuition.

7.3 TIME-VARYING CHANNEL ESTIMATION

Most of the described ICI mitigation techniques assume that the receiver knows the LTV channel. In rapidly time-varying scenarios, the channel estimation task is rather cumbersome because the CIR is not constant within the OFDM block. As a consequence, multiple parameter estimation is necessary for each channel path, and the estimation has to be repeated (or updated) for each OFDM block. Therefore, we include an overview of LTV channel estimation for OFDM (this issue is treated in more detail within Chapter 4). In this section, we first review the basis expansion model (BEM), which is one of the most popular channel models used for LTV channel estimation. Then, we describe some pilot-aided and data-aided channel estimation methods. An iterative channel estimation method based on turbo processing will also be considered.

7.3.1 Basis Expansion Model of LTV Channels

Since in the kth block the discrete-time LTV channel is expressed by $h[kN + L_{\mathrm{CP}} + n, m]$, for $0 \leq n \leq L-1$ and $0 \leq m \leq M-1$, it is clear that ML parameters are required to represent the time-varying CIR, i.e., L values for each of the M channel paths. To reduce the number of parameters, each channel path can be modeled by a BEM in the kth OFDM block. With reference to the windowed channel, the BEM for the mth channel path is expressed by (Tsatsanis & Giannakis, 1996)

$$\Delta[n]h[kN + L_{\mathrm{CP}} + n, m] = \sum_{i=0}^{I-1} c_{i,k}[m]u_i[n], \tag{7.37}$$

where $\Delta[n] \triangleq [\mathbf{\Delta}]_{n,n}$ is the nth window coefficient, I is the number of basis functions, $u_i[n]$ represents the ith basis function, which models the time variability, and $c_{i,k}[m]$ is the ith coefficient for the mth channel tap in the kth block. This approach reduces the number of channel parameters from ML to MI per block. By (7.37), the windowed time-domain channel matrix, which is obtained by multiplying the diagonal windowing matrix $\mathbf{\Delta}$ by $\mathbf{H}_{\mathrm{T}}[k]$, can be expressed as

$$\mathbf{\Delta H}_{\mathrm{T}}[k] = \sum_{m=0}^{M-1}\sum_{i=0}^{I-1} c_{i,k}[m]\mathbf{U}_{\mathrm{d},i}\mathbf{Z}_{\mathrm{c}}[m] = \sum_{i=0}^{I-1} \mathbf{U}_{\mathrm{d},i}\mathbf{C}_{\mathrm{c},i}[k], \tag{7.38}$$

where $\mathbf{U}_{\mathrm{d},i}$ is a diagonal matrix defined by $\left[\mathbf{U}_{\mathrm{d},i}\right]_{n,n} \triangleq u_i[n]$, $\mathbf{Z}_{\mathrm{c}}[m]$ is the $L \times L$ cyclic-shift matrix with ones in the mth lower diagonal and in the $(L-m)$th upper diagonal, and zeros elsewhere, and $\mathbf{C}_{\mathrm{c},i}[k] \triangleq \sum_{m=0}^{M-1} c_{i,k}[m]\mathbf{Z}_{\mathrm{c}}[m]$. Consequently, the $L \times L$ windowed frequency-domain channel matrix $\mathbf{H}_{\mathrm{W}}[k]$ in (7.29) can be expressed as

$$\mathbf{H}_{\mathrm{W}}[k] = \sum_{m=0}^{M-1}\sum_{i=0}^{I-1} c_{i,k}[m]\mathbf{U}_{\mathrm{c},i}\mathbf{Z}_{\mathrm{d}}[m] = \sum_{i=0}^{I-1} \mathbf{U}_{\mathrm{c},i}\mathbf{C}_{\mathrm{d},i}[k]. \tag{7.39}$$

In (7.39), $\mathbf{U}_{\mathrm{c},i} \triangleq \mathbf{W}\mathbf{U}_{\mathrm{d},i}\mathbf{W}^H$ is a circulant matrix that contains the (shifted version of the) discrete Doppler spectrum associated with the ith basis function, $\mathbf{Z}_{\mathrm{d}}[m] \triangleq \mathbf{W}\mathbf{Z}_{\mathrm{c}}[m]\mathbf{W}^H$ is diagonal with elements $[\mathbf{Z}_{\mathrm{d}}[m]]_{n,n} = \mathrm{e}^{j2\pi mn/L}$ that represent the frequency shift associated with the mth delay lag, and $\mathbf{C}_{\mathrm{d},i}[k] \triangleq \sum_{m=0}^{M-1} c_{i,k}[m]\mathbf{Z}_{\mathrm{d}}[m] = \mathbf{W}\mathbf{C}_{\mathrm{c},i}[k]\mathbf{W}^H$ models the frequency selectivity associated with the projection of the multipath channel onto the ith basis function. Note that (7.38) and (7.39) are equivalent to (4.9) and (4.10), respectively, in Chapter 4. The specific structure of the channel matrix $\mathbf{H}_{\mathrm{W}}[k]$ in (7.39) depends on the chosen basis. In any case, the structured model (7.39) opens the way to low-complexity algorithms for channel estimation and equalization.

Regarding the type of basis functions, a popular choice is to adopt $I = 2D+1$ orthogonal critically sampled complex exponential (CCE) functions $u_i[n] = \mathrm{e}^{j2\pi(i-D)n/L}$, $i = 0, \ldots, I-1$ (Tsatsanis & Giannakis, 1996) (see also (1.64) in Chapter 1), i.e., each function corresponds to a discrete Doppler frequency shift of i subcarriers. Thus, $\mathbf{U}_{\mathrm{c},i} = \mathbf{Z}_{\mathrm{c}}[i-D]$, and hence, the basis functions produce a perfectly cyclically banded $\mathbf{H}_{\mathrm{W}}[k]$ with $I = 2D+1$ diagonals. In other words, the ICI support of CCE-BEM is finite (Tang, Cannizzaro, Leus, & Banelli, 2007). However, the CCE functions are periodic, with period equal to one OFDM block, while the time variability of the channel path is not periodic. As a consequence, the modeling error can be significant, especially at the edges of the block.

Alternatively, other types of basis functions can be employed. Another intuitive choice is to model the time variability by means of polynomials $u_i[n] = (n - \frac{L}{2})^i$ (Borah & Hart, 1999) (see also (1.66) in Chapter 1), which arise from the power series model of an LTV channel. In this case, the modeling error is negligible when the normalized maximum Doppler shift is low (Gorokhov & Linnartz, 2004; Tang et al., 2007). Other possible basis functions include oversampled complex exponential functions (Leus, 2004) (see also (1.65) in Chapter 1), discrete prolate spheroidal sequences (Zemen & Mecklenbräuker, 2005) (see also Section 1.6.1.3), and discrete Karhunen–Loève functions (Teo & Ohno, 2005, November/ December). The discrete Karhunen–Loève basis minimizes the MSE of the channel modeling, but it requires full statistical information about the Doppler power profile. On the other hand, the discrete prolate spheroidal basis assumes a flat Doppler power profile, and hence, it only requires knowledge of the maximum Doppler frequency. A comparison of the error produced by different basis functions has been performed by Zemen & Mecklenbräuker (2005) and Tang et al. (2007).

7.3.2 Training-Based Channel Estimation

When the LTV channel is represented by a BEM, the channel estimation problem reduces to the estimation of the BEM coefficients $c_i[m]$ from the received vector $\mathbf{y}_W[k]$ (and known pilots). With reference to the kth OFDM block, since $\mathbf{x}[k] = \mathbf{a}[k] + \mathbf{p}[k]$, (7.29) can be rewritten as

$$\mathbf{y}_W[k] = \mathbf{H}_W[k]\mathbf{x}[k] + \mathbf{z}_W[k] = \mathbf{H}_W[k]\mathbf{p}[k] + \mathbf{i}_W[k] + \mathbf{z}_W[k], \tag{7.40}$$

where $\mathbf{p}[k]$ is the known pilot or training vector, while $\mathbf{i}_W[k] \triangleq \mathbf{H}_W[k]\mathbf{a}[k]$ is the "interference" produced by the still unknown data vector $\mathbf{a}[k]$. From (7.39), the windowed channel matrix can be expressed by

$$\mathbf{H}_W[k] = \sum_{m=0}^{M-1}\sum_{i=0}^{I-1} c_{i,k}[m]\mathbf{U}_{c,i}\mathbf{Z}_d[m] = \boldsymbol{\Omega}(\mathbf{c}[k] \otimes \mathbf{I}_L), \tag{7.41}$$

where the BEM coefficients are collected in a single vector, as expressed by

$$\mathbf{c}[k] \triangleq (c_{0,k}[0] \cdots c_{I-1,k}[0] \cdots c_{0,k}[M-1] \cdots c_{I-1,k}[M-1])^T, \tag{7.42}$$

and $\boldsymbol{\Omega} \triangleq (\boldsymbol{\Omega}_{0,0} \cdots \boldsymbol{\Omega}_{I-1,0} \cdots \boldsymbol{\Omega}_{0,M-1} \cdots \boldsymbol{\Omega}_{I-1,M-1})$ with $\boldsymbol{\Omega}_{i,m} \triangleq \mathbf{U}_{c,i}\mathbf{Z}_d[m]$. Note that the vector $\mathbf{c}[k]$ defined in (7.42) is a permuted version of the vector defined in Chapter 4 after (4.8). From (7.40) and (7.41), using the identity $(\mathbf{c}[k] \otimes \mathbf{I}_L)\mathbf{p}[k] = (\mathbf{I}_{MI} \otimes \mathbf{p}[k])\mathbf{c}[k]$, we obtain

$$\mathbf{y}_W[k] = \boldsymbol{\Omega}\,(\mathbf{I}_{MI} \otimes \mathbf{p}[k])\,\mathbf{c}[k] + \mathbf{i}_W[k] + \mathbf{z}_W[k], \tag{7.43}$$

which clearly reveals the linear relationship between the received vector $\mathbf{y}_W[k]$ and the BEM coefficient vector $\mathbf{c}[k]$ to be estimated. Note that in general, the unknown vector $\mathbf{c}[k]$ also appears in the data-dependent term $\mathbf{i}_W[k]$, which can be rewritten as $\mathbf{i}_W[k] = \boldsymbol{\Omega}(\mathbf{c}[k] \otimes \mathbf{I}_L)\mathbf{a}[k] = \boldsymbol{\Omega}(\mathbf{I}_{MI} \otimes \mathbf{a}[k])\mathbf{c}[k]$.

In order to estimate the BEM coefficients, to reduce complexity, we can select only ω elements of $\mathbf{y}_W[k]$. Denoting by $\mathbf{S}[k]$ the $\omega \times L$ observation selection matrix constructed by selecting ω rows of $\mathbf{I}_L$, we obtain from (7.43)

$$\mathbf{y}_S[k] \triangleq \mathbf{S}[k]\mathbf{y}_W[k] = \mathbf{S}[k]\boldsymbol{\Omega}(\mathbf{I}_{MI} \otimes \mathbf{p}[k])\mathbf{c}[k] + \mathbf{i}_S[k] + \mathbf{z}_S[k], \tag{7.44}$$

where $\mathbf{i}_S[k] \triangleq \mathbf{S}[k]\mathbf{i}_W[k] = \mathbf{S}[k]\boldsymbol{\Omega}(\mathbf{c}[k] \otimes \mathbf{I}_L)\mathbf{a}[k]$ and $\mathbf{z}_S[k] \triangleq \mathbf{S}[k]\mathbf{z}_W[k]$. Obviously, the choice of the ω elements depends on the training pattern. Intuitively, we should select those elements of $\mathbf{y}_W[k]$ that mainly depend on the pilot vector $\mathbf{p}[k]$ and exclude those elements that mainly depend on the unknown data vector $\mathbf{a}[k]$.

In the following, we first review some LTV channel estimation techniques for OFDM when the MSE-optimal training for CCE-BEM channels is employed. Then, we describe some low-complexity channel estimation techniques for OFDM that use a suboptimal frequency-domain training without zero pilots, such as the DVB-T/H pilot pattern. Besides, we consider an adaptive frequency-domain approach based on both pilots and data.

7.3.2.1 *Estimation with Optimal Frequency-Domain Training*

Differently from LTI channels and from single-path LTV channels, where the MSE-optimal training is known (Tong, Sadler, & Dong, 2004) (see also Chapter 5), the MSE-optimal training for multipath LTV channels is known only for a specific channel model, i.e., CCE-BEM (Kannu & Schniter, 2008; Ma, Giannakis, & Ohno, 2003). For CCE-BEM LTV channels, the optimal training, known as frequency-domain Kronecker delta training, consists of M equispaced clusters placed in the frequency domain (Kannu & Schniter, 2008). Each cluster contains a single pilot subcarrier, surrounded by $2D$ zero subcarriers on each side (Kannu & Schniter, 2008). Therefore, the total number of training symbols in each OFDM block is $P = M(4D+1) = M(2I-1)$, while the other $A = L - P$ subcarriers can be used for the data symbols. Basically, the zero subcarriers should separate the data from the pilot clusters, in such a way that data and pilots remain orthogonal after the Doppler dispersion caused by the LTV channel. Indeed, a CCE-BEM channel of order $I = 2D+1$ corresponds to a perfectly banded channel matrix that produces a maximum Doppler dispersion of $\pm D$ subcarriers. Therefore, a separation of $2D$ subcarriers between data and pilot clusters is the minimum amount that avoids pilot-data overlapping after the LTV channel. As a consequence, with this training pattern, a convenient choice is $\omega = MI$. This is obtained by selecting, for each of the M pilot clusters, only the $I = 2D+1$ central elements. In this case, the interference $\mathbf{i}_S[k]$ is virtually absent, and (7.44) becomes

$$\mathbf{y}_S[k] \approx \mathbf{P}_S[k]\mathbf{c}[k] + \mathbf{z}_S[k], \tag{7.45}$$

where the system matrix $\mathbf{P}_S[k] = \mathbf{S}[k]\boldsymbol{\Omega}\,(\mathbf{I}_{MI} \otimes \mathbf{p}[k])$ is square. Hence, the BEM coefficients $\mathbf{c}[k]$ can be estimated by a deterministic approach, such as the LS estimator, expressed by (Kay, 1993)

$$\hat{\mathbf{c}}[k] = \mathbf{P}_S^{\#}[k]\mathbf{y}_S[k],$$

where $^{\#}$ denotes pseudoinverse. Alternatively, a stochastic approach can be used, such as the LMMSE estimator expressed by (Kay, 1993)

$$\hat{\mathbf{c}}[k] = \mathbf{C}_{\mathbf{cc}}\mathbf{P}_S^H[k](\mathbf{P}_S[k]\mathbf{C}_{\mathbf{cc}}\mathbf{P}_S^H[k] + \mathbf{C}_{\mathbf{z}_S\mathbf{z}_S}[k])^{-1}\mathbf{y}_S[k],$$

where $\mathbf{C}_{\mathbf{z}_S\mathbf{z}_S}[k] = \gamma\mathbf{S}[k]\boldsymbol{\Gamma}\boldsymbol{\Gamma}^H\mathbf{S}^T[k]$ is the noise covariance and $\mathbf{C}_{\mathbf{cc}}$ is the covariance matrix of the BEM coefficients, which is calculated as a function of the time correlation of the channel paths (Tang et al., 2007). In both the LS and LMMSE cases, the computational complexity of channel estimation is $\mathcal{O}(M^2I^2)$ per OFDM block.

To enhance the channel estimation performance, more observations $\omega > MI$ should be collected. In this case, differently from (7.45), the data-dependent interference cannot be neglected, and

$$\mathbf{y}_S[k] = \mathbf{P}_S[k]\mathbf{c}[k] + \mathbf{i}_S[k] + \mathbf{z}_S[k]. \tag{7.46}$$

Consequently, stochastic approaches should include the covariance $\mathbf{C}_{\mathbf{i}_S\mathbf{i}_S}[k]$ of the data-dependent interference $\mathbf{i}_S[k] = \mathbf{S}[k]\boldsymbol{\Omega}\,(\mathbf{c}[k] \otimes \mathbf{I}_L)\,\mathbf{a}[k]$, which however depends on the unknown $\mathbf{c}[k]$. Tang et al. (2007) have solved this problem by employing an iterative best linear unbiased estimator (BLUE), as expressed by

$$\hat{\mathbf{c}}^{(\kappa)}[k] = \left[\mathbf{P}_S^H[k](\hat{\mathbf{C}}_{\mathbf{i}_S\mathbf{i}_S}^{(\kappa)}[k] + \mathbf{C}_{\mathbf{z}_S\mathbf{z}_S}[k])^{-1}\mathbf{P}_S[k]\right]^{-1}\mathbf{P}_S^H[k](\hat{\mathbf{C}}_{\mathbf{i}_S\mathbf{i}_S}^{(\kappa)}[k] + \mathbf{C}_{\mathbf{z}_S\mathbf{z}_S}[k])^{-1}\mathbf{y}_S[k],$$

where κ is the iteration index, and

$$\hat{\mathbf{C}}_{\mathbf{i}_S\mathbf{i}_S}^{(\kappa)}[k] = \mathbf{S}[k]\boldsymbol{\Omega}(\hat{\mathbf{c}}^{(\kappa-1)}[k] \otimes \mathbf{I}_L)\mathbf{C}_{\mathbf{aa}}(\hat{\mathbf{c}}^{(\kappa-1)}[k] \otimes \mathbf{I}_L)^H\boldsymbol{\Omega}^H\mathbf{S}^T[k], \tag{7.47}$$

where $\hat{\mathbf{c}}^{(0)}[k] = \mathbf{0}_{MI\times 1}$ at the first iteration. As far as performance is concerned, the iterative BLUE produces a lower channel estimation MSE than an LMMSE estimator that neglects the covariance matrix of the interference (Tang et al., 2007). Iterative decision-feedback estimators are also possible: alternatively to (7.47), we can exploit a data-aided covariance estimate, as expressed by

$$\hat{\mathbf{C}}_{\mathbf{i}_S\mathbf{i}_S}^{(\kappa)}[k] = \mathbf{S}[k]\boldsymbol{\Omega}(\mathbf{I}_{MI} \otimes \hat{\mathbf{a}}^{(\kappa-1)}[k])\mathbf{C}_{\mathbf{cc}}(\mathbf{I}_{MI} \otimes \hat{\mathbf{a}}^{(\kappa-1)}[k])^H\boldsymbol{\Omega}^H\mathbf{S}^T[k],$$

which is based on the expression $\mathbf{i}_S[k] = \mathbf{S}[k]\boldsymbol{\Omega}(\mathbf{I}_{MI} \otimes \mathbf{a}[k])\mathbf{c}[k]$. Another option is to employ a data-dependent interference canceller that subtracts from (7.46) the term

$$\hat{\mathbf{i}}_S[k] = \mathbf{S}[k]\boldsymbol{\Omega}(\mathbf{I}_{MI} \otimes \hat{\mathbf{a}}^{(\kappa-1)}[k])\hat{\mathbf{c}}^{(\kappa-1)}[k],$$

where $\hat{\mathbf{a}}^{(\kappa-1)}[k]$ is a (soft or hard) data estimate obtained by a channel equalizer at the previous iteration.

A different approach relies on adaptive channel estimation. As an example, Kalman filtering and recursive least-squares methods, discussed by Cannizzaro, Banelli, & Leus (2006), exploit the time correlation of the channel over successive OFDM blocks. In addition, adaptive techniques are able to track the variability of the channel statistics and are therefore suitable for nonstationary environments. In case of frequent changes of the channel statistics, a robust approach based on H_∞ filtering may be appropriate (Banelli & Rugini, 2010).

7.3.2.2 *Estimation with Suboptimal Training*

Current OFDM-based systems, such as DVB-T/H, IEEE 802.11a, and IEEE 802.16e (WiMAX) (ETSI, 2004, 2005; IEEE, 1999, 2006), are designed for LTI or slowly time-varying channels, and therefore, their training patterns are different from the frequency-domain Kronecker delta pattern discussed so far. Usually, packet-based systems such as IEEE 802.11a employ a time-domain preamble, mainly used for time synchronization and channel estimation, and few frequency-domain pilots, mainly used for CFO compensation and tracking. On the other hand, continuous systems such as DVB-T/H usually employ only frequency-domain pilots, whose locations change with the OFDM block index. In these cases, the frequency-domain pilots are adjacent to the data symbols, and therefore, significant ICI is present when the channel is rapidly time varying.

In order to estimate LTV channels by means of (nonzero-guarded) frequency-domain pilots, such as in DVB-T/H, many techniques have been proposed. Usually, these techniques first collect information from consecutive OFDM blocks; then, for the pilot positions, they estimate a sampled version of the LTV channel in the joint time-frequency domain; and finally, for the data positions, they use interpolation techniques to recover the whole LTV channel. To keep the complexity low, the channel is sometimes modeled as constant within a single OFDM block. This assumption actually leads to the estimation of the *time-averaged CIR* (expressed inside the square brackets in (7.7)). Since the ICI is neglected, the time variation of the channel within the OFDM block is ignored. However, the time variation can subsequently be reconstructed by interpolating the estimated channel values corresponding to consecutive OFDM blocks.

Among the proposed techniques, we can distinguish between batch algorithms (Hoeher, Kaiser, & Robertson, 1997; Hutter, Hasholzner, & Hammerschmidt, 1999; Lai & Chiueh, 2006) and adaptive algorithms (Sanzi & Speidel, 2000; Schafhuber & Matz, 2005; Schafhuber, Matz, & Hlawatsch, 2003, April). In addition, the two-dimensional interpolation can be performed either jointly in the time-frequency domain (Hoeher et al., 1997; Schafhuber & Matz, 2005; Schafhuber et al., 2003, April) or separately, as a cascade of two one-dimensional interpolators (Hutter et al., 1999; Lai & Chiueh, 2006; Sanzi & Speidel, 2000), as summarized in the following:

- Two-dimensional batch interpolators (Hoeher et al., 1997);
- Successive one-dimensional batch interpolators (Hutter et al., 1999; Lai & Chiueh, 2006);
- Two-dimensional adaptive interpolators (Schafhuber & Matz, 2005; Schafhuber et al., 2003, April);
- Successive one-dimensional adaptive interpolators (Sanzi & Speidel, 2000).

Among the interpolation methods, Wiener filtering is a popular choice, since it can exploit the available information about the time-frequency correlation of the channel (Hoeher et al., 1997; Hutter et al., 1999; Sanzi & Speidel, 2000; Schafhuber & Matz, 2005; Schafhuber et al., 2003, April). A Wiener filter approach has been proposed also by Sgraja & Lindner (2003), without neglecting the ICI produced by the channel time variation within the OFDM block. Alternatively, an LS interpolation approach can be used, as done by Fertl & Matz (2006), who incorporated a trigonometric polynomial model of the time-frequency channel, and proposed a conjugate gradient algorithm with early termination to deal with the severe ill conditioning generated by fast LTV channels.

When the channel time variation is rapid, comb-type pilot placement schemes suffer from significant ICI. Besides zero-guarded pilots, another possible solution is to employ clustered pilot schemes, where groups of adjacent subcarriers are employed as pilots. This issue has been investigated in Shin, Andrews, & Powers (2007), where simulation results show that the MSE-optimal number of clusters is again equal to the number M of discrete-time channel paths. Moreover, as in the zero-guarded case, the M clusters should be equispaced.

7.3.2.3 Data-Aided Channel Estimation and Tracking

In the two previous subsections, we have considered the channel estimation task for two types of frequency-domain training: the zero-guarded clusters of Kannu & Schniter (2008), and the classical (nonzero-guarded) comb pilot pattern used in DVB-T/H (ETSI, 2004). The first type, being MSE-optimal for CCE-BEM channels, obviously allows for good performance, at the price of a non-negligible rate reduction. On the other hand, focusing on a single OFDM block, the comb pattern of DVB-T/H minimizes the rate reduction due to training, at the price of a significant performance degradation caused by the unmodeled ICI. To overcome these drawbacks, a different type of training pattern should be adopted. Banelli, Cannizzaro, & Rugini (2007) used the zero-guarded clusters only for K_1 consecutive OFDM blocks, while no pilots are transmitted in the K_2 successive OFDM blocks. There are two key ideas in this approach. First, the time variability of the BEM coefficients over successive OFDM blocks is modeled as a first-order Gauss-Markov process, as expressed by

$$\mathbf{c}[k+1] = \tilde{\mathbf{A}}\mathbf{c}[k] + \boldsymbol{\nu}[k],$$

where $\tilde{\mathbf{A}}$ characterizes the BEM evolution and $\boldsymbol{\nu}[k]$ is the BEM innovation. This model permits LTV channel prediction from block to block during the K_2 blocks without pilots, by means of a Kalman filter (Banelli et al., 2007). Second, the same Kalman filter structure is exploited in a data-aided approach to refine the channel estimation over the K_2 blocks without pilots. Specifically, for each of the K_2 blocks, M equispaced data clusters are identified as virtual pilots, using a reliability-based selection metric. Using these virtual pilots, the BEM coefficients of the current block are iteratively estimated in a decision-directed mode, to obtain a better channel estimate and then more reliable data re-estimates. Successively, the BEM coefficients of the next block are predicted, and new virtual pilots are identified. For the virtual pilot selection, different reliability-based metrics are possible, based on the soft decisions and the MSE of the symbol estimates (Banelli et al., 2007).

The Kalman filter used by Banelli et al. (2007) is based on frequency-domain vector observations. To reduce complexity, a Kalman filter based on time-domain scalar observations can be employed (Muralidhar & Li, 2009). Indeed, the Kalman filter of Banelli et al. (2007) tracks a single vector that collects the BEM coefficients of all the paths, while the Kalman filter of Muralidhar & Li (2009) separately tracks each subvector related to the corresponding channel path, in a parallel way.

7.3.3 Iterative Channel Estimation and Turbo Equalization

Most of the ICI mitigation techniques investigated in the technical literature have two separate steps for channel estimation and channel equalization. However, joint channel estimation and equalization is a valid alternative, as shown by the data-aided approaches of Banelli et al. (2007) and Fang, Rugini, & Leus (2008, April). Additional performance improvement can be obtained by incorporating the channel decoder into the iterative scheme that performs joint channel estimation and ICI cancelation (Tomasin et al., 2005). In this section, we briefly describe the joint approach of Fang, Rugini, & Leus (2008, April), which combines the channel estimator of Tang et al. (2007) with the turbo equalizer of Fang et al. (2008). Specifically, the channel estimator can be modified in order to exploit the LLR values of the data produced by the turbo equalizer. These LLR values can be used to update the a priori symbol mean vector $\hat{\boldsymbol{\mu}}_a^{(\kappa)}[k]$, where κ is the iteration index. Since $\hat{\boldsymbol{\mu}}_a^{(\kappa)}[k]$ represents our current knowledge about the data vector $\mathbf{a}[k]$, an improved channel estimator can be designed by updating the model in (7.43), which becomes

$$\mathbf{y}_{\mathrm{W}}^{(\kappa)}[k] = \boldsymbol{\Omega}\left[\mathbf{I}_{MI} \otimes (\mathbf{p}[k] + \hat{\boldsymbol{\mu}}_a^{(\kappa)}[k])\right]\mathbf{c}[k] + \tilde{\mathbf{i}}_{\mathrm{W}}^{(\kappa)}[k] + \mathbf{z}_{\mathrm{W}}[k], \tag{7.48}$$

where $\tilde{\mathbf{i}}_{\mathrm{W}}^{(\kappa)}[k] \triangleq \mathbf{H}_{\mathrm{W}}[k](\mathbf{a}[k] - \hat{\boldsymbol{\mu}}_a^{(\kappa)}[k])$ is the updated data-dependent interference. From (7.48), it is clear that $\mathbf{p}[k] + \hat{\boldsymbol{\mu}}_a^{(\kappa)}[k]$ can be considered as a new pilot vector, which incorporates our soft knowledge $\hat{\boldsymbol{\mu}}_a^{(\kappa)}[k]$ about the data $\mathbf{a}[k]$, while $\tilde{\mathbf{i}}_{\mathrm{W}}^{(\kappa)}[k]$ represents the interference coming from the unknown part of the data. Since the interference power is reduced, the channel estimator exhibits improved performance, which helps the turbo equalizer to refine the estimate $\hat{\boldsymbol{\mu}}_a^{(\kappa)}[k]$ used at the next iteration. This iterative channel estimation and turbo equalization approach yields improved performance not only in the presence of zero-guarded pilot clusters but also when (nonzero-guarded) comb pilots are used as training pattern (Fang et al., 2008).

Alternatively, iterative channel estimation can be performed using both a time-domain preamble and (nonzero-guarded) frequency-domain pilots and exploiting the soft data outputs of the channel decoder (Zhao, Shi, & Reed, 2008). Specifically, the iterative channel estimator of Zhao et al. (2008) neglects the time variation of the channel within the OFDM block, and therefore, the ICI is considered as an additional noise.

7.3.4 Impact of Channel Estimation on BER Performance

The aim of this section is to understand how much performance loss is introduced by channel estimation errors at the receiver side. We consider an OFDM system with $L = 128$ subcarriers, of which $P = 21$ are dedicated to zero-guarded pilot symbols and $A = L - P = 107$ to QPSK data symbols. The $P = 21$ pilot symbols are divided into 7 clusters of length 3, i.e., each nonzero pilot is surrounded by a single zero on each side. The average pilot power of each cluster is equal to the data power. We consider a WSSUS Rayleigh fading channel with uniform delay power profile and channel length $M = 6$. The CP length is set to $L_{\mathrm{CP}} = 5$. Regarding the time variation of the channel, we assume a Jakes' Doppler power profile with $\vartheta_{\max} = 0.12$. The receiver window used for both channel estimation and equalization is MBAE-SOE with $D_{\mathrm{b}} = 3$. Each channel path is modeled using a discrete Karhunen–Loève BEM with $I = 5$ basis functions. To estimate the LTV channel, all pilot and data locations are exploited, i.e., the number of observations is $\omega = L = 128$. The LMMSE channel estimator described by Tang et al. (2007) is employed in order to provide the CSI to the iterative banded block MMSE equalizer of Fang et al. (2008). For $E_{\mathrm{s}}/N_0 > 25$ dB, Tikhonov regularization is adopted at the receiver. Figure 7.9 compares the BER performance of the turbo banded block MMSE equalizer (Fang et al., 2008) with perfect CSI and estimated CSI. The performance gap is moderate, and it is noteworthy that in both cases the BER curve does not evidence any error floor for $E_{\mathrm{b}}/N_0 < 30$ dB.

7.3.5 Channel Estimation in MIMO-OFDM

Compared to single-antenna OFDM systems, channel estimation for MIMO-OFDM systems is more difficult for two reasons. First, since multiple channels have to be estimated, additional pilot symbols should be inserted with respect to the single-channel case. The second reason is the presence of IAI arising from multiple transmit antennas, which adds to the ICI. Similarly to the ICI mitigation issue, LTV channel estimation in MIMO-OFDM systems is quite recent, and relatively few techniques have been proposed so far. One of these techniques has been discussed in Section 4.5.3. Here, we briefly list the main features of other proposed schemes, which can be used for both spatial multiplexing and space-time-frequency coded OFDM.

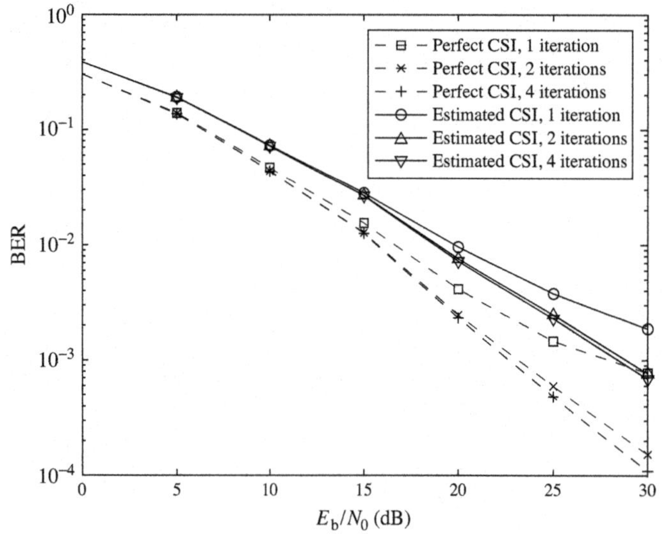

FIGURE 7.9

Effect of channel estimation on the BER performance.

Schafhuber, Rupp, Matz, & Hlawatsch (2003), have proposed a least-mean-squares-based adaptive approach to estimate a subsampled version of the MIMO frequency-domain channel matrix. The pilot symbols of each transmit antenna, which are orthogonal to those of the other antennas, are scattered comb-type nonzero-guarded pseudo-noise sequences. The optimum adaptation constants are obtained by automatic tuning (Schafhuber et al., 2003, June).

Salvo Rossi & Müller (2008) have used discrete prolate spheroidal sequences as basis functions to represent the MIMO channel transfer function. Pilot symbols are sparsely placed to sample the two-dimensional time-frequency channel. The linear MIMO channel estimators considered by Salvo Rossi & Müller (2008) counteract the IAI while ignoring the ICI.

Gao & Liu (2008) have proposed an expectation-maximization (EM) approach for maximum a posteriori (MAP) channel estimation. First, an LS estimator is designed for LTI channels, assuming the optimal pilot pattern presented by Barhumi, Leus, & Moonen (2003). Then, parallel ICI cancelation is included in the estimation process to deal with LTV channels. The proposed EM estimator also applies a low-rank approximation that avoids large matrix inversion, thereby reducing complexity.

Next, Li et al. (2008) have presented an LS estimator based on a linear model of the channel time variation. Orthogonal pilot clusters are assumed. The channel estimation task is performed before the data detection, neglecting the ICI coming from the data symbols.

Another MAP-based channel estimation scheme has been presented by Kim & Lim (2008). Also in this case, the ICI coming from the data symbols is neglected. A distinctive feature of the channel estimator of Kim & Lim (2008) is that $h^{(i,j)}[n,m]$ is obtained as a linear combination of the $\tilde{N}$ previous values $h^{(i,j)}[n-1,m],\ldots,h^{(i,j)}[n-\tilde{N},m]$, where $\tilde{N} < L$, and i (j) is the index of the receive (transmit) antenna. In addition, to reduce the number of parameters, Kim & Lim (2008) assumes that all the $MM_{\mathrm{T}}M_{\mathrm{R}}$ MIMO channel paths have the same statistics and hence share the same set of linear combining coefficients.

Similarly to Zhao et al. (2008), the iterative channel estimator of Zhao, Shi, & Reed (2007) uses a time-domain preamble, frequency-domain pilots, and soft-decoded data. To deal with the MIMO scenario, space-time processing and IAI cancelation are included. Since the estimation process is carried out on a per-subcarrier basis, the estimation quality can be improved by a low-pass filter that exploits the frequency-domain channel correlation (Zhao et al., 2007).

7.4 CONCLUDING REMARKS

We now summarize some important issues about multicarrier systems transmitting over LTV channels. First, we stress that when the scattering function is separable, the ICI power only depends on the shape of the Doppler power profile and on the normalized maximum Doppler shift $\vartheta_{\max}$. On the other hand, the ICI power depends neither on the delay power profile of the channel (provided that the CP is sufficiently long) nor on the absolute maximum Doppler shift $\nu_{\max}$.

Second, we have shown that many options are available for channel estimation and ICI mitigation, with different performance levels and complexity requirements. In most cases, the use of statistical CSI (e.g., in window-aided receivers) guarantees a performance gain with negligible additional complexity. Therefore, statistical CSI should be exploited as much as possible. Note that the statistical CSI usually does not change as fast as the CIR.

Third, although we have focused on OFDM, most of the results discussed in this chapter can be extended to multiuser multicarrier systems, such as multicarrier CDMA (MC-CDMA) or orthogonal frequency-division multiple-access (OFDMA). For instance, one-tap MMSE equalizers and banded MMSE equalizers for MC-CDMA downlink systems have been investigated by Linnartz (2001) and by Rugini, Banelli, & Leus (2005, March), respectively, while an iterative joint channel estimator and multiuser detector for MC-CDMA uplink systems has been proposed by Zemen, Mecklenbräuker, Wehinger, & Müller (2006). Its extension for multiuser MIMO communications will be discussed in Chapter 8.

In the following, we highlight some topics related to OFDM system and application aspects, and then, we conclude this chapter by discussing some open issues.

7.4.1 System and Application Aspects

The first application aspect we consider is related to OFDM broadcasting systems designed as a single-frequency network (SFN). In a broadcasting SFN, many fixed transmitters (one for each geographical cell) transmit the same OFDM signal, using the same carrier frequency, so that a mobile receiver can avoid handovers. Therefore, the multiple signals are seen by the receiver as a single signal with different multipath components, which are beneficial from the frequency-diversity viewpoint. A drawback arises when two (or more) different multipath components have a significant relative delay so that the composite CIR seen by the receiver, i.e., the sum of the delayed CIRs, becomes very long. For instance, in DVB-T/H 8k, the length of a Typical Urban CIR is $M \approx 64$, while, due to the relative delays of the CIRs, the length of a composite CIR can be, e.g., $M_{\mathrm{comp}} \approx 750$. Here, the problem is not the CP length (assuming $L_{\mathrm{CP}} = L/8 = 1024$, the ISI is avoided), but the lack of enough DVB-T/H pilot symbols for estimating the channel. Indeed, since the DVB-T/H pilot spacing is 12 times the subcarrier spacing, the maximum tolerated length of the composite CIR is $M_{\mathrm{max}} \approx 8192/12 \approx 683$. Therefore, when a composite channel with CIR length $M_{\mathrm{comp}} \approx 750$ is estimated using only the pilots of a single OFDM block, a delay-domain CIR aliasing arises, and the whole system performance degrades considerably. For LTI channels, this delay-domain aliasing is circumvented by exploiting the DVB-T/H scattered pilots of four OFDM blocks, thereby reducing the effective pilot spacing to $12/4 = 3$ and increasing M_{max} to $M_{\mathrm{max}} \approx 8192/3 \approx 2971$, which is much greater than the limit given by the CP length $L_{\mathrm{CP}} = 1024$. However, for fast LTV channels, the composite CIR changes from block to block, and therefore, the methods used for LTI channels are not appropriate.

Poggioni, Rugini, & Banelli (2009, June), avoided the delay-domain aliasing by means of an iterative algorithm that performs data-aided channel estimation. Basically, some data symbols are first estimated and successively used as virtual pilots to estimate the other data symbols. To reduce complexity, the positions of the virtual pilots are chosen to be equidistant from those of the pilot subcarriers. This way, the effective pilot spacing reduces to $12/2 = 6$ times the subcarrier spacing, and hence $M_{\mathrm{max}} \approx 8192/6 \approx 1365$, which is beyond the CP limit $L_{\mathrm{CP}} = 1024$. In the presence of a reduced number of pilots, other emerging techniques, such as compressed sensing, can also be employed to take advantage of the channel sparsity in the delay-Doppler domain (Tauböck, Hlawatsch, Eiwen, & Rauhut, 2010).

Another system aspect is the compatibility of the mentioned techniques with current OFDM standards, such as DVB-T/H. Clearly, the LTV equalization algorithms are fully compatible, since they only act at the receiver side. On the other extreme, pulse-shaping OFDM algorithms require a different transmitter and hence a moderate modification of current standards, due to the overlapping of different OFDM blocks. Besides, some channel estimation algorithms are not compatible, because they assume zero-guarded pilot patterns, differently from current standards. We identify three cases:

- Maximum compatibility: the receiver is designed assuming the current standard, at the price of a possibly low performance or a possibly high complexity;
- Moderate compatibility: the transmitter is slightly modified in order to simplify the receiver algorithms or to enhance the performance;
- No compatibility: in this case, an entirely new system design is performed, taking into account the requirements caused by fast LTV channels.

Multiuser systems can be specifically designed to deterministically reduce the ICI and the multiple-access interference (MAI): for instance, the multiuser multicarrier system of Leus, Zhou, & Giannakis (2003) completely removes both the ICI and the MAI for CCE-BEM channels, while an approximately MAI-free precoded multiuser OFDM system has been investigated by Tadjpour, Tsai, & Kuo (2007).

7.4.2 Open Issues

Even though OFDM systems with fast LTV channels have been largely investigated, there are still many questions that call for a definitive answer. One issue is related to the performance-complexity trade-off of structured equalizers, such as banded equalizers. Although the computational complexity of banded equalizers is well known, their theoretical performance has been scarcely investigated. A semianalytical approach that predicts the BER performance of banded linear block MMSE equalizers has been investigated by Rugini & Banelli (2007). However, little is known about the theoretical performance of nonlinear equalizers, such as ICI cancellers and turbo approaches. For instance, for DFEs, only some performance bounds are available (Rugini & Banelli, 2007). Moreover, it is not completely clear which is the best structure (i.e., the best BEM basis) for channel estimation purposes. Simulation results of Tang et al. (2007) have shown that the polynomial BEM presents a reduced modeling error at low Doppler spread, while the modeling error of an oversampled complex exponential BEM and discrete Karhunen–Loève BEM is quite low at high Doppler spread. For BEMs based on statistical CSI, an aspect that deserves a deeper investigation is the robustness to mismatches of the Doppler power profile (Zemen & Mecklenbräuker, 2005). In addition, the BEM choice can affect both the equalization performance and the rate reduction caused by the number of pilots, as shown by Tang & Leus (2008) for CCE-BEM and polynomial BEM. Channel estimation by means of superimposed training is another interesting topic for further investigation (Zhang, Dai, Li, & Ye, 2009). In addition, it is still unknown whether it is advantageous to bypass the channel estimation step, for instance, by using noncoherent schemes (Hwang & Schniter, 2008; Wetz, Periša, Teich, & Lindner, 2008) (see also Section 6.4.5).

Undoubtedly, MIMO-OFDM systems present even more open issues than single-antenna OFDM systems. A significant issue is the pilot design (Dai, 2007), and the related channel estimation, which could exploit multiple OFDM blocks to reduce the pilot overhead, as done by Tang, Leus, & Banelli (2006) for single-antenna OFDM systems (see also Section 4.4). In addition, space-time-frequency codes could be specifically designed in order to counteract (or, better, to exploit) the rapid time variation of the channel. Last but not least, in MIMO-OFDM, time and frequency synchronization issues are important. In this case, synchronization algorithms should explicitly take into account the time variation of the channel, as done for single-antenna OFDM systems in Lv, Li, & Chen (2005), Mostofi & Cox (2007), Nguyen-Le & Le-Ngoc (2009) Lottici, Reggiannini, & Carta (2010).

References

Alamouti, S. M. (1998). A simple transmit diversity scheme for wireless communications. *IEEE Journal of Selected Areas in Communication*, *16*(10), 1415–1458.

Al-Gharabally, M., & Das, P. (2006, June). On the performance of OFDM systems in time varying channels with channel estimation error. In *Proceedings of the IEEE ICC-06*: (pp. 5180–5185). Istanbul, Turkey.

Amini, P., & Farhang-Boroujeny, B. (2009, June). Per-tone equalizer design and analysis of filtered multitone communication systems over time-varying frequency-selective channels. In *Proceedings of the IEEE ICC-09*, Dresden, Germany.

Armstrong, J. (1999). Analysis of new and existing methods of reducing intercarrier interference due to carrier frequency offset in OFDM. *IEEE Transactions on Communications*, *47*(3), 365–369.

Banelli, P., & Rugini, L. (2010, June). An H-infinity filtering approach for robust tracking of OFDM doubly-selective channels. In *Proceedings of the IEEE SPAWC-10*, Marrakech, Morocco.

Banelli, P., Cannizzaro, R. C., & Rugini, L. (2007, April). Data-aided Kalman tracking for channel estimation in Doppler-affected OFDM systems. In *Proceedings of the IEEE ICASSP-07*: Vol. 3, (pp. 133–136). Honolulu, HI.

Barbarossa, S., & Torti, R. (2001, May). Chirped-OFDM for transmissions over time-varying channels with linear delay/Doppler spreading. In *Proceedings of the IEEE ICASSP-01*: (pp. 2377–2380). Salt Lake City, UT.

Barhumi, I., Leus, G., & Moonen, M. (2003). Optimal training design for MIMO OFDM systems in mobile wireless channels. *IEEE Transactions on Signal Processing*, *51*(6), 1615–1624.

Barhumi, I., Leus, G., & Moonen, M. (2004). Time-domain and frequency-domain per-tone equalization for OFDM in doubly-selective channels. *Signal Processing*, *84*(11), 2055–2066.

Bello, P. A. (1963). Characterization of randomly time-variant linear channels. *IEEE Transactions on Communications Systems*, *11*(4), 360–393.

Bingham, J. A. C. (1990). Multicarrier modulation for data transmission: An idea whose time has come. *IEEE Communications Magazine*, *28*(5), 5–14.

Bölcskei, H. (2003). Orthogonal frequency division multiplexing based on offset QAM. In H. G. Feichtinger & T. Strohmer (Eds.), *Advances in Gabor analysis*, (pp. 321–352). Boston: Birkhäuser.

Borah, D. K., & Hart, B. D. (1999). Frequency-selective fading channel estimation with a polynomial time-varying channel model. *IEEE Transactions on Communications*, *47*(6), 862–873.

Cai, X., & Giannakis, G. B. (2003). Bounding performance and suppressing intercarrier interference in wireless mobile OFDM. *IEEE Transactions on Communications*, *51*(12), 2047–2056.

Cannizzaro, R. C., Banelli, P., & Leus, G. (2006, August). Adaptive channel estimation for OFDM systems with Doppler spread. In *Proceedings of the IEEE SPAWC-06*, Cannes, France.

Chang, K., Han, Y., Ha, J., & Kim, Y. (2006, May). Cancellation of ICI by Doppler effect in OFDM systems. In *Proceedings of the IEEE VTC-06 Spring*: (pp. 1411–1415). Melbourne, Australia.

Chen, S., & Yao, T. (2004). Intercarrier interference suppression and channel estimation for OFDM systems in time-varying frequency-selective fading channels. *IEEE Transactions on Consumer Electronics*, *50*(2), 429–435.

Chiavaccini, E., & Vitetta, G. M. (2000). Error performance of OFDM signaling over doubly-selective Rayleigh fading channels. *IEEE Communications Letters*, *4*(7), 328–330.

Choi, Y.-S., Voltz, P. J., & Cassara, F. A. (2001). On channel estimation and detection for multicarrier signals in fast and selective Rayleigh fading channels. *IEEE Transactions on Communications*, *49*(8), 1375–1387.

Chow, J., & Jeremic, D. (2006, November). Diversity benefits of OFDM in fast fading. In *Proceedings of the IEEE GLOBECOM-06*, San Francisco, CA.

Cimini, L. J., Jr. (1985). Analysis and simulation of a digital mobile channel using orthogonal frequency division multiplexing. *IEEE Transactions on Communications*, *33*(7), 665–675.

Cui, T., & Tellambura, C. (2007). Blind receiver design for OFDM systems over doubly-selective channels. *IEEE Transactions on Communications*, *55*(5), 906–917.

Dai, X. (2007). Optimal training design for linearly time-varying MIMO/OFDM channels modelled by a complex exponential basis expansion. *IET Communications*, *1*(5), 945–953.

Dao, D. N., & Tellambura, C. (2005). Intercarrier interference self-cancellation space-frequency codes for MIMO-OFDM. *IEEE Transactions on Vehicular Technology*, *54*(5), 1729–1738.

Das, S., & Schniter, P. (2007). Max-SINR ISI/ICI-shaping multicarrier communication over the doubly dispersive channel. *IEEE Transactions on Signal Processing*, *55*(12), 5782–5795.

Erseghe, T., Laurenti, N., & Cellini, V. (2005). A multicarrier architecture based upon the affine Fourier transform. *IEEE Transactions on Communications*, *53*(5), 853–862.

ETSI (2004, November). *Digital video broadcasting (DVB); framing structure, channel coding and modulation for digital terrestrial television.* EN 300 744, V1.5.1. Sophia Antipolis, France: European Telecommunications Standards Institute.

ETSI (2005, July). *Digital audio broadcasting (DAB); data broadcasting – MPEG-2 TS streaming.* TS 102 427, V1.1.1. Sophia Antipolis, France: European Telecommunications Standards Institute.

Fang, K., Leus, G., & Rugini, L. (2006, November-December). Alamouti space-time coded OFDM systems in time- and frequency-selective channels. In *Proceedings of the IEEE GLOBECOM-06*, San Francisco, CA.

Fang, K., Rugini, L., & Leus, G. (2008, April). Iterative channel estimation and turbo equalization for time-varying OFDM systems. In *Proceedings of the IEEE ICASSP-08*: (pp. 2909–2912). Las Vegas, NV.

Fang, K., Rugini, L., & Leus, G. (2008). Low-complexity block turbo equalization for OFDM systems in time-varying channels. *IEEE Transactions on Signal Processing*, *56*(11), 5555–5566.

Feng, S., Minn, H., Yan, L., & Jinhui, L. (2010). PIC-based iterative SDR detector for OFDM systems in doubly-selective fading channels. *IEEE Transactions on Wireless Communications*, *9*(1), 86–91.

Fertl, P., & Matz, G. (2006, October). Efficient OFDM channel estimation in mobile environments based on irregular sampling. In *Proceedings of the IEEE Asilomar-06*: (pp. 1777–1781). Pacific Grove, CA.

Fu, Y., Tellambura, C., & Krzymien, W. A. (2007). Transmitter precoding for ICI reduction in closed-loop MIMO OFDM systems. *IEEE Transactions on Vehicular Technology*, *56*(1), 115–125.

Gao, J., & Liu, H. (2008). Low-complexity MAP channel estimation for mobile MIMO-OFDM systems. *IEEE Transactions on Wireless Communications*, *7*(3), 774–780.

Gorokhov, A., & Linnartz, J.-P. (2004). Robust OFDM receivers for dispersive time-varying channels: Equalization and channel acquisition. *IEEE Transactions on Communications*, *52*(4), 572–583.

Gradshteyn, I. S., & Ryzhik, I. M. (1994). *Table of integrals, series, and products.* (5th ed.). Florida: Academic Press.

Haas, R., & Belfiore, J.-C. (1997). A time-frequency well-localized pulse for multiple carrier transmission. *Wireless Personal Communications*, *5*(1), 1–18.

Hampejs, M., Švač, P., Tauböck, G., Gröchenig, K., Hlawatsch, F., & Matz, G. (2009, June). Sequential LSQR-based ICI equalization and decision-feedback ISI cancellation in pulse-shaped multicarrier systems. In *Proceedings of the IEEE SPAWC-09*: (pp. 1–5). Perugia, Italy.

Harris, F. J. (1978, January). On the use of windows for harmonic analysis with the discrete Fourier transform. *Proceedings of the IEEE*, *66*(1), 51–83.

Hoeher, P., Kaiser, S., & Robertson, P. (1997, April). Two-dimensional pilot-symbol-aided channel estimation by Wiener filtering. In *Proceedings of the IEEE ICASSP-97*: (pp. 1845–1848). Munich, Germany.

How, W.-S., & Chen, B.-S. (2005). ICI cancellation for OFDM communication systems in time-varying multipath fading channels. *IEEE Transactions on Wireless Communications*, *4*(9), 2100–2110.

Hrycak, T., & Matz, G. (2006, October). Low-complexity time-domain ICI equalization for OFDM communications over rapidly varying channels. In *Proceedings of the IEEE Asilomar-06*: (pp. 1767–1771). Pacific Grove, CA.

Huang, D., Letaief, K. B., & Lu, J. (2005). Bit-interleaved time-frequency coded modulation for OFDM systems over time-varying channels. *IEEE Transactions on Communications*, *53*(7), 1191–1199.

Hunziker, T., & Dahlhaus, D. (2003). Iterative detection for multicarrier transmission employing time-frequency concentrated pulses. *IEEE Transactions on Communications*, *51*(4), 641–651.

Hutter, A. A., Hasholzner, R., & Hammerschmidt, J. S. (1999, September). Channel estimation for mobile OFDM systems. In *Proceedings of the IEEE VTC-99 Fall*: (pp. 305–309). Amsterdam, The Netherlands.

Hwang, S.-J., & Schniter, P. (2006). Efficient sequence detection of multi-carrier transmissions over doubly dispersive channels. *EURASIP Journal on Applied Signal Processing*, article ID 93638.

Hwang, S.-J., & Schniter, P. (2008). Efficient multicarrier communication for highly spread underwater acoustic channels. *IEEE Journal on Selected Areas in Communications*, *26*(9), 1674–1683.

IEEE (1999). *Part 11: Wireless LAN medium access control (MAC) and physical layer (PHY) specifications: High-speed physical layer in the 5 GHz band.* New York: Institute of Electrical and Electronics Engineers.

IEEE (2006). *Part 16: Air interface for fixed and mobile broadband wireless access systems.* New York: Institute of Electrical and Electronics Engineers.

Jeon, W. G., Chang, K. H., & Cho, Y. S. (1999). An equalization technique for orthogonal frequency-division multiplexing systems in time-variant multipath channels. *IEEE Transactions on Communications*, *47*(1), 27–32.

Jung, P., & Wunder, G. (2007). The WSSUS pulse design problem in multicarrier transmission. *IEEE Transactions on Communications*, *55*(10), 1918–1928.

Kannu, A. P., & Schniter, P. (2008). Design and analysis of MMSE pilot-aided cyclic-prefixed block transmission for doubly selective channels. *IEEE Transactions on Signal Processing*, *56*(3), 1148–1160.

Kay, S. M. (1993). *Fundamentals of statistical signal processing, Vol. I: Estimation Theory.* (1st ed.), Upper Saddle River, NJ, Prentice Hall.

Keller, T., & Hanzo, L. (2000). Adaptive multicarrier modulation: A convenient framework for time-frequency processing in wireless communications. *Proceedings of the IEEE*, *88*(5), 611–640.

Kim, J., Heath, R. W., Jr & Powers, E. J. (2005). Receiver designs for Alamouti coded OFDM systems in fast fading channels. *IEEE Transactions on Wireless Communications*, *4*(2), 550–559.

Kim, J.-G., & Lim, J.-T. (2008). MAP-based channel estimation for MIMO-OFDM over fast Rayleigh fading channels. *IEEE Transactions on Vehicular Technology*, *57*(3), 1963–1968.

Kim, K., & Park, H. (2006, May). A low complexity ICI cancellation method for high mobility OFDM systems. In *Proceedings of the IEEE VTC-06 Spring*: (pp. 2528–2532), Melbourne, Australia.

Kou, Y. J., Lu, W.-S., & Antoniou, A. (2005, August). An iterative intercarrier-interference reduction algorithm for OFDM systems. In *Proceedings of the IEEE PACRIM-05*: (pp. 538–541), Victoria, Canada.

Kozek, W., & Molisch, A. F. (1998). Nonorthogonal pulseshapes for multicarrier communications in doubly dispersive channels. *IEEE Journal on Selected Areas in Communications*, *16*(8), 1579–1589.

Lai, I-W., & Chiueh, T.-D. (2006, May). One-dimensional interpolation based channel estimation for mobile DVB-H reception. In *Proceedings of the IEEE ISCAS-06*: (pp. 5207–5210). Kos, Greece.

Le Floch, B., Alard, M., & Berrou, C. (1995). Coded orthogonal frequency division multiplex. *Proceedings of the IEEE*, *83*(6), 982–996.

Leung, E., & Ho, P. (1998, June). A successive interference cancellation scheme for an OFDM system. In *Proceedings of the IEEE ICC-98*: (pp. 375–379), Atlanta, GA.

Leus, G. (2004, September). On the estimation of rapidly time-varying channels. In *Proceedings of the EUSIPCO-04*. (pp. 2227–2230), Vienna, Austria.

Leus, G., Zhou, S., & Giannakis, G. B. (2003). Orthogonal multiple access over time- and frequency-selective channels. *IEEE Transactions on Information Theory*, *49*(8), 1942–1950.

Li, G., Yang, H., Cai, L., & Gui, L. (2003, October). A low-complexity equalization technique for OFDM system in time-variant multipath channels. In *Proceedings of the IEEE VTC-03 Fall*: (pp. 2466–2470), Orlando, FL.

Li, R., Li, Y., & Vucetic, B. (2008, March-April). Iterative receiver for MIMO-OFDM systems with joint ICI cancellation and channel estimation. In *Proceedings of the IEEE WCNC-08*: (pp. 7–12), Las Vegas, NV.

Li, Y., & Cimini, L. J., Jr. (2001). Bounds on the interchannel interference of OFDM in time-varying impairments. *IEEE Transactions on Communications*, *49*(3), 401–404.

Lin, D.-B., Chiang, P.-H., & Li, H.-J. (2005). Performance analysis of two-branch transmit diversity block-coded OFDM systems in time-varying multipath Rayleigh-fading channels. *IEEE Transactions on Vehicular Technology*, *54*(1), 136–148.

Linnartz, J.-P. M. G. (2001). Performance analysis of synchronous MC-CDMA in mobile Rayleigh channel with both delay and Doppler spreads. *IEEE Transactions on Vehicular Technology*, *50*(6), 1375–1387.

Liu, D. N., & Fitz, M. P. (2008, September). Turbo MIMO equalization and decoding in fast fading mobile coded OFDM. In *Proceedings of the 2008 Allerton Conference*: (pp. 977–982), Monticello, IL.

Liu, K., Kadous, T., & Sayeed, A. M. (2004). Orthogonal time-frequency signaling over doubly dispersive channels. *IEEE Transactions on Information Theory*, *50*(11), 2583–2603.

Lottici, V., Reggiannini, R., & Carta, M. (2010). Pilot-aided carrier frequency estimation for filter-bank multicarrier wireless communications on doubly-selective channels. *IEEE Transactions on Signal Processing*, *58*(5), 2783–2794.

Lu, S., Kalbasi, R., & Al-Dhahir, N. (2006, September). OFDM interference mitigation algorithms for doubly-selective channels. In *Proceedings of the IEEE VTC-06 Fall*, Montreal, Canada.

Lv, T., Li, H., & Chen, J. (2005). Joint estimation of symbol timing and carrier frequency offset of OFDM signals over fast time-varying multipath channels. *IEEE Transactions on Signal Processing*, *53*(12), 4526–4535.

Ma, X., Giannakis, G. B., & Ohno, S. (2003). Optimal training for block transmissions over doubly-selective wireless fading channels. *IEEE Transactions on Signal Processing*, *51*(5), 1351–1366.

Martone, M. (2001) A multicarrier system based on the fractional Fourier transform for time-frequency-selective channels. *IEEE Transactions on Communications*, *49*(6), 1011–1020.

Matz, G., Schafhuber, D., Gröchenig, K., Hartmann, M., & Hlawatsch, F. (2007). Analysis, optimization, and implementation of low-interference wireless multicarrier systems. *IEEE Transactions on Wireless Communications*, *6*(5), 1921–1931.

Molisch, A. F., Toeltsch, M., & Vermani, S. (2007). Iterative methods for cancellation of intercarrier interference in OFDM systems. *IEEE Transactions on Vehicular Technology*, *56*(4), 2158–2167.

Mostofi, Y., & Cox, D. C. (2005). ICI mitigation for pilot-aided OFDM mobile systems. *IEEE Transactions on Wireless Communications*, *4*(3), 765–774.

Mostofi, Y., & Cox, D. C. (2007). A robust timing synchronization design in OFDM systems – Part 2: High-mobility cases. *IEEE Transactions on Wireless Communications*, *6*(12), 4340–4348.

Muralidhar, K., & Li, K. H. (2009). A low-complexity Kalman approach for channel estimation in doubly-selective OFDM systems. *IEEE Signal Processing Letters*, *16*(7), 632–635.

Muschallik, C. (1996). Improving an OFDM reception using an adaptive Nyquist windowing. *IEEE Transactions on Consumer Electronics*, *42*(3), 259–269.

Nguyen-Le, H., & Le-Ngoc, T. (2009, June). Joint synchronization and channel estimation for OFDM transmissions over doubly selective channels. In *Proceedings of the IEEE ICC-09*, Dresden, Germany.

Ohno, S. (2005, March). Maximum likelihood inter-carrier interference suppression for wireless OFDM with null subcarriers. In *Proceedings of the IEEE ICASSP-05*, (pp. 849–852), Philadelphia, PA.

Ozdemir, M. K., & Arslan, H. (2007). Channel estimation for wireless OFDM systems. *IEEE Communications Surveys and Tutorials*, *9*(2), 18–48.

Park, K. W., & Cho, Y. S. (2005). An MIMO-OFDM technique for high-speed mobile channels. *IEEE Communications Letters*, *9*(7), 604–606.

Peiker, E., Dominicus, J., Teich, W., & Lindner, J. (2008, December). Improved performance of OFDM systems for fast time-varying channels. In *Proceedings of the IEEE ICSPCS-08*, Gold Coast, Australia.

Poggioni, M., Rugini, L., & Banelli, P. (2008). A novel simulation model for coded OFDM in Doppler scenarios. *IEEE Transactions on Vehicle Technology*, *57*(5), 2969–2980.

Poggioni, M., Rugini, L., & Banelli, P. (2009, June). Multistage decoding-aided channel estimation and equalization for DVB-H in single-frequency networks. In *Proceedings of the IEEE SPAWC-09*: (pp. 181–185), Perugia, Italy.

Poggioni, M., Rugini, L., & Banelli, P. (2009). DVB-T/H and T-DMB: Physical layer performance comparison in fast mobile channels. *IEEE Transactions on Broadcasting*, *55*(4), 719–730.

Robertson, P., & Kaiser, S. (1999, September). The effects of Doppler spreads in OFDM(A) mobile radio systems. In *Proceedings of the IEEE VTC-99 Fall*: (pp. 329–333), Amsterdam, The Netherlands.

Rugini, L., & Banelli, P. (2006, September). Banded equalizers for MIMO-OFDM in fast time-varying channels. In *Proceedings of the EUSIPCO-06*, Florence, Italy.

Rugini, L., & Banelli, P. (2007, June). Performance analysis of banded equalizers for OFDM systems in time-varying channels. In *Proceedings of the IEEE SPAWC-07*, Helsinki, Finland.

Rugini, L., Banelli, P., & Leus, G. (2005, March). Reduced-complexity equalization for MC-CDMA systems over time-varying channels. In *Proceedings of the IEEE ICASSP-05*, Vol. 3, (pp. 473–476), Philadelphia, PA.

Rugini, L., Banelli, P., & Leus, G. (2005). Simple equalization of time-varying channels for OFDM. *IEEE Communications Letters*, *9*(7), 619–621.

Rugini, L., Banelli, P., & Leus, G. (2006). Low-complexity banded equalizers for OFDM systems in Doppler spread channels. *EURASIP Journal on Applied Signal Processing*, article ID 67404, (pp. 13).

Rugini, L., Banelli, P., Fang, K., & Leus, G. (2009, June). Enhanced turbo MMSE equalization for MIMO-OFDM over rapidly time-varying frequency-selective channels. In *Proceedings of the IEEE SPAWC-09*: (pp. 36–40), Perugia, Italy.

Russell, M., & Stüber, G. L. (1995, July). Interchannel interference analysis of OFDM in a mobile environment. In *Proceedings of the IEEE VTC-95*: (pp. 820–824), Chicago, IL.

Salvo Rossi, P., & Müller, R. R. (2008). Slepian-based two-dimensional estimation of time-frequency variant MIMO-OFDM channels. *IEEE Signal Processing Letters*, *15*(1), 21–24.

Sanzi, F., & Speidel, J. (2000). An adaptive two-dimensional channel estimator for wireless OFDM with application to mobile DVB-T. *IEEE Transactions on Broadcasting*, *46*(2), 128–133.

Sari, H., Karam, G., & Jeanclaude, I. (1995). Transmission techniques for digital terrestrial TV broadcasting. *IEEE Communications Magazine*, *33*(2), 100–109.

Schafhuber, D., & Matz, G. (2005). MMSE and adaptive prediction of time-varying channels for OFDM systems. *IEEE Transactions on Wireless Communications*, *4*(2), 593–602.

Schafhuber, D., Matz, G., & Hlawatsch, F. (2003, April). Adaptive Wiener filters for time-varying channel estimation in wireless OFDM systems. In *Proceedings of the IEEE ICASSP-03*: (pp. 688–691), Hong Kong.

Schafhuber, D., Rupp, M., Matz, G., & Hlawatsch, F. (2003, June). Adaptive identification and tracking of doubly selective fading channels for wireless MIMO-OFDM systems. In *Proceedings of the IEEE SPAWC-03*: (pp. 417–421), Rome, Italy.

Schniter, P. (2004). Low-complexity equalization of OFDM in doubly selective channels. *IEEE Transactions on Signal Processing*, *52*(4), 1002–1011.

Seyedi, A., & Saulnier, G. J. (2005). General ICI self-cancellation scheme for OFDM systems. *IEEE Transactions on Vehicular Technology*, *54*(1), 198–210.

Sgraja, C., & Lindner, J. (2003, May). Estimation of rapid time-variant channels for OFDM using Wiener filtering. In *Proceedings of the IEEE ICC-03*: (pp. 2390–2395), Anchorage, AK.

Shin, C., Andrews, J. G., & Powers, E. J. (2007). An efficient design of doubly selective channel estimation for OFDM systems. *IEEE Transactions on Wireless Communications*, *6*(10), 3790–3802.

Song, H., Kim, J., Nam, S., Yu, T., & Hong, D. (2008). Joint Doppler-frequency diversity for OFDM systems using hybrid interference cancellation in time-varying multipath fading channels. *IEEE Transactions on Vehicular Technology*, 57(1), 635–641.

Song, W. G., & Lim, J. T. (2006). Channel estimation and signal detection for MIMO-OFDM with time varying channels. *IEEE Communications Letters*, *10*(7), 540–542.

Stamoulis, A., Diggavi, S. N., & Al-Dhahir, N. (2002). Intercarrier interference in MIMO OFDM. *IEEE Transactions on Signal Processing*, *50*(10), 2451–2464.

Stantchev, B., & Fettweis, G. (2000). Time-variant distortions in OFDM. *IEEE Communications Letters*, *4*(9), 312–314.

Strohmer, T., & Beaver, S. (2003). Optimal OFDM design for time-frequency dispersive channels. *IEEE Transactions on Communications*, *51*(7), 1111–1122.

Tadjpour, L., Tsai, S. H., & Kuo, C. C. J. (2007). An approximately MAI-free multiaccess OFDM system in fast time-varying channels. *IEEE Transactions on Signal Processing*, *55*(7), 3787–3799.

Tan, P., & Beaulieu, N. C. (2004). Reduced ICI in OFDM systems using the "better than" raised-cosine pulse. *IEEE Communications Letters*, *8*(3), 135–137.

Tang, Z., & Leus, G. (2008). A novel receiver architecture for single-carrier transmission over time-varying channels. *IEEE Journal on Selected Areas in Communications*, *26*(2), 366–377.

Tang, Z., Leus, G., & Banelli, P. (2006, August). Pilot-assisted time-varying OFDM channel estimation based on multiple OFDM symbols. In *Proceedings of the IEEE SPAWC-06*, Cannes, France.

Tang, Z., Cannizzaro, R. C., Leus, G., & Banelli, P. (2007). Pilot-assisted time-varying channel estimation for OFDM systems. *IEEE Transactions on Signal Processing*, *55*(5), 2226–2238.

Tauböck, G., Hampejs, M., Matz, G., Hlawatsch, F., & Gröchenig, K. (2007, June). LSQR-based ICI equalization for multicarrier communications in strongly dispersive and highly mobile environments. In *Proceedings of the IEEE SPAWC-07*, Helsinki, Finland.

Tauböck, G., Hlawatsch, F., Eiwen, D., & Rauhut, H. (2010). Compressive estimation of doubly selective channels in multicarrier systems: Leakage effects and sparsity-enhancing processing. *IEEE Journal on Selected Topics in Signal Processing*, *4*(2), 255–271.

ten Brink, S. (2001). Convergence behavior of iteratively decoded parallel concatenated codes. *IEEE Transactions on Communications*, *49*(10), 1727–1737.

Teo, K. A. D., & Ohno, S. (2005, November/December). Optimal MMSE finite parameter model for doubly-selective channels. In *Proceedings of the IEEE GLOBECOM-05*: (pp. 3503–3507), St. Louis, MO.

Tomasin, S., Gorokhov, A., Yang, H., & Linnartz, J. P. (2005). Iterative interference cancellation and channel estimation for mobile OFDM. *IEEE Transactions on Wireless Communications*, *4*(1), 238–245.

Tonello, A. M., & Pecile, F. (2008). Analytical results about the robustness of FMT modulation with several prototype pulses in time-frequency selective fading channels. *IEEE Transactions on Wireless Communications*, *7*(5), 1634–1645.

Tong, L., Sadler, B. M., & Dong, M. (2004). Pilot-aided wireless transmissions: General model, design criteria, and signal processing. *IEEE Signal Processing Magazine*, *21*(6), 12–25.

Tran, X. N., & Fujino, T. (2005, December). Groupwise successive ICI cancellation for OFDM systems in time-varying channels. In *Proceedings of the IEEE ISSPIT-05*: (pp. 489–494), Athens, Greece.

Tsatsanis, M. K., & Giannakis, G. B. (1996). Modeling and equalization of rapidly fading channels. *International Journal of Adaptive Control and Signal Processing*, *10*(2/3), 159–176.

Vahlin, A., & Holte, N. (1995). Optimal finite duration pulses for OFDM. *IEEE Transactions on Communications*, *44*(1), 10–14.

Vogeler, S., Brötje, L., Klenner, P., Kühn, V., & Kammeyer, K. D. (2004, September). Intercarrier interference suppression for OFDM transmission at very high velocities. In *Proceedings of the International OFDM-Workshop*, Dresden, Germany.

Wang, T., Proakis, J. G., Masry, E., & Zeidler, J. R. (2006). Performance degradation of OFDM systems due to Doppler spreading. *IEEE Transactions on Wireless Communications*, *5*(6), 1422–1432.

Wang, T., Proakis, J. G., & Zeidler, J. R. (2007). Interference analysis of filtered multitone modulation over time-varying frequency-selective fading channels. *IEEE Transactions on Communications*, *55*(4), 717–727.

Wang, Z., & Giannakis, G. B. (2000). Wireless multicarrier communications: Where Fourier meets Shannon. *IEEE Signal Processing Magazine*, *17*(3), 29–48.

Wang, Z., & Giannakis, G. B. (2003). Complex-field coding for OFDM over fading wireless channels. *IEEE Transactions on Information Theory*, *49*(3), 707–720.

Wetz, M., Periša, I., Teich, W., & Lindner, J. (2008). Robust transmission over fast fading channels on the basis of OFDM-MFSK. *Wireless Personal Communications*, *47*(1), 113–123.

Yun, J. Y., Chung, S. Y., & Lee, Y. H. (2007). Design of ICI canceling codes for OFDM systems based on capacity maximization. *IEEE Signal Processing Letters*, *14*(3), 169–172.

Zhang, H., & Li, Y. (2003). Optimum frequency-domain partial response encoding in OFDM system. *IEEE Transactions on Communications*, *51*(7), 1064–1068.

Zhang, H., Dai, X., Li, D., & Ye, S. (2009). Linearly time-varying channel estimation and symbol detection for OFDMA uplink using superimposed training. *EURASIP Journal on Wireless Communications and Networking*, article ID 307375, (pp. 11).

Zhang, Y., & Liu, H. (2006). Frequency-domain correlative coding for MIMO-OFDM systems over fast fading channels. *IEEE Communications Letters*, *10*(5), 347–349.

Zhang, Y. H., Lu, W. S., & Gulliver, T. A. (2007, June). A successive intercarrier interference reduction algorithm for OFDM systems. In *Proceedings of the IEEE ICC-07*: (pp. 2936–2941), Glasgow, UK.

Zhao, Y., & Häggman, S. G. (2001). Intercarrier interference self-cancellation scheme for OFDM mobile communication systems. *IEEE Transactions on Communications*, *49*(7), 1185–1191.

Zhao, Y., Leclercq, J. D., & Häggman, S. G. (1998). Intercarrier interference compression in OFDM communication systems by using correlative coding. *IEEE Communications Letters*, *2*(8), 214–216.

Zemen, T., & Mecklenbräuker, C. F. (2005). Time-variant channel estimation using discrete prolate spheroidal sequences. *IEEE Transactions on Signal Processing*, *53*(9), 3597–3607.

Zemen, T., Mecklenbräuker, C. F., Wehinger, J., & Müller, R. R. (2006). Iterative joint time-variant channel estimation and multi-user detection for MC-CDMA. *IEEE Transactions on Wireless Communications*, *5*(6), 1469–1478.

Zhao, M., Shi, Z., & Reed, M. C. (2007, November). An iterative receiver with channel estimation for MIMO-OFDM over a time and frequency dispersive fading channel. In *Proceedings of the IEEE GLOBECOM-07*: (pp. 4155–4159), Washington, DC.

Zhao, M., Shi, Z., & Reed, M. C. (2008). Iterative turbo channel estimation for OFDM system over rapid dispersive fading channel. *IEEE Transactions on Wireless Communications*, *7*(8), 3174–3184.

Zhu, J., Wen, S., & Du, Q. (2008). Space-frequency-Doppler coded OFDM over the time-varying Doppler fading channels. *International Journal of Electronics and Communications, (AEÜ)*, *62*, 307–315.

Zou, W. Y., & Wu, Y. (1995). COFDM: An overview. *IEEE Transactions on Broadcasting*, *41*(1), 1–8.

CHAPTER

Multiuser MIMO Receiver Processing for Time-Varying Channels

8

Charlotte Dumard[1], Joakim Jaldén[2], Thomas Zemen[1]

[1]*FTW Forschungszentrum Telekommunikation Wien, Vienna, Austria*

[2]*Royal Institute of Technology (KTH), Stockholm, Sweden*

8.1 INTRODUCTION

Wireless broadband communications for mobile users at vehicular speed is the cornerstone of future fourth-generation systems. This chapter deals with joint iterative channel estimation and multiuser detection for the uplink of a multicarrier (MC) code division multiple access (CDMA) system. MC-CDMA is based on orthogonal frequency division multiplexing (OFDM) and employs spreading sequences in the frequency domain (Kaiser, 1998). Both the mobile stations and the base station employ multiple antennas; hence, we deal with a multiuser multiple-input multiple-output (MIMO) receiver.

So far, most research on multiuser detection has dealt with block-fading frequency-selective channels, where the channel state is assumed to stay constant for the duration of a single data block of K data symbols. Even so, the optimal maximum a posteriori (MAP) detector for such a system is prohibitively complex although it can be approximated using iterative linear minimum mean-square error (LMMSE) multiuser detection and parallel interference cancelation (Zemen, Mecklenbräuker, Wehinger, & Müller, 2006).

This work deals with mobile users where the MIMO channels are time and frequency selective. Due to the rapid time variation of the MIMO channel, the computational complexity of conventional multiuser receivers, based on channel estimation, parallel interference cancelation, multiuser detection, and iterative decoding, increases drastically since the multiuser detection filters need to be recalculated for each data symbol individually.

In this chapter, we address this complexity issue by trading accuracy for efficiency. As a starting point, we adopt a joint iterative structure based on LMMSE multiuser detection and channel estimation. The decoding stage, implemented by the BCJR algorithm (Bahl, Cocke, Jelinek, & Raviv, 1974), supplies extrinsic probabilities (EXT) and a posteriori probabilities (APP) on the code symbols. This APP and EXT information is fed back for enhanced channel estimation and multiuser detection, respectively (Mecklenbräuker, Wehinger, Zemen, Artés, & Hlawatsch, 2006; Zemen et al., 2006).

The remainder of this chapter is organized as follows: in Section 8.2, the signal model is established, and in Section 8.3, key ideas for complexity reduction are introduced. For complexity reduction, we

- approximate the MAP detector using an iterative receiver structure;
- establish a low-dimensional reduced-rank model of the time-varying MIMO channel;

- apply the Krylov subspace method for reducing the complexity of LMMSE filters; and
- investigate an alternative nonlinear detector using the sphere decoding algorithm.

The rest of this chapter builds on these four tools. We present time-varying multiuser MIMO channel estimation in Section 8.4. In Section 8.5, we adapt the parallel interference cancelation in user space such that it can be combined with the Krylov subspace method for data detection. A sphere decoding algorithm exploiting the reduced-rank model of time-varying channels is presented in Section 8.6. In Section 8.7, simulation results are used to compare the bit error rate (BER) versus signal-to-noise ratio (SNR) performance of the presented algorithms as well as their computational complexity.

8.2 MULTIUSER MIMO SYSTEMS

We consider a multiuser MIMO uplink system. U users, having M_{T} transmit antennas each, transmit to a receiver with M_{R} receive antennas (see Fig. 8.1). The transmission is based on OFDM with L subcarriers.

We uniquely identify the data symbols sent from antenna $t \in \{1,\ldots,M_{\mathrm{T}}\}$ of user $u \in \{1,\ldots,U\}$ by a random spreading sequence $\mathbf{s}_{u,t} \in \mathbb{C}^L$. The elements of each spreading sequence are independent identically distributed from the set $\{\pm 1 \pm j\}/\sqrt{2L}$ (Paulraj, Nabar, & Gore, 2003). We refer to transmit antenna t of user u as transmit antenna or virtual user (u,t).

We consider block-based transmission. Each data block transmitted from a single antenna consists of $K-J$ OFDM data symbols and J pilot symbols. The data symbols $d_{u,t}[k]$ at discrete time $k \in \{0,\ldots,K-1\}$ result from the binary information sequence $b_k[i]$ of length $2(K-J)R_{\mathrm{c}}$ by convolutional encoding with code rate R_{c}, random interleaving, and quadrature phase shift keying (QPSK) modulation.

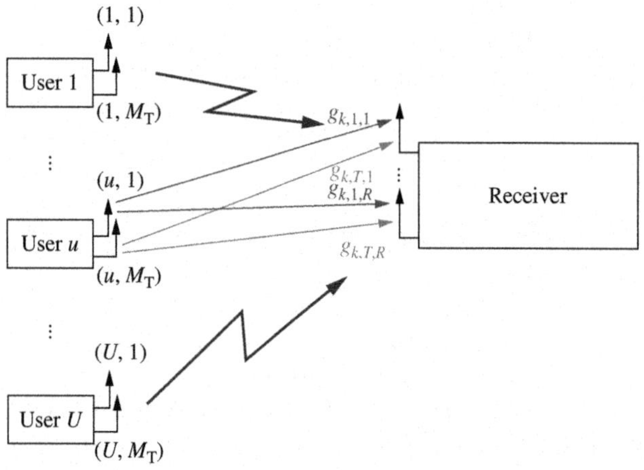

FIGURE 8.1

Multiuser MIMO uplink.

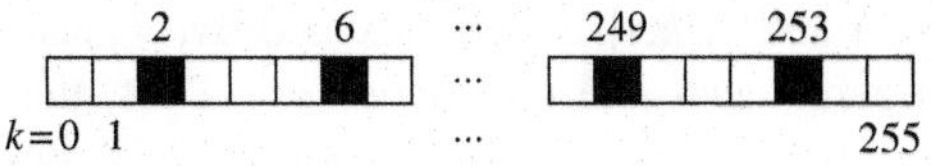

FIGURE 8.2

Pilot placement for $J = 60$ pilots among $K = 256$ symbols.

We jointly encode, interleave, and map M_T data blocks of one user. Then, the $M_T(K-J)$ coded symbols are split into M_T coded symbol blocks (Wolniansky, Foschini, Golden, & Valenzuela, 1998). The $K-J$ data symbols are distributed over a block of length K such that

$$d_{u,t}[k] \in \begin{cases} \{\pm 1 \pm j\}/\sqrt{2} & \text{for} \quad k \notin \mathcal{P}, \\ 0 & \text{for} \quad k \in \mathcal{P}, \end{cases}$$

allowing for pilot symbol insertion. The pilot placement is defined through the index set

$$\mathcal{P} = \left\{ \left\lfloor i\frac{K}{J} + \frac{K}{2J} \right\rfloor \middle| \, i \in \{0, \ldots, J-1\} \right\}.$$

Figure 8.2 gives an illustration of the pilot placement.

Each data symbol is independently spread as follows: each element $d_{u,t}[k]$ of the symbol block $\mathbf{d}_{u,t} \in \mathbb{C}^{K-J}$ is spread over the L OFDM subcarriers using the spreading sequence $\mathbf{s}_{u,t}$ defined above, leading to the OFDM data symbol at discrete time k

$$\mathbf{s}_{u,t} d_{u,t}[k] \in \mathbb{C}^L.$$

For $k \in \mathcal{P}$, the pilot vector $\mathbf{p}_{u,t}[k] \in \mathbb{C}^L$ has elements chosen randomly from the QPSK symbol set $\{\pm 1 \pm \mathrm{j}\}/\sqrt{2L}$. Otherwise, $\mathbf{p}_{u,t}[k] = \mathbf{0}_L$ for $k \notin \mathcal{P}$. After spreading, the pilot symbols $\mathbf{p}_{u,t}[k] \in \mathbb{C}^L$ are added as

$$\mathbf{a}_{u,t}[k] = \mathbf{s}_{u,t} d_{u,t}[k] + \mathbf{p}_{u,t}[k]. \tag{8.1}$$

Each transmit antenna (u,t) transmits data symbols $d_{u,t}[k]$ with symbol rate $1/T$, where T denotes symbol duration. At each transmit antenna, an L-point inverse discrete Fourier transform (DFT) is performed and a cyclic prefix of length G is inserted. A single OFDM symbol together with the cyclic prefix has length $P = L + G$ chips. After parallel-to-serial conversion, the chip stream with chip rate $1/T_c = P/T$ is transmitted over a time-varying multipath fading channel with M resolvable paths.

The transmission of UM_T symbols at time instant k is done over U independent $M_T \times M_R$ MIMO channels. The UM_TM_R component channels are assumed uncorrelated. Thus, the receiver treats the UM_T antennas in the same way as if they were UM_T independent users having one transmit antenna each.

The receiver is equipped with $M_R \geq 1$ receive antennas. At each receive antenna $r \in \{1, \ldots, M_R\}$, the signals of all UM_T transmit antennas add up. At each of the M_R receive antennas, cyclic prefix removal and a DFT are performed on the corresponding received signal. After these two operations, the received signal at receive antenna $r \in \{1, \ldots, M_R\}$ for subcarrier l and time instant k is given by

$$y_r[k,l] = \sum_{u=1}^{U} \sum_{t=1}^{M_T} g_{u,t,r}[k,l] a_{u,t}[k,l] + z_r[k,l], \tag{8.2}$$

where $a_{u,t}[k,l]$ is the lth element of $\mathbf{a}_{u,t}[k]$ defined in (8.1) and $z_r[k,l]$ is complex additive Gaussian noise that is white with respect to r, k, and l, with zero mean and covariance σ_z^2. Thus, the noise is assumed to be spatially and temporally white. The time-varying frequency response between transmit antenna (u,t) and receive antenna r for discrete time-index $k \in \{0,\ldots,K-1\}$ and subcarrier $l \in \{0,\ldots,L-1\}$ is denoted by $g_{u,t,r}[k,l]$.

The variation of the wireless channel $g_{u,t,r}[k,l]$ over the duration of a long coded data block is caused by user mobility and multipath propagation. The Doppler shifts on the individual paths depend on the user's direction, its velocity υ, the carrier frequency f_c, and the scattering environment. The maximum variation in time of the wireless channel is upper bounded by the maximum (one-sided) normalized Doppler bandwidth

$$\upsilon_{\max} = \frac{\upsilon_{\max} f_\mathrm{c}}{c} T,$$

where $\upsilon_{\max}$ is the maximum supported velocity, T is the symbol duration, and c denotes the speed of light. The parameter $\upsilon_{\max}$ can be considered a system parameter that is valid for all channels.

Under the assumption of small inter-carrier interference, each time-varying frequency-flat channel is fully described by a sequence of complex scalars at the OFDM symbol rate $1/T$. This sequence is band-limited by $\nu_{\max}$. In order to perform coherent multiuser detection, we need to estimate a time-limited snapshot of this band-limited sequence at the receiver side. The length of these snapshots is equal to the length K of a data block consisting of OFDM data symbols with interleaved OFDM pilot symbols.

In matrix-vector notation, we can compactly represent the joint received signal vector of the multiuser MIMO system at time k as (Dumard & Zemen, 2006; Hanly & Tse, 2001)

$$\mathbf{y}[k] = \tilde{\mathbf{S}}[k]\mathbf{d}[k] + \mathbf{z}[k] \quad \text{for} \quad k \notin \mathcal{P}, \tag{8.3}$$

where

$$\mathbf{y}[k] = \begin{pmatrix} \mathbf{y}_1[k] \\ \vdots \\ \mathbf{y}_{M_\mathrm{R}}[k] \end{pmatrix}, \quad \tilde{\mathbf{S}}[k] = \begin{pmatrix} \tilde{\mathbf{S}}_1[k] \\ \vdots \\ \tilde{\mathbf{S}}_{M_\mathrm{R}}[k] \end{pmatrix}, \quad \text{and} \quad \mathbf{z}[k] = \begin{pmatrix} \mathbf{z}_1[k] \\ \vdots \\ \mathbf{z}_{M_\mathrm{R}}[k] \end{pmatrix}.$$

In the above equation, $\mathbf{y}_r[k] = (y_r[k,0] \cdots y_r[k,L-1])^T$ is the received chip sequence at antenna r, and $\mathbf{z}_r[k]$ denotes the additive complex white Gaussian noise vector at the receiver defined similarly. The data symbols for the UM_T transmit antennas are contained in $\mathbf{d}[k]$. The elements of the time-varying effective spreading matrix are given by

$$\tilde{\mathbf{S}}_r[k] \triangleq \Big(\underbrace{\tilde{\mathbf{s}}_{1,1,r}[k] \cdots \tilde{\mathbf{s}}_{1,M_\mathrm{T},r}[k]}_{\text{user } 1} \cdots \underbrace{\tilde{\mathbf{s}}_{U,1,r}[k] \cdots \tilde{\mathbf{s}}_{U,M_\mathrm{T},r}[k]}_{\text{user } U} \Big) \in \mathbb{C}^{L \times UM_\mathrm{T}}$$

with

$$\tilde{\mathbf{s}}_{u,t,r}[k] \triangleq \operatorname{diag}\left\{\mathbf{g}_{u,t,r}[k]\right\} \mathbf{s}_{u,t} \in \mathbb{C}^L, \tag{8.4}$$

where $\mathbf{g}_{u,t,r}[k] \in \mathbb{C}^L$ is the time-varying frequency response of the channel between transmit antenna (u,t) and receive antenna r. The notation $\mathrm{diag}\{\mathbf{v}\}$ denotes the diagonal matrix whose elements are the elements of the vector $\mathbf{v}$.

To make the dependence on the time-varying channel more explicit, we can write (8.3) equivalently as

$$\mathbf{y}[k] = \underbrace{(\mathbf{G}[k] \odot \mathbf{S})}_{\tilde{\mathbf{S}}[k]}\mathbf{d}[k] + \mathbf{z}[k] \quad \text{for} \quad k \notin \mathcal{P},$$

where $\mathbf{S} \in \mathbb{C}^{LM_\mathrm{R} \times UM_\mathrm{T}}$ collects all (time-invariant) spreading sequences $\mathbf{s}_{u,t,r}$ in the same ordering as used for $\tilde{\mathbf{S}}[k]$ in (8.3), i.e.,

$$\mathbf{S}_r \triangleq \begin{pmatrix} \mathbf{s}_{1,1,1} & \cdots & \mathbf{s}_{1,M_\mathrm{T},1} & \cdots & \mathbf{s}_{U,1,1} & \cdots & \mathbf{s}_{U,M_\mathrm{T},1} \\ \vdots & \ddots & \vdots & \ddots & \vdots & \ddots & \vdots \\ \mathbf{s}_{1,1,M_\mathrm{R}} & \cdots & \mathbf{s}_{1,M_\mathrm{T},M_\mathrm{R}} & \cdots & \mathbf{s}_{U,1,M_\mathrm{R}} & \cdots & \mathbf{s}_{U,M_\mathrm{T},M_\mathrm{R}} \end{pmatrix},$$

$\mathbf{G}[k] \in \mathbb{C}^{LM_\mathrm{R} \times UM_\mathrm{T}}$ collects all time-varying channels $\mathbf{g}_{u,t,r}[k]$ according to

$$\mathbf{G}[k] \triangleq \begin{pmatrix} \mathbf{g}_{1,1,1}[k] & \cdots & \mathbf{g}_{1,M_\mathrm{T},1}[k] & \cdots & \mathbf{g}_{U,1,1}[k] & \cdots & \mathbf{g}_{U,M_\mathrm{T},1}[k] \\ \vdots & \ddots & \vdots & \ddots & \vdots & \ddots & \vdots \\ \mathbf{g}_{1,1,M_\mathrm{R}}[k] & \cdots & \mathbf{g}_{1,M_\mathrm{T},M_\mathrm{R}}[k] & \cdots & \mathbf{g}_{U,1,M_\mathrm{R}}[k] & \cdots & \mathbf{g}_{U,M_\mathrm{T},M_\mathrm{R}}[k] \end{pmatrix},$$

and $\odot$ denotes the element-wise Hadamard product.

8.3 TOOLS FOR COMPLEXITY REDUCTION

Mathematically exact solutions in the context of multiuser detection require high computational complexity. Often, this accuracy is simply not needed due to a low SNR or due to reduced-rank properties of the system of equations to be solved. In the following, four key ideas for multiuser detection in time-varying MIMO channels are outlined that allow trading accuracy for efficiency. These ideas are applied and tested on the multiuser MIMO receiver in later sections.

8.3.1 Iterative Approximation of the MAP Detector

Assuming (for a moment) perfect channel knowledge, the MAP sequence detector is given by (Kühn, 2006)

$$\mathbf{d}'_\mathrm{MAP} = \arg\min_{\mathbf{d}' \in \mathcal{D}} p(\mathbf{y}'|\mathbf{d}')p(\mathbf{d}'),$$

where

$$\mathbf{y}' \triangleq \begin{pmatrix} \mathbf{y}[0] \\ \vdots \\ \mathbf{y}[K-1] \end{pmatrix}, \quad \mathbf{d}' \triangleq \begin{pmatrix} \mathbf{d}[0] \\ \vdots \\ \mathbf{d}[K-1] \end{pmatrix},$$

and $\mathcal{D}$ represents the constraint imposed by coding. Specializing the metric to (8.3), using the Gaussianity and whiteness of the noise, and assuming that all code words are a priori equally likely yields

$$\mathbf{d}'_{\mathrm{MAP}} = \arg\min_{\mathbf{d}' \in \mathcal{D}} \sum_{k \notin \mathcal{P}} \left\| \mathbf{y}[k] - \tilde{\mathbf{S}}[k]\mathbf{d}[k] \right\|^2. \tag{8.5}$$

The optimization problem in (8.5) is a quadratic problem in $U(K-J)M_{\mathrm{T}}$ variables constrained to a set of $2^{U(K-J)M_{\mathrm{T}}R_{\mathrm{C}}}$ possible code words. Solving (8.5) directly is therefore unfeasible for any realistic system. Applying the Viterbi algorithm to (8.5) would still require a complexity of $\mathcal{O}((K-J)2^{UM_{\mathrm{T}}\gamma})$ where γ is the constraint length of the code, under the best case assumption that there is no interleaver. In short, one cannot realistically expect to obtain the exact solution to (8.5). Taking the channel uncertainty into account will further complicate the MAP metric and the subsequent optimization problem. This motivates suboptimal approximations to (8.5).

Virtually all suboptimal approaches are iterative in nature. The iterative approach contains two steps where in the first step the metric in (8.5) is minimized while neglecting the code constraints. The code constraint is then enforced in the second step. The first step is referred to as the multiuser detection step (see Fig. 8.3). Soft information is passed between the two steps, indicating the reliability of (or belief in) candidate solutions to (8.5). The iterative approach also has the benefit of being able to nicely incorporate updates of the channel estimates as a third step in the iteration. While the iterative procedure has traditionally been viewed as ad hoc, recent theoretical justification based on divergence minimization was given in Hu, Land, Rasmussen, Piton, & Fleury (2008). In the iterative approach considered herein, the code constraint is enforced using the BCJR algorithm, which has a complexity of $\mathcal{O}(U(K-J)M_{\mathrm{T}}2^{\gamma})$. The focus of this chapter is, however, on the multiuser detection and channel estimation. A schematic overview of the approximated MAP receiver is given in Fig. 8.3.

The key gain of neglecting the code constraint in (8.5) is that the optimization metric decouples into $K-J$ separate terms, i.e., each term in the sum can be treated separately by the detector. In effect, the multiuser detector takes soft information in the form of bit probabilities (or log-likelihood ratios) from the output of the BCJR decoder and updates these based on the signal model in (8.3), for each $k \notin \mathcal{P}$ separately. This alone reduces the number of variables that need to be simultaneously considered by a factor of $K-J$.

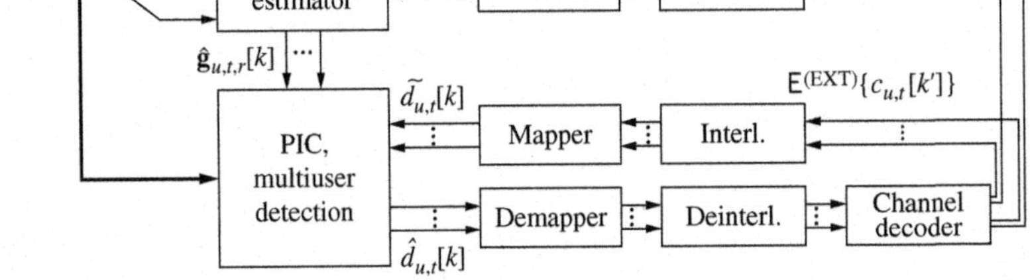

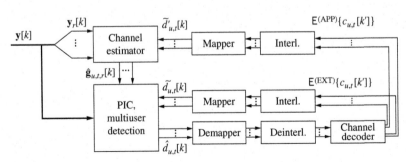

FIGURE 8.3

Iterative multiuser MIMO receiver for MC-CDMA.

Still, the signal model in (8.5) contains UM_{T} data symbols and a large number of channel parameters. For example, to construct an LMMSE filter with the aim of only retrieving the data symbols requires a complexity of $\mathcal{O}((UM_{\text{T}})^3)$, and due to the time-varying nature of the channel, this construction must be done separately for the $K-J$ OFDM symbols containing data. When the number of users or antennas is large, this may still be prohibitive. It is therefore crucial to exploit the structural properties of the time-varying channel in order to further reduce the complexity and improve performance. In this chapter, several key ideas that accomplish this goal are presented.

8.3.2 Reduced-Rank Model for the Time-Varying Channel

Starting from (8.2), it is in principle possible to construct an LMMSE filter for estimating all UM_{T} time-varying frequency responses $\mathbf{g}_{u,t,r}$ at receive antenna r. However, the following practical problems arise:

- The second-order statistics of all UM_{T} channels are unknown. These statistics are usually difficult to estimate in a multiuser system due to interference and limited stationarity time of the fading process (Matz, 2005).
- The block length is upper bounded by the maximum round trip delay; hence, the number of pilots J for estimating all UM_{T} channels is limited as well.

The performance of the iterative receiver depends on the channel estimates for the time-varying frequency response $\mathbf{g}_{u,t,r}[k]$ since the effective spreading sequence $\tilde{\mathbf{s}}_{u,t,r}[k]$ in (8.4) directly depends on the current channel realization. We take advantage of Slepian's basic result that time-limited parts (snapshots) of band-limited sequences span a low-dimensional subspace (Slepian, 1978). A convenient set of orthonormal basis functions for this subspace is given by the discrete prolate spheroidal (DPS) sequences. Using these results from the theory of time-concentrated and band-limited sequences, we represent a time-varying subcarrier through a Slepian basis expansion of low dimensionality (Zemen & Mecklenbräuker, 2005). Projecting a vector on an incomplete set of orthonormal basis functions always incurs a nonzero quadratic bias. It was shown in Zemen & Mecklenbräuker (2005) that the square bias of the Slepian basis expansion is more than one order of magnitude smaller compared to the square bias of the Fourier basis expansion (Sayeed, Sendonaris, & Aazhang, 1998) (i.e., a truncated DFT). The reason for this improvement is that the frequency leakage effect of the Fourier transform is significantly reduced. Channel estimation for a single-user OFDM system using DPS sequences was also investigated independently in Javaudin, Lacroix, & Rouxel (2003).

The DPS sequences $\{u_i[k]\}$ are defined as the solution of the eigenvalue problem (Slepian, 1978)

$$\sum_{\ell=0}^{K-1} \frac{\sin(2\pi\nu_{\max}(\ell-k))}{\pi(\ell-k)} u_i[\ell] = \lambda_i u_i[k].$$

The sequences $\{u_i[k]\}$ are doubly orthogonal over the infinite set $\{-\infty,\ldots,\infty\}$ and the finite set $\mathcal{I}_K = \{0,\ldots,K-1\}$, band-limited by $\nu_{\max}$, and maximally energy-concentrated on $\mathcal{I}_K$.

We are interested in describing the time variation of the frequency-selective channel $\mathbf{g}_{u,t,r}[k]$ for the duration of a single data block $\mathcal{I}_K$. For $k \in \mathcal{I}_K$, we model $\mathbf{g}_{u,t,r}[k]$ using the Slepian basis expansion (Zemen & Mecklenbräuker, 2005). The Slepian basis functions $\mathbf{u}_i = (u_i[0] \cdots u_i[K-1])^T$ for $i \in \{0,\ldots,D-1\}$ are the time-limited DPS sequences. The time-varying channel $\mathbf{g}_{u,t,r}[k] \in \mathbb{C}^L$ is projected onto the subspace spanned by the first D Slepian sequences and is approximated as

$$g_{u,t,r}[k,l] \approx \tilde{g}_{u,t,r}[k,l] \triangleq \sum_{i=0}^{D-1} \psi_{u,t,r}[i,l] u_i[k], \tag{8.6}$$

for $k \in \mathcal{I}_K$ and $l \in \{0,\ldots,L-1\}$. In the following, we distinguish the exact channel, the projected channel, and the estimated channel, which are denoted $g_{u,t,r}[k,l]$, $\tilde{g}_{u,t,r}[k,l]$, and $\hat{g}_{u,t,r}[k,l]$, respectively.

In Zemen, Mecklenbräuker, Fleury, & Kaltenberger (2007), it is shown that from the Karhunen–Loève identity, it follows that the basis vectors $\mathbf{u}_i$ minimize the mean square reconstruction error per subcarrier

$$\mathrm{MSE}_{u,t,r}[l] = \frac{1}{K} \sum_{k=0}^{K-1} \mathsf{E}\left\{ \left| g_{u,t,r}[k,l] - \hat{g}_{u,t,r}[k,l] \right|^2 \right\}$$

for a fading process with constant Doppler spectrum with support $(-\nu_{\max}, \nu_{\max})$. The subspace dimension minimizing the mean square error per observation interval for a given SNR is found to be (Scharf & Tufts, 1987)

$$D = \operatorname*{argmin}_{D' \in \{1,\ldots,K\}} \left(\frac{1}{2\nu_{\max} K} \sum_{i=D'}^{K-1} \lambda_i + \frac{D'}{K} \sigma_z^2 \right). \tag{8.7}$$

We implicitly assume here that the eigenvalues are ranked as $\lambda_0 \geq \lambda_1 \geq \cdots \geq \lambda_{K-1}$. By optimizing D, we achieve a square bias-variance compromise.

In Fig. 8.4, we plot the subspace dimension D versus E_a/σ_z^2, where E_a denotes the symbol energy. Due to the exponential decay of the λ_i for $i \geq \lfloor 2\nu_{\max} K \rfloor + 1$ (Slepian, 1978), the optimum number of basis functions, $D(\sigma_z^2)$ in (8.7) stays constant for large intervals of E_a/σ_z^2 (see Fig. 8.4). The near-optimum solution would be to select D based on the actual E_a/σ_z^2 according to Fig. 8.4. For practical mobile communication systems, $D \leq 5$ for $K = 256$ and $J = 64$ (see Zemen & Mecklenbräuker, 2005).

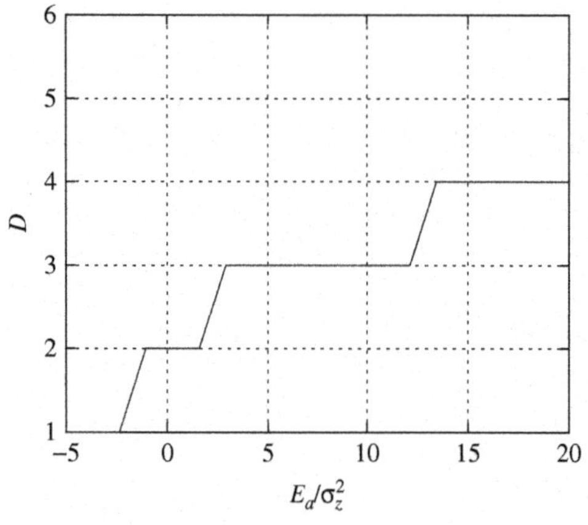

FIGURE 8.4

Subspace dimension D versus E_a/σ_z^2.

Substituting the basis expansion (8.6) for the time-varying subcarrier coefficients $g_{u,t,r}[k,l]$ into the system model (8.2), we obtain

$$y_r[k,l] = \sum_{u=1}^{U} \sum_{t=1}^{M_T} \sum_{i=0}^{D-1} \psi_{u,t,r}[i,l] u_i[k] a_{u,t}[k,l] + z_r[k,l].$$

This reduced-rank channel model will be used for low-complexity channel estimation in Section 8.4 and for a convenient implementation of sphere decoding in Section 8.6.

8.3.3 The Krylov Subspace Method

In Zemen et al. (2006), two LMMSE filters are used for multiuser detection and time-varying channel estimation (see Fig. 8.3). The computational complexity of obtaining these filters is substantial, and suboptimal approximations are thus of interest. One such method is based on projections on Krylov subspaces and has been extensively used in the literature. Details on the Krylov subspace approximation can be found in Kailath & Sayed (1999), Moon & Stirling (2000), Saad (2003); and van der Vorst (2003). Krylov subspaces have been shown to define a framework in which the Wiener–Hopf equation can be solved, leading to the multistage nested Wiener filter (Joham & Zoltowski, 2000). Applications range from beamforming (Kecicioglu & Torlak, 2004; Kirsteins & Ge, 2006) to detection (Cottatellucci, Müller, & Debbah, 2006; Dietl & Utschick, 2007), where a substantial reduction of computational complexity has been shown.

For multiuser detection, the authors of Cottatellucci et al. (2006) make use of the self-averaging properties of random matrices modeling the channel to compute the necessary projection weights. However, they do not consider an iterative scheme using interference cancelation. In such a case, the projection computations are not common to all users anymore, and no computational complexity reduction can be achieved this way.

In Dietl & Utschick (2007), the authors use the Lanczos algorithm to approximate the Wiener filter in an iterative receiver for a *single-user* single-input multiple-output system. They use an adjusted mean of the signal based on a priori information to cancel the multipath interference and obtain a computational complexity that scales quadratically with the length of the observation vector.

The LMMSE filters for channel estimation and multiuser detection have the general form (Dumard, Jaldén, & Zemen, 2008; Wehinger, 2005)

$$\mathbf{f} = \mathbf{N}^{-1}\mathbf{n}, \quad \text{with} \quad \mathbf{N} \triangleq \mathbf{D}_1 + \mathbf{M}\mathbf{D}_2\mathbf{M}^H, \quad \mathbf{n} \triangleq \mathbf{M}\mathbf{v}. \tag{8.8}$$

Here, $\mathbf{M} \in \mathbb{C}^{a\times b}$ and $\mathbf{v} \in \mathbb{C}^{b}$, and $\mathbf{D}_1 \in \mathbb{C}^{a\times a}$ and $\mathbf{D}_2 \in \mathbb{C}^{b\times b}$ are diagonal. The filter $\mathbf{f}$ is applied on the received signal $\mathbf{y}$ to estimate either the channel or the symbols using $\mathbf{f}^H\mathbf{y}$.

The LMMSE filter $\mathbf{f}$ in (8.8) is the exact solution of a linear system. The computational complexity of its calculation is given by

$$C_{\text{LMMSE}} = 8a^2\left(\frac{a}{3} + b\right) \quad \text{flops.}$$

We define a *flop* as a floating point operation, as given in Golub & Van Loan (1996). A flop is an addition, subtraction, multiplication, division, or square root operation in the *real* domain. Thus, one complex multiplication requires 4 real multiplications and 2 additions, leading to 6 flops. Similarly, one complex addition requires 2 flops.

Computing the exact LMMSE filter using (8.8) is expensive. Trading complexity for accuracy, we can employ the Krylov subspace method to approximate it, as described below. Considering the linear system

$$\mathbf{N}\mathbf{f} = \mathbf{n},$$

we define the Krylov subspace of dimension s defined by $\mathbf{N}$ and $\mathbf{n}$ as

$$\mathcal{K}_s = \mathrm{span}\{\mathbf{n}, \mathbf{N}\mathbf{n}, \ldots, \mathbf{N}^{s-1}\mathbf{n}\}.$$

8.3.3.1 Basic Idea

The Krylov subspace-based algorithm computes iteratively an approximation of the solution $\mathbf{f}$, starting from an initial guess $\mathbf{f}_0$, using projections onto the Krylov subspaces $\mathcal{K}_s$. Based on the Cayley–Hamilton theorem (Atiyah, 1994) for an invertible matrix $\mathbf{N}$, we can state that there is a minimum polynomial $\mathbf{K}(\cdot)$ of degree $d \le a$ such that $\{\mathbf{I}_a, \mathbf{N}, \ldots, \mathbf{N}^{b-1}\}$ are linearly independent and $\mathbf{K}(\mathbf{N}) = \mathbf{0}$. This leads to

$$\mathbf{N}^{-1} = \sum_{i=1}^{d} \alpha_i \mathbf{N}^{i-1},$$

where $\alpha_1, \ldots, \alpha_{d-1} \in \mathbb{C}$ are computed from $\mathbf{K}(\cdot)$. It follows that

$$\mathbf{f} = \mathbf{N}^{-1}\mathbf{n} = \sum_{i=1}^{d} \alpha_i \mathbf{N}^{i-1}\mathbf{n} \in \mathcal{K}_d.$$

The idea behind Krylov subspace methods is to iteratively compute an approximation $\mathbf{f}_s$ of $\mathbf{f}$ for s incrementing up to d, using projections of the corresponding residual vector $\mathbf{t}_s = \mathbf{n} - \mathbf{N}\mathbf{f}_s$ onto the Krylov subspace $\mathcal{K}_s$. To limit the complexity of such a method, which is proportional to the number of iterations performed by the algorithm, a given maximum dimension S is fixed such that the output

$$\sum_{i=1}^{S} \alpha_i \mathbf{N}^{i-1}\mathbf{n} \in \mathcal{K}_S$$

approximates f.

8.3.3.2 The Algorithm

The iterative process begins with some initial value $\mathbf{f}_0$. At step s, the approximation of $\mathbf{f}$ is given by $\mathbf{f}_s$. We define for each step the residual vector $\mathbf{t}_s$ according to

$$\mathbf{t}_s = \mathbf{n} - \mathbf{N}\mathbf{f}_s.$$

The iteration is done until $s = S$ by constructing $\mathbf{f}_s$ such that

$$\mathbf{f}_s \in \mathbf{f}_0 + \mathrm{span}\left\{\mathbf{t}_0, \mathbf{N}\mathbf{t}_0, \ldots, \mathbf{N}^{s-1}\mathbf{t}_0\right\} \quad \text{and} \quad \mathbf{t}_s = \mathbf{n} - \mathbf{N}\mathbf{f}_s \perp \mathcal{K}_s. \tag{8.9}$$

In the previous equation, the notation $\mathbf{f}_0 + \mathcal{S}$ denotes the subspace $\mathcal{S}$ shifted by $\mathbf{f}_0$. Thus, $\mathbf{f}_s \in \mathbf{f}_0 + \mathcal{S}$ is equivalent to $(\mathbf{f}_s - \mathbf{f}_0) \in \mathcal{S}$.

It can be easily proven that the error $\boldsymbol{\epsilon}_s = \mathbf{f} - \mathbf{f}_s$ satisfies

$$\mathbf{N}\boldsymbol{\epsilon}_s = \mathbf{t}_s.$$

The above orthogonality condition on $\mathbf{t}_s$ (8.9) is the basis for the so-called Ritz–Galerkin approach (van der Vorst, 2003). Another possible condition is that of minimum residual norm, i.e., computing $\mathbf{f}_s$ such that the ℓ_2 norm of $\mathbf{t}_s$ is minimized. We will not consider this second approach in this work.

Note that for $s \leq S$, $\mathbf{f}_s - \mathbf{f}_0$ is an element of $\mathcal{K}_s$ and can as such be written

$$\mathbf{f}_s - \mathbf{f}_0 = \mathbf{W}_s \mathbf{v}_s,$$

for any basis

$$\mathbf{W}_s = (\mathbf{w}_1 \cdots \mathbf{w}_s)$$

of $\mathcal{K}_s$ and $\mathbf{v}_s \in \mathbb{C}^s$. An orthonormal basis $\mathbf{W}_s$ can be computed by applying the Gram–Schmidt method (Golub & Van Loan, 1996) on the Krylov basis $\mathbf{B}_s = \left(\mathbf{t}_0 \mathbf{N} \mathbf{t}_0 \cdots \mathbf{N}^{s-1} \mathbf{t}_0\right)$. The Ritz–Galerkin condition $\mathbf{t}_s \perp \mathcal{K}_s$ can be expressed as

$$\mathbf{W}_s^H \mathbf{t}_s = \mathbf{0} \Leftrightarrow \mathbf{W}_s^H \mathbf{t}_0 = \mathbf{W}_s^H \mathbf{N} \mathbf{W}_s \mathbf{v}_s. \tag{8.10}$$

Furthermore, the vectors $\mathbf{w}_i$ for $i \in \{1, \ldots, s\}$ are such that $\mathbf{N}\mathbf{w}_i \in \mathcal{K}_{i+1}$, leading by construction to the property

$$\mathbf{w}_\ell^H \mathbf{N} \mathbf{w}_i = 0 \quad \text{if} \quad s \geq \ell > i+1.$$

The matrix $\mathbf{T}^{(s)} \triangleq \mathbf{W}_s^H \mathbf{N} \mathbf{W}_s$ has elements $[\mathbf{T}^{(s)}]_{i,\ell} = \mathbf{w}_\ell^H \mathbf{N} \mathbf{w}_i$, where the superscript (s) indicates the increasing subspace dimension s. Thus, it is an upper Hessenberg matrix (i.e., the elements below the first subdiagonal are zeros). For $\mathbf{N}$ being symmetric, $\mathbf{T}^{(s)}$ will be tridiagonal symmetric (i.e., all elements outside the three main diagonals are zeros), and we denote its elements on the main diagonal as $\mu_i \in \mathbb{C}$ and on the secondary diagonals as $\nu_i \in (0, +\infty)$. Finally, we know by construction of the orthonormal basis $\mathbf{W}_s$ that $\mathbf{t}_0 = \|\mathbf{t}_0\| \mathbf{w}_1$.

Inserting these results into (8.10), we now need to solve

$$\mathbf{v}_s = \left(\mathbf{T}^{(s)}\right)^{-1} \|\mathbf{t}_0\| \mathbf{e}_s,$$

where $\mathbf{e}_s = (1\,0 \cdots 0)^T$ has length s. To compute $\mathbf{v}_s$, the first column of $\left(\mathbf{T}^{(s)}\right)^{-1}$ is needed only. We apply the matrix inversion lemma for partitioned matrices (Moon & Stirling, 2000) to obtain the recursive relation

$$\mathbf{T}_s = \begin{pmatrix} \mathbf{T}_{s-1} & \nu_s \tilde{\mathbf{e}}_{s-1} \\ \nu_s \tilde{\mathbf{e}}_{s-1}^T & \mu_s \end{pmatrix},$$

where $\tilde{\mathbf{e}}_s = (0 \cdots 0\,1)^T$ has length s for $s \in \{1, \ldots, S\}$. This gives the following set of recursive equations

$$\mathbf{c}_{\text{first}}^{(s)} = \begin{pmatrix} \mathbf{c}_{\text{first}}^{(s-1)} \\ 0 \end{pmatrix} + \beta_s^{-1} \left[\mathbf{c}_{\text{last}}^{(s-1)}\right]_1^* \begin{pmatrix} \nu_s^2 \mathbf{c}_{\text{last}}^{(s-1)} \\ -\nu_s \end{pmatrix}$$

$$\mathbf{c}_{\text{last}}^{(s)} = \beta_s^{-1} \begin{pmatrix} -\nu_s \mathbf{c}_{\text{last}}^{(s-1)} \\ 1 \end{pmatrix},$$

where $\mathbf{c}_{\text{first}}^{(s)}$ and $\mathbf{c}_{\text{last}}^{(s)}$ denote respectively the first and last columns of $\left(\mathbf{T}^{(s)}\right)^{-1}$, and

$$\beta_s = \mu_s - \nu_s^2 \left[\mathbf{c}_{\text{last}}^{(s-1)}\right]_{s-1}$$

is a scalar. Note that $\mathbf{c}_{\text{first}}^{(s)}$ and $\mathbf{c}_{\text{last}}^{(s)}$ have dimension s at iteration s of the algorithm.

Algorithm 8.1 Krylov Subspace Algorithm for a Hermitian Matrix

1	input $\mathbf{N}, \mathbf{n}, \mathbf{f}_0, S$	11	$\mathbf{u} = \mathbf{N}\mathbf{w}_s$
2	$\mathbf{t}_0 = \mathbf{n} - \mathbf{N}\mathbf{f}_0$	12	$\mu = \mathbf{w}_s^H \mathbf{u}$
3	$\mathbf{w}_1 = \mathbf{t}_0 / \|\mathbf{t}_0\|$	13	$\beta = \mu - \nu^2 \mathbf{c}_{last}^{(s-1)}$
4	$\mathbf{u} = \mathbf{N}\mathbf{w}_1$	14	$\mathbf{c} = \beta^{-1} \left(-\nu \mathbf{c}_{last}^T \ 1 \right)^T$
5	$\mu = \mathbf{w}_1^H \mathbf{u}$	15	$\mathbf{c}_{first} = \left(\mathbf{c}_{first}^T \ 0 \right)^T - [\mathbf{c}_{last}]_1^* \nu \mathbf{c}$
6	$\mathbf{c}_{first} = \mathbf{c}_{last} = 1/\mu$	16	$\mathbf{c}_{last} = \mathbf{c}$
7	$\mathbf{w} = \mathbf{u} - \mu \mathbf{w}_1$	17	$\mathbf{w} = \mathbf{u} - \mu \mathbf{w}_s - \nu \mathbf{w}_{s-1}$
8	for $s = 2, \ldots, S$	18	end
9	$\nu = \|\mathbf{w}\|$	19	$\mathbf{W}_S = (\mathbf{w}_1 \cdots \mathbf{w}_S)$
10	$\mathbf{w}_s = \mathbf{w}/\nu$	20	output $\mathbf{f}_S = \|\mathbf{t}_0\| \mathbf{W}_S \mathbf{c}_{first} + \mathbf{f}_0$

The algorithm is detailed in Algorithm 8.1. Note that by using an appropriate initial value $\mathbf{f}_0$, the Krylov subspace algorithm may converge faster. In other words, a smaller subspace dimension s would be needed to achieve a given error ϵ

$$\left\| \mathbf{N}^{-1}\mathbf{n} - \mathbf{f}_s \right\| \leq \epsilon.$$

The reduction of computational complexity achieved by using the Krylov subspace method is given by

$$\gamma \triangleq \frac{C_K}{C_{LMMSE}} \leq \mathcal{O}\left(\frac{2S+3}{a} \right), \tag{8.11}$$

where C_K and C_{LMMSE} denote the computational complexity of calculating the approximate Krylov subspace-based filter and the exact LMMSE filter, respectively. A more detailed complexity calculation can be found in Dumard et al. (2008). Due to the symmetric structure of $\mathbf{f}$, we can set $\min\{a,b\} = a$ without loss of generality. Assuming $S \ll a$, the computational complexity reduction achieved by the Krylov subspace method is substantial.

As can be noticed, these kinds of iterative methods do not compute the matrix $\mathbf{N}^{-1}$ explicitly, but give an approximation of the solution $\mathbf{N}^{-1}\mathbf{n}$ of the linear system for a given $\mathbf{n}$. Thus, if we need to solve several linear systems with the same matrix $\mathbf{N}$ but different vectors $\mathbf{n}$, then the Krylov subspace method will have to be performed for each $\mathbf{n}$. This will be an important aspect for our discussion of computational complexity in Section 8.5.1.

8.3.4 Sphere Decoding

Linear multiuser detectors solving (8.5) ignore the constraint imposed by the code, thus allowing the symbol vector $\mathbf{d}[k]$ to take any value in $\mathbb{C}^{UM_T}$. While this generally leads to a less complex optimization problem, it also generally entails a large performance loss. The performance may often be significantly improved by enforcing the constraints imposed by the finite symbol alphabet, while still neglecting the code constraint. This entails solving optimization problems of the form

$$\hat{\mathbf{d}}[k] = \underset{\mathbf{d}[k] \in \mathcal{A}^{UM_T}}{\arg\min} \ \|\mathbf{y}[k] - \tilde{\mathbf{S}}[k]\mathbf{d}[k]\|^2, \tag{8.12}$$

where $\mathcal{A} = \{\pm 1 \pm \mathrm{j}\}/\sqrt{2}$ denotes the finite symbol alphabet. Problems of this form are also encountered when calculating soft symbol and bit estimates (this will be further discussed in Section 8.6.2).

The problem in (8.12) is unfortunately NP-hard for general $\mathbf{y}[k]$ and $\tilde{\mathbf{S}}[k]$ (Verdú, 1989), implying that there are no efficient (i.e., polynomial-time) algorithms for its solution. Nevertheless, for moderately sized problems, a feasible approach is offered by the sphere decoding algorithm.

8.3.4.1 Dimensionality Reduction

It has been shown that the expected number of nodes visited in the search phase still grows exponentially with UM_{T} (Jalden & Ottersten, 2005). While this is not a problem for moderately sized systems, it does however pose a serious problem for systems of the size considered herein. A suitable compromise between performance and complexity is to only perform nonlinear detection in a lower-dimensional subspace spanned by the effective spreading sequences corresponding to a particular user. The multiuser interference in this subspace is mitigated by (linear) parallel interference cancelation (see Sections 8.5.1 and 8.5.2).

The contribution of user u in (8.3) is given by

$$\mathbf{y}^{(u)}[k] = \tilde{\mathbf{S}}^{(u)}[k]\mathbf{d}^{(u)}[k] + \mathbf{z}^{(u)}[k], \tag{8.13}$$

where

$$\tilde{\mathbf{S}}^{(u)}[k] = \begin{pmatrix} \tilde{\mathbf{s}}_{u,1,1}[k] & \cdots & \tilde{\mathbf{s}}_{u,M_{\mathrm{T}},1}[k] \\ \vdots & \ddots & \vdots \\ \tilde{\mathbf{s}}_{u,1,M_{\mathrm{R}}}[k] & \cdots & \tilde{\mathbf{s}}_{u,M_{\mathrm{T}},M_{\mathrm{R}}}[k] \end{pmatrix} \in \mathbb{C}^{LM_{\mathrm{R}} \times M_{\mathrm{T}}} \tag{8.14}$$

contains the effective spreading sequences from all transmit antennas of user u to all receive antennas. The number of possible data vectors $\mathbf{d}^{(u)}[k] = \left(d_{u,1}[k] \cdots d_{u,M_{\mathrm{T}}}[k]\right)^T$ transmitted by user u is $|\mathcal{A}|^{M_{\mathrm{T}}}$. By definition, the contribution of user u is also given by

$$\mathbf{y}^{(u)}[k] = \mathbf{y}[k] - \sum_{u' \neq u} \mathbf{y}^{(u')}[k]. \tag{8.15}$$

Parallel interference cancelation is performed by approximating the contribution of the other users in (8.15) according to

$$\begin{aligned} \mathbf{y}^{(u)}[k] &\approx \mathbf{y}[k] - \sum_{u' \neq u} \left(\tilde{\mathbf{S}}^{(u')}[k]\mathbf{d}^{(u')}[k] + \mathbf{z}^{(u')}[k]\right) \\ &\approx \mathbf{y}[k] - \tilde{\mathbf{S}}[k]\tilde{\mathbf{d}}[k] + \tilde{\mathbf{S}}^{(u)}[k]\tilde{\mathbf{d}}^{(u)}[k] \triangleq \tilde{\mathbf{y}}^{(u)}[k]. \end{aligned} \tag{8.16}$$

The soft symbol estimates $\tilde{d}_{u,t}[k]$ for $k \in \{0,\ldots,M-1\}$ corresponding to $d_{u,t}[k]$ are calculated from the extrinsic probabilities after detection, $\Pr^{(\mathrm{EXT})}\left\{c_{u,t}[k]\right\}$, as follows:

$$\tilde{d}_{u,t}[k] = \frac{2\mathrm{E}^{(\mathrm{EXT})}\{c_{u,t}[2k]\} - 1}{\sqrt{2}} + j\frac{2\mathrm{E}^{(\mathrm{EXT})}\{c_{u,t}[2k+1]\} - 1}{\sqrt{2}}, \tag{8.17}$$

where for $k' \in \{0,\ldots,2M-1\}$

$$\begin{aligned}\mathsf{E}^{(\mathrm{EXT})}\{c_{u,t}[k']\} &= \Pr^{(\mathrm{EXT})}\{c_{u,t}[k'] = +1\} - \Pr^{(\mathrm{EXT})}\{c_{u,t}[k'] = -1\} \\ &= 2\Pr^{(\mathrm{EXT})}\{c_{u,t}[k'] = +1\} - 1\end{aligned} \tag{8.18}$$

calculates the expectation over the alphabet of c which is $\{-1,+1\}$. The notation $\Pr^{(\mathrm{EXT})}$ denotes extrinsic probabilities supplied by the BCJR decoder, and the notation $\mathsf{E}^{(\mathrm{EXT})}$ is chosen to explicitly show that extrinsic probabilities are used for the calculation of the expectation. In the next chapter on channel estimation, we will use soft symbols based on a posteriori probabilities; the corresponding expectation will be indicated by the notation $\mathsf{E}^{(\mathrm{APP})}$.

The maximum likelihood detector for (8.13) searches for $\mathbf{d}^{(u)}[k] \in \mathcal{A}^{M_\mathrm{T}}$ which minimizes the distance $\|\mathbf{y}^{(u)}[k] - \tilde{\mathbf{S}}^{(u)}[k]\mathbf{d}^{(u)}[k]\|^2$. This is expressed as

$$\hat{\mathbf{d}}^{(u)}[k] = \underset{\mathbf{d}^{(u)}[k]\in\mathcal{A}^{M_\mathrm{T}}}{\operatorname{argmin}} \|\mathbf{y}^{(u)}[k] - \tilde{\mathbf{S}}^{(u)}[k]\mathbf{d}^{(u)}[k]\|^2.$$

8.3.4.2 *Sphere Decoding*

The sphere decoder is based on an algorithm first proposed in Fincke & Pohst (1985). The algorithm was used in the communications context already in Mow (1994) and gained mainstream recognition with Viterbo & Boutros (1999), where also the term sphere decoder was used. There are several different implementations of the algorithm. For a full description of the sphere decoder and possible implementations, the reader is referred to the semitutorial papers (Agrell, Eriksson, Vardy, & Zeger, 2002; Damen, El Gamal, & Caire, 2003; Murugan, El Gamal, Damen, & Caire, 2006).

In the remainder of this section, we drop the user superscript (u) in (8.13) to simplify the notation. The sphere decoder operates on a triangularized system model, obtained by taking the *thin* QR decomposition (Golub & Van Loan, 1996) of $\tilde{\mathbf{S}}[k]$. Let $\mathbf{Q}[k]\mathbf{R}[k] = \tilde{\mathbf{S}}[k]$ denote the QR decomposition of $\tilde{\mathbf{S}}[k]$, i.e., $\mathbf{Q}[k]$ is a matrix with orthonormal columns and $\mathbf{R}[k]$ is square upper triangular. By the rotational invariance of the Euclidean norm, it follows that (8.12) is equivalently given by

$$\hat{\mathbf{d}}[k] = \underset{\mathbf{d}[k]\in\mathcal{A}^{M_\mathrm{T}}}{\operatorname{argmin}} \|\mathbf{w}[k] - \mathbf{R}[k]\mathbf{d}[k]\|^2, \tag{8.19}$$

where $\mathbf{w}[k] \triangleq \mathbf{Q}^H[k]\mathbf{y}[k]$. Note that the search space in (8.19) is now reduced to $\mathcal{A}^{M_\mathrm{T}}$ due to parallel interference cancelation. The sphere decoder solves (8.19) by enumerating symbol vectors within a sphere of radius ρ, centered at $\mathbf{w}[k]$, i.e., symbol vectors $\mathbf{d}[k]$ that satisfy

$$\|\mathbf{w}[k] - \mathbf{R}[k]\mathbf{d}[k]\|^2 \le \rho^2. \tag{8.20}$$

Naturally, if the condition in (8.20) were to be verified for every vector $\mathbf{d}[k]$, the algorithm would be a low-complexity implementation of an exhaustive search, i.e., enumerating all solutions. Choosing a suitable value for ρ allows reducing the search space and, as a consequence, the overall complexity.

By exploiting the triangular structure of (8.19), the sphere decoder verifies (8.20) recursively. To this end, let $\mathbf{d}[k] = \left(d_1[k]\cdots d_{M_\mathrm{T}}[k]\right)^T$, $\mathbf{w}[k] = \left(w_1[k]\cdots w_{M_\mathrm{T}}[k]\right)^T$, and

$$\mathbf{R}[k] = \begin{pmatrix} R_{1,1}[k] & \cdots & R_{1,M_\mathrm{T}}[k] \\ 0 & \ddots & \vdots \\ 0 & 0 & R_{M_\mathrm{T},M_\mathrm{T}}[k] \end{pmatrix}.$$

Due to the upper triangular structure of $\mathbf{R}[k]$, the metric in (8.20) may be expressed according to

$$\|\mathbf{w}[k] - \mathbf{R}[k]\mathbf{d}[k]\|^2 = \delta(1,k),$$

where

$$\delta(t,k) \triangleq \sum_{i=t}^{M_{\mathrm{T}}} \left| w_i[k] - \sum_{j=i}^{M_{\mathrm{T}}} R_{i,j}[k] d_j[k] \right|^2, \quad t \in \{1,\ldots,M_{\mathrm{T}}\} \tag{8.21}$$

is referred to as the tth partial distance. As a sum of positive elements, $\delta(t,k)$ is increasing when t decreases. If

$$\delta(t,k) > \rho^2 \tag{8.22}$$

for some t and $k \notin \mathcal{P}$, then (8.20) cannot be satisfied. (Recall that $\mathcal{P}$ denotes the pilot index set.) This allows for a simple recursive implementation of the sphere decoder: the distance $\delta(t,k)$ is computed for t decreasing and starting from $t = M_{\mathrm{T}}$. As soon as some t is reached such that (8.22) is satisfied, all vectors $\mathbf{d}[k]$ having the partial vector $\mathbf{d}^{(t)}[k] = (d_t[k] \cdots d_{M_{\mathrm{T}}}[k])^T$ that leads to (8.22) are discarded since they lie outside the sphere radius ρ.

The sphere decoder search is effectively visualized as a tree search over a regular tree with M_{T} levels and with $|\mathcal{A}|$ children per node (Murugan et al., 2006). A node at layer $M_{\mathrm{T}} - t + 1$ (with layer 0 corresponding to the root node and layer M_{T} corresponding to the leaf nodes) is represented by the partial data vector $\mathbf{d}^{(t)}[k]$. A node is pruned if it violates (8.20) and the set of leaf nodes visited corresponds to valid solutions to (8.20). This process is illustrated for a four-element alphabet in Fig. 8.5.

At this stage, several strategies exist for traversing the search tree and for selecting the search radius ρ (Murugan et al., 2006). The currently most popular strategy is known as the Schnorr–Euchner strategy (Agrell et al., 2002). The Schnorr–Euchner strategy updates the sphere radius such that (8.20) is satisfied with equality once a valid leaf node is visited. This ensures that once a candidate solution to (8.19) is found, the sphere decoder does not consider candidates that would yield a larger metric. It also has the benefit that ρ may be set infinitely large at the beginning of the search, thus alleviating the

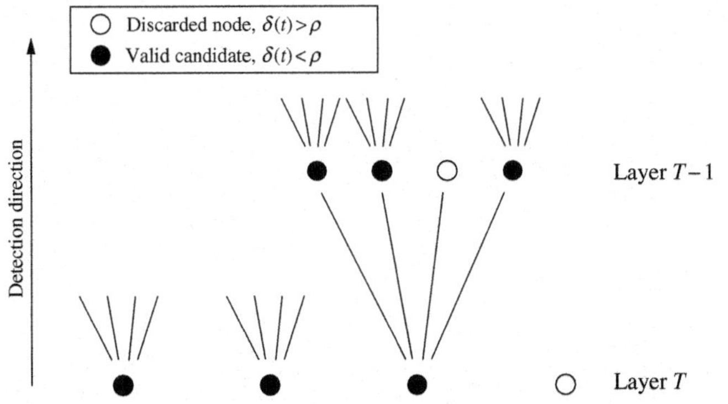

FIGURE 8.5

Illustration of sphere decoding using tree pruning, for a four-element alphabet.

potential problem when no nodes satisfy (8.20). It can be shown that the first candidate found by the Schnorr–Euchner sphere decoder corresponds to the zero-forcing decision-feedback estimate of $\mathbf{d}[k]$ (Agrell et al., 2002).

In order to apply the sphere decoder, the QR factorization of $\tilde{\mathbf{S}}^{(u)}[k]$ must be computed for every user u and time instant k. This results in a significant overhead complexity. Section 8.6.1 explores how the structure of the time-varying channel may be exploited to reduce the complexity of this step.

8.4 ITERATIVE MULTIUSER MIMO TIME-VARYING CHANNEL ESTIMATION

In this section, we describe the channel estimator using the reduced-rank channel model described in Section 8.3.2. For channel estimation, J pilot symbols in (8.1) are known. The remaining $K-J$ symbols are not known. We replace them by soft symbols that are calculated from the a posteriori probabilities (APP) obtained in the previous iteration from the BCJR decoder output. The soft symbols are computed as

$$\tilde{d}'_{u,t}[k] = \frac{2\mathsf{E}^{(\mathrm{APP})}\{c_{u,t}[2k]\}-1}{\sqrt{2}} + j\frac{2\mathsf{E}^{(\mathrm{APP})}\{c_{u,t}[2k+1]\}-1}{\sqrt{2}}, \tag{8.23}$$

where for $k' \in \{0,\ldots,2M-1\}$

$$\begin{aligned}\mathsf{E}^{(\mathrm{APP})}\{c_{u,t}[k']\} &= \Pr^{(\mathrm{APP})}\{c_{u,t}[k']=+1\} - \Pr^{(\mathrm{APP})}\{c_{u,t}[k']=-1\}\\ &= 2\Pr^{(\mathrm{APP})}\{c_{u,t}[k']=+1\}-1,\end{aligned}$$

similarly as in (8.17) and (8.18). (The notation $'$ in $\tilde{d}'_{u,t}[k]$ is used here to distinguish these soft symbols obtained from a posteriori probabilities from those obtained from extrinsic probabilities in (8.17).) This allows us to obtain refined channel estimates if the soft symbols get more reliable from iteration to iteration. The $UM_{\mathrm{T}}M_{\mathrm{R}}$ channels are assumed uncorrelated and thus, channel estimation can be performed for every receive antenna r independently, without loss of information. For clarity, we drop the receive antenna subscript r for the rest of the section.

8.4.1 Signal Model

At each receive antenna r, the basis expansion coefficients $\psi_{u,t}[i,l]$ can be obtained jointly for all UM_{T} virtual users but individually for every subcarrier l. We define the vector

$$\boldsymbol{\psi}_{l,i} \triangleq (\psi_{1,1}[i,l]\cdots\psi_{1,M_{\mathrm{T}}}[i,l]\cdots\psi_{u,t}[i,l]\cdots\psi_{U,1}[i,l]\cdots\psi_{U,M_{\mathrm{T}}}[i,l])^T \in \mathbb{C}^{UM_{\mathrm{T}}} \tag{8.24}$$

for $i \in \{0,\ldots,D-1\}$ and the vector

$$\boldsymbol{\phi}_l \triangleq \left(\boldsymbol{\psi}_{l,0}^T\cdots\boldsymbol{\psi}_{l,D-1}^T\right)^T \in \mathbb{C}^{UM_{\mathrm{T}}D} \tag{8.25}$$

containing the basis expansion coefficients of all UM_{T} virtual users for subcarrier l. The received symbol sequence associated with each single data block on subcarrier l is given by

$$\mathbf{y}_l = (y[0,l]\cdots y[K-1,l])^T \in \mathbb{C}^K.$$

Using these definitions, we write

$$\mathbf{y}_l = \mathbf{U}_l \boldsymbol{\phi}_l + \mathbf{z}_l, \tag{8.26}$$

where

$$\mathbf{U}_l \triangleq \left(\operatorname{diag}(\mathbf{u}_0)\,\tilde{\mathbf{A}}_l \cdots \operatorname{diag}(\mathbf{u}_{D-1})\,\tilde{\mathbf{A}}_l\right) \in \mathbb{C}^{K \times UM_\mathrm{T}D}.$$

The matrix $\tilde{\mathbf{A}}_l \in \mathbb{C}^{K\times UM_\mathrm{T}}$ contains all the transmitted symbols at all time instants $k \in \mathcal{I}_K$ on subcarrier l:

$$\tilde{\mathbf{A}}_l = \begin{pmatrix} \tilde{a}'_{1,1}[0,l] & \cdots & \tilde{a}'_{U,M_\mathrm{T}}[0,l] \\ \vdots & \ddots & \vdots \\ \tilde{a}'_{1,1}[K-1,l] & \cdots & \tilde{a}'_{U,M_\mathrm{T}}[K-1,l] \end{pmatrix}.$$

Here,

$$\tilde{a}'_{u,t}[k,l] \triangleq s_{u,t}[l]\tilde{d}'_{u,t}[k] + p_{u,t}[k,l]$$

are the symbols calculated using the APP-based soft data symbols $\tilde{d}'_{u,t}[k]$ provided by the decoding stage (8.23).

8.4.2 Reduced-Rank LMMSE Channel Estimator

Based on (8.26), the LMMSE estimator for channel estimation can be expressed as (Mecklenbräuker et al., 2006; Zemen, 2004; Zemen et al., 2006),

$$\hat{\boldsymbol{\phi}}_l = \left(\mathbf{U}_l^H \boldsymbol{\Delta}^{-1} \mathbf{U}_l + \boldsymbol{\Phi}^{-1}\right)^{-1} \mathbf{U}_l^H \boldsymbol{\Delta}^{-1} \mathbf{y}_l. \tag{8.27}$$

Here, $\boldsymbol{\Delta} \triangleq \boldsymbol{\Lambda} + \sigma_z^2 \mathbf{I}_K \in \mathbb{C}^{K\times K}$, where the elements of the diagonal matrix $\boldsymbol{\Lambda} \in \mathbb{C}^{K\times K}$ are defined as

$$[\boldsymbol{\Lambda}]_{kk} = \frac{1}{2L\nu_\mathrm{max}} \sum_{i=0}^{D-1} \lambda_i u_i^2[k] \sum_{u=1}^{U} \sum_{t=1}^{M_\mathrm{T}} \left(1 - |\tilde{d}'_{u,t}[k]|^2\right).$$

The diagonal covariance matrix $\boldsymbol{\Phi}$ for $\boldsymbol{\phi}_l$ is given by

$$\boldsymbol{\Phi} = \frac{1}{2\nu_\mathrm{max}} \mathbf{I}_{UM_\mathrm{T}} \otimes \operatorname{diag}\{(\lambda_0 \cdots \lambda_{D-1})\},$$

where $\otimes$ denotes the Kronecker matrix product. We note that $\boldsymbol{\Phi}$ does not depend on the subcarrier l. After estimating $\boldsymbol{\phi}_l$ and using (8.24) and (8.25), an estimate of the time-varying frequency response is given by

$$\hat{g}'_{u,t}[k,l] = \sum_{i=0}^{D-1} u_i[k] \hat{\psi}_{u,t}[i,l]. \tag{8.28}$$

A further noise suppression is achieved if we exploit the correlation between the subcarriers by projecting $\hat{g}'_{u,t}[k,l]$ according to

$$\hat{\mathbf{g}}_{u,t}[k] = \mathbf{F}_{L\times K}\mathbf{F}^{H}_{L\times K}\hat{\mathbf{g}}'_{u,t}[k], \tag{8.29}$$

where $[\mathbf{F}_{L\times K}]_{l,k} = \frac{1}{\sqrt{L}}e^{\frac{-j2\pi lk}{L}}$ for $(l,k) \in \{0,\ldots,L-1\} \times \{0,\ldots,K-1\}$.

8.4.3 Comparison of the Slepian and Fourier Bases

In order to estimate the time-varying and frequency-selective channel in the previous section, we employed two sets of basis function. We use a Slepian basis in the time domain (cf. (8.26)) and a Fourier basis in the frequency domain (cf. (8.29)). One can ask why such a mixed approach is appropriate. A projection onto a Slepian basis in the time domain and onto another Slepian basis in the frequency domain is described in Zemen, Hofstetter, & Steinböck (2005) (see also Salvo Rossi and Müller, 2008).

Clearly, the Slepian basis achieves a smaller square-bias compared to the Fourier basis and allows for a low-complexity implementation as long as the subspace dimension is a small single-digit number (Zemen & Mecklenbräuker, 2005). This condition is fulfilled in the time-domain (cf. (8.7)); however, in the frequency domain, the situation is different: the required subspace dimension is much larger, and the gain in square-bias compared to the Fourier basis is reduced. Furthermore, the projection onto the Fourier basis can be implemented using a low-complexity DFT implementation.

In order to demonstrate the different situations in the time and frequency domain, Fig. 8.6 shows the mean square error obtained when estimating a general random process with spectral support

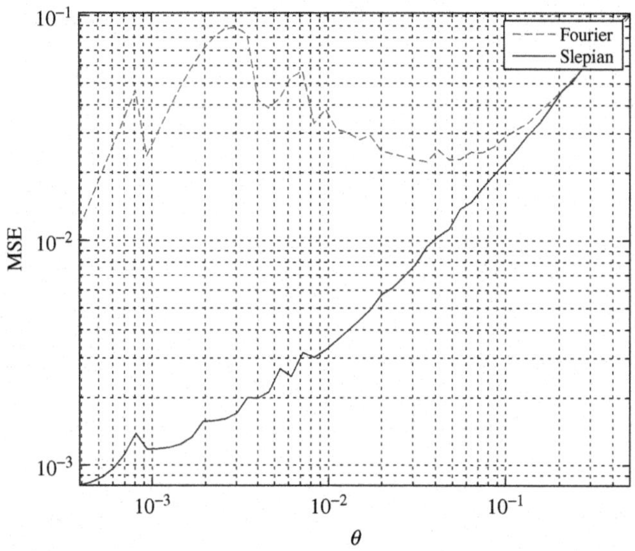

FIGURE 8.6

Mean square error for projection onto a subspace spanned by Fourier or DPS basis functions. The signal-to-noise ratio is fixed at 10dB, and the normalized frequency is varied in the interval $0.1/K \leq \theta \leq 1/2$.

in $[-\theta,\theta]$ using K observations. We use (25) and (32) in Zemen & Mecklenbräuker (2005) to calculate the mean square error versus θ for the Fourier and Slepian subspaces for a fixed SNR of 10 dB.

The left side of Fig. 8.6 (i.e., for small normalized frequencies) is representative of time-varying channel estimation of a frequency-flat subcarrier. Here, we set $\theta = \nu_{\max}$. Hence, in this case, the Slepian basis expansion provides a considerable gain compared to the Fourier basis expansion.

The right side of Fig. 8.6 is representative of noise suppression in the frequency domain. For this case, we set $\theta = \tau_{\max}/T_c L$ equal to the maximum normalized path delay. It becomes clear that for values of the normalized frequency $\theta \geq 0.2$, a low-complexity DFT-based implementation is preferable, because the mean square error gain provided by the Slepian subspace tends to zero for $\theta \geq 0.2$. For the OFDM-based system considered, θ is given by the ratio of cyclic prefix length G to number of subcarriers L. Hence, $\theta = G/L = 15/64 = 0.23$ (this value will be used in the numeric simulation provided in Section 8.7).

8.4.4 Krylov Approximation of the Reduced-Rank LMMSE Channel Estimator

It is possible to implement the LMMSE channel estimator (8.27) using the Krylov subspace method. By matching the LMMSE filter structure (8.27) with the general structure in (8.8) and using (8.11), we can calculate the complexity reduction for the multiuser MIMO channel estimator employing the Krylov subspace algorithm (Dumard et al., 2008),

$$\gamma = \frac{C_{\mathrm{K}}}{C_{\mathrm{LMMSE}}} \leq \mathcal{O}\left(\frac{2S'+3}{K}\right).$$

For the selection of a sufficient Krylov subspace dimension S', we resort to numerical simulation (see Section 8.7).

8.5 LINEAR JOINT ANTENNA MULTIUSER DETECTION

In this section, we describe several methods to perform linear multiuser detection for each time instant $k \in \{0,\ldots,K-1\}$. For each method, we discuss its implementation using the Krylov subspace method. To simplify the notation, we omit the time index k for the remainder of this section.

8.5.1 Multiuser Detection in Chip Space

After parallel interference cancelation for user (u,t) (cf. (8.16)), the received signal (8.3) becomes

$$\mathbf{y}_{u,t} = \mathbf{y} - \tilde{\mathbf{S}}\tilde{\mathbf{d}} + \tilde{\mathbf{s}}_{u,t}\tilde{d}_{u,t} = \tilde{\mathbf{S}}(\mathbf{d}-\tilde{\mathbf{d}}) + \tilde{\mathbf{s}}_{u,t}\tilde{d}_{u,t} + \mathbf{z}. \tag{8.30}$$

Multiuser detection is performed using the soft symbol estimates $\tilde{d}_{u,t}$, computed from the extrinsic probabilities as in (8.17). We define

$$\tilde{\mathbf{d}} \triangleq \left(\tilde{d}_{1,1}\cdots\tilde{d}_{1,M_{\mathrm{T}}}\cdots\tilde{d}_{U,1}\cdots\tilde{d}_{U,M_{\mathrm{T}}}\right)^T \in \mathbb{C}^{UM_{\mathrm{T}}},$$

and the error covariance matrix of the soft symbols

$$\mathbf{V} \triangleq \tilde{\mathrm{E}}\left\{(\mathbf{d}-\tilde{\mathbf{d}})(\mathbf{d}-\tilde{\mathbf{d}})^H\right\}$$

with diagonal elements

$$V_{u,t} = \tilde{\mathrm{E}}\left\{1 - |\tilde{d}_{u,t}|^2\right\} \tag{8.31}$$

that are constant during one iteration of the receiver (i.e., independent of the time instant k). The expectation operator in (8.31) is implemented as an empirical mean according to

$$\tilde{\mathrm{E}}\left\{1 - |\tilde{d}_{u,t}|^2\right\} = \frac{1}{K}\sum_{k=0}^{K-1}\left(1 - |\tilde{d}_{u,t}[k]|^2\right).$$

The symbols $\mathbf{d}$ and $\tilde{\mathbf{d}}$ are supposed to be independent, and the other elements of $\mathbf{V}$ are assumed to be zero.

The corresponding unbiased LMMSE filter (Müller & Caire, 2001; Wehinger, Müller, Lončar, & Mecklenbräuker, 2002) is given by

$$\mathbf{f}_{u,t}^H = \frac{\tilde{\mathbf{s}}_{u,t}^H\left(\sigma_z^2\mathbf{I}_{LM_\mathrm{R}} + \tilde{\mathbf{S}}\mathbf{V}\tilde{\mathbf{S}}^H\right)^{-1}}{\tilde{\mathbf{s}}_{u,t}^H\left(\sigma_z^2\mathbf{I}_{LM_\mathrm{R}} + \tilde{\mathbf{S}}\mathbf{V}\tilde{\mathbf{S}}^H\right)^{-1}\tilde{\mathbf{s}}_{u,t}}, \tag{8.32}$$

and the estimate of $d_{u,t}$ is obtained as

$$\hat{d}_{u,t} = \mathbf{f}_{u,t}^H\mathbf{y}_{u,t}.$$

The $UM_\mathrm{T}(K-J)$ estimates $\hat{d}_{u,t}$ are then demapped, deinterleaved, and decoded using a BCJR decoder (Bahl et al., 1974).

In this situation, each virtual user (u,t) requires its own filter, while the UM_T filters (8.32) have a common matrix inverse. However, an approximate algorithm such as the Krylov subspace method has to be performed *per (virtual) user*. This adds a multiplicative factor UM_T in the global computational complexity using the Krylov subspace method.

For a nonoverloaded system (i.e., $UM_\mathrm{T} \leq LM_\mathrm{R}$),

$$\gamma = \frac{C_\mathrm{K}}{C_\mathrm{LMMSE}} \leq \mathcal{O}(2S+3);$$

see Dumard et al. (2008) for more details. Hence, the complexity reduction achieved by using the Krylov subspace method is neutralized by the multiplicative factor UM_T, and no complexity reduction is achieved. However, parallelization of the computations in UM_T branches is possible (Dumard & Zemen, 2006), thus reducing latency time by a factor of order $UM_\mathrm{T}/(2S+3)$.

8.5.2 Multiuser Detection in User Space

Ultimately, we want to define an LMMSE filter that allows joint detection of all users using only one filter such that we can apply the Krylov subspace method optimally. Without loss of information

(Verdú, 1998), we can apply a matched filter on the received signal (8.3):

$$\mathbf{x} = \begin{pmatrix} x_{1,1} \\ \vdots \\ x_{u,t} \\ \vdots \\ x_{U,M_\mathrm{T}} \end{pmatrix} \triangleq \tilde{\mathbf{S}}^H \mathbf{y} = \tilde{\mathbf{S}}^H \tilde{\mathbf{S}} \mathbf{b} + \tilde{\mathbf{S}}^H \mathbf{z}.$$

Performing interference cancelation for user (u,t) in a mathematically identical way as in (8.30), we obtain

$$\mathbf{x}_{u,t} = \tilde{\mathbf{S}}^H \mathbf{y} - \tilde{\mathbf{S}}^H \tilde{\mathbf{S}} \tilde{\mathbf{d}} + \tilde{\mathbf{S}}^H \tilde{\mathbf{s}}_{u,t} \tilde{d}_{u,t}.$$

In this equation, the element

$$x_{u,t} \triangleq \left[\mathbf{x}_{u,t}\right]_{M_\mathrm{T}(u-1)+t}$$

contains the most information on the specific user (u,t). In all other elements of $\mathbf{x}_{u,t}$, the information about user (u,t) consists of interference which is mostly canceled by the parallel interference cancelation. From now on, we set these error terms to zero. This way the received signal after parallel interference cancelation for user (u,t) becomes

$$\hat{x}_{u,t} \triangleq x_{u,t} - \tilde{\mathbf{s}}_{u,t}^H \tilde{\mathbf{S}} \tilde{\mathbf{d}} + \tilde{\mathbf{s}}_{u,t}^H \tilde{\mathbf{s}}_{u,t} \tilde{d}_{u,t}.$$

Combining these UM_T values into a vector $\hat{\mathbf{x}}$ leads to

$$\hat{\mathbf{x}} = \tilde{\mathbf{S}}^H \mathbf{y} - (\tilde{\mathbf{S}}^H \tilde{\mathbf{S}} - \mathbf{D}) \tilde{\mathbf{d}}, \tag{8.33}$$

where $\mathbf{D} \in \mathbb{C}^{UM_\mathrm{T} \times UM_\mathrm{T}}$ is defined as the diagonal matrix whose diagonal elements equal those of the covariance matrix $\mathbf{R} \triangleq \tilde{\mathbf{S}}^H \tilde{\mathbf{S}}$, i.e.,

$$[\mathbf{D}]_{i,i} = [\mathbf{R}]_{i,i} \quad \text{for} \quad i \in \{1, \ldots, UM_\mathrm{T}\}.$$

Performing parallel interference cancelation in user space as described in (8.33) allows joint detection of all users *using one filter only*. Some information is lost since we have neglected some terms, and thus, a slight loss in performance can be expected.

The LMMSE filter $\mathbf{F}$ for parallel interference cancelation in user space is defined by

$$\mathbf{F} = \underset{\mathbf{F}'}{\operatorname{argmin}}\, \mathrm{E}\left\{\|\mathbf{F}'^H \hat{\mathbf{x}} - \mathbf{d}\|^2\right\}$$

and leads to the symbol estimates

$$\hat{\mathbf{d}} = \mathbf{F}^H \hat{\mathbf{x}}.$$

As shown in Dumard et al. (2008), it can be expressed as

$$\mathbf{F} = \left[\mathbf{RVR} + \sigma_z^2 \mathbf{R} + \mathbf{D}\left(\mathbf{I}_{UM_\mathrm{T}} - \mathbf{V}\right)\mathbf{D}\right]^{-1} (\mathbf{RV} - \mathbf{DV} + \mathbf{D}).$$

The $UM_\mathrm{T}K$ estimates in $\hat{\mathbf{d}}$ are demapped, deinterleaved, and decoded by a BCJR decoder.

Although the LMMSE filter in this case is more complex than the one in chip space, the product $\mathbf{F}^H\hat{\mathbf{x}}$ needs to be computed only once to detect all users simultaneously. The complexity reduction achieved by the Krylov subspace method is given by

$$\gamma = \frac{C_{\mathrm{K}}}{C_{\mathrm{LMMSE}}} \leq \mathcal{O}\left(\frac{6S+9}{UM_{\mathrm{T}}}\right).$$

Hence, depending on the Krylov subspace dimension S, a considerable complexity reduction can be achieved. Again, the selection of the sufficient subspace dimension S will be obtained by numerical simulations (see Section 8.7).

8.6 NONLINEAR DETECTION

In Section 8.3.4, the basic method for sphere decoding in a multiuser MIMO system was introduced. However, for a time-varying MIMO channel, the QR decomposition must be calculated for every time instant k.

The reduced-rank channel model from Section 8.3.2 describes the time-varying frequency-selective channel $\mathbf{g}_{u,t,r}[k] \in \mathbb{C}^L$ for the duration of a single data block $k \in \mathcal{I}_K = \{0,\ldots,K-1\}$. In this section, we will show how such a reduced-rank channel model can be combined with sphere decoding for complexity reduction.

8.6.1 Exploiting the Reduced-Rank Channel Model

For $k \in \mathcal{I}_K$, we write $\mathbf{g}_{u,t,r}[k]$ as a linear superposition of the first D DPS sequences index-limited to the time interval $\mathcal{I}_K$, as in (8.6). Accordingly, the channel estimates $\hat{g}_{u,t,r}[k,l]$ in (8.28) can also be written as a linear combination of the DPS sequences:

$$\hat{\mathbf{g}}_{u,t,r}[k] = \boldsymbol{\Gamma}_{u,t,r}\mathbf{u}[k] \quad \text{for} \quad k \in \mathcal{I}_K, \tag{8.34}$$

where

$$\mathbf{u}[k] = \begin{pmatrix} u_1[k] \\ \vdots \\ u_{D-1}[k] \end{pmatrix} \in \mathbb{C}^D,$$

and $\boldsymbol{\Gamma}_{u,t,r}$ contains the projection coefficients $\psi_{u,t,r}[i,l]$, i.e.,

$$\boldsymbol{\Gamma}_{u,t,r} = \begin{pmatrix} \psi_{u,t,r}[0,0] & \cdots & \psi_{u,t,r}[D-1,0] \\ \vdots & \ddots & \vdots \\ \psi_{u,t,r}[0,L-1] & \cdots & \psi_{u,t,r}[D-1,L-1] \end{pmatrix} \in \mathbb{C}^{L\times D}.$$

Using (8.34) for estimating (8.4), we obtain

$$\tilde{\mathbf{s}}_{u,t,r}[k] = \operatorname{diag}\{\mathbf{s}_{u,t}\}\boldsymbol{\Gamma}_{u,t,r}\mathbf{u}[k]. \tag{8.35}$$

Inserting (8.35) into (8.14) leads to

$$\tilde{\mathbf{S}}^{(u)}[k] = \boldsymbol{\Gamma}^{(u)}\mathbf{U}[k],$$

where

$$\mathbf{\Gamma}^{(u)} \triangleq \begin{pmatrix} \operatorname{diag}\{\mathbf{s}_{u,1}\}\mathbf{\Gamma}_{u,1,1} & \cdots & \operatorname{diag}\{\mathbf{s}_{u,M_\mathrm{T}}\}\mathbf{\Gamma}_{u,M_\mathrm{T},1} \\ \vdots & \ddots & \vdots \\ \operatorname{diag}\{\mathbf{s}_{u,1}\}\mathbf{\Gamma}_{u,1,M_\mathrm{R}} & \cdots & \operatorname{diag}\{\mathbf{s}_{u,M_\mathrm{T}}\}\mathbf{\Gamma}_{u,M_\mathrm{T},M_\mathrm{R}} \end{pmatrix}$$

and

$$\mathbf{U}[k] \triangleq \begin{pmatrix} \mathbf{u}[k] & \mathbf{0} & \mathbf{0} \\ \mathbf{0} & \ddots & \mathbf{0} \\ \mathbf{0} & \mathbf{0} & \mathbf{u}[k] \end{pmatrix} \in \mathbb{C}^{DM_\mathrm{T} \times M_\mathrm{T}}. \tag{8.36}$$

The received signal of user u after parallel interference cancelation (8.15) finally becomes

$$\tilde{\mathbf{y}}^{(u)}[k] = \mathbf{\Gamma}^{(u)}\mathbf{U}[k]\mathbf{d}^{(u)}[k] + \mathbf{z}^{(u)}[k].$$

Note that $\mathbf{\Gamma}^{(u)}$ is time-invariant but user-dependent, while $\mathbf{U}[k]$ is common to all users but time-varying. This specific structure is essential to reduce the computational complexity of the sphere decoder preprocessing for time-varying MIMO channels.

Dropping index (u), we write

$$\tilde{\mathbf{y}}[k] = \mathbf{\Gamma}\mathbf{U}[k]\mathbf{d}[k] + \mathbf{z}[k]. \tag{8.37}$$

We consider the *thin* QR factorization (Golub & Van Loan, 1996) of $\mathbf{\Gamma}$, i.e., $\mathbf{\Gamma} = \mathbf{QR}$ where $\mathbf{Q} \in \mathbb{C}^{LM_\mathrm{R} \times DM_\mathrm{T}}$ is unitary and $\mathbf{R} \in \mathbb{C}^{DM_\mathrm{T} \times DM_\mathrm{T}}$ is upper triangular. Multiplying (8.37) from the left side by $\mathbf{Q}^H$, we obtain

$$\mathbf{w}[k] \triangleq \mathbf{Q}^H\tilde{\mathbf{y}}[k] = \mathbf{RU}[k]\mathbf{d}[k] + \mathbf{Q}^H\mathbf{z}[k].$$

We rewrite the maximum likelihood detector after QR factorization of $\mathbf{\Gamma}$ as

$$\hat{\mathbf{d}}[k] = \underset{\mathbf{d}[k] \in \mathcal{A}^{M_\mathrm{T}}}{\operatorname{argmin}} \|\mathbf{w}[k] - \mathbf{RU}[k]\mathbf{d}[k]\|^2. \tag{8.38}$$

To make use of the block diagonal structure of $\mathbf{U}[k]$ in (8.36), we decompose the matrix $\mathbf{R}$ into blocks of size $D \times D$:

$$\mathbf{R} = \begin{pmatrix} \mathbf{R}_{1,1} & \mathbf{R}_{1,2} & \cdots & \mathbf{R}_{1,M_\mathrm{T}} \\ \mathbf{0} & \mathbf{R}_{2,2} & \cdots & \mathbf{R}_{2,M_\mathrm{T}} \\ \vdots & \ddots & \ddots & \vdots \\ \mathbf{0} & \cdots & \mathbf{0} & \mathbf{R}_{M_\mathrm{T},M_\mathrm{T}} \end{pmatrix}, \tag{8.39}$$

where $\mathbf{R}_{t,t} \in \mathbb{C}^{D \times D}$ is upper triangular and $\mathbf{R}_{t,t'>t} \in \mathbb{C}^{D \times D}$ is a full matrix.

Due to the reduced-rank channel model (8.34) and the resulting signal model for user u shown in (8.37), we need to compute the QR factorization only once per data block, which yields a complexity reduction. However, the tree search still has to be performed for each time instant $k \notin \mathcal{P}$.

The sphere decoder can be adapted to use the block QR factorization (see (8.39)), which is not time-dependent, by replacing the partial distance $\delta(t,k)$ in (8.21) with the time-dependent vector metric defined by

$$\delta(t,k) \triangleq \sum_{i=t}^{M_{\mathrm{T}}} \left\| \mathbf{w}_i[k] - \sum_{j=i}^{M_{\mathrm{T}}} \mathbf{R}_{i,j}\mathbf{u}[k]d_j[k] \right\|^2 = \delta(t+1,k) + \left\| \mathbf{w}_t[k] - \sum_{j=t}^{M_{\mathrm{T}}} \mathbf{R}_{t,j}\mathbf{u}[k]d_j[k] \right\|^2, \tag{8.40}$$

where $\mathbf{w}_t[k] \triangleq \left(w_{D(t-1)+1}[k] \cdots w_{Dt}[k]\right)^T$. Recursively, this leads to the metric

$$\delta(1) = \sum_{t=1}^{M_{\mathrm{T}}} \left\| \mathbf{w}_t[k] - \sum_{j=t}^{M_{\mathrm{T}}} \mathbf{R}_{t,j}\mathbf{u}[k]d_j[k] \right\|^2,$$

which can be rewritten as

$$\delta(1,k) = \sum_{t=1}^{M_{\mathrm{T}}} \left\| \begin{pmatrix} w_{D(t-1)+1}[k] \\ \vdots \\ w_{Dt}[k] \end{pmatrix} - \begin{pmatrix} \mathbf{R}_{t,t} & \cdots & \mathbf{R}_{t,M_{\mathrm{T}}} \end{pmatrix} \begin{pmatrix} \mathbf{u}[k] & \mathbf{0} & \mathbf{0} \\ \mathbf{0} & \ddots & \mathbf{0} \\ \mathbf{0} & \mathbf{0} & \mathbf{u}[k] \end{pmatrix} \begin{pmatrix} d_t[k] \\ \vdots \\ d_{M_{\mathrm{T}}}[k] \end{pmatrix} \right\|^2. \tag{8.41}$$

Equation (8.41) corresponds to the following block computations:

$$\delta(1,k) = \left\| \begin{pmatrix} w_1[k] \\ \vdots \\ \boxed{\begin{matrix} w_{D(t-1)+1}[k] \\ \vdots \\ w_{Dt}[k] \end{matrix}} \\ \vdots \\ w_{DM_{\mathrm{T}}}[k] \end{pmatrix} - \begin{pmatrix} \mathbf{R}_{1,1} & \mathbf{R}_{1,2} & \cdots\cdots & \mathbf{R}_{1,M_{\mathrm{T}}} \\ \vdots & \ddots & \ddots\ \ddots & \vdots \\ \mathbf{0} & \ddots & \boxed{\mathbf{R}_{t,t} \ \cdots \ \mathbf{R}_{t,M_{\mathrm{T}}}} \\ \vdots & \ddots & \ddots\ \ddots & \vdots \\ \mathbf{0} & \cdots & \cdots\ \mathbf{0} & \mathbf{R}_{M_{\mathrm{T}},M_{\mathrm{T}}} \end{pmatrix} \begin{pmatrix} \mathbf{u}[k] & \mathbf{0} & \cdots\cdots & \mathbf{0} \\ \vdots & \ddots & \ddots\ \ddots & \vdots \\ \mathbf{0} & \ddots & \boxed{\begin{matrix} \mathbf{u}[k] & \cdots & \mathbf{0} \\ \ddots & \ddots & \vdots \\ \cdots & \mathbf{0} & \mathbf{u}[k] \end{matrix}} \\ \vdots & \ddots & & \\ \mathbf{0} & \cdots & & \end{pmatrix} \begin{pmatrix} d_1[k] \\ \vdots \\ \boxed{\begin{matrix} d_t[k] \\ \vdots \\ d_{M_{\mathrm{T}}}[k] \end{matrix}} \end{pmatrix} \right\|^2$$

which follow from the structure of $\mathbf{R}$ and $\mathbf{U}[k]$. As

$$\delta(1,k) = \|\mathbf{w}[k] - \mathbf{R}\mathbf{U}[k]\mathbf{d}[k]\|^2,$$

the tree search procedure used in the original sphere decoder may also be used to solve (8.38), albeit with (8.40) in place of (8.21), thus circumventing the need to recompute the QR factorization for each time instant k. All search strategies available for the classical sphere decoder (such as the Schnorr–Euchner strategy (Dumard et al., 2008)) extend naturally to this case.

Sphere decoding after an initial parallel interference cancelation step allows for substantial complexity reduction if the sphere decoder exploits the reduced-rank channel model. However, in comparison to the previously described LMMSE strategies, a performance loss in terms of BER versus SNR is incurred, mainly due to the hard decision supplied by the sphere decoder (see the simulation results in Section 8.7). In the next section, we demonstrate that a soft-output max-log sphere decoder (Dumard et al., 2008) can exploit the reduced-rank channel model as well.

8.6.2 Soft Sphere Decoding

In Section 8.6.1, we developed an algorithm for solving (8.38) that takes the block structure of the basis expansion matrix $\mathbf{U}[k]$ into account. This method is less complex than LMMSE detection, but it introduces a substantial loss in performance. This loss is partly due to the fact that the hard sphere decoder assumes perfect interference cancelation and thus does not take residual interference into account. An additional loss is due to the fact that the BCJR decoder receives hard inputs. To solve this latter problem, it is necessary to compute soft symbols to feed the BCJR decoder. Wang & Giannakis (2004) present a method for computing log-likelihood ratios using sphere decoding to reduce the complexity. We will recall the method in the following and apply our subspace-based implementation that yields a further complexity reduction for time-varying channels. For reasons of computational complexity, soft sphere decoding is performed on a per-user basis similar to hard sphere decoding in Section 8.6.1. The user index u is again omitted.

So far, all quantities have been complex-valued; in particular, the transmitted symbols $\mathbf{d}[k]$ stem from a QPSK symbol alphabet $\mathcal{A} \in \{\pm 1 \pm j\}/\sqrt{2}$. However, the soft sphere decoder requires computations of log-likelihood ratios, which is somewhat simpler to perform in the real domain. In order to convert the signal model into the real domain, we use the superscripts (r) and (i) denoting, respectively, the real and imaginary part of a complex vector or matrix. We thus define the following stacked real-valued variables:

$$\bar{\mathbf{d}}[k] \triangleq \sqrt{2}\begin{pmatrix} \mathbf{d}^{(r)}[k] \\ \mathbf{d}^{(i)}[k] \end{pmatrix}, \quad \bar{\mathbf{y}}[k] \triangleq \begin{pmatrix} \tilde{\mathbf{y}}^{(r)}[k] \\ \tilde{\mathbf{y}}^{(i)}[k] \end{pmatrix}, \quad \bar{\mathbf{z}}[k] \triangleq \begin{pmatrix} \mathbf{z}^{(r)}[k] \\ \mathbf{z}^{(i)}[k] \end{pmatrix},$$

$$\bar{\mathbf{\Gamma}} \triangleq \begin{pmatrix} \mathbf{\Gamma}^{(r)} & -\mathbf{\Gamma}^{(i)} \\ \mathbf{\Gamma}^{(i)} & \mathbf{\Gamma}^{(r)} \end{pmatrix}, \quad \text{and} \quad \bar{\mathbf{U}}[k] \triangleq \begin{pmatrix} \mathbf{U}_0[k] & \mathbf{0} \\ \mathbf{0} & \mathbf{U}_0[k] \end{pmatrix},$$

where $\mathbf{U}_0[k]$ contains the first DM_T rows and the first M_T columns of $\mathbf{U}[k]$. Note that we scale the input vector such that we consider entries for $\bar{\mathbf{d}}[k]$ in $\{\pm 1\}$. The signal model (8.37) can be rewritten as

$$\bar{\mathbf{y}}[k] = \frac{1}{\sqrt{2}}\bar{\mathbf{\Gamma}}\bar{\mathbf{U}}[k]\bar{\mathbf{d}}[k] + \bar{\mathbf{z}}[k]. \tag{8.42}$$

Using the signal model (8.42), the maximum likelihood detector reads

$$\hat{\mathbf{d}}_\mathrm{ML}[k] = \underset{\bar{\mathbf{d}}[k]}{\operatorname{argmin}} \left\| \bar{\mathbf{y}}[k] - \frac{1}{\sqrt{2}}\bar{\mathbf{\Gamma}}\bar{\mathbf{U}}[k]\bar{\mathbf{d}}[k] \right\|^2.$$

After QR-decomposition $\bar{\mathbf{\Gamma}} = \bar{\mathbf{Q}}\bar{\mathbf{R}}$ and left multiplication by $\bar{\mathbf{Q}}^T$, this becomes

$$\hat{\mathbf{d}}_\mathrm{ML}[k] = \underset{\bar{\mathbf{d}}[k]}{\operatorname{argmin}} \|\bar{\mathbf{w}}[k] - \bar{\mathbf{R}}\bar{\mathbf{U}}[k]\bar{\mathbf{d}}[k]\|^2,$$

where

$$\bar{\mathbf{w}}[k] \triangleq \sqrt{2}\bar{\mathbf{Q}}^T\bar{\mathbf{y}}[k],$$

$\bar{\mathbf{Q}}$ is unitary, and $\bar{\mathbf{R}}$ is square upper triangular.

8.6.2.1 Definition of Log-Likelihood Ratios

Let us denote by $\lambda_{\text{PRIOR}}(\bar{d}_t)$, $\lambda_{\text{POST}}(\bar{d}_t)$, and $\lambda_{\text{EXT}}(\bar{d}_t)$ the a priori, a posteriori, and extrinsic probabilities of the $2M_{\text{T}}$ real-valued elements of $\bar{\mathbf{d}}$, respectively. They are defined as

$$\begin{aligned}\lambda_{\text{PRIOR}}(\bar{d}_t) &\triangleq \ln\left(\frac{p(\bar{d}_t=+1)}{p(\bar{d}_t=-1)}\right),\\ \lambda_{\text{POST}}(\bar{d}_t) &\triangleq \ln\left(\frac{p(\bar{d}_t=+1|\bar{\mathbf{w}})}{p(\bar{d}_t=-1|\bar{\mathbf{w}})}\right),\\ \lambda_{\text{EXT}}(\bar{d}_t) &\triangleq \lambda_{\text{POST}}(\bar{d}_t)-\lambda_{\text{PRIOR}}(\bar{d}_t),\end{aligned} \tag{8.43}$$

where $p(\bar{d}_t)$ denotes the probability mass function of $\bar{d}_t \in \{-1,+1\}$ and, by abuse of notation, $t \in \{1,\ldots,2M_{\text{T}}\}$ due to the notation in the real domain.

8.6.2.2 Explicit Computation of Extrinsic Probabilities

As shown in the Appendix, it is possible to approximate the extrinsic probability of $\bar{d}_t$ as

$$\lambda_{\text{EXT}}(\bar{d}_t) = -\frac{\bar{d}_{\text{SD},t}}{\sigma_z^2}\|\bar{\mathbf{w}}+\mathbf{v}-\bar{\mathbf{R}}\bar{\mathbf{U}}\bar{\mathbf{d}}_{\text{SD}}\|^2+\frac{\bar{d}_{\text{SD},t}}{\sigma_z^2}\|\bar{\mathbf{w}}+\mathbf{v}-\bar{\mathbf{R}}\bar{\mathbf{U}}\bar{\mathbf{d}}_{\text{SD}}^{(t)}\|^2-\lambda_{\text{PRIOR}}(\bar{d}_t). \tag{8.44}$$

Here, vector $\bar{\mathbf{d}}_{\text{SD}}$ can be computed using hard sphere decoding, solving

$$\bar{\mathbf{d}}_{\text{SD}} \triangleq \underset{\bar{\mathbf{d}}\in\mathcal{A}^{M_{\text{T}}}}{\operatorname{argmin}}\,\|\bar{\mathbf{w}}+\mathbf{v}-\bar{\mathbf{R}}\bar{\mathbf{U}}\bar{\mathbf{d}}\|^2$$

Vector $\bar{\mathbf{d}}_{\text{SD}}^{(t)}$ contain $\bar{\mathbf{d}}_{\text{SD}}$ except element $\bar{d}_{\text{SD},t}$. Vector $\mathbf{v}$ is defined as

$$\mathbf{v} = \frac{\sigma_z^2}{2u_{D-1}}\left(\mathbf{v}_1^T\cdots\mathbf{v}_{M_{\text{T}}}^T\right)^T$$

with

$$\mathbf{v}_t = \left(0\cdots0\ \lambda_{\text{PRIOR}}(\bar{d}_t)\right)^T \in \mathbb{R}^{2D} \quad \text{for} \quad t\in\{1,\ldots,M_{\text{T}}\}.$$

By abuse, computing soft outputs by means of a sphere decoder is generally referred to as a soft sphere decoder.

8.6.3 Computational Complexity

In this section, we will discuss complexity aspects of the hard and soft sphere decoders. We denote by $q_t \leq Q^{M_{\text{T}}-t+1}$ (resp. $q_t \leq Q^{2M_{\text{T}}-t+1}$) the number of candidates after the step t of the hard sphere decoder (resp. soft sphere decoder) (see Dumard & Zemen, 2007 for more details). q_t is a random variable since it depends on the realization of $\mathbf{d}$, $\tilde{\mathbf{S}}$, and $\mathbf{z}$. $Q=|\mathcal{A}|$ is the size of the alphabet.

8.6.3.1 Hard Sphere Decoder

As shown in Dumard & Zemen (2007), the computational complexity for a block of length $K-J$ using hard sphere decoding requires

- one *thin* complex QR factorization of size $LM_R \times DM_T$, with complexity (Golub & Van Loan, 1996)

$$c_{QR} = 8(DM_T)^2\left(LM_R - \frac{DM_T}{3}\right) \text{ flops};$$

- $K-J$ runs of the hard sphere decoder, with complexity for a given k according to

$$c_{\text{hard}}[k] = 2D\sum_{t=2}^{M_T}[(M_T-t)(4D+3)+2D+4]\,q_t[k]+4(M_T-1)DQ(D+2) \text{ flops}.$$

An upper bound on this expression can be computed using $q_t = Q^{M_T-t+1}$, which corresponds to the case where all symbol vectors remain valid candidates at all steps. This is equivalent to an exhaustive search using the sphere decoder implementation and an infinite radius:

$$c_{\text{hard}}[k] \leq 2D\left[(4D+3)\sum_{t=1}^{M_T-1} tQ^t-(2D-1)\sum_{t=1}^{M_T-1} Q^t\right]+4(M_T-1)DQ(D+2) \text{ flops}. \tag{8.45}$$

The complexity for a data block of length $K-J$ is given by

$$C_{\text{hard}} = c_{QR} + \sum_{k\notin\mathcal{P}} c_{\text{hard}}[k]$$

and can be upper bounded using (8.45).

8.6.3.2 Soft Sphere Decoder

The complexity computations for the soft sphere decoder are very similar to the hard sphere decoder, except that they are done in the real domain. Thus, dimensions are multiplied by a factor 2. However, we have now real multiplications and additions, equivalent to one flop each. Furthermore, the alphabet size is reduced to $Q/2$. The computations required are

- one *thin* real QR factorization of size $2LM_R \times 2DM_T$, with complexity (Golub & Van Loan, 1996)

$$c_{QR} = 64(DM_T)^2\left(LM_R - \frac{DM_T}{3}\right) \text{ flops};$$

- $(K-J)(4M_T+1)$ runs of the subspace-based sphere decoder, with the complexity for a given k according to

$$c_{\text{soft}}[k] = D\sum_{t=1}^{2M_T-1}[(2M_T-t)(2D+1)-D]\,q_{t+1}[k]+\frac{(2M_T-1)DQ(D+1)}{2} \text{ flops}.$$

As previously, an upper bound on this expression can be computed using $q_t = \left(\frac{Q}{2}\right)^{2M_T-t+1}$:

$$c_{\text{soft}}[k] \leq D(2D+1)\sum_{t=1}^{2M_T-1} t\left(\frac{Q}{2}\right)^t - D\sum_{t=1}^{2M_T-1}\left(\frac{Q}{2}\right)^t+\frac{(2M_T-1)DQ(D+1)}{2} \text{ flops}. \tag{8.46}$$

The complexity for a data block of length $K-J$ is given by

$$C_{\text{soft}} = c_{\text{QR}} + (4M_{\text{T}}+1)\sum_{k\notin\mathcal{P}} c_{\text{soft}}[k]$$

and can be upper bounded using (8.46).

8.7 SIMULATION RESULTS

To compare the performance of the presented multiuser detection algorithms, we resort to numeric simulations. Realizations of the time-varying frequency-selective channel $h_{u,t,r}[l,m]$, sampled at the chip rate $1/T_{\text{c}}$, are generated using an exponentially decaying power delay profile

$$\eta^2[k] = \frac{\mathrm{e}^{-\frac{k}{4}}}{\sum_{k'=0}^{K-1} \mathrm{e}^{-\frac{k'}{4}}}$$

with root mean square delay spread $T_{\text{d}} = 4T_{\text{c}} = 1\,\mu\text{s}$ for a chip rate of $1/T_{\text{c}} = 3.84\cdot 10^6\,\text{s}^{-1}$ (Correia, 2001). We assume $M = 15$ resolvable paths. The autocorrelation for every channel tap is given by the classical Clarke spectrum (Clarke, 1968).

The system operates at carrier frequency $f_{\text{c}} = 2$ GHz, and there are $U = 32$ users moving with velocity $v = 70$ km/h. This gives a Doppler bandwidth of $B_{\text{D}} = 126$ Hz. We use $M_{\text{T}} = 4$ transmit antennas per user and $M_{\text{R}} = 4$ receive antennas at the base station. The number of subcarriers is $L = 64$, and the OFDM symbol with cyclic prefix has length $P = L+G = 79$. The data block consists of $K = 256$ OFDM symbols including $J = 60$ pilot symbols. The system is designed for $v_{\text{max}} = 102.5$ km/h, which results in a dimension of $D = 3$ for the Slepian basis expansion. In order to analyze the diversity gain of the receiver only, the MIMO channel taps are normalized according to

$$\mathsf{E}\left\{\sum_{r=1}^{M_{\text{R}}}\sum_{t=1}^{M_{\text{T}}}\sum_{m=0}^{M-1} |h_{u,t,r}[l,m]|^2\right\} = 1.$$

No antenna gain is present due to this normalization.

For data transmission, a convolutional, nonsystematic, nonrecursive, 4 state, rate $R_{\text{c}} = 1/2$ code with code generators [101] and [111] (denoted $(5,7)_8$ in octal notation) is used (Hanzo, Liew, & Yeap, 2002). All results shown are obtained by averaging over 100 independent channel realizations. The QPSK symbol energy is normalized to 1. The SNR is defined as

$$\frac{E_b}{N_0} = \frac{1}{2R_{\text{c}}\sigma_z^2}\frac{P}{L}\frac{K}{K-J},$$

which takes into account the loss due to coding, pilots, and cyclic prefix. The noise variance σ_z^2 is assumed to be known at the receiver.

8.7.1 Bit Error Rate Comparison

In Fig. 8.7, we compare the BER performance of all presented multiuser detection methods. In addition, we show the performance obtained with LMMSE multiuser detection and perfect channel knowledge

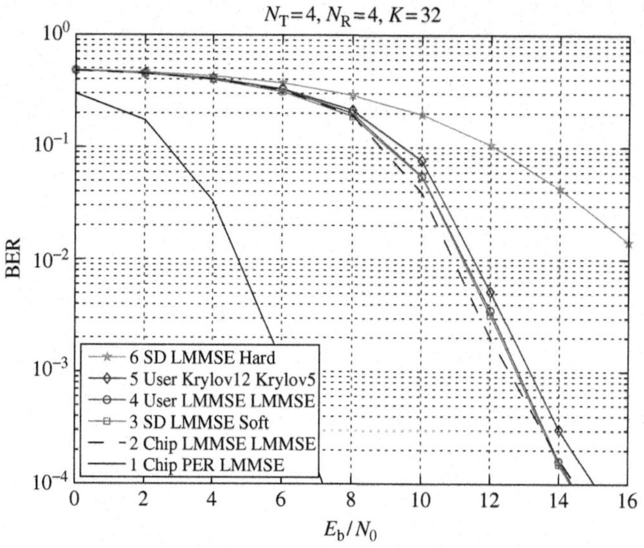

FIGURE 8.7

Comparison of detection methods: BER versus SNR after receiver iteration 4 for $U = 32$ users. We compare the results obtained with the following receivers: parallel interference cancelation in chip and user space using LMMSE and Krylov subspace method for multiuser detection, and hard- and soft-sphere decoder (SD). The third and fourth columns in the legend indicate the channel estimation method and the multiuser detector, respectively. "PER" denotes perfect channel knowledge, and "Krylov 12" denotes the Krylov subspace method with 12 iterations.

as a performance benchmark. The ultimate performance limit is shown for perfect channel knowledge and LMMSE multiuser detection.

First, let us consider parallel interference cancelation and multiuser detection in chip space and in user space, where channel estimation and multiuser detection use the exact LMMSE filters (curves 2 and 4). The receiver performs four iterations. We see that parallel interference cancelation in user space results in a slight increase in BER compared to parallel interference cancelation and multiuser detection in chip space. In the next step, we consider the implementation of the Krylov subspace method to approximate the LMMSE filters with Krylov dimensions $S = 5$ and $S' = 12$ for multiuser detection and channel estimation, respectively, for parallel interference cancelation in user space (curve 5). We see that using the approximation introduces again a slight loss of BER. Finally, we look at the sphere decoder for multiuser detection. The hard sphere decoder (curve 6) exhibits a large performance loss. This is mainly due to two facts: on the one hand, for feasibility reasons, the sphere decoder has to be implemented on a per-user basis. On the other hand, the soft symbols provided by linear detection to the BCJR decoder were lost when implementing the hard sphere decoder, providing only hard values. For this reason, we implemented the soft sphere decoder, computing log-likelihood ratios using sphere decoding (curve 3). We see that the soft sphere decoder outperforms Krylov subspace-based detection and reaches very close to the exact LMMSE detection and channel estimation in chip space.

In order to discuss the dependence of the performance of the multiuser detector on the Krylov subspace dimension, we focus now on the joint antenna detector with parallel interference cancelation

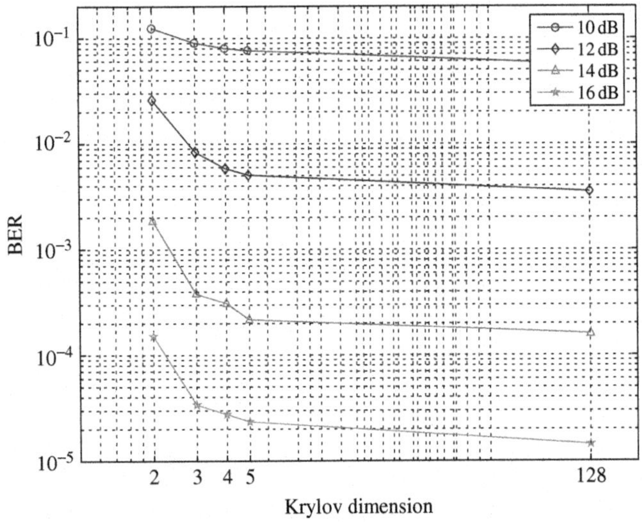

FIGURE 8.8

BER for varying Krylov subspace dimension S parameterized by the SNR. The multiuser detector employs the Krylov subspace method, and channel estimates are obtained by means of the LMMSE filter (8.27).

in user space, using the LMMSE channel estimator (8.27). Figure 8.8 shows the BER for varying Krylov subspace dimension S parameterized by the SNR. We see that the BER decreases strongly up to $S = 5$. The decay up to the full LMMSE solution with dimension $S = UM_{\mathrm{T}} = 32 \cdot 4 = 128$ is only marginal.

To assess the optimal Krylov subspace dimension for channel estimation, we keep $S = 5$ constant for multiuser detection with parallel interference cancelation in user space and vary the Krylov subspace dimension S' for channel estimation. The results are shown in Fig. 8.9. Again, we see that the BER decays up to a dimension $S' = 12$. The additional gain achieved by using the full LMMSE solution with dimension $S' = UM_{\mathrm{T}}D = 32 \cdot 4 \cdot 3 = 384$ is small. Hence, a Krylov dimension of $S' = 12$ allows for a good trade-off between computational complexity and performance.

8.7.2 Computational Complexity Comparison

To assess the complexity reduction achieved with the methods introduced in this chapter, we evaluate the computational complexity for the simulation scenario presented above. The results presented are given in millions of flops per user and take into account the whole block of $K = 256$ symbols and all $L = 64$ subcarriers.

The complexity savings obtained with the Krylov subspace method for channel estimation with $S' = 12$ are shown in Fig. 8.10. It is seen that one order of magnitude can be saved by the Krylov subspace methods for channel estimation.

In Fig. 8.11, we show upper bounds on the complexity of the multiuser detector over a block of size $K - J$. We compare all scenarios analyzed previously, i.e., parallel interference cancelation in chip space followed by LMMSE or Krylov subspace-based multiuser detection, parallel interference cancelation in user space followed by LMMSE or Krylov subspace-based multiuser detection, and

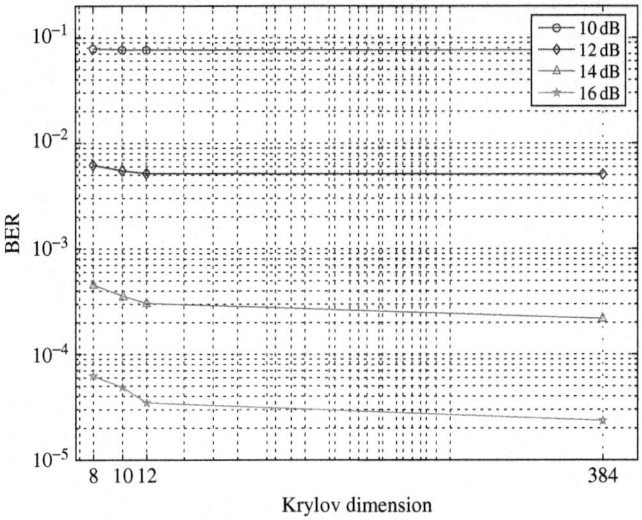

FIGURE 8.9

BER for varying Krylov subspace dimension S' for channel estimation parameterized by the SNR. Both the multiuser detector and the channel estimator utilize the Krylov subspace method.

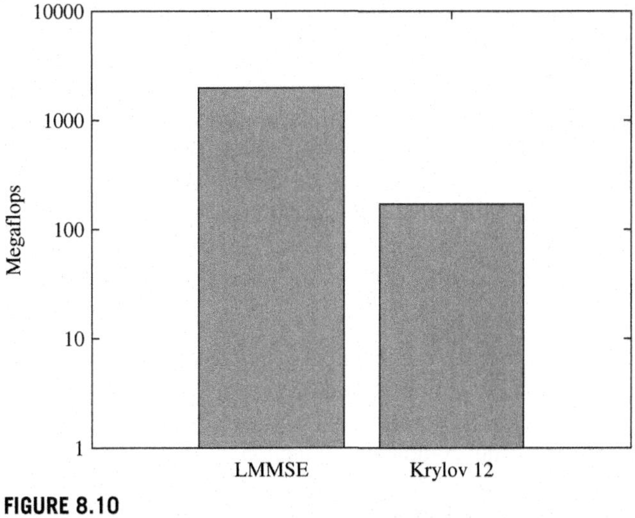

FIGURE 8.10

Computational complexity of the channel estimator. Channel estimation is performed using either an exact LMMSE filter or the Krylov subspace method with subspace dimension $S' = 12$.

finally, hard and soft sphere decoding. We can see that the soft sphere decoder is about one order of magnitude more complex than the hard sphere decoder and one order of magnitude less complex than LMMSE detection.

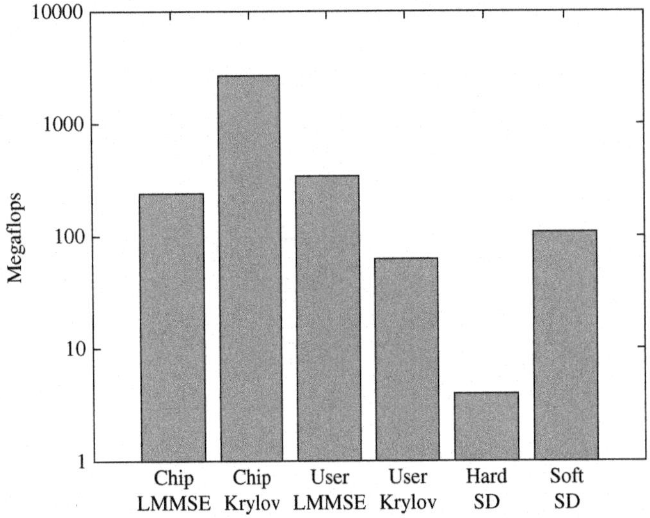

FIGURE 8.11

Computational complexity of the multiuser detector. We compare the complexity using (1) exact LMMSE filtering or Krylov subspace approximation with dimension $S = 5$, (2) parallel interference cancelation in chip space or user space, and (3) sphere decoding ("SD") with hard or soft outputs.

8.8 CONCLUSIONS

Multiuser detection in time-varying channels is a demanding task from the computational complexity viewpoint. In this chapter, we traded accuracy for efficiency by introducing several complexity reduction methods enabling a real-world low-complexity implementation. In particular, we used an iterative approximation of the MAP detector in combination with a reduced-rank model for the time-varying channel. This reduced-rank channel model projects the time-varying channel onto a subspace spanned by band-limited and time-concentrated prolate spheroidal sequences. We investigated two different multiuser detection methods:

- First, we used the Krylov subspace method to reduce the complexity of multiuser detection using LMMSE filtering. We showed that for chip-space linear multiuser detection, *no* complexity reduction due to the Krylov subspace method can be achieved. However, for a user-space LMMSE filter, the Krylov subspace method yields a significant complexity reduction.
- Second, we investigated sphere decoding and developed a sphere decoder that exploits the reduced-rank channel model for complexity reduction. Furthermore, we showed that a soft-output sphere decoder can be devised using the same approach.

We provided simulation results and complexity comparisons for all presented low-complexity architectures. The complexity of channel estimation and multiuser detection could be reduced by more than one magnitude while retaining the good performance of exact LMMSE filters.

Acknowledgments

This research was supported by the FTW projects "Cooperative Communications for Traffic Telematics" (COCOMINT) and "Future Mobile Communications Systems" (Math+MIMO), both funded by the Vienna Science and Technology Fund (WWTF), by the Austrian Science Fund (FWF) under grant S10607, and by the EC under the FP7 Network of Excellence project NEWCOM++ and the STREP MASCOT. The Telecommunications Research Center Vienna (FTW) is supported by the Austrian Government and the City of Vienna within the competence center program COMET.

Appendix

8.A COMPUTATION OF THE LOG-LIKELIHOOD RATIO (8.44)

Let $\mathbb{B}_t^+$ and $\mathbb{B}_t^-$ be the subsets of $\{-1,+1\}^{2M_T}$ such that $\bar{d}_t = +1$ or $\bar{d}_t = -1$, respectively. We have

$$
\begin{aligned}
p\left(\bar{d}_t = \pm 1 | \bar{\mathbf{w}}\right) &= \sum_{\bar{\mathbf{d}} \in \mathbb{B}_t^{\pm}} p\left(\bar{\mathbf{d}} | \bar{\mathbf{z}}\right) = \sum_{\bar{\mathbf{d}} \in \mathbb{B}_t^{\pm}} \frac{p\left(\bar{\mathbf{w}} | \bar{\mathbf{d}}\right) p(\bar{\mathbf{d}})}{p(\bar{\mathbf{w}})} \\
&= \frac{p\left(\bar{d}_t = \pm 1\right)}{p(\bar{\mathbf{w}})} \sum_{\bar{\mathbf{d}} \in \mathbb{B}_t^{\pm}} p\left(\bar{\mathbf{w}} | \bar{\mathbf{d}}\right) \prod_{t' \neq t} p\left(\bar{d}_{t'}\right),
\end{aligned}
$$

leading to

$$
\lambda_{\text{POST}}\left(\bar{d}_t\right) = \lambda_{\text{PRIOR}}\left(\bar{d}_t\right) + \ln \left(\frac{\sum_{\bar{\mathbf{d}} \in \mathbb{B}_t^{+}} p\left(\bar{\mathbf{w}} | \bar{\mathbf{d}}\right) \prod_{t' \neq t} p\left(\bar{d}_{t'}\right)}{\sum_{\bar{\mathbf{d}} \in \mathbb{B}_t^{-}} p\left(\bar{\mathbf{w}} | \bar{\mathbf{d}}\right) \prod_{t' \neq t} p\left(\bar{d}_{t'}\right)} \right).
$$

From the definition of the a priori probability in (8.43), we obtain

$$
\exp\left(\pm \lambda_{\text{PRIOR}}\left(\bar{d}_{t'}\right)\right) = \frac{p\left(\bar{d}_{t'} = \pm 1\right)}{p\left(\bar{d}_{t'} = \mp 1\right)} = \frac{p\left(\bar{d}_{t'} = \pm 1\right)}{1 - p\left(\bar{d}_{t'} = \pm 1\right)},
$$

which yields

$$
\begin{aligned}
p\left(\bar{d}_{t'} = \pm 1\right) &= \frac{\exp\left(\pm \lambda_{\text{PRIOR}}\left(\bar{d}_{t'}\right)\right)}{1 + \exp\left(\pm \lambda_{\text{PRIOR}}\left(\bar{d}_{t'}\right)\right)} \\
&= \frac{\exp\left(-\lambda_{\text{PRIOR}}\left(\bar{d}_{t'}\right)/2\right)}{1 + \exp\left(-\lambda_{\text{PRIOR}}\left(\bar{d}_{t'}\right)\right)} \exp\left(\pm \frac{\lambda_{\text{PRIOR}}\left(\bar{d}_{t'}\right)}{2}\right) \\
&= A_{t'} \exp\left(\bar{d}_{t'} \frac{\lambda_{\text{PRIOR}}\left(\bar{d}_{t'}\right)}{2}\right),
\end{aligned}
$$

where

$$A_{t'} \triangleq \frac{\exp\left(-\lambda_{\text{PRIOR}}(\bar{d}_{t'})/2\right)}{1+\exp\left(-\lambda_{\text{PRIOR}}\left(\bar{d}_{t'}\right)\right)}$$

does not depend on the sign of $\bar{d}_{t'}$. It follows that

$$\prod_{t'\neq t} p(\bar{d}_{t'}) = \left(\prod_{t'\neq t} A_{t'}\right)\exp\left(\sum_{t'\neq t}\frac{\bar{d}_{t'}\lambda_{\text{PRIOR}}\left(\bar{d}_{t'}\right)}{2}\right)$$

$$= \mathcal{A}_t \exp\left(\frac{\bar{\mathbf{d}}_t^T \boldsymbol{\lambda}_{\text{PRIOR},t}}{2}\right),$$

with

$$\mathcal{A}_t \triangleq \prod_{t'\neq t} A_{t'},$$

and where $\bar{\mathbf{d}}_t$ contains $\bar{\mathbf{d}}$ except $\bar{d}_t$ and $\boldsymbol{\lambda}_{\text{PRIOR},t}$ contains the a priori probabilities of the elements of $\bar{\mathbf{d}}_t$. Finally, after simplification, we obtain (see (8.43))

$$\lambda_{\text{EXT}}(\bar{d}_t) = \boldsymbol{\lambda}_{\text{POST}}\left(\bar{d}_t\right) - \boldsymbol{\lambda}_{\text{PRIOR}}\left(\bar{d}_t\right) = \ln\left(\frac{\sum_{\bar{\mathbf{d}}\in\mathbb{B}_t^+} p(\bar{\mathbf{w}}|\bar{\mathbf{d}})\exp\left(\frac{\bar{\mathbf{d}}_t^T\boldsymbol{\lambda}_{\text{PRIOR},t}}{2}\right)}{\sum_{\bar{\mathbf{d}}\in\mathbb{B}_t^-} p(\bar{\mathbf{w}}|\bar{\mathbf{d}})\exp\left(\frac{\bar{\mathbf{d}}_t^T\boldsymbol{\lambda}_{\text{PRIOR},t}}{2}\right)}\right).$$

In what follows, let $\boldsymbol{\lambda}_{\text{PRIOR}}$ denote the vector containing the a priori probabilities of $\bar{\mathbf{d}}$, which are given by the soft outputs from the iterative receiver. Knowing

$$\bar{\mathbf{d}}_t^T\boldsymbol{\lambda}_{\text{PRIOR},t} = \bar{\mathbf{d}}^T\boldsymbol{\lambda}_{\text{PRIOR}} - \bar{d}_t\lambda_{\text{PRIOR}}(\bar{d}_t)$$

and

$$p(\bar{\mathbf{w}}|\bar{\mathbf{d}}) \propto \exp\left(-\frac{1}{\sigma^2}\left\|\bar{\mathbf{w}}-\bar{\mathbf{R}}\bar{\mathbf{U}}\bar{\mathbf{d}}\right\|^2\right),$$

and using the max-log approximation (Hochwald & ten Brink, 2003), the extrinsic probability becomes

$$\begin{aligned}\lambda_{\text{EXT}}(\bar{d}_t) \approx &\max_{\bar{\mathbf{d}}\in\mathbb{B}_t^+}\left(-\frac{1}{\sigma_z^2}\left\|\bar{\mathbf{w}}-\bar{\mathbf{R}}\bar{\mathbf{U}}\bar{\mathbf{d}}\right\|^2+\frac{\bar{\mathbf{d}}^T\boldsymbol{\lambda}_{\text{PRIOR}}}{2}\right)\\ &-\max_{\bar{\mathbf{d}}\in\mathbb{B}_t^-}\left(-\frac{1}{\sigma_z^2}\left\|\bar{\mathbf{w}}-\bar{\mathbf{R}}\bar{\mathbf{U}}\bar{\mathbf{d}}\right\|^2+\frac{\bar{\mathbf{d}}^T\boldsymbol{\lambda}_{\text{PRIOR}}}{2}\right)\\ &-\lambda_{\text{PRIOR}}(\bar{d}_t).\end{aligned} \tag{A.1}$$

To maximize the two expressions in (A.1), it is useful to find a vector $\mathbf{v}$ satisfying $4(\bar{\mathbf{R}}\bar{\mathbf{U}})^T\mathbf{v} = \sigma_z^2\boldsymbol{\lambda}_{\mathrm{PRIOR}}$. This allows us to write

$$\max\left\{-\frac{1}{\sigma_z^2}\left\|\bar{\mathbf{w}}-\bar{\mathbf{R}}\bar{\mathbf{U}}\bar{\mathbf{d}}\right\|^2+\frac{\bar{\mathbf{d}}^T\boldsymbol{\lambda}_{\mathrm{PRIOR}}}{2}\right\} = \max\left\{-\frac{1}{\sigma_z^2}\left\|\bar{\mathbf{w}}-\bar{\mathbf{R}}\bar{\mathbf{U}}\bar{\mathbf{d}}\right\|^2+2\frac{(\bar{\mathbf{R}}\bar{\mathbf{U}}\bar{\mathbf{d}})^T\mathbf{v}}{\sigma_z^2}\right\}$$

$$= -\frac{1}{\sigma_z^2}\min\left\|\bar{\mathbf{w}}+\mathbf{v}-\bar{\mathbf{R}}\bar{\mathbf{U}}\bar{\mathbf{d}}\right\|^2+c,$$

where c is not dependent on $\bar{\mathbf{d}}$. For instance, it can be easily checked that a valid solution is

$$\mathbf{v} = \frac{\sigma_z^2}{2u_{D-1}}\left(\mathbf{v}_1^T\cdots\mathbf{v}_{M_\mathrm{T}}^T\right)^T$$

with

$$\mathbf{v}_t = \left(0\cdots 0\ \lambda_{\mathrm{PRIOR}}(\bar{d}_t)\right)^T \in \mathbb{R}^{2D} \quad \text{for} \quad t\in\{1,\ldots,M_\mathrm{T}\}.$$

Let us now define

$$\bar{\mathbf{d}}_{\mathrm{SD}} \triangleq \underset{\bar{\mathbf{d}}\in\mathcal{A}^{M_\mathrm{T}}}{\operatorname{argmin}}\left\|\bar{\mathbf{w}}+\mathbf{v}-\bar{\mathbf{R}}\bar{\mathbf{U}}\bar{\mathbf{d}}\right\|^2,$$

and let s denote the sign of the tth element $\bar{d}_{\mathrm{SD},t}$. The extrinsic probability in (A.1) is then simplified to

$$\lambda_{\mathrm{EXT}}(\bar{d}_t) = -\frac{\bar{d}_{\mathrm{SD},t}}{\sigma_z^2}\left\|\bar{\mathbf{w}}+\mathbf{v}-\bar{\mathbf{R}}\bar{\mathbf{U}}\bar{\mathbf{d}}_{\mathrm{SD}}\right\|^2+\frac{\bar{d}_{\mathrm{SD},t}}{\sigma_z^2}\min_{\bar{\mathbf{d}}\in\mathbb{B}_t^{-s}}\left\|\bar{\mathbf{w}}+\mathbf{v}-\bar{\mathbf{R}}\bar{\mathbf{U}}\bar{\mathbf{d}}\right\|^2-\lambda_{\mathrm{PRIOR}}(\bar{d}_t),$$

where $\mathbb{B}_t^{-s}\in\{\mathbb{B}_t^+,\mathbb{B}_t^-\}$ depending on s. Denoting

$$\bar{\mathbf{d}}_{\mathrm{SD}}^{(t)} \triangleq \underset{\bar{\mathbf{d}}\in\mathbb{B}_t^{-s}}{\operatorname{argmin}}\left\|\bar{\mathbf{w}}+\mathbf{v}-\bar{\mathbf{R}}\bar{\mathbf{U}}\bar{\mathbf{d}}\right\|^2,$$

we obtain

$$\lambda_{\mathrm{EXT}}(\bar{d}_t) = -\frac{\bar{d}_{\mathrm{SD},t}}{\sigma_z^2}\left\|\bar{\mathbf{w}}+\mathbf{v}-\bar{\mathbf{R}}\bar{\mathbf{U}}\bar{\mathbf{d}}_{\mathrm{SD}}\right\|^2+\frac{\bar{d}_{\mathrm{SD},t}}{\sigma_z^2}\left\|\bar{\mathbf{w}}+\mathbf{v}-\bar{\mathbf{R}}\bar{\mathbf{U}}\bar{\mathbf{d}}_{\mathrm{SD}}^{(t)}\right\|^2-\lambda_{\mathrm{PRIOR}}(\bar{d}_t),$$

which is (8.44).

References

Agrell, E., Eriksson, T., Vardy, A., & Zeger, K. (2002). Closest point search in lattices. *IEEE Transactions on Information Theory*, *48*(8), 2201–2214.

Atiyah, M. (1994). *Introduction to commutative algebra*. Boulder, CO: Westview Press.

Bahl, L. R., Cocke, J., Jelinek, F., & Raviv, J. (1974). Optimal decoding of linear codes for minimizing symbol error rate. *IEEE Transactions on Information Theory*, *20*(2), 284–287.

Clarke, R. H. (1968). A statistical theory of mobile-radio reception. *Bell System Technical Journal*, *47*, 957–1000.

Correia, L. M. (2001). *Wireless flexible personalised communications*. New York: Wiley.

Cottatellucci, L., Müller, R. R., & Debbah, M. (2006, September). Linear detectors for multiuser systems with correlated spatial diversity. In *Proceedings of the Fourteenth European Signal Processing Conference (EUSIPCO)*, Florence, Italy.

Damen, M. O., El Gamal, H., & Caire, G. (2003). On maximum-likelihood detection and the search for the closest lattice point. *IEEE Transactions on Information Theory*, *49*(10), 2389–2402.

Dietl, G., & Utschick, W. (2007). Complexity reduction of iterative receivers using low-rank equalization. *IEEE Transactions on Signal Processing*, *55*(3), 1035–1046.

Dumard, C., & Zemen, T. (2006, September). Krylov subspace method based low-complexity MIMO multiuser receiver for time-variant channels. In *Proceedings of the Seventeenth IEEE International Symposium on Personal, Indoor and Mobile Radio Communication (PIMRC)*, Helsinki, Finland.

Dumard, C., & Zemen, T. (2007, September). Subspace based sphere decoder for MC-CDMA in time-varying MIMO channels. In *Proceedings of the Eighteenth IEEE Workshop on Personal, Indoor and Mobile Radio Communciations (PIMRC)*, Athens, Greece.

Dumard, C., Jaldén, J., & Zemen, T. (2008, August). Soft sphere decoder for an iterative receiver in time-varying channels. In *Proceedings of the Sixteenth European Signal Processing Conference (EUSIPCO)*, Lausanne, Switzerland.

Fincke, U., & Pohst, M. (1985). Improved methods for calculating vectors of short length in a lattice, including a complex analysis. *Mathematics of Computation*, *44*(169–170), 463–471.

Golub, G. H., & Van Loan, C. F. (1996). *Matrix Computations*. (3rd ed.). Baltimore, MD: Johns Hopkins University Press.

Hanly, S. V., & Tse, D. N. C. (2001). Resource pooling and effective bandwidths in CDMA networks with multiuser receivers and spatial diversity. *IEEE Transactions on Information Theory*, *47*(4), 1328–1351.

Hanzo, L., Liew, T. H., & Yeap, B. L. (2002). *Turbo coding, turbo equalization and space-time coding for transmission over fading channels*. New York: Wiley.

Hochwald, B. M., & ten Brink, S. (2003). Achieving near-capacity on a multiple-antenna channel. *IEEE Transactions on Communications*, *51*(3), 389–399.

Hu, B., Land, I., Rasmussen, L., Piton, R., & Fleury, B. H. (2008). A divergence minimization approach to joint multiuser decoding for coded CDMA. *IEEE Journal on Selected Areas in Communications*, *26*(3), 432–445.

Jalden, J., & Ottersten, B. (2005). On the complexity of sphere decoding in digital communications. *IEEE Transactions on Signal Processing*, *53*(4), 1474–1484.

Javaudin, J. P., Lacroix, D., & Rouxel, A. (2003, April). Pilot-aided channel estimation for OFDM/OQAM. In *Vehicular Technology Conference (VTC) 2003-Spring*: Vol. 3. (pp. 1581–1585).

Joham, M., & Zoltowski, M.D. (2000). *Interpretation of the multi-stage nested Wiener filter in the Krylov subspace framework*. Tech. Rep. TUM-LNS-TR-00-6, Munich University of Technology.

Kailath, T., & Sayed, A. H. (1999). *Fast reliable algorithms for matrices with structure*. Philadelphia: SIAM.

Kaiser, S. (1998). Multi-carrier CDMA mobile radio systems – Analysis and optimization of detection, decoding, and channel estimation. Vol. 10. *Fortschritts-Berichte VDI Reihe*. Düsseldorf, Germany: VDI Verlag GmbH.

Kecicioglu, B., & Torlak, M. (2004). Reduced rank beamforming methods for SDMA/OFDM communications. In *Proceedings of the Thirty-eighth Asilomar Conference on Signals, Systems and Computers*: Vol. 2 (pp. 1973–1977).

Kirsteins, I. P., & Ge, H. (2006, July). Performance analysis of Krylov space adaptive beamformers. In *Proceedings of IEEE Workshop on Sensor Array and Multichannel Signal Processing*: Vol. 3. (pp. 16–20).

Kühn, V. (2006). *Wireless communications over MIMO channels*. New York: John Wiley & Sons.

Matz, G. (2005). On non-WSSUS wireless fading channels. *IEEE Transactions on Wireless Communications*, *4*(5), 2465–2478.

Mecklenbräuker, C. F., Wehinger, J., Zemen, T., Artés, H., & Hlawatsch, F. (2006). Multiuser MIMO channel equalization. In T. Kaiser, A. Bourdoux, H. Boche, J. R. Fonollosa, J. Bach Andersen, and W. Utschick (Eds.), *Smart antennas – state-of-the-art*, EURASIP book series on signal processing and communications, (chap. 1.4, pp. 53–76). New York: Hindawi.

Moon, T. K., & Stirling, W. C. (2000). *Mathematical methods and algorithms for signal processing*. Upper Saddle River, NJ: Prentice Hall.

Mow, W. H. (1994). Maximum likelihood sequence estimation from the lattice viewpoint. *IEEE Transactions on Information Theory*, *40*(5), 1591–1600.

Müller, R. R., & Caire, G. (2001, August). The optimal received power distribution for IC-based iterative multiuser joint decoders. In *Proceedings of the Thirty-ninth Annual Allerton Conference on Communication, Control and Computing*, Monticello (IL), USA.

Murugan, A. D., El Gamal, H., Damen, M. O., & Caire, G. (2006). A unified framework for tree search decoding: rediscovering the sequential decoder. *IEEE Transactions on Information Theory*, *52*(3), 933–953.

Paulraj, A., Nabar, R., & Gore, D. (2003). *Introduction to space-time wireless communications*. Cambridge, UK: Cambridge University Press.

Saad, Y. (2003). *Iterative methods for sparse linear systems*. (2nd ed.). Philadelphia: SIAM.

Salvo Rossi, P., & Müller, R. (2008). Slepian-based two-dimensional estimation of time-frequency variant MIMO-OFDM channels. *IEEE Signal Processing Letters*, *15*(1), 21–24.

Sayeed, A. M., Sendonaris, A., & Aazhang, B. (1998). Multiuser detection in fast-fading multipath environment. *IEEE Journal on Selected Areas in Communication*, *16*(9), 1691–1701.

Scharf, L. L., & Tufts, D. W. (1987). Rank reduction for modeling stationary signals. *IEEE Transactions on Acoustics, Speech, and Signal Processing*, ASSP-*35*(3), 350–355.

Slepian, D. (1978). Prolate spheroidal wave functions, Fourier analysis, and uncertainty – V: The discrete case. *Bell System Technical Journal*, *57*(5), 1371–1430.

van der Vorst, H. (2003). *Iterative Krylov methods for large linear systems*. Cambridge, UK: Cambridge University Press.

Verdú, S. (1989). Maximum likelihood sequence estimation from the lattice viewpoint. *Algorithmica*, *4*(1–4), 303–312.

Verdú, S. (1998). *Multiuser detection*. New York: Cambridge University Press.

Viterbo, E., & Boutros, J. (1999). A universal lattice code decoder for fading channels. *IEEE Transactions on Information Theory*, *45*, 1639–1642.

Wang, R., & Giannakis, G. B. (2004, March). Approaching MIMO channel capacity with reduced-complexity soft sphere decoding. In *Proceedings of the Wireless Communications and Networking Conference (WCNC)*, Atlanta, Georgia.

Wehinger, J. (2005). *Iterative multi-user receivers for CDMA systems*. Unpublished doctoral dissertation, Vienna University of Technology, Vienna, Austria.

Wehinger, J., Müller, R. R., Lončar, M., & Mecklenbräuker, C. F. (2002). Performance of iterative CDMA receivers with channel estimation in multipath environments. In *Proceedings of the Thirty-sixth Asilomar Conference on Signals, Systems and Computers*, Vol. 2, (pp. 1439–1443). Pacific Grove CA.

Wolniansky, P. W., Foschini, G. J., Golden, G. D., & Valenzuela, R. A. (1998). V–BLAST: An architecture for achieving very high data rates over rich-scattering wireless channels. In *Proceedings of the International Symposium on Signals, Systems, and Electronics (ISSSE–98)*, Pisa, Italy.

Zemen, T. (2004). *OFDM multi-user communication over time-variant channels*. Unpublished doctoral dissertation, Vienna University of Technology, Vienna, Austria.

Zemen, T., & Mecklenbräuker, C. F. (2005). Time-variant channel estimation using discrete prolate spheroidal sequences. *IEEE Transactions on Signal Processing*, *53*(9), 3597–3607.

Zemen, T., Hofstetter, H., & Steinböck, G. (2005, June). Successive Slepian subspace projection in time and frequency for time-variant channel estimation. In *Proceedings of the Fourteenth IST Mobile and Wireless Communication Summit*, Dresden, Germany.

Zemen, T., Mecklenbräuker, C. F., Wehinger, J., & Müller, R. R. (2006). Iterative joint time-variant channel estimation and multi-user detection for MC-CDMA. *IEEE Transactions on Wireless Communications*, *5*(6), 1469–1478.

Zemen, T., Mecklenbräuker, C. F., Fleury, B. H., & Kaltenberger, F. (2007). Minimum-energy band-limited predictor with dynamic subspace selection for time-variant flat-fading channels. *IEEE Transactions on Signal Processing*, *55*(9), 4534–4548.

CHAPTER

9

Time-Scale and Dispersive Processing for Wideband Time-Varying Channels

Antonia Papandreou-Suppappola[1], Cornel Ioana[2], Jun Jason Zhang[1]

[1]*Arizona State University, Tempe, AZ, USA*

[2]*Grenoble Institute of Technology, Grenoble, France*

9.1 INTRODUCTION

9.1.1 Need for Wideband Channel Characterizations

Linear time-varying (LTV) channel characterizations have found many successes in mobile wireless communications as they can be effectively used to detect, estimate, and diversify the communication process. As discussed in earlier chapters, one such useful characterization of the channel output is in terms of time shifts and Doppler (frequency) shifts on the transmitted signal; these shifts can be due to multipath propagation and time dispersion and due to motion or carrier frequency offsets, respectively (Bello, 1963; Giannakis & Tepedelenlioglu, 1998; Molisch, 2005; Proakis, 2001; Sayeed & Aazhang, 1999). Although such a representation can be used to describe any LTV channel, it is not well matched to all possible mobile communication channels. A very rapidly varying wideband communication channel can cause Doppler (scale) changes on the transmitted signal; the scale changes cannot be approximated by frequency shifts (Margetts, Schniter, & Swami, 2007; Ye & Papandreou-Suppappola, 2006). For example, a shallow underwater communication channel can cause different frequencies to be shifted in time by different amounts (Iem, Papandreou-Suppappola, & Boudreaux-Bartels, 2002; Stojanovic, 2003; Ye & Papandreou-Suppappola, 2007). For these types of channels, a matched characterization is expected to yield more effective processing than the aforementioned channel characterization that is matched to time-frequency shifts. A higher processing performance is also expected when using a matched channel model instead of adopting techniques that simply compensate for wideband or nonlinear channel transformations.

The narrowband LTV channel model represents the channel output in terms of time-frequency-shifted versions of the transmitted signal. However, when the relative motion between the transmitter, the receiver, and the scatterers in the propagation channel becomes fast, and the fractional bandwidth (ratio of bandwidth over carrier frequency) of the signal is large, then the signal is scaled (expanded or compressed) during transmission (Davies, Pointer, & Dunn, 2000; Shenoy & Parks, 1995; Weiss, 1996). Under these conditions, the resulting time variation due to Doppler scaling effects, coupled with dispersive scattering due to multipath propagation, can severely limit the receiver performance. Thus, a wideband channel model needs to be considered at the receiver to improve performance.

Doppler scaling was taken into consideration in high-speed underwater acoustic communications (Davies et al., 2000; Freitag et al., 1998; Freitag, Stojanovic, Singh, & Johnson, 2001; Freitag, Stojanovic, Kilfoyle, & Preisig, 2004; Johnson, Freitag, & Stojanovic, 1997; Kim & Lu, 2000; Li, Zhou, Stojanovic, Freitag, & Willett, 2007; Stojanovic, 1996; Stojanovic, Catipovic, & Proakis, 1994; Sharif, Neasham, Hinton, & Adams, 2000), high data rate wireless communications (Bircan, Tekinay, & Akansu, 1998; Zhang, Fan, & Lindsey, 2001), and emerging ultra-wideband communications operating at high fractional bandwidths (He, Wang, Zuo, & Zhang, 2008; Scholtz & Lee, 2002; Win & Scholtz, 2000, 2002). A Doppler preprocessing procedure, which uses interpolation via polyphase filters to provide a computationally efficient method for adjusting the time scale of the data, was proposed in Johnson et al. (1997) for high-speed underwater acoustic communications. Based on this preprocessing, an acoustic modem for bidirectional communication with an unmanned underwater vehicle was developed and tested in the field (Freitag et al., 1998). This Doppler correction method was also used in Freitag et al. (2001) to establish a direct-sequence, frequency-hopped spread-spectrum acoustic communication. In Li et al. (2007), a two-step approach was proposed to mitigate frequency-dependent Doppler drifts in zero-padded OFDM transmissions over fast-varying channels. A time-scale compensation mechanism for underwater acoustic communications was proposed in Sharif et al. (2000) based on a time-frequency-based Doppler estimation method. In particular, the predicted and simulated performance of a prototype communication system was successfully compared, and communication was demonstrated at 16 kb/s with a transmitting platform moving up to $\pm 2.6\,\text{m/s}$. Following Sharif et al. (2000), a similar system was used for OFDM in Kim & Lu (2000). The effect of the Doppler scale variability was studied in Davies et al. (2000) to discuss the benefits of robust open- and closed-loop Doppler compensation for both high and low data rate modulation methods. In He et al. (2008), an ultra-wideband data link for micro air vehicles was used for low probability of intercept and transmission of digital voice. Also, without compensating for Doppler scaling, a scale-lag rake receiver was used in Margetts, Schniter, & Swami (2007).

Although the aforementioned studies took into consideration the Doppler scale effect, the underwater communication performance could be further improved by the use of highly matched channel characterizations. This is because these studies assumed only one significant Doppler-scale component in the channel. However, underwater environments can have multiple causes for different dense Doppler-scales, such as reflections from the sea surface and the rough seabed (Eggen, Baggeroer, & Preisig, 2000, 2001). Also, these studies did not consider any potential diversity that could be exploited in a time-scale channel characterization to increase communication performance.

Dispersive LTV channels can cause different frequencies to be time-shifted by different amounts, depending on the nonlinear change caused on the phase of the transmitted signal (Iem et al., 2002; Papandreou-Suppappola, Murray, Iem, & Boudreaux-Bartels, 2001). Many real-life applications require processing using dispersive LTV channel models. For example, shallow water can induce nonlinear time-frequency dispersion to acoustic propagation. In recent years, there has been a growing interest in underwater acoustic communications in various applications such as telemetry, military operations, remote control, and data transmission. A large class of underwater channels can be characterized as rapidly varying time-dispersive channels. This follows from the time-varying multipath and Doppler-induced changes on the transmitted signal when propagating over an underwater channel (Stojanovic, 2003). Time-varying multipath corresponds to the propagation of the transmitted signal along several paths, where each path is characterized by a given delay and attenuation. The time-varying multipath can be thought of as being materialized by very long channel responses spanning up to several tens of symbol intervals, resulting in intersymbol interference (ISI) and thus making channel equalization extremely difficult. Although there are several techniques to reduce ISI, such as differential phase-shift-keying and linear equalization (Hinton, Howe, & Adams, 1992) or direct sequence spread spectrum (Fischer, Bennett, Reible, Cafarella, & Yao, 1992), the main problem is the recovery of the carrier in the presence of strong multipath. In Stojanovic (2004), spatio-temporal processing was performed by splitting equalization between the transmitter and receiver and reducing ISI by transmitting a time-reversed version of a probe signal received from the source position. Note, however, that the data rate was also reduced. As a result, when a dispersively matched channel representation is not considered, the time-varying multipath and Doppler remain a challenging topic for underwater communications.

For both time-scale and dispersive signal transformations, the constant frequency shift assumption of the narrowband model fails, and wideband[1] time-varying models are required for processing. This chapter presents wideband time-varying channel models and their application towards improving wireless communication performance. Before discussing the new channel models, and in order to further demonstrate the need for such models, we first present some underwater communication examples.

9.1.2 Examples of Wideband Dispersive Channel Characteristics

Figure 9.1 illustrates a mobile underwater acoustic communication channel. The transmitted sound waveform uses a very low carrier frequency and is shown scattered at different ranges and propagating at different velocities. When the depth of transmission is also low, the communication channel can be dispersive; that is, it can produce different delays in spectral components such that higher frequencies are less delayed than lower frequencies (Frisk, 1994; Jensen, Kuperman, Porter, & Schmidt, 1994; Tolstoy & Clay, 1966). As a result, the Doppler effect must be considered when designing the communication receiver. The key idea is to take advantage of the natural dispersive changes that appear at very low frequencies (typically, frequencies lower than 300–400 Hz).

In order to demonstrate the possible dispersive changes on a signal transmitted in an underwater communication channel, we consider an isovelocity normal mode channel model with constant sound profile (see Section 9.5 for more details on the channel model) that is simulated using the *KRAKEN*

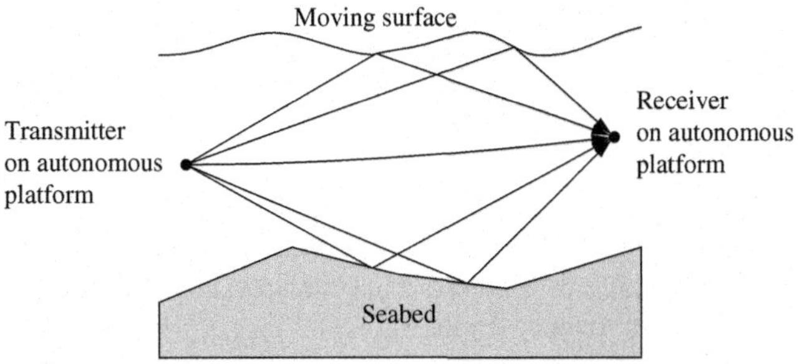

FIGURE 9.1

Mobile underwater acoustic communication channel.

[1]Note that in Federal Communications Commission (2002), a wideband communication system is defined as a system whose fractional bandwidth exceeds 0.2; the fractional bandwidth is given by the ratio of the single-sided bandwidth to the center frequency.

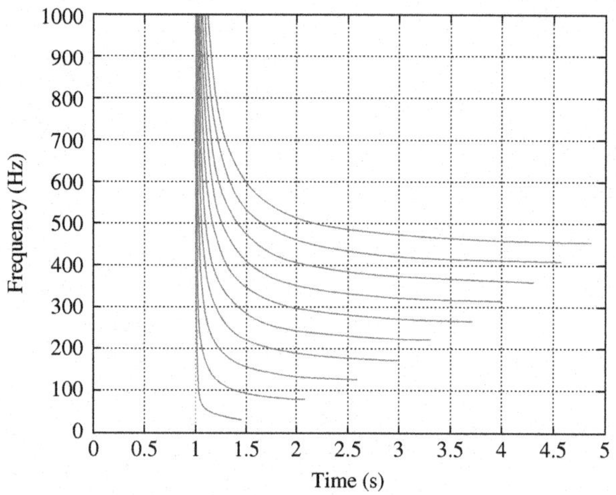

FIGURE 9.2

Time-frequency representation of the impulse response of an isovelocity dispersive channel.

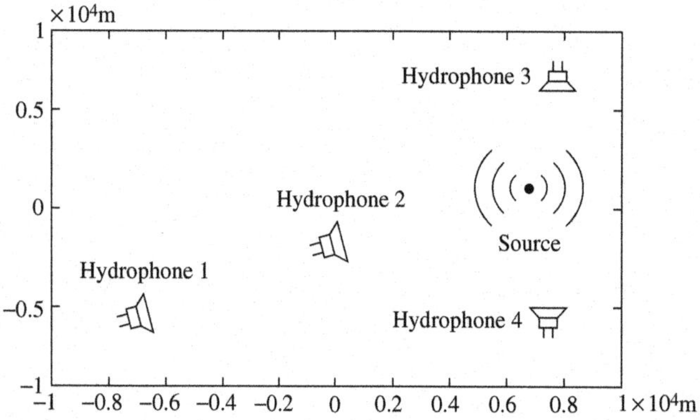

FIGURE 9.3

Positions of the hydrophones and the presumed position of the blue whale (the source of the sound).

software (Porter, 1991). For the channel's physical parameters, we choose the ocean depth to be 16 m, the transmitter and receiver depth to be 4 m, and the range between the transmitter and receiver to be 2 km. Figure 9.2 demonstrates that the simulated impulse response of the channel exhibits dispersive time-frequency characteristics, especially in the low-frequency region.

The dispersive channel characteristics can also be demonstrated using experimental ocean data. The experiment we consider employed four hydrophones placed in the ocean to receive impulse signals from a blue whale in Saint Lawrence bay[2] (Gervaise, Vallez, Stephan, & Simard, 2008). The positions of the hydrophones and the presumed position of the blue whale are depicted in Fig. 9.3.

[2]The real dataset is courtesy of C. Gervaise (École Nationale Supérieure des Ingénieurs, Brest, France) and Y. Simard (Institut des Sciences de la Mer de Rimouski, Rimouski, Québec, Canada).

The received data follow a more complicated sound speed profile than the isovelocity model. This is seen from the dispersive time-frequency structure of the spectrogram time-frequency representations as well as the extracted time-frequency modes of the signals received by the hydrophones; the extracted time-frequency modes are shown superimposed on the spectrograms in Fig. 9.4. Specifically, the figure demonstrates that the dispersion effect is more pronounced at the low-frequency components than at the high-frequency components of the received signals. Note that investigating the dispersive characteristics of underwater channels can also help increase communication data rates. One possible way to exploit dispersive characteristics is by using processing techniques that match the time-frequency structures of the signals; such methods can be similar to the ones we used (Zhang, Papandreou-Suppappola, Gottin, & Ioana, 2009) to extract the time-frequency modes in Fig. 9.4.

We further demonstrate the dispersive nature of the underwater channel in the following scenario, which we will reconsider in Section 9.6 to show how time-dispersion diversity can be exploited to improve communication. We consider a multiuser underwater communication scenario with two communication channels, as shown in Fig. 9.5. The first channel enables communication between a

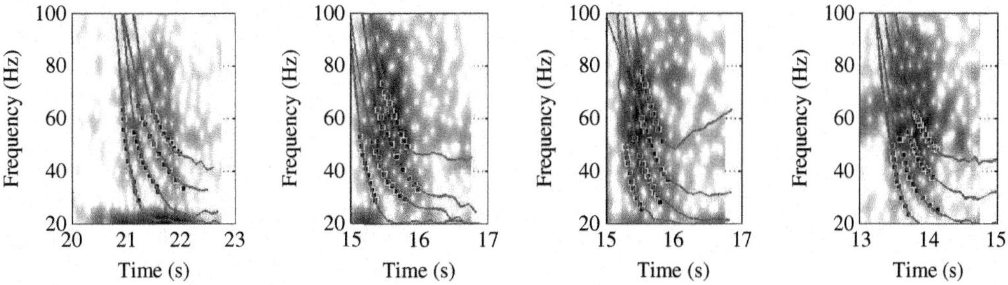

FIGURE 9.4

Spectrograms of experimental data, demonstrating the dispersive nature of the underwater channel. Each plot shows the spectrogram of the signal received by one of the four hydrophones.

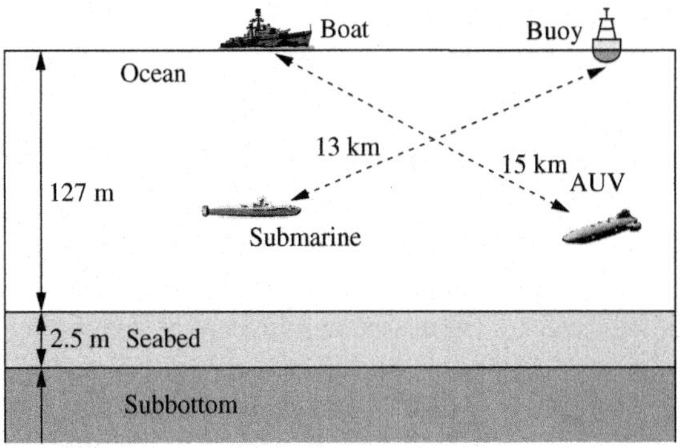

FIGURE 9.5

Configuration of the two underwater communication channels used in the simulation.

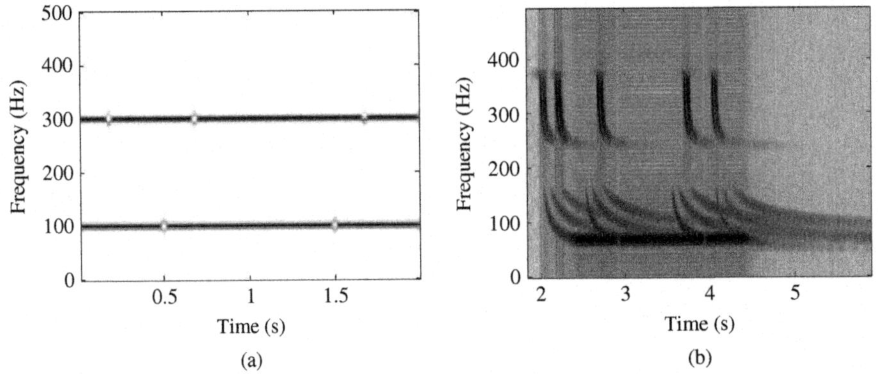

FIGURE 9.6

Superimposed spectrograms of (a) the transmitted signals and (b) the received signals. The spectrograms are plotted in dB scale.

boat and an autonomous underwater vehicle (AUV) that are 15 km apart; the second channel enables communication between a submarine and a surface buoy that are 13 km apart. We consider an ideal sound velocity profile, where the sound speed is 1500 m/s for both channels. We assume that the water column and the bottom have a structure that is similar to the real configuration characterizing the North Elba Sea (Gingras, 1994). Specifically, the ocean depth is assumed to be 127 m; the sediment of the seabed is 2.5 m deep; the sound speed from the top to the bottom of the sediment is assumed to increase linearly from 1520 m/s to 1580 m/s; the density of the seabed is $1.75\,\mathrm{g/cm^3}$; and the sound attenuation coefficient of the seabed is $0.13\,\mathrm{dB}$ per wavelength. The subbottom has a constant sound velocity of 1600 m/s, a density of $1.8\,\mathrm{g/m^3}$, and a sound attenuation coefficient of 0.15 dB per wavelength. Both communication channels use binary phase-shift-keying (BPSK) modulation, and the corresponding carrier frequencies are 100 and 300 Hz. The signals are also corrupted by white Gaussian noise with a signal-to-noise ratio (SNR) of 5 dB. Figure 9.6(a) shows two spectrograms, each corresponding to a transmitted signal, that are overlaid on the same time-frequency plot. Figure 9.6(b) shows the superimposed spectrograms of the corresponding received signals (simulated using *KRAKEN*). All the spectrograms are plotted in dB scale. The multiple nonlinear time-frequency components in the received signals show the dispersive nature of the channel. Thus, if we can understand the dispersion characteristics of the channel and design methods to extract the nonlinear time-frequency components, then we can use them to our advantage to improve communication performance.

9.1.3 Chapter Organization

The rest of this chapter is organized as follows. In Section 9.2, we review the narrowband channel characterization and its discretized model that is essential in processing applications. The delay-scale wideband channel is discussed in Section 9.3, and its use in wireless communications exploiting multipath-scale diversity is provided in Section 9.3.6. In Section 9.4, we discuss the wideband dispersive channel characterization and its equivalence to the narrowband model through time-warping processing. Its application to underwater communications is presented in Section 9.5. Finally, an application for underwater communications is provided in Section 9.6 to demonstrate the increase in performance obtained when the matched channel model is used for processing.

9.2 NARROWBAND CHANNEL CHARACTERIZATION

9.2.1 Narrowband Spreading Function

An LTV channel **H** can be characterized by a kernel representation that is based on the time-varying impulse response $h(t,\tau)$ of the channel, as discussed in Chapter 1 (Kailath, 1959; Zadeh & Desoer, 1963). A narrowband LTV channel can cause time shifts τ and frequency or Doppler shifts ν on the transmitted signal $s(t)$. This time-frequency spreading of the signal can be quantified using the narrowband spreading function (SF) based representation, that has been widely used, specifically for radar signal design (Vakman, 1968), pulse shaping in multicarrier communications (Bölcskei, Duhamel, & Hleiss, 1999; Haas & Belfiore, 1997; Kozek & Molisch, 1998; Matz, Schafhuber, Gröchenig, Hartmann, & Hlawatsch, 2007), and wireless channel modeling (Giannakis & Tepedelenlioglu, 1998; Sayeed & Aazhang, 1999; Ye & Papandreou-Suppappola, 2006). Using the narrowband SF, $S_{\mathbf{H}}(\tau,\nu)$, the (noise free) received signal,

$$r(t) = (\mathbf{H}s)(t) = \int_{-\infty}^{\infty}\int_{-\infty}^{\infty} S_{\mathbf{H}}(\tau,\nu)s(t-\tau)e^{j2\pi\nu t}\,d\tau\,d\nu \tag{9.1}$$

can be physically interpreted in terms of the time-shift operation $s(t-\tau) = (\mathbf{S}_\tau s)(t)$ and the frequency-shift operation $s(t)e^{j2\pi\nu t} = (\mathbf{M}_\nu s)(t)$, which are the effects of the channel on the transmitted signal (Bello, 1963; Bhashyam, Sayeed, & Aazhang, 2000; Eggen et al., 2000; Sayeed & Aazhang, 1999; Sostrand, 1968; Ziomek, 1985).

9.2.2 Discrete Channel Characterization and Finite Approximations

Although Eqn (9.1) can completely characterize LTV channels, it is not very practical to use in simulations as it is based on continuous formulations of the signal transformations. Using some physical assumptions on the channel, an equivalent discrete representation has been derived in terms of sampled time shifts and sampled frequency shifts (Bello, 1963; Kailath, 1959; Sayeed & Aazhang, 1999). The structure of this discrete canonical formulation can be exploited in order to analyze the channel and improve receiver performance. This has already been demonstrated by the tapped-delay line model, which is a representation in terms of discrete time shifts. The tapped-delay line model decomposes frequency-selective channels into multiple independent flat fading channels, leading to the well-known rake receiver and multipath diversity (Kennedy, 1969; Proakis, 2001).

The canonical time-frequency channel model provides a discrete representation of the narrowband channel **H** as a weighted summation of discrete time shifts m and discrete frequency shifts d (Bhashyam et al., 2000; Sayeed & Aazhang, 1999). This model decomposes a time-frequency selective channel into independent flat fading channels, and it can be used to obtain joint multipath-Doppler diversity by extending the rake receiver to a time-frequency rake receiver formulation (Bello, 1963; Bhashyam et al., 2000; Sayeed & Aazhang, 1999). The baseband transmitted signal $s(t)$ in (9.1) is assumed band-limited to $[-B/2, B/2]$, and without loss of generality, the reception of the first symbol during $[0,T]$ is considered. We also consider a doubly selective channel with maximum multipath spread $\tau_{\max} \ll T$ and maximum Doppler spread $\nu_{\max}$. Then, the representation in (9.1) can be decomposed as

$$r(t) = \frac{1}{TB}\sum_{m=0}^{M-1}\sum_{d=-D}^{D} \tilde{S}_{\mathbf{H}}\left(\frac{m}{B},\frac{d}{T}\right)s\left(t-\frac{m}{B}\right)e^{j2\pi(d/T)t}, \tag{9.2}$$

where

$$\tilde{S}_{\mathbf{H}}(\tau,\nu) = TB \int_{0}^{\tau_{\max}} \int_{-\nu_{\max}}^{\nu_{\max}} S_{\mathbf{H}}(\tau',\nu') e^{-j2\pi(\nu-\nu')T} \operatorname{sinc}(\pi(\nu-\nu')T) \operatorname{sinc}(\pi(\tau-\tau')B) \, d\nu' \, d\tau'$$

are two-dimensional (2D) samples of the smoothed SF, sampled at the uniform grid $\tau = m/B$ and $\nu = d/T$. The number of summation terms in (9.2) is determined from $M = \lceil \tau_{\max} B \rceil$ and $D = \lceil \nu_{\max} T \rceil$ (where $\lceil a \rceil$ is the smallest integer not less than a). Following the (effectively) time-limited and band-limited nature of the transmitted signal, the representation in (9.2) decomposes the channel into $M(2D+1)$ slow, flat fading subchannels by appropriately sampling the multipath-Doppler (or time-frequency lag) plane. The sampling intervals are $1/B$ along the multipath or time-lag axis and $1/T$ along the Doppler or frequency-lag axis (as demonstrated in Fig. 9.7(a)).

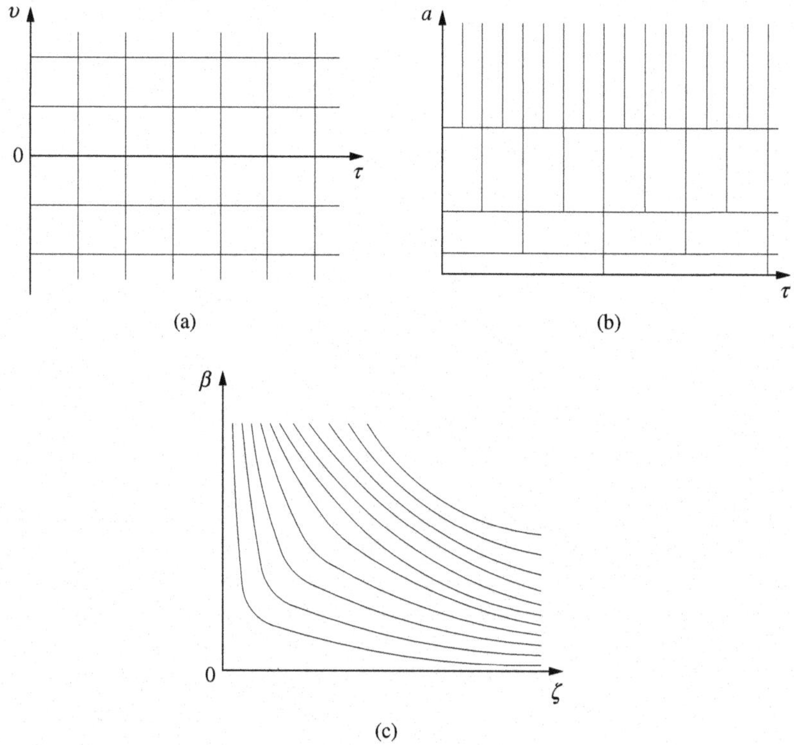

FIGURE 9.7

Spreading function representations for LTV systems in different types of environments: (a) Narrowband SF sampling in the time-frequency lag plane according to Eqn (9.2). (b) Wideband SF sampling in the delay-scale plane according to Eqn (9.9). (c) Dispersive SF sampling according to Eqn (9.22), showing shifts in instantaneous frequency $\beta\eta'(t/t_r)$.

For the wide-sense stationary uncorrelated scatterer (WSSUS) channel model, the correlation of the samples $\tilde{S}_{\mathsf{H}}(m/B, d/T)$ in (9.2) can be represented in terms of a double integral involving the scattering function $C_{\mathsf{H}}(\tau, \nu)$ (Ye & Papandreou-Suppappola, 2006),

$$\mathsf{E}\left\{\tilde{S}_{\mathsf{H}}\left(\frac{m}{B},\frac{d}{T}\right)\tilde{S}^{*}_{\mathsf{H}}\left(\frac{m'}{B},\frac{d'}{T}\right)\right\} = \frac{T^2B^2}{e^{j\pi(d-d')}}\int_{0}^{\tau_{\max}}\int_{-\nu_{\max}}^{\nu_{\max}} C_{\mathsf{H}}(\tau,\nu)\,\mathrm{sinc}\left(\pi\left(d-\nu T\right)\right)$$
$$\times\,\mathrm{sinc}\left(\pi\left(d'-\nu T\right)\right)\mathrm{sinc}\left(\pi\left(m-\tau B\right)\right)\mathrm{sinc}\left(\pi\left(m'-\tau B\right)\right)d\nu\, d\tau.$$

This shows that when the scattering function varies slowly with respect to τ (on the order of $1/B$) and with respect to ν (on the order of $1/T$), the coefficients are approximately uncorrelated across different time shifts and frequency shifts, that is,

$$\mathsf{E}\left\{\tilde{S}_{\mathsf{H}}\left(\frac{m}{B},\frac{d}{T}\right)\tilde{S}^{*}_{\mathsf{H}}\left(\frac{m'}{B},\frac{d'}{T}\right)\right\} \approx TB\, C_{\mathsf{H}}\left(\frac{m}{B},\frac{d}{T}\right)\delta_{mm'}\,\delta_{dd'},$$

where δ_{ij} is the Kronecker delta (defined to be 1 when $i = j$ and 0 when $i \neq j$).

9.3 WIDEBAND DELAY-SCALE CHANNEL CHARACTERIZATION

9.3.1 Narrowband and Wideband Conditions

Just like all LTV channels, if the channel has wideband characteristics, it can still be represented in terms of the channel's time-varying impulse response or using the SF representation in (9.1). However, this representation is in terms of time-frequency shifts and thus better matched to narrowband channels. For wideband LTV channels characterized by time delays and time-scale changes that describe the physical effect of the channel on the transmitted signal, better matched representations in terms of the wideband SF are needed (Iem et al., 2002; Margetts and Schniter, 2004; Shenoy & Parks, 1995; Sibul, Weiss, & Dixon, 1994; Weiss, 1994; Ye & Papandreou-Suppappola, 2003; Young, 1993).

The combined effects of multipath and mobility on transmitted signals are modeled differently in narrowband channels than in wideband channels. Specifically, in narrowband channels, when a dense ring of scatterers surrounds the receiver, then mobility causes a spreading of the signal spectrum that results in a constant frequency shift. On the other hand, in wideband channels, mobility results in time-domain and frequency-domain scale spreading. In general, movement of the scatterers in a communication channel causes Doppler scaling (compression or expansion) on the transmitted signal. So, although in many cases the Doppler effect can be approximated as a frequency shift, it actually corresponds to a time scale and frequency scale change (compression or dilation) on the transmitted signal. The Doppler scale $\alpha \in \mathbb{R}$ is directly related to the radial velocity υ of the scatterer and the velocity c of the signal in the propagation medium according to the relationship $\alpha = (c-\upsilon)/(c+\upsilon)$ (Weiss, 1994). The *narrowband condition* approximates the Doppler scale by a Doppler shift or frequency shift that is given by

$$f_d \approx (\alpha - 1) f_c,$$

where f_c is the carrier frequency of the signal. This approximation is valid if two conditions hold. The first condition requires the signal to have a small fractional bandwidth (ratio of the signal bandwidth

B to the center frequency) (Margetts, Schniter, & Swami, 2007). For example, for ultra-wideband (UWB) systems, the fractional bandwidth is typically greater than 0.25 (Win & Scholtz, 2000). The second condition requires the target to move slowly such that its position does not change by much when compared to the positional resolution of the signal. In particular, the second condition is given by

$$TB \ll \frac{c}{2v},$$

where TB is the time-bandwidth product of the signal. When both conditions do not hold (Weiss, 1994), then even a very small scale value can make a difference when computing the effect of the channel on the transmitted signal. Thus, a wideband channel would be more appropriately characterized in terms of Doppler scale changes together with multipath delays.

9.3.2 Wideband Spreading Function

For a wideband LTV system $\mathbf{H}$ defined on $L^2(\mathbb{R})$ with an input signal $s(t)$, the output $r(t)$ can be characterized by the superposition of time shifts τ and scale changes α, weighted by the wideband or delay-scale SF, $F_{\mathbf{H}}(\tau,\alpha)$. Specifically,

$$r(t) = \int_0^\infty \int_{-\infty}^\infty F_{\mathbf{H}}(\tau,\alpha)\,\sqrt{\alpha}s(\alpha(t-\tau))\,d\tau\,d\alpha. \tag{9.3}$$

Thus, reflections off fast moving point scatterers result in a wideband signal that is a time-delayed and time-scaled version of the transmitted signal. For most cases $|v| < c$, thus $\alpha > 0$ in (9.3). When $\alpha < 1$, the scatterer is moving away from the receiver (resulting in a time expansion); when $\alpha > 1$, the scatterer is approaching the receiver (resulting in a time compression). The wideband SF in (9.3) represents the reflection strength of the scatterers.

When the channel is randomly varying, a wideband scattering function, $B_{\mathbf{H}}(\tau,\alpha)$, can be used to measure the second-order statistics of the wideband SF (Balan, Poor, Rickard, & Verdu, 2004; Margetts & Schniter, 2004; Ye & Papandreou-Suppappola, 2003, 2005). If the transmitted energy is scattered uncorrelatedly in the time-scale domain, it can be shown that

$$\mathsf{E}\{F_{\mathbf{H}}(\tau,\alpha)F^*_{\mathbf{H}}(\tau',\alpha')\} = B_{\mathbf{H}}(\tau,\alpha)\delta(\alpha-\alpha')\delta(\tau-\tau'), \tag{9.4}$$

where $\delta(\cdot)$ is the Dirac impulse.

9.3.3 Discrete Delay-Scale Channel Characterization

Just as the discrete time-frequency model can efficiently improve narrowband processing (as discussed in Section 9.2), a discrete delay-scale or wideband channel characterization can be useful in analyzing and processing wideband channel outputs and in providing increased performance by model-inherent diversity paths. Such a characterization decomposes a wideband LTV channel output into discrete time shifts and time-scale changes on the input signal, weighted by a smoothed and sampled version of the wideband SF.

There are two discretization methods for wideband system characterizations developed independently in the literature. The first method is transform-based, and it uses the Mellin transform that is inherently matched to scale changes (Bertrand, Bertrand, & Ovarlez, 1996) to geometrically sample the scale parameter and the Fourier transform (FT) to arithmetically sample the time delay parameter (Ye & Papandreou-Suppappola, 2003, 2005, 2006). The second method uses an operator approach, where it was shown that by time-shifting with a band-limiting projection and time-scaling with a scale-limiting projection, a bounded wideband operator can be decomposed into a weighted sum of operators (Balan et al., 2004; Rickard, 2003). In this section, we concentrate on the transform-based discretization approach as it provides an intuitive physical interpretation of the sampled signal transformations, given physical constraints on the wideband channel. This leads to a finite expansion that, with an appropriately designed signal reception scheme, can be used to exploit multipath-scale diversity.

For the transform-based approach, we first note that the choice of the Mellin transform conforms with the nature of the wideband Doppler scale transformation. In the same way that the FT is invariant to time shifts (up to a phase shift), the Mellin transform is invariant to time-scale changes (up to a phase shift) (Altes & Titlebaum, 1970; Bertrand et al., 1996; Flandrin, 1999). As a result, the Mellin transform has been widely used in applications involving signal compression or dilation. For example, it has been used to estimate wideband signal parameters (Ovarlez, 1993) and to compute affine (time-scale) time-frequency representations (Ovarlez, Bertrand, & Bertrand, 1992). The Mellin transform of a signal $s(t)$ is defined as (Bertrand et al., 1996; Papandreou, Hlawatsch, & Boudreaux-Bartels, 1993)

$$\mathscr{M}_s(q) = \int_0^\infty \frac{1}{\sqrt{t}} s(t) \exp(j2\pi q \ln(t/t_r)) dt, \tag{9.5}$$

where $q \in \mathbb{R}$ and $t_r > 0$ is a reference time. The aforementioned scale invariance property of the Mellin transform, up to a phase shift, states that if a signal is time-scaled, $\sqrt{\alpha}\, s(\alpha t)$, $\alpha > 0$, then the Mellin transform of the scaled signal is given by $e^{-j2\pi \ln \alpha} \mathscr{M}_s(q)$. Another important property that is used in the development of the discrete wideband channel model is the multiplicative convolution (Bertrand et al., 1996; Cohen, 1993) given by

$$s_1(t) \circledast s_2(t) \triangleq \frac{1}{\sqrt{t_r}} \int_0^\infty s_1(t\,t'/t_r)\, s_2^*(t')\, dt' \quad \leftrightarrow \quad \mathscr{M}_{s_1}(q)\, \mathscr{M}_{s_2}^*(q).$$

This property states that the multiplicative convolution between two signals corresponds to the product of their Mellin transforms in the Mellin domain.

For the discrete time-scale channel characterization, we sample the delay-scale parameters in (9.3) by assuming finite bounds in the Fourier and Mellin domains. Thus, we assume that the FT $S(f)$ of the transmitted signal $s(t)$ is supported within $f \in [-B/2, B/2]$ and that the Mellin transform $\mathscr{M}_s(q)$ of $s(t)$ is supported within $q \in [-q_0/2, q_0/2]$ (and thus localized in the time-frequency plane, as shown in Bertrand et al., 1996). Using these support assumptions, the 2D integral representation in (9.3) can be decomposed into the following 2D summation

$$r(t) = \sum_{\kappa \in \mathbb{Z}} \sum_{m \in \mathbb{Z}} \tilde{F}_{\mathbf{H}}\left(\frac{m}{\alpha_0^\kappa B}, \alpha_0^\kappa\right) \alpha_0^{\kappa/2} s\left(\alpha_0^\kappa t - \frac{m}{B}\right), \tag{9.6}$$

where the basic scaling factor is $\alpha_0 = e^{1/q_0}$ and $\tilde{F}_{\mathbf{H}}(\tau, \alpha)$ is a 2D smoothed version of the wideband SF $F_{\mathbf{H}}(\tau, \alpha)$. This smoothed wideband SF is given by

$$\tilde{F}_{\mathbf{H}}(\tau, \alpha) = \int_0^\infty \int_{-\infty}^\infty F_{\mathbf{H}}(\tau', \alpha') \operatorname{sinc}(\pi \alpha B(\tau - \tau')) \operatorname{sinc}\left(\pi \frac{\ln \alpha - \ln \alpha'}{\ln \alpha_0}\right) d\tau' d\alpha'. \tag{9.7}$$

Note that the weighting coefficients $\tilde{F}_{\mathbf{H}}(m/(\alpha_0^\kappa B),\alpha_0^\kappa)$ in (9.6) are delay-scale samples of the smoothed wideband SF $\tilde{F}_{\mathbf{H}}(\tau,\alpha)$. Specifically, the time-scale parameter is geometrically sampled as $\alpha=\alpha_0^\kappa$, $\kappa \in \mathbb{Z}$, and for a fixed κ, the delay parameter is uniformly sampled as $\tau = m/(\alpha_0^\kappa B)$, $m \in \mathbb{Z}$. The detailed steps of the geometric sampling using the Mellin transform can be found in Ye & Papandreou-Suppappola (2006). Summarized, the steps include taking the Mellin transform of an appropriately defined intermediate function, applying the multiplicative convolution property of the Mellin transform and the inverse Mellin transform to the result, and then expanding the Mellin transform using a Fourier series expansion. The Fourier series expansion coefficients, within the finite Mellin support domain $[-q_0/2, q_0/2]$, lead to the geometric sampling $\alpha=\alpha_0^\kappa$, where κ is an integer and $\alpha_0 = \mathrm{e}^{1/q_0} > 1$ is the basic scaling factor. The delay parameter is uniformly sampled using a similar procedure but based on the FT (Ye & Papandreou-Suppappola, 2006).

9.3.4 Finite Approximations

In real scenarios, channels have various physical restrictions such as path loss or velocity limits. As a result, we assume that the wideband SF has a compact support, i.e.,

$$F_{\mathbf{H}}(\tau,\alpha)=0, \quad \forall\,(\tau,\alpha)\notin[0,\tau_{\max}]\times[\alpha_1,\alpha_2].$$

Here, $\tau_{\max}$ is the maximum delay, and $[\alpha_1,\alpha_2]$ is the range of possible time-scale values. With this compact support, (9.3) is more realistically written as

$$r(t)=\int_{\alpha_1}^{\alpha_2}\int_0^{\tau_{\max}} F_{\mathbf{H}}(\tau,\alpha)\,\sqrt{\alpha}s(\alpha(t-\tau))\,\mathrm{d}\tau\,\mathrm{d}\alpha.$$

As a result, the infinite summation in (9.6) can be approximated by

$$r(t)\approx\sum_{\kappa=\kappa_1}^{\kappa_2}\sum_{m=0}^{M(\kappa)}\tilde{F}_{\mathbf{H}}\left(\frac{m}{\alpha_0^\kappa B},\alpha^\kappa\right)\alpha_0^{\kappa/2}\,s\left(\alpha_0^\kappa\,t-\frac{m}{B}\right)=\sum_{\kappa=\kappa_1}^{\kappa_2}\sum_{m=0}^{M(\kappa)}\tilde{F}_{\mathbf{H}}[m,\kappa]s_{m,\kappa}(t), \tag{9.8}$$

where the summation limits are determined as $\kappa_1=\lfloor\ln\alpha_1/\ln\alpha_0\rfloor$, $\kappa_2=\lceil\ln\alpha_2/\ln\alpha_0\rceil$, and $M(\kappa)=\lceil\alpha_0^\kappa\;B\;\tau_{\max}\rceil$. In (9.8), the sampled smoothed wideband SF (as demonstrated in Fig. 9.7(b)) is given by

$$\tilde{F}_{\mathbf{H}}[m,\kappa]\triangleq\tilde{F}_{\mathbf{H}}\left(\frac{m}{\alpha_0^\kappa B},\alpha^\kappa\right),\quad \kappa=\kappa_1,\ldots,\kappa_2,\ m=0,1,\ldots,M(\kappa), \tag{9.9}$$

and the transmitted signal with the sampled wideband transformations is given by

$$s_{m,\kappa}(t)=\alpha_0^{\kappa/2}s\left(\alpha_0^\kappa\,t-\frac{m}{B}\right),\quad \kappa=\kappa_1,\ldots,\kappa_2,\ m=0,1,\ldots,M(\kappa). \tag{9.10}$$

The resulting discrete time-scale model corresponding to (9.8) is demonstrated in Fig. 9.8. The output $r(t)$ can be viewed as the superposition of $(\kappa_2-\kappa_1+1)$ parallel tapped-delay lines. The tapped-delay line corresponding to the κth sampled scale $\alpha=\alpha_0^\kappa$, $\kappa=\kappa_1,\ldots,\kappa_2$, consists of $M(\kappa)$

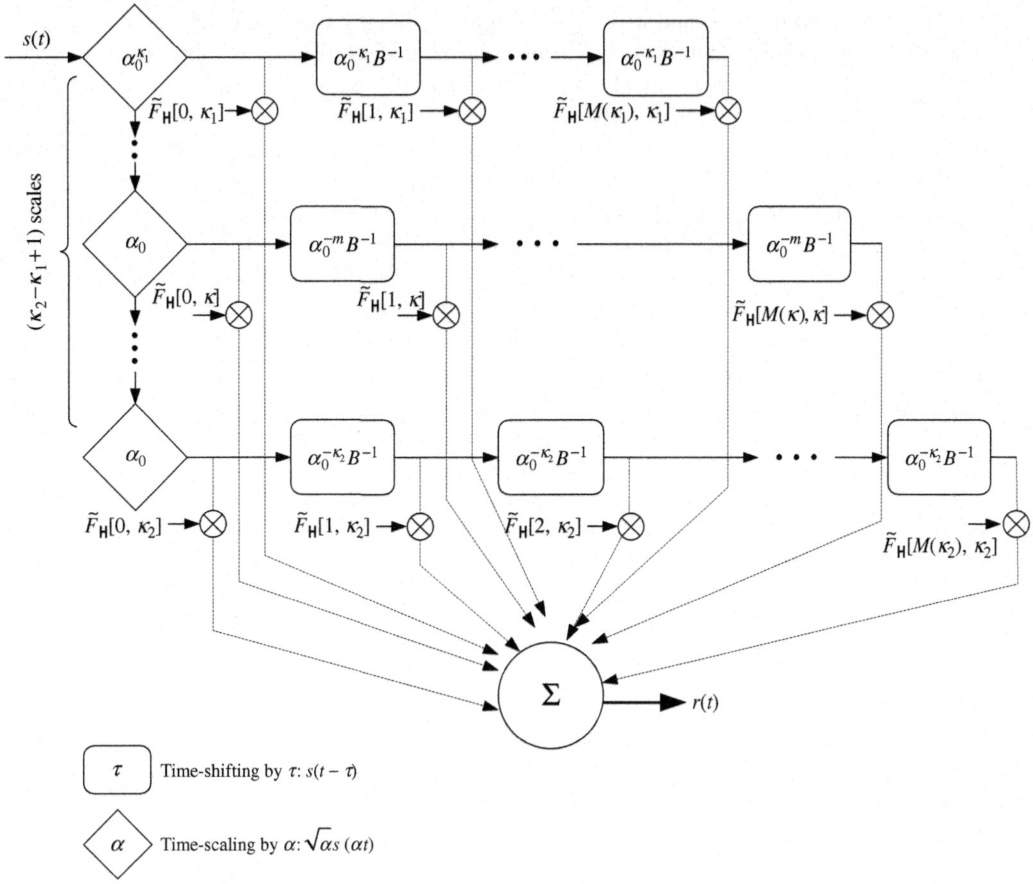

FIGURE 9.8

The transmitted signal $s(t)$ is scaled $(\kappa_2 - \kappa_1 + 1)$ times, and the α_0^κ scale, $\kappa = \kappa_1, \ldots, \kappa_2$, undergoes $M(\kappa)$ time shifts when propagating over the wideband channel model in Eqn (9.8). Each path is weighted by the wideband SF sample $\tilde{F}_{\mathsf{H}}[m,\kappa]$, $m = 0, \ldots, M(\kappa)$.

sampled time shifts, $\tau = m/(\alpha_0^\kappa B)$, $m = 1, \ldots, M(\kappa)$. The corresponding weighting coefficient is $\tilde{F}_{\mathsf{H}}[m,\kappa]$ in (9.9).

9.3.5 Multipath-Scale Diversity

With the discretized wideband channel representation in (9.8) now available, new opportunities arise in exploiting diversity when the narrowband condition does not hold. For example, for underwater communications based on high-speed moving platforms, where time-varying scattering and multipath propagation cause fading degradations (Stojanovic, 1996), the time-frequency model of multipath and

Doppler will not provide the assumed diversity. Currently, in many existing underwater communication systems, Doppler scale changes and multipath are treated as distortions rather than potential diversity sources. As a result, different techniques are used to compensate for the assumed distortion after estimation (Johnson et al., 1997; Sharif et al., 2000) or to suppress the distortion by adaptive equalization (Stojanovic et al., 1994). However, instead of trying to eliminate these inherent wideband channel effects, we can exploit them in a joint multipath-scale diversity, as discussed next. We note that the application of the discrete wideband channel model was used in Margetts and Schniter (2004) to obtain a diversity gain based on a direct sequence spread spectrum ultra-wideband system.

We now derive the diversity order that can be achieved using the discrete wideband model. We consider a single user transmitting a baseband signal with binary phase-shift keying (BPSK) modulation, $\sum_i b[i]\, s(t-iT)$, where $b[i] = \pm 1$ is the ith information-bearing symbol and T is the symbol duration. We choose $T \gg \tau_{\max}$, where $\tau_{\max}$ is the maximum channel delay, to avoid intersymbol interference so that each symbol can be decoded independently. As a result, we can continue our analysis with only one symbol b, and we can choose the transmission signal to be $bs(t)$. We also assume that the FT of $s(t)$ is supported within $[-B/2, B/2]$ and that the Mellin transform $\mathscr{M}_s(q)$ of $s(t)$ is supported within $[-q_0/2, q_0/2]$. For a wideband channel with physical constraints of a maximum delay $\tau_{\max}$ and Doppler scale spread $\kappa_2 - \kappa_1$, and using the decomposition in Eqn (9.8), the corresponding noisy received signal can be expressed as

$$r(t) \approx b \sum_{\kappa=\kappa_1}^{\kappa_2} \sum_{m=0}^{M(\kappa)} \tilde{F}_{\mathsf{H}}[m,\kappa]\, s_{m,\kappa}(t) + w(t). \tag{9.11}$$

From our previous analysis of the multipath sampling and its dependence on the scale sampling, given the signal bandwidth B, the multipath resolution in a wideband channel is $1/(\alpha B)$ when the signal is scaled by α; in the narrowband case, the resolution would be the fixed value $1/B$.

Under the wideband WSSUS assumption in (9.4), and following (9.7), it can be shown that

$$\begin{aligned} \mathsf{E}\{\tilde{F}_{\mathsf{H}}[m,\kappa], \tilde{F}^*_{\mathsf{H}}[m',\kappa']\} = \int_{\ln\kappa_1}^{\ln\kappa_2} \int_0^{\tau_{\max}} B_{\mathsf{H}}(\tau, e^{\gamma})\, e^{2\gamma}\, \mathrm{sinc}\left(\pi\left(\frac{\gamma}{\ln\alpha_0} - \kappa\right)\right) \\ \times\ \mathrm{sinc}\left(\pi\left(\frac{\gamma}{\ln\alpha_0} - \kappa'\right)\right) \mathrm{sinc}\left(\pi(m - \alpha_0^{\kappa} B\tau)\right) \mathrm{sinc}\left(\pi(m' - \alpha_0^{\kappa'} B\tau)\right) d\tau\, d\gamma, \end{aligned}$$

where $B_{\mathsf{H}}(\tau,\alpha)$ is the wideband scattering function. If $B_{\mathsf{H}}(\tau, e^{\gamma})\, e^{2\gamma}$ in the above equation is sufficiently smooth and varies slowly with changes in γ (on the order of $\ln\alpha_0$), and if it also varies slowly with changes in τ (on the order of $1/(\alpha_0^{\kappa} B)$) for a fixed κ, then the channel coefficients can be assumed mutually uncorrelated over different sampled delays and scales. Specifically, we can obtain

$$\mathsf{E}\{\tilde{F}_{\mathsf{H}}[m,\kappa] \tilde{F}^*_{\mathsf{H}}[m',\kappa']\} \approx B_{\mathsf{H}}\left(\frac{m}{\alpha_0^{\kappa} B}, \alpha_0^{\kappa}\right) \alpha_0^{2\kappa}\, \delta_{mm'} \delta_{\kappa\kappa'}.$$

Using the central limit theorem, when we have a very large number of scatterers, then the discrete channel coefficient random variables $\tilde{F}_{\mathsf{H}}[m,\kappa]$ can be assumed Gaussian and therefore independent

(Kennedy, 1969). Thus, (9.11) results in effectively decomposing the wideband channel into

$$\Upsilon \triangleq \sum_{\kappa=\kappa_1}^{\kappa_2} [M(\kappa)+1] \tag{9.12}$$

independent, flat-fading subchannels. These Υ subchannels have the potential to yield a joint multipath-scale diversity of order Υ that can be exploited to increase system performance. The time-scale rake receiver described next achieves this diversity if the subchannels are properly combined. Note that when $\kappa = 0$ in (9.12), then no scaling occurs, and the time-scale rake receiver simplifies to the conventional rake receiver (Ye & Papandreou-Suppappola, 2006).

9.3.6 Time-Scale Rake Receiver

In order to achieve the aforementioned time-scale diversity, the transmitted signal must be designed such that the signals $s_{m,\kappa}(t)$ in (9.10) are orthonormal. For the design, because the time-scaling effect in (9.10) is equal to the transformation underlying dyadic wavelet bases, we use an orthonormal wavelet-based design scheme (Ye & Papandreou-Suppappola, 2003) with dyadic scalings (basis $\alpha_0 = 2$) given by $\alpha = 2^\kappa$, $\kappa \in \mathbb{Z}$. In particular, we let $\psi(t)$ be a wavelet function (Flandrin, 1999) and

$$\psi_{m,\kappa}(t) = 2^{\kappa/2}\psi(2^\kappa t - m), \quad m,\kappa \in \mathbb{Z}$$

constitute an orthonormal basis for $L^2(\mathbb{R})$, i.e., $\{\psi_{m,\kappa}(t)\}$ is complete in $L^2(\mathbb{R})$ and

$$\int_{-\infty}^{\infty} \psi_{m,\kappa}(t)\psi^*_{m',\kappa'}(t)\,\mathrm{d}t = \delta_{mm'}\delta_{\kappa\kappa'}, \quad \forall m,m',\kappa,\kappa' \in \mathbb{Z}.$$

We design the transmit signal to be a scaled version of the wavelet function, given by

$$s(t) = \frac{1}{\sqrt{T_w}}\,\psi\left(\frac{t}{T_w}\right). \tag{9.13}$$

As a result, assuming $B \approx 1/T_w$ and fixing $\alpha_0 = 2$ in (9.10), then it can be shown that $s(t)$ in (9.13) induces an orthonormal basis since

$$\int_{-\infty}^{\infty} s_{m,\kappa}(t)s^*_{m',\kappa'}(t)\mathrm{d}t = \frac{1}{T_w}\int_{-\infty}^{\infty} \psi_{m,\kappa}\left(\frac{t}{T_w}\right)\psi^*_{m',\kappa'}\left(\frac{t}{T_w}\right)\mathrm{d}t = \delta_{mm'}\delta_{\kappa\kappa'}, \quad \forall m,m',\kappa,\kappa' \in \mathbb{Z}.$$

The dyadic sampling with $\alpha_0 = 2$ in (9.6) is explicitly demonstrated in Fig. 9.9. As we have assumed that $s(t)$ is bounded in the Mellin transform domain within $q \in [-q_0/2, q_0/2]$ and $\alpha_0 = 2 = \mathrm{e}^{1/q_0}$, then for dyadic sampling, $s(t)$ and $\psi(t)$ in (9.13) must both be bounded in the Mellin domain within $q \in [-0.5/\ln 2, 0.5/\ln 2]$. This condition is satisfied, for example, when $\psi(t)$ is the Haar wavelet (Flandrin, 1999).

Using maximum ratio combining, we can compute the coherent detector of the time-scale rake receiver (Ye & Papandreou-Suppappola, 2003, 2006). Specifically, for the mth delay and κth scale diversity component, the detection statistic $\lambda_{m,\kappa}$ can be obtained as the inner product between the received signal $r(t)$ and $s_{m,\kappa}(t)$ in (9.10) and corresponds to the dyadic time-scale sampled wavelet

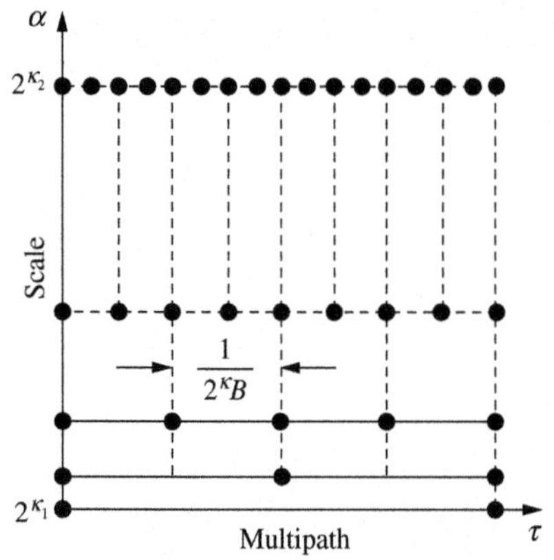

FIGURE 9.9

Dyadic sampling of the smoothed wideband SF, $\tilde{F}_{\mathbf{H}}(m/(2^{\kappa}B), 2^{\kappa})$, in the time-scale plane with $\alpha = 2^{\kappa}$ scale resolution and $1/(2^{\kappa}B)$ multipath resolution for fixed κ.

transform. That is, $\lambda_{m,\kappa} = W_{rs}(m/(2^{\kappa}B), 2^{\kappa})$, where the cross wavelet transform of $r(t)$ and $s(t) \in L^2(\mathbb{R})$ is defined as (Flandrin, 1999)

$$W_{rs}(\tau,\alpha) = \int_{-\infty}^{\infty} r(t')\, \sqrt{\alpha} s^*(\alpha(\tau - t'))\, \mathrm{d}t', \quad \alpha > 0.$$

The implementation of $\lambda_{m,\kappa}$ is demonstrated in Fig. 9.10 using a bank of matched filters obtained by time-shifting and time-scaling the transmit signal $s(t)$. For signal detection, the channel coefficients can be combined coherently to obtain an estimate of the transmitted information symbol b as

$$\hat{b} = \text{sign}\left(\text{Re}\left\{ \sum_{\kappa=\kappa_1}^{\kappa_2} \sum_{m=0}^{M(\kappa)} \lambda_{m,\kappa}\, \tilde{F}^*_{\mathbf{H}}[m,\kappa] \right\} \right).$$

In order to demonstrate the importance of the discrete delay-scale model when the channel is wideband, we next compare the performance of the time-scale rake receiver and the time-frequency rake receiver using simulations. We use a transmitted signal $s(t)$ with bandwidth $B = 2.5\,\text{kHz}$ that is designed based on the Haar wavelet according to Eqn (9.13). The signal is assumed to be transmitted in an underwater acoustic mobile channel, where the sound velocity in water is $c = 1500\,\text{m/s}$, and rapidly moving platforms such as autonomous underwater vehicles act as scatterers and move with velocity $v = 15\,\text{m/s}$ (Sharif et al., 2000). The maximum delay of the channel is $\tau_{\max} = 0.8\,\text{ms}$, and the Doppler scale spread is $A_s = \alpha_2 - \alpha_1$. From this relation, we can deduce that, given the scale spread, $\alpha_1 = 1 + A_s$ and $\alpha_2 = 1 - A_s$. For the discretization, we use the dyadic scale basis $\alpha_0 = 2$ in (9.8). In determining the total number of possible diversity paths, we need to consider that the κth scale path, $\kappa = \kappa_1, \ldots, \kappa_2$, results in $M(\kappa) + 1$ delay paths, where $M(\kappa) = \lceil \alpha_0^{\kappa}\, B\, \tau_{\max} \rceil$, and there are $\kappa_2 - \kappa_1 + 1$ scale paths.

As an example, if $\alpha_1 = 0.98$ and $\alpha_2 = 1.02$ (so that $A_s = 0.04$), then $\kappa_1 = \lfloor \ln\alpha_1 / \ln\alpha_0 \rfloor = -1$, $\kappa_2 = \lceil \ln\alpha_2 / \ln\alpha_0 \rceil = 1$, and $M(\kappa) = \lceil \alpha_0^{\kappa}\, B\, \tau_{\max} \rceil = 2^{\kappa+1}$, $\kappa = -1, 0, 1$. The $\kappa = -1$ scale results in the $m =$

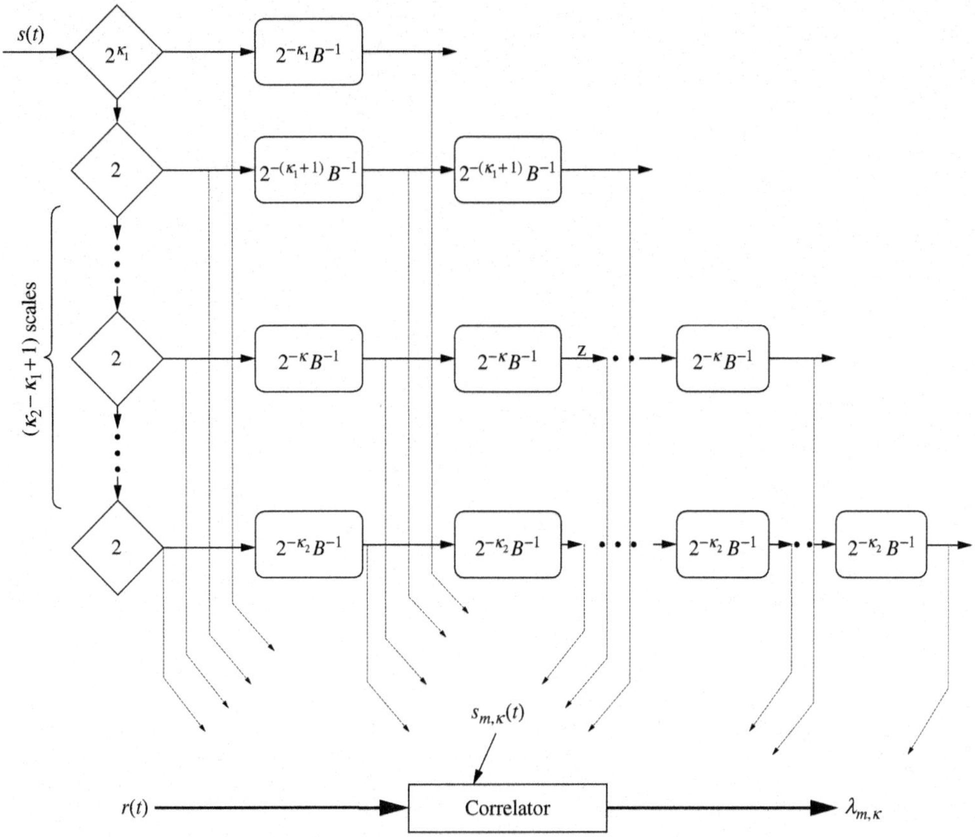

FIGURE 9.10

Bank of matched filters used to compute the detection statistic $\lambda_{m,\kappa}$ for the mth delay and κth scale diversity component.

$0,1$ delays since $M(-1) = 2^0 = 1$; the $\kappa = 0$ scale results in the $m = 0,1,2$ delays since $M(0) = 2^1 = 2$; and the $\kappa = 1$ scale results in the $m = 0,1,2,3,4$ delays since $M(1) = 2^2 = 4$. Thus, for this example, the possible time-scale diversity order is $\Upsilon = 10$; this is obtained by counting all the paths or by using (9.12). Note that the conventional rake receiver assumes that the transmit signal experiences only multipath changes and that there is no scale spread. The assumption of $A_s = 0$ results in $\alpha_1 = \alpha_2 = 1$, $\kappa_1 = \kappa_2 = 0$, $M(0) = 2$ and thus in a $\Upsilon = 3$ diversity order in (9.12).

Figure 9.11 shows the bit-error-rate (BER) performance improvement obtained for varying average signal-to-noise ratio (SNR) for different scale spread values (but constant diversity order) with coherent maximum ratio combining reception. As we can observe, as the scale spread increases, the SNR gains also increase. Specifically, when $A_s = 1$, corresponding to about a third of the total power in the $\kappa = \pm 1$ Doppler scale paths, the conventional rake receiver would require 25 dB more SNR to achieve the same 10^{-5} BER as the time-scale rake receiver. For the aforementioned example with a scale spread $A_s = 0.04$, an 8 dB SNR gain is achieved by the time-scale rake receiver over the conventional rake receiver.

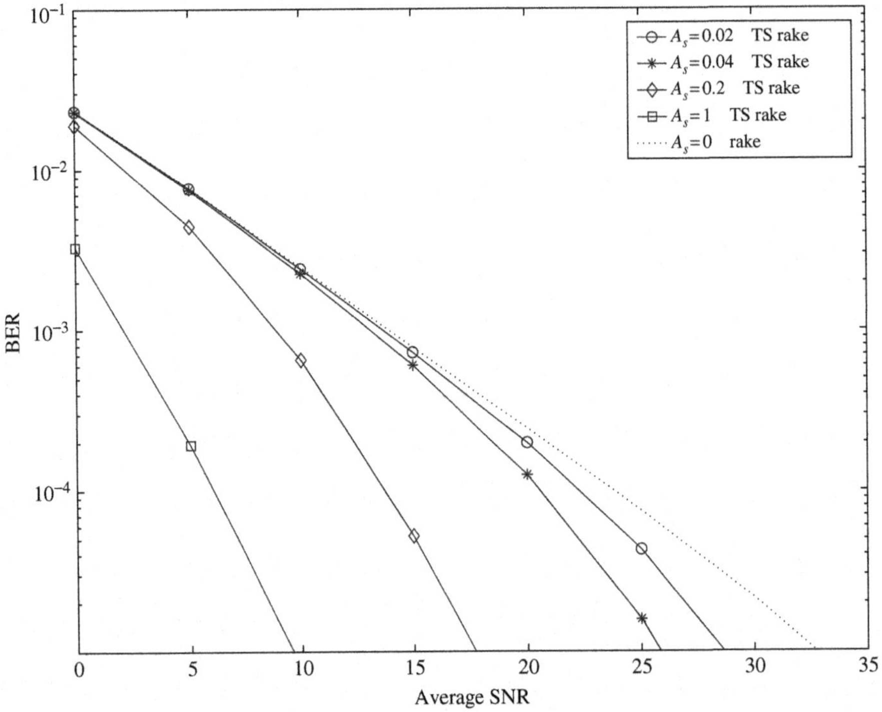

FIGURE 9.11

BER performance for varying average SNR, demonstrating the diversity gains of the time-scale (TS) rake receiver over the conventional rake receiver. For the time-scale rake receivers, the scale spread A_s takes values from $\{0.02, 0.04, 0.2, 1\}$, and all yield a diversity order $\Upsilon = 10$. For the conventional rake receiver, $A_s = 0$ and the diversity order is $\Upsilon = 3$.

9.4 WIDEBAND DISPERSIVE CHANNEL CHARACTERIZATION

9.4.1 Dispersive Time-Frequency Structures

In nature, there exist many channels with dispersive time-frequency structures that are not well-matched to the aforementioned narrowband and wideband channel models. Such dispersive signal transformations are specific to the nature of the channel or environment that the signal propagates in. These transformations can be characterized by the nonlinear change $\eta(t/t_r)$ in the phase function of the signal, where $t_r > 0$ is a time reference point. Some examples of such dispersive changes can be found in shallow water acoustic environments, where the frequency of the signal is shifted nonlinearly with time (Chen, James, Boudreaux-Bartels, Gopu, & Colin, 2003; Tolstoy & Clay, 1966), and in the dispersive propagation of a shock wave in a steel beam that causes changes in the phase function by $\eta(t/t_r) = |t/t_r|^{1/2}$ (Newland, 1998). Other examples of dispersive systems include dielectric media (e.g., white light through a prism), optical channels (Agazzi, Hueda, Carrer, & Crivelli, 2005), ultrasonic testing systems, radio waves in the ionosphere (Freeman, Dunham, & Qian, 1995), and acoustic waves reflected from underwater spherical shells (Sessarego, Sageloli, Flandrin, & Zakharia, 1989). In such scenarios, characterizations using the narrowband or wideband SFs do not provide accurate representations of dispersive channels. Instead, matched SFs that depend on $\eta(t/t_r)$ need to be employed to provide a pointwise measure of how the transmission energy spreads in the time-frequency plane (Iem et al., 2002). As a result, the dispersive channel output needs to be modeled as a superposition of dispersive signal transformations, weighted by this matched dispersive SF. As it can be shown, the SF used for characterizing narrowband LTV channels can be extended to dispersive channels through a unitary transformation that depends on this characteristic phase function (Iem, 1998; Iem et al., 2002; Ye & Papandreou-Suppappola, May 2006, 2007).

9.4.2 Dispersive Spreading Function and Unitary Warping Relations

When a signal $s(t)$ is transmitted over a dispersive channel $\mathbf{Z}$, its phase function can be changed by a nonlinear time-varying amount $\eta(t/t_r)$. This change in phase results in a (possibly) nonlinear shift in the instantaneous frequency of the signal that is given by $\nu(t) \triangleq \frac{\mathrm{d}}{\mathrm{d}t}\eta(t/t_r)$. The dispersive SF, $D_{\mathbf{Z}}(\zeta,\beta)$, can be used to characterize a dispersive channel in terms of the nonlinear instantaneous frequency shifts (Iem et al., 2002). Specifically, the channel output $r(t)$ can be written as

$$r(t) = (\mathbf{Z}s)(t) = \int_{-\infty}^{\infty}\int_{-\infty}^{\infty} D_{\mathbf{Z}}(\zeta,\beta)\left(\mathbf{D}_{\beta}^{(\eta)}\mathbf{G}_{\zeta}^{(\eta)}s\right)(t)\,\mathrm{d}\zeta\,\mathrm{d}\beta, \tag{9.14}$$

where the transformation parameters ζ and β are dimensionless. The dispersive frequency-shift operation in (9.14) is defined as

$$\left(\mathbf{D}_{\beta}^{(\eta)}s\right)(t) = \left(\mathbf{U}_{\eta}^{-1}\mathbf{M}_{\beta/t_r}\mathbf{U}_{\eta}s\right)(t) = s(t)\,\mathrm{e}^{j2\pi\beta\eta(t/t_r)}, \tag{9.15}$$

where $(\mathbf{M}_{\nu}s)(t) = s(t)\,\mathrm{e}^{j2\pi\nu t}$ is the frequency-shift operation in (9.1). The generalized time-shift operation in (9.14) is given by

$$\left(\mathbf{G}_{\zeta}^{(\eta)}s\right)(t) = \left(\mathbf{U}_{\eta}^{-1}\mathbf{S}_{t_r\zeta}\mathbf{U}_{\eta}s\right)(t) \tag{9.16}$$

$$= \left|\frac{\nu(t)}{\nu\left(t_r\eta^{-1}\left(\eta\left(\frac{t}{t_r}\right)-\zeta\right)\right)}\right|^{1/2} s\left(t_r\eta^{-1}\left(\eta\left(\frac{t}{t_r}\right)-\zeta\right)\right), \tag{9.17}$$

where $(\mathbf{S}_{\tau}s)(t) = s(t-\tau)$ is the time-shift operation in (9.1). The unitary warping operator $\mathbf{U}_{\eta}$ in (9.15) and (9.17) is defined as

$$(\mathbf{U}_{\eta}s)(t) = \left|t_r\nu\left(t_r\eta^{-1}\left(\frac{t}{t_r}\right)\right)\right|^{-1/2} s\left(t_r\eta^{-1}\left(\frac{t}{t_r}\right)\right), \tag{9.18}$$

where $(\mathbf{U}_{\eta}^{-1}\mathbf{U}_{\eta}s)(t) = s(t)$. The warping relationship assumes that $\eta(t/t_r)$ is a one-to-one function with $\eta^{-1}(\eta(t/t_r)) = t/t_r$.

From (9.14), it is seen that a dispersive channel $\mathbf{Z}$ is uniquely characterized by its specific function $\eta(t/t_r)$, which in turn defines the unitary warping operator in (9.18), and by its dispersive SF $D_{\mathbf{Z}}(\zeta,\beta)$. The unitary warping operator in (9.18) plays an important role in relating the narrowband and dispersive channel characterizations. As shown in (9.15) and (9.16), to obtain the generalized frequency-shift and generalized time-shift operations, the frequency-shift and time-shift operations are warped using the warping operator in (9.18); the operator directly depends on the channel's characteristic function $\eta(t/t_r)$. It was shown in Iem et al. (2002) that the dispersive SF is related to the narrowband SF using the same warping. Specifically, if $\mathbf{Z}$ represents a dispersive channel, then $\mathbf{U}_\eta \mathbf{Z} \mathbf{U}_\eta^{-1}$ represents an equivalent narrowband channel, and the dispersive SF of $\mathbf{Z}$ is equivalent to the narrowband spreading function of $\mathbf{U}_\eta \mathbf{Z} \mathbf{U}_\eta^{-1}$. This relationship can be expressed as (Iem et al., 2002)

$$D_{\mathbf{Z}}(\zeta,\beta) = S_{\mathbf{U}_\eta \mathbf{Z} \mathbf{U}_\eta^{-1}}(t_r\zeta,\beta/t_r). \tag{9.19}$$

The generalized frequency-shift or instantaneous frequency-shift operation in (9.15) depends on the channel over which the transmitted signal propagates. For example, if the channel is such that the signal undergoes hyperbolic instantaneous frequency shifts $e^{j2\pi\beta\ln(t/t_r)}s(t)$, then the corresponding generalized time-shift operation in (9.17) is a scaling transformation.

9.4.3 Discrete Dispersive Channel with Physical Limitations

We consider next the discretization of the dispersive channel representation for processing. For the generalized dispersive channel model in (9.14), we follow the transform-based discretization approach, similar to the methods used for the narrowband and wideband delay-scale models. For these two models, the transforms used were chosen to match the transformations caused by the channel on the transmitted signal. For example, as narrowband channels cause constant frequency shifts, the transform used to sample the narrowband channel parameters is the Fourier transform. On the other hand, the Mellin transform was used for delay-scale channels that cause a transmit signal to be compressed or dilated. The corresponding transform that is used for discretizing generalized dispersive channels will be referred to as the generalized or matched transform. This transform can be considered a linear expansion of a signal $s(t)$ into nonlinear frequency-modulated (FM) signals, in the same way that the Fourier transform is a linear expansion of a signal $s(t)$ into single-frequency or sinusoidal signals. The generalized transform is given by (Hlawatsch, Papandreou-Suppappola, & Boudreaux-Bartels, 1999; Papandreou-Suppappola, Hlawatsch, & Boudreaux-Bartels, 1998; Papandreou-Suppappola et al., 2001)

$$\aleph_s^{(\eta)}(q) \triangleq \int_{t\in\wp} |\nu(t)|^{1/2} s(t) e^{-j2\pi\, q\,\eta(t/t_r)}\, dt, \tag{9.20}$$

with $\nu(t) = \frac{d}{dt}\eta(t/t_r)$ and $q \in \mathbb{R}$, for a given monotonic function $\eta(t/t_r)$ with domain $\wp$. If a dispersive environment causes a change in phase by $\eta(t/t_r)$ to the transmitted signal, then the transform in (9.20) that depends on the same function $\eta(t/t_r)$ is ideally matched to the characterization and discretization of the channel model. For example, when $\eta(t/t_r) = t/t_r$, this is simply the Fourier transform that is matched to narrowband signal transformations such as simple frequency shifts, as seen in Section 9.2. On the other hand, when $\eta(t/t_r) = \ln(t/t_r)$, $t > 0$, the corresponding transform is the Mellin transform in (9.5) that is matched to wideband scale changes, as we have seen in Section 9.3.

Using the unitary warping operator $\mathbf{U}_\eta$ in (9.18), it can be shown that the generalized transform is a warped version of the Fourier transform, i.e.,

$$\aleph_s^{(\eta)}(f t_r) = (\mathbf{F}\mathbf{U}_\eta s)(f),$$

where $\mathbf{F}$ is the Fourier transform operator. Using this relation, together with the relation between the narrowband SF and the dispersive SF in (9.19), we can first apply warping on the received signals and then use the narrowband model to discretize the dispersive channel (Ye & Papandreou-Suppappola, May 2006, 2007; Ye, Shen, & Papandreou-Suppappola, 2006).

We assume that the generalized transform $\aleph_s^{(\eta)}(q)$ of the transmitted signal $s(t)$ in (9.20) is supported within $q \in [q_1, q_2]$, with a width $Q \triangleq q_2 - q_1$. We also assume that the time-warped signal $(\mathbf{U}_\eta \mathbf{Z} s)(t)$ has a bounded time support $t \in [t_r\gamma_1, t_r\gamma_2]$, with a normalized duration $\Gamma \triangleq \gamma_2 - \gamma_1$. Within this support, the discrete version of the dispersive model (9.14) is

$$r(t) = \sum_{x\in\mathbb{Z}} \sum_{y\in\mathbb{Z}} \tilde{D}_{\mathbf{Z}}\left(\frac{x}{Q}, \frac{y}{\Gamma}\right) \left(\mathbf{D}_{y/\Gamma}^{(\eta)}\, \mathbf{G}_{x/Q}^{(\eta)} s\right)(t), \tag{9.21}$$

where the operators $\mathbf{D}_\beta^{(\eta)}$ and $\mathbf{G}_\zeta^{(\eta)}$ are defined in (9.15) and (9.17), respectively. The smoothed dispersive SF in (9.21) is related to the smoothed narrowband SF in (9.2) using the warping relation in (9.19).

For a real channel that admits a finite approximation due to physical constraints such that the dispersive SF is nonzero only when $(\zeta, \beta) \in [\zeta_1, \zeta_2] \times [\beta_1, \beta_2]$, the summation limits in (9.21) become finite. Specifically, the smoothed dispersive SF coefficients $\tilde{D}_{\mathbf{Z}}(x/Q, y/\Gamma)$ are significantly nonzero only when the function's mainlobe supports, $[(x-1)/Q, (x+1)/Q]$ and $[(y-1)/\Gamma, (y+1)/\Gamma]$, are effectively overlapped with $[\zeta_1, \zeta_2]$ and $[\beta_1, \beta_2]$, respectively. The resulting summation limits and discretized model are thus given by

$$r(t) = \sum_{x=\lfloor \zeta_1 Q \rfloor}^{\lceil \zeta_2 Q \rceil} \sum_{y=\lfloor \beta_1 \Gamma \rfloor}^{\lceil \beta_2 \Gamma \rceil} \tilde{D}_{\mathbf{Z}}[x, y] \left(\mathbf{D}_{y/\Gamma}^{(\eta)}\, \mathbf{G}_{x/Q}^{(\eta)} s\right)(t), \tag{9.22}$$

where

$$\tilde{D}_{\mathbf{Z}}[x, y] \triangleq \tilde{D}_{\mathbf{Z}}\left(\frac{x}{Q}, \frac{y}{\Gamma}\right).$$

9.4.4 Generalized Time-Frequency Rake Receiver

The dispersive nature of channels, such as shallow water, would normally require the communication system designer to attempt to compensate for the dispersion effects. However, when the matched discrete model in (9.22) is used and the transmitted signal is appropriately designed, then the performance will actually improve and dispersion will not be a nuisance. In order to make use of the discrete channel model in (9.22) to achieve performance gains, we need to identify the characteristic function $\eta(t/t_r)$ of the channel. In Ye & Papandreou-Suppappola (2007), this was specifically demonstrated for a shallow water acoustic communication channel (Chen et al., 2003; Tolstoy & Clay, 1966). The signaling was chosen to be direct sequence spread spectrum (DSSS), and the transformations of the DSSS signal performed by the channel can be approximated by dispersive frequency shift and time delay. When propagating through the dispersive channel, the delayed-dispersed signal $s_{x,y}(t) = (\mathbf{D}^{(\eta)}_{y/\Gamma}\,\mathbf{G}^{(\eta)}_{x/Q}\,s)(t)$ can be seen as a summation of weighted orthogonal basis functions. Thus, a rake receiver, analogous to the time-scale rake receiver described in Section 9.3.6, can resolve different diversity components and achieve performance gains. Specifically, for coherent detection, maximum ratio combining can be used, where the entries of the sufficient statistics vector can be combined in an optimal way to maximize the output SNR.

9.5 UNDERWATER WIRELESS COMMUNICATION CHANNELS

Models for underwater acoustic propagation environments are well-documented in the literature (Tolstoy & Clay, 1966; Urick, 1975; Ziomek, 1985, 1999). However, communication systems in these environments are often being designed by considering dispersion as a distortion that needs to be compensated for. The dispersive effect can severely limit the performance of underwater acoustic communications unless the channel is appropriately taken into consideration (Beaujean & LeBlanc, 2004; Morozov, Preisig, & Papp, 2008; Morozov, Preisig, & Papp, 2008, December; Preisig, 2003, 2005; Preisig & Johnson, 2001; Stojanovic & Preisig, 2009). In this section, we consider in detail a specific communication channel with wideband dispersive characteristics that extensively benefits from the matched dispersive models we developed.

Note that the wideband dispersive channel model in Section 9.4 is in terms of the time-domain transmitted signal. As a result, the considered dispersion is with respect to the instantaneous frequency of the signal. As the underwater acoustic propagation models were developed in the frequency domain, in this section, we will consider corresponding dispersive channel models in terms of the Fourier transform of the transmitted signal. Thus, the dispersion in this section will be with respect to the group delay of the signal.

9.5.1 Shallow Water Environment Model

In underwater acoustics, various waveguide models were developed to describe the effects of the ocean environment on transmitted signals (Tolstoy & Clay, 1966; Urick, 1975; Ziomek, 1985, 1999). When the environment corresponds to shallow water, the model finds the response of an omnidirectional point source $s(t)$, with Fourier transform $S(f)$, in the environment. Models of different complexity can be found in the literature. In particular, the isovelocity model assumes a perfect waveguide and models the ocean surface as an ideal pressure release boundary and the ocean bottom as an ideal rigid boundary (Ziomek, 1999). A more realistic scenario would require a more advanced model such as the Pekeris model (Zhang, Gottin, Papandreou-Suppappola, & Ioana, 2007; Zhang et al., 2009) that characterizes the shallow water environment with a pressure-release surface and a fluid seabed (Deane, 1997; Frisk, 1994; Ziomek, 1999).

As depicted in Fig. 9.12, we consider a scenario with both the transmitter and the receiver under shallow water and a coordinate system consisting of range r and depth z. We assume that the depths of the ocean, surface, transmitter, and receiver are z_{ocean}, 0, z_{tx}, and z_{rx}, respectively. The ranges of the transmitter and receiver are 0 and r_{rx}, respectively. Following the isovelocity model, the received signal spectrum $R(f)$ at location $(r_{\text{rx}}, z_{\text{rx}})$ is a linear combination of $\mathcal{N}$ mode transformations of the

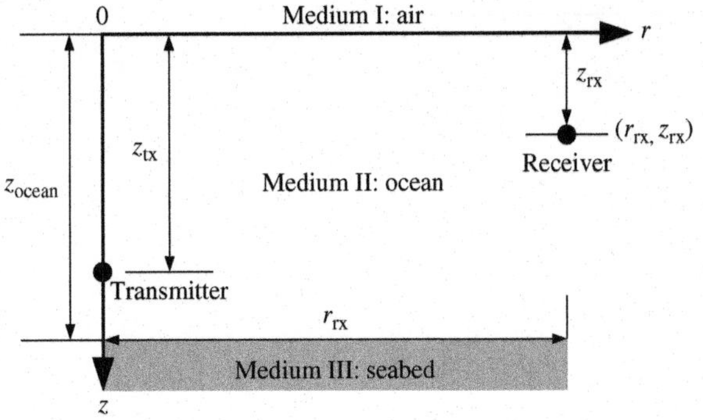

FIGURE 9.12

Shallow water isovelocity acoustic model when a signal is transmitted from a point source.

transmitted signal spectrum $S(f)$, and it is given by (Ziomek, 1999)

$$R(f) = \sum_{\varsigma=1}^{\mathcal{N}} U_\varsigma(f) = \sum_{\varsigma=1}^{\mathcal{N}} S(f)\Theta_\varsigma(f) \tag{9.23}$$

$$= \sum_{\varsigma=1}^{\mathcal{N}} S(f)\rho_\varsigma\left(\frac{r_{rx}}{c} f_r \vartheta_\varsigma\left(\frac{f}{f_r}\right)\right)^{-\frac{1}{2}} e^{-j2\pi \frac{r_{rx}}{c} f_r \vartheta_\varsigma\left(\frac{f}{f_r}\right)}. \tag{9.24}$$

Here, $U_\varsigma(f) \triangleq S(f)\Theta_\varsigma(f)$ represents the ςth mode of the received signal spectrum, $f_r > 0$ is a normalization frequency, c is the constant speed of sound in the ocean, and

$$\Theta_\varsigma(f) = \rho_\varsigma\left(\frac{r_{rx}}{c} f_r \vartheta_\varsigma\left(\frac{f}{f_r}\right)\right)^{-\frac{1}{2}} e^{-j2\pi \frac{r_{rx}}{c} f_r \vartheta_\varsigma\left(\frac{f}{f_r}\right)} \tag{9.25}$$

is the ςth dispersive change on the transmitted signal spectrum, which includes both a dispersive amplitude modulation term and a dispersive phase change. The ςth phase change is given by

$$\vartheta_\varsigma\left(\frac{f}{f_r}\right) \triangleq \begin{cases} \left[\left(\frac{f}{f_r}\right)^2 - \left(\frac{f_\varsigma}{f_r}\right)^2\right]^{\frac{1}{2}}, & f > f_\varsigma \\ 0, & \text{otherwise} \end{cases} \tag{9.26}$$

where

$$f_\varsigma = \frac{(2\varsigma - 1)c}{4 z_{ocean}} \tag{9.27}$$

is the cutoff frequency of the ςth mode,

$$\rho_\varsigma = e^{-j\frac{\pi}{4}} \frac{z_r}{z_{ocean}} \sin\left(\frac{(2\varsigma - 1)\pi z_{tx}}{2 z_{ocean}}\right) \sin\left(\frac{(2\varsigma - 1)\pi z_{rx}}{2 z_{ocean}}\right)$$

is constant for the ςth mode, and $z_r > 0$ is a normalization constant.

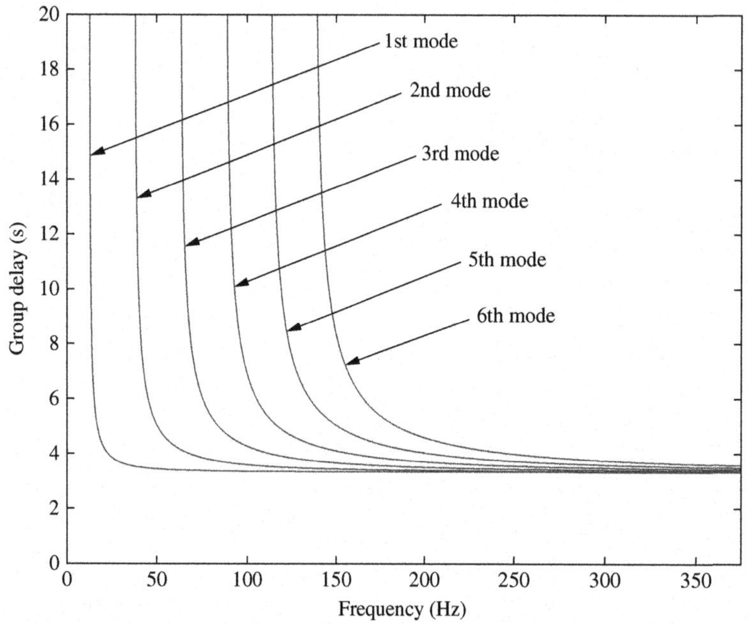

FIGURE 9.13

Group delay shifts in (9.28) for $\mathcal{N} = 6$ modes, following the isovelocity model.

9.5.2 Time-Frequency Characteristics of Shallow Water Environments

The transformation at the receiver in (9.24) corresponds to a linear combination of $\mathcal{N}$ nonlinear phase changes of the transmitted signal spectrum $S(f)$, caused by the ςth, $\varsigma = 1,\ldots,\mathcal{N}$, mode propagation components in the horizontal radial direction; equivalently, these correspond to a linear combination of $\mathcal{N}$ group delay shifts of the transmitted signal $s(t)$. Thus, each of the $\mathcal{N}$ mode components in the acoustic shallow water environment model in (9.24) introduces a different dispersive transformation by causing a different shift in the group delay of the transmitted signal. The group delay shift depends on the cutoff frequency of the ς th mode in (9.27), and it is given by the derivative of the phase change in (9.26) as

$$\xi_\varsigma(f) = \frac{\mathrm{d}}{\mathrm{d}f}\vartheta_\varsigma\left(\frac{f}{f_r}\right) = \frac{f}{\left(f^2 - f_\varsigma^2\right)^{1/2}} \tag{9.28}$$

for $f > f_\varsigma$ and $\varsigma = 1,\ldots,\mathcal{N}$, and zero otherwise.

The effect of the dispersive channel is depicted for $\mathcal{N} = 6$ modes in Fig. 9.13. From (9.28), $\mathcal{N} = 6$ different modal frequencies are shifted by different amounts in time as the transmitted signal is scattered at different speeds. This results in $\mathcal{N} = 6$ dispersive curves in the time-frequency plane, each of which is characterized by f_ς, $\varsigma = 1,\ldots,6$, in (9.27).

9.5.3 Mode Separation in Time-Frequency

If we want to use the shallow water channel model for processing wireless communication signals, it is important to be able to separate the different modes in (9.24). For example, the mode separation would be needed to achieve dispersion diversity and improve communication performance. Our objective is thus to design the transmitted signal $s(t)$ in such a way as to yield $\mathcal{N}$ distinct dispersive curves in the time-frequency plane at the receiver, corresponding to the $\mathcal{N}$ modes of the shallow water model in (9.24).

Many techniques have been published in the literature on how to separate sinusoidal signal components in the time-frequency plane. Based on this, we want to design the transmitted signal spectrum $S(f)$ such that, after transmission, the received ςth mode component of the received signal spectrum corresponds to a frequency-domain sinusoid whose modulation rate depends on the ςth mode parameters in (9.24). Following this idea, we design the transmitted signal spectrum to be $S(f) = \sqrt{f/f_r}$, for $f > 0, f_r > 0$; using this transmit signal, the resulting noiseless received signal spectrum $R(f)$, after propagation over the shallow water channel, is given by (9.24).

In order to match the ςth mode and extract the ςth sinusoid, we apply the unitary warping operator $\mathbf{J}_\varsigma$ to the received signal. The warping operation is defined by the resulting signal transformation

$$\begin{aligned}\mathcal{R}_\varsigma(f) &= (\mathbf{J}_\varsigma R)(f) = \mathcal{G}_\varsigma(f) R\left(f_r \vartheta_\varsigma^{-1}\left(\frac{f}{f_r}\right)\right) = \mathcal{G}_\varsigma(f) \sum_{\varsigma'=1}^{\mathcal{N}} U_{\varsigma'}\left(f_r \vartheta_\varsigma^{-1}\left(\frac{f}{f_r}\right)\right) \\ &= \mathcal{G}_\varsigma(f) U_\varsigma\left(f_r \vartheta_\varsigma^{-1}\left(\frac{f}{f_r}\right)\right) + \sum_{\substack{\varsigma'=1 \\ \varsigma' \neq \varsigma}}^{\mathcal{N}} (\mathbf{J}_\varsigma U_{\varsigma'})(f),\end{aligned} \tag{9.29}$$

where $\mathcal{G}_\varsigma(f) = \left| f_r \xi_\varsigma\left(f_r \vartheta_\varsigma^{-1}(f/f_r)\right)\right|^{-1/2}$, $\vartheta_\varsigma(f/f_r)$ is defined in (9.26), and its inverse function is given by $\vartheta_\varsigma^{-1}(f/f_r) = \left((f/f_r)^2 + (f_\varsigma/f_r)^2\right)^{1/2}$. Replacing (9.24) with $S(f) = \sqrt{f}, f > 0$, in (9.29), we obtain

$$\begin{aligned}\mathcal{R}_\varsigma(f) &= \mathcal{G}_\varsigma(f) \rho_\varsigma S\left(f_r \vartheta_\varsigma^{-1}\left(\frac{f}{f_r}\right)\right) \Theta_\varsigma\left(f_r \vartheta_\varsigma^{-1}\left(\frac{f}{f_r}\right)\right) + \sum_{\substack{\varsigma'=1 \\ \varsigma' \neq \varsigma}}^{\mathcal{N}} (\mathbf{J}_\varsigma U_{\varsigma'})(f) \\ &= \mathcal{G}_\varsigma(f) \rho_\varsigma \left|\vartheta_\varsigma^{-1}\left(\frac{f}{f_r}\right)\right|^{\frac{1}{2}} \Theta_\varsigma\left(f_r \vartheta_\varsigma^{-1}\left(\frac{f}{f_r}\right)\right) + \sum_{\substack{\varsigma'=1 \\ \varsigma' \neq \varsigma}}^{\mathcal{N}} (\mathbf{J}_\varsigma U_{\varsigma'})(f).\end{aligned}$$

Using $\xi_\varsigma(f)$ in (9.28) to compute $\mathcal{G}_\varsigma(f)$ and replacing $\Theta_\varsigma(f)$ with (9.25), we can further simplify $\mathcal{R}_\varsigma(f)$ to obtain

$$\mathcal{R}_\varsigma(f) = \rho_\varsigma \sqrt{c/(f_r r_{\mathrm{rx}})}\, \mathrm{e}^{-j2\pi \frac{r_{\mathrm{rx}}}{c} f} + \sum_{\substack{\varsigma'=1 \\ \varsigma' \neq \varsigma}}^{\mathcal{N}} (\mathbf{J}_\varsigma U_{\varsigma'})(f). \tag{9.30}$$

The warped received signal will yield a single frequency-domain sinusoid if the warping operator $\mathbf{J}_\varsigma$ in (9.29) matches the ςth shallow water mode; the sinusoid will then have modulation rate r_{rx}/c, where c is the speed of sound in water and r_{rx} is the range of the receiver. The warped received signal will also consist of dispersive signal terms, corresponding to the remaining $\mathcal{N} - 1$ warped signal modes. However, these terms have highly nonlinear group delay functions, and only the matched ςth mode will appear as a line in the time-frequency plane at $t = r_{rx}/c$. In order to separate all $\mathcal{N}$ mode components, successive warping operations, for $\varsigma = 1, \ldots, \mathcal{N}$, would need to be applied.

In real channels, a signal propagating through a shallow water environment will also experience fluctuations from the ocean surface and roughness from the ocean bottom that can affect the signal reflections. We can include these distortion effects in the isovelocity model in (9.24) by introducing the random fading coefficient H_ς. We also include zero-mean, additive white Gaussian noise (AWGN) $W(f)$, with variance σ_w^2, to account for random disturbances in the ocean environment. The received signal is thus more realistically given by $R(f) = \sum_{\varsigma=1}^{\mathcal{N}} H_\varsigma S(f)\Theta_\varsigma(f) + W(f)$ where $\Theta_\varsigma(f)$ is defined in (9.25).

We simulated data using the *KRAKEN* software (Porter, 1991; Porter & Duncan, 2002) to demonstrate the dispersive time-frequency signatures of the modes and the warping effect in (9.30). This software was written to model underwater acoustic propagation for a given transmitted signal and specific environmental parameters (similar to those in Fig. 9.12). Shown in Fig. 9.14(a) is the spectrogram time-frequency representation (TFR) of the received (noiseless) signal obtained using the *KRAKEN* software when the transmitted signal is $S(f) = \sqrt{f/f_r}$, for $f > 0$ and $f_r = 1$ Hz. The speed of sound in water was fixed to $c = 1500\,\text{m/s}$, and we assumed that the signal was dispersed over $\mathcal{N} = 3$ modes, following the isovelocity channel model in (9.24). Based on that model, lower frequencies are shifted in time by larger amounts than higher frequencies. Each mode is represented as a dispersive (nonlinear) curve in the time-frequency plane, and the modes are not easily separable. In Figs 9.14(b)–9.14(d), we computed the spectrograms of the warped received signals $\mathscr{R}_1(f)$, $\mathscr{R}_2(f)$, and $\mathscr{R}_3(f)$, respectively, in (9.30). In each one of these three plots, we can distinctly see a frequency-domain sinusoid that behaves as an impulse function, or a wideband pulse, in the time domain. Each one of these impulses is thus highly localized at r_{rx}/c s, according to (9.30). Note, also, that when the ςth warping, $\varsigma = 1,2,3$, is applied to obtain the ςth sinusoid, the other two modes, $\varsigma' = 1,2,3$, $\varsigma' \neq \varsigma$, remain dispersive, especially at low frequencies. At high frequencies, even without the warping in Fig. 9.14(a), the modes behave as wideband pulses since the phase becomes $(f^2 - f_\varsigma^2)^{1/2} \approx f$ provided $f \gg f_\varsigma$. This is the case, for example, when the signal is transmitted in a high-frequency band, and the resulting modes are closer to each other in the time-frequency plane. As a result, the modes are almost impossible to separate at high frequencies, and the warping operation is best applied in the low-frequency region (Gottin, Zhang, Papandreou-Suppappola, & Ioana, 2007; Zhang et al., 2009). Also, the modes are easier to separate if they are more dispersive, and they are more dispersive if the transmission distance is longer.

9.5.4 Dispersion Diversity Receiver Design

The shallow water channel model with the multiple dispersive time-frequency mode characteristics in (9.24) provides a natural diversity path that can be exploited for communication. We specifically design a receiver to detect a binary antipodal symbol, $b = \pm 1$, after it is modulated with an appropriate signal $S(f)$ and transmitted over a shallow water channel. The transmit waveform is designed to be $S(f) = \sqrt{f/f_r}$, for $f > 0$. The corresponding received signal is given by

$$R(f) = \sum_{\varsigma=1}^{\mathcal{N}} H_\varsigma b S(f)\Theta_\varsigma(f) + W(f). \tag{9.31}$$

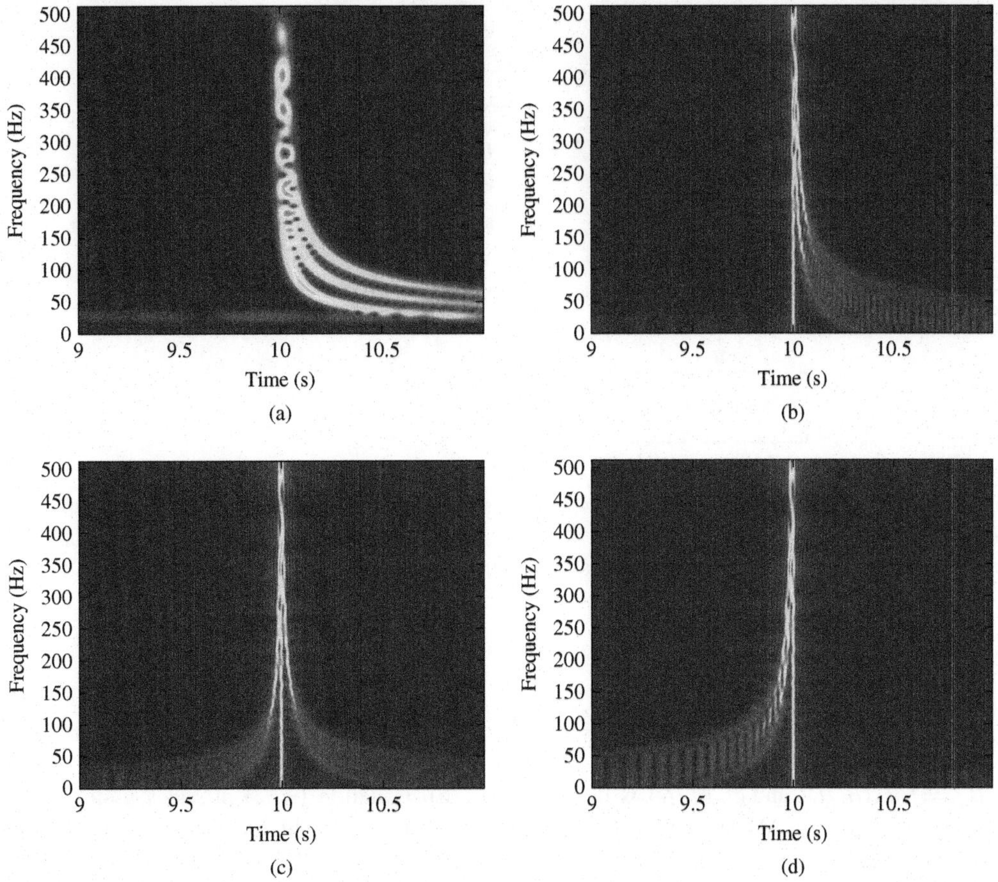

FIGURE 9.14

Spectrogram of (a) a received signal $R(f)$ that was simulated using the KRAKEN software with the isovelocity shallow water model in (9.24), (b) warped signal $\mathscr{R}_1(f)$, (c) warped signal $\mathscr{R}_2(f)$, and (d) warped signal $\mathscr{R}_3(f)$ in (9.30) with $\varsigma = 1,2,3$.

At the receiver, in order to determine the information symbol b, we first matched-filter (MF) the received signal $R(f)$ with $\mathcal{N}$ dispersive mode components and then combine the MF outputs in the minimum-error-probability (MEP) sense.

The MEP receiver algorithm is depicted in Fig. 9.15 and described next in more detail (Zhang & Papandreou-Suppappola, 2007). We first assume that the channel characteristics and the transmitted signal, up to the unknown information bit, are known at the receiver. This implies that the ςth component, $U_\varsigma(f) = S(f)\Theta_\varsigma(f)$ in (9.23), is also known at the receiver and corresponds to the ςth expected received component, without fading and AWGN, assuming that $b = +1$ was transmitted. With this assumption, we form the ςth MF output as $z_\varsigma = \langle R, U_\varsigma \rangle = \int_{B_r} R(f) U_\varsigma^*(f) \mathrm{d}f$. Here, B_r is the effective bandwidth of the received signal spectrum. Due to the unitarity of the warping operator in (9.29), the

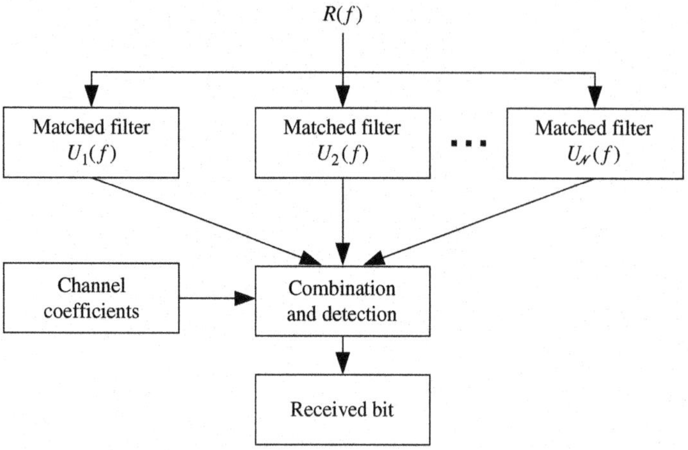

FIGURE 9.15

Receiver design for the shallow water channel: the received signal spectrum $R(f)$ is matched filtered with each of the dispersive mode components $U_\varsigma, \varsigma = 1,\ldots,\mathcal{N}$, in (9.23); the resulting $\mathcal{N}$ matched filtered outputs are then combined, in the minimum-error-probability sense, to make a decision on the received bit.

ςth MF output can also be computed using the warped received signal spectrum as

$$z_\varsigma = \langle R, U_\varsigma \rangle = \langle \mathbf{J}_\varsigma R, \mathbf{J}_\varsigma U_\varsigma \rangle = \int_{B_{wr}} (\mathbf{J}_\varsigma R)(f)\, \rho_\varsigma \sqrt{c/(f_r r_{\mathrm{rx}})}\, e^{j2\pi \frac{r_{\mathrm{rx}}}{c} f}\, df, \tag{9.32}$$

where $B_{wr} = f_r \vartheta_\varsigma(B_r/f_r)$ and $\vartheta_\varsigma(f/f_r)$ is defined in (9.26). Equation (9.32) provides a simplified implementation of the ςth MF since the ςth warped mode $(\mathbf{J}_\varsigma U_\varsigma)(f)$, which corresponds to the first term on the right-hand side of (9.30), is a frequency-domain sinusoid with modulation rate r_{rx}/c, i.e., $(\mathbf{J}_\varsigma U_\varsigma)(f) = \rho_\varsigma \sqrt{c/r_{\mathrm{rx}}}\, e^{j2\pi (r_{\mathrm{rx}}/c) f}$. Since processing now involves sinusoidal signals, (9.32) can also be implemented as a short-time Fourier transform TFR of the warped received signal $(\mathbf{J}_\varsigma R)(f)$ using a frequency-domain rectangular window with amplitude $c/(f_r r_{\mathrm{rx}})$.

For the MEP detector formulation, we use the following vector notation. We form the $\mathcal{N} \times 1$ vector $\mathbf{u}(f) \triangleq (U_1(f) \cdots U_{\mathcal{N}}(f))^T$ by concatenating the $U_\varsigma(f)$ components in (9.23); here, T denotes vector transpose. Combining the MF outputs z_ς in (9.32), we obtain the $\mathcal{N} \times 1$ MF vector $\mathbf{z} \triangleq (z_1 \cdots z_{\mathcal{N}})^T$. We also consider the information vector $b\mathbf{1}$, where $\mathbf{1}$ is the $\mathcal{N} \times 1$ vector of ones, and the $\mathcal{N} \times \mathcal{N}$ diagonal matrix $\mathbf{H} \triangleq \mathrm{diag}\{H_1,\ldots,H_{\mathcal{N}}\}$, whose diagonal elements are the random channel coefficients H_ς in (9.31). The received signal can thus be written as

$$R(f) = \mathbf{u}^T(f)\, \mathbf{H}\, b\mathbf{1} + W(f). \tag{9.33}$$

Using this notation, the output of the MF can be given as $\mathbf{z} = \mathbf{P}\mathbf{H}b\mathbf{1} + \mathbf{w}$, where the $\mathcal{N} \times \mathcal{N}$ matrix $\mathbf{P} \triangleq \int_{B_r} \mathbf{u}^*(f)\,\mathbf{u}^T(f)\, df$ contains the correlation values between the different modes, and $\mathbf{w} \triangleq \int_{B_r} \mathbf{u}^*(f)\, W(f)\, df$ is the noise at the MF outputs with covariance $\sigma_w^2 \mathbf{P}$. If we denote $\mathbf{d} \triangleq \mathbf{H}\, b\mathbf{1}$ in (9.33), then $\mathbf{z} = \mathbf{P}\mathbf{d} + \mathbf{w}$. As $\mathbf{w}$ is not AWGN, we prewhiten it using the unitary matrix $\mathbf{Q}$ whose columns are the eigenvalues of the Hermitian matrix $\mathbf{P}$ (Kay, 1998). Specifically, $\mathbf{P} = \mathbf{Q}^H \mathbf{\Lambda} \mathbf{Q}$, where

$\boldsymbol{\Lambda} = \mathrm{diag}\{\lambda_1, \lambda_2, \ldots, \lambda_{\mathcal{N}}\}$ is the eigenvalue matrix of $\mathbf{P}$ and H denotes conjugate transpose. Using prewhitening, the modified MF output vector is given by $\breve{\mathbf{z}} \triangleq \mathbf{Q}\mathbf{z} = \boldsymbol{\Lambda}\breve{\mathbf{d}} + \breve{\mathbf{w}}$, where $\breve{\mathbf{d}} \triangleq \mathbf{Q}\mathbf{d}$, and $\breve{\mathbf{w}} \triangleq \mathbf{Q}\mathbf{w}$ can be shown to be AWGN with covariance matrix $\sigma_w^2 \boldsymbol{\Lambda}$.

If $\mathbf{P}$ has N_r nonzero eigenvalues, ordered such that $\lambda_1 \geq \lambda_2 \geq \cdots \lambda_{N_r} > 0$, then the Neyman–Pearson detector (Kay, 1998) decides that $b = +1$ was transmitted if

$$\frac{\prod_{\varsigma=1}^{N_r} \exp\left(- |\breve{z}_\varsigma - \lambda_\varsigma \breve{d}_\varsigma|^2/(2\sigma_w^2 \lambda_\varsigma)\right)}{\prod_{\varsigma=1}^{N_r} \exp\left(-|\breve{z}_\varsigma + \lambda_\varsigma \breve{d}_\varsigma|^2/(2\sigma_w^2 \lambda_\varsigma)\right)} > \gamma, \tag{9.34}$$

where $\breve{z}_\varsigma$ is the ςth element of vector $\breve{\mathbf{z}}$ and $\breve{d}_\varsigma$ is the ςth element of vector $\breve{\mathbf{d}} = \mathbf{H1}$. Note that the detection threshold γ is computed based on a desired detection performance. If we use the Bayesian approach to minimize the BER in the received symbol, assuming equal probability in transmitting $+1$ and -1, then $\gamma = 1$. In that case, we can form $\mathbf{Q}_r$ using the first N_r rows of $\mathbf{Q}$ to simplify the minimum BER detector. Specifically, we decide that $b = +1$ was transmitted if

$$\mathrm{Re}\left\{\breve{\mathbf{d}}^H \mathbf{Q}_r^H \mathbf{Q}_r \mathbf{z}\right\} > 0. \tag{9.35}$$

The aforementioned receiver design scheme transforms the $\mathcal{N}$ correlated MF outputs into N_r independent signals that can be expressed as

$$\breve{z}_\varsigma = \lambda_\varsigma \mathbf{q}_\varsigma^T \mathbf{d} + \breve{w}_\varsigma, \quad \varsigma = 1, \ldots, N_r,$$

where $\mathbf{q}_\varsigma^T$ is the $1 \times \mathcal{N}$ ςth row of the unitary matrix $\mathbf{Q}$ and $\breve{w}_\varsigma$ is the ςth element of the noise vector $\breve{\mathbf{w}}$. Using the minimum BER detector rule in (9.35), we can compute the SNR of the prewhitened ςth MF output $\breve{z}_\varsigma$ and then use it to obtain the average BER (Veeravalli, 2001) (for details, see Zhang et al. (2009)). This calculation shows that the potential diversity order of the shallow water receiver is N_r, which is the rank of $\mathbf{P}$ (Zhang & Papandreou-Suppappola, 2007; Zhang et al., 2009).

9.5.5 Numerical Simulations for Shallow Water Communications

In order to demonstrate the BER and diversity performance for the shallow water channel, we consider next some numerical simulation results. In the previous section, the dispersion diversity level depended on the rank of matrix $\mathbf{P}$. However, $\mathbf{P}$ may not be full rank (that is, N_r may not be equal to the number of modes $\mathcal{N}$) since the normal modes are not mutually orthogonal, and thus, the matched filters for corresponding modes are not orthogonal. In fact, $\mathbf{P}$ was formed from the correlation values between the different modes, and as a result, it depends on the environment parameters and on the transmission frequency band. Figure 9.16 shows the amplitude of the correlation between any two modes for a shallow water channel with $\mathcal{N} = 10$ modes. For this particular example, we chose the depth of the ocean to be $z_{\mathrm{ocean}} = 100\,\mathrm{m}$, the transmission distance to be $r_{\mathrm{rx}} = 15{,}000\,\mathrm{m}$, and $c = 1500\,\mathrm{m/s}$. We considered four different transmission frequency bands: 1–2, 5–6, 6–7, and 7–8 kHz. For the lowest frequency band in Fig. 9.16(a), $\mathbf{P}$ is almost full rank and similar to an identity matrix. However, as the position of the frequency band increases, the rank of $\mathbf{P}$ decreases and the correlation matrix begins to spread since the modes become more correlated. In Fig. 9.16(d), the rank of $\mathbf{P}$ is reduced to 7. In general, for the same set of environment parameters, the modes are more difficult to discriminate in higher frequency, and thus, the rank of $\mathbf{P}$ decreases as the position of the transmission frequency

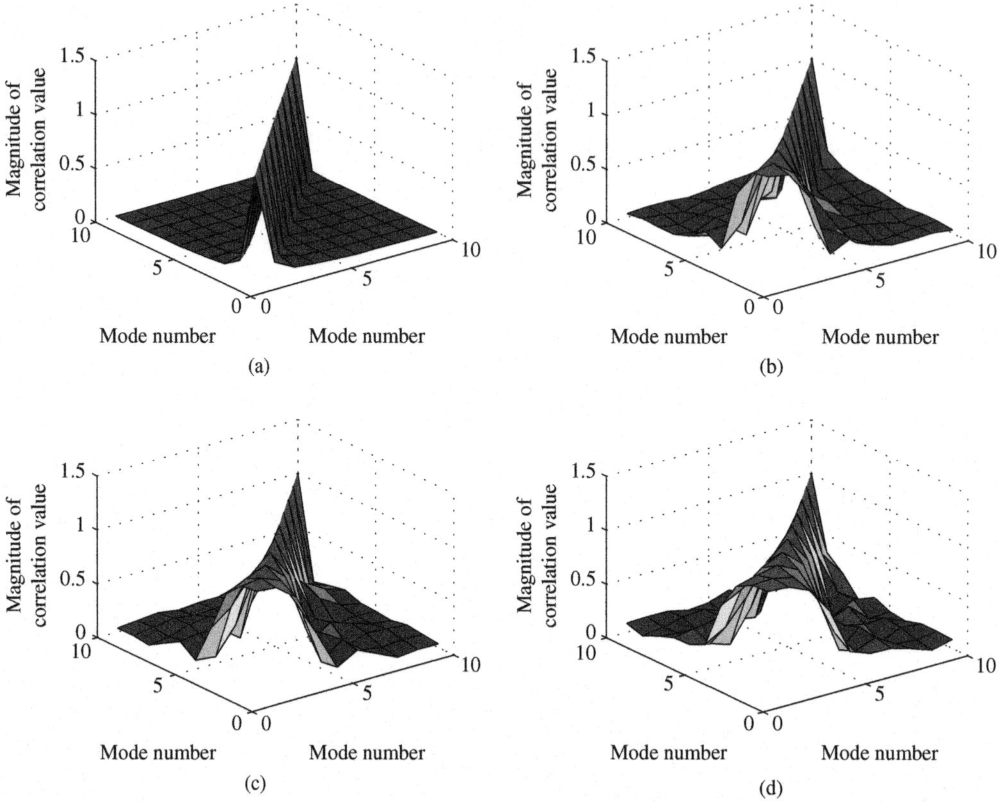

FIGURE 9.16

Correlation matrices (magnitudes of matrix elements) for different shallow water modes over different frequency bands: (a) 1–2 kHz frequency band (correlation matrix rank 10), (b) 5–6 kHz frequency band (correlation matrix rank ≈ 9), (c) 6–7 kHz frequency band (correlation matrix rank ≈ 8), and (d) 7–8 kHz frequency band (correlation matrix rank ≈ 7).

band increases. Using the shallow water environment parameters of $z_{\text{ocean}} = 50\,\text{m}$, $r_{\text{rx}} = 15{,}000\,\text{m}$, and $c = 1500\,\text{m/s}$, we simulated a received signal and obtained the BER results and approximate diversity order performances for three different frequency bands. The results are shown in Fig. 9.17 for SNR values between 0 and 30 dB. As expected, as the position of the frequency band increases, the BER performance deteriorates and the diversity order decreases. The diversity order was 9.8 for the 1–2 kHz frequency band, 7.9 for the 5–6 kHz band, and 5.2 for the 7–8 kHz band.

In real channel applications, the mode parameters of the shallow water channel are not known and need to be estimated. In such cases, warping techniques to separate the dispersive time-frequency modes at the receiver have been investigated using the formulation in (9.29) (Zhang & Papandreou-Suppappola, 2007; Zhang et al., 2009). Time-frequency separation was achieved by adaptively computing and extracting subsequent group delay curves corresponding to different modes and then using the curves to design the receiver and achieve dispersion diversity for improved communication performance.

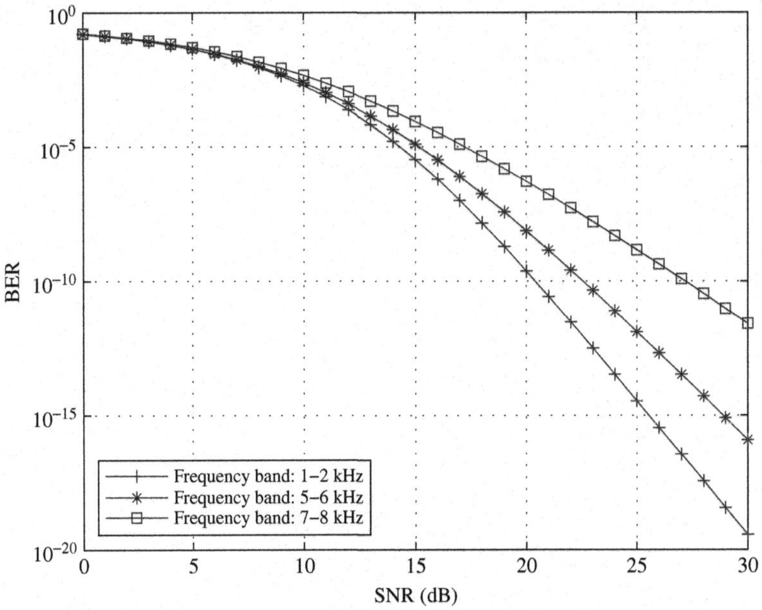

FIGURE 9.17

BER and diversity performance (slope of the BER curve at high SNR) for different transmission bands using the isovelocity shallow water model.

9.6 APPLICATION EXAMPLE

We provide an application example of underwater acoustic communications that is based on the multichannel communications scenario and ocean environment described in Fig. 9.5 in Section 9.1.2. With this example, we demonstrate the advantage of processing techniques that make use of the dispersive channel model. In particular, we show that an effective method of dealing with time-varying multipath in a multiuser underwater communication is to use carriers with very low frequencies, lower than 300–400 Hz, in order to exploit the natural dispersive properties of the channel at these frequencies. This method is potentially useful in reducing underwater time-varying multipath and increasing multiuser capabilities. The diversity can be successfully exploited, and the BER can be decreased, by estimating the corresponding dispersive curves in the time-frequency plane (for example, for the transmitted signal in Fig. 9.6(b)). In order to demonstrate the effectiveness of this method, we will compare it with communication at higher carrier frequencies and the use of equalization techniques to improve communication performance.

For the example in Fig. 9.5, we will refer to the communication channel between the boat and the AUV as *Channel 1* and the communication channel between the submarine and the surface buoy as *Channel 2*. We consider a transmitted signal of duration 2 s using BPSK modulation (with 10 kHz sampling frequency) for each of the channels, as depicted in Figs 9.18(a) and (c). We start by first choosing a 1 kHz carrier frequency for Channel 1 and a 3 kHz carrier frequency for Channel 2. The spectrograms

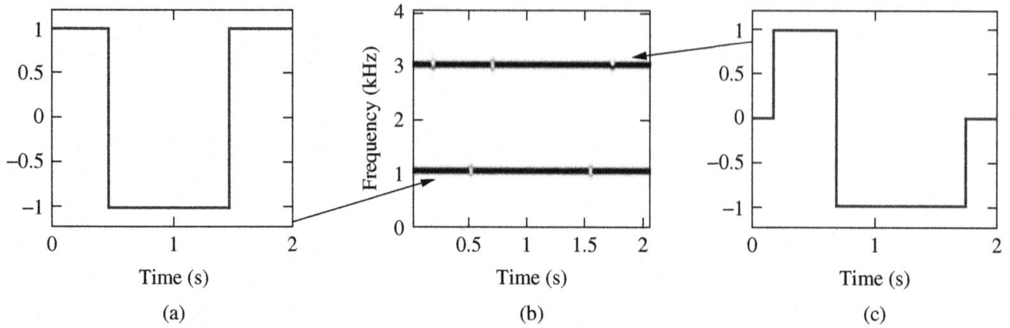

FIGURE 9.18

(a) BPSK modulated signal transmitted over Channel 1, (b) spectrogram of signals transmitted by both channels, and (c) BPSK modulated signal transmitted over Channel 2.

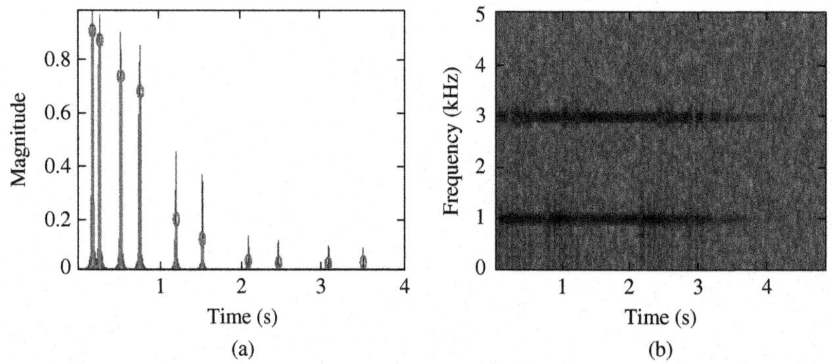

FIGURE 9.19

(a) Theoretical (solid) and estimated (circles) channel impulse response for Channel 1 and (b) spectrogram of received signals plotted in dB scale.

of the transmitted signals of both channels are shown superimposed in Fig. 9.18(b). Taking into account the high carrier frequencies and the physical parameters of the channels, the propagation can be modeled using ray theory (Jensen et al., 1994; Pedersen & Gordon, 1972), and the impulse response of the channels can be approximated using ray tracing as shown in Fig. 9.19(a) (Porter & Bucker, 1987). The received signals are also corrupted by white Gaussian noise with an SNR of 10 dB. Their spectrograms are plotted in dB scale in Fig. 9.19(b). This figure also shows the presence of several undesired spikes which correspond to the delayed versions of informative phase transients that are part of the transmitted signals.

The instantaneous frequency of the transmitted and received signals is illustrated in Fig. 9.20(a) for Channel 1 and in Fig. 9.20(b) for Channel 2. As the instantaneous frequency is the derivative of a signal's time-varying phase, sudden changes (or spikes) in the instantaneous frequency carry the binary message information in the case of BPSK; one such instantaneous frequency spike corresponds to a change in the binary state. From Figs 9.20(a) and 9.20(b), we can observe the big difference between

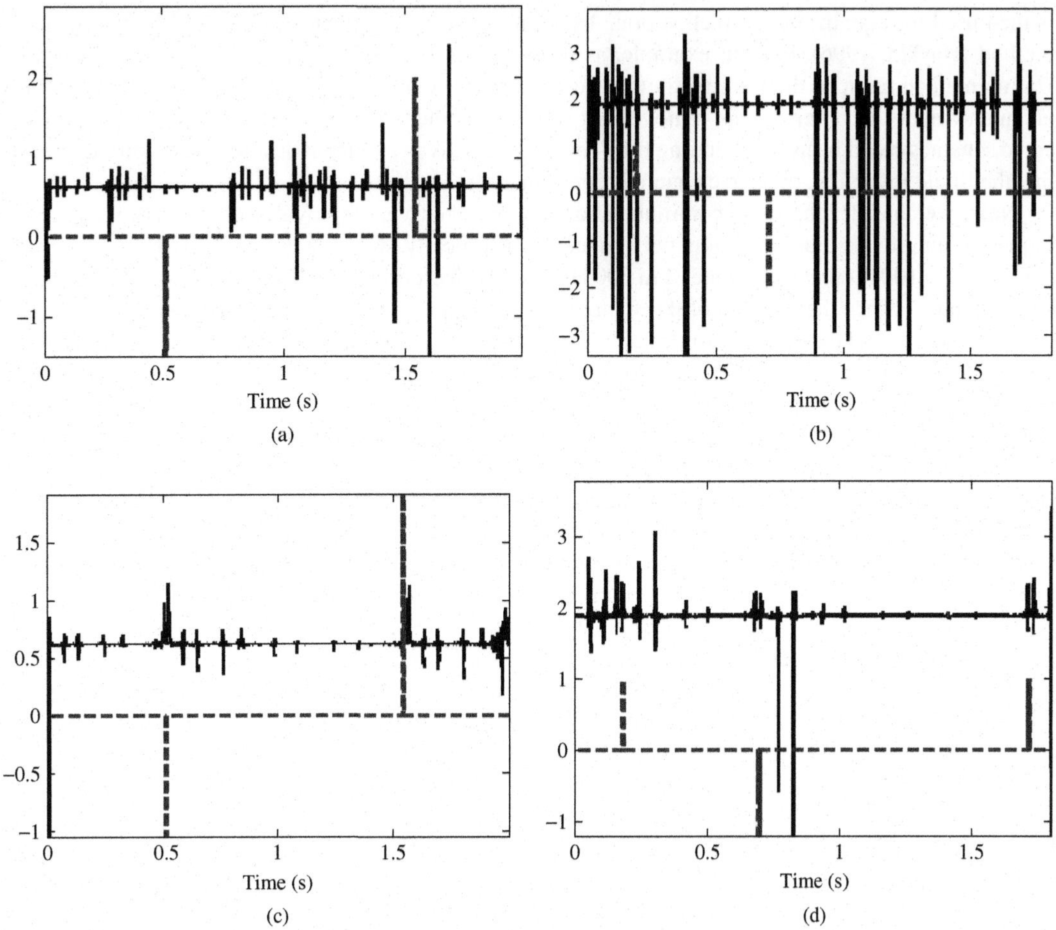

FIGURE 9.20

Comparison of instantaneous frequencies of transmitted (dashed) and received (solid) signals before and after equalization for Channels 1 and 2: (a) Instantaneous frequency of transmitted signal (dashed) and received signal without equalization (solid) for Channel 1, (b) instantaneous frequency of transmitted signal (dashed) and received signal without equalization (solid) for Channel 2, (c) instantaneous frequency of received signal with equalization (solid) for Channel 1, and (d) instantaneous frequency of received signal with equalization (solid) for Channel 2.

the instantaneous frequencies of the transmitted and received signals in each of the channels. As a result, we can clearly see the need for equalization to improve communication performance.

The widely used method proposed in Stojanovic (2003) is employed for equalization, and the results are shown in Figs 9.20(c) and 9.20(d). This method first estimates the channel's impulse response using a linear chirp probe signal whose duration is 0.5 s and whose frequency band spans 1–3 kHz. Figure 9.19(a) presents the results of channel impulse response estimation for Channel 1 and shows how well the theoretical ray model matches the true impulse response. The estimated impulse response is then used for equalization, which is done by means of the least-mean-square (LMS) adaptive filter method (Haykin, 1996). For our example, as illustrated in Figs 9.20(c) and (d), although the equalization method reduced the multipath effect (especially for Channel 1, as shown in Fig. 9.20(c)), not all intersymbol interference was removed. In order to further improve performance, more sophisticated schemes could be used, including time-reversal, adaptive waveform design (for the probe signal), matched-field inversion, or sparse channel representation.

Next, we consider the same communication problem, but we employ dispersive processing techniques to improve communication performance. Specifically, we reduce the carrier frequency of Channel 1 to 100 Hz and of Channel 2 to 300 Hz, thereby increasing the dispersion effect in the underwater channel. Since the position of the signal frequency band is now decreased by an order of 10, the normal mode model can be used to describe the underwater environment, and the *KRAKEN* software can be used for numerical simulations (Porter, 1991). The spectrograms of the received signals at 5 dB SNR are illustrated in Fig. 9.6(b). We can observe that the propagation effects for the two channels are quite different even though the two carrier frequencies are close in value. The transmitted symbols are now characterized by groups of dispersive curves in the time-frequency plane in different frequency bands as a result of the different mode components of the channel. We also observe that for each channel, there are two additional groups of dispersive curves that correspond to the modulation start and end points. For example, in Fig. 9.6(b), we observe five groups of dispersive curves around the carrier frequency of 300 Hz of Channel 2; the groups correspond to the three transmitted symbols, the modulation start point, and the modulation end point. Therefore, the dispersion curve phenomenon could be interpreted as a natural modulator that can be used to identify a channel according to the time-frequency characteristics of the received signal. While the natural dispersive phenomena are unique for a given transmitter-channel-receiver configuration, the propagation modes form a specific group of dispersive curves in the time-frequency plane that could be used to increase the multiuser capabilities of a channel. In addition, the diversity that can be obtained from the dispersion can be exploited to improve communication performance provided the dispersive curves in the time-frequency plane can be estimated.

The block diagram for the system we consider for achieving dispersion diversity is shown in Fig. 9.21. The system consists of U users. The uth user, $u = 1, \ldots, U$, is transmitting signal $s_u(t)$ that carries symbol a_u with BPSK modulation. The underwater channel is dispersive as low carrier

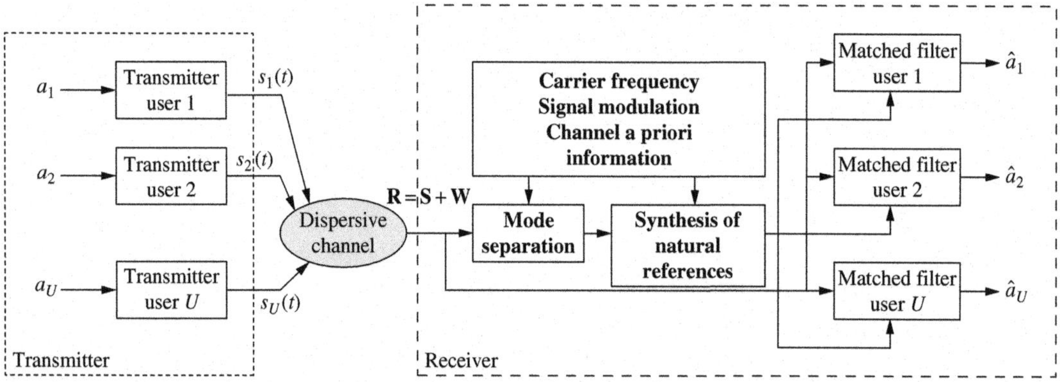

FIGURE 9.21

Communication system exploiting dispersion diversity.

frequencies are used. At the receiver, mode separation is first performed by estimating the characteristics of the dispersive curves (Gottin et al., 2007; Ioana, Stankovic, Quinquis, & Stankovic, 2005; Jarrot, Ioana, Gervaise, & Quinquis, 2007; Zhang et al., 2009). Note that the initialization of the time-frequency extraction algorithm requires a priori information, including the bandwidth of the transmitted signals, the carrier frequencies, the number of users, and the symbol duration. After mode extraction, similar time-frequency curves corresponding to the same user are grouped together. These groups, as well as additional information such as carrier frequency and signal duration, allow synthesizing the natural references associated with each user, which are then used to perform matched filtering in order to extract the signal associated with each user.

Figure 9.22(a) shows the extracted time-frequency curves of the normal modes corresponding to the propagation of the phase transitions of both BPSK signals. Note that each arrival is characterized by three modes. Comparing the extracted time-frequency modal curves in Fig. 9.22(a) and the spectrogram of the received signals in Fig. 9.22(b), we can see the difference in time-frequency structure between the two different channels. This is a result of the difference in the dispersion effect, which

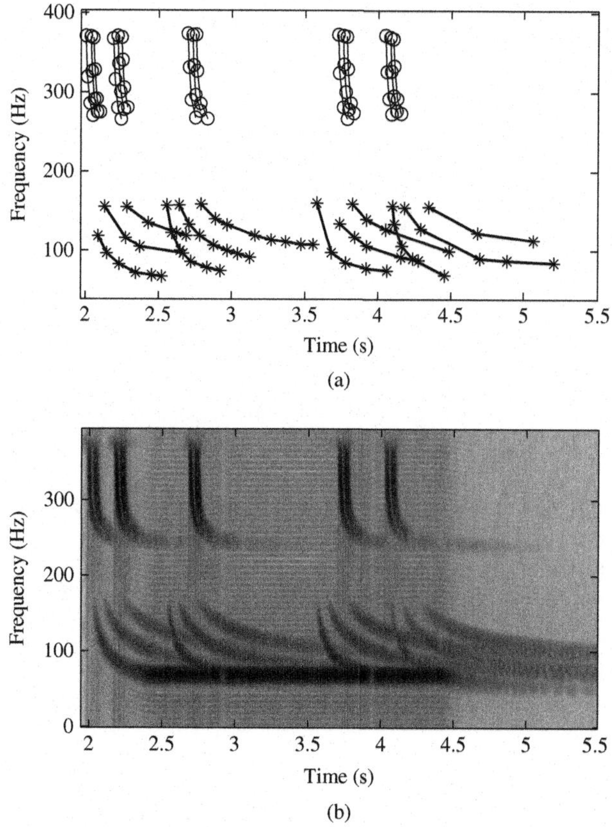

FIGURE 9.22

(a) Extracted time-frequency dispersive curves and (b) superimposed spectrograms of received signals of Channel 1 and Channel 2, plotted in dB scale.

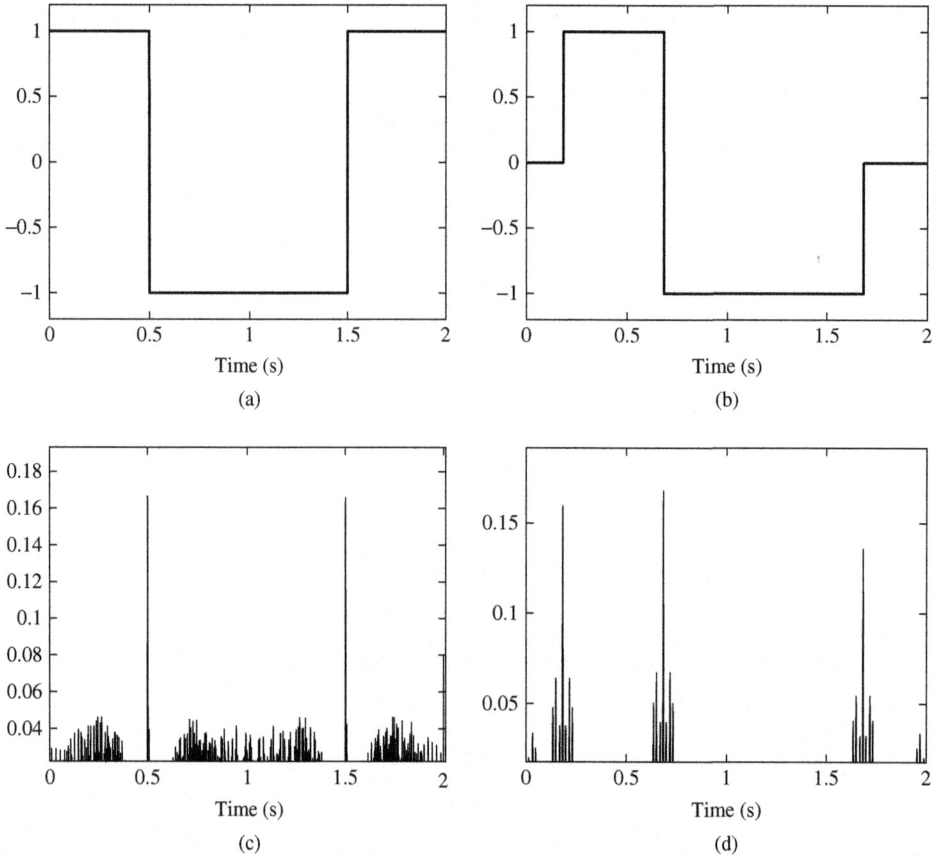

FIGURE 9.23

Symbols transmitted in (a) Channel 1 and (b) Channel 2, and estimated symbols using dispersion diversity in (c) Channel 1 and (d) Channel 2.

depends on the locations of the transmitter and receiver and on the signal's carrier frequency and bandwidth. Because a transmitter-receiver channel configuration has a unique time-frequency signature, a multiuser identification is feasible, especially when different carrier frequencies are used. Using the estimated dispersive curves and matched filtering, the signals are extracted as shown in Fig. 9.23 for both Channel 1 and Channel 2. We notice that the spikes corresponding to changes in the binary state are correctly detected and the false alarms are substantially reduced. Additional examples when there is relative motion between the transmitter and receiver can be found in Josso, Ioana, Gervaise, & Mars (2008).

9.7 CONCLUSIONS

Wireless communication systems can be greatly challenged by the characteristics of the propagation environment. Specifically, when we consider linear time-varying (LTV) channels with mobile transmitter, mobile receiver, and/or mobile reflecting scatterers, the effect of the channel on the transmitted signal may vary considerably. This variation depends on relative velocities and reflectivity properties, and it can affect the time-frequency signature of the transmitted signal. As a result, the LTV channel model needs to be versatile in order to allow for different signal transformations. It also needs to allow for discretization of the various signal transformations so that the channel can be decomposed into multiple independent, flat fading subchannels, leading to diversity in the transformation space. Based on these considerations, we investigated two types of wideband channels: the wideband delay-scale channel that causes multipath and Doppler scaling on the transmitted signal and the wideband dispersive channel that causes nonlinear dispersive changes on the transmitted signal.

For the wideband delay-scale channel, we provided the conditions under which Doppler changes in the signal cannot be approximated as Doppler shifts and have to be considered dilations. The corresponding channel representation is based on the wideband spreading function. We demonstrated that the Doppler scale parameter of this representation can be sampled geometrically using the Mellin transform, and for each discrete scale, the multipath parameter can be uniformly sampled. We furthermore demonstrated that the delay-scale framework can be used to improve wideband channel communication performance due to its inherent multipath-scale diversity, which is captured by the discrete time-scale model. This is achieved by using a wavelet signaling scheme to dyadically decompose the channel into independent subchannels. The adoption of wavelet signaling also facilitates the efficient implementation of a wideband time-scale rake receiver using wavelet transform techniques.

The wideband dispersive channel generalizes the narrowband channel characterization. This generalization framework provides an important foundation to exploit dispersive characteristics and improve system performance. We based our discretization of the wideband dispersive channel on a unitary warping relation between the narrowband and dispersive spreading functions. This relation uses the dispersive spreading function to characterize nonlinear phase changes on the signal caused by the dispersive channel. Thus, the discrete model decomposes the dispersive channel output as a discrete representation of uniformly sampled dispersive frequency shifts and generalized time shifts, weighted by a smoothed and sampled version of the dispersive spreading function.

Shallow water acoustic communications using low frequencies provide an example of a dispersive channel. Most current communications are limited to short distances because they are constrained by the use of the narrowband model. However, use of the dispersive normal-mode model can allow for long-distance shallow water communications. We specifically investigated the isovelocity normal-mode model for shallow water environments. Based on this model, we designed the transmission waveform and constructed a matched filter bank receiver structure that matched the channel's time-frequency dispersive characteristics and could be exploited to improve diversity performance. We furthermore demonstrated the advantage of using the inherent dispersive normal modes at low frequencies in order to improve the performance of a multichannel underwater communication system.

References

Agazzi, O. E., Hueda, M. R., Carrer, H. S., & Crivelli, E. E. (2005). Maximum-likelihood sequence estimation in dispersive optical channels. *Journal of Lightwave Technology*, *23*(2), 749–763.

Altes, R. A., & Titlebaum, E. L. (1970). Bat signals as optimally Doppler tolerant waveforms. *Journal of the Acoustical Society of America*, *48*, 1014–1020.

Balan, R., Poor, H. V., Rickard, S., & Verdu, S. (2004, September). Time-frequency and time-scale canonical representations of doubly spread channels. In *Proceedings of EUSIPCO 2004*, (pp. 445–448). Vienna, Austria.

Beaujean, P.-P. J., & LeBlanc, L. R. (2004). Adaptive array processing for high-speed acoustic communication in shallow water. *IEEE Journal of Oceanic Engineering*, *29*(3), 807–823.

Bello, P. A. (1963). Characterization of randomly time-variant linear channels. *IEEE Transactions on Communication Systems*, *11*, 360–393.

Bertrand, J., Bertrand, P., & Ovarlez, J. P. (1996). The Mellin transform. In A. D. Poularikas (Ed.), *The transforms and applications handbook*, chap. 11, (pp. 829–885). Boca Raton, FL: CRC Press.

Bhashyam, S., Sayeed, A. M., & Aazhang, B. (2000). Time-selective signaling and reception for communication over multipath fading channels. *IEEE Transactions on Communications*, *48*(1), 83–94.

Bircan, A., Tekinay, S., & Akansu, A. N. (1998, October). Time-frequency and time-scale representation of wireless communication channels. In *Proceedings of the IEEE International Symposium on Time-Frequency/Time-Scale Analysis*, (pp. 373–376). Pittsburgh, PA.

Bölcskei, H., Duhamel, P., & Hleiss, R. (1999, June). Design of pulse shaping OFDM/OQAM system for high data-rate transmission over wireless channels. In *Proceedings of IEEE ICC 1999*: (pp. 559–564). Vancouver, Canada.

Chen, C., James, H. M., Boudreaux-Bartels, G. F., Gopu, R. P., & Colin, J. L. (2003, September). Time-frequency representations for wideband acoustic signals in shallow water. In *Proceedings of IEEE OCEANS 2003*, Vol. 5, (pp. 2903–2907).

Cohen, L. (1993). The scale representation. *IEEE Transactions on Signal Processing*, *41*(12), 3275–3292.

Davies, J. J., Pointer, S. A., & Dunn, S. M. (2000, September). Wideband acoustic communications dispelling narrowband myths. In *Proceedings of MTS/IEEE OCEANS 2000*, Vol. 1, (pp. 377–384). Providence, RI.

Deane, G. B. (1997). Internal friction and boundary conditions in lossy fluid seabeds. *Journal of the Acoustical Society of America*, *101*(1), 233–240.

Eggen, T. H., Baggeroer, A. B., & Presig, J. C. (2000). Communication over Doppler spread channels, Part I: Channel and receiver presentation. *IEEE Journal of Oceanic Engineering*, *25*(1), 62–71.

Eggen, T. H., Baggeroer, A. B., & Presig, J. C. (2001). Communication over Doppler spread channels, Part II: Receiver characterization and practical results. *IEEE Journal of Oceanic Engineering*, *26*(4), 612–621.

Federal Communications Commission. (2002). *Revision of part 15 of the Commission's rules regarding ultra-wideband transmission systems, First report and order*, ET Docket 98-153, FCC 02-48, (pp. 1–118).

Fischer, J. H., Bennett, K. R., Reible, S. A., Cafarella, J. H., & Yao, I. (1992, October). A high data rate, underwater acoustic data-communications transceiver. In *Proceedings of MTS/IEEE OCEANS 1992*, Vol. 2, (pp. 571–576). Newport, RI.

Flandrin, P. (1999). *Time-frequency/time-scale analysis*. San Diego, CA: Academic Press.

Freeman, M. J., Dunham, M. E., & Qian, S. (1995, August). Trans-ionospheric signal detection by time-scale representation. In *Proceedings of the IEEE UK Symposium on Applications of Time-Frequency and Time-Scale Methods*, (pp. 152–158). University of Warwick, Coventry, UK.

Freitag, L., Grund, M., Singh, S., Smith, S., Christenson, R., Marquis, L., & Catipovic, J. (1998, September). A bidirectional coherent acoustic communication system for underwater vehicles. In *Proceedings of IEEE OCEANS 1998*, Vol. 1, (pp. 482–486). Nice, France.

Freitag, L., Stojanovic, M., Singh, S., & Johnson, M. (2001). Analysis of channel effects on direct-sequence and frequency-hopped spread-spectrum acoustic communication. *IEEE Journal of Oceanic Engineering*, *26*(4), 586–593.

Freitag, L., Stojanovic, M., Kilfoyle, D., & Preisig, J. (2004, July). High-rate phase coherent acoustic communication: A review of a decade of research and a perspective on future challenges. In *Proceedings of the European Conference on Underwater Acoustics*, Delft, The Netherlands.

Frisk, G. V. (1994). *Ocean and seabed acoustics*. Englewood Cliffs: Prentice-Hall.

Gervaise, C., Vallez, S., Stephan, Y., & Simard, Y. (2008). Robust 2D localization of low-frequency calls in shallow waters using modal propagation modelling. *Canadian Acoustics*, *36*(1), 153–159.

Giannakis, G. B., & Tepedelenlioglu, C. (1998). Basis expansion models and diversity techniques for blind identification and equalization of time-varying channels. *Proceedings of the IEEE*, *86*(10), 1969–1986.

Gingras, D. F. (1994). *North Elba Sea trial summary*. Tech. Rep. NATO SACLANTCEN, La Spezia, Italy.

Gottin, B., Zhang, J., Papandreou-Suppappola, A., & Ioana, C. (2007). Diversity in shallow water environments using blind time-frequency separation techniques. In *Proceedings of the 41st Asilomar Conference on Signals, Systems, and Computers*: (pp. 1159–1163). Pacific Grove, CA.

Haas, R., & Belfiore, J. (1997). A time-frequency well-localized pulse for multiple carrier transmission. *Wireless Personal Communications*, *5*, 1–18.

Haykin, S. (1996). *Adaptive filter theory*. Upper Saddle River, NJ: Prentice-Hall.

He, Y., Wang, Y., Zuo, X., & Zhang, Y. (2008, April). Ultra wideband technology for micro air vehicles data links systems. In *Proceedings of the International Conference on Microwave and Millimeter Wave Technology*, Vol. 1, (pp. 108–111). Nanjing, China.

Hinton, O., Howe, G., & Adams, A. (1992). An adaptive, high bit rate, sub-sea communications system. In *Proceedings of the European Conference on Underwater Acoustics*, (pp. 75–79). Brussels, Belgium.

Hlawatsch, F., Papandreou-Suppappola, A., & Boudreaux-Bartels, G. F. (1999). The power classes—Quadratic time-frequency representations with scale covariance and dispersive time-shift covariance. *IEEE Transactions on Signal Processing*, *47*, 3067–3083.

Iem, B. G. (1998). *Generalization of the Weyl symbol and the spreading function via time-frequency warpings: Theory and application*. Unpublished doctoral dissertation, University of Rhode Island, Kingston, RI.

Iem, B. G., Papandreou-Suppappola, A., & Boudreaux-Bartels, G. F. (2002). Wideband Weyl symbols for dispersive time-varying processing of systems and random signals. *IEEE Transactions on Signal Processing*, *50*, 1077–1090.

Ioana, C., Stankovic, S., Quinquis, A., & Stankovic, L. (2005, March). Modelling of signal's time-frequency content using warped complex-time distributions. In *Proceedings of IEEE ICASSP 2005*, Vol. 4, (pp. 477–480). Philadelphia, PA.

Jarrot, A., Ioana, C., Gervaise, C., & Quinquis, A. (2007, April). A time-frequency characterization framework for signals issued from underwater dispersive environments. In *Proceedings of IEEE ICASSP 2007*, Vol. 3, (pp. 1145–1148). Honolulu, HI.

Jensen, F. B., Kuperman, W. A., Porter, M. B., & Schmidt, H. (1994). *Computational ocean acoustics*. New York: AIP Press.

Johnson, M., Freitag, L., & Stojanovic, M. (1997, April). Improved Doppler tracking and correction for underwater acoustic communications. In *Proceedings of IEEE ICASSP 1997*, Vol. 1, (pp. 575–578), Munich, Germany.

Josso, N., Ioana, C., Gervaise, C., & Mars, J. I. (2008, June/July). On the consideration of motion effects in underwater geoacoustic inversion. In *Proceedings of Acoustics '08*, Vol. 123, (pp. 4779–4784). Paris, France.

Kailath, T. (1959). *Sampling models for linear time-variant filters*. (Tech. Rep. No. 352). M.I.T. Research Laboratory of Electronics, Cambridge, MA.

Kay, S. M. (1998). *Fundamentals of statistical signal processing: Detection theory*. Upper Saddle River, NJ: Prentice Hall.

Kennedy, R. S. (1969). *Fading dispersive communication channels*. New York: Wiley.

Kim, B. C., & Lu, I-T. (2000). Parameter study of OFDM underwater communications system. In *Proceedings of MTS/IEEE OCEANS 2000*, Vol. 2, (pp. 1251–1255). Providence, RI.

Kozek, W., & Molisch, A. F. (1998). Nonorthogonal pulse shapes for multicarrier communication in doubly dispersive channels. *IEEE Journal on Selected Areas in Communications*, *16*(8), 1579–1589.

Li, B., Zhou, S., Stojanovic, M., Freitag, L., & Willett, P. (2007, June). Non-uniform Doppler compensation for zero-padded OFDM over fast-varying underwater acoustic channels. In *Proceedings of IEEE OCEANS 2007*, (pp. 1–6). Aberdeen, Scotland.

Margetts, A. R., & Schniter, P. (2004, November). Joint scale-lag diversity in mobile ultra-wideband systems. In *Proceedings of the 38th Asilomar Conference on Signals, Systems and Computers*, (pp. 1496–1500). Pacific Grove, CA.

Margetts, A. R., Schniter, P., & Swami, A. (2007). Joint scale-lag diversity in wideband mobile direct sequence spread spectrum systems. *IEEE Transactions on Wireless Communications*, *6*(12), 4308–4319.

Matz, G., Schafhuber, D., Gröchenig, K., Hartmann, M., & Hlawatsch, F. (2007). Analysis, optimization, and implementation of low-interference wireless multicarrier systems. *IEEE Transactions on Wireless Communications*, *6*(5), 1921–1931.

Molisch, A. F. (2005). *Wireless communications*. Chichester, UK: Wiley.

Morozov, A. K., Preisig, J. C., & Papp, J. C. (2008). Investigation of modal processing for low frequency acoustic communications in shallow water. In *Proceedings of Meetings on Acoustics*, Vol. 4, Issue 1, (pp. 070005–070005-16).

Morozov, A. K., Preisig, J. C., & Papp, J. C. (2008). Modal processing for acoustic communications in shallow water experiment. *Journal of the Acoustical Society of America*, *124*(3), 177–181.

Newland, D. E. (1998). Time-frequency and time-scale analysis by harmonic wavelets. In A. Prochazka (Ed.), *Signal analysis and prediction*, chap. 1. Boston, MA: Birkhäuser.

Ovarlez, J.-P. (1993, April). Cramer Rao bound computation for velocity estimation in the broad-band case using the Mellin transform. In *Proceedings of IEEE ICASSP 1993*, Vol. 1, (pp. 273–276). Minneapolis, MN.

Ovarlez, J.-P., Bertrand, J., & Bertrand, P. (1992, March). Computation of affine time-frequency distributions using the fast Mellin transform. In *Proceedings of IEEE ICASSP 1992*, Vol. 5, (pp. 117–120). San Francisco, CA.

Papandreou, A., Hlawatsch, F., & Boudreaux-Bartels, G. F. (1993). The hyperbolic class of quadratic time-frequency representations, part I: Constant-Q warping, the hyperbolic paradigm, properties, and members. *IEEE Transactions on Signal Processing*, *41*(12), 3425–3444.

Papandreou-Suppappola, A., Hlawatsch, F., & Boudreaux-Bartels, G. F. (1998). Quadratic time-frequency representations with scale covariance and generalized time-shift covariance: A unified framework for the affine, hyperbolic and power classes. *Digital Signal Processing: A Review Journal*, *8*, 3–48.

Papandreou-Suppappola, A., Murray, R. L., Iem, B. G., & Boudreaux-Bartels, G. F. (2001). Group delay shift covariant quadratic time-frequency representations. *IEEE Transactions on Signal Processing*, *49*, 2549–2564.

Pedersen, M. A., & Gordon, D. F. (1972). Normal-mode and ray theory applied to underwater acoustic conditions of extreme downward refraction. *Journal of the Acoustical Society of America*, *51*(1), 323–368.

Porter, M. B. (1991). The KRAKEN normal mode program. Saclantcen memorandum sm-245, SACLANT Undersea Research Centre, La Spezia, Italy. (1997 updated document at http://oalib.hlsresearch.com/Modes/AcousticsToolbox/manual_html/kraken.html).

Porter, M. B., & Bucker, H. P. (1987). Gaussian beam tracing for computing ocean acoustic fields. *Journal of the Acoustical Society of America*, *82*(4), 1348–1359.

Porter, M., & Duncan, A. (2002). Underwater acoustic propagation modelling software - AcTUI version 1.6. http://www.cmst.curtin.edu.au/products/actoolbox/.

Preisig, J. C. (2003). The impact of bubbles on underwater acoustic communications in shallow water environments. *Journal of the Acoustical Society of America*, *114*(4), 2370.

Preisig, J. C. (2005). Performance analysis of adaptive equalization for coherent acoustic communications in the time-varying ocean environment. *Journal of the Acoustical Society of America*, *118*(1), 263–278.

Preisig, J. C., & Johnson, M. P. (2001). Signal detection for communications in the underwater acoustic environment. *IEEE Journal of Oceanic Engineering*, *26*(4), 572–585.

Proakis, J. G. (2001). *Digital communications*. 4th ed. New York: McGraw-Hill.

Rickard, S. (2003). *Time-frequency and time-scale representations of doubly spread channels*. Unpublished doctoral dissertation, Princeton University.

Sayeed, A. M., & Aazhang, B. (1999). Joint multipath-Doppler diversity in mobile wireless communications. *IEEE Transactions on Communications*, *47*, 123–132.

Scholtz, R. A., & Lee, J.-Y. (2002, November). Problems in modeling UWB channels. In *Proceedings of the 36th Asilomar Conference on Signals, Systems and Computers*, Vol. 1, (pp. 706–711). Pacific Grove, CA.

Sessarego, J. P., Sageloli, J., Flandrin, P., & Zakharia, M. (1989). Time-frequency Wigner-Ville analysis of echoes scattered by a spherical shell. In J. M. Combes, A. Grossman, & P. Tchamitchian (Eds.), *Wavelets, time-frequency methods and phase space*, (pp. 147–153). Berlin: Springer-Verlag.

Sharif, B. S., Neasham, J., Hinton, O. R., & Adams, A. E. (2000). A computationally efficient Doppler compensation system for underwater acoustic communications. *IEEE Journal of Oceanic Engineering*, *25*(1), 52–61.

Shenoy, R. G., & Parks, T. W. (1995). Wide-band ambiguity functions and affine Wigner distributions. *Signal Processing*, *41*, 339–363.

Sibul, L. H., Weiss, L. G., & Dixon, T. L. (1994). Characterization of stochastic propagation and scattering via Gabor and wavelet transforms. *Journal of Computational Acoustics*, *2*(3), 345–369.

Sostrand, K. A. (1968). Mathematics of the time-varying channel. In *NATO Advanced Study Institution on Signal Processing with Emphasis on Underwater Acoustics*, Vol. 2, (pp. 1–20). Enschede, The Netherlands.

Stojanovic, M. (1996). Recent advances in high-speed underwater acoustic communications. *IEEE Journal of Oceanic Engineering*, *21*(2), 125–136.

Stojanovic, M. (2003). Acoustic underwater communications. In J. G. Proakis (Ed.). *Encyclopedia of telecommunications*. Hoboken, NJ: Wiley.

Stojanovic, M. (2004). Spatio-temporal focusing for elimination of multipath effects in high rate acoustic communications. In M. Porter, M. Siderious, & W. Kupermann (Eds.), *High frequency ocean acoustics*, (pp. 65–73). New York: American Institute of Physics.

Stojanovic, M., & Preisig, J. C. (2009). Underwater acoustic communication channels: Propagation models and statistical characterization. *IEEE Communications Magazine*, *47*(1), 84–89.

Stojanovic, M., Catipovic, J. A., & Proakis, J. G. (1994). Phase-coherent digital communications for underwater acoustic channels. *IEEE Journal of Oceanic Engineering*, *19*(1), 100–111.

Tolstoy, I., & Clay, C. S. (1966). *Ocean acoustics: Theory and experiment in underwater sound.* New York: McGraw-Hill.

Urick, R. J. (1975). *Principles of underwater sound.* 2nd ed. New York: McGraw-Hill.

Vakman, D. E. (1968). *Sophisticated signals and the uncertainty principle in radar*, chap. 4, (pp. 126–218). New York: Springer-Verlag, translated by K. N. Trirogoff.

Veeravalli, V. V. (2001). On performance analysis for signaling on correlated fading channels. *IEEE Transactions on Communications*, *49*(11), 1879–1883.

Weiss, L. G. (1994). Wavelets and wideband correlation processing. *IEEE Signal Processing Magazine*, *11*(1), 13–32.

Weiss, L. G. (1996). Time-varying system characterization for wideband input signals. *Signal Processing*, *55*, 295–304.

Win, M. Z., & Scholtz, R. A. (2000). Ultra-wide bandwidth time-hopping spread-spectrum impulse radio for wireless multiple-access communications. *IEEE Transactions on Communications*, *48*(4), 679–689.

Win, M. Z., & Scholtz, R. A. (2002). Characterization of ultra-wide bandwidth wireless indoor channels: A communication-theoretic view. *IEEE Journal on Selected Areas in Communications*, *20*(9), 1613–1627.

Ye, J., & Papandreou-Suppappola, A. (2003, September). Characterization of wideband time-varying channels with multipath-scale diversity. In *Proceedings of the IEEE Statistical Signal Processing Workshop*, (pp. 50–53). St. Louis, MO.

Ye, J., & Papandreou-Suppappola, A. (2005, March). Time-scale canonical model for wideband system characterization. In *Proceedings of IEEE ICASSP 2005*, Vol. 4, (pp. 281–284). Philadelphia, PA.

Ye, J., & Papandreou-Suppappola, A. (2006, May). Discrete time-frequency models of generalized dispersive systems. In *Proceedings of IEEE ICASSP 2006*, Vol. 3, (pp. 349–352). Toulouse, France.

Ye, J., & Papandreou-Suppappola, A. (2006). Discrete time-scale characterization of wideband time-varying systems. *IEEE Transactions on Signal Processing*, *54*, 1364–1375.

Ye, J., & Papandreou-Suppappola, A. (2007). Discrete time-frequency characterizations of dispersive time-varying systems. *IEEE Transactions on Signal Processing*, *55*(5), 2066–2076.

Ye, J., Shen, H., & Papandreou-Suppappola, A. (2006, January). Characterization of shallow water environments and waveform design for diversity. In *Proceedings of the International Waveform Diversity and Design Conference*, Kauai, HI.

Young, R. K. (1993). *Wavelet theory and its applications*. Boston: Kluwer.

Zadeh, L. A., & Desoer, C. A. (1963). *Linear system theory: The state space approach*. New York: McGraw-Hill.

Zhang, H., Fan, H. H., & Lindsey, A. (2001, March). A wavelet packet based model for time-varying wireless communication channels. In *Proceedings of the IEEE Workshop on Signal Processing Advances in Wireless Communications*, (pp. 50–53). Taoyuan, Taiwan, R.O.C.

Zhang, J., & Papandreou-Suppappola, A. (2007, April). Time-frequency based waveform and receiver design for shallow water communications. In *Proceedings of IEEE ICASSP 2007*, Vol. 3, (pp. 1149–1152). Honolulu, HI.

Zhang, J., Gottin, B., Papandreou-Suppappola, A., & Ioana, C. (2007, August). Time-frequency modeling of shallow water environments: Rigid vs. fluid seabed. In *Proceedings of the IEEE Workshop on Statistical Signal Processing*, (pp. 740–744). Madison, WI.

Zhang, J., Papandreou-Suppappola, A., Gottin, B., & Ioana, C. (2009). Time-frequency characterization and receiver waveform design for shallow water environments. *IEEE Transactions on Signal Processing*, *57*(8), 2973–2985.

Ziomek, L. J. (1985). *Underwater acoustics: A linear systems theory approach*. Orlando, FL: Academic Press.

Ziomek, L. J. (1999). *Fundamentals of acoustic field theory and space-time signal processing*. Boca Raton, FL: CRC Press.

Index

A

B

C

D

G

H

I

J

K

L

M

N

O

P

Q

R

S

T

U

V

W

Y

Z